S0-BLM-003

Column Handbook for Size Exclusion Chromatography

Column Handbook for Size Exclusion Chromatography

Edited by

Chi-san Wu

International Specialty Products
Wayne, New Jersey

ACADEMIC PRESS

San Diego London Boston New York Sydney Tokyo Toronto

This book is printed on acid-free paper. ∞

Academic Press
a division of Harcourt Brace & Company
525 B Street, Suite 1900, San Diego, California 92101-4495, USA
http://www.apnet.com

Academic Press
24-28 Oval Road, London NW1 7DX, UK
http://www.hbuk.co.uk/ap/

Library of Congress Catalog Card Number: 98-86760

International Standard Book Number: 0-12-765555-7

PRINTED IN THE UNITED STATES OF AMERICA
99 00 01 02 03 04 QW 9 8 7 6 5 4 3 2 1

CONTENTS

5 American Polymer Standards Columns for Size Exclusion Chromatography

JOHN E. ARMONAS AND BRIAN H. PEABODY

6 Shodex Columns for Size Exclusion Chromatography

HIROSHI SUZUKI AND SADAO MORI

7 Size Exclusion Chromatography on Fractogel EMD BioSEC

LOTHAR R. JACOB AND LOTHAR BRITSCH

8 Size Exclusion High-Performance Liquid Chromatography of Small Solutes

ANDREW J. ALPERT

9 Design, Properties, and Testing of Polymer Standards Service Size Exclusion Chromatography (SEC) Columns and Optimization of SEC Separations

P. KILZ

10 SynChropak Size Exclusion Columns

KAREN M. GOODING

11 Waters Columns for Size Exclusion Chromatography

UWE DIETER NEUE

12 Size Exclusion Chromatography Columns from Polymer Laboratories

ELIZABETH MEEHAN

13 Jordi Gel Columns for Size Exclusion Chromatography

HOWARD JORDI

II CONTRIBUTIONS FROM USERS OF SIZE EXCLUSION CHROMATOGRAPHY AND GEL-PERMEATION CHROMATOGRAPHY

14 General Characterization of Gel-Permeation Chromatography Columns

RUDOLPH BRUESSAU

18 Application of Size Exclusion–High-Performance Liquid Chromatography for Biopharmaceutical Protein and Peptide Therapeutics

DAVID P. ALLEN

19 Column Selection and Related Issues for Acrylic Acid and Acrylate Ester Polymers

MICHAEL T. BENDER AND DANIEL A. SAUCY

20 Applications and Uses of Columns for Aqueous Size Exclusion Chromatography of Water-Soluble Polymers

DENNIS J. NAGY

21 Quality Control of Columns for High-Temperature Gel-Permeation Chromatography

A. WILLEM deGROOT

III COLUMNS FOR OTHER RELATED POLYMER SEPARATION OR FRACTIONATION TECHNIQUES

CONTRIBUTORS

Numbers in parentheses indicate the pages on which the authors' contributions begin.

David P. Allen (531), Biopharmaceutical Product Development, Eli Lilly and Company, Indianapolis, Indiana 46285

Andrew J. Alpert (249), PolyLC Inc., Columbia, Maryland 21045

John E. Armonas (159), American Polymer Standards Corporation, Mentor, Ohio 44060

Michael T. Bender (539), Analytical Research Department, Rohm and Haas Company, Spring House, Pennsylvania 19477

Dušan Berek (445), Polymer Institute of the Slovak Academy of Sciences, 842 36 Bratislava, Slovakia

Lothar Britsch (219), Merck KGaA, 64271 Darmstadt, Germany

Rudolf Bruessau (429), BASF Aktiengessellschaft, ZKM/A-B1, Kunstofflaboratorium, D-67056 Ludwigshafen, Germany

Tom M. H. Cheng (499), International Specialty Products, Wayne, New Jersey 07470

A. Willem deGroot (583), The Dow Chemical Company, Freeport, Texas 77541

Jon Fisher (93), TosoHaas, Montgomeryville, Pennsylvania 18936

Karen M. Gooding (305), Formerly of SynChrom, Inc., Lafayette, Indiana 47901 and MICRA Scientific, Northbrook, Illinois 60062

Lars Hagel (27), Amersham Pharmacia Biotech AB, SE-751 82 Uppsala, Sweden

Shyhchang S. Huang (597), Advanced Technology Group, The BFGoodrich Company, Brecksville, Ohio 44141

Anton Huber (459), Institut für Physikalische Chemie, KF-Universität, A-8010 Graz, Austria

Lothar R. Jacob (219), Merck KGaA, 64271 Darmstadt, Germany

Howard Jordi (367), Jordi Associates, Inc., Bellingham, Massachusetts 02019

Yoshio Kato (93), TosoHaas, Montgomeryville, Pennsylvania 18936

P. Kilz (267), PSS Polymer Standards Service, D-55023 Mainz, Germany

Michael J. Lu (3), BioChrom Labs, Inc., Terre Haute, Indiana 47808

Elizabeth Meehan (349), Polymer Laboratories, Shropshire SY6 6AX, United Kingdom

Robert T. Moody (75), MAC-MOD Analytical, Inc., Chadds Ford, Pennsylvania 19317

Sadao Mori (171), Faculty of Engineering, Department of Industrial Chemistry, Mie University, Tsu Mie 514, Japan

Dennis J. Nagy (559), Analytical Technology Ctr., Air Products and Chemicals, Inc., Allentown, Pennsylvania 18195

Uwe Dieter Neue (325), Waters Corporation, Milford, Massachusetts 01757

J. Kevin O'Donnell (93), TosoHaas, Montgomeryville, Pennsylvania 18936

Donna Osborne (499), International Specialty Products, Wayne, New Jersey 07470

Brian H. Peabody (159), American Polymer Standards Corporation, Mentor, Ohio 44060

Werner Praznik (459), IFA Tulln, Analytikzentrum, A-3430 Tulln, Austria, and Institut für Chemie, Universität für Bodenkultur, A-1190 Vienna, Austria

Daniel A. Saucy (539), Analytical Research Department, Rohm and Haas Company, Springhouse, Pennsylvania 19477

Larry Senak (499), International Specialty Products, Wayne, New Jersey 07470

Hiroshi Suzuki (171), Shodex Group, Showa Denko K.K. 13-9, Minato-ku, Tokyo 105-8518, Japan

Iwao Teraoka (611), Department of Chemical Engineering, Chemistry, and Materials Science, Polytechnic University, Brooklyn, New York 11201

Alan Williams (27), Amersham Pharmacia Biotech Inc., Piscataway, New Jersey 08855

Chi-san Wu (499), International Specialty Products, Wayne, New Jersey 07470

PREFACE

Size exclusion chromatography (SEC) [also known as gel-permeation chromatography (GPC) or gel-filtration chromatography (GFC)] is the most practical method of characterizing the molecular weight (MW) and molecular weight distribution (MWD) of polymers. The column is the heart of the SEC or GFC processes because it determines the separation of the polymer into its specific molecular weight fractions. Many excellent books on SEC or GFC have been published during the past 30 years. Most of them, however, deal more with subjects other than columns, such as absolute molecular weight detectors, calibrations, band broadening, and chromatographic theories. No book has been devoted solely to the technology, application, characterization, evaluation, quality control, and maintenance of commercially available columns for SEC or GFC.

Literature on commercial columns from manufacturers and distributors is scattered and sparse. Catalogs from most manufacturers contain only basic information, such as the plate count, exclusion limit, and a calibration curve, which are necessary but are often not sufficient to help users make intelligent decisions as to which is the best column for a specific application. To further complicate the problem, many column manufacturers use nonexclusive distributors who cannot provide expert advice on the columns they sell. It is very difficult and time consuming for practitioners to gather the needed information on columns. At the time of this writing, a column typically costs more than $1000. Thus, it is very easy to make an expensive purchase, only to find out later that the column does not provide the expected result.

Particle technology and column technology have advanced rapidly in the past decade. Template polymerization has complemented suspension polymerization in obtaining monodisperse particles without laborious refining or classification. The particle size of packing materials and overall column dimensions have both decreased. Due to its much improved capability, the linear or mixed-bed column (a single column that is packed with particles with two or more pore sizes) has become more popular than a column bank (a series of columns with different pore sizes). Linear columns designed for different molecular weight ranges are also available. Columns for high-molecular-weight polymers designed to prevent shear degradation have been offered by several manufacturers. There are many more choices available now for SEC or GFC columns in solvents such as methanol, HFIP, DMSO, DMAC, water, and mixed mobile phases. The temperature stability and solvent compatibility of columns have improved significantly. A column has also been introduced specifically to separate the system peaks from the low-molecular-weight tail of the polymer peak. This type of column is needed due to the regulatory requirement that manufacturers provide information on the percentage of materials below 1000 or 500 Da in a new product. However, even with these advancements, there are relatively few review articles on columns in the literature.

This book is intended to fill the information void on columns for SEC or GFC. Virtually every major SEC or GFC column manufacturer worldwide has specially prepared a chapter for it. Many manufacturers reveal unprecedented details on the technology, application, maintenance, and quality control of their columns. This will enable readers to develop a better understanding of the different approaches used by manufacturers to design, produce, and perform quality control of columns for use in various applications. For example, four different types of columns have been introduced for SEC in HFIP: conventional PS/DVB packing, modified silica gels, monodisperse packing obtained by template polymerization, and DVB packing. Extensive coverage of SEC or GFC columns for all polymers (synthetic, bio-, or natural) is included. Column manufacturers have provided several hundred excellent examples of how to separate synthetic, bio-, or natural polymers. Another important feature of this book is that expert column users share their valuable experiences with characterization, evaluation, maintenance, selection, and application of columns.

Column characteristics can change with time, and sometimes a little maintenance can restore performance. Therefore, we include very useful and systematic maintenance procedures for columns. Many experienced column users may have noticed the difference in interactive properties of commercial columns with polymers. A systematic way to characterize the differences in interactive properties among commercial columns is also discussed.

There has been very little discussion on quality control of columns in the literature from either the manufacturer's or the user's perspective. SEC or GFC has become an ever important quality control tool for polymer manufacturers and users. The establishment of a quality control program for columns is critical for obtaining consistent MW and MWD on a long-term basis. Readers will be able to find useful information not only on the quality control of columns, but also on how to set up a successful quality control program for columns between manufacturers and users.

SEC, GPC, and GFC, which separate polymers entropically, are used interchangeably in this book by the contributors. However, the contributors also address columns for two other techniques: hydrodynamic chromatography and high osmotic pressure chromatography, which are based on different separation mechanisms. These techniques are included because they are closely related to SEC, GPC, and GFC in the separation of polymers by size or MW, and they also can supplement SEC, GPC, or GFC.

Despite advancements made in the past decade, the columns currently available for SEC and GFC are still not ideal. This book is intended to promote proactive communication between column users and column manufacturers, which will stimulate the advancement in column technology. We hope that the usefulness of the information will encourage column manufacturers and column users to soon begin working on the second edition.

The preparation of this book was an experiment by itself in which both column manufacturers and experienced column users have participated. The credit belongs to all of them for their enthusiastic support throughout all stages of production and also for their sacrifices of personal time from their busy schedules to prepare the respective chapters. They are true professionals who have significantly helped in many other ways, such as reviewing manuscripts and providing advice.

Dr. David Packer, Senior Acquisitions Editor at Academic Press, is credited for his vision to support this book since its inception and for providing valuable guidance during the different stages of production. The editor also thanks Dr. Edward G. Malawer, Director of the Analytical Department, and the management team of International Specialty Products (ISP) for permission to take on the challenge of editing this book.

Chi-San Wu

I

CONTRIBUTIONS FROM SIZE EXCLUSION CHROMATOGRAPHY AND GEL-FILTRATION CHROMATOGRAPHY COLUMN MANUFACTURERS

1 PREPARATION OF BEADED ORGANIC POLYMERS AND THEIR APPLICATIONS IN SIZE EXCLUSION CHROMATOGRAPHY

MICHAEL J. LU

BioChrom Labs, Inc., Terre Haute, Indiana 47808

I. INTRODUCTION

Beaded polymeric supports are widely used as packing materials in various chromatography techniques such as size exclusion, ion-exchange, reversed-phase, hydrophobic interaction, and affinity chromatography. Microspherical polymeric supports are also employed in a number of related applications, including diagnosis, water treatment, extractions of precious metals, solid-phase peptide synthesis, oligonucleotide synthesis, catalysis, and other chemical applications. The demands on packing material for high-performance liquid chromatography are rigidity, chemical stability toward solvents and pH changes, high load capacities, no nonspecific interactions, and no hindrance of solute diffusion. Compared to silica particles, which are the most commonly used packing materials in chromatography, the advantages of the synthetic polymeric supports are a much increased chemical stability and sophisticated surface modification.

Beaded polymeric supports are produced by numerous manufacturing processes and technologies, the details of which are not frequently reported in the literature. The systematic coverage of the preparation of these materials is necessary. The manufacturing parameters are controlled to obtain the beaded products with specific physical and chemical properties such as particle-size distribution, pore-size distribution, and particle morphologies. The products obtained may be hydrophilic, hydrophobic, or amphiphilic, gelatinous, or rigid. In addition, they may have high or low porosity and surface area, depending on the chemicals, degree of cross-linkage, stabilizer used, porogens, mechanic

Column Handbook for Size Exclusion Chromatography

design and forces, and reaction conditions. Following production of the polymer, the microbeads may be altered by surface modification to change their hydrophobicity or may be derivatized and activated by chemical modification to introduce specific functional groups.

This chapter is intended to provide a general picture and a special design of the manufacturing aspects of beaded polymeric supports and also discusses various polymerization and particle formation processes employed in the manufacture of synthetic polymeric microbeads. This chapter covers the preparation of major synthetic organic polymeric packing materials, including polystyrene–divinylbenzene, polystyrene, polyvinyl alcohol, polymethacrylates, and composites. Chemical structures and the control of physicochemical criteria such as particle size, pore size, porosity volume, porosity distribution, surface area, and swelling behavior are discussed as well. This chapter provides a useful guided source of information for those interested in preparing synthetic, polymeric microbeads. For this purpose, detailed coverage of beaded organic polymers and synthetic methodologies is provided.

II. BASIC FEATURES OF CONVENTIONAL SUSPENSION POLYMERIZATION

Beaded polymeric support, whether polystyrene–divinylbenzene, polymethacrylate, or polyvinyl alcohol, is conventionally produced by different variations of a two-phase suspension polymerization process, in which liquid microdroplets are converted to the corresponding solid microbeads (1).

Suspension polymerization of water-insoluble monomers (e.g., styrene and divinylbenzene) involves the formation of an oil droplet suspension of the monomer in water with direct conversions of individual monomer droplets into the corresponding polymer beads. Preparation of beaded polymers from water-soluble monomers (e.g., acrylamide) is similar, except that an aqueous solution of monomers is dispersed in oil to form a water-in-oil (w/o) droplet suspension. Subsequent polymerization of the monomer droplets produces the corresponding swollen hydrophilic polyacrylamide beads. These processes are often referred to as inverse suspension polymerization.

Among the various suspension systems mentioned, the details of oil-in-water (o/w) suspension polymerizations are fully known. The criteria of droplet formation, droplet stabilization, and droplet hardening, as will be discussed for the o/w suspension system, can apply equally to the preparation of beaded polymer particles in w/o systems.

A. Droplet Formation

The most important feature of o/w suspension polymerization is the formation of an oil droplet suspension of the monomer in the water and the maintenance of the individual droplets throughout the polymerization process. Droplet formation in an oil-in-water mixture is accomplished and controlled by two major factors: mechanical stirring and the volume ratio of the monomer phase to water. The stirring speed is a key factor in controlling the size of oil droplets and the final size of the polymers. The stirring speed usually needs to be over

2000 rpm to produce droplets of oil less than 10 μm when a stabilizer, such as a surfactant or a seed particle, is not used. For most cases, the volume ratio of the monomer phase to water must be between 1 : 10 and 1 : 2 to produce stable oil droplets.

In the suspension system, the suspended droplets collide with each other, coalesce into large oil droplets, and might redivide into smaller ones. However, redivision of the coalesced larger monomer droplets becomes gradually more difficult as a result of polymerization and increased viscosity. This means that at a certain stage of polymerization (sticky period), redivision of the partially polymerized droplets becomes almost impossible, and continued droplet coalescence leads to coagulation of the entire bulk of the monomer phase. The sticky period is usually observed between 25 and 75% conversion, depending on the composition of the monomer mixture. Individuality of partially polymerized droplets can be maintained by one means or another. Progress of the polymerization reaction leads to gradual hardening of the droplets. However, at the end of the sticky period, the hardened droplets will no longer coalesce in the event of a collision.

B. Droplet Stabilization

Coagulation during the sticky period can be prevented by reducing the surface tension of droplets and minimizing the force with which they collide. The latter is controlled by proper reactor design and stirring force. The force with which the droplets collide can be reduced by decreasing the stirring speed, but the stirring speed must be kept high enough to prevent the droplets from aggregating and separating in the sticky period. Reducing surface tension is achieved by using a small amount of a suitable droplet stabilizer as a coagulation inhibitor. In o/w suspension polymerization, a highly effective droplet stabilizer is a small amount of water-insoluble inorganic salt or organic polymer that is insoluble in the monomer droplets and has relatively low solubility in the suspension medium to the suspension system. Organic polymers are usually preferred over insoluble inorganic salts because they are removed more easily from the surface of the beads. Examples of inorganic droplet stabilizers used for suspension polymerization include calcium sulfate, calcium phosphate, and benzonite. Among the most commonly used organic stabilizers for suspension systems are polyvinyl alcohols (75–98% hydrolyzed) and polyvinylpyrrolidone. A wide range of other water-soluble polymers such as methylcellulose, gelatin, and other natural gums are also used. In general, a relatively low concentration of the stabilizer (0.15 to 1%) is sufficient to maintain a stable suspension system under constant stirring conditions. The droplet stabilization is a surface phenomenon. Therefore, the minimum stabilization concentration required for a full monolayer coverage of droplets increases with decreasing particle size.

C. Particle Size and Pore Size

Microspherical polymer beads are widely used as packing materials for chromatography and a variety of other applications. Size exclusion chromatography is based on pore size and pore-size distribution of microbeads to separate molecules of different sizes. Particle size and pore size also play a very important role in ion-exchange and affinity chromatography and in the polymeric supports for catalyst. In all cases, the particle size, pore size, surface area of microbeads, swelling behavior, and chemical structure of the polymer's backbone strongly influence the overall performance of the polymer supports in the described applications (2).

Beaded polymeric supports are produced by a two-phase suspension polymerization in which microdrops of a monomer solution are directly converted to the corresponding microbeads. The size of a microdroplet is usually determined by a number of interrelated manufacturing parameters, which include the reactor design, the rate of stirring, the ratio of the monomer phase to water, the viscosity of both phases, and the type and concentration of the droplet stabilizer.

The size distribution of the polymer beads obtained with conventional two-phase suspension polymerization depends mainly on the configuration of the reactor, the shape of the stirrer, and the stirring speed. It is possible to obtain relatively uniform beads in which the deiviation from the average size is not greater than about 100%. Among various factors influencing particle size, stirring speed provides a relatively convenient means of particle-size control for most purposes.

Two-phase suspension systems produce beaded products with broader particle-size distribution (e.g., 1–50 μm). The microspherical particles usually need to be classified repeatedly to reduce the particle-size distribution in order to improve the resolution and efficiency in the separation for use in chromatography. The actual classification process depends on the size range involved, the nature of the beaded product, and its intended applications. Relatively large (>50 μm) and mechanically stable particles can be sieved easily in the dry state, whereas small particles are processed more conveniently in the wet state. For very fine particles (<20 μm), classification is accomplished by wet sedimentation, countflow setting, countflow centrifugation, or air classification.

The pore size, the pore-size distribution, and the surface area of organic polymeric supports can be controlled easily during production by precipitation processes that take place during the conversion of liquid microdroplets to solid microbeads. For example, polystyrene beads produced without cross-linked agents or diluent are nonporous or contain very small pores. However, by using high divinylbenzene (DVB) concentrations and monomer diluents, polymer beads with wide porosities and pore sizes can be produced, depending on the proportion of DVB and monomer diluent. Control of porosity by means of monomer diluent has been extensively studied for polystyrene (3–6) and polymethacrylate (7–10).

The porosity of polymer beads is controlled by the ratio of diluents (porogen) to monomers in the organic phase. The increase in the ratio of diluents to monomer in the monomer mixture increases the porosity of polymer beads. The pore size can be manipulated by adjusting the ratio of nonsolvating and solvating diluents in the monomer mixture. The increase in the ratio of nonsolvating diluent (precipitant) in the monomer mixture increases the pore sizes and vice versa.

Macroporous polymer beads are usually produced by using inert linear organic polymers such as polystyrene, alkylcellulose, polyalkylvinyl ether, and inorganic polymers (such as silica microbeads) as the porogen in the monomer mixture to control macropores. After polymerization, these large porogens are removed by solvent extraction or hydrolysis with a strong alkali to form macropores in the polymeric beads. The pore-size distribution is controlled by these processes and polymerization methods. Conventional suspension polymerization has a tendency to produce narrow pore-size distribution in the polymeric particles. However, the activated swelling and polymerization method produces polymeric beads with wide pore-size distribution.

Porosity and surface area are routinely measured by nitrogen absorption–desorption, mercury intrusion, and low-angle X ray. The electron microscope (EM) provides direct visual evidence of pore size and pore-size distribution. Thus, a combination of EM and conventional methods of pore-size measurement should provide reliable information on the pore structure of polymers.

III. PREPARATION OF ORGANIC POLYMERIC BEADS BY CONVENTIONAL SUSPENSION POLYMERIZATION

Synthetic organic polymers, which are used as polymeric supports for chromatography, as catalysts, as solid-phase supports for peptide and oligonucleotide synthesis, and for diagnosis, are based mainly on polystyrene, polystyrene–divinylbenzene, polyacrylamide, polymethacrylates, and polyvinyl alcohols. A conventional suspension of polymerization is usually used to produce these organic polymeric supports, especially in large-scale industrial production.

A. Polystyrene/Polystyrene–Divinylbenzene

Styrene-based polymer supports are produced by o/w suspension polymerization of styrene and divinylbenzene. Suspension polymerization is usually carried out by using a monomer-soluble initiator such as benzoperoxide (BPO) or 2,2-azo-bis-isobutylnitrile (AIBN) at a temperature of 55–85°C (19). A relatively high initiator concentration of 1–5% (w/w) based on the monomer is used. The time required for complete monomer conversion must be determined by preliminary experiments and is usually between 5 and 20 h, depending on the initiator concentration, the temperature, and the exact composition of the monomer mixture (11–18).

In addition to monomers and the initiator, an inert liquid (diluent) must be added to the monomer phase to influence the pore structure and swelling behavior of the beaded resin. The monomer diluent is usually a hydrophobic liquid such as toluene, heptane, or pentanol. It is noteworthy that the nature and the percentage of the monomer diluent also influence the rate of polymerization. This may be mainly a concentration or precipitation effect, depending on whether the diluent is a solvent or precipitant for the polymer. For example, when the diluent is a good solvent such as toluene to polystyrene, the polymerizations proceed at a correspondingly slow rate, whereas with a nonsolvent such as pentanol to polystyrene the opposite is true.

Following the completion of the polymerization process, the beaded polymer is recovered from the suspension mixture and freed from the stabilizer, diluents, and traces of monomers and initiators. For laboratory and small-scale preparation, repeated washings with water, methanol, or acetone are appropriate. Complete removal of the monomer diluent, solvents, and initiator, especially from macroporous resin, may require a long equilibration time with warm methanol or acetone. In industry, this is usually accomplished by stream stripping.

Macroporous resins are prepared by copolymerizing an aqueous suspension of styrene and divinylbenzene in the presence of a saturated aliphatic carboxylic acid as the porogenic agent, which is relatively water insoluble ($<2\%$) and has at least five carbon atoms (20). The use of a saturated aliphatic carboxylic acid as the porogenic agent has numerous advantages over the use of porogenic agents such as alcohol, hydrocarbon, and ether. The saturated aliphatic acids are relatively nonvolatile and may be removed from resin merely by treatment with a base solution, which renders the acid soluble in the water of dispersion.

Producing a polystyrene (PS)–DVB copolymer of increasing porosity has been accomplished by dissolving 50–80% styrene, 10–50% divinylbenzene, and 30–70% of an inert organic liquid. Toluene is a solvent for the monomer but is a nonsolvent for the polymerized polymer. The monomer solution is then incorporated into water to form a dispersion of oil droplets followed by the polymerization of the suspended oil droplets from the aqueous medium into the polymer (21).

A macroporous polystyrene–divinylbenzene copolymer is produced by a suspension polymerization of a mixture of monomers in the presence of water as a precipitant. This is substantially immiscible with the monomer mixture but is solubilized with a monomer mixture by micelle-forming mechanisms in the presence of the surfactant sodium bis(2-ethylhexylsulfosuccinate) (22). The porosity of percentage void volume of macroporous resin particles is related to percentage weight of the composite (50% precipitant, 50% solvent) in the monomer mixture.

A porous copolymer containing a cyano group is obtained by the polymerization of acrylnitrile and divinylbenzene in the presence of an organic solvent, such as toluene, as a swelling agent for the resulting polymers and with a linear polystyrene as a porogen. The polymer is large in surface area, high in porosity, and shows high adsorption for polar substances (23).

A porous polystyrene–divinylbenzene gel is produced by suspension polymerization in an aqueous system with incorporation of more than 5 mol% initiator to a total amount of styrene and divinylbenzene with an inert organic solvent as diluent and porogen (24).

A macroporous polystyrene–divinylbenzene copolymer, produced by copolymerizing a mixture of styrene and divinylbenzene, is dissolved in an organic liquid such as *t*-amyl alcohol or isooctane, which is a solvent for monomers. This solvent is unable to substantially swell the resulting copolymer. Macroporous cation-exchange beads are also produced from these macroporous copolymers (25,26).

Macroporous divinylbenzene gels having a high concentration of chemically and accessibly free vinyl groups can be produced. The gels are prepared by polymerization of divinylbenzene in a mixture of nonpolar solvents such as toluene and polar solvents such as tetrahydrofuran (THF), in the presence of the polymerization initiator butyllithium (27).

B. Polyacrylamide–Polymethacrylate

Beaded acrylamide resins (28) are generally produced by w/o inverse-suspension polymerization. This involves the dispersion of an aqueous solution of the monomer and an initiator (e.g., ammonium peroxodisulfates) with a droplet stabilizer such as carboxymethylcellulose or cellulose acetate butyrate in an immiscible liquid (the oil phase), such as 1,2-dichloroethane, toluene, or a liquid paraffin. A polymerization catalyst, usually tetramethylethylenediamine, may also be added to the monomer mixture. The polymerization of beaded acrylamide resin is carried out at relatively low temperatures (20–50°C), and the polymerization is complete within a relatively short period (1–5 hr). The polymerization of most acrylamides proceeds at a substantially faster rate than that of styrene in o/w suspension polymerization. The problem with droplet coagulation during the synthesis of beaded polyacrylamide by w/o suspension polymerization is usually less critical than that with a styrene-based resin.

Beaded methacrylate polymers, poly(hydroxyethylmethacrylate), Spheron, Separon (29), and poly(glycidylmethacrylate), Eupergin (30,31), are studied extensively at the Czechoslovak Academy of Macromolecular Sciences. An addition to this type of support is poly(oxyethylene–dimethacrylate) (32). Heitz *et al.* (33) described the preparation of beaded poly(methylacrylates) cross-linked with ethanedimethacrylates.

A polymethacrylate copolymer has been produced by suspension polymerization of ethylene glycol dimethacrylate and glycidyl methacrylate. The copolymer is then subjected to chemical modification with fluorine compounds such as 2,2,2-trifluoroethanol and 2,2,3,3,4,4,4-heptafluorobutanol and glycidol to introduce the fluorine-containing group and hydrophilic-containing group. The copolymer has been used in gel-permeation chromatography and has a porosity in the range of 0.5 to 3.0 ml/g and a particle size in the range of 1 to 200 μm (34).

A polymethacrylate copolymer is modified by successive reaction with epichlorohydrin, *m*-aminophenylboric acid, and nitric acid to introduce a 1-amino-(2′-nitrophenyl-5′-boric acid)-2-hydroxyl-3-*o*-propyl group. The modified polymethacrylates are used as chromatographic support materials and can be used to analyze biological materials without prior deproteinization (35).

C. Polyvinyl Alcohol

Heitz *et al.* (33) also described the preparation of polyvinyl acetate cross-linked with butanediol divinyl ether. The polymer is the base of the Merckogel series of size exclusion chromatography packings, and its hydrolyzed derivative, polyvinyl alcohol, is marketed as Fractogel and Toyopearls.

A process for the preparation of porous polyvinyl alcohol gels in three steps is: (1) suspension polymerization of vinyl acetate with diethylene glycol dimethacrylate in the presence of a diluent as porogen, (2) saponifying of the resulting porous polyvinyl acetate gel with an alkali, and then (3) subjecting of the resulting porous polyvinyl alcohol to post-cross-linking with epichlorohydrin (36).

A weak cation-exchange resin is obtained by reaction of glyoxylic acid and a cross-linked polyvinyl alcohol. The polyvinyl alcohol is cross-linked with glutaraldehyde in the presence of hydrochloric acid. The cation-exchange resin has an exchange capacity of 3 meq/g or greater and a swelling volume of 10 ml/g or smaller (37–38).

IV. SPECIAL POLYMERIC SUPPORTS

Some special polymeric supports have been designed and developed by several different groups. These supports are not produced by conventional suspension polymerization or activated swelling, and polymerization methods require predetermined shape, size, pore structure, and composition. In general, these supports are produced by filling monomers or monomer mixtures into special containers, cartridges, or pores of inorganic and organic porous materials, and then polymerizing them to convert oil droplets into solid particles.

A. Templated Polymeric Supports

New templated polymer support materials have been developed for use as reversed-phase packing materials. Pore size and particle size have not usually been precisely controlled by conventional suspension polymerization. A templated polymerization is used to obtain controllable pore size and particle-size distribution. In this technique, hydrophilic monomers and divinylbenzene are formulated and filled into pores in templated silica material, at room temperature. After polymerization, the templated silica material is removed by base hydrolysis. The surface of the polymer may be modified in various ways to obtain the desired functionality. The particles are useful in chromatography, adsorption, and ion exchange and as polymeric supports of catalysts (39,40).

Alternative processes in templated polymerization have been developed. The pores of inorganic silica particles are filled with a blowing agent, which is a substance capable of being decomposed under appropriate conditions. As the pores of the inorganic particles are filled during the polymerization reaction, the organic polymer that forms the beads surrounds, rather than penetrates, the inorganic particles. After the polymerization is complete, organic, roughly spherical beads, containing a multiplicity of the inorganic porogens within their volume, are formed. The inorganic particles are insulated from each other by a small amount of organic polymer that surrounds the organic particles. The blowing agent is then activated, causing the formation of channels between inorganic particles, facilitating the dissolution of these inorganic particles in an appropriate solution, and creating a network of channels through the organic beads (41).

B. Pore–Matrix Composite Supports

A composite separating agent has been produced by filling soft gels such as dextran and a agarose, which have poor mechanical strength but effectively separate biomolecules, into pores of an organic polymer substrate such as a polymethacrylate, which has a high mechanic strength. These are then cross-linked with a cross-linking agent, epichlorohydrin, to stabilize the filling of soft gel inside the pores of polymeric supports (42).

The composite separating agents have two major advantages. They show the mechanical strength of rigid polymer supports and can be operated under a higher pressure and flow rate during the separation. They also have the high-loading capacity of a soft gel (4). In fact, pore–matrix composite supports can be developed by using highly rigid supports such as polystyrene–divinylbenzene beads or ceramic, in which the surface of the materials is modified with hydrophilic surface modification agents.

C. Dual Functionality of Polymeric Supports

Pore-size-specific functionality has been developed by Frechetz and colleagues (43–46). The concept of pore-size-specific modification of porous materials relies on the use of a catalyst with defined molecular volumes and able to perform a chemical modification only in those pores large enough to allow this access. The same principles apply as with size exclusion chromatography. It is assumed that internal surfaces of porous beads that have a relatively broad pore-size distribution are lined with reactive functional groups. When a macromolecular catalyst is added to such beads in solution, it will penetrate only the large pores of the beads. Any reactive group that comes into contact with the catalyst is modified into another functionality (first functional group). In contrast, the hydrodynamic volume of the soluble polymer does not allow it to penetrate into small pores, which therefore retain their functional groups. In a subsequent reaction step, a second reagent or catalyst with a smaller hydrodynamic volume than the previous one may be used to penetrate the smaller pores and to modify the reactive group into the second functionality (second functional group).

Separation media, with bimodal chemistry, are generally designed for the complete separation of complex samples, such as blood plasma serum, that typically contain molecules differing in properties such as size, charge, and polarity. The major principle of bifunctional separation relies on the pore size and functional difference in the media. For example, a polymer bead with hydrophilic large pores and hydrophobic small pores will not interact with and retain large molecules such as proteins, but will interact with and retain small molecules such as drugs and metabolites.

In a combination of the hydrophobic interaction and reverse-phase chromatography (HIC–RPC) approach, the large pores of the beads provide a low density of phenyl functional groups to the hydrophilic surface, whereas a higher surface density of hydrophobic phenyl group is introduced into the small pores. The modification process involves four steps: (1) reaction of groups in all pores with phenol under mild conditions, (2) reactivation of the newly created hydroxy groups with epichlorohydrin, (3) pore-size-specific hydrolysis of the epoxide groups in the large pores, and (4) additional hydrophobization of the small pores with phenol under strong conditions (44). See Fig. 1.1.

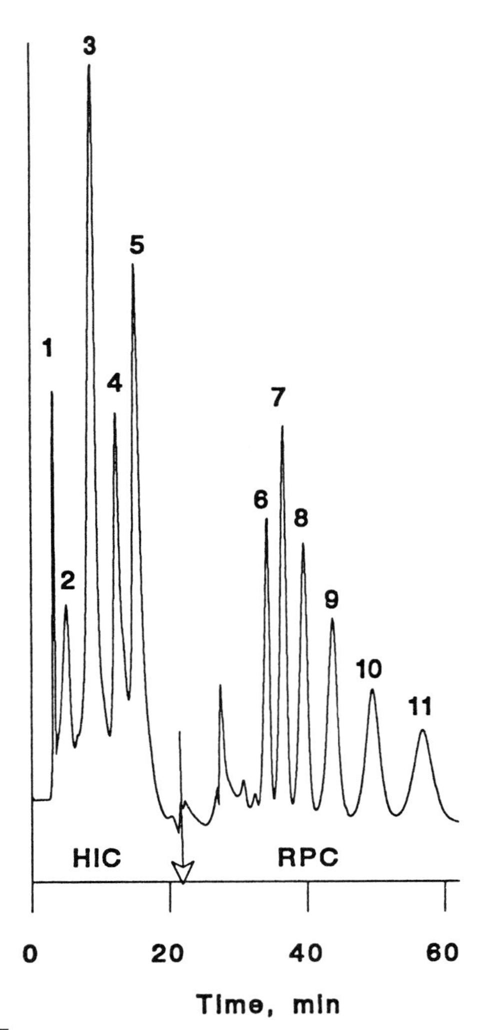

FIGURE 1.1 Hydrophobic interaction and reversed-phase chromatography (HIC–RPC). Two-dimensional separation of proteins and alkylbenzenes in consecutive HIC and RPC modes. Column: 100 × 8 mm i.d. HIC: mobile phase, gradient decreasing from 1.7 to 0 mol/liter ammonium sulfate in 0.02 mol/liter phosphate buffer solution (pH 7) in 15 min. RPC: mobile phase, 0.02 mol/liter phosphate buffer solution (pH 7); acetonitrile (65:35 vol/vol):flow rate, 1 ml/min; UV detection 254 nm. Peaks: (1) cytochrome c, (2) ribonuclease A, (3) conalbumin, (4) lysozyme, (5) soybean trypsin inhibitor, (6) benzene, (7) toluene, (8) ethylbenzene, (9) propylbenzene, (10) butylbenzene, and (11) amylbenzene. [Reprinted from J. M. J. Frechet (1996). Pore-size specific modification as an approach to a separation media for single-column, two-dimensional HPLC, *Am. Lab.* **28,** 18, p. 31. Copyright 1996 by International Scientific Communications, Inc., Shelton, CT.]

The separation medium for the combination of ion-exchange (IEC)/RPC modes can be obtained in a process consisting of the following four steps: (1) Hydrolysis of a copolymer of glycidyl methacrylate and ethylene dimethacrylate by poly(styrenesulfonic acid) (PSSA) introduces vinyl hydroxyl groups into pores large enough to accommodate PSSA (2). The remaining epoxide groups react with octadecylamine in dioxane to yield beads with C-18 functionalities. These beads can be used immediately for the separation of mixtures using a combination of size exclusion chromatography (SEC) and RPC and for direct-injection chromatography in a clinical laboratory. (3) The hydroxyl groups are activated again by the reaction with epichlorohydrin in the presence of a base catalyst to form the active epoxide functional groups. (4) The aminolysis of the activated beads with diethylamine yields beads containing ion exchanges or diethylamino (DEAE) groups. The chromatogram shown in Fig. 1.2 illustrates the separation obtained with a column packed with dual-mode

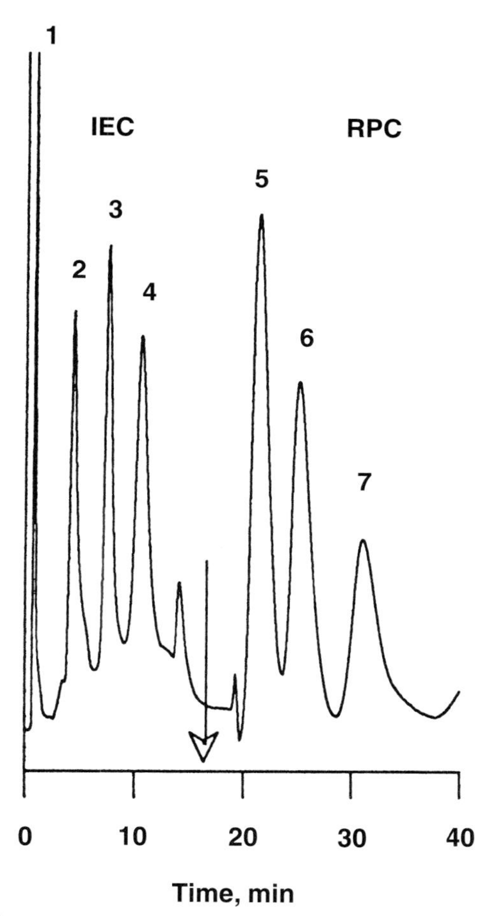

FIGURE 1.2 Ion-exchange and reversed-phase chromatography (IEC–RPC). Separation of proteins and alkylbenzenes in consecutive IEC and RPC modes in a column packed with beads containing segregated chemistries. Column: 50 × 8 mm i.d.; mobile phase, gradient from 0 to 0.5 mol/liter NaCl in 0.01 mol/liter Tris–HCl buffer (pH 7.6) within 30 min followed by buffer acetonitrile (80:20 vol/vol); flow rate, 1 ml/min; UV detection, 254 nm. Peaks: (1) myoglobin, (2) cytochrome c, (3) bovine serum albumin, (4) soybean trypsin inhibitor, (5) toluene, (6) ethylbenzene, and (7) propylbenzene. [Reprinted from J. M. J. Frechet (1996). Pore-size specific modification as an approach to a separation media for single-column, two-dimensional HPLC, *Am. Lab.* **28,** 18, p. 31. Copyright 1996 by International Scientific Communications, Inc., Shelton, CT.]

IEC–RPC beads and used first in the IEC mode for the separation of protein, followed by the separation of small hydrophobic molecules in the RPC mode.

D. Continuous Rods of Macroporous Polymeric Supports

Continuous porous polymer rods have been prepared by an *in situ* polymerization within the confines of a chromatographic column. The column is filled with glycidyl methacrylate and ethylene dimethacrylate monomer mixtures, cyclohexanol and dodecanol diluents, and AIBN initiator. They are then purged with nitrogen, stopped, and closed with a silicon rubber septum. The polymerization is allowed to proceed for 6 hr at 70°C with the column acting as a mold (47).

A macroporous polystyrene–divinylbenzene rod has been prepared by a free-radical polymerization of a mixture containing monomers, initiator, and porogenic solvents, in the confines of a chromatographic column, and is then used in the very fast reversed-phase HPLC of proteins (48). Characterization of the pore structures of continuous rods by mercury intrusion porosimetry reveals a large volume of pores with a diameter of about 1 μm to pores below 100 nm. A scanning electron micrograph of a cross section of the rod reveals clusters of globules separated by large pores. The characteristic of continuous rods of macroporous polymer is the likely absence of an interparticle void volume in a chromatography column, in comparison with a conventional column. This characteristic may result in several advantages. It contributes to better column space utilization compared to a packed column of the same size. The column contains more separation medium in which the actual separation takes place. The absence of empty space in the column forces all of the mobile phase to flow through the separation medium rather than around it.

The effects of the polymerization time, the porogenic solvents, and the polymerization temperature on the pore size and pore-size distribution of the molded rods have been more thoroughly investigated (49). A continuous rod of porous polyglycidyl methacrylate–*co*-ethylene dimethacrylate has been prepared by a free-radical polymerization within the confines of a chromatographic column. The epoxide groups of rods have been modified by a reaction with diethylamine. This reaction step introduces the ionizable functionalities required for the ion-exchange chromatography mode.

E. Porous Perfusion Particles

A chromatography matrix composed of first and second interconnected sets of pores and a high surface area for solute interaction in fluid communication with the number of second sets of pores has been developed (51,52). The first and second sets of pores are embodied as the interstices among particles and through pores within the particles. The pores are dimensioned so that, at achievable high fluid flow rates, convective flow occurs majorly in the first pore set, and the convective flow rate exceeds the rate of solute diffusion in the second pore set. This approach couples convective and diffusive mass transport to and from the active surface and permits an increase in fluid velocity without the normally expected band spreading. Disclosed in the patent are chromatography methods and matrix geometries that permit high-flow-rate separations of mixtures of solutes, particularly biological materials.

V. BASIC FEATURES OF MONOSIZED POLYMERIC PARTICLES PREPARED BY ACTIVATED SWELLING AND POLYMERIZATION

Starting with monosized seed particles, the activated swelling and polymerization method can be used to prepare monosized particles in the range of 1 to 100 μm with a standard deviation of particle diameter of less than 1% when the method is properly performed, an equal swelling of each particle is secured, and the monodispersity throughout the swelling and polymerization processes is maintained. The methods are applicable to the production of monosized polymeric particles from a number of different monomer and monomer mixtures. An important feature of this process is that all the necessary monomer, initiator, and solvent porogens are absorbed in the highly swollen seed particles before polymerization is started. The process is especially suitable for the preparation of monosized cross-linked polymeric-particles and for the production of monosized macroporous polymeric particles (53).

A. Monosized Seed Particles

The method of activated swelling requires the use of small, highly monosized particles as seed. Highly monosized particles of up to 0.5 μm are prepared by emulsion polymerization with mixed ionic and nonionic emulsifiers (54). The general aspect of the preparation of monodispersed particles in emulsion polymerization is the same as that for other monodispersed colloids. The essential condition is the supply of a fixed number of nuclei that subsequently grow to much larger without any nucleation occurring during the growing period. In the case of emulsion polymerization, further growth may result in a self-sharpening of the particle size. This fact has led to the development of seed techniques for the preparation of monodispersed particles (55,56). By this method, monosized particles with diameters up to 2 μm may be formed in limited yield from small rather polydisperse particles of about 0.1 μm by a series of build-up steps.

During performance of these continuous steps, the amount of the emulsifier must be large enough to prevent the coagulation of particles during polymerization and small enough to prevent the formation of new particles. Methods for the production of monosized particles without the application of emulsifiers have also been developed (57). The emulsifier-free emulsion polymerizations have been reported to produce particles over 1 μm, when the effects of initiator concentration, ion strength of polymerization medium, polymerization temperature, and monomer/water ratio are optimized. In the case of methyl methacrylate and glycidyl methacrylate, only much smaller particles, not exceeding about 0.7 μm in size, can be obtained (58).

Monosized polystyrene particles in the size range of 2–10 μm have been obtained by dispersion polymerization of styrene in polar solvents such as ethyl alcohol or mixtures of alcohol with water in the presence of a suitable steric stabilizer (59–62). Dispersion polymerization may be looked upon as a special type of precipitation polymerization and was originally meant to be an alternative to emulsion polymerization. The components of a dispersion polymerization include monomers, initiator, steric stabilizer, and the dispersion medium in which all the components are soluble and in one phase. The polymer formed is insoluble in the dispersion medium and is stabilized against coagulation by a steric stabilizer. Most of the work published so far has dealt with styrene, acrylates, and the mixture of these. The reasons that monodispersed beads may be obtained in dispersion polymerization are poorly understood. The experimental conditions that may be varied include temperature; the solvency of the dispersion medium for the polymer initially formed; and the concentration and type of monomers, initiators, and stabilizers. These parameters may have significant effects on the particle size, the molecular weight, and the kinetics of the polymerization. Interrelationships of these variables appear to be quite complicated. A general observation is that any change in the system that enhances the solubility of the polymer initially formed increases the polymer size. This may be achieved by changing to a better solvent for the polymer formed or by increasing the monomer concentration. An increase in the initiator concentration results in the formation of molecules with shorter chain lengths that are more soluble and, consequently, in an increase in the particle size. An increase in temperature during polymerization is expected to favor the formation of large particles for two reasons. The ratio of radical formation increases, which in turn leads to the formation of a polymer with a shorter chain length. At the same time, the solubility of the polymer formed is enhanced due to the temperature effect.

Only particles of linear or very slightly cross-linked (<0.6%) polymers may be produced by dispersion polymerization. Obviously, dispersion polymerization may be used for the production of monosized seed particles, which, after transfer to aqueous conditions, are used for the production of different cross-linked and macroporous particles by the activated swelling and polymerization method.

B. Activated Seed Particles

Two different methods have been used for the incorporation of the activating ogliomer (or monomer) in the seed particles. The first method involves the application of a small organic chemical, such as chloroundecane or dibutyl phthalate, which is incorporated into the particles in the first swelling step. In the second method, an ogliomer compound is formed by polymerization of monomers that are absorbed inside the seed particles.

1. The Two-Step Swelling Method

This method involves swelling of the seed particles (phase a) in a first step with a Y compound (phase b). Because the Y compound is highly water insoluble, special precautions must be taken to facilitate the transfer of the Y compound from phase b to phase a. The use of small-radius oil droplets of liquid Y compound (i.e., applying finely dispersed droplets of Y compound) is essential for the swelling of the particles in the first step. The molar volume of the Y compound is also large in comparison with the molar volume of seed particles in the first step. However, the rate of swelling in the first step can be further increased by addition of a water-soluble organic solvent such as acetone or a low-molecular-weight alcohol, which is also a solvent for the liquid, the Y compound, and initiator and results in a higher solubility of the Y compound in the continuous aqueous phase. This solvent is removed by evaporation or dilution with water before the addition of the monomer mixture in the next step. Y compounds that have been applied include highly water-insoluble initiators such as dioctyl peroxide, which serve a dual role, acting both as swelling agents and as initiators.

2. Preparation of Oligomer–Polymer Particles

Two methods have been used for the production of oligomers by polymerization inside particles. In the first method, the seed particles are swollen with a mixture of monomer and a high concentration oil-soluble initiator. The type and the amount of initiator and reaction temperature are chosen so that the polymerization gives an oligomer of appropriate chain-length for the subsequent swelling with monomer in the second step. The second method requires that seed particles are swollen with a mixture of monomer, initiator, and a chain-transfer agent. The type and amount of chain-transfer agent are chosen so that an oligomer of appropriate chain length for the subsequent swelling with a monomer mixture in the second step is subjected to polymerization.

C. Swelling of the Activated Seed Particles with Monomers or a Mixture of Monomers

The second step in the production of monodispersed polymer particles involves the swelling of activated particles with a monomer or a mixture of monomers, diluents, and porogens, and the shape of the swollen oil droplets must be maintained in the continuous aqueous phase. The monomer or the mixture of monomers may be added in bulk form, preferably as an aqueous dispersion to increase the rate of swelling, especially in the case of relatively water-insoluble monomers.

In the two-step swelling method, for the process to function, there is an absolute requirement that no Y compound be left outside the particles in the continuous phase when a large amount of Y compound is used in the first step to swell seed particles. Likewise, there should not be any transport of Y compound out of the particles during the second-step swelling with the monomer or the mixture of monomers and diluents. During the addition of the monomer or the mixture of monomers to the activated swelling seed particles, the shape of particles must be maintained and the Y compound prevented from being transported out of the seed particles during the swelling process. Preferably, all the monomers are allowed to swell the activated particles before polymerization is started. The activated swelling is the only method known at present that will allow the production of spheres of monosized, highly cross-linked polymeric particles.

D. Polymerization of Multistep Swelling Particles

The maintenance of monodispersed swelling seed particles through polymerization is a key step in producing monodispersed particles. Coagulation during the sticky period of polymerization increases particle size, increases the aggregation of swelling particles, and decreases monodispersity of the swelling particles.

Several major parameters affect coagulation during polymerization of the swelling particles. These parameters must be controlled properly to maintain monodispersity of the swelling particles and to convert these swelling particles into microspherical polymer particles. The parameters with the greatest impact on the polymerization processes are reaction temperature, stabilizer type and amount, reactor design and stirring speed, and volume ratio of the monomer phase to the water phase.

E. Preparation of Monosized Macroporous Particles

The method of activated swelling is especially suitable for the preparation of monosized cross-linked polymeric particles and for the production of monosized macroporous particles. This method has been used to produce macroporous polymeric particles from a number of different vinyl monomers with a very high percentage of polyfunctional vinyl monomers as the cross-linked agents. In this case, the swelling is carried out by applying inert solvents as the pore-forming and precipitate agents, in addition to monomers and cross-linked agents. The pore volume is determined by the amount of inert solvent used relative to the vinyl compounds. The pore size and pore-size distribution may be varied by adjusting relative amounts of typical swelling and typical precipitating agents in the mixture. Inert solvents with a solubility parameter value that varies greatly from that of the matrix polymer tend to increase the pore size. On the other hand, the use of inert solvents with solubility parameters close to those of the polymer results in particles with smaller pore sizes. The activated swelling method allows the production of nearly perfect spheres of macroporous particles in the size range of 1–100 μm and with pore volumes of up to 80%.

The molar volume of seed particles in the activated swelling and polymerization method is usually kept at less than 1% of the molar volume of the final polymer beads to ensure the chemical and physical properties of the final particles. However, the recent development of monodispersed macroporous particles by Frechet's group (64,65,70) and Vanderhoff's group (66,67) has changed the approach. Oligo–polymer particles of up to 5–9 μm are used as shape templates and linear polymer porogens with monomer mixtures and low-molecular-weight porogens are used for the second step of swelling. After polymerization, the linear polymer porogens are removed by solvent extraction to form macroporous structures of polymer beads in about 7.4 and 10 μm.

Monodispersed poly(methyl methacrylate–ethyleneglycol dimethacrylate) is prepared by a multistep swelling and polymerization method. When a good solvent such as toluene is applied as a porogen, the seed polymer severely affects the pore structure, whereas no effects are observed with poor solvents, such as cyclohexanol, as a porogen, in comparison with the conventional suspension polymerization (68,69).

In the multistep swelling and polymerization method, polymerization time, temperature, and initiator concentration have marginal effects only. The ratio of monovinyl to divinyl monomer in the polymerization mixture, along with the composition of the porogenic systems, is known to be the most important factor in controlling the ultimate macroporous structure. The average molecular weight of the polymer porogens has great impact on the pore-size distribution. Data show that a decrease in the molecular weight of the polymeric porogen results in a decrease in macropores (>50 nm) and an increase in the mesopores (2–50 nm). However, the macroporous PS–DVB beads formed in the presence of high molecular weight polystyrene do not possess the mechanical strength required for its application in HPLC (70).

The effects of the concentration of divinylbenzene on pore-size distribution and surface areas of micropores, mesopores, and macropores in monosized PS–DVB beads prepared in the presence of linear polymeric porogens have been studied (65). While the total surface area is clearly determined by the content of divinylbenzene, the sum of pore volumes for mesoforms and macropores, as well as their pore-size distribution, do not change within a broad range of DVB concentrations. However, the more cross-linked the beads, the better the mechanical and hydrodynamic properties.

Information about the particle size, distribution, sphericity, and morphology of the porous monosized polymer has been obtained by scanning and transmission electron microscopy. Scanning electron micrographs of a highly porous and monosized 15-μm chromatographic packing material are shown in Fig. 1.3. The narrow particle-size distribution and sphericity are illustrated in Fig. 1.3a, whereas Fig. 1.3b shows a close-up of one particle revealing the surface pore structure.

The advantages of monosized chromatographic supports are as follows: a uniform column packing, uniform flow velocity profile, low back pressure, high resolution, and high-speed separation compared with the materials of broad size distribution. Optical micrographs of 20-μm monosized macroporous particles and a commercial chromatography resin of size 12–28 μm are shown in Fig. 1.4. There is a clear difference in the size distribution between the monodispersed particles and the traditional column material (87).

VI. APPLICATION OF POLYMERIC BEADS AS SIZE EXCLUSION CHROMATOGRAPHY SUPPORTS

Chromatographic separations of macromolecular substances were reported in 1959 (72). The earliest work was essentially done using procedures of modern gel-permeation chromatography (GPC), which was referred to as gel filtration, and using soft gels such as dextran for column packing with an aqueous buffer as the eluent. In general, separation times were long because the column packings were soft and compressible; therefore, the operating pressure was severely restricted. Moore's (73,74) introduction of rigid macroporous cross–linked polystyrene–divinylbenzene resin allowed the rapid analysis of synthetic organic soluble polymers in 1960. Since then, the technology of GPC has progressed steadily as successive advancements in polystyrene–divinylbenzene column packings have been made. The latest high-resolution SEC columns employ monodispersed 3-, 5-, and 10-μm polystyrene–divinylbenzene particles, which allow even faster analysis. The use of high-performance GPC columns, packed with very small particles, was introduced in advance of other modes of liquid chromatography (75,76). Columns packed with a mixture of sorbents

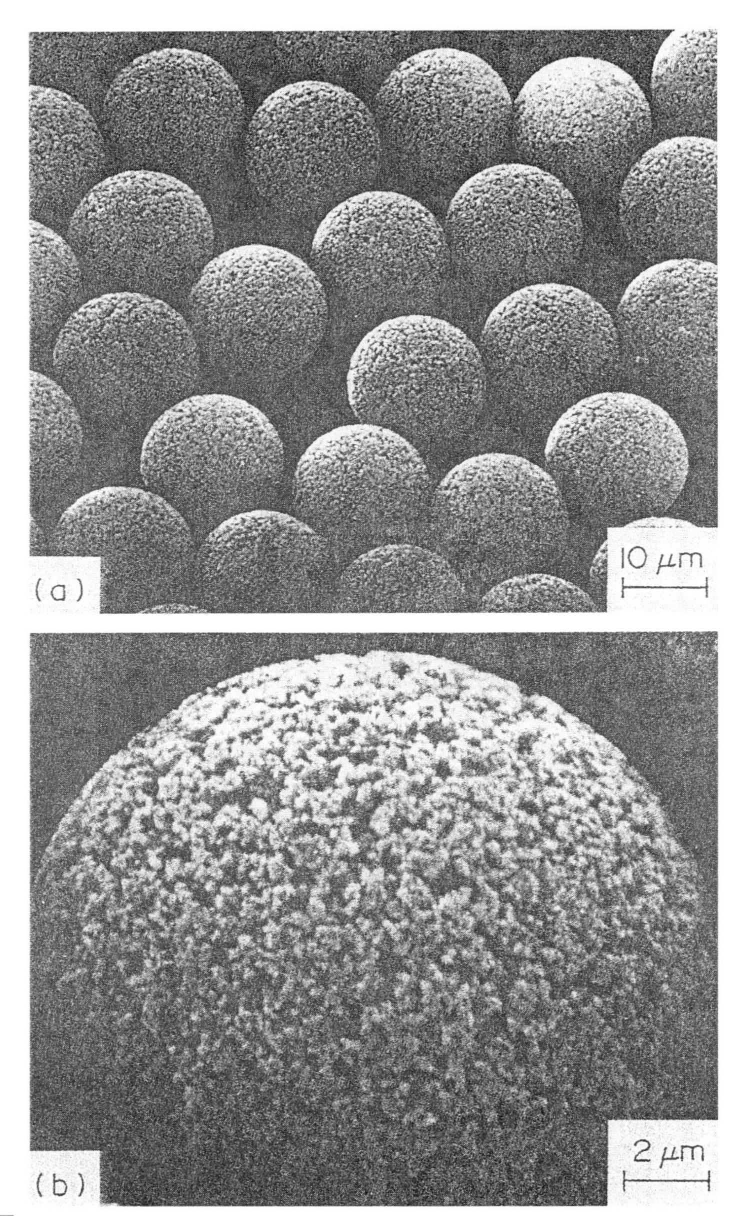

FIGURE 1.3 15-μm macroporous packing material. (a) Narrow particle-size distribution. (b) Close-up of one particle to reveal the pore structure, porosity 75%, surface area (BET) 494 m^2/g. [Reprinted from J. Ugelstad (1992). Preparation and Application of New Monosized Polymer Particles. *Prog. Polym. Sci.* **17,** 87–161; with kind permission from Elsevier Science Ltd., Kidlington OX5 1GB UK.]

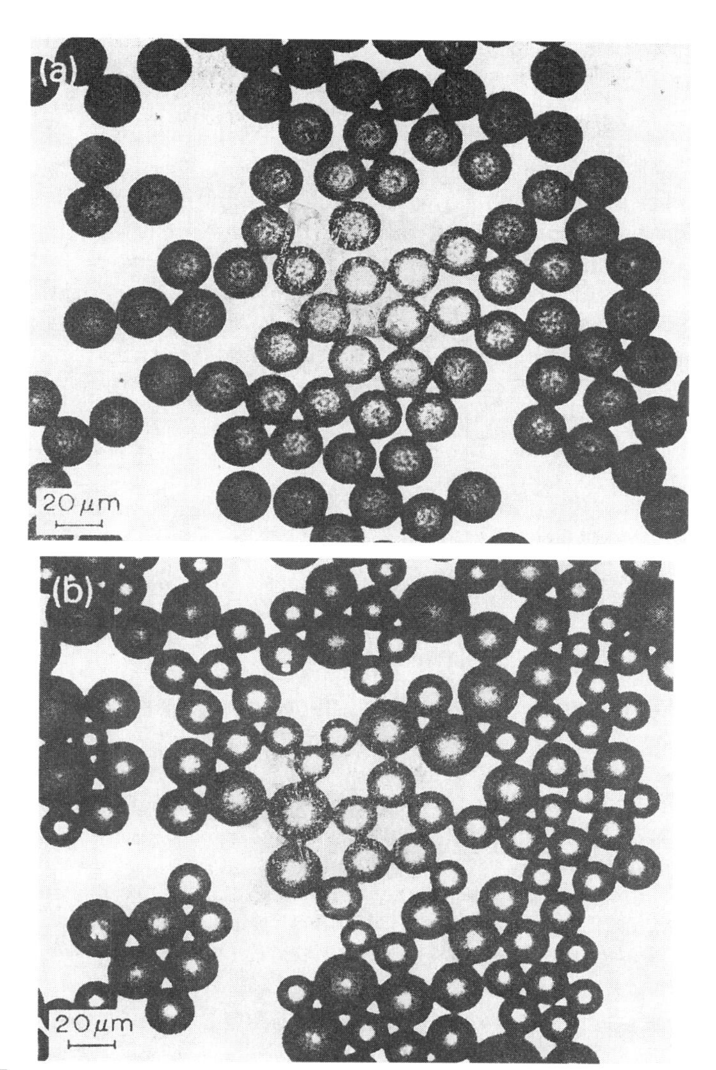

FIGURE 1.4 Optical micrograph of macroporous chromatographic column materials. (a) Monosized particles of 20 μm. (b) Commercial column filling of 12–28 μm. [Reprinted from T. Ellingsen *et al.* (1990). Monosized stationary phases for chromatography. *J. Chromatogr.* **535,** 147–161; with kind permission from Elsevier Science-NL, Amsterdam, The Netherlands.]

differing in pore sizes were also developed at early stages (76). Such columns have wide separation ranges and hence are very versatile.

A. Polystyrene–Divinylbenzene-Based Resin

A number of polystyrene resins with significant residual double-bond contents have been used for grafting of polar hydrophilic polymethacrylate–polyethylene glycol macromolecules in an attempt to produce a thin, uniform coating on the resin surface for use in aqueous GPC packing materials. The attempt was only partially successful. Most of the modified resins obviously showed the aggregated and poorly distributed coating, as well as showing the significant hydrophobic adsorption in water eluent (77).

A novel cross-linked polystyrene–divinylbenzene copolymer has been produced from suspension polymerization with toluene as a diluent, having an average particle size of 2 to 50 μm, with an exclusive molecular weight for the polystyrene standard from about 500 to 20,000 in gel-permeation chromatography. A process for preparing the PS–DVB copolymer by suspension polymerization in the presence of at least one free-radical polymerization initiator, such as 2,2′-azo-bis (2,4-dimethylvaleronitrile) with a half-life of about 2 to 60 min at 70°C, has been disclosed (78).

The packing material for liquid chromatography is produced from styrene and divinylbenzene dissolved in 50 to 300% by weight of organic solvent to both monomers. The constitution of divinylbenzene in the monomer mixture is not less than 60% by weight. In gel-permeation chromatography, the exclusive molecular weight is not less than 1×10^3 in terms of standard polystyrene (79).

B. Polyacrylamide- and Polymethacrylate-Based Gels

Cross-linked microspherical polyacrylamides, which are macroporous and rigid, have been examined as column packings for aqueous, high-performance size-exclusion chromatography. These column packings are suitable for high-resolution separation of oligosaccharides and polysaccharides in water. Peak resolution may be varied by changing eluent flow rate and temperature. A mixed-bed column has been evaluated for use in gel-permeation chromatography of the water-soluble polymers. It has a wide separation range, and the molecular weight calibration curve for polyethylene glycol is almost linear over the range of 10^2–10^6 (80). A variety of water-soluble polymers have been fractionated according to molecular size, with no evidence of adsorption providing that an appropriate salt or salt and organic eluent were employed (81).

An aqueous ethanol solution of acrylamide, 2,2′-methylenebisacrylamide as cross-linked agent and third acrylamide derivative, is dispersed in an *n*-alkane. Then three monomers are polymerized to spherical porous gels. The effect of the composition of the third monomer on the exclusion limits of the gel in size-exclusion chromatography has been investigated (82).

Cross-linked polyacrylamides attached with morpholine pendent at some repeating unit of the backbone chains have been prepared and used for the separation of discrete chemical compounds by gel-permeation chromatography (83).

C. Polyvinyl Alcohol-Based Gels

A polyvinyl alcohol is obtained by suspension polymerization of vinyl acetate and the cross-linking agent, triallyl isocyanurate, with a triazine ring followed by alkali hydrolysis. The polyvinyl alcohol gel is used as packing for gel-permeation chromatography. The exclusion limit for spherical proteins is about 1×10^5 (84).

Macroporous polyvinyl alcohol particles with a molecular weight cutoff of ca. 8×10^5 in gel-permeation chromatography have been prepared. The particles are produced by first dispersing an aqueous solution of polyvinyl alcohol in an organic solvent to make spheres of polyvinyl alcohol solution. Holding the dispersion in such a state that a gel will then form spontaneously will cause the gel to react with glutaraldehyde in the presence of an acidic catalyst (85).

D. Application of Monosized Polymeric Particles in Size Exclusion Chromatography

A trend in chromatography has been to use monosized particles as supports for ion-exchange and size-exclusion chromatography and to minimize the column size, such as using a 15 × 4.6-mm column packed with 3-μm polymer particles for size exclusion chromatography. The more efficient and lower back pressure of monosized particles is applied in the separation.

Monosized macroporous polystyrene–divinylbenzene particles have been prepared in a multistep swelling process, in which particles of different sizes

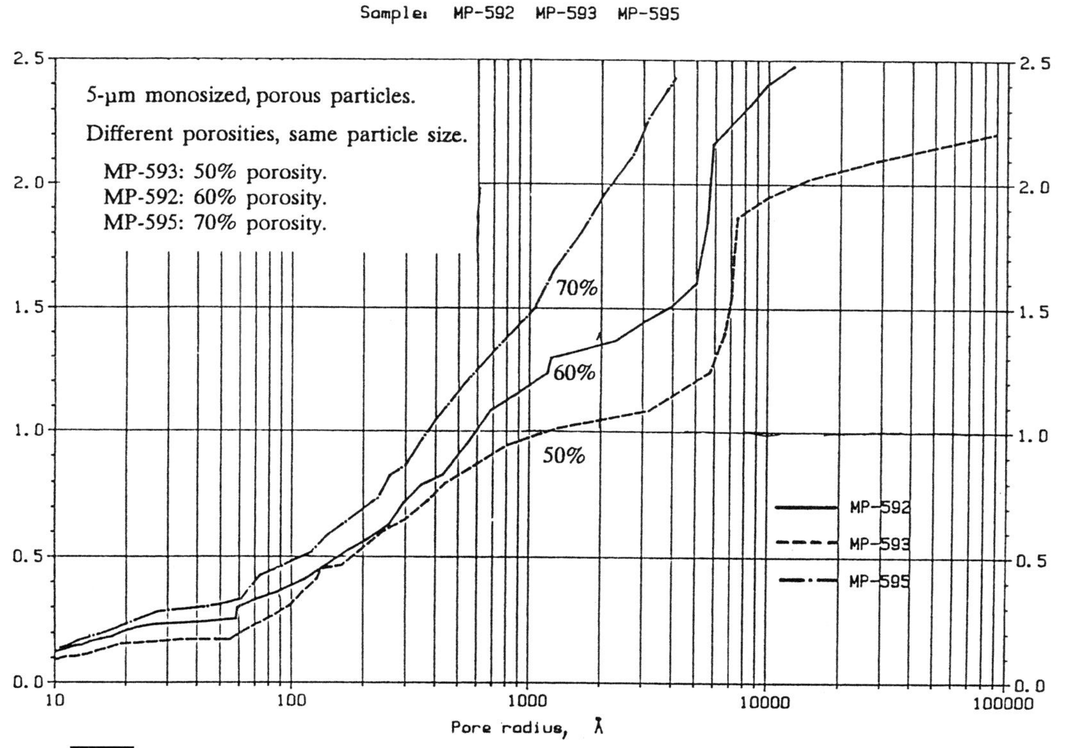

FIGURE 1.5 Cumulative pore volume curves of 5-μm monosized porous particles. [Reprinted from T. Ellingsen *et al.* (1990). Monosized stationary phases for chromatography. *J. Chromatogr.* **535,** 147–161; with kind permission from Elsevier Science-NL, Amsterdam, The Netherlands.]

FIGURE 1.6 SEM of 5-μm porous particles of 50% porosity. [Reprinted from T. Ellingsen *et al.* (1990). Monosized stationary phases for chromatography. *J. Chromatogr.* **535,** 147–161; with kind permission from Elsevier Science-NL, Amsterdam, The Netherlands.]

(5, 10, and 20 μm) with similar wide-pore-size distribution are prepared. All particle sizes yield columns with graphically similar calibration. This is expected because of their nearly identical pore distributions. With 5-μm particles, more than 50,000 theoretical plates can be obtained in a 30-cm column (HETP $\approx$ 0.006 mm). The resolution is ca. 4.40. The pressure drop for the standard size-exclusion column, 30 $\times$ 7.8 mm, is about 975 psi at a 1ml/min flow rate with toluene as eluent. The calibration curves are linear in the range of 20–350 kDa (86).

Figure 1.5 shows the cumulative pore volume curve for 5-μm monosized porous PS–DVB particles with 50, 60, and 70% porosity. The curves were drawn by overlapping the measurements from nitrogen adsorption–desorption and mercury intrusion. A scanning electron micrograph of 5-μm monosized particles with 50% porosity is shown in Fig. 1.6 (87).

REFERENCES

1. Arshady, R. (1991). *J. Chromatogr.* **586,** 181–197.
2. Arshady, R. (1991). *J. Chromatogr.* **586,** 199–216.
3. Galina, H., Colaz, W. B., Wiezorek, P. P., and Wojsznska, M., Sr. (1985). *Polym. J.* **17,** 215.
4. Heitz, W. (1977). *Adv. Polym. Sci.* **23,** 1.
5. Guyot, A., and Bartholin, M. (1982). *Prog. Polym. Sci.* **8,** 277.
6. Moore, J. C. (1969). *J. Polym. Sci., Part A-2,* 835.

7. Horak, D., Pelzbauer, Z., Bleha, M., Zlavskky, M., Svec, F., and Klal, J. (1980). *J. Appl. Polym. Sci.* **26**, 411.
8. Horak, D., Svec, F., Bleha, M., and Klal, J. (1981). *Angew. Makromol. Chem.* **95**, 109.
9. Horak, D., Svec, F., Bleha, M., and Klal, J. (1981). *Angw. Makromol. Chem.* **95**, 117.
10. Coupek, J., Krivakova, M., and Pokorny, S. (1973). *J. Polym. Sci., Polym Symp.* **42**, 185.
11. Arshady, R. (1988). *Chim. Ind. (Milan)* **70**(9), 70.
12. Wolf, F., and Ecket, S. (1971). *Plaste Kautsch.* **18**, 650 and 890.
12a. Wolf, F., and Ecket, S. (1972). *Plaste Kautsch.* **19**, 26.
13. Ahmed, S. M. (1984). *J. Dispersion Sci. Technol.* **5**, 421.
14. Jacobelli, H., Bartholin, M., and Guyot, A. (1979). *J. Appl. Polym. Sci.* **23**, 927.
15. Sederel, W. L., and de Jong, G. J. (1973). *J. Appl. Polym. Sci.* **17**, 2835.
16. Tomoi, M., and Ford, W. T. (1981). *J. Am. Chem. Soc.* **103**, 821.
17. Guyot, A. (1988). *Pure Appl. Chem.* **60**, 365.
18. Seidl, J., Malinsky, J., Dusek, K., and Heitz, W. (1967). *Adv. Polym. Sci.* **5**, 113.
19. Standard Suspension Polymerization Techniques, Appendix (1980). *In* "Polymer-Supported Reaction in Organic Synthesis" (P. Hodge and D. C. Sherington eds.), Wiley: Chichester.
20. Werotte, L. E., and Grammont, P. D. U.S. Patent 3,418,262, December 24, 1968.
21. Mindick, M., and Svarg, J. J. U.S. Patent 3,549,562, December 22, 1970.
22. Morse, L. D., Calmon, C., and Grundner, W. T. U.S. Patent 3,627,708, December 14, 1971.
23. Fuchiwaki, Y., *et al.* U.S. Patent 3,791,999, February 12, 1974.
24. Kido *et al.* U.S. Patent 4,174,430, November 13, 1979.
25. Meitzner, E. F., and Oline, J. A. U.S. Patent 4,256,840, March 17, 1981.
26. Meitzner, E. F., and Oline, J. A. U.S. Patent 4,297,220, October 27, 1981.
27. Bates, F. S., and Cohen, R. E. U.S. Patent 4,485,207, November 27, 1984.
28. Dawkins, J. V. (1981). *Polymer* **22**, 291.
29. Coupek, J., Krivakeva, M., and Pokorny, S. (1973). *J. Polym. Sci., Polym. Symp.* **42**, 185.
30. Svec, F., Hradil, J., Coupek, J., and Kalal, J. (1975). *Angew. Makromol. Chem.* **48**, 135.
31. Horak, D., Svec, F., Bleha, M., and Kalal, J. (1981). *Angew. Makromol. Chem.* **95**, 109.
32. Trijasson, P., Ferere, Y., and Gramain, P. (1990). *Makromol. Chem. Rapid. Commun.* **11**, 235.
33. Heitz, W., Ulliner, H., and Hoeker, H. (1966). *Makromol. Chem.* **98**, 42.
34. Itagaki, T., Kusano, H., and Kubota, H. U.S. Patent 4,696,745, September 29, 1987.
35. Boos, K. S., Wilmers, B., Sauerbrey, R., and Schlimme, E. U.S. Patent 4,767,529, August 30, 1988.
36. Kido *et al.* U.S. Patent 4,104,208, August 1, 1978.
37. Itagaki, T., and Ouchi, H. U.S. Patent 4,306,031, December 15, 1981.
38. Itagaki, T., and Ouchi, H., U.S. Patent 4,350,773, September 21, 1982.
39. Feibush, B., and Li, N. U.S. Patent No. 4,933,372, June 12, 1990.
40. Knox, J. H., and Gilbert, M. T. U.S. Patent No. 4,263,268, April 21, 1981.
41. Li, N., and Mazid, M. A. U.S. Patent No. 5,168,104, December 1, 1992.
42. Kusano, H., *et al.* U.S. Patent 5,114,577, May 19, 1992.
43. Svec, F., and Frechet, J. M. J. *Am. Lab.* 25–34. December 1996.
44. Smigol, V., Svec, F., and Frechet, J. M. J. (1993). *Macromolecules* **26**, 5615–5620.
45. Smigol, V., Svec, F., and Frechet, J. M. J. (1994). *Anal. Chem.* **66**, 4308–4315.
46. Smigol, V., Svec, F., and Frechet, J. M. J. (1994). *J. Liq. Chromatogr.* **17**(4), 891–911.
47. Svec, F., and Frechet, J. M. J. (1992). *Anal. Chem.* **64**(7), 820–822.
48. Wang, O. C., Svec, F., and Frechet, J. M. J. (1993). *Anal. Chem.* 2243–2248.
49. Svec, F., and Frechet, J. M. J. (1995). *Chem. Mater.* **7**, 707–715.
50. Svec, F., and Frechet, J. M. J. (1995). *J. Chromatogr. A.* **702**, 89–95.
51. Afeyan, N. B., Regnier, F. E., and Dean, R. C. U.S. Patent 5,019,270, May 28, 1991.
52. Afeyan, N. B., Regnier, F. E., and Dean, R. C. U.S. Patent 5,552,041, September 3, 1996.
53. Ugelstad, J. (1992). *Prog. Polym. Sci.* **17**, 87–161.
54. Woods, M. E., Dodge, J. S., Krieger, I. M., and Pierce, P. (1968). *J. Paint Technol.* **40**(527), 541.
55. Bradford, E. B., and Vanderhott, J. W. (1955). *J. Appl. Phys.* **26**, 864.
56. Vanderhoff, J. W., Bradford, E. B., Takowski, H. L., and Wilkinson, B. W. (1961). *J. Polym. Sci.* **50**, 265.
57. Goodwin, J. W., Hearn, J., Ho, C. C., and Ottewill, R. H. (1974). *Colloid Polym. Sci.* **252**, 464.

58. Smigol, V., Svec, F., Hosoya, K., Wang, O., and Frechet, J. M. J. (1992). *Angew. Makromol. Chem.* **195**, 151–164.
59. Almog, Y., Reich, S., and Levy, M. (1982). *Br. Polym. J.* **131.**
60. Almog, Y., and Levy, M. (1982). *J. Polym. Sci. Polym. Chem. Ed.* **20**, 417.
61. Ober, C. K., and Hair, M. L. (1987). *J. Polym. Sci. A* **25**, 1395.
62. Lok, K. P., and Ober, C. K. (1985). *Can. J. Chem.* **63**, 209.
63. Kulin, L., Flodin, P., Ellingsem, T., and Ugelstad, J. (1990). *J. Chromatogr.* **514**, 1–9.
64. Wang, O. C., Hosoya, K., Svea, F., and Frechet, J. M. J. (1992). *Anal. Chem.* **64**, 1232–1238.
65. Wang, O. C., Svea, F., and Frechet, J. M. J. (1992). *Polym. Bull.* **28**, 569–575.
66. Cheng, C. M., Micale, F. J., Vanderhoff, J. W., and El-aasser, M. S. (1992). *J. Polym. Sci., A* **30**, 235–244.
67. Cheng, C. M., Vanderhoff, J. W., and El-aasser, M. S. (1992). *J. Polym. Sci. A* **30**, 245–256.
68. Hosoya, K., *et al.* (1992). *Chem. Lett.* 1145–1148.
69. Hosoya, K., and Frechet, J. M. J. (1993). *J. Polym. Sci. A* **31**, 2129–2141.
70. Wang, O. C., Svea, F., and Frechet, J. M. J. (1994). *J. Polym. Sci. A* **32**, 2577–2588.
71. Galia, M., Svec, F., and Frechet, J. M. J. (1994). *J. Polym. Sci. A* **32**, 2169–2175.
72. Parath, J., and Flodin, P. (1959). *Nature (London)* **183**, 1657.
73. Moore, J. C. (1964). *J. Polym. Sci. A* **2**, 835.
74. Moore, J. C. U.S. Patent 3,326,875, June 20, 1967.
75. Kato, Y., *et al.* (1973). *J. Polym. Sci. Polym. Phys. Ed.* **11**, 2329.
76. Kato, Y., *et al.* (1974). *J. Polym. Sci. Polym. Phys. Ed.* **12**, 1339.
77. Hefferman, J. G. (1984). *J. Appl. Polym. Sci.* **29**, 3013–3025.
78. Tanaka, Y., Takeda, J., and Noguchi, K. U.S. Patent 4,338,404, July 6, 1982.
79. Tokunaga, K., and Hushimoto, T. U.S. Patent 4,686,269, August 11, 1987.
80. Dawkin, J. V., *et al.* (1986). *J. Chromatogr.* **371**, 283–291.
81. Kato, Y., *et al.* (1985). *J. Chromatogr.* **332**, 39–46.
82. Suzuki, K., *et al.* (1990). *J. Chromatogr.* **535**, 173–180.
83. Epton, R., *et al.* U.S. Patent 3,896,092, July 22, 1975.
84. Murayama, N., and Sakagami, T. U.S. Patent 4,314,032, February 2, 1982.
85. Itagaki, T., Kussno, H., Miyata, E., and Tashiro, T. U.S. Patent 4,863,972, September 5, 1989.
86. Kulin, L., Flodin, P., Ellingsem, T., and Ugelstad, J. (1990). *J. Chromatogr.* **514**, 1–9.
87. Ellingsen, T., Aune, O., Ugelstad, J., and Hagen, S. (1990). *J. Chromatogr.* **535**, 147–161.

2

SIZE EXCLUSION FOR ANALYSIS AND PURIFICATION OF AQUEOUS MACROMOLECULES

ALAN WILLIAMS

Amersham Pharmacia Biotech Inc., Piscataway, New Jersey 08855

LARS HAGEL

Amersham Pharmacia Biotech AB, SE-751 82 Uppsala, Sweden

I. INTRODUCTION: DEVELOPMENT OF SIZE EXCLUSION CHROMATOGRAPHY (SEC)

Size exclusion was first noted in the late fifties when separations of proteins on columns packed with swollen maize starch were observed (Lindqvist and Storgårds, 1955; Lathe and Ruthven, 1956). The run time was typically 48 hr. With the advent of a commercial material for size separation of molecules, a gel of cross-linked dextran, researchers were given a purposely made material for size exclusion, or gel filtration, of solutes as described in the classical work by Porath and Flodin (1959). The material, named Sephadex, was made available commercially by Pharmacia in 1959. This promoted a rapid development of the technique and it was soon applied to the separation of proteins and aqueous polymers. The work by Porath and Flodin promoted Moore (1964) to apply the technique to size separation, gel permeation chromatography of organic molecules on gels of lightly cross-linked polystyrene (i.e., Styragel).

Classical gels had a low degree of cross-linkage and were of a large particle size. This resulted in that modest flow rates could only be applied and the separation time was typically 10 hr, which at that time was perfectly acceptable, keeping in mind that preparation of the column could take up to 2 days or more. After the introduction of Sephadex, new materials have been introduced continuously on the market, and still, 30 years after the introduction of the first commercial material, new media are still introduced, also from the originators of Sephadex. What are the driving forces behind this development and what are the features of these new media?

First line of development addressed the need for media of different separation ranges. Because the separation range is determined solely by the pore size distribution of the media, different pore size distributions (e.g., achieved by different degrees of cross-linkage of Sephadex) are needed for the separation of small solutes, medium sized solutes, and large solutes. At the same time, a narrow pore size distribution will yield a high selectivity factor and therefore a variety of media of different pore sizes have been introduced for various applications. This also encouraged researchers to test different polymers in order to achieve media of unique pore structures. However, as been concluded, size exclusion is rather insensitive to the pore structure of the material (Hagel, 1988). Not all media survived the rigorous test of nonspecific adsorption—a property that is essential to a gel filtration medium—or did not offer advantages over existing commercially available media, and the majority of media are still supplied from a handful of companies.

The second line of development was to decrease the separation time for analytical applications. This was predominantly achieved by decreasing the particle size. However, small particle-sized media, e.g., smaller that 15 μm, are generally more difficult to pack and therefore prepacked columns for size exclusion chromatography were introduced. Smaller particle size media yield higher pressure drops and therefore the rigidity of the materials were increased. This was achieved by using rigid packings, e.g., silica-based or heavily cross-linked polymers, and a characteristic of these media are a higher matrix volume, in turn leading to a smaller separating volume than classical media. Thus, the peak capacity of some of these small particle-sized media is therefore of the same order as for traditional media in longer columns (Hagel, 1992). However, the gain in reduced separation time for these columns is considerable and separation times down to 30 min or less are achieved regularly.

The third line of development was to increase the selectivity in order to achieve the highest possible resolution to address difficult separations. This may be achieved by a very narrow pore size distribution of the media, e.g., such as achieved by porous silica microspheres (PSM) or by modifying the porous phase by a composite material, e.g., as for Superdex. In practice, this material shows a maximum selectivity over the separation range (e.g., see Fig. 2.2).

Irrespective of the development of media, many of the traditional media are successfully defending their position. This is due to their hydrophilic nature, preserving biological function of the separated molecules, but also the fact that columns may be prepared easily and, finally, some of the classical media, e.g., Sephadex, have a selectivity that is so far unsurpassed and therefore very fit for use. Intersting enough, Sephadex is still the premiere gel filtration medium for desalting due to the optimal pore size and particle size of this medium (see Section II,C).

The evolution of media covering aqueous and nonaqueous systems on the one hand and analytical as well as microscale and macroscale preparative applications on the other hand has resulted in an arbitrarily nomenclature within the field. Thus the current practice is to refer to the separation principle based on solute size as size exclusion chromatography (SEC) whereas the application in aqueous systems is traditionally referred to as gel filtration (GF) and the application in nonaqueous systems is designated gel-permeation chromatography (GPC). The term high-performance size exclusion may be misleading as many traditional media fulfill the (arbitrarily) definition of high-performance systems (Hagel and Janson, 1992). Thus, this chapter uses the term gel filtration to describe aqueous size exclusion chromatography of standard media, fast media, and process media.

Size exclusion is basically employed for three different applications. One is the original one, i.e., desalting or separation of solutes differing more than a decade in molecular size. Another application is fractionation where the size differences are smaller, i.e., typically a factor of two to five. The third application is the determination of the molecular weight(s) of a sample.

There are different requirements on the media for the different application areas which is the reason why so many media exist and why some of the pioneering media still are the preferred choice for some applications, e.g., the use of Sephadex G-25 for desalting purposes.

II. APPLICATION AREAS OF SEC

The separation of various molecules based on size can be applied readily to both analytical and preparative applications. Design considerations that distinguish analytical from preparative applications are more than just the presence or absence of a fraction collector at the outlet of the system but rather requires establishing specific priorities for the separation at hand. Analytical applications such as molecular weight determinations or purity determinations typically place greater emphasis on resolution and/or minimizing the applicable sample mass in comparison to preparative applications such as group separations (i.e., desalting) and buffer exchange, which may place a premium on throughput or yield. However, in practice, many SEC applications, such as high-resolution fractionation of complex mixtures, must balance resolution and throughput to reach the desired goal. In the design of all SEC applications, careful consideration of the physical and chemical attributes of both the column and media, the nature of the sample, and the methodological approach to be used is required to achieve the desired results.

Numerous application examples are related to gel filtration (e.g., see Hagel, 1989; Hagel and Janson, 1992; Pharmacia, 1991). A selected number of applications are discussed with respect to their goals and to types of methods and SEC media attributes, which impact the selection and or construction of a suitable SEC column. Specific examples of these various applications types are given later under Sections II,C and III. The optimization of running conditions to achieve the desired results are discussed in Section VI.

A. Determination of Molecular Size and Sample Purity

The observation of molecular size or polydispersity and the subsequent determination of relative molecular mass, (M_r) or molecular mass (weight) distribution (MWD), is the most common analytical application of SEC. The goal of these types of experiments is to either observe the solvated size of one or more molecular species or to observe the distribution of sizes present in a mixture (see Fig. 2.1). Based on the shape of the solvated molecule(s) and its respective retention volume, V_R, during SEC, an estimation of molecular size can be made relative to calibration data collected from the retention volume of appropriate standards. The operational attributes of the SEC separation to attain these goals should primarily focus on keeping high resolution throughout the desired molecular size range by selecting media with the correct selectivity and using appropriate column dimensions (see Section VI) . However, in such determinations it is critical that the solute does not interact with the matrix (i.e., be adsorbed to or repelled from the matrix), which will effect the retention volume and hence the estimation of size. In testing environments where large numbers of samples are processed, economic factors such as reducing the total run time, the solvent volume used per run, and the sample mass and/or sample volume consumed may become of high priority.

The resolution required in any analytical SEC procedure, e.g., to detect sample impurities, is primarily based on the nature of the sample components with respect to their shape, the relative size differences of species contained in the sample, and the minimal size difference to be resolved. These sample attributes, in addition to the range of sizes to be examined, determine the required selectivity. Earlier work has shown that the limit of resolvability in SEC of molecules [i.e., the ability to completely resolve solutes of different sizes as a function of (1) plate number, (2) different solute shapes, and (3) media pore volumes] ranges from close to 20% for the molecular mass difference required to resolve spherical solutes down to near a 10% difference in molecular mass required for the separation of rod-shaped molecules (Hagel, 1993). To approach these limits, a SEC medium and a system with appropriate selectivity and efficiency must be employed.

The highest selectivity achievable for any specific solute size range in SEC separations requires a porous particle with a very narrow distribution of pore sizes or, ideally, a single pore size. The theoretical limit of selectivity based on a single pore size of 60 Å in comparison to Superdex 75 HR is shown in Fig. 2.2. It is readily apparent that the theoretical limit in selectivity is available commercially in certain separation ranges. Figure 2.3 presents a series of the selectivity curves for various analytical SEC columns from Amersham Pharmacia Biotech. The maximal size selectivity of a support is obtained at the inflection point of the selectivity curve (Hagel, 1988), typically located near $K_{av} = 0.5$. An inherent limitation with any size-based separation is that porous media designed for high selectivity necessarily yield a narrower separation range (i.e., typically spanning one decade in solute size) than media with broader pore size distributions. This limitation suggests that high-resolution analysis of broad molecular weight distributions over 2 or 3 decades of size may require the use of several different selectivities with overlapping separation ranges (Hagel, 1988). However, when initially examining the molecular weight distribution of any given sample, it may be desirable to first use a medium with a broad separation range to ensure inclusion of the entire range of solute sizes and to establish the specific separation range of interest. From results of an initial scouting run using media with a broad separation range, a medium with higher selectivty in the separation range of interest may be subsequently selected. It is therefore of critical importance to have a range of SEC media

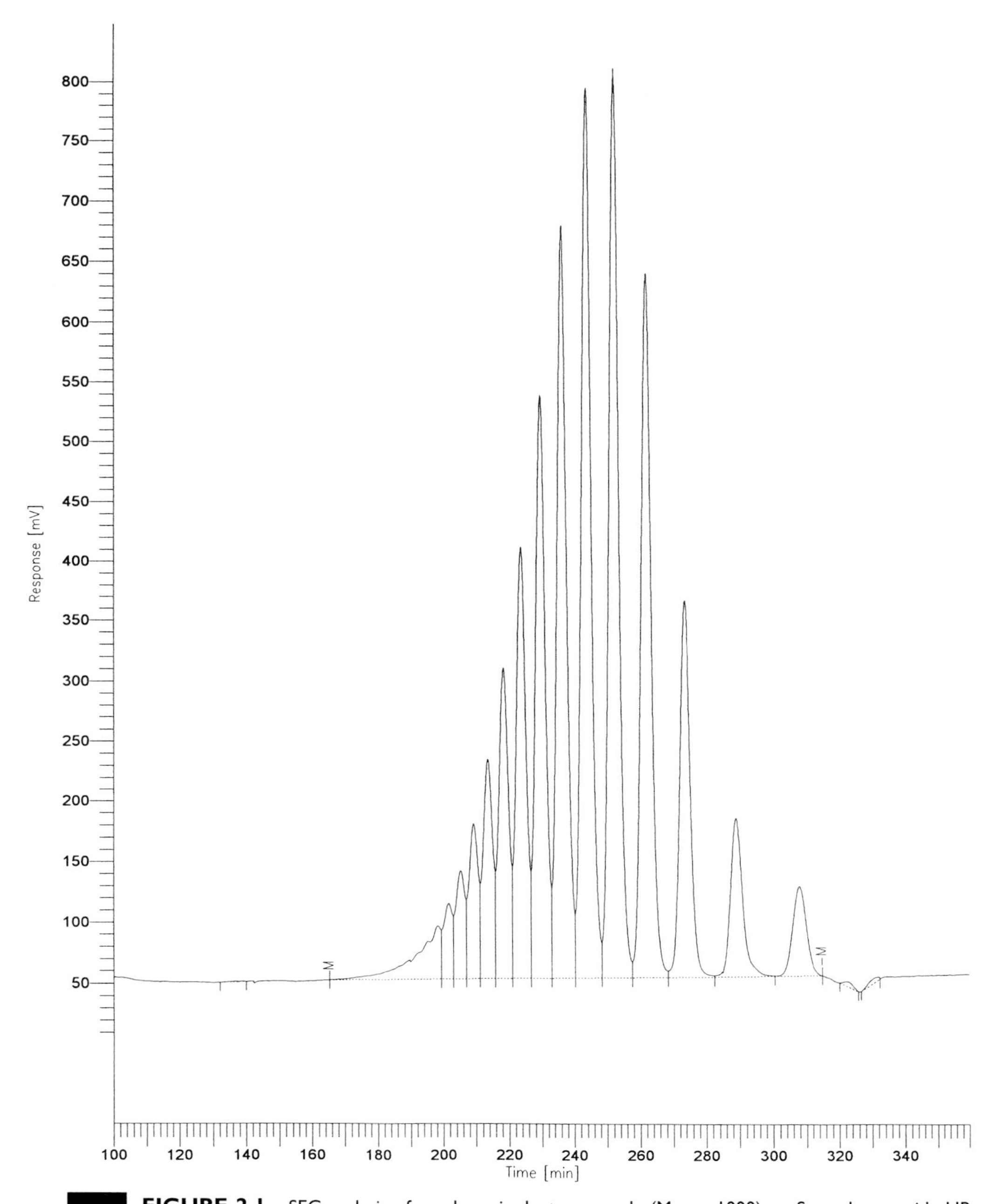

FIGURE 2.1 SEC analysis of a polymeric dextran sample (Mw = 1000) on Superdex peptide HR 10/50. The very high resolution between individual components of the sample is obtained by using two columns in series. Courtesy of T. Andersson. (Reproduced with permission from Amersham Pharmacia Biotech.)

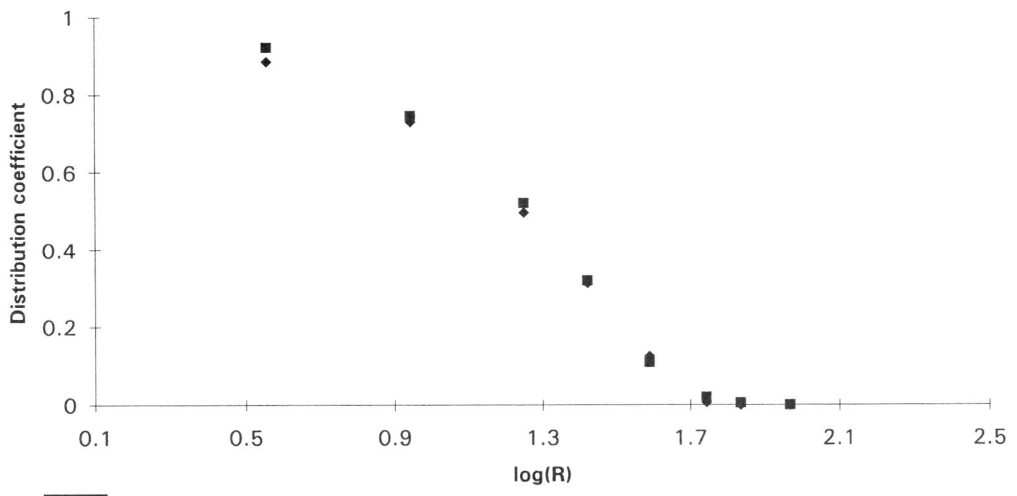

FIGURE 2.2 Selectivity curve of Superdex 75, HR 10/30, as compared to a hypothetical single pore-size support. ■, experimental data from dextran fractions; ◆, calculated for a SEC medium having a single pore radius of 60 Å. [Reproduced from Hagel (1996), with permission.]

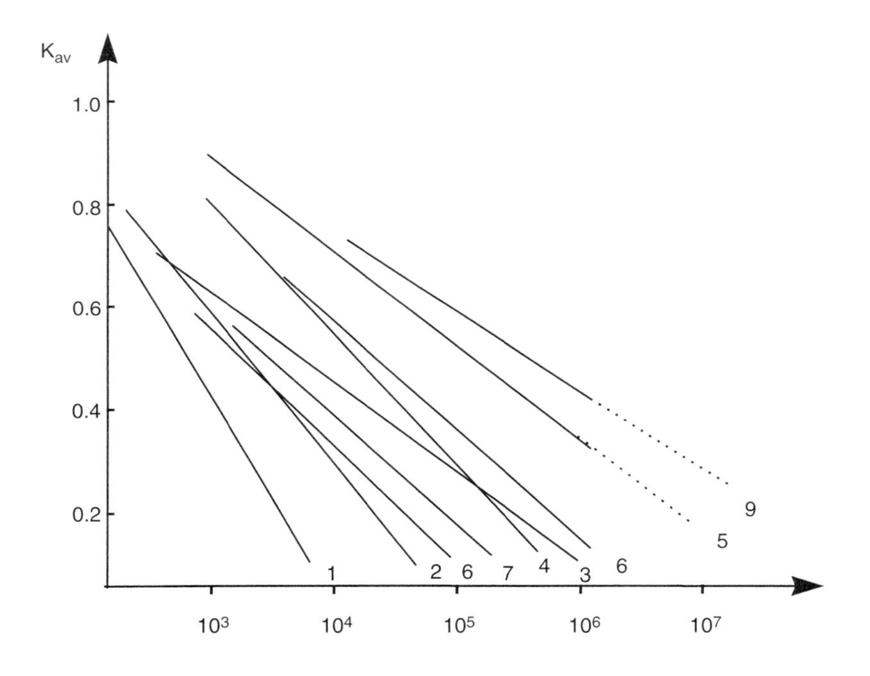

FIGURE 2.3 Selectivity curves of various modern SEC media. (1) Superdex peptide, (2) Superdex 75, (3) Superdex 200, (4) Superose 12, (5) Superose 6, (6) Sephacryl S-100 HR, (7) Sephacryl S-200 HR, (8) Sephacryl S-300 HR, and (9) Sephacryl S-400 HR. (Reproduced with permission of Amersham Pharmacia Biotech.)

available with a variety of selectivities and separation ranges to choose from. If high selectivity SEC medium is not available in the separation range required, another approach to increasing resolution is to increase efficiency of the separation (typically by selecting a smaller particle size) or to use a longer column, but this is never as effective as having the correct selectivity at hand.

Smaller particle sizes have been employed to increase separation efficiency and have reduced the column lengths, the amount of sample, and the total run times required for many SEC-based separations. However, the reduction in particle size has placed higher demands on the SEC chromatographic equipment. Smaller particles have increased the operation pressures required to push the liquid through the bed. Second, the reduction in the sample volumes with smaller columns has increased the demand on the system to minimize extra column effects such as pre- and postcolumn dead volumes, which cause band broadening and a loss of resolution (Huber, 1978; Hagel, 1989).Third, reduced sample mass requires more sensitive detectors and lower signal-to-noise ratios. These three factors elucidate the need for higher performance SEC systems to maintain the resolution gained by increasing efficiency with decreased particle size.

Two methodological approaches can be used to achieve molecular size and/or molecular weight distribution data. The first and most common approach is to use a calibrated system. The term "calibrated system" refers to establishing the retention volume for various standards and presumes a fixed volume from injector to detector, a precise and reproducible flow rate (especially if time will be used to report elution results), and the use of standards with similar shape and behavior in solution as that of the sample to be studied. This method usually requires running multiple samples followed by plotting the molecular mass versus retention volume and data fitting to produce a calibration curve. This classical approach relies on the availability of appropriate standards (i.e., homologous in size and shape relative to the unknown) to calibrate the SEC system. The quality of the standards will ultimately determine the accuracy of the determination (Yau *et al.*, 1979). A modification of this approach described previously (Hagel *et al.*, 1993) uses a polydisperse polymer mix with a well-characterized molecular size distribution for an integral calibration of the system.

When applied to the SEC column, the calibrated polydisperse polymer solution provides a large number of data points in a single run. Use of a standard with a molecular size distribution that encompasses the full separation range for the column allows the entire separation range to be calibrated in a single run (Fig. 2.4).

The second methodological approach to determining molecular size and molecular mass from SEC separations is to use on-line size-sensitive and/or mass detectors or to fractionate the eluent and use off-line analysis. Direct size and/or mass monitoring precludes the need for column calibration and alleviates concerns about nonideal SEC effects such as matrix/solute interactions affecting the retention volume. Liquid chromatography with in-line mass spectrometry (LC-MS) systems have been constructed for the analysis of low molecular weight materials from SEC columns (e.g., See Nylander *et al.*, 1995). Low angle and multiangle laser light scattering monitors (MALLS) are available commercially for monitoring much larger molecular sizes (e.g., see Jackson *et*

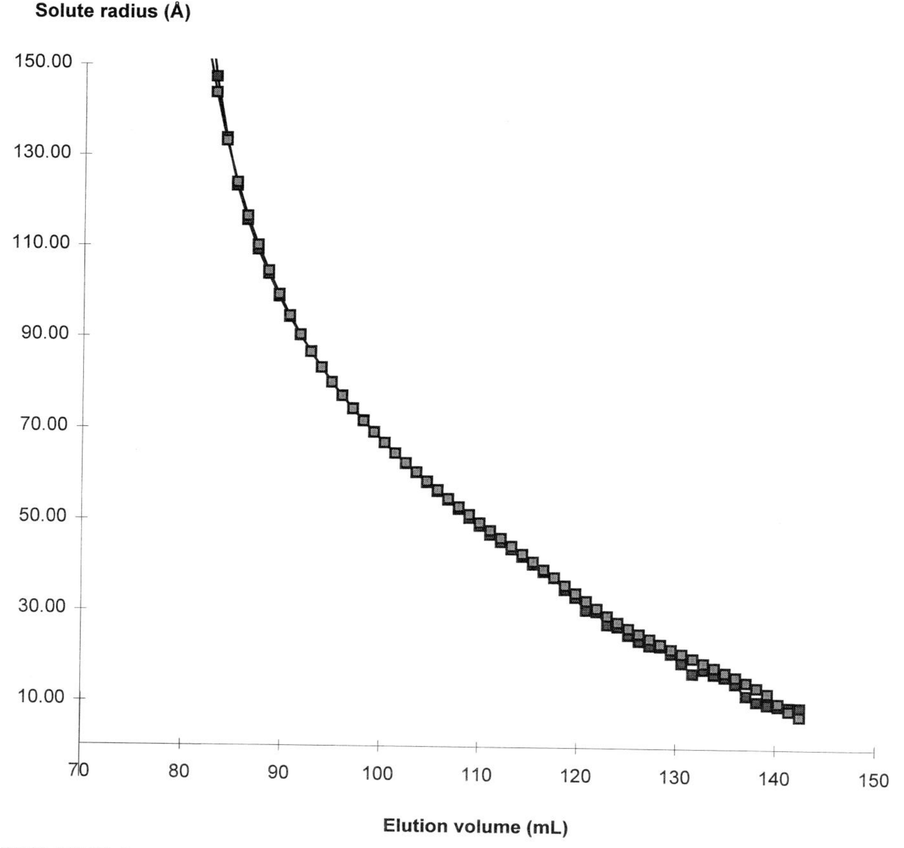

FIGURE 2.4 Calibration curve of dextran on Sephacryl S-300 SF. Calibration curves were calculated from one chromatogram of a broad MWD reference sample using data for the molecular mass distribution as obtained by a calibrated gel filtration column (□, upper curve) and on-line MALLS (■). The calibration curve was found useful for estimating the size of globular proteins. [Reproduced from Hagel *et al.* (1993), with permission.]

al., 1989), and differential viscosity detectors are useful for the characterization of polymer samples (Yau, 1990).

Accuracy and precision achieved either during molecular weight determinations or during the characterization of molecular weight distributions by either the calibrated system and the direct monitoring approach are often dependent on the equipment and technique employed. Developing robust methodologies must take into account the potential band-broadening effects that may result from differences in sample mass or sample volume, loss of column efficiency over time, or solute/solute interactions from sample components or contaminants that may build up in the system over time. To reduce the potential for such problems it is recommended to determine sample mass and volume effects over a series of test experiments, as this will also allow the quantitation of sample components in addition to the desired qualification. In both single and multicomponent samples the possibility of a concentration-dependent interaction of sample components or change in effective molecular size should be examined by establishing a range of concentrations suitable for sample analysis. The removal of particulates from the sample prior to introduction to the system, the use of high-quality solvents, which have also been filtered and degassed, and the development of routine cleaning in place (CIP) and storage protocols should be established to prevent the buildup of contaminants from samples, solvents, or the environment. It is usually easier and less expensive to prevent a buildup of contaminants than to remove them. In laboratory environments where large numbers of samples will be evaluated, periodic testing and calibration of system components and efficiency testing of the column are good preventative measures to ensure quality. The frequency of such testing must be determined for each particular experimental system, but daily calibration is a good starting frequency.

The analytical capability of a SEC column is sometimes judged by the peak capacity, which is the number of unique species that can be resolved on any given SEC column. This number will increase with decreased particle size, increased column length, and increased pore volume. Because small particle-sized medium generally has a lower pore volume and a shorter column length, peak capacities of ca. 13 for fully resolved peaks can be expected for high-resolution modern media as well as traditional media, (see Fig. 2.5). It was found that SEC columns differ widely in pore volume, which affects the effective peak capacity (Hagel, 1992).

Traditionally, Sephadex has been used for molecular weight analyses of proteins (Determan and Brewer, 1975) and clinical dextran (Granath and Kvist, 1967). Today, Sepharose, Sephacryl, or Superose is used for the assay of broad MWDs. Superdex is the premiere choice for attaining the highest resolution of components of similar size (see Fig. 2.1).

B. High-Resolution Fractionation

The most demanding application in preparative SEC is the purification of a single target molecule from similar-sized contaminants while allowing for a high sample load. This high-resolution fractionation application of SEC always requires a balance between load and resolution. Preparations of protein(s) or peptide(s) from complex biological mixtures are representative examples. SEC is normally relegated to the later steps of purification due to the limited sample volume that may be applied (e.g., as opposed to adsorptive purification processes). An exception to this is desalting, which may be used as a first conditioning step. Initial steps in a purification strategy typically require the use of adsorptive techniques to capture the target from a crude extract followed by one or more intermediate absorptive steps that exploit different physical/chemical attributes of the target molecule relative to the impurity. The effect of these early purification steps is to reduce the total sample volume and mass while enriching the sample with respect to the target molecule concentration. However, the remaining impurities will be more similar to the target molecule with each subsequent purification step. While reduction in sample mass and sample volume will reduce the size of the column required, the need for resolution can be great.

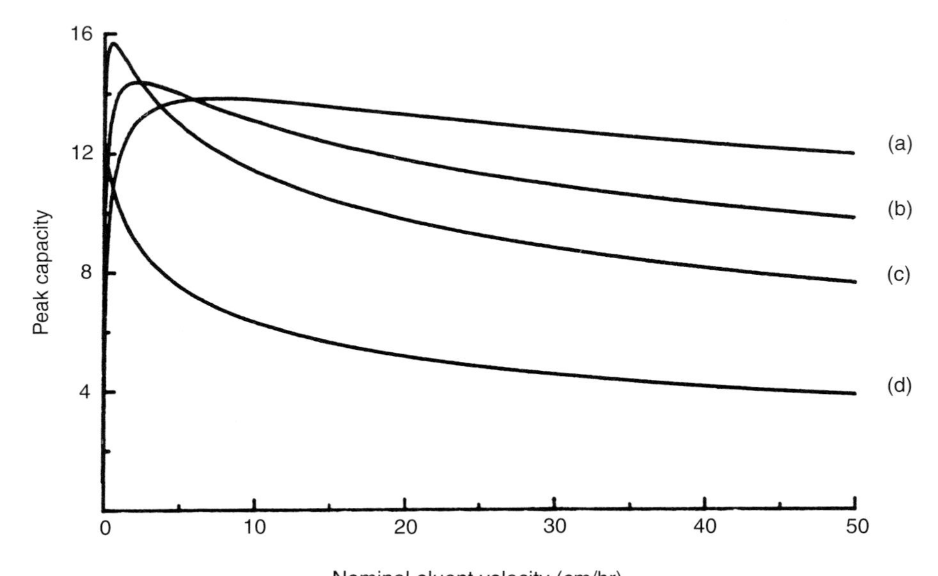

FIGURE 2.5 Theoretical peak capacity of different columns for SEC.

	d_p (μm)	V_p/V_o	L (cm)
a.	4	0.78	25
b.	10	1.30	30
c.	30	2.20	60
d.	100	2.25	100

The theoretical values were confirmed by testing commercial columns. [Reproduced from Hagel (1992) with permission.]

The smallest size difference that can be resolved is related to the pore volume, the solute shape, and the efficiency of the column (see Fig. 2.6). However, this is at very low loadings. At higher loadings the sample volume will contribute to zone broadening and may, in some cases, be the dominating factor for resolution. Thus, for fractionation, an optimum exists with respect to column efficiency (represented by the flow rate as operational parameter) and sample volume for processing a particular volume of feed per unit time. As a rule of thumb this optimum can be found at a relative sample volume of 2–5% of the column volume (Hagel *et al.*, 1989).

For the purpose of high-resolution fractionation, the gel medium must be tailor made to cope with different separation ranges. The Superdex family is designed for the high resolution of peptides and proteins having a molecular mass of 500 to 100,000. Also, Sephacryl media have found wide applicability as a final polishing step in process scale SEC (see Section III,C).

C. Desalting (Group Fractionation)

Separation of solutes having more than a decade of difference in size between the larger target species and the smaller contaminant species may be performed

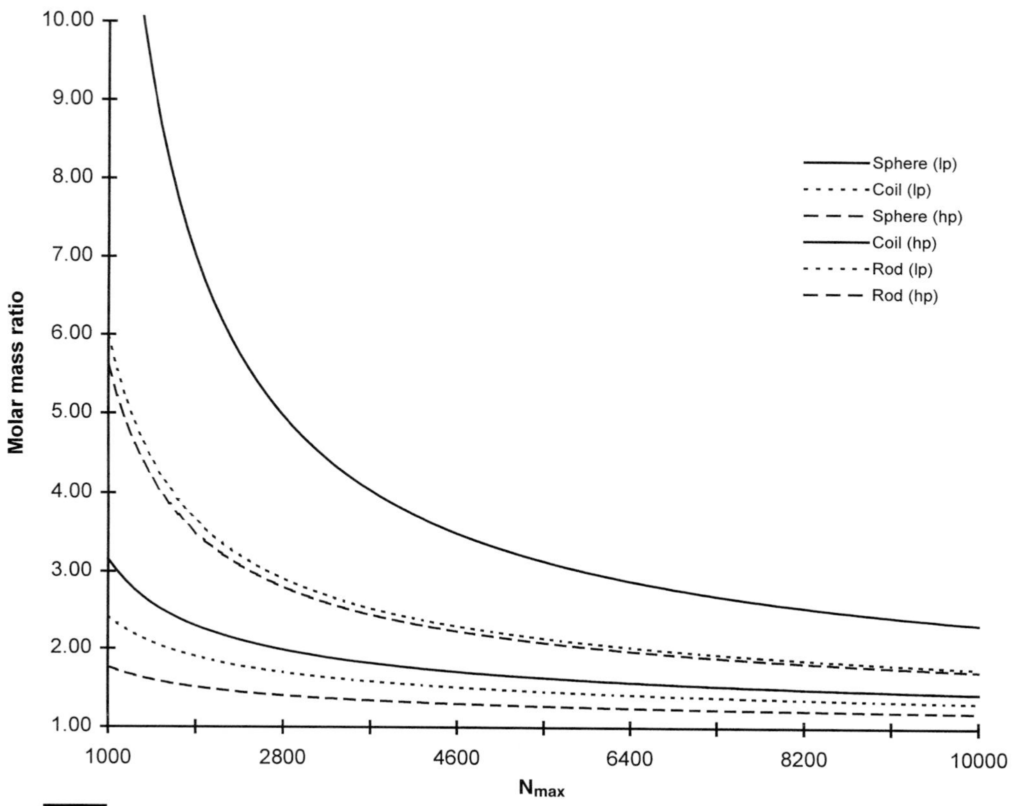

FIGURE 2.6 Resolvability of SEC. The molecular mass ratio needed for the complete resolution of solutes of various shapes as a function of column plate number. The influence of pore volume is given by the designations (lp), which stands for low porous for which V_p/V_o is 0.75, and (hp), which stands for highly porous for which V_p/V_o is typically 2.0. [Reproduced from Hagel (1993), with permission.]

by desalting or group fractionation. Applications such as buffer exchange to remove residual salts or solvents after precipitation of the target molecule, removal of unincorporated radio nucleotides after DNA-labeling reactions, and the removal of unreacted monomeric species after synthetic polymerization reactions are typical examples of group separations. The scale of operation for desalting applications may range from small laboratory-scale applications with only a few microliters of sample to process scale operations requiring throughput of hundreds of liters per day. Irrespective of scale, the primary operational goals during group separation are high yield and high throughput. The matrix for desalting is selected such that the pore size will just exclude the target molecule while allowing all contaminant species to permeate the matrix. Selection of a suitable matrix for desalting should be based primarily on the exclusion limit of the matrix (i.e., the smallest size molecule that will be excluded from the pores of the matrix) relative to the size of the target molecule. If the target molecule can permeate the matrix (e.g., the exlusion limit is too large), then flow-dependent zone broadening will occur. Matrixes with exclusion limits smaller than required may reduce the maximum sample volume, which can be applied due to a reduction in pore volume from a higher matrix volume, as well as cause impurities to be eluted prior to the total volume. However, the higher matrix volume typically imparts greater mechanical rigidity, and hence the selection of a matrix must always balance the gains in pore volume at the expense of attainable pressures and flow rates. However, because particle size has a minor effect on resolution, as compared to other modes of SEC, it is possible to use large particle-sized media for desalting, allowing high flow rates to be used (e.g., see Table 2.1).

The yield from desalting applications is typically 100% for proteins and media based on natural polymers (also, the surface area sensed by the nonpermeating solutes is very low). Because the target molecule is excluded from the pore volume, there is little sample dilution. With only smaller, faster diffusing contaminant species needing to permeate the matrix, flow-dependent band broadening of the chromatographic zone containing the larger target molecules is avoided, allowing for rapid separations.

Throughput in desalting, the amount of sample processed per unit time, depends on the volume and concentration of sample applied and the flow rate. The theoretical limit or maximum volume that can be applied is equal to the pore volume of the column. Hence, low matrix volumes and high pore volumes are essential criteria for selecting the appropriate desalting media. The maximum flow rate is typically a function of the particle size and the mechanical strength of the matrix or of flow capabilities of the system components. As long as the matrix is not compressed (which will reduce the pore volume), the flow may be increased or the viscosity of the mobile phase from the sample concentration may be increased. Linear velocities up to 200 cm/hr are readily attainable and quite reasonable when using columns packed to a bed height of 10–20 cm with Sephadex G-25.

Desalting applications are less dependent on particle size than high-resolution applications, which may require higher efficiency to achieve the desired resolution. Because the target molecule need not permeate the pore structure, larger particle sizes, which allow higher flow rates at lower pressures, are preferred. Similarly, because the pressure drop across a column is proportional to column length, short columns may be operated at higher flow rates. Columns around 10 cm in length are typical for laboratory desalting applications. Columns for process scale desalting may be 60–100 cm long to give beds with the required amount of gel (e.g., 2500 liters).

The pore size and matrix rigidity make Sephadex very suitable for desalting or buffer exchange. This is especially true for Sephadex G-25, which still is the premiere choice for many desalting operations, in micropreparative scale as well as in process scale. This is illustrated by the desalting of 50 μl reduction/alkylation mixture on a 10 × 0.32-cm i.d. column of Sephadex G-25 Superfine prior to sequence analysis (Hellman *et al.,* 1990) and the process scale desalting of 875 liters of a crude enzyme preparation on a 100 × 180-cm i.d. stainless-steel column packed with coarse Sephadex G-25 (Horton, 1972). If the group separation goal is to purify larger substances, e.g., DNA from proteins, then a medium of larger pore size, e.g., Sepharose, may be needed.

III. CHARACTERISTICS OF GEL-FILTRATION/SIZE EXCLUSION MEDIA

The development and optimization of porous supports for chromatographic applications have depended on advances in polymer chemistry together with advances in particle manufacturing technology. Separations that required hours or days to achieve with early SEC supports are now typically performed in less than an hour. The following description of various types of media is presented in the approximate chronological order in which the various media were developed.

A. Sephadex

Sephadex, the first commercially available gel-filtration media, is formed from cross-linking dextran with epichlorohydrin and is available as a dry powder or in various prepacked column types for desalting. The various Sephadex G types listed in Table 2.1 are bead-formed gels prepared by varying the amount of cross-linking to produce different selectivities. Increasing the degree of cross-linking per gram dry gel increases the mechanical stability and decreases the porosity. Increasing the amount of cross-linking per gram dry gel also reduces the amount of free hydroxyl groups, which limits water regain, and results in differential swelling characteristics of the various Sephadex G types (Table 2.1). The various grades (e.g., bead sizes) of Sephadex G types listed in Table 2.1 are produced by sieving the dry Sephadex preparations. Characteristics of commercially available columns prepacked with Sephadex are given in Table 2.2.

Sephadex is insoluble in all solvents that do not chemically degrade the cross-linked dextran. However, the gel may shrink or swell in the presence of organic solvents, which will impact the porosity of the gel and hence the selectivity. Shrinkage of Sephadex in ethanol is used for producing stable beds by a special packing procedure. Dextran, like most carbohydrates, may be subject to chemical modifications such as methylation, amination, carboxylation, or sulfonation at the reducing end of the polymer or at various hydroxyl groups along the polymer. Although many of the reactions have been used to introduce various ligands to Sephadex for adsorptive chromatographic applications, the strongly oxidative conditions required to derivitize dextrans should be avoided when using Sephadex. Although stable over the pH range of 2 to 10, the prolonged exposure of Sephadex to strongly acidic or basic conditions may cause hydrolysis of the gel by disruption of the glycosidic linkages or ring opening of the component glucose residues. However, exposure of Sephadex to 0.1 *M* HCl or 0.1 *M* NaOH for 2 hr or storage in 0.02 *M* HCl for 6 months showed no effect on chromatographic performance. Sephadex may be used at elevated temperatures and may even be autoclaved at 120°C for 30 min without affecting chromatographic performance. Above 120°C, Sephadex may begin to caramelize. Freezing and thawing of hydrated Sephadex may result in disruption of the bead structure and should be avoided.

The maximal flow rates and operational pressures that can be used with Sephadex G types depend primarily on the concentration of dextran and the degree of cross-linking used to stabilize the particle. Sephadex G-10, G-15, G-25, and G-50 are quite highly cross-linked, and the first three types are consid-

TABLE 2.1 Characteristics of Sephadex G Types

Sephadex type	Grade	Dry bead diameter (μm)	Fractionation range peptides and proteins (g/mol)	Fractionation range dextrans (g/mol)	Swelling factor (ml/g dry Sephadex)	Maximum operating pressure[a] (cm H_2O)	Permeability K_0	Maximum linear velocity[a] (cm/hr)	Swelling time (h) 20°C	Swelling time (h) 90°C
G-10		40–120	−700	−700	2–3		19	>100[b]	3	1
G-15		40–120	−1,500	−1,500	2.5–3.5		18	>100[b]	3	1
G-25	Coarse	100–300	1,000–5,000	1,000–5,000	4–6		290	>100[b]	3	1
	Medium	50–150					80		3	1
	Fine	20–80					30		3	1
	Superfine	10–40					9		3	1
G-50	Coarse	100–300	1,500–30,000	500–10,000	9–11		400	>100[b]	3	1
	Medium	50–150					145		3	1
	Fine	20–80					36		3	1
	Superfine	10–40					13.5		3	1
G-75		40–120	3,000–80,000	1,000–50,000	12–15	160		77	24	3
	Superfine	10–40	3,000–70,000			160		18	24	3
G-100		40–120	4,000–150,000	1,000–100,000	15–20	96		50	72	5
	Superfine	10–40	4,000–100,000			96		12	72	5
G-150		40–120	5,000–300,000	1,000–150,000	20–30	36		23	72	5
	Superfine	10–40	5,000–150,000		18–22	36		6	72	5
G-200		40–120	5,000–600,000	1,000–200,000	30–40	16		12	72	5
	Superfine	10–40	5,000–250,000		20–25	16		3	72	5

[a]Maximum linear velocities and operating pressures were determined in 2.5-cm-diameter columns packed to a bed height of ca. 30 cm.

[b]Sephadex G-10–G-50 behave as rigid spheres and the maximum linear velocity may be calculated by Darcy's law where $U = K_o\,(\Delta p/L)$; U is the linear velocity (cm/hr); K_o is the permeability; Δp is the pressure drop over the column (cm H_2O); and L is the bed height (cm).

ered rigid particles such that the linear velocity of eluent achieved through the column is linearly proportional to the pressure applied, i.e., which obeys the simple relationship (Darcy's law)

$$u = K\ (\Delta p/L), \tag{1}$$

where u is the fluid velocity (cm/hr), Δp is the pressure drop over the bed (cm H_2O), L is the bed height (cm), and K is a constant that depends on the particle diameter of the bed material, the viscosity of the eluent, and the void fraction of the bed.

Assuming an eluent viscosity of 1 cP, K can be read from Table 2.1 and the theoretical linear velocity of an eluent at any given pressure can be calculated. For the less rigid Sephadex G types, the maximum operating pressures at which the relation between superficial velocity and applied pressure is still linear are given in Table 2.1. Exceeding the pressures listed will result in bead compression, a reduction in pore volume, and a decreased flow rate.

Sephadex is supplied as a free-flowing dry powder that must be hydrated and degassed prior to use. Swelling and degassing of the Sephadex slurry can be accomplished by simply autoclaving the Sephadex in an excess of aqueous solvent for 30 min at 120°C. Swelling at lower temperatures requires an increase in swelling times, as indicated in Table 2.1. After swelling Sephadex G-100, G-150, and G-200, any fines present in solution may be removed by swirling a 50% gel slurry in a flask and allowing the gel to settle for 1 hr and then decanting the excess liquid containing the fines. In its hydrated form, Sephadex should be stored near neutral pH in a suitable antimicrobial agent. Sephadex in its dry form is stable for more than 10 years at room temperature.

Carbohydrates, and hence Sephadex, may contain a small number of residual carboxyl groups in addition to the numerous hydroxyl groups. The carboxyl groups may cause ionic interactions with cations in solution or may cause exclusion of anionic solutes from the beads. These effects are readily overcome by maintaining an ionic strength above 20 m*M* or operating at a pH were the solute of interest is uncharged. High ionic strengths may promote hydrophobic interactions of small organic hydrophobic molecules with Sephadex (Yano, 1980, Haglund and Marsden, 1980). Presumably, the alkyl backbone of the cross-linker used in Sephadex G types is involved for the hydrophobic retention mechanism. These interactions, which are more prevalent in highly cross-linked Sephadex G types, can often be eliminated by lowering the ionic strength of the eluent or by reducing the polarity of the mobile phase by including small amounts of ethanol or methanol in the eluent. The hydroxyl groups in Sephadex have the potential to hydrogen bond with some solutes. Hydrogen bonding may be eliminated by the inclusion of chaotrophic agents such as urea or guanidine HCl in the eluent without affecting selectivity.

B. Sepharose

The family of agarose-based gels, Sepharose , Sepharose CL, and Sepharose Fast Flow, are bead-formed gels prepared from 2, 4, or 6% agarose solutions. The matrix porosity decreases and rigidity of the bead structure increases with increasing agarose concentrations. The open pore structure and broad

TABLE 2.2 Characteristics of Commercially Available Perpacked Sephadex Columns

Column	Sephadex type	Column type[a]		Bed volume (ml)	Column dimensions		Column fitting		Maximum pressure/flow (mPa)/ml/min	Supplied in	Recommended use
		Tube	Frit		i.d. (mm)	L (mm)	Top	Bottom			
Fast desalting column HR 10/10	G-25 SF	Glass	PE	8	10	100	M6	M6	1.2 mPa (2–6 ml/min)	20% EtOH	For FPLC or HPLC systems of 0.5- to 2-ml samples
Fast desalting column PC 3.2/10[b]	G-25 SF	Glass	Peek	0.8	3.2	100	M6	M6	1.2 mPa (0.01–0.9 ml/min)	20% EtOH	HPLC or SMART system ≤0.2 ml samples
HiTrap desalting	G-25 SF	PP	PE	5	16	25	M6	M6	0.3 mPa (1–10 ml/min)	20% EtOH	Any LC system or syringe 0.5- to 1.5-ml samples
Gravity operated											
PD-10	G-25 M	PP	PE	9.1	15	54	Open	Leur Taper	Gravity Flow	0.15% Kathon CG	≤2.5 ml of samples
NAP 5	G-25 M[c]	PP	PE	1.5	9	28	Open	Leur Taper	Gravity Flow	0.15% Kathon CG	DNA purification (>10 mers) 0.5-ml sample
NAP 10	G-25 M[c]	PP	PE	3.6	13	27	Open	Leur Taper	Gravity Flow	0.15% Kathon CG	DNA purification (>10 mers) 1-ml sample

NAP 25	G-25 M[c]	PP	PE	9	15	50	Open	Leur Taper	Gravity Flow	0.15% Kathon CG	DNA purification (>10 mers) 2.5-ml sample
NICK	G-50 M[c]	PP	PE	1.2	9	19	Open	Leur Taper	Gravity Flow	0.15% Kathon CG	DNA purification (>20 mers) <0.1-ml sample
Centrifuge operated											
Nick spin[d]	G-50 F[c]	PP	PE	1.1	7	28	Open	Leur Taper	Gravity Flow	0.15% Kathon CG	DNA purification (>20 mers) 0.075- to 0.150-ml samples
MicroSpin G-50[d]	G-50 F[c]	PP	PE	0.5	NA[e]	NA[e]	Open	Open	Gravity Flow	0.15% Kathon CG	25- to 50-μl samples
MicroSpin G-25[d]	G-25 F[c]	PP	PE	0.5	NA[e]	NA[e]	Open	Open	Gravity Flow	0.15% Kathon CG	25- to 50-μl samples

[a]PP, polypropylene; PE, polyethylene.
[b]Requires column holder unless used with SMART system.
[c]DNA grade Sephadex: Tested for nonspecific adsorption of DNA.
[d]Columns are designed for centrifuge operation.
[e]Bed shape is conical and depends on rotor angles and *g* force applied.
Bold type indicates male-type fitting.

fractionation ranges of the Sepharose gels, introduced in the 1960s, enabled the preparation and characterization of large biomolecules and polymer distributions by SEC. The original noncross-linked Sepharoses (2B, 4B, and 6B) are formed by cooling hot solutions of agarose, whereby individual agarose chains come together to form double helices that subsequently aggregate to form bundles and the stable polymer network (Låås, 1975). The agarose gel is stabilized primarily by hydrogen bonding between individual polysaccharide chains and readily dissociates in the presence of chaotropic agents and melts above 40°C. Cross-linked Sepharose CL (CL-2B, CL-4B, and CL-6B) gels are prepared from the corresponding noncross-linked Sepharose gels using 2,3-dibromopropanol as a cross-linking agent. Sepharose CL gels have substantially the same porosity as noncross-linked gels but have superior thermal and chemical stability. The Sepharose CL series of gels can be used in the presence of chaotropic agents and may also be autoclaved. For this reason the noncross-linked gels are no longer recommended for use in SEC. More recently, improvements in the cross-linking of Sepharose gels have led to the introduction of Sepharose Fast Flow medium, which has superior flow performance characteristics relative to the Sepharose CL series. Characteristics of the Sepharose-based gels are given in Table 2.3.

Cross-linked Sepharose gels (e.g., Sepharose CL and Sepharose Fast Flow) are stable to a broad range of pH (Table 2.3) and are insensitive to the presence of chaotropic agents. Prolonged exposed to strong oxidizing agents may cause limited hydrolysis of the polysaccharide chains and should be avoided. However, the stability in alkaline pH is quite good and even exposure to 2 *M* NaOH at 70°C for several hours has no apparent effect on the chromatographic performance of cross-linked gels. The structure of the cross-linked gels is quite stable to many organic solvents and may be used with ethanol, acetonitrile, dichloroethane, chloroform, tetrahydrofuran, or dimethyl sulfoxide with no significant change in pore size. Sepharose gels may be sterilized repeatedly at pH 7 at 120°C without significant changes in porosity or rigidity. Freezing and thawing of Sepharose-based gels may result in disruption of the bead structure and should be avoided.

Sepharose gels are supplied in 20% ethanol and are not available in prepacked columns. Sepharose should be stored in the presence of an antimicrobial agent (i.e., 20% ethanol or 10 m*M* NaOH) and is stable for more than 10 years.

The broad pore size distribution of Sepharose makes it well apt for the analysis of broad molecular mass distributions of large molecules. One example is given by the method for determination of MWD of clinical dextran suggested in the Nordic Pharmacopea (Nilsson and Nilsson, 1974). Because Superose 6 has the same type of pore size distributions as Sepharose 6, many analytical applications performed earlier on Sepharose have been transformed to Superose in order to decrease analysis time. However, Sepharose is suitable as a first "try out" when no information about the composition of the sample, in terms of size, is available.

C. Sephacryl

Sephacryl, a gel formed by cross-linking allyl dextran with *N,N'*-methylenebisacrylamide, was introduced in the 1970s. The original Sephacryl

TABLE 2.3 Characteristics of Sepharose-Based Gels

Type	Bead diameter (μm)	Composition (% agarose)	Size exclusion pore dimension[a] (nm)	Fractionation range		Maximum operating pressure[b] (kPa)	Maximum operating linear velocity[b] (cm/hr)	pH stability
				Proteins (M_r)	Dextrans (M_r)			
Sepharose 2B	60–200	2		7×10^4–4×10^7	1×10^5–2×10^7	4	10	3–11
Sepharose 4B	45–165	4		6×10^4–2×10^7	3×10^4–5×10^6	8	11	3–11
Sepharose 6B	45–165	6		1×10^4–1×10^6	1×10^4–1×10^6	20	14	3–11
Sepharose CL-2B	60–200	2, cross-linked	75	7×10^4–4×10^7	1×10^5–2×10^7	5	15	3–13
Sepharose CL-4B	45–165	4, cross-linked	42	6×10^4–2×10^7	3×10^4–5×10^6	12	26	3–13
Sepharose CL-6B	45–165	6, cross-linked	24	1×10^4–1×10^6	1×10^4–1×10^6	20	30	3–13
Sepharose 4 Fast Flow	45–165	4, cross-linked	45	6×10^4–2×10^7	3×10^4–5×10^6	80	240	2–12
Sepharose 6 Fast Flow	45–165	6, cross-linked	29	1×10^4–1×10^6	1×10^4–1×10^6	100	300	2–12

[a]Data reproduced from Hagel *et al.* (1996).

[b]Maximum operating linear velocities and pressures were determined in 2.5 × 30-cm beds for Sepharose and Sepharose CL media and with 5 × 15-cm beds for Sepharose fast flow gels, assuming pure/water as mobile phase.

Superfine (SF) gels offered a narrower bead size distribution (45–105 μm, d_{50} = 70 μm) and a more rigid particle extended the separation ranges to smaller solutes relative to the available agarose-based gels. The smaller, more rigid particle provided higher efficiency separations and reduced separation times by a factor of 2–3. In the early 1980s improved Sephacryl high resolution (HR) gels were introduced. From the optimization of particle formation, the Sephacryl HR series of gels offer a smaller particle diameter and size distribution (25–75 μm, d_{50} = 50 μm), and from optimization in cross-linking the original selectivities were unchanged and a smaller average pore size gel, Sephacryl S-100, was introduced. Characteristics of the current Sephacryl gels are presented in Table 2.4 and available prepacked columns in Table 2.5. Columns packed with Sephacryl HR typically exhibit a packing efficiency (reduced plate height, *h*, see Section V) of 1.9 to 2.1, a matrix volume that ranges from near 8% of the total bed volume for Sephacryl S-500, increasing to 15% for Sephacryl S-100, and the void fraction is typically 35% of bed volume (Hagel *et al.*, 1989).

Sephacyl is stable to a broad range of solvents, including 1 *M* acetic acid, 8 *M* urea, formamide, dimethyl sulfoxide and acetonitrile. There are negligible changes in the swollen volume of Sephacryl from changes in pH or ionic strength. Some shrinking or swelling may occur in changing between various solvent systems (Table 2.6). Sephacryl HR media are stable for prolonged exposure of pH in the range of 3–11. However, a short duration exposure for CIP may be performed from pH 2 to 13 with no effect on the chromatographic performance. Sephacryl may be autoclaved repeatedly at pH 7, 120°C without significant changes in porosity or rigidity. Freezing and thawing of Sephacryl-based gels may result in disruption of the bead structure and should be avoided. Sephacryl is supplied in 20% ethanol.

Sephacryl offers a matrix that is suitable for laboratory as well as large-scale applications. Sephacryl was used in an inexpensive setup for the analysis of dextran MWD of clinical samples. The column, Sephacryl S-300, was calibrated with one broad fraction of dextran covering the molecular weight range of interest. The elution profiles of dextran were determined from anthrone analysis of collected fractions (Fig. 2.7, page 49), the molecular weights were converted to size, and the clearance of the glomerular barrier plotted (see Fig. 2.8, page 50).

A process scale application of Sephacryl S-100 is demonstrated by the polishing of a pharmaceutical in a system of three BPSS columns giving a final bed volume of 2500 liters (Fig. 2.9, page 51).

D. Superose

In the mid-1980s, advances in agarose particle technology gave rise to smaller and highly cross-linked agarose particles known as Superose (Andersson *et al.*, 1995). Superose 6 and Superose 12 are agarose-based gels formed from 6 and 12% agarose solutions, respectively. The decreased particle size allowed for high efficiency separations and reduced run times. The increased cross-linking and particle construction maintained the chemical stability and increased the physical stability in comparison to other agarose-based particles. Characteristics of Superose and the larger particle size Superose prep grade gels (for scaleup of preparative separations) are presented in Table 2.7 (page

TABLE 2.4 Characteristics of Sephacryl

Type	Bead diameter (μm)	Size exclusion pore dimension[a] (nm)	Fractionation range		Maximum operating pressure/velocity[b] (kPa)/(cm/hr)	Recommended operating linear velocity (cm/hr)	pH stability (short term)
			Proteins (M_r)	Dextran (M_r)			
Sephacryl S-100 HR	25–75	6.6	1×10^3–1×10^5	1×10^3	3/60	20–50	2–13
Sephacryl S-200 HR	25–75	7.7	5×10^3–2.5×10^5	1×10^3–8×10^4	3/60	20–50	2–13
Sephacryl S-300 HR	25–75	13	1×10^4–1.5×10^6	2×10^3–4×10^5	3/60	20–50	2–13
Sephacryl S-400 HR	25–75	31	2×10^4–8×10^6	1×10^4–2×10^6	3/60	20–50	2–13
Sephacryl S-500 HR	25–75			4×10^4–4×10^7	3/60	20–50	2–13
Sephacryl S-1000 SF	40–105			5×10^5–$>10^8$	2/50	10–40	3–11

[a]Data reproduced from Hagel *et al.* (1996).
[b]Maximum pressure and linear velocity were determined in 1.6 × 60-cm beds at 25°C with H_2O.

TABLE 2.5 Characteristics of Prepacked Sephacryl Columns

Column	Bed volume (ml)	Column dimensions [diameter (cm) × length (cm)]	Column fittings	Column materials[a] (Tube/Frit)	Maximum operating pressure (kPa)	Maximum linear velocity[b] (cm/hr)
HiPrep 16/60 Sephacryl S-100	120	1.6 × 60	M6	PP/Nylon	3	30
HiPrep 26/60 Sephacryl S-100	300	2.6 × 60	M6	PP/Nylon	3	30
HiPrep 16/60 Sephacryl S-200	120	1.6 × 60	M6	PP/Nylon	3	30
HiPrep 26/60 Sephacryl S-200	300	2.6 × 60	M6	PP/Nylon	3	30
HiPrep 16/60 Sephacryl S-300	120	1.6 × 60	M6	PP/Nylon	3	30
HiPrep 26/60 Sephacryl S-300	300	2.6 × 60	M6	PP/Nylon	3	30

[a]PP, polypropylene (surgical grade).
[b]Maximum linear velocities were determined with H_2O at 25°C.

51). Characteristics of columns prepacked with Superose are presented in Table 2.8 (page 52).

Ionic interactions between solutes and Superose are negligible at ionic strengths above 50 m*M*. However, some hydrophobic interactions have been observed with small hydrophobic peptides, membrane proteins, and lipopro-

TABLE 2.6 Swelling Characteristics of Sephacryl HR in Various Solvents[a]

Solvent	Bed volume (ml)				
	S-100 HR	S-200 HR	S-300 HR	S-400 HR	S-500 HR
Water	100	100	100	100	100
Formamide	110	115	100	100	100
Dimethyl sulfoxide	100	110	90	90	95
Methanol	100	100	100	100	100
Ethanol	100	100	95	95	100
Acetone	85	85	85	85	90

[a]Reproduced with permission of Pharmacia Biotech.

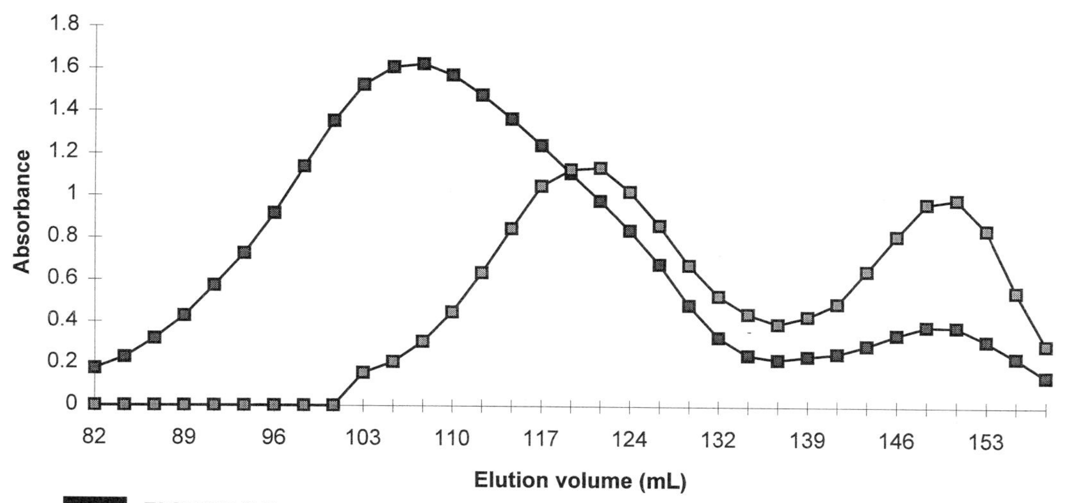

FIGURE 2.7 SEC elution profiles of dextran in clinical samples, serum (■) and urine (□). The first peak represent dextran and the second peak inulin (used as a reference for clearance). The content of carbohydrates was determined in collected fractions with the anthrone method. [Reproduced from Hagel *et al.* (1993), with permission.]

teins that elute later than predicted. The degree of interaction is less for the Superose prep grade than observed with the prepacked Superose 6 columns, which exhibit weaker effects than the Superose 12 columns (Pharmacia, 1991). Superose 6 and Superose 12 columns are extremely robust and have been shown to be unaffected by 1000 repetitive serum injections at basic pH (Johansson and Ellström, 1985).

Superose-based media are stable for the prolonged exposure of pH in the range of 3–12. However, short duration exposure for CIP may be performed from pH 1 to 14 with no effect on the chromatographic performance. Superose may be autoclaved repeatedly at pH 7, 120°C without significant changes in porosity or rigidity. Freezing and thawing of Superose-based gels may result in disruption of the bead structure and should be avoided. Superose is supplied in 20% ethanol.

Superose offers the same "gentle" agarose base matrix as Sepharose. Smaller particle size yields faster analysis (e.g., see Andersson *et al.*, 1985). Desalting large molecules from medium-sized molecules may require a pore size that is larger than that for Sephadex G-25. Superose 6 prep grade was used to selectively exclude plasmid DNA from the mixture of proteins and RNA. Gel filtration was found to be a favorable alternative to CsCl density gradient centrifugation for this application (McClung and Gonzales, 1989).

E. Superdex

Superdex is a composite media of agarose and dextran. The base particle is formed of cross-linked agarose to which dextran is then covalently attached. The resulting family of Superdex-based gels has been constructed to provide

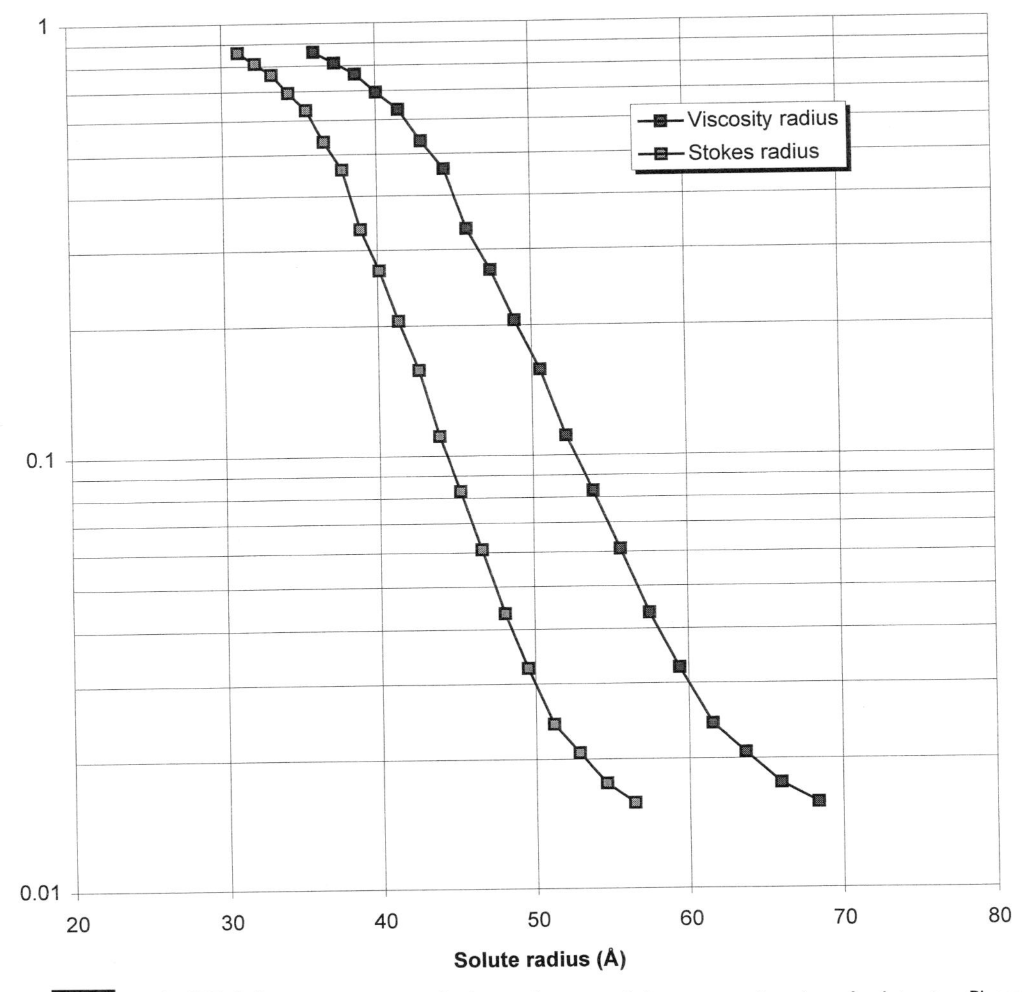

FIGURE 2.8 Determination of relative clearance of dextran as a function of solute size. Please note that Stokes radius (left curve) will yield too low a value for the renal clearance barrier. [Reproduced from Hagel *et al.* (1993), with permission.]

high-resolution separations over relatively narrow fractionation ranges compared to traditional agarose media. The highest selectivity achievable for any specific size range in SEC separations requires a porous particle with a very narrow distribution of pore sizes or ideally a single pore size. The theoretical limit of selectivity based on a single pore size of 60 Å in comparison to Superdex 75 is shown in Fig. 2.2. It is readily apparent that the theoretical limit in selectivity is available commercially in certain separation ranges. Characteristics of Superdex media are presented in Table 2.9. Characteristics of columns prepacked with Superdex are presented in Table 2.10.

Superdex is stable to the same conditions as cross-linked Sepharose media (see earlier). The chemical stability in acidic and basic solutions, as well as ionic and hydrophobic interactions of various molecules, has been studied

FIGURE 2.9 Industrial fractionation by SEC using Sephacryl S-100 in three BPSS columns (L = 30 cm, i.d. = 140 cm) to give a total bed volume of 1500 liters. Courtesy of R. Hersberg. (Reproduced by permission of Amersham Pharmacia Biotech.)

extensively (Hellberg *et al.*, 1996). Like Sephadex, small amounts of dextran can leach from the Superdex particle under prolonged exposure to strong acids (100 m*M* HCl) or bases (1 *M* NaOH). There is also a corresponding increase in the available pore volume with the loss of dextran (Drevin and Johansson, 1991). Residual carboxyl groups are present in most carbohydrates and Superdex contains typically less than 1 μmol/ml gel. Ionic interactions are readily overcome by maintaining an ionic strength above 50 m*M* in the eluent. The hydrophobic interaction of aliphatic alcohols with Superdex has been shown to increase as a function of the amount of dextran coupled to the Superdex (the smaller the exclusion limit, the higher the dextran concentration used) and is enhanced by increasing the ionic strength of the eluent (Hellberg *et al.*, 1996).

TABLE 2.7 Characteristics of Superose

Type	Bead size d_{50} (μm)	Size exclusion pore dimension[a] (nm)	Exclusion limit protein (M_r)	Fractionation range protein (M_r)	pH stability (long term/short term)
Superose 12	10	13	2×10^6	1×10^3–3×10^5	3–12/1–14
Superose 12 prep grade	34	not determined	2×10^6	1×10^3–3×10^5	3–12/1–14
Superose 6	13	25	4×10^7	5×10^3–5×10^6	3–12/1–14
Superose 6 prep grade	34	21	4×10^7	5×10^3–5×10^6	3–12/1–14

[a]Data reproduced from Hagel *et al.* (1996).

TABLE 2.8 Characteristics of Prepacked Superose Columns

Column	Column dimension (i.d. × L) (cm × cm)	Bed volume (ml)	Column materials[a] (Tube/Frit)	Column fittings[b] (inlet/outlet)	Theoretical plates (N/m)	Maximum operating pressure/flow rate (kPa)/(ml/hr)
Superose 12 PC 3.2/30	0.32 × 30	2.4	Glass/PP	M6/M6	>40,000	12/0.1
Superose 12 HR 10/30	1.0 × 30	24	Glass/PP	**M6/M6**	>40,000	15/1.0
BioPilot Superose 12 prep grade 35/600	3.5 × 60	600	Glass/Nylon	**M6/M6**	>10,000	8.0/6
BioPilot Superose 12 prep grade 60/600	6.0 × 60	1700	Glass/Nylon	**M6/M6**	>10,000	24/6
Superose 6 PC 3.2/30	0.32 × 30	2.4	Glass/PP	M6/M6	>30,000	24/0.1
Superose 6 HR 10/30	1.0 × 30	24	Glass/PP	**M6/M6**	>30,000	30/1.5

[a]PP, polypropylene.
[b]**Bold** lettering denotes male fittings.

Superdex and prepacked Superdex columns are supplied in 20% ethanol. All Superdex may be autoclaved repeatedly at pH 7, 120°C without significant changes in porosity or rigidity. Freezing and thawing of Superdex-based gels may result in disruption of the bead structure and should be avoided.

TABLE 2.9 Characteristics of Superdex

Type	Bead size d_{50} (μm)	Size exclusion pore dimension[a] (nm)	Exclusion limit peptide (M_r)	Fractionation range		pH stability
				Peptide (M_r)	PEG[b] (M_r)	
Superdex peptide	13	3.3	20,000	100–7,000	100–4,000	1–14
Superdex 30 prep grade	34	3.8	30,000	3,000–10,000	400–7,000	3–12
					Dextran (M_r)	
Superdex 75	13	6.0	100,000	3,000–70,000	500–30,000	3–12
Superdex 75 prep grade	34		100,000	3,000–70,000	500–30,000	3–12
Superdex 200	13	13	1,300,000	10,000–600,000	1,000–100,000	3–12
Superdex 200 prep grade	34		1,300,000	10,000–600,000	1,000–100,000	3–12

[a]Data reproduced from Hagel *et al.* (1996).
[b]PEG, polyethylene glycol.

TABLE 2.10 Characteristics of Prepacked Superdex Columns

Column type	Bed volume (ml)	Column dimension (i.d. × L) (cm × cm)	Column material[a] (tube/frit)	Column fittings[b] (inlet/outlet)	Theoretical plates (N/m)	Maximum operating pressure/flow rate (kPa)/(ml/min)
Superdex peptide PC 3.2/30	2.4	0.32 × 30	Glass/PP	M6/M6[c]	>30,000	20/0.150
Superdex 75 PC 3.2/30	2.4	0.23 × 30	Glass/PP	M6/M6[c]	>30,000	24/0.100
Superdex 200 PC 3.2/30	2.4	0.32 × 30	Glass/PP	M6/M6[c]	>30,000	12/0.100
Superdex peptide PE 7.5/300	13	0.75 × 30	PEEK/PEEK	1/16/1/16	>30,000	15/0.70
Superdex peptide HR 10/30	24	1.0 × 30	Glass/PP	**M6/M6**	>30,000	15/1.2
Superdex 75 HR 10/30	24	1.0 × 30	Glass/PP	**M6/M6**	>30,000	18/1.5
Superdex 200 HR 10/30	24	1.0 × 30	Glass/PP	**M6/M6**	>30,000	15/1.0
HiLoad 16/60 Superdex 30 pg	120	1.6 × 60	Glass/nylon	**M6/M6**	>13,000	3/2.5
HiLoad 26/60 Superdex 30 pg	300	2.6 × 60	Glass/nylon	**M6/M6**	>13,000	3/6.0
HiLoad 16/60 Superdex 75 pg	120	1.6 × 60	Glass/nylon	**M6/M6**	>13,000	3/2.5
HiLoad 26/60 Superdex 75 pg	300	2.6 × 60	Glass/nylon	**M6/M6**	>13,000	3/6.0
HiLoad 16/60 Superdex 200 pg	120	1.6 × 60	Glass/nylon	**M6/M6**	>13,000	3/2.5
HiLoad 26/60 Superdex 200 pg	300	2.6 × 60	Glass/nylon	**M6/M6**	>13,000	3/6.0
BioPilot Superdex 75 pg 35/600	600	3.5 × 60	Glass/nylon	**M6/M6**	>13,000	3/8
BioPilot Superdex 75 pg 60/600	1700	6.0 × 60	Glass/nylon	**M6/M6**	>13,000	3/23
BioPilot Superdex 200 pg 35/600	600	3.5 × 60	Glass/nylon	**M6/M6**	>13,000	3/8
BioPilot Superdex 200 pg 60/600	1700	6.0 × 60	Glass/nylon	**M6/M6**	>13,000	3/23

[a]PP, polypropylene.
[b]**Bold** lettering denotes male-type fittings.
[c]Requires column holder if not used with SMART system.

Superdex offers a very high resolving power for peptides and medium-sized proteins and polymeric species of corresponding size. The high resolving power may be used for purity check, as demonstrated in Fig. 2.10. The same properties may of course be used advantageously for polishing and then preferentially using the larger sized bulk material (to allow for the preparation of process columns). The high resolvability of Superdex 200 was used for the separation of DNA and DNA-bound transcription factor USF (Bresnick and Felsenfeldt, 1994). A quite different application of SEC was described by Werner and co-workers (1994), who used gel filtration on Superdex 75 HR 10/30 to refold recombinant proteins produced as insoluble inclusion bodies. The possibility of separating solutes of small size makes the gel Superdex peptide very suitable for rapid analysis of low molecular weight solutes (see Fig. 2.11).

IV. COLUMNS FOR PACKING SEC MEDIA

Four column systems are available from Amersham Pharmacia Biotech that can be used to pack SEC media for various applications at the laboratory scale. These include C, XK, SR, and HR column systems. All of the laboratory-scale columns are constructed with borosilicate glass tubes. Columns for larger scale process applications include INdEX, BPG, FineLINE, BPSS, and Stack columns. The larger scale columns are constructed to meet stringent validation requirements for the production of biopharmaceuticals. Each of the column types are described.

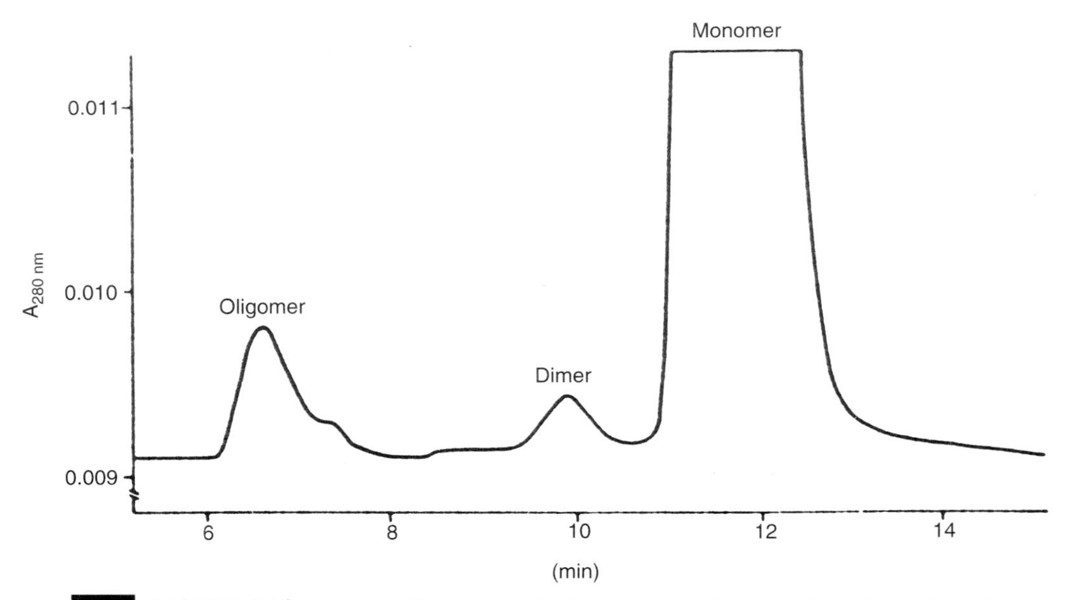

FIGURE 2.10 Analytical SEC for determination of aggregates in preparations of recombinant human growth hormone using Superdex 75 HR 10/30. [Reproduced from Hagel (1993), with permission.]

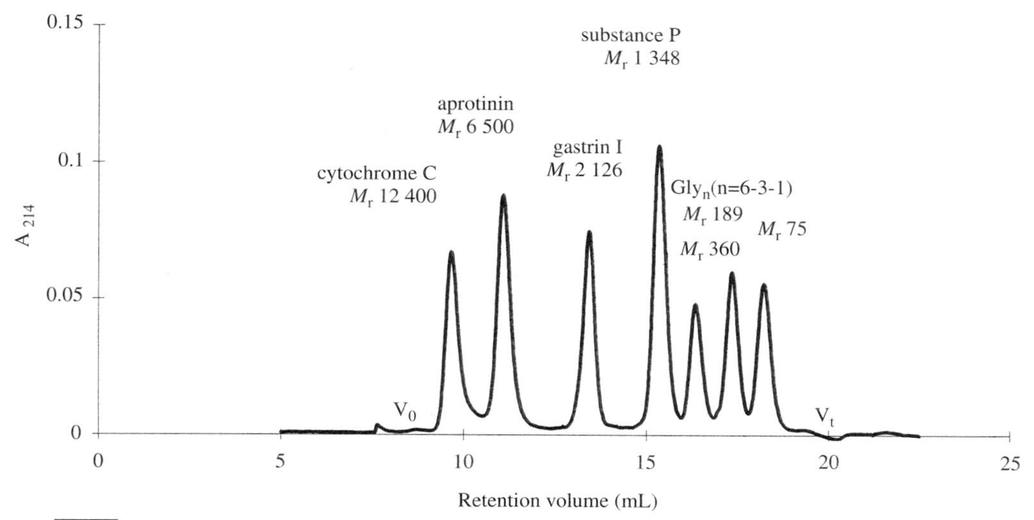

FIGURE 2.11 Separation of a low molecular mass sample mixture on Superdex peptide HR 10/30 in 20 min. (Reproduced with permission from Amersham Pharmacia Biotech.)

A. C Columns

The C column system is a modular system composed of the basic C column, flow adaptors for accommodating variable bed heights, packing reservoirs, and thermostatic jackets. The basic C column is designed for use at fixed bed heights and is suitable for operating pressures up to 1 bar (0.1 MPa) and with a thermostatic jacket may be used up to 3 bar (0.3 MPa). The wetted materials include EPDM, TEFZEL, superpolyoxymethylene, and fluororubber. The columns use nylon nets of 10-, 20-, or 80-μm mesh size and may be used with most SEC media with particle diameters above 20, 40, and 120 μm, respectively. Columns are intended for use with aqueous solutions and are not suitable for use with ketones, chlorinated hydrocarbons, aliphatic esters, or phenol. Columns may be used with mild acids and bases, including NaOH (<2 *M*), HCl ($<10\%$), and acetic acid ($<5\%$). Strong mineral acids should be avoided. C columns are stable over the pH range of 1–14, may be used from 4 to 60°C, and may be sterilized with ethylene oxide. The columns use low-pressure compression fittings, which will accommodate 1.8-mm o.d. tubings. The available column sizes and specifications for the C column series are given in Table 2.11.

B. XK Columns

The XK column system is a medium-pressure jacketed glass column system designed for operating pressures up to 5 bar (0.5 MPa). Column dead volumes are less than 0.1% of the total column volume. Wetted materials include EPDM, TEFZEL, superpolyoxymethylene, and flurorubber. Columns use nylon nets of 10-μm mesh size and may be used with most SEC media with particle diameters >20 μm. Columns are intended for use with aqueous solutions and

TABLE 2.11 C Columns

			No adaptor		One adaptor		Two adaptors	
Column	i.d. (mm)	Length (cm)	Volume (ml)	Bed height (cm)	Volume (ml)	Bed height (cm)	Volume (ml)	Bed height (cm)
C 10/10	10	10	7	8	0–7	0–8	0–7	0–8
C 10/20	10	20	14	18	7–14	9–18	0–14	0–18
C 10/40	10	40	30	38	23–30	29–38	16–30	20–38
C 16/20	16	20	36	18	6–36	3–18	0–36	0–18
C 16/40	16	40	77	38	47–77	23–38	8–77	4–38
C 16/70	16	70	137	68	107–137	53–68	69–137	34–68
C 16/100	16	100	197	98	167–197	83–98	129–197	64–98
C 26/40	26	40	202	38	122–202	23–38	21–202	4–38
C 26/70	26	70	361	68	281–361	53–68	181–361	34–68
C 26/100	26	100	520	98	441–520	83–98	340–520	64–98

are not suitable for use with ketones, chlorinated hydrocarbons, aliphatic esters, or phenol. Columns may be used with mild acids and bases, including NaOH (<2 *M*), HCl (<10%), and acetic acid (<5%). Strong mineral acids should be avoided. XK columns are stable over the pH range of 1–14, may be used from 4 to 60°C, and may be sterilized with ethylene oxide. The columns are supplied with one flow adaptor and a fixed bottom end piece. XK columns use M6 female fittings and 1.2-mm i.d., 1.8-mm o.d. Tefzel tubing. Adaptors with larger tubings (1.8 mm i.d., 2.9 mm o.d.) are available. The available column sizes and specifications for the XK column series are given in Table 2.12.

C. SR Columns

The SR (solvent resistant) column series is designed for use with organic solvents. Wetted materials include borosilicate glass, PTFE, fluropolymer, and stainless steel (SR25) or titanium (SR10). SR columns are stable in most solvents, including alcohols, pyridine, toluene, dimethyl sulfoxide, formamide, chloroform, and ethylene chloride. SR columns are stable over the pH range of 1–14, may be used from 4 to 60°C, and may be sterilized with ethylene oxide. Columns are supplied with two flow adaptors and PTFE nets with a mesh size of 10 (SR10) or 42 (SR25) μm. SR 10 columns may use pressures up to 10 bar (1 MPa) and SR 25 columns up to 3 bar (0.3 MPa). The column uses internal compression fitting designed for use with 1.8-mm o.d. PTFE tubing. The available column sizes and specifications for the SR column series are given in Table 2.13.

D. HR Columns

The HR column series was designed for high-resolution applications at higher pressures than the C, XK, or SR column series. Wetted materials include

TABLE 2.12 XK Columns

Column	i.d. (mm)	Length (cm)	One adaptor		Two adaptors	
			Volume (ml)	Bed height (cm)	Volume (ml)	Bed height (cm)
XK 16/20	16	20	2–34	1–17	0–34	0–17
XK 16/40	16	40	42–74	21–37	8–74	4–37
XK 16/70	16	70	102–35	61–67	68–135	34–67
XK 16/100	16	100	163–195	81–97	129–195	64–97
XK 26/20	26	20	0–80	0–15	0–80	0–15
XK 26/40	26	40	122–196	23–37	32–196	6–37
XK 26/70	26	70	281–356	53–67	191–356	36–67
XK 26/100	26	100	440–515	83–97	350–515	66–97
XK 50/20	50	20	0–275	0–14	0–275	0–14
XK 50/30	50	30	330–510	15–26	0–510	0–26
XK 50/60	50	60	785–1099	40–56	471–1099	24–56
XK 50/100	50	100	1570–1884	80–96	1256–1884	64–96

borosilicate glass, polypropylene, and polyvinyl chloride. HR columns may be used at pressures from 30 to 100 bar (3–10 MPa), depending on the diameter of the column (See Table 2.4). The column may be used with most solvents except chlorinated hydrocarbons or aromatic solvents. Columns may be used with SEC media >5 μm. HR columns are supplied with one fixed end piece, a single flow adaptor, and 1-μm polypropylene frits. The column uses M6 female fittings. The available column sizes and specifications for the HR column series are given in Table 2.14.

E. INdEX Columns

The INdEX column series was designed for use in process development and small-scale production. This column series uses a unique hydraulic adaptor for

TABLE 2.13 SR Columns

Column	i.d. (mm)	Length (cm)	Two adaptors	
			Volume (ml)	Bed height (cm)
SR 10/50	10	50	16–39	20–50
SR 10/50 J[a]	10	50	16–39	20–50
SR 25/45	25	45	73–220	15–45
SR 25/100	25	100	343–490	70–100

[a]Column is supplied fitted with a glass thermostatic jacket.

TABLE 2.14 HR Columns

Column	i.d. (mm)	Length (cm)	Pressure limit (MPa)	One adaptor		Two adaptors	
				Volume (ml)	Bed height (cm)	Volume (ml)	Bed height (cm)
HR 5/2	5	2	10	0.20–0.59	1.0–3.0	0–0.65	0.0–3.3
HR 5/5	5	5	10	0.8–1.2	4.0–6.0	0.45–1.25	2.3–6.3
HR 5/10	5	10	10	1.8–2.2	9.0–11.0	1.40–2.20	7.3–11.3
HR 5/20	5	20	10	3.7–4.1	19.0–21.0	3.40–4.20	17.2–21.2
HR 10/10	10	10	5	6.4–8.7	8.1–11.1	4.7–9.4	6.0–12.0
HR 10/30	10	30	5	22.1–24.4	17.2–21.2	20.4–25.2	26.0–32.0
HR 16/10	16	10	3	14.3–22.3	7.1–11.1	6.2–22.3	3.1–11.1
HR 16/50	16	50	3	95–103	47.5–51.5	87.0–103	33.5–51.5

fast and easy column packing. Columns may be used at pressures up to 3 bar (0.3 MPa). Polyamide 10-μm nets or Polypropylene 23-μm nets are available for all column diameters. Wetted materials include the borosilicate glass column tube, polypropylene, stainless steel, nitrile, and EPDM. Index columns are supplied with one fixed end piece and one hydraulic flow adaptor. Column fittings are three-piece sanitary fittings with 6 mm i.d. The available column sizes and specifications for the INdEX column series are given in Table 2.15. Columns stands and packing equipment are available.

F. FineLINE Columns

The FineLINE series of columns was designed to meet the stringent demands of hygiene required in the production of biopharmaceuticals. These stainless-steel-based columns employ a hydraulic flow adaptor and may be operated at

TABLE 2.15 INdEX Columns

Column	i.d. (mm)	Length (cm)	Pressure limit (bar)	Volume[a] (liter)	Bed height (cm)
INdEX 70/500	70	50	3	0.0–1.6	3–41
INdEX 70/950	70	95	3	1.9–3.0	48–79
INdEX 100/500	100	50	3	0.2–3.2	3–41
INdEX 100/950	100	95	3	3.8–4.8	48–79
INdEX 140/500	140	50	3	0.5–46.3	3–41
INdEX 140/950	140	95	3	7.4–912.2	48–79
INdEX 200/500	200	50	3	0.9–12.9	3–41
INdEX 200/950	200	95	3	15.1–24.8	48–79

[a]Maximum bed volumes require packing equipment.

pressures up to 10 bar (1 MPa). Wetted parts include stainless steel and EPDM. The nets are 2 μm stainless steel. All materials conform to the biological activity tests described in the U.S. Pharmacopoeia (USP XXII). FineLINE columns are supplied with one fixed end piece and one hydraulic flow adaptor. Column fittings are three-piece sanitary fittings with 6 mm i.d. The available column sizes and specifications for the FineLINE column series are given in Table 2.16. Columns stands and packing equipment are available. These columns are autoclavable at 121°C for 30 min.

G. BPG Columns

The BPG series of columns was designed to meet the stringent demands of hygiene required in the production of biopharmaceuticals. These borosilicate glass-based columns employ a single-screw adaptor and may be operated at pressures between 4 and 8 bar, depending on the column diameter. Wetted parts include stainless steel, polypropylene, PTFE, polyamide, acetal plastic, polyurethane, and EPDM. Nets for BPG columns are available in 10-, 12-, and 23-μm mesh sizes. All materials conform to the biological activity tests described in the U.S. Pharmacopoeia (USP XXII). BPG columns are supplied with one fixed end piece and one flow adaptor. Column fittings are three-piece sanitary fittings. The available column sizes and specifications for the BPG column series are given in Table 2.17. Columns stands and packing equipment are available.

H. BioProcess Stainless-Steel Columns

BioProcess stainless-steel columns are fixed bed height columns designed for the most stringent requirements in the routine production of biopharmaceuticals. Wetted materials include stainless steel, polypropylene, and EPDM. The BPSS series may be operated at pressures up to 3 bar (0.3 MPa) and are supplied with sanitary fittings of 10 or 22 mm i.d. The available column sizes and specifications for the BPSS column series are given in Table 2.18.

I. Stack Columns

Stack columns are a modular process system that allows several column units to be stacked in series. This column system is well suited for packing soft compressible gels in short segments that can then be hooked in series. The

TABLE 2.16 FineLINE Columns

Column	i.d. (mm)	Length (cm)	Volume (liter)	Bed height (cm)
FineLINE 100	100	35	236–1178	3–15
	100	70	393–2355	5–30
FineLINE 200	200	35	942–4710	3–15
	200	70	1570–9420	5–30

TABLE 2.17 BPG Columns

Column	i.d. (mm)	Length (cm)	Pressure limit (bar)	Volume[a] (liter)	Bed height[a] (cm)
BPG 100/500	100	50	8	0–3.8	0–48
BPG 100/750	100	75	8	2–5.2	25–66
BPG 100/950	100	95	8	3.5–6.6	45–83
BPG 140/500	140	50	6	0–7.4	0–48
BPG 140/950	140	95	6	6.9–12.8	45–83
BPG 200/500	200	50	6	0–15.1	0–48
BPG 200/750	200	75	6	7.3–20.7	25–66
BPG 200/950	200	95	6	14.1–26.1	45–83
BPG 300/500	296	50	4	0–33	0–48
BPG 300/750	296	75	4	17.2–45.4	25–66
BPG 300/950	296	95	4	31–57.1	45–83

[a]Maximum bed volumes and bed heights require packing equipment.

columns are constructed of polymethylpentene and are supplied with fixed end pieces. The stack columns can be operated up to 1 bar (0.1 MPa) and are ideal for desalting applications with Sephadex. The stack column specifications are shown in Table 2.19.

TABLE 2.18 BPSS Columns

Column	i.d. (mm)	Length (cm)	Volume (liter)
BPSS 400/150	40	15	18.8
BPSS 400/300	40	30	37.7
BPSS 400/600	40	60	75.4
BPSS 400/1000	40	100	126.0
BPSS 600/150	60	15	42.4
BPSS 600/300	60	30	84.8
BPSS 600/600	60	60	169.6
BPSS 600/1000	60	100	282.7
BPSS 800/150	80	15	75.4
BPSS 800/300	80	30	150.8
BPSS 800/600	80	60	301.6
BPSS 800/1000	80	100	502.6
BPSS 1000/150	100	15	117.8
BPSS 1000/300	100	30	235.6
BPSS 1000/600	100	60	471.2
BPSS 1200/150	120	15	169.6
BPSS 1200/300	120	30	339.3

TABLE 2.19 Stack Column Characteristics

Column	Number of sections	i.d. (mm)	Length (cm)	Pressure limit (bar)	Volume (L)	Bed height (cm)
PS370	1	370	15	1	16.1	15
	2	370	15	1	32.2	30
	3	370	15	1	48.3	45
	4	370	15	1	64.4	60
	5	370	15	1	80.5	75
	6	370	15	1	96.6	90

V. PREPARATION OF THE SIZE EXCLUSION COLUMN

Based on the requirements of the separation, media of suitable pore size, particle size, and surface properties are selected as well as column dimensions and column material. In some cases a suitable combination of media type and column dimensions may be available as a prepacked column. In most cases, this is a more expensive alternative to preparing the column yourself but will provide a consistent quality as assured by the manufacturing and testing procedures of the vendor. The consistent quality may be critical in obtaining reproducible results and may thus be a cost-effective solution. Also, the fact that smaller particle-sized media are more difficult to pack and require special, and expensive, equipment has resulted in that gel filtration media of small particle size, e.g. smaller than 15 μm, are predominantly supplied as prepacked columns.

Often, media of similar characteristics as the one used for prepacked columns (i.e., except for particle size) are obtainable, e.g., as preparative grade material, having particle sizes of typically 30–50 μm. This is required for large-scale gel filtration where the sample capacity of the prepacked columns is often insufficient.

The selection of column characteristics is determined by solvent resistance, the need to visually inspect the bed, the pressure rating of the system, and the dimensions [column inner diameter (i.d.) and length (L)] required from productivity considerations. Productivity considerations will vary if the requirement is based on the amount of information per unit time (analytical gel filtration) or the amount of substance per unit time (preparative gel filtration).

A. Selection Criteria for Empty Columns

Gel filtration separations are performed in an aqueous environment and the requirement on chemical resistance is therefore modest. Most column tubes and accessories (e.g., support net, sealings) withstand the solvent mixtures sometimes used in aqueous SEC such as 20% ethanol, 3 *M* guanidinium hydrochloride, 6 *M* urea, or 0.1 *M* HCl and 1 *M* NaOH, the last two being used for cleaning the packed column (see Section V).

The length of the column is determined by the resolution that is needed to obtain the purity of the peak(s) that is required. The resolution is proportional to the square root of the column length. However, the pressure drop also increases with column length. Because the resolution is also affected by the flow rate, the comparatively lower resolution of a short column may be compensated for by using a lower flow rate (i.e., keeping the separation time constant). The following column lengths have been found to be a good first choice for different particle sizes; L = 25–30 cm for d_p = 5–15 μm, L = 50–60 cm for d_p = 30–50 μm, and L = 60–100 cm for d_p = 70–150 μm.

The diameter of the column is selected from the volume of sample that is to be processed. As a rule of thumb the maximum productivity is obtained at a sample volume of 2–6% of the bed volume in preparative gel filtration on a 50-μm chromatographic medium (Hagel *et al.*, 1989). Thus, the required column diameter is calculated from the bed volume needed to cope with the sample volume and the column length needed to give the resolution desired.

A third parameter to consider is the column construction. Thus the sample applicator should provide optimal sample application to give the most performance possible out of the packed bed. Constructions should also allow simple, fast, and reproducible packing of the column. Because costs for repacking of columns are a substantial operating cost item in industrial chromatography, the selection of column construction from this point of view is also important. Some novel column constructions allow very simple procedures both for laboratory and for industrial scale (e.g., INdEX columns, see Section V).

Finally, the construction should allow very flexible handling to cope with the various packing instructions recommended by the suppliers of the gel filtration medium.

B. General Packing Methodology

The packing method supplied by the manufacturer of the gel filtration medium may need to be revised according to the column being selected. It is therefore important to have an understanding about the basic principles governing the packing of chromatographic beds.

The first condition to be fulfilled is to obtain a good dispersion of the particles in the slurry used for packing. Thus, additives may need to be used in order to prevent particle–particle interactions. The choice of suitable additives is regulated by the surface properties of the chromatographic medium (e.g., salt to prevent ionic interaction and ethanol to prevent hydrophobic interaction or surfactants such as Tween). Different types of chromatographic material will require different additives. The concentration of particles in the slurry should generally not exceed 70% to prevent particle aggregation and trapped air when pouring the slurry into the column tube.

The second condition is that the well-dispersed slurry forms a homogeneous bed by formation of the bed under well-controlled conditions. This is achieved by a two-step procedure where the bed is formed using constant velocity of the mobile phase and then stabilizing the bed at a constant pressure (Hagel, 1989). The rationale for the first step at constant velocity is that this will create uniform drag forces from the flowing liquid on the gel particles and thus prevent compaction of the particles at the bottom part of the column. The stabilization at constant pressure will ascertain that the bed will withstand different conditions (e.g., solvents of different viscosity) with the pressure drop as the limiting condition. Deviations from this general procedure may be found, e.g., for rigid materials that will not compact the first step may also be performed at constant pressure (although the flow rate and thus impact velocity will vary with the continuously growing bed height during packing). However, a stabilization step is recommended, also for rigid materials such as silica.

In some cases the column construction used does not support a movable adapter, e.g., the popular stack column (see Section V) where a stable bed is achieved by shrinking the gel slightly in an ethanol–water mixture prior to packing. After the column is packed and the column has been closed the gel is allowed to expand by equilibrating the column in distilled water.

Often, the maximum eluent velocity that is applicable to the size exclusion of macromolecules is restricted by the loss in resolution at higher velocities (see Section VI), and therefore column packing is traditionally performed at a flow rate slightly higher (e.g., 20–50%) than that used for the separations. However, an optimally packed bed, i.e., homogeneous and with the lowest possible void fraction, will be formed using a high eluent velocity, e.g., 70% of the maximum applicable, and then stabilizing the bed at the pressure given by that flow rate (see Fig. 2.12). The compaction of nonrigid beds and the

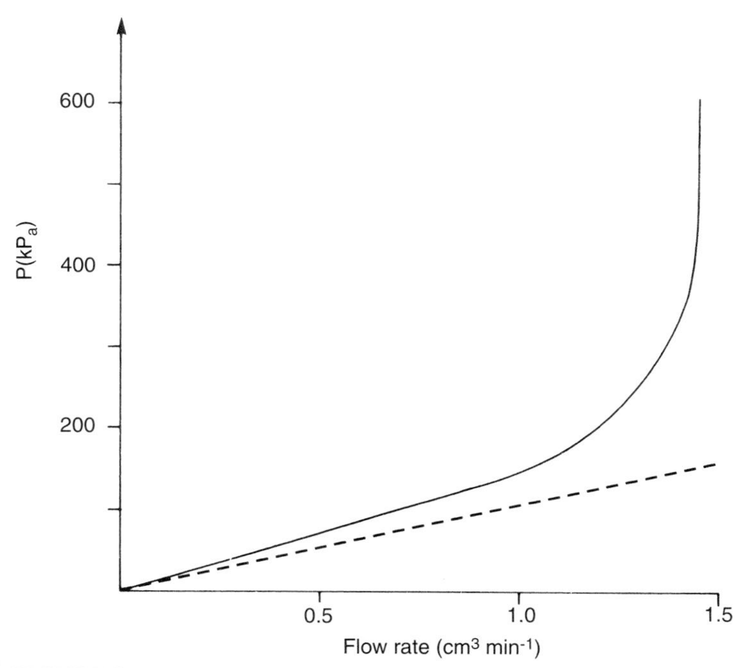

FIGURE 2.12 Pressure drop as a function of liquid flow through a packed bed of Superose 6 prep grade packed in a HR 10/30 column. At high velocities of the mobile phase the beads are compressed and the void channels reduced, which leads to a high pressure drop. If this happens, the material can be resuspended and packed at a lower flow rate. [Reproduced from Hagel and Andersson (1984), with permission.]

collapse of rigid materials are prevented by supportive forces from the column tube. This supportive effect is more pronounced for narrow tubes and will gradually decrease to be negligible at a column-to-particle diameter ratio of 1000 (Hagel, 1989).

The popularity of traditional media irrespective of their sometimes lower rigidity is partly due to the ease with which beds of good performance are prepared. The reason for this is probably the elasticity of the polymer beads to allow for a dynamic rearrangement of particles (e.g., by slipping over) to form a dense bed as compared to irregular silica for which the particles are believed to be locked in an initial position with little chance for rearrangement unless the forces are considerable (e.g., as in a stabilizing step). This phenomenon can be detected by differences in the void fraction of Superose (polymer) of 0.33 as compared to Waters I-125 (irregular silica) having a void fraction of 0.5 (Hagel, 1989).

C. Qualification of Column Packing

If the quality of the packed bed has a significant influence on the results (e.g., if it is not used for simple laboratory-scale desalting operations), it needs to be checked prior to use. This is mostly done by running a pulse of a small solute, e.g., acetone, salt, azide, or sodium nitrate, through the column and determining the zone broadening and symmetry of the peak (see Fig. 2.13). As a rule of thumb the minimal reduced plate height is approximately 2 for a well-packed column and should not exceed 5 for an acceptable column. The asymmetry factor should be between 0.8 and 1.2. Sometimes it is necessary to try several solutes to avoid solute–matrix interactions and misleading results

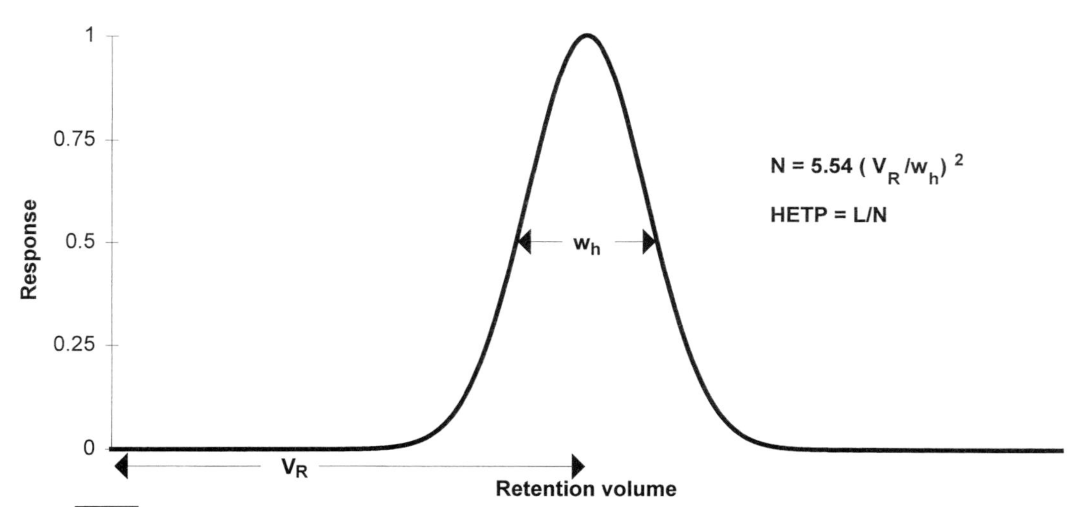

FIGURE 2.13 From measurements of the retention volume, V_R, and the peak width at half peak height, w_h, of a gaussian peak, an estimate of column efficiency N and relative efficiency, HETP, may be calculated. The last figure is for very well packed columns close to $2 \times d_p$. [Reproduced from Sofer and Hagel (1997), with permission.]

(these small solutes will sense a larger portion of the surface area than the macromolecules to be separated and may thus be susceptible to interactions not otherwise noticed). In addition to this test it is advisable to run a test with a sample mixture that is relevant for the separation problem at hand, i.e., even though the plate height or asymmetry factor of the column is slightly worse than for an earlier column, this may not be critical to the separation; however, large deviations in plate height and asymmetry numbers indicate that the packing procedure should be revised. The sample volume needs to be low in order not to add extra column effects to the measured reduced plate height. The effect is more serious for columns of inherent higher plate counts, as seen in Fig. 2.14.

The quality of the packed bed may also be determined by frontal analysis where the sample is applied until it reaches a plateau to give the residence time function and then the solution is momentarily switched to wash to give the washout function. The latter is used to calculate the plate height of the column

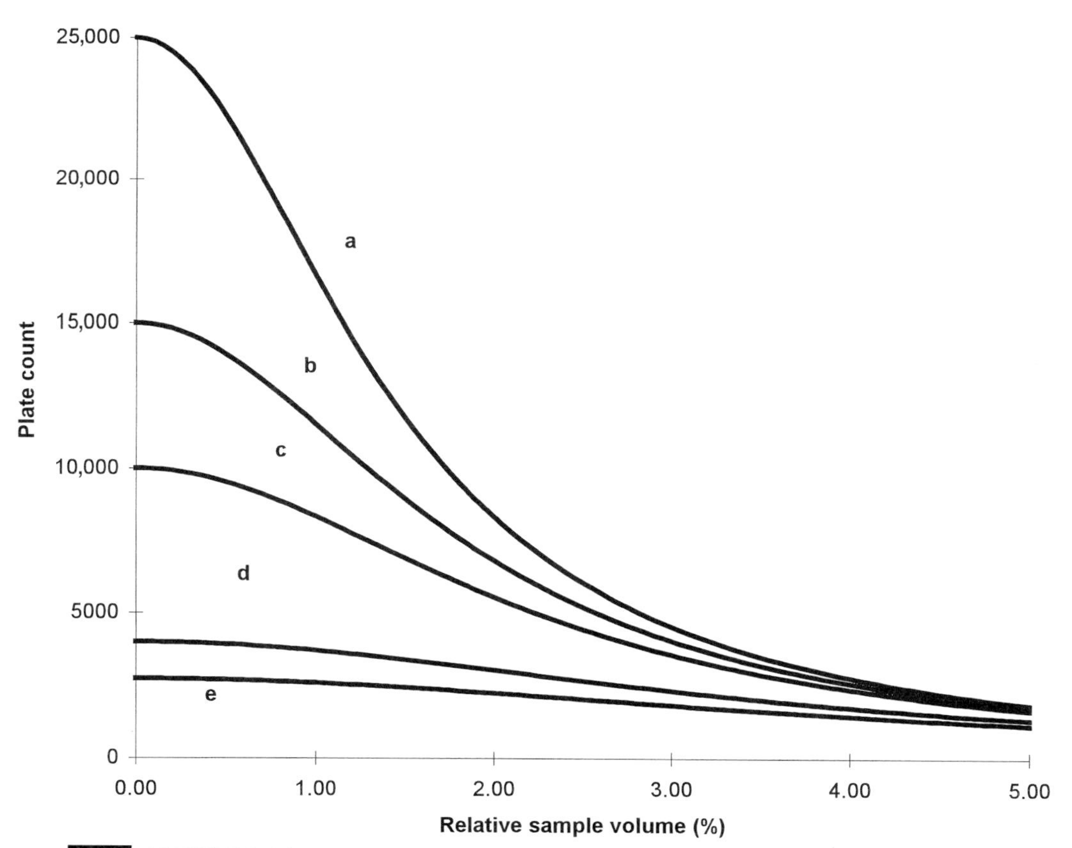

FIGURE 2.14 Influence of sample volume (as percentage of bed volume) on the calculated column efficiency. Conditions: (a) $d_p = 5$ μm and L = 25 cm, (b) $d_p = 10$ μm and L = 30 cm, (c) $d_p = 30$ μm and L = 60 cm, (d) $d_p = 75$ μm and L = 60 cm, and (e) $d_p = 110$ μm and L = 60 cm. [Reproduced from Sofer and Hagel (1997), with permission.]

(Sofer and Hagel, 1997). The result should, in theory, be identical to the pulse experiment. This method is used frequently in industrial laboratories.

The qualification just stated should also preferentially be performed on prepacked columns to ensure that the column fulfills the users' own specifications. Furthermore, it is good practice to run the qualification periodically and after cleaning and stand still.

D. Typical Results for Self-Packed Columns

Requirements on the quality of the column packing vary according to the application at hand. Therefore, "typical results" may only serve as a guideline of what to expect; furthermore, the examples given here are in most cases results achieved by persons having experience in column packing. Nevertheless, the results are not extraordinary and should be quite possible to achieve and even excel.

1. Laboratory Columns

The uncomplicated packing procedure of traditional media to yield efficient columns is demonstrated by the reduced plate heights of 2 or slightly below as reported for Sephadex and for the qualification of Sepharose columns for the determination of MWDs.

The preparation of high-resolution columns of Sephacryl HR, where the second step was performed at a constant high flow rate and the packing direction was toward the inlet adapter to assure a homogeneous zone at the inlet part of the column, was reported to give reduced plate heights of better than 2.1 and asymmetry factors in the range of 0.9–1.1 (Hagel *et al.*, 1989). Preparation of columns of various dimensions was reported, and high plate counts, i.e., 12,000 to 14,000 plates per meter, were also found for columns of 5 cm i.d. (XK 50/100).

2. Process Columns

Traditionally, a slightly higher reduced plate height has been accepted for process scale columns. This may be due to that handling of the slurry, column construction, or zone broadening in the systems limits the plate height that may be achieved. However, with an optimal column construction and reducing the influence of extra column effects to zone broadening during the qualification of the column (e.g., by assuring the shortest possible piping between sample introduction to column and from column to detector and by keeping the sample volume below 1% of bed volume; see Section VI), reduced plate heights of close to 2 may also be obtained with process scale columns.

However, it must be stressed that a reduced plate number of 2.0 and an asymmetry factor of 1.0 are not goals per se, i.e., for other reasons, such as convenient handling, a less optimal construction from packing quality point of view may be selected and the characteristics of the bed being well apt for the separation problem at hand. For instance, spending a lot of time to optimize a packing procedure to give a reduced plate height of 2 instead of 5 for a desalting column will certainly be a waste of time.

VI. OPTIMIZATION OF GEL FILTRATION/SIZE EXCLUSION SEPARATIONS

The optimization of chromatographic separations can generally be seen as a compromise between speed, i.e., to produce the largest possible amount of data or substance per unit time, and resolution, i.e., to produce the highest possible quality of data or purity of substance. Obviously the goal for optimization differs according to the purpose of the separation and also between scale of operation. Therefore, different parameters are critical for different situations. Still, some basic rules for optimization may be applied.

Generally, optimizing the selectivity by choosing a gel medium of suitable pore size and pore size distribution is the single most important parameter. Examples of the effect of pore size on the separation of a protein mixture are given in Fig. 2.15. The gain in selectivity may then be traded for speed and/or sample load. However, if the selectivity is limited, other parameters such as eluent velocity, column length, and sample load need to be optimized to yield the separation required.

The resolution in size exclusion may be calculated from the following equation showing the influence of various parameters (Hagel, 1989):

$$R_s = \frac{2(V_{R_2} - V_{R_1})}{w_{b_2} + w_{b_1}} = \frac{1}{4} \cdot \log \frac{R_2}{R_1} \cdot \frac{dK_D/d\log R}{\frac{V_0}{V_p} + K_D} \cdot \frac{\sqrt{L}}{\sqrt{H}}. \tag{2}$$

The important parameters to consider are the selectivity ($dK_D/d\log R$), the ratio of pore volume, V_p, over void volume, V_0, the plate height, H, and the column length, L. The distribution coefficient, K_D, has a slight effect on resolution (with an optimum at K_D 0.3–0.5). In addition to this, extra column effects, such as sample volume, may also contribute to the resolution.

A. Selectivity

The selectivity of a gel, defined by the incremental increase in distribution coefficient for an incremental decrease in solute size, is related to the width of the pore size distribution of the gel. A narrow pore size distribution will typically have a separation range of one decade in solute size, which corresponds to roughly three decades in protein molecular mass (Hagel, 1988). However, the largest selectivity obtainable is the one where the solute of interest is either totally excluded (which is achieved when the solute size is of the same order as the pore size) or totally included (as for a very small solute) and the impurities differ more than a decade in size from the target solute. In this case, a gel of suitable pore size may be found and the separation carried out as a desalting step. This is very favorable from an operational point of view (see later).

B. Separating Volume

Because the separation in size exclusion chromatography takes place only in the range of the pore volume, this parameter has a great impact on the resolution. A large pore volume will result in that peaks are separated further apart. In

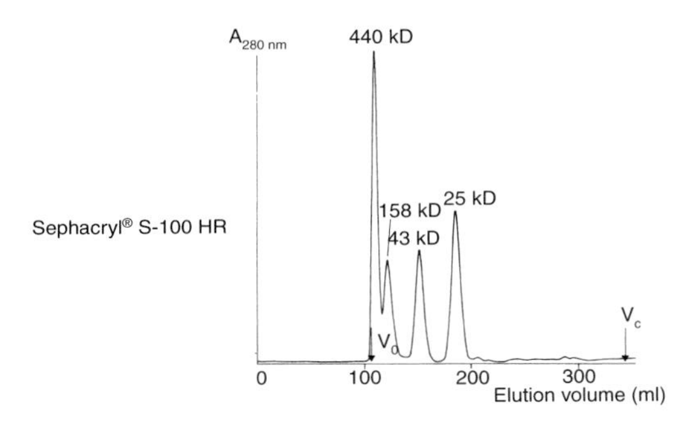

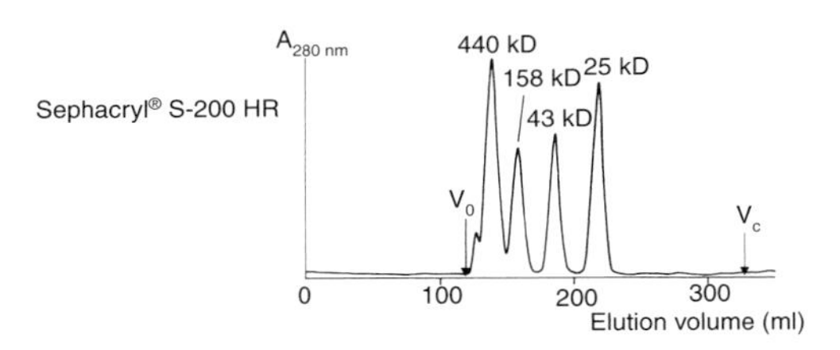

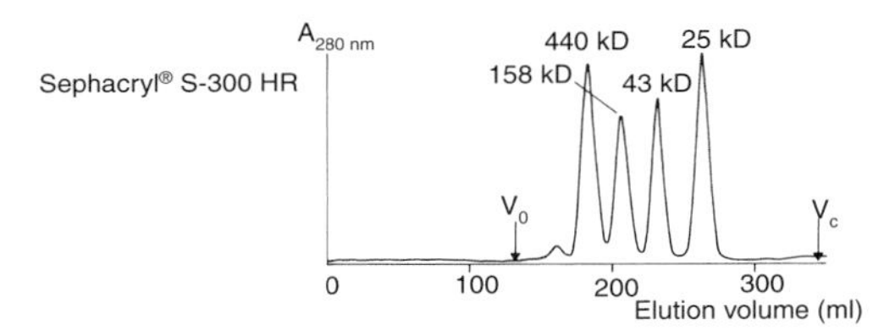

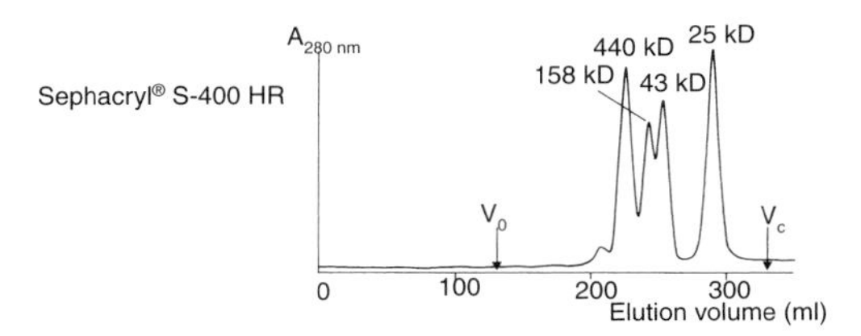

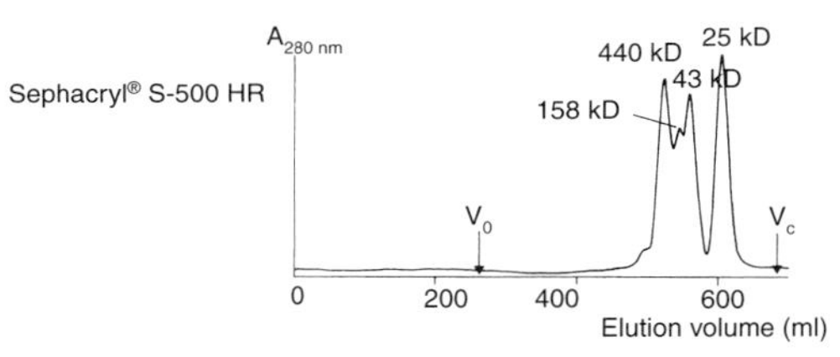

FIGURE 2.15 Influence of the pore size of Sephacryl HR on the separation of proteins of various molecular mass. The protein mixture is composed of ferritin, aldolase, ovalbumin, and chymotrypsinogen A. [Reproduced from Hagel *et al.* (1989), with permission.]

addition to this, a small void fraction will reduce the nonseparating volume. Therefore, a medium with as high a pore volume as possible giving beds of low void volume is favorable from a resolution point of view (however, a lower void fraction is covariant with a higher pressure drop over the bed). The importance of a high pore volume has been illustrated by many of the figures in the preceding sections. The influence is normally calculated from the pore fraction, V_p/V_o (i.e, pore volume over void volume). The effect may be illustrated by the fact that a gel medium having $V_p/V_o = 1.2$ will require twice as many plates per column to give the same resolution as a gel medium having $V_p/V_o = 2.0$ (Chang *et al.*, 1976).

C. Plate Height

If the selectivity and the pore volume are not sufficient to separate the target component from impurities, then the resolution needs to be increased further. This can be accomplished by decreasing the widths of the peaks by decreasing the plate height. This is achieved by using media of small particle size or running the separation at a velocity corresponding to Hagel (1989):

$$F = A_c \cdot \varepsilon \cdot 65 \cdot K_D \cdot \frac{D_M}{d_p} \cdot 60, \tag{3}$$

where F is the flow rate (ml/min), A_c is the column cross-sectional area (cm^2), ε is the void fraction of the column, D_M is the diffusion coefficient of the solute (cm^2/s), and d_p is the particle diameter (cm) (60 being the conversion factor between seconds and minutes). Thus for large solutes the velocity needs to be low (e.g., 5–10 cm/min) and for small solutes the velocity should be kept higher (e.g., >50 cm/min).

D. Column Length

The resolution is theoretically proportional to the square root of the bed length. This is also sometimes encountered in practice (Hagel *et al.*, 1989). Therefore, adding column in series may yield the resolution that is needed. However, some resolution is generally lost in the tubings, fittings, and distribution systems when connecting columns and therefore it may not be reasonable to expect to achieve more than 80% of the theoretically calculated improvement in resolution.

E. Sample Load

Sample load is primarily a concern in preparative gel filtration. In analytical applications the only precaution is to ascertain that the sample volume is sufficiently low as not to contribute to peak widths (and thus decrease the quality of the information) (Hagel, 1985). The concentration of the sample should not exceed 30 mg/ml for globular proteins or 5 mg/ml for polymers and DNA (Hagel and Janson, 1992).

The influence of the sample volume on the zone broadening is more pronounced for media of smaller particle size (due to their inherent low zone broadening) and therefore there is no reason for employing small particle-sized media for separating large sample volumes (i.e., the resulting sample zone will be affected by the sample volume and not by the column efficiency). Therefore, medium-sized particles (e.g., of 30–70 μm) of high pore volume (e.g., $V_p/V_o > 2$) that can be packed readily into optional column configurations is preferred for large-scale preparative work. As a rule of thumb, approximately 4% of the bed volume can be applied for fractionation work and 30% for desalting operations. The sample concentrations may be as high as 70 mg/ml for globular proteins of intermediate size (e.g., molecular mass of 50,000 g/mol). At higher concentrations the sample zone will get too viscous relative to an aqueous solvent and distortion of the rear part of the sample zone will result from "viscous fingering." The influence of the sample volume in SEC on the resolution is illustrated by Fig. 2.16.

F. Analytical Gel Filtration

Gel filtration is very suitable for the purity check of protein preparations, especially if these have been purified by adsorptive techniques. It can be expected that high-resolution gel filtration columns will easily separate dimeric forms from monomeric forms to reveal heterogeneities of the preparations. However, a size difference of less than 20% will not result in total resolution of the peaks (although the chromatogram may be used for a qualitative judgment of the

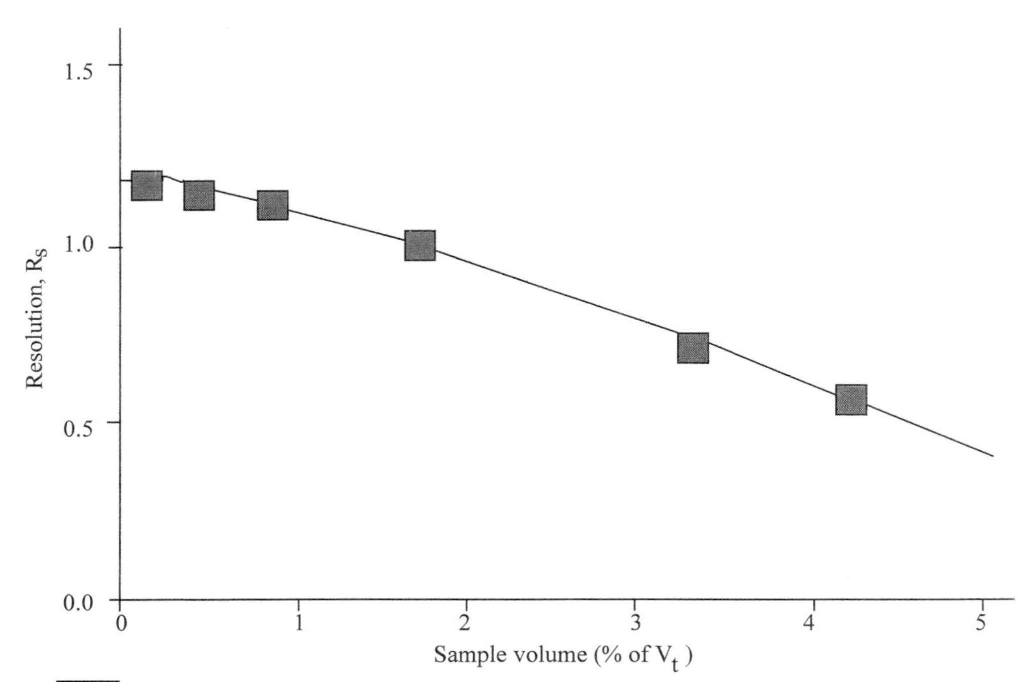

FIGURE 2.16 Influence of sample volume on the resolution of a 1 : 1 (w/w) mixture of transferrin (M_r, 81,000) and IgG (M_r, 160,000) on Superdex 200 prep grade. A sample concentration of 8 mg/ml was applied at 30 cm/hr to a 1.6 × 70-cm column in 50 m*M* sodium acetate, 20 m*M* sodium phosphate, 100 m*M* sodium chloride, pH 7.2. (Reproduced with permission of Amersham Pharmacia Biotech.)

purity of the preparation). In purity check the highest resolvability is needed and therefore media of very high selectivity (e.g., Superdex) with a pore size tailored to the substance class in question, having a particle size of less than 15 μm and packed in columns of 25–50 cm, should be optimal. The flow rate should be close to or less than the one given by Eq. (3), e.g., an eluent velocity (cm/sec) of roughly 40 times D_m/d_p, which for $D_m = 10^{-6}$ cm^2/sec, d_p = 10 μm, and column i.d = 1 cm (and a void fraction of 0.33) corresponds to a flow rate of 0.6 ml/min. To further enhance the separation the flow rate may be reduced to the minimum of the van Deemter equation (Deemter and Klinkenberg, 1956), which is obtained at five times D_m/d_p (e.g., 0.1 ml/min in the case cited). It may be noted that the peak capacity of self-packed and prepacked columns does not differ substantially (as the larger particle size of the former is compensated by a longer bed height) and is typically 11–13 (Hagel, 1992).

Conditions for the determination of molecular mass or molecular mass distributions need not be as stringent as for purity check. Because the molecular size is given by the retention volume (as given by the first moment of the distribution), the concern is that peaks should not overlap to significantly affect this estimate. The flow rate will not affect the retention volume, but may affect the peak width and thus "hide" smaller peaks. The net contribution from smaller peaks to the retention volume of larger peaks may be neglected in most cases. Thus, if the sample is relatively pure the requirements on the operating conditions are modest and the run may be done at high flow rates. However, if the sample is composed of several species, as for a polymer sample, the estimates of mass averages will be affected by the entire distribution and thus the broadening of the peak (e.g., the number average molecular mass is heavily weighted by the low molecular portion). However, a modest zone broadening is not likely to severely affect the estimates (unless for samples of narrow MWDs) and the running conditions are not critical, i.e., running at a velocity of 40 times D_m/d_p should be sufficient (corresponding to 0.8 ml/min for the case cited earlier). For a column of 30 cm length, this will result in a separation time of 30 min. This time may be seen as the maximum required to give high-quality information. It may be further reduced if only a rough estimate of the size and size distribution is the goal, and separations times as low as a few seconds have been reported. However, at these extreme eluent velocities the risk of polymer degradation by shear forces must be taken into account.

Analytical gel filtration may also be performed by employing the size exclusion column as a separation device and assaying the effluent, on line or by fractions, by a selective method. Thus, the use of mass spectrometrometers, viscosity detectors, and light-scattering detectors have all been used for the on-line characterization of proteins and polymers (e.g., see Nylander *et al.*, 1995; Yau, 1990; Jackson *et al.*, 1989). One advantage to the use of specific detectors (as opposed to ultraviolet and refractive index detectors) is that sorptive effects or shape effects will not be misinterpreted for molecular weight effects.

G. Preparative Gel Filtration

In preparative gel filtration the maximum throughput at a predefined level of purity is the goal of the optimization. Whereas the maximum sample concentration is determined by the viscosity of the sample solution as compared to that of the solvent (i.e., limiting the total protein concentration to approximately 70 mg/ml sample or the relative viscosity to 1.5), the sample volume and flow rate need to be optimized in order to maximize the productivity.

In fractionation of solutes having a similar size (e.g., monomer from oligomers) the pore size of the gel should be selected as to give a distribution coefficient of the lower molecular weight fraction around 0.6 (if low molecular weight impurities such as salt are not present a higher K_D may be chosen). This will in most cases allow the elution of higher molecular weight impurities at a K_D below 0.4 if the selectivity of the material is appropriate. A high pore volume of the gel filtration medium is very advantageous in preparative purifications. The optimum running conditions will balance the detrimental effects of a large sample volume (i.e., running the feed in few cycles) and the zone broadening at running at high flow rates (i.e., split the feed into many cycles). A guidance to the optimal sample volume is given by Hagel *et al.* (1989):

$$V_{\mathrm{sample}} \approx \left(\frac{V_{\mathrm{feed}} \cdot K_{\mathrm{inj}} \cdot V_{\mathrm{c}} \cdot V_{\mathrm{p}} \cdot d_{\mathrm{p}}^{2}}{15 \cdot D_{M}} \right)^{1/3}, \tag{4}$$

where V_{feed} ml sample is to be processed per hour. The beneficial influence of a large column volume, V_{c}, and pore volume, V_{p}, is obvious. The influence from the particle size, d_{p}, and the solute diffusivity, D_M, may at a first glance seem strange. However, what the equation states is that if the particle size is small or the diffusivity is large then the optimal sample volume will be found at smaller sample volumes and the flow rate should be increased. Equation (4) was found to support the general rule of processing a volume equal to 2–6% of the column volume each cycle at cycle times of respectively 5 to 1 hr (Hagel *et al.*, 1989).

H. Desalting

In desalting the size differences between solute and impurities are larger than one decade. In this instance, selecting a medium that will selectively exclude the macromolecule while including the impurities is the optimal choice. This may be seen as an inverse filtration operation, hence the original designation gel filtration. In desalting the entire pore volume may in principle be utilized for the purification. The zone broadening caused by excessive flow rates or the use of large particles for the operation contributes only slightly to the total peak width. Therefore, desalting may be carried out using large particle sizes (which is beneficial due to the low back pressure generated) and also at high flow rates to give rapid purifications. The maximum sample volume is restricted by the pore volume, but the influence from zone broadening in the column and connectors yields a rule of thumb of not applying more than 80% of the pore volume. For a traditional medium such as Sephadex G-25, this corresponds to roughly 40% of the bed volume. In contrast to general expectations, gel filtration in the desalting mode provides a high productivity (e.g., 64 g/liter of chromatography medium an hour for the deethanolization of albumin solutions on coarse Sephadex G-25 was calculated), which competes favorably with the productivity of adsorptive techniques (Sofer and Hagel, 1997).

ACKNOWLEDGMENT

Sephadex, Sepharose, Sephacryl, Superose, Superdex, FPLC, SMART, HiTrap, HiPrep, and INdEX are trademarks owned by Amersham Pharmacia Biotech AB.

REFERENCES

Andersson, T., Carlsson, M., Hagel, L., Pernemalm, P.-Å., and Janson, J.-C. (1985). *J. Chromatogr.* **326**, 33.

Chang, S. H., Gooding, K. M., and Reguier, F. E. (1976). *J. Chromatogr.*, **125**, 103.

Determan, H., and Brewer, J. E. (1975). *In* "Chromatography" (E. Heftmann, ed.), 3rd Ed., p. 385. Van Nostrand-Reinhold, New York.

Drevin, I., and Johansson, B.-L. (1991). *J. Chromatogr.* **547**, 21.

Granath, K. A., and Kvist, B. E. (1967). *J. Chromatogr.* **99**, 425.

Haglund, A. C., and Marsden, N. V. B. (1980). *J. Polym. Sci. Polym. Lett. Edn.* **18**, 271.

Hagel, L. (1985). *J. Chromatogr.* **324**, 422.

Hagel, L. (1988). Pore size distributions. *In* "Aqueous Size-Exclusion Chromatography" (P. Dubin, ed.), pp. 119–150. Elsevier, Amsterdam.

Hagel, L. (1989). *Gel filtration. In* "Protein Purification, Principles, High Resolution Methods, and Applications (J.-C. Janson, and L. Rydén, eds.), pp. 63–106. VCH Publishers, New York.

Hagel L. (1992). *J. Chromatogr.* **591**, 47.

Hagel, L. (1993). *J. Chromatogr.* **648**, 19.

Hagel, L. (1996). Characteristics of modern media for aqueous size exclusion chromatography. *In* "Strategies in Size Exclusion Chromatography" (M. Potschka and P. Dubin, eds.), ACS Symposium Series 635, pp. 225–248. Am. Chem. Society, Washington, DC.

Hagel, L., Hartmann, A., and Lund, K. (1993). *J. Chromatogr.* **641**, 63.

Hagel, L., and Andersson, T. (1984). *J. Chromatogr.*, **285**, 295.

Hagel, L., and Janson, J.-C. (1992). Size-exclusion chromatography. *In* "Chromatography" (E. Heftmann ed.), 5th Ed., pp. A267–A307. Elsevier, Amsterdam.

Hagel, L., Lundström, H., Andersson, T., and Lindblom, H. (1989). *J. Chromatogr.* **476**, 329.

Hagel, L., Östberg, M., and Andersson, T. (1996). *J. Chromatogr.*, A, **743**, 33.

Hellberg, U., Ivarsson, J.-P., and Johansson, B.-L. (1996). *Process Biochem.* **31**(2), 163.

Hellman, U., Wiksell, E., and Karlsson, B.-M. (1990). A new approach to micropreparative desalting exemplified by desalting a reduction/alkylation mixture, presented at Eight International Conference on Methods in Protein Sequence Analysis, Kiruna, Sweden, July 1–6.

Horton, T. (1972). *Amer. Lab.*, **May.**

Huber, J. F. K. (1978). *In* "Instrumentation for High-Performance Liquid Chromatography," pp. 1–9, Elsevier, Amsterdam.

Jackson, C., Nilsson, L. M., and Wyatt, P. J. (1989). *J. Appl. Polym. Sci.*, **43**, 99.

Johansson, B-L., and Ellström, C. (1985). *J. Chromatogr.* **330**, 360.

Lathe, G. H. and Ruthven, C. R., J. (1956). *Biochem. J.*, **62**, 665–674.

Lindqvist, B. and Storgårds, T. (1955). *Nature (London)*, **175**, 511.

Låås, T. (1975). Doctoral Thesis Acta Universitatis Upsaliensis.

McClung, J. K. and Gonzales, R. A. (1989). *Anal. Biochem.*, **117**, 378.

Moore, J., C. (1964). *J. Polym. Sci.*, Part A, **2**, 835.

Nilsson, G. And Nilsson, K. (1974). *J. Chromatogr.* **101**, 137.

Nylander, I., Tan-No, K., and Winter, A. (1995). *Life Sci.*, **57**, 123.

Porath, J. and Flodin, P. (1959). *Nature* **183**, 1657.

Pharmacia Biotech, Gel Filtration: Principles and Methods 5th ed. 1991.

Sofer, G. and Hagel L. (1997). *Handbook of Process Chromatography, A Guide to Optimization, Scale-up and Validation,* Academic Press, London.

van Deemter, J. J., Zuiderweg, F. J., and Klinkenberg, A. (1956). *Chem. Eng. Sci.*, **5**, 271.

Werner, M. H., Clore, G. M., Gronenborn, A. M., Kondoh, A., and Fisher, R. J. (1994). *FEBS Lett.*, **345**, 125.
Yau, W. W, Kirkland, J. J. and Bly, D. D. (1979). *Modern Size Exclusion Liquid Chromatography*, p. 116, Wiley, NY.
Yau, W. W. (1990). *Chemtr.-Macromol. Chem.*, **1**, 1.
Yano, Y. (1980). *J. Chromatogr.* **200**, 125.

3 ZORBAX POROUS SILICA MICROSPHERE COLUMNS FOR HIGH-PERFORMANCE SIZE EXCLUSION CHROMATOGRAPHY

ROBERT T. MOODY

MAC-MOD Analytical, Inc., Chadds Ford, Pennsylvania 19317

I. BACKGROUND INFORMATION

A. Development of Silica Particles for High-Performance Size Exclusion Chromatography (HPSEC)

In 1959, Porath and Flodin first reported using cross-linked poldextran gels to separate water-soluble macromolecules (1). This technique was referred to as gel-filtration chromatography (GFC). In 1964, Moore at Dow Chemical reported using cross-linked polystyrene gels to determine the molecular weight and molecular weight distribution of synthetic polymers via liquid chromatography (2). This chromatographic technique was referred to as gel-permeation chromatography (GPC) because it was based on selective permeation of porous gels by polymer molecules. Both GFC and GPC are types of size exclusion chromatography (SEC).

Modern SEC columns are packed with material other than polystyrene gels, such as porous silica particles or highly cross-linked styrene–divinylbenzene copolymers. Because of improvements in speed and resolution, the term SEC is sometimes replaced by the term high-performance size-exclusion chromatography (HPSEC).

In 1972, Kirkland at E. I. du Pont de Nemours patented porous silica microspheres (PSM) specifically for high-performance liquid chromatography (HPLC) applications (3). Prior to this development, silica particles used for chromatographic applications were simply adapted from some other use. In the 1970s, Kirkland showed that porous silica particles could be used for size-based separation of polymers (4). In 1978, Yau, Ginnard, and Kirkland at du Pont reported the use of porous silica microspheres for SEC, and du Pont commercialized porous silica materials as SEC packings (5). Columns packed with porous silica packings quickly became popular with polymer chemists because they offered improved stability over polymer gel columns and were compatible with a wider variety of solvents.

B. Advantages and Disadvantages of Porous Silica Packings for HPSEC

Porous silica packings are rigid and mechanically stable. Because of its mechanical strength, porous silica can usually be packed into columns at high pressures, resulting in better long-term stability. Porous silica packings can be used with a wide variety of organic and aqueous mobile phases. Because they do not swell in solvents, porous silica packings equilibrate rapidly with changes in the mobile phase; thus, they offer greater versatility and convenience than polymer gels. They have proved particularly valuable for HPSEC applications that require extremely high temperatures, such as molecular weight analysis of polyolefins requiring column temperatures as high as 150°C.

Porous silica packings do, however, sometimes suffer from adsorption between the sample and silanol groups on the silica surface. This interaction can interfere with the size exclusion experiment and yield erroneous information. In many cases, this problem is easily overcome by selecting mobile phases that eliminate these interactions. In addition, the surface of porous silica packings is routinely modified in order to reduce these undesirable interactions. Trimethylsilane modified packing is typically used with synthetic polymers. Diol modified packing is typically used with proteins and peptides.

C. Manufacture of Zorbax Porous Silica Microspheres (PSM)

Zorbax PSM particles are made from small (80–2000 Å), extremely uniform colloidal silica sol beads. In a patented polymerization process, these beads are agglutinated to form spherical particles. The size of the Zorbax PSM particles is controlled by the polymerization process, and the pore size is determined by the size of the silica sol beads. After polymerization, the silica is heated to remove the organic polymer and sinter the particles. The result is a spherical, porous, mechanically stable, pure silica particle that provides excellent chromatographic performance (Fig. 3.1).

The patented process used to produce Zorbax PSM packings is unique in its ability to carefully control both the particle size and the pore size. The polymerization process that "builds" the particle is used to produce a packing material of a desired particle size. Through careful control of this polymerization process, and other proprietary control techniques, packing material with an extremely narrow particle-size distribution can be produced. In addition, the size of the silica sol beads used in the manufacturing process is controlled to produce a particle with a pore size that will provide separation over a specific molecular weight range. These narrow size packings allow for homogeneous packed columns that have excellent bed stability, high-efficiency performance, and moderate back pressures.

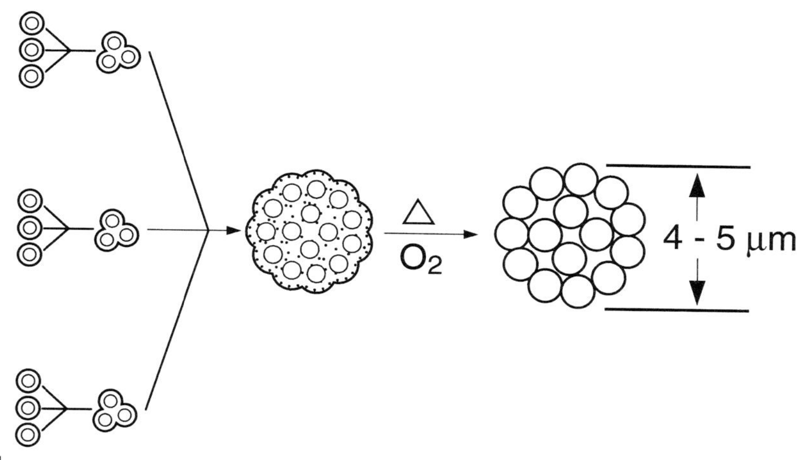

FIGURE 3.1 Formation of Zorbax porous silica microspheres.

Zorbax PSM packings are produced in three forms: unmodified, trimethylsilane modified, and diol modified. Modified Zorbax PSM packings are produced by chemically bonding a layer on the silica surface through siloxane bonds (Table 3.1). Silanized Zorbax PSM packings suppress adsorption effects and are the preferred choice when the mobile phase contains organic solvents. Unsilanized and diol modified Zorbax PSM packings should be used when the mobile phase consists of aqueous solvents.

D. Separation Mechanism

The mechanism of separation is the same for Zorbax PSM columns as it is for other types of SEC columns. As the mobile phase flows through the column, large molecules are forced down the column at faster rates than small molecules because the large molecules have less access to the column volume inside the pores. Consequently, molecules that are too large to permeate any of the pore

TABLE 3.1 Characteristics of Zorbax PSM Column Packings

Particle	Porous spherical silica
Particle size	5 μm
Pore size	
PSM 60	60 Å
PSM 300	300 Å
PSM 1000	1000 Å
PSM 3000	3000 Å
Bonded phase	
Silanized packing	Trimethylsilane
Unsilanized packing	No bonded phase

volume will elute from the column first (exclusion volume); molecules that are small enough to completely permeate the pore volume will elute last (permeation volume); and molecules of a size that allows partial access to the volume inside the pores will elute from the column in between. The elution volume of a molecule will depend on its hydrodynamic size (which may be related to the molecular weight of the molecule) and the pore size of the column packing.

II. USING AND SELECTING ZORBAX PSM COLUMNS AND MOBILE PHASES

A. Conducting HPSEC Experiments with Zorbax PSM Columns

Running an HPSEC experiment with Zorbax PSM columns is similar to running an experiment with any other GPC or HPSEC column. The process is summarized as follows:

1. Select the appropriate chromatographic column or columns. Choose a column packing with a pore size that will resolve the molecular size range of the sample.
2. Select suitable mobile phase condition. Choose a mobile phase that will solubilize the sample and will be compatible with the column packing material.
3. Select the detector. To acquire molecular weight distribution data, use a general detector such as a refractive index detector. To acquire structural or compositional information, employ a more selective detector such as an ultraviolet (UV) or infrared (IR) detector. Viscometric and light-scattering detectors facilitate more accurate molecular weight measurement when appropriate standards are not available.
4. Set the mobile phase flow rate. Use a flow rate of 1 to 2 ml/min (Table 3.2).
5. Calibrate the "system." Use narrowly dispersed molecular weight standards of the polymer of interest to construct a calibration curve of log molecular weight versus elution volume (Fig. 3.2). If a more sophisticated software system is available, a broad molecular weight standard may be used to calibrate the system.
6. Analyze the sample. Inject the dissolved sample onto the column.
7. Apply the results. Correlate molecular weight and molecular weight distribution data ($\overline{M}_n$, $\overline{M}_w$, $\overline{M}_z$, etc.) with desirable or undesirable properties of polymers.

TABLE 3.2 Characteristics of Zorbax PSM Columns

Dimensions (i.d. × L)	6.2 × 250 mm
Recommended flow rate	1 to 2 ml/min
Maximum operating pressure	3000 psi
Maximum operating temperature	150°C

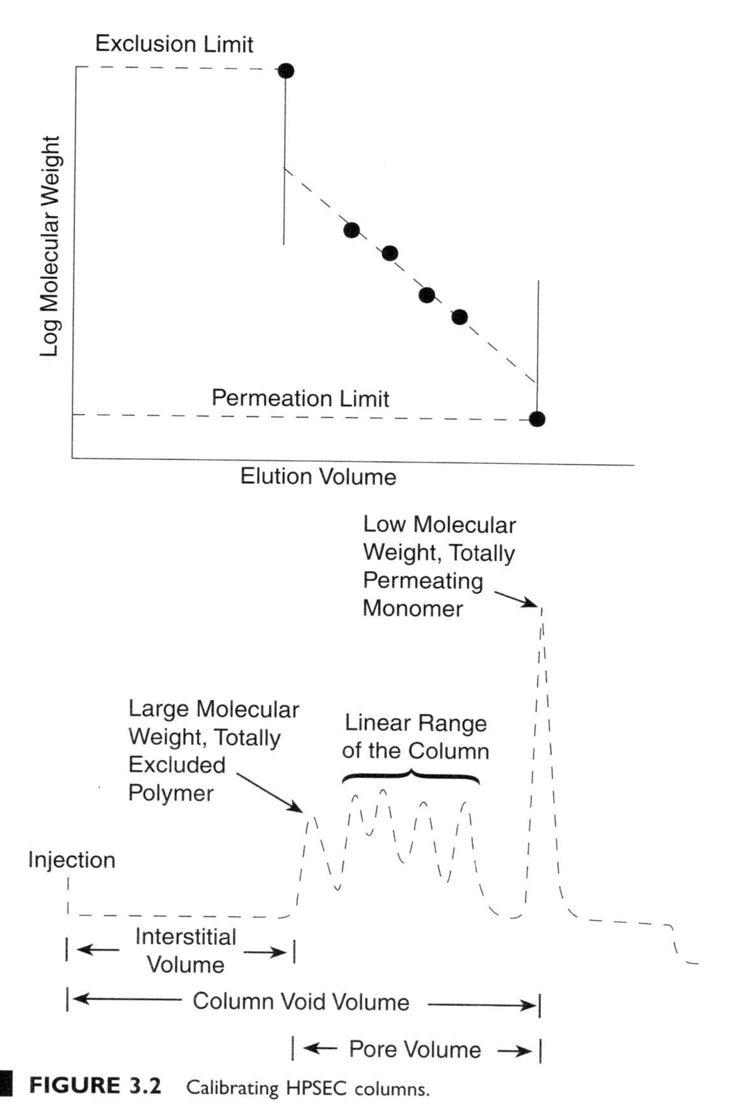

FIGURE 3.2 Calibrating HPSEC columns.

B. Selecting Columns and Mobile Phases for HPSEC

1. Selecting Zorbax PSM Columns for HPSEC

The elution volume of a molecule in HPSEC is determined by its hydrodynamic size and the pore size of the column packing. In setting up an HPSEC experiment, the chromatographer must match the pore size of the column to the molecular size range of the sample.

Figure 3.3 shows the calibration plots for Zorbax PSM columns. (The calibration plots for silanized and unsilanized columns are comparable.) These calibration plots allow the chromatographer to select the appropriate columns for samples. For example, the Zorbax PSM 60 column provides resolution of

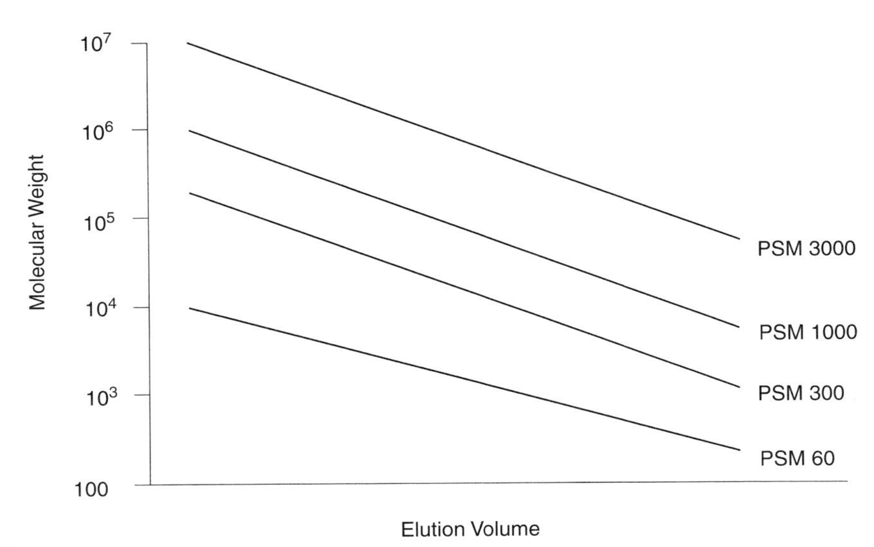

FIGURE 3.3 Calibration plots for Zorbax PSM columns.

molecular weight from 500 to 10,000, and the Zorbax PSM 1000 has a molecular weight range of 10,000 to 1,000,000. The molecular weight range data in Fig. 3.3 and Table 3.3 are based on polystyrene in tetrahydrofuran (THF) and do not necessarily apply to other polymers or solvent conditions. If the molecular weight range of a sample extends beyond the resolving range of any one packing pore size, a mixture of pore sizes must be used. This is achieved by combining columns packed with different pore size packings or selecting columns that are packed with a mixture of pore sizes.

TABLE 3.3 Separation Ranges of Zorbax PSM Columns

Column	Molecular weight range[a]
PSM 60	500 to 10,000
PSM 300	3000 to 300,000
PSM 1000	10,000 to 1,000,000
PSM 3000	100,000 to 9,000,000
Bimodal[b]	500 to 1,000,000
Trimodal[c]	500 to 9,000,000

[a]Data are based on polystyrene in THF and do not necessarily apply to other polymers or solvents.

[b]Contains PSM 60 and PSM 1000 particles.

[c]Contains PSM 60, PSM 300, and PSM 3000 particles.

2. Coupling Columns or Using Mixed-Bed Columns to Extend the Molecular Weight Range

Several important considerations are involved in selecting combinations of columns for molecular weight analysis. First, select only columns that cover the molecular weight range of the sample. Columns that separate outside the molecular weight range of the sample do not increase the resolving power of the system, but increase analysis time, back pressure, and band broadening. Second, use combined columns that provide a linear plot of log molecular weight versus elution volume over the required molecular weight range. Linear calibration plots are easier to use and provide improved accuracy in the molecular weight data; nonlinear calibration plots require expensive software.

To achieve linear performance, combine columns that cover the desired molecular weight range, but do not include columns that have overlapping molecular weight ranges. In addition, the pore volumes of the columns should be nearly identical because the slope of the calibration curve is directly related to the column pore volume. The combination of Zorbax PSM 60 and Zorbax PSM 1000 columns yields good linearity over the molecular weight range of 500 to 1,000,000 because they meet these criteria.

Zorbax PSM Bimodal and Trimodal columns are packed with mixed pore-size packing to achieve linear size separations over a broad molecular weight range (Table 3.3). Zorbax PSM Bimodal columns are packed with PSM 60 and PSM 1000 particles, and Trimodal columns contain PSM 60, PSM 300, and PSM 3000 particles (Fig. 3.4). Carefully selecting and mixing different pore-size particles in columns provide much better linearity than coupling columns that are each packed with single pore-size particles.

Resolution increases when columns of the same pore size are used in series; however, increasing the number of columns in series also increases the total

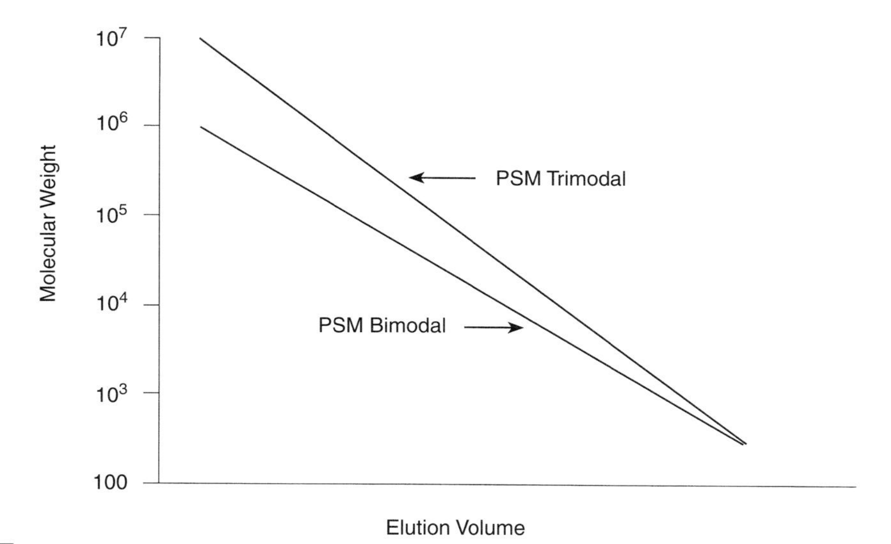

FIGURE 3.4 Calibration plots for Zorbax PSM Bimodal and Trimodal columns.

column volume and adds to band broadening. Analysis time and system back pressure also increase as column length increases. Thus, there is a limit to the number of columns that should be used in series. For most HPSEC columns, two columns in series provide an acceptable compromise for most applications.

3. Selecting Mobile Phases for HPSEC

Select mobile phases for HPSEC based on their ability to dissolve the sample and their compatibility with the column. Zorbax PSM columns are compatible with a wide variety of organic and aqueous mobile phases (Table 3.4), but analysts should avoid aqueous mobile phases with a pH greater than 8.5. As mentioned earlier, select mobile phases that minimize adsorption between samples and silica-based packings. Sample elution from the column after the permeation volume indicates that adsorption has occurred. If adsorption is observed or suspected, select a mobile phase that will be more strongly adsorbed onto the silica surface than the sample. For example, *N,N*-dimethylformamide (DMF) is often used for polyurethanes and polyacrylonitrile because it eliminates adsorption and dissolves the polymers. When aqueous mobile phases are required, highly polar macromolecules such as Carbowax can be used to coat the silica surface and eliminate adsorption. Table 3.5 provides a list of recommended mobile-phase conditions for some common polymers.

III. HPSEC TROUBLESHOOTING

Errors in the molecular weight data from HPSEC are usually due to improperly prepared samples, column dispersity, or flow rate variations. The sample to be analyzed should be completely dissolved in the mobile phase and filtered prior to injection onto the column. A plugged column inlet frit will invalidate results. In addition, do not load the column with excess sample. Column overloading affects the accuracy of data by broadening peaks, reducing resolution, and increasing elution volume. For best results, the concentration of the injected sample should be as low as possible while still providing adequate

TABLE 3.4 Examples of Acceptable Solvents for Zorbax PSM Columns

Acetone	Hexafluoroisopropanol
Benzene	Isopropanol
Buffers, aqueous (pH <8.5)	Methanol
Carbon tetrachloride	Methylene chloride
Chloroform	*N*-Methylpyrrolidone
m-Cresol	Tetrahydrofuran
o-Dichlorobenzene	Toluene
N,N-Dimethylformamide	Trichlorobenzene
Dimethyl sulfoxide	Water

TABLE 3.5 Recommended Mobile Phases for HPSEC Analyses of Polymers

Polymer	Mobile phase
ABS	DMF at 135°C
Adhesives, hot melt	TCB at 135°C
Alkyd resin	THF
Ashphaltenes	THF
Carboxylated polybutadiene	THF
Carboxylated methylcullulose	Water at 80°C
Cellulose	DMSO at 80°C
Cellulose esters	THF
Cellulose nitrate	THF
Coal tars	TCB at 135°C
Elastomers	Toluene at 80°C
EPDM	TCB at 135°C
Epoxy resin	THF
EPR	TCB at 135°C
Melamine resin	DMF at 80°C or HFIP at 40°C
Nylon	*m*-Cresol at 25–135°C or HFIP at 40°C
Phenolic resin	THF at 50°C or DMF at 50°C
Polyacrylamides	Water at 85°C
Polyacrylates	Water at 80°C or DMF at 100°C
Polyacrylic acid	Water at 80°C
Polyacrylonitrile	DMF at 100°C
Polyamides	*m*-Cresol at 100°C or HFIP at 40°C
Polybutadiene	Toluene at 85°C or THF at 40°C
Polycarbonate	THF
Polydextrans	Water at 70°C
Polydimethyl siloxane	ODCB at 90–140°C
Polyester polyol	THF
Polyesters	*m*-Cresol at 100°C or HFIP at 40°C
Polyethylene	ODCB at 145°C
Polyethylene oxide	Water at 70°C
Polyethylene terephthalate	*m*-Cresol at 135°C or HFIP at 40°C
Polyglycols	Water at 70°C
Polyisocyanate	THF
Polyisoprene	THF at 80°C or toluene at 80°C
Polymethylmethacrylate	THF
Polypropylene	ODCB at 145°C
Polystyrene	THF
Polyurethanes	DMF at 100°C
Polyvinylacetate	THF
Polyvinylchloride	THF
Polyvinylpyrrolidone	*N*-Methylpyrrolidone at 80°C
SBR	Toluene at 80°C or THF at 40°C
Silicones	Toluene or THF
Starches	Water at 80°C or DMF at 100°C
Urea–formaldehyde resins	HFIP

TABLE 3.6 Guidelines for Injected Sample Concentration[a]

Molecular weight	Sample concentration (weight %)
Less than 20,000	0.25
34,000 to 200,000	0.10
400,000 to 2,000,000	0.05
Greater than 2,000,000	0.01

[a]For best results, sample injection volume should not exceed 100 μl.

detection. Table 3.6 provides guidelines for selecting sample concentration based on the molecular weight of the polymer.

Column dispersity (band spreading) causes the measured molecular weight distribution to be broader than the true molecular weight distribution (Fig. 3.5). Because Zorbax PSM columns exhibit very low band-spreading characteristics, these columns have historically provided better molecular weight distribution accuracy than many gel-type columns.

Most modern HPLC pumping systems provide excellent flow rate control and have sufficient precision for HPSEC applications. Because molecular weight determinations are based on a linear relationship between log molecular weight and elution volume, and because many systems relate elution time to elution volume, even small errors in flow rate can lead to substantial errors in molecular weight data. Random changes in flow rate may not only lead to inaccurate molecular weight data, but also lead to problems with precision between runs. Depending on the magnitude of the drift, flow rate drift can cause even larger problems in determining molecular weight data. Even if all other factors are constant, a mere 1% change in flow rate can cause an error of 20% in molecular weight average (Fig. 3.6). Flow rate disparity creates substantial problems when molecular weight distribution data from two different HPSEC systems are compared. Even if the flow rates of the calibrated pumps are only slightly

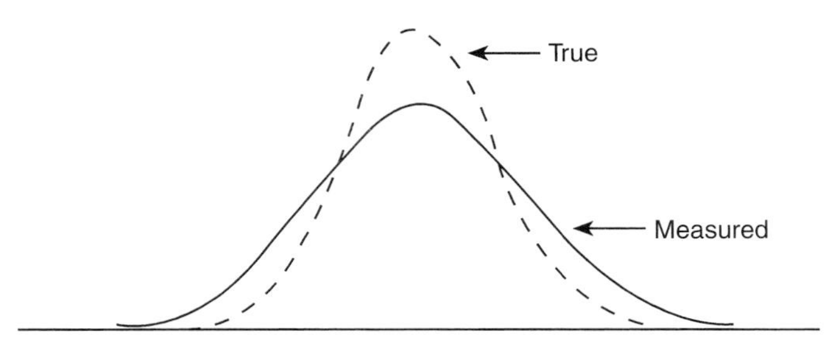

FIGURE 3.5 Measured molecular weight distribution (by HPSEC) is broader than the true molecular weight distribution.

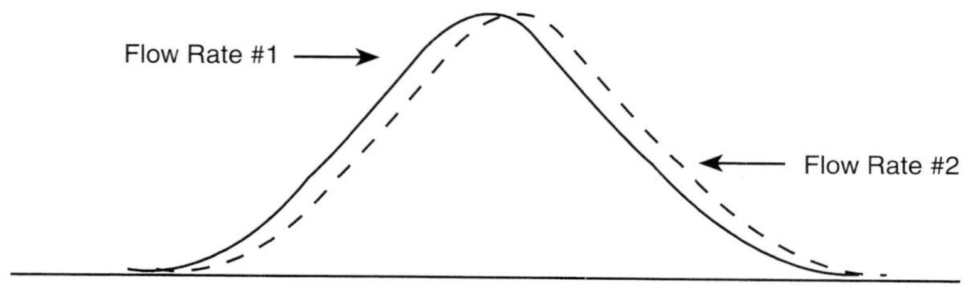

% Deviation in Flow Rate	Relative _% Error in Mw
+0.5	+9.7
−0.5	−8.9
+1.0	+20.1
−1.0	−17.0

FIGURE 3.6 Small differences in flow rate can have a significant effect on the molecular weight measurement.

different, the reported molecular weight data from the two systems may not agree.

IV. APPLICATIONS

A. General Applications

HPSEC is commonly used to compare the molecular weight distribution of polymers. This comparison is sometimes made by simple visual inspection of the HPSEC "curves." However, quantitative moments of the molecular weight distribution ($\overline{M}_n$, $\overline{M}_w$, $\overline{M}_z$) are typically used because they are sensitive to different areas of the molecular weight distribution and provide useful information about specific properties of polymers. The number average molecular weight ($\overline{M}_n$) is more sensitive to smaller molecules and gives information about a polymer's impact resistance and plasticity. The weight average molecular weight ($\overline{M}_w$) is more sensitive to larger molecules and often provides information about the general strength of a polymer. The z average molecular weight ($\overline{M}_z$) provides information on the largest molecules and can be used to gain information on creep. Dispersity ($\overline{M}_w$ / $\overline{M}_n$) often provides information about a polymer's processability.

Figure 3.7 shows HPSEC data that identify good and bad polyurethane. The good polyurethane has larger $\overline{M}_n$, $\overline{M}_w$, and $\overline{M}_z$ values, indicating a larger amount of high-molecular-weight components. This can even be seen in a visual inspection of the overlapping HPSEC curves.

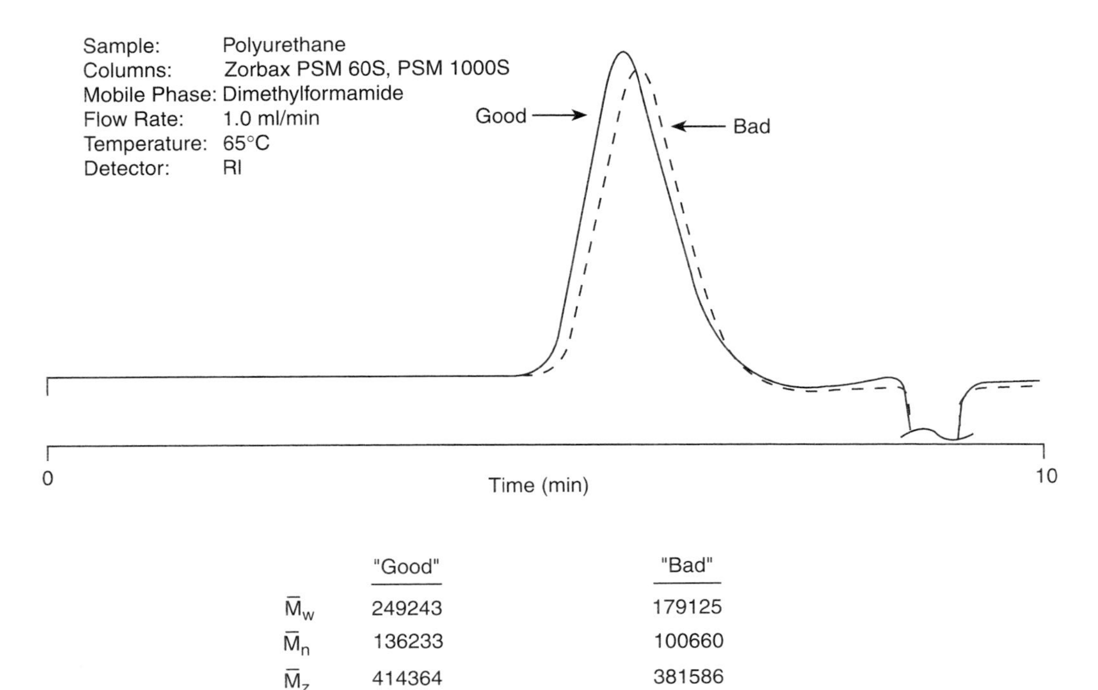

	"Good"	"Bad"
$\bar{M}_w$	249243	179125
$\bar{M}_n$	136233	100660
$\bar{M}_z$	414364	381586

FIGURE 3.7 "Good" and "bad" performance characteristics of polymers can often be identified by molecular weight distribution data.

Many high molecular weight synthetic polymers, such as polyethylene and polypropylene, have a large percentage of their molecules in the crystalline state. Prior to dissolution, these polymers must usually be heated almost to their melting points to break up the crystalline forces. Orthodichlorobenzene (ODCB) is a typical mobile phase for these polymers at 150°C. The accuracy and stability of the Zorbax PSM columns under such harsh conditions make them ideal for these analyses (Fig. 3.8).

Variations in the composition of a copolymer can cause substantial differences in the properties of the copolymer. Compositional information about copolymers may be acquired using selective detectors. Figure 3.9 shows the separation of an ethylene–vinyl acetate (EVA) copolymer by HPSEC using IR detectors. One IR detector monitors the vinyl acetate carbonyl at 5.75 μm, and the other IR detector monitors the total alkyl absorbance at 3.4 μm.

B. Special Applications: Proteins

The size separation of proteins has been routinely called gel filtration because of the historic use of cross-linked gels for this application. Specially modified Zorbax PSM columns, Zorbax GF-250 and Zorbax GF-450, are used for separating proteins by size. These columns are packed with porous silica micro-

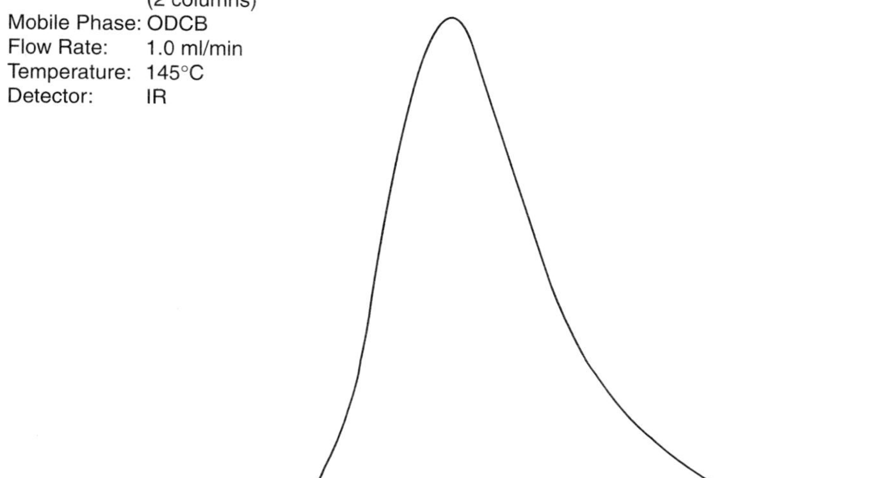

	$\overline{M}_w$	$\overline{M}_n$
Known Value (NBS standard)	53,100	18,300
HPSEC Calculated Value	53,176	18,942

FIGURE 3.8 The stability of Zorbax PSM columns makes them ideal for applications that require extremely harsh conditions, such as high temperature HPSEC of polyolefins.

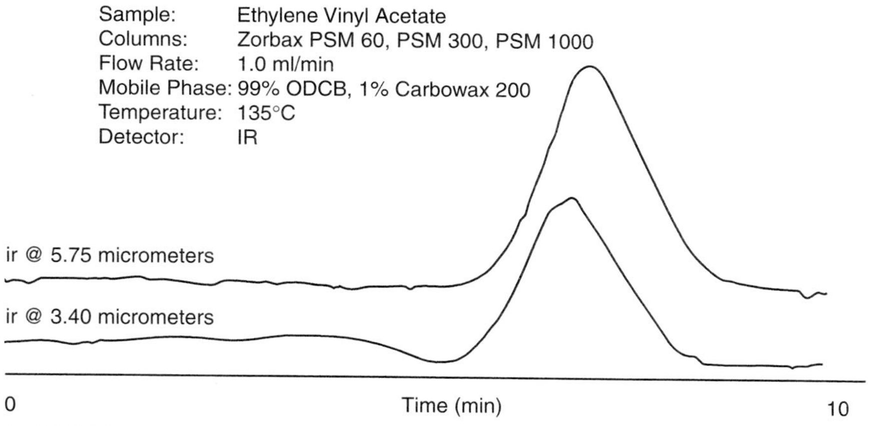

FIGURE 3.9 Selective detectors can be used to provide structural information about copolymers.

TABLE 3.7 Characteristics of Zorbax GF Column Packings

	GF-250	GF-450
Particle	Porous spherical silica	Porous spherical silica
Particle size	4 μm	6 μm
Pore size	150 Å	300 Å
Surface area	140 m^2/g	50 m^2/g
Surface modification	Zirconium-stabilized	Zirconium-stabilized
Bonded phase	Diol	Diol
pH range	3 to 8.5	3 to 8.5
Separation range (approximate)	4000 to 400,000 Da	10,000 to 1,000,000 Da

TABLE 3.8 Recovery of Purified Proteins on Zorbax GF-250 Columns

Protein	% recovered
Ovalbumin	100.4
Rnase	99.5
α-Chymotrypsin	97.3
β-Amylase	93.8
Thyroglobulin	92.3
Carbonic anhydrase	91.5
Cytochrome c	91.1
Myoglobin	85.6

TABLE 3.9 Characteristics of Zorbax GF Columns

Dimensions (i.d. × L)	4.6 × 250 mm 9.4 × 250 mm 21.2 × 250 mm
Maximum operating pressure	5000 psi
Maximum operating temperature	80°C

spheres that have a unique surface stabilized by zirconia. These zirconia-stabilized phases permit the Zorbax GF columns to be used over a much wider pH range than other surface-modified, silica-based SEC columns. A mobile phase pH of 3.0 to 8.5 can be safely used (Table 3.7). In addition, a diol phase is bonded to the silica surface to produce a hydrophilic phase that minimizes protein adsorption and yields high protein recovery. Typical protein recovery on a Zorbax GF column is more than 90% at loads of 5 μg or greater (Table 3.8). Table 3.9 provides characteristics of Zorbax GF columns.

Proteins are separated on Zorbax GF columns based on their hydrodynamic size, which may be related to the proteins' molecular weights (Fig. 3.10). Under ideal conditions, two proteins whose molecular sizes differ by a factor of 2 can be baseline separated.

Zorbax GF columns can be used for size-separation applications such as estimating molecular weight; purifying complex mixtures; monitoring reac-

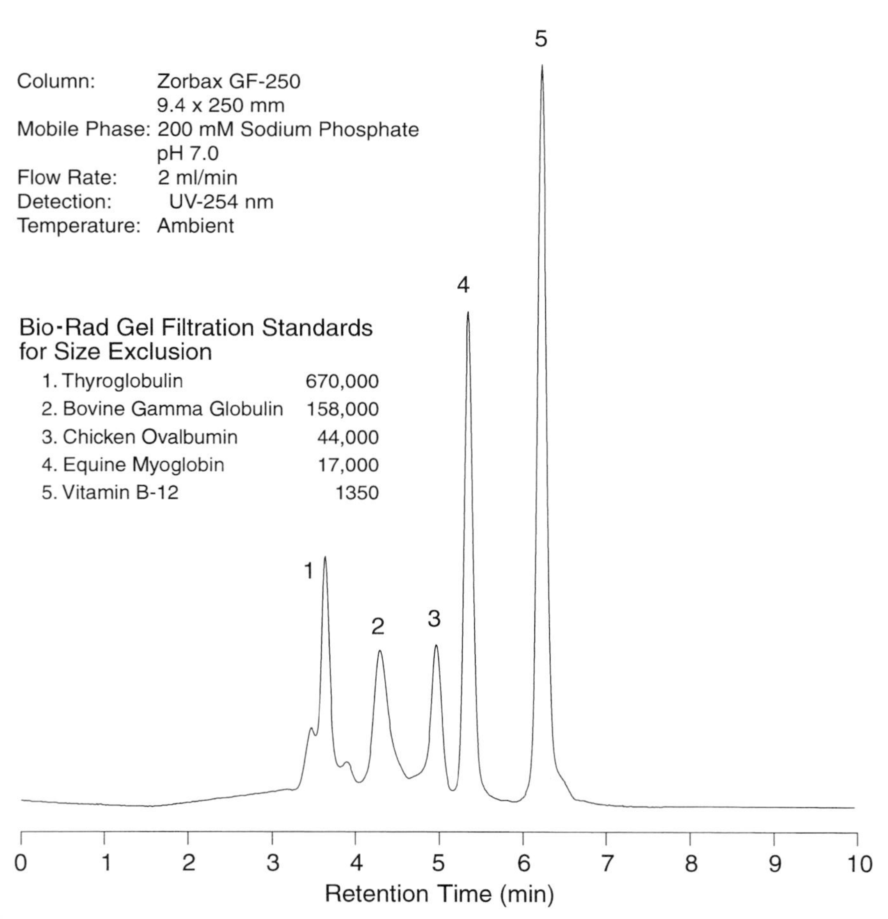

FIGURE 3.10 Under ideal conditions, proteins are separated on Zorbax GF columns based on their hydrodynamic size.

tions, conformational changes, and binding reactions (Fig. 3.11); and desalting samples. For accurate and reproducible results, electrostatic and hydrophobic interaction of the protein with the packing must be eliminated or at least minimized and controlled. The selection of the proper mobile phase ionic strength is the key to achieving reproducible results. Table 3.10 gives relative elution values for several proteins with mobile phases having different ionic strengths. (The elution value is the elution volume of the protein divided by the total permeation volume of the column.) For best results, choose an ionic strength at which small changes in the ionic strength will not affect the elution volume of the protein.

Tailing peaks or longer than expected elution volumes are sometimes caused by low solubility of the protein in the mobile phase. Using a trial-and-error process, select the proper pH and ionic strength to address this problem. Detergents such as sodium dodecyl sulfate (SDS) are sometimes helpful but, because they change the conformation of many proteins and are difficult to remove from the column should be used only if other methods fail.

As with other size-exclusion techniques, the pore size of the selected Zorbax GF column should provide resolution over the molecular size range of the proteins that are to be separated. The Zorbax GF-250 column separates proteins in the range of 4000 to 400,000 Da. The Zorbax GF-450 provides separation over the range of 10,000 to 1,000,000 Da. When these two columns are coupled, they can be used to separate proteins with molecular weights of 4000 to 1,000,000.

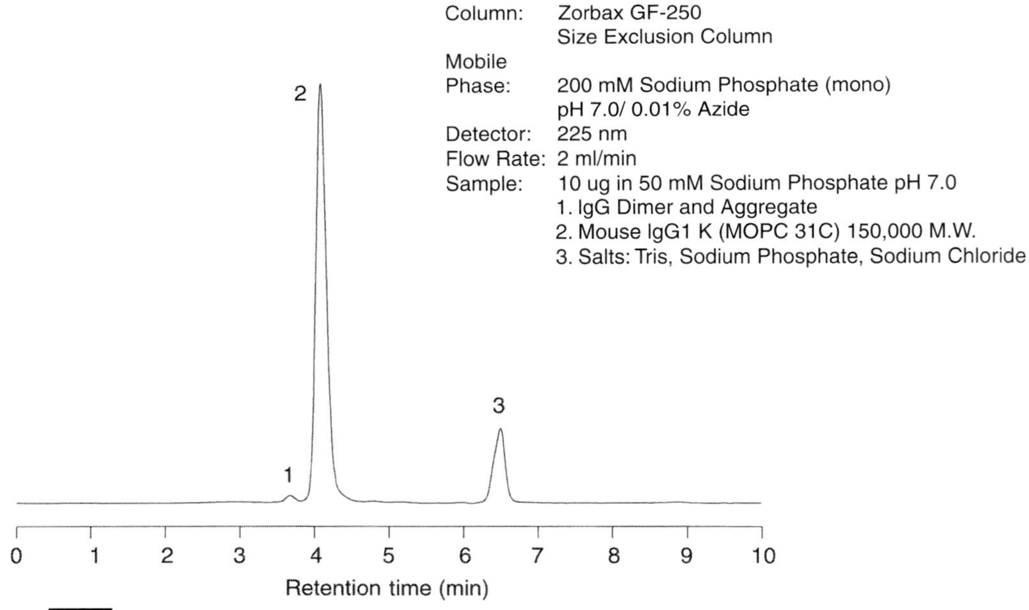

FIGURE 3.11 Zorbax GF columns are routinely used to quantitate or isolate different forms (monomer, dimer, aggregates) of biomolecules.

TABLE 3.10 Relative Elution Value (K^*)[a] of Samples at Various Ionic Strengths[b]

Sample	pI	MW	K* of sample at listed concentration of pH 7.0 sodium phosphate				
			80 mM	100 mM	150 mM	200 mM	400 mM
Thyroglobulin	5.1	669,000	0.54	0.54	0.54	0.53	0.55
Ferritin	4.2	440,000	0.64	0.65	0.64	0.64	0.67
Catalase	5.4	250,000	0.68	0.69	0.68	0.68	0.71
β-Amylase	5.4	200,000	0.63	0.64	0.64	0.64	0.65
Alcohol dehydrogenase	6.8	150,000	0.66	0.67	0.67	0.67	0.67
BSA dimer	5.4	132,000	0.64	0.64	0.64	0.64	0.65
BSA	5.1	66,430	0.70	0.70	0.70	0.70	0.72
Ovalbumin	4.6	44,000	0.75	0.75	0.75	0.75	0.78
Peroxidase	9.0	44,000	0.73	0.73	0.73	0.73	0.73
Pepsin	2.0	35,000	0.81	0.80	0.81	0.81	0.92
Carbonic anhydrase	5.9	29,000	0.85	0.83	0.82	0.82	0.83
Chymotrypsinogen A	9.3	25,000	1.00[c]	0.88	0.84	0.83	0.83
Trypsinogen	9.3	23,970	0.83	0.82	0.80	0.82	0.83
Trypsin	10.5	23,280	0.87	0.85	0.83	0.83	0.85
α-Chymotrypsin	8.8	21,600	0.96	0.86	0.82	0.82	0.83
Myoglobin	7.2	17,600	0.82	0.82	0.82	0.82	0.83
Lysozyme	10.0	14,300	1.33[c]	1.02[c]	0.94	0.95	0.95
Ribonuclease A	7.8	13,700	1.82[c]	1.78[c]	0.96	0.85	0.86
Cytochrome c	9.6	12,400	4.26[c]	3.72[c]	1.20[c]	0.92	0.85
Aprotinin	10.0	6,500	1.72[c]	1.11[c]	0.96	0.94	0.93
Insulin	5.7	6,000	0.97	0.95	0.96	0.97	1.09[c]
Vitamin B_{12}	—	1,350	0.95	0.95	0.96	0.99	1.04[c]
Uridine	—	240	0.95	0.94	0.95	0.96	0.97
Uracil	—	120	0.99	0.98	0.99	1.00	1.01[b]
Sodium Azide	—	60	1.00	1.00	1.00	1.00	1.01[b]

[a]$K^* = V_R/V_m$, where V_R is the retention volume of sample and V_M is the total permeation volume of the column.

[b]From Ref. 6.

[c]Nonideal behavior.

For best results, use the flow rate, injection volume, and column sample capacity guidelines in Table 3.11. Conditions outside these guidelines may be used, but poor resolution between proteins may result from extensive deviations from these guidelines.

TABLE 3.11 Operating Guidelines for Zorbax GF Columns

Column i.d. × L (mm)	Flow rate (ml/min)	Injection volume (μl)	Sample mass (mg)
4.6 × 250	0.5–1.0	10	0.25
9.4 × 250	1.0–3.0	50	1.0
21.2 × 250	5.0–10.0	250	5.0

ACKNOWLEDGMENTS

I thank J. J. Kirkland and J. J. DeStefano of Hewlett-Packard for reviewing the technical information presented in this chapter and for their helpful advice. I also thank L. Amos and P. E. Antle of Writers, Inc., for their assistance in preparing the text and tables. Zorbax is a registered trademark of the DuPont Company.

REFERENCES

1. Porath, J., and Flodin, P. (1959). *Nature* **183,** 1657.
2. Moore, J. C. (1964). *J. Polym. Sci. A* **2,** 835.
3. Kirkland, J. J. (1972). *J. Chromatogr. Sci.* **10,** 593.
4. Kirkland, J. J. (1976). *J. Chromatogr.* **125,** 231.
5. Yau, W. W., Ginnard, C. R., Kirkland, J. J. (1978). *J. Chromatogr.* **149.**
6. Ricker, R. D., Sandoval, L. A., Justice, J. D., and Geiser, F. O. (1995). *J. Chromatogr. A,* **691,** 67.

BIBLIOGRAPHY

GPD*HPSEC Column Selection Guide. (1988). MAC-MOD Analytical, Inc., Chadds Ford, PA.

Kirkland, J. J. (1973). *J. Chromatogr. Sci.* **83.**

Theodore Provder, Editor. (1987). *Detection and Data Analysis in Size Exclusion Chromatography,* ACS Symposium Series.

Yau, W. W., Kirkland, J. J., and Bly, D. D. (1979). *Modern Size-Exclusion Liquid Chromatography,* John Wiley & Sons.

Zorbax GF-250 and GF-450 Size Exclusion Columns. (1996). Rockland Technologies, Inc.

Zorbax HPLC Column Catalog. (1988). MAC-MOD Analytical, Inc., Chadds Ford, PA.

Zorbax HPLC Columns for Analytical Biochemistry. (1995). MAC-MOD Analytical, Inc., Chadds Ford, PA.

4

SIZE EXCLUSION CHROMATOGRAPHY USING TSK-GEL COLUMNS AND TOYOPEARL RESINS

YOSHIO KATO
J. KEVIN O'DONNELL
JON FISHER

TosoHaas, Montgomeryville, Pennsylvania 18936

I. SILICA-BASED TSK-GEL SW/SW$_{XL}$ COLUMNS FOR GEL-FILTRATION CHROMATOGRAPHY (GFC)

A. Description

The separation of biomolecules by size exclusion chromatography (SEC) in aqueous systems is referred to as gel-filtration chromatography (GFC). Silica-based TSK-GEL SW packings are designed for the GFC of proteins and peptides. Many molecular weight estimations in the literature have been performed with TSK-GEL SW columns. TSK-GEL SW type packings are composed of rigid spherical silica gel chemically bonded with hydrophilic compounds that have primary alcohols on the surface (1,2). These packings have advantages due to low adsorption and well-defined pore size distribution, which are required for high-performance SEC. Particles having three different pore sizes are available packed as TSK-GEL G2000SW, TSK-GEL G3000SW, and TSK-GEL G4000SW columns. The properties of these packings are summarized in Table 4.1.

The high-performance TSK-GEL SW$_{XL}$ versions are packed with 5-μm particles (8-μm particles in G4000SW$_{XL}$ columns) whereas standard TSK-GEL SW columns are packed with 10-μm particles (13-μm particles in G4000SW columns). TSK-GEL SW and TSK-GEL SW$_{XL}$ analytical packings are available in both glass and stainless-steel column formats. Semipreparative, 21.5-mm i.d. columns are packed with 13-μm TSK-GEL SW particles (17-μm particles in G4000SW columns), and preparative 55-mm, 108-mm, and larger inside

Column Handbook for Size Exclusion Chromatography

TABLE 4.1 Properties and Separation Ranges for TSK-GEL SW Type Packings[a]

TSK-GEL packing	Particle size (μm)	Pore size (Å)	Molecular weight of sample: Globular protein	Dextran	Polyethylene glycol/oxide
Super SW2000	4	125	5000–150,000	1000–30,000	500–15,000
G2000SW$_{XL}$	5	125	5000–150,000	1000–30,000	500–15,000
G2000SW	10, 13, 20	125	5000–100,000	1000–30,000	500–15,000
Super SW3000	4	250	10,000–500,000	2000–70,000	1000–35,000
G3000SW$_{XL}$	5	250	10,000–500,000	2000–70,000	1000–35,000
G3000SW	10, 13, 20	250	10,000–500,000	2000–70,000	1000–35,000
G4000SW$_{XL}$	8	450	20,000–10,000,000	4000–500,000	2000–250,000
G4000SW	13, 17	450	20,000–7,000,000	4000–500,000	2000–250,000

[a]Column: Two 5-μm, 7.8 mm × 30 cm TSK-GEL SW$_{XL}$ columns in series; two 10-μm, 7.5 mm × 60 cm TSK-GEL SW columns in series.

Elution: Proteins: 0.3 *M* NaCl in 0.1 *M* (0.05 *M* for SW$_{XL}$ columns) phosphate buffer, pH 7; dextrans and PEOs: distilled water.

diameter columns (G2000SW and G3000SW only) are packed with 20-μm TSK-GEL SW particles. Appropriate guard columns are available for all columns. The separation of a mixture of protein standards is compared on each TSK-GEL SW and TSK-GEL SW$_{XL}$ analytical column in Fig. 4.1.

The availability of different pore sizes for the three TSK-GEL SW packings results in different exclusion limits for several sample series. These are shown by the calibration curves in Fig. 4.2. From these data, recommended separation ranges for different sample types can be made for each column (Table 4.1). In Fig. 4.3, the calibration curves for protein standards are compared using the three high-performance, 5-μm TSK-GEL SW$_{XL}$ packings. The separation ranges listed in Table 4.1 for the TSK-GEL$_{XL}$ packing are slightly different from those of the 10-μm TSK-GEL SW packings due to slight variations in manufacturing the smaller particle size silica and in column packing techniques.

B. Conditions

1. Choosing the Right Column

The right chromatography column should separate the sample sufficiently to enable identification or quantitative measurement of the components within a reasonable period of time. The resolution factor (Rs) for two sample components is determined by the width of the two peaks and the distance between the peak maxima. In general, Rs values of 1.0 are required for good qualitative or quantitative work, whereas Rs values >1.5 indicate baseline resolution for two components (3).

The different pore sizes and exclusion limits of each TSK-GEL SW column will have a substantial effect on the resolution of a biomolecule mixture. G2000SW packing, which has the smallest pores, provides the best resolution for smaller proteins such as myoglobin and cytochrome c (16,900 and 12,400 Da, respectively, Rs = 1.01). Resolution of the proteins myoglobin

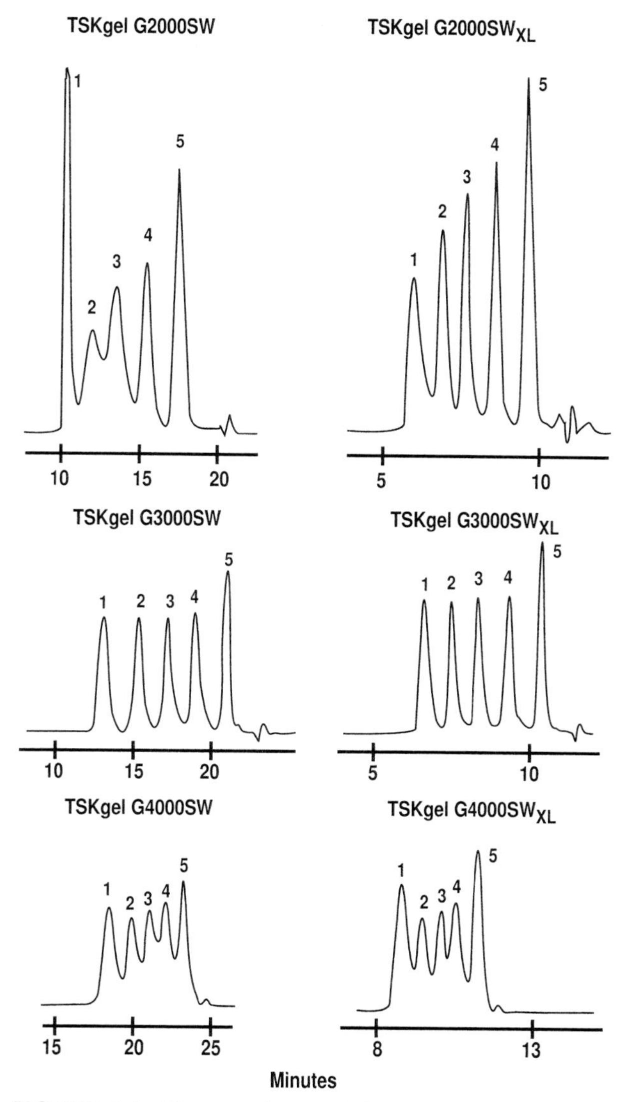

FIGURE 4.1 Higher resolution with 5-μm TSK-GEL SW_{XL} columns compared to 10-μm TSK-GEL SW columns. Columns: (left) TSK-GEL SW, two 10-μm, 7.5 mm × 30 cm columns in series and (right) TSK-GEL SW_{XL}, two 5-μm, 7.8 mm × 30 cm columns in series. Sample: (1) glutamate dehydrogenase, (2) lactate dehydrogenase, (3) enolase, (4) adenylate kinase, and (5) cytochrome c. Elution: 0.3 *M* NaCl in 0.05 *M* phosphate buffer, pH 6.9. Flow rate: 1.0 ml/min. Detection: UV at 220 nm.

and β-lactoglobulin (35,000 Da) and bovine serum albumin (67,000 Da) is better on the larger pore G3000SW packing (Rs = 0.55 and 2.74) compared to the G2000SW packing (Rs = 0.30 and 1.78). The very large protein thyroglobulin (660,000 Da) was excluded from both the G2000SW and the G3000SW packings, but was retained on the G4000SW packing, which has the largest pores of the three packings. Because most proteins are between

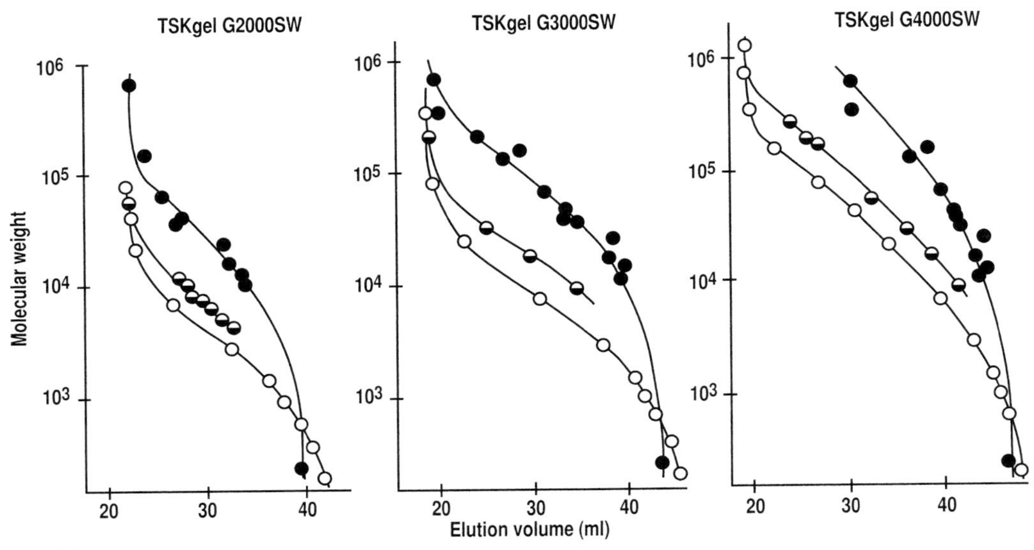

FIGURE 4.2 Polyethylene oxide, dextran, and protein calibration curves for TSK-GEL SW Columns. Column: TSK-GEL SW, two 7.5 mm × 60 cm columns in series. Sample: ●, proteins; ◒, polyethylene oxides; O, dextrans. Elution: dextrans and polyethylene oxides: distilled water; proteins: 0.3 *M* NaCl in 0.1 *M* phosphate buffer, ph 7. Flow rate: 1.0 ml/min. Detection: UV at 220 nm and RI.

20,000 and 200,000 Da, the TSK-GEL G3000SW column is most often recommended and is the most widely used. Resolution factors for various proteins on 5-, 8-, and 10-μm TSK-GEL SW columns are compared in Table 4.2.

2. Mobile Phase

a. Ionic Strength

Proper selection of elution conditions is necessary to maximize molecular sieving mechanisms and to minimize secondary effects such as ionic and hydrophobic interactions between the sample and the column packing material. Under conditions of high ionic strength (>1.0 *M*), hydrophobic interactions may occur. Under low ionic strength (<0.1 *M*), ionic interactions are more likely to occur. Secondary interactions take place more often with small solutes as the residual silanol sites of TSK-GEL SW packings are mainly located in small pores, which are accessible to small solutes. In general, the use of relatively high ionic strength buffers can be used to overcome secondary interactions. A neutral salt, such as sodium sulfate, is often added to increase buffer ionic strength.

While the extent of secondary interactions is not as great for proteins as for smaller molecules, complex interactions can occur. For each protein, there will be an optimum buffer type and concentration, which results in the highest resolution and recovery. As shown in Table 4.3, resolution among protein samples varies according to buffer type and strength. Although each salt used has the same molar concentration in the mobile phase, each provides a different ionic strength, from the lowest in strength, sodium chloride, to the highest, ammonium phosphate.

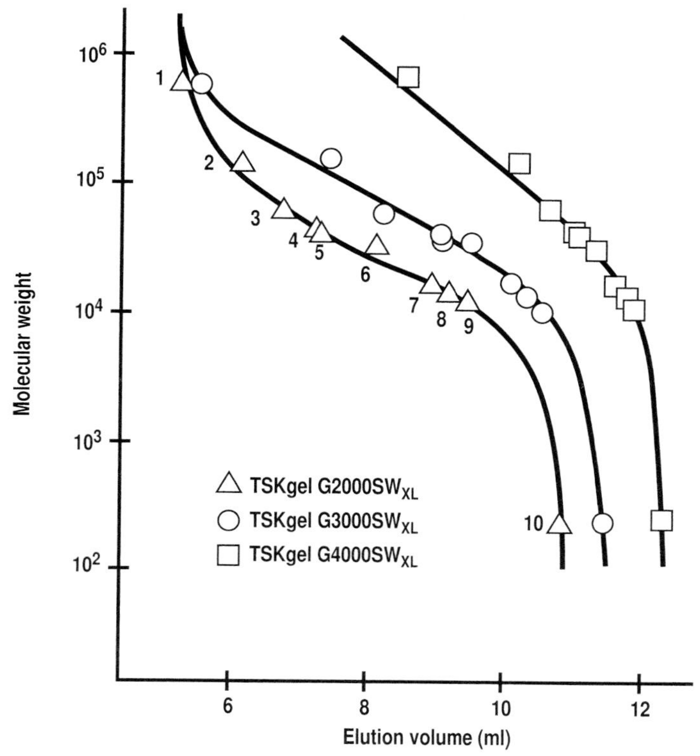

FIGURE 4.3 Protein calibration curves for TSK-GEL SW$_{XL}$ columns. Column: TSK-GEL SW$_{XL}$ columns, 5 μm, 7.8 mm $\times$ 30 cm. Sample: (1) thyroglobulin (660, 000 Da), (2) IgG (156,000 Da), (3) bovine serum albumin (67,000 Da), (4) ovalbumin (43,000 Da), (5) peroxidase (40,200 Da), (6) β-lactoglobulin (35,000 Da), (7) myoglobin (16,900 Da), (8) ribonuclease A (13,700 Da), (9) cytochrome c (12,400 Da), and (10) glycine tetramer (246 Da). Elution: 0.3 *M* NaCl in 0.1 *M* sodium phosphate buffer, pH 7. Detection: UV at 220 nm.

b. Ionic Species

The ionic species of the mobile phase will also affect the separation. This is shown in Table 4.3 by the difference in resolution values for magnesium chloride buffer compared to sodium sulfate buffer. In addition, calibration curves for proteins in potassium phosphate buffers are shallower than those generated in sodium phosphate buffers. The slope of the curve in Sorenson buffer (containing both Na^+ and K^+) is midway between the slopes generated with either cation alone (1). Table 4.4 illustrates the impact of different buffer conditions on mass recovery for six sample proteins. In this case, the mass recovery of proteins (1,4) is higher with sodium or potassium phosphate buffers (pH 6.9) than with Tris–HCl buffers (pH 7.8).

c. pH

The optimal buffer pH for a separation depends on the protein isoelectric point (p*I*) and the stability of the packing. In general, the use of a buffer at a

TABLE 4.2 Comparison of Protein Resolution Factors (Rs) on TSK-GEL SW Columns[a]

		G2000SW		G3000SW		G4000SW	
Protein	**Da**	**5 μm**	**10 μm**	**5 μm**	**10 μm**	**8 μm**	**13 μm**
Thyroglobulin	660,000						
		2.4	1.6	4.1	4.4	3.2	2.8
IgG	156,000						
		3.1	2.2	3.7	2.3	1.5	1.3
BSA	67,000						
		6.4	2.9	7.1	4.2	3.3	2.4
Peroxidase	40,200						
		9.1	5.8	8.3	5.7	3.0	2.8
Myoglobin	16,900						
		13.0	5.2	8.5	4.3	3.3	2.4
Cytochrome c	12,400						
		2.9	1.5	2.7	1.3	0.7	0.7
Glycine tetramer	246						

[a]Column: TSK-GEL G2000SW and G3000SW: 5 μm, 7.8 mm × 30 cm; 10 μm, 7.5 mm × 30 cm
TSK-GEL G4000SW: 8 μm, 7.8 mm × 30 cm; 13 μm, 7.5 mm × 30 cm
Elution: 0.3 *M* NaCl in 0.1 *M* phosphate buffer, pH 7.0.
Note: $^*Rs = 2(V_2 - V_1)/[(W_1 + W_2)]$.

TABLE 4.3 Resolution Factors (Rs) of Proteins with Various Ionic Species Added to the Elution Buffer[a]

		0.2 *M*		0.2 *M*		
	Da	**$(NH_4)_2HPO_4$**	**0.2 *M* Na_2SO_4**	**$(NH_4)_2SO_4$**	**0.2 *M* $MgCl_2$**	**0.2 *M* NaCl**
Glutamate dehyd.	280,000					
		1.4	1.1	1.5	1.3	1.5
Alcohol dehyd.	150,000					
		0.4	0.5	0.2	0.5	0.6
Glutathione reduct.	113,000					
		0.9	1.1	0.9	0.9	1.2
Enolase	67,000					
		1.1	0.8	1.3	0.8	0.4
Ovalbumin	43,000					
		1.3	1.3	1.4	1.9	2.2
Trypsinogen	24,000					
		1.0	0.5	0.2	0.8	0.9
Cytochrome c	12,400					

[a]Column: TSK-GEL G3000SW, 7.5 mm × 30 cm.
Elution: 0.05 *M* GTA buffer, pH 7.0 (G, 3,3-dimethylglutaric acid; T, trishydroxyaminomethane; A, 2-amino-2-methyl-1,3-propanediol) plus indicated salt.
Detection: UV at 220 nm.

TABLE 4.4 Effect of Buffer Composition on Mass Recovery of Proteins from TSK-GEL G3000SW[a]

	Recovery (%)		
Protein	**A. $NaPO_4$**	**B. KPO_4**	**C. Tris–HCl**
Cytochrome c	98	101	92
Lysozyme	92	96	75
α-Chymotrypsinogen	95	98	90
IgG	95	98	88
Thyroglobulin	94	94	85
Ovalbumin	96	92	66

[a]Column: TSK-GEL G3000SW, 7.5 mm × 60 cm
Elution: (A) 0.2 *M* NaH_2PO_4 and 0.2 *M* Na_2HPO_4, pH 6.9; (B) 0.2 *M* KH_2PO_4 and 0.2 *M* K_2HPO_4, pH 6.9; and (C) 0.2 *M* NaCl and 0.05 *M* Tris–HCl, pH 7.8.
Flow rate: 1.0 ml/min.
Detection: UV at 220 nm.

pH that is different from the p*I* of the protein is recommended, so the protein will carry an overall net charge. The pH stability of TSK-GEL SW packings is shown in Fig. 4.4. If the separation must be performed under basic conditions, pH-stable, polymer-based TSK-GEL PW columns should be used.

3. Sample Load

In all modes of chromatography, high sample loads distort peak shapes and cause an overall decrease in efficiency due to column overload. Sample loads may be increased by using organic solvents to enhance the solubility of the sample or by using higher column temperatures to lower the viscosity of

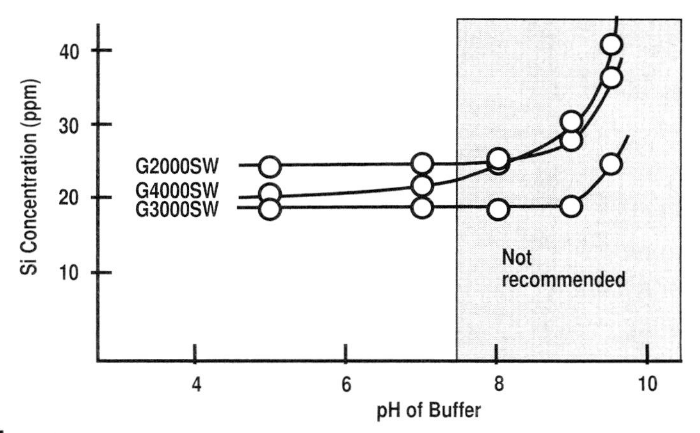

FIGURE 4.4 pH stability of TSK-GEL SW packings. Conditions: 1 g of each TSK-GEL SW packing was suspended in 10 ml of buffer at indicated pH for 1 month. The concentration of silica dissolved in each buffer was then determined.

the mobile phase. When recovering protein activity, however, these techniques usually are not desirable.

The sample load limitations of TSK-GEL SW and TSK-GEL SW_{XL} analytical and semipreparative columns are shown in Fig. 4.5 (1). At protein injections higher than 1 mg/100 μl, HETP increases rapidly and column efficiency decreases with TSK-GEL SW and TSK-GEL SW_{XL} analytical columns (Fig. 4.5A). Sample concentrations in the range of 1 to 20 mg/ml are recommended, but proteins can be loaded at higher concentrations and higher total loads than synthetic macromolecules. For example, a TSK-GEL G3000SW semipreparative column can provide high efficiency for 100 mg of bovine serum albumin (Fig. 4.5B). For a synthetic, linear polyethylene glycol (7500 Da), sample loads higher than 20 mg cause a rapid increase in HETP. The high sample mass

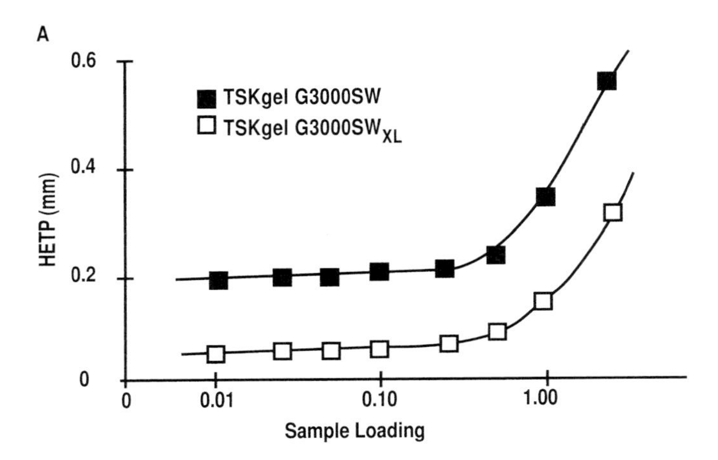

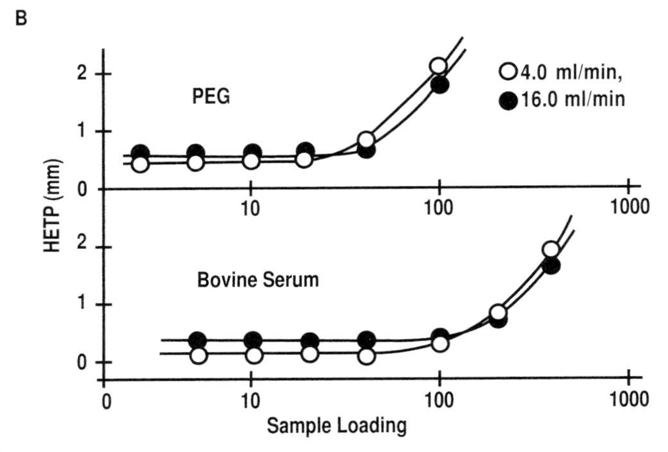

FIGURE 4.5 Effect of sample load on efficiency. Column: (A) TSK-GEL G3000SW_{XL}, 5 μm, 7.8 mm × 30 cm; TSK-GEL G3000SW, 10 μm, 7.5 mm × 30 cm, and (B) TSK-GEL G3000SW, 13 μm, 21.5 mm × 60 cm, two columns in series. Sample: (A) bovine serum albumin in 100 μl and (B) PEG 7500 or bovine serum albumin in 4 ml. Elution: (A) 0.3 *M* NaCl in 0.05 *M* phosphate buffer, pH 7 and (B) 0.3 *M* NaCl in 0.1 *M* phosphate buffer pH 7. Flow rate: (A) 1.0 ml/min and (B) 4.0 and 16.0 ml/min. Detection: (A) UV at 220 nm and (B) RI for PEG 7500, UV at 220 nm for BSA.

TABLE 4.5 Mass Recovery as a Function of Protein Sample Load[a]

	Mass recovery (%)				
Sample loading (μg)	1	5	10	50	100
Ribonuclease A	104	106	103	103	94
IgG	91	90	107	97	104
Thyroglobulin	78	90	91	102	101

[a]Column: TSK-GEL G4000SW$_{XL}$, 7.8 mm × 30 cm.
Elution: 0.3 *M* NaCl in 0.05 *M* phosphate buffer, pH 7.0.
Flow rate: 1.0 ml/min.
Detection: UV at 220 nm.

recoveries observed for 1- to 100-μg loads of small and large proteins on a TSK-GEL G4000SW$_{XL}$ column (Table 4.5) are representative of the protein mass recoveries expected for all TSK-GEL SW analytical and preparative columns. High mass recoveries are attributed to the minimal adsorption properties of the TSK-GEL SW silica matrix.

4. Flow Rate

Flow rate determines the separation time and can significantly affect resolution and efficiency. The effect of flow rate on HETP for TSK-GEL SW and TSK-GEL SW$_{XL}$ analytical columns is shown in Fig. 4.6. Resolution is typically higher at slower flow rates, although results shown in Fig. 5B indicate that, with increasing sample load, the faster flow rates can give higher resolution.

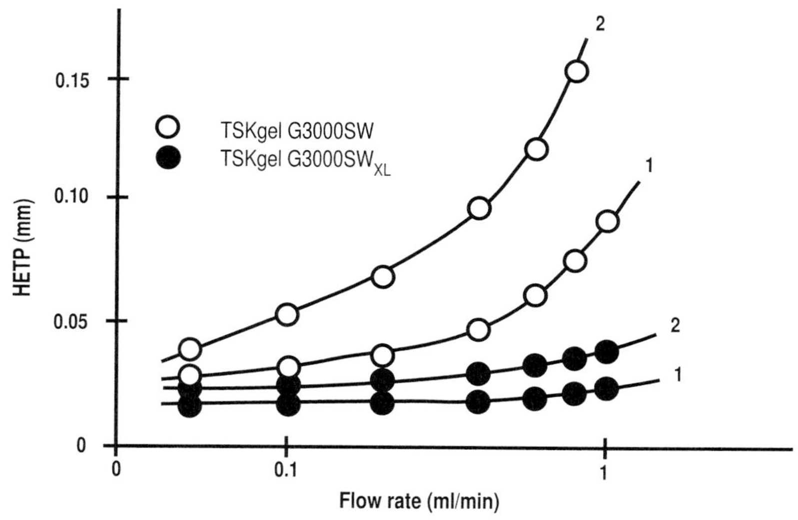

FIGURE 4.6 Flow rate has greater effect on efficiency for larger packing particles. Column: TSK-GEL G3000SW$_{XL}$, 5 μm, 7.8 mm × 30 cm; TSK-GEL G3000SW, 10 μm, 7.5 mm × 30 cm. Sample: (1) myoglobin and (2) bovine serum albumin. Elution: 0.3 *M* NaCl in 0.05 *M* phosphate buffer, pH 7. Detection: UV at 220 nm.

The recommended flow rates of 0.5 to 1.0 ml/min for stainless steel and 0.4 to 0.8 ml/min for glass analytical columns will provide adequate resolution of most protein mixtures.

C. Applications

1. Peptides

Peptides with a molecular mass of only a few hundred daltons can be separated using aqueous GFC. When GFC is coupled with multiangle laser light scattering (MALLS) it is a powerful technique for determining the absolute molecular weights of various biomolecules. Figure 4.7 shows the separation of four peptides on a TSK-GEL G2000SW$_{XL}$ column with a Wyatt Technologies miniDAWN 90 light-scattering detector (5).

2. Enzymes

A principal feature of GFC is its minimal interaction with the sample. This gentle technique allows for high recovery of enzymatic activity. For example,

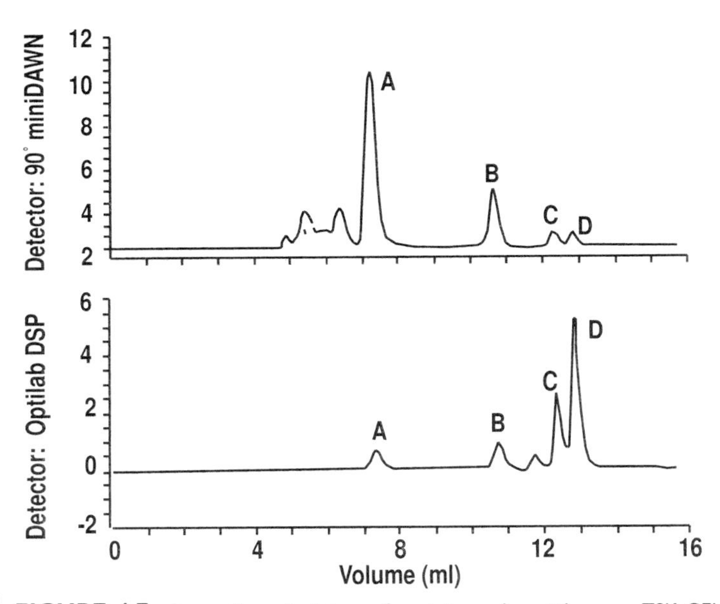

FIGURE 4.7 Separation of mixture of proteins and peptides on a TSK-GEL G2000SW$_{XL}$ column using Wyatt/Optilab DSP and miniDawn detectors.

Peak	Protein	Molar Mass (Da)	
		Sequence	miniDAWN
A	BSA	67,000	64,300 ± 700
B	Lysozyme	14,300	14,600 ± 300
C	Bradykinin	1,060	1,090 ± 10
D	Leucine-enkephalin	556	592 ± 6

crude samples of peroxidase and glutathione *S*-transferase were separated in only 15 min on a TSK-GEL G3000SW$_{XL}$ column, and activity recovery was 98 and 89%, respectively. The elution profiles of the separations in Fig. 4.8 show that all of the activity eluted in a narrow band of about 1.5 ml.

3. Antibodies

A therapeutic solution of intravenous IgG will contain albumin as a stabilizer, and both proteins must be quantified following manufacture. Although literature reports describe the separation of these two proteins by many other chromatographic methods, long analysis times and complex gradient elutions are required. A method developed on TSK-GEL G3000SW$_{XL}$ provides quantitative separation of the two proteins in 15 min with a simple, isocratic elution system (6). As shown in Fig. 4.9, albumin can be separated from a 20-fold excess of IgG, and quantified, using the optimized elution buffer method. This simple separation method can be applied to the isolation of other IgGs, such as monoclonal antibodies from albumin in ascites fluid or supernatants.

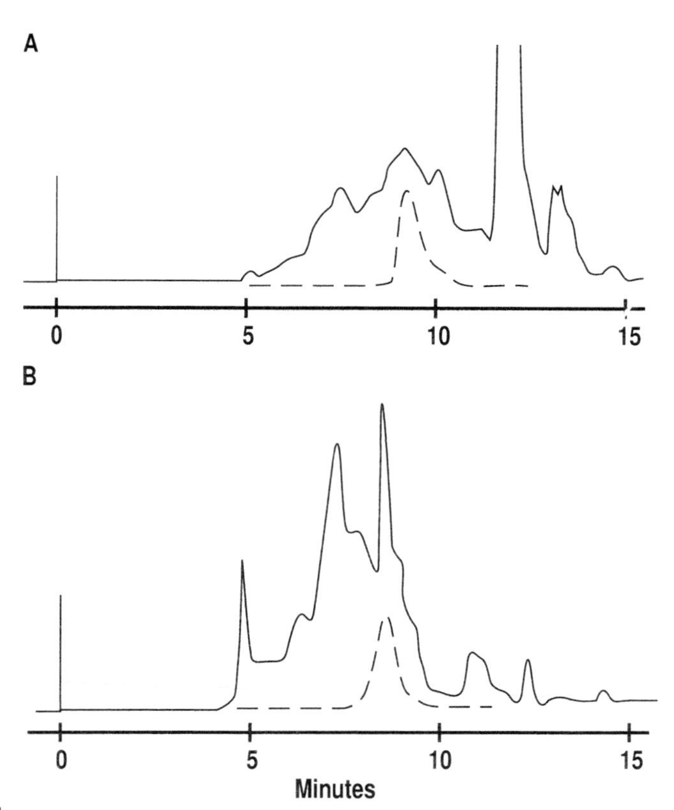

FIGURE 4.8 Separation of crude protein samples on TSK-GEL G3000SW$_{XL}$. Column: TSK-GEL G3000SW$_{XL}$, 5 μm, 7.8 mm × 30 cm. Sample: (A) crude peroxidase from Japanese radish, 0.15 mg in 0.1 ml and (B) crude glutathione *S*-transferase from guinea pig liver extract, 0.7 mg in 0.1 ml. Elution: 0.3 *M* NaCl in 0.05 *M* phosphate buffer, pH 7. Flow rate: 1.0 ml/min. Detection: UV at 220 nm and enzyme assay tests (----). Recovery: Enzymatic activity recovered was 98% in A and 89% in B.

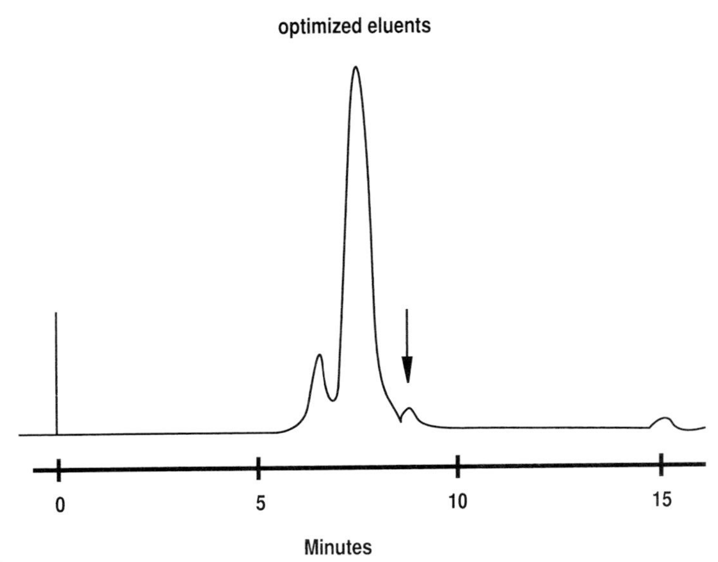

FIGURE 4.9 QC test for albumin in intravenous human IgG. Column: TSK-GEL G3000SW$_{XL}$, 5 μm, 7.8 mm × 30 cm. Sample: 5 μl of Venilon containing 237.5 mg of IgG and 12.5 mg of albumin. Elution: 0.1 *M* Na_2SO_4 in 0.05 *M* sodium phosphate buffer, pH 5.0. Flow rate: 1.0 ml/min. Detection: UV at 280 nm.

TSK-GEL SW columns have also been used to isolate heavy and light chains from immunoglobulins. Kast *et al.* (7) reported how these chains were desalted and isolated on a 21.5-mm i.d. × 60-cm TSK-GEL G4000SW semipreparative column using a 0.1 trifluoroacetic acid (TFA)/40% acetonitrile mobile phase.

4. DNA Restriction Fragments

Although DNA fragments are usually separated by gel electrophoresis, liquid chromatography provides an alternative method that is highly reproducible and capable of handling large sample volumes. Furthermore, with GFC, sample recovery is usually quantitative, isocratic elution systems allow simplified method development, and scale-up is easy and straightforward. The elution positions of sample peaks enable easy estimation of chain length, and these conditions are reproducible. Figure 4.10 shows elution patterns for the cleavage of two common cloning vectors with the restriction endonuclease *Hae*III. In Fig. 4.10A, fragments ranging from 100 to 300 bp are resolved on a 5-μm TSK-GEL G4000SW$_{XL}$ 30-cm column. For higher resolution of the same base pair range, two 10-μm TSK-GEL G4000SW 60-cm columns can be connected in series (8), as shown in Fig. 4.10B. Alternatively, two 5-μm TSK-GEL SW$_{XL}$ 30-cm columns can be connected in series.

D. Preparative TSK-GEL SW Columns

Also available are semipreparative, 21.5-mm i.d. and preparative 55-, 108-, 158-, and 210-mm i.d. stainless-steel columns packed with TSK-GEL SW parti-

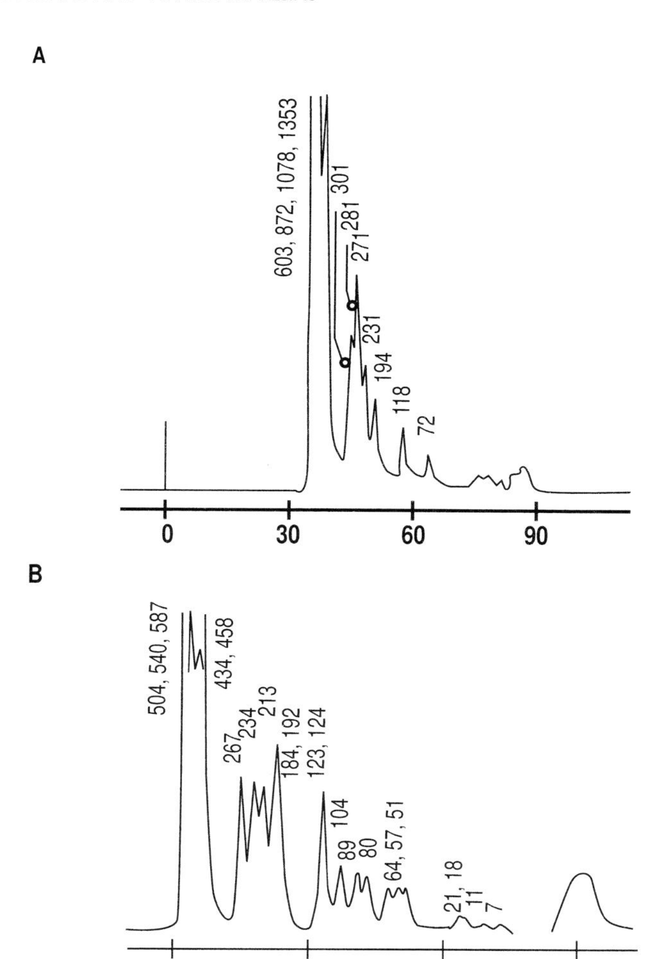

FIGURE 4.10 Separation of *Hae*III-cleaved DNA samples. Column: (A) TSK-GEL G4000SW$_{XL}$, 5 μm, 7.8 mm $\times$ 30 cm and (B) TSK-GEL G4000SW, 10 μm, 7.5 mm $\times$ 60 cm, two in series. Sample: (A) *Hae*III-cleaved ϕX 174 DNA, 4.5 μg in 50 μl and (B) *Hae*III-cleaved pBR 322 DNA, 0.1 ml of 0.015% solution. Elution: 0.3 *M* NaCl in 0.05 *M* phosphate buffer, pH 7, plus 1 m*M* EDTA. Flow rate: (A) 0.15 ml/min and (B) 0.33 ml/min. Detection: UV at 260 nm.

cles for precise scaleup to the commercial production of therapeutic proteins and other biopharmaceuticals. These TSK-GEL SW particles have larger diameters and are appropriate for process-scale columns. They have the same pore sizes and provide the same selectivity as their analytical TSK-GEL SW counterparts. Therefore, high sample resolution can be achieved in the same separation ranges as listed in Table 4.1. Semipreparative TSK-GEL G2000SW and TSK-GEL G3000SW columns are packed with 13-μm particles, whereas TSK-GEL G4000SW columns are packed with 17-μm particles. Preparative columns are packed with 20-μm TSK-GEL G2000SW or TSK-GEL G3000SW particles. In a commercial application, more than 20 kg of product have been purified yearly on a large diameter (108 mm × 60 cm) TSK-GEL G3000SW column. Because scale-up from analytical columns is relatively straightforward, preparative TSK-GEL SW columns are an economical route for the rapid commercial production of biomolecules.

II. POLYMER-BASED TSK-GEL PW/PW$_{XL}$ COLUMNS FOR AQUEOUS GFC

A. Description

Polymeric TSK-GEL PW and TSK-GEL PW$_{XL}$ columns are designed for aqueous GFC of proteins, peptides, polysaccharides, oligosaccharides, DNA, RNA, water-soluble organic polymers, and other large water-soluble samples. TSK-GEL PW column packings are hydrophilic, rigid, spherical, porous polymeric beads that have excellent chemical and mechanical stability. TSK-GEL PW and TSK-GEL PW$_{XL}$ columns can be used with eluents in the pH range of 2 to 12, with up to 50% organic cosolvent (9) and at temperatures up to 80°C (up to 50°C for TSK-GEL G-DNA-PW). Although most TSK-GEL PW packings bear a slight negative charge, from 5 to 18 μeq/ml, the separation of nonionic, anionic, and most cationic samples is not affected by the addition of small amounts of salt.

TSK-GEL PW type columns are commonly used for the separation of synthetic water-soluble polymers because they exhibit a much larger separation range, better linearity of calibration curves, and much lower adsorption effects than TSK-GEL SW columns (10). While TSK-GEL SW columns are suitable for separating monodisperse biopolymers, such as proteins, TSK-GEL PW columns are recommended for separating polydisperse compounds, such as polysaccharides and synthetic polymers.

TSK-GEL PW analytical columns are available in seven pore sizes, in either 10- or 17-μm particle sizes. Resins having the same seven pore sizes with 17-, 22-, and 25-μm particles are available in semipreparative 21.5 mm × 60 cm columns. Resins of select pore sizes are available packed in 55- and 108-mm i.d. preparative columns. Resins packed in TSK-GEL PW$_{XL}$ columns are 6-, 10-, or 13-μm high-performance versions of the TSK-GEL PW resins and are available in five different pore sizes. Specialty resin-based columns include the mixed-bed TSK-GEL GMPW and TSK-GEL GMPW$_{XL}$ columns for samples with broad molecular weight range and TSK-GEL G-Oligo-PW columns and TSK-GEL G-DNA-PW columns for oligosaccharides and for DNA or RNA, respectively. The properties and molecular weight separation ranges for all TSK-GEL PW columns are summarized in Table 4.6.

B. Conditions

1. Choosing the Right Column

The range of pore sizes in which TSK-GEL PW and TSK-GEL PW$_{XL}$ columns are available permits a wide spectrum of water-soluble substances to be analyzed. Calibration curves for polyethylene glycols chromatographed on

TABLE 4.6 Properties and Molecular Weight Separation Ranges for TSK-GEL PW Type Packings[a]

TSK-GEL column	Particle size[b] (μm)	Average pore size (Å)	Molecular weight of sample: Polyethylene oxides/glycols	Dextrans[c]	Globular proteins[c]
G1000PW	10, 17	<100	Up to 1000	—	<2000
G2000PW	10, 17, 20	125	Up to 2000	—	<5000
G2500PW$_{XL}$	6	<200	Up to 3000	—	<8000
G2500PW	10, 17, 20				
G3000PW$_{XL}$	6	200	Up to 50,000	Up to 60,000	500–800,000
G3000PW	10, 17, 20				
G4000PW$_{XL}$	10	500	2000–300,000	1000–700,000	10,000–1,500,000
G4000PW	17, 22				
G5000PW$_{XL}$	10	1000	4000–1,000,000	50,000–7,000,000	<10,000,000
G5000PW	17, 20, 22				
G6000PW$_{XL}$	13	>1000	40,000–8,000,000	500,000–50,000,000	<200,000,000
G6000PW	17, 25				
GMPW$_{XL}$	13	<100–1000	500–8,000,000	<50,000,000	<200,000,000
GMPW	17				
G-Oligo-PW	6	125	Up to 3000	—	<3000
G-DNA-PW	10	4000	40,000–8,000,000	—	<200,000,000

[a]Column: TSK-GEL PW columns, 7.5 mm × 60 cm; TSK-GEL PWXL, G-Oligo-PW and G-DNA-PW, 7.8 mm × 30 cm.
Elution: Polyethylene glycols and oxides: distilled water, dextrans and proteins: 0.2 *M* phosphate buffer, pH 6.8.
Flow rate: 1.0 ml/min.

[b]Larger particle sizes of each group are for 21.5 mm × 60 cm semipreparative and 55 or 108 mm × 60 cm preparative columns.

[c]Maximum separation range determined from estimated exclusion limits.

TSK-GEL PW and TSK-GEL PW_{XL} columns are shown in Fig. 4.11. Although many methods for polymer analysis have been developed satisfactorily on TSK-GEL PW columns, higher resolution can often be achieved with a TSK-GEL PW_{XL} column. The smaller particle sizes of the resins packed in TSK-GEL PW_{XL} columns provide almost 2.5 times the resolution of their TSK-GEL PW counterparts. In addition, with shorter TSK-GEL PW_{XL} columns, higher resolution separations are possible in less than half the time, as shown in Fig. 4.12.

For analytical purposes, TSK-GEL PW_{XL} columns are preferred. For preparative work, or for other cases in which large amounts of sample must be used, TSK-GEL PW columns are recommended because of their larger loading capacity. To select the proper TSK-GEL PW type column for a particular sample, consult the separation ranges listed in Table 4.6 or the calibration curves in Fig. 4.11.

When the molecular weight range of the sample is very broad, or is unknown, the two mixed-bed columns, TSK-GEL GMPW and TSK-GEL $GMPW_{XL}$, are recommended for analysis. The TSK-GEL GMPW column and its high resolution counterpart, TSK-GEL $GMPW_{XL}$, are packed with the G2500, G3000, and G6000 PW or corresponding PW_{XL} resins and offer a broad molecular weight separation range. As shown in Fig. 4.11, the calibration curve for polyethylene glycols and oxides on these mixed-bed columns is fairly flat and is linear over the range from 100 to 1,000,000 Da (11).

This TSK-GEL G-Oligo-PW is designed for high-resolution separations of nonionic and cationic oligomers (12). Figure 4.13 demonstrates excellent

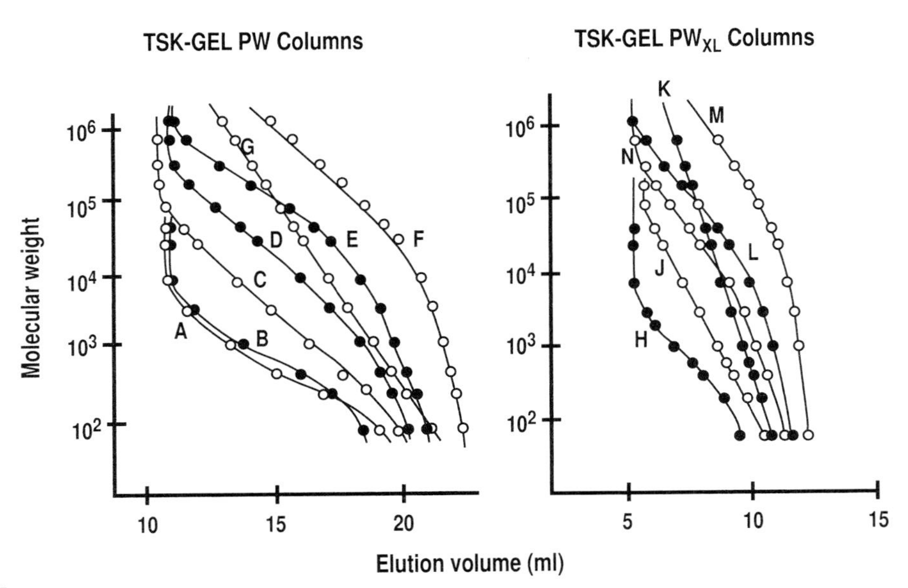

FIGURE 4.11 Polyethylene glycol and oxide calibration curves for TSK-GEL PW and TSK-GEL PW_{XL} columns. Column: TSK-GEL PW columns: (A) G2000PW, (B) G2500PW, (C) G3000PW, (D) G4000PW, (E) G5000PW, (F) G6000PW, (G) GMPW, all 7.5 mm × 60 cm. TSK-GEL PW_{XL} columns: (H) G2500PW_{XL}, (J) G3000PW_{XL}, (K) G4000PW_{XL}, (L) G5000PW_{XL}, (M) G6000PW_{XL}, (N) $GMPW_{XL}$, all 7.8 mm × 30 cm. Elution: Distilled water. Flow rate: 1.0 ml/min. Detection: RI.

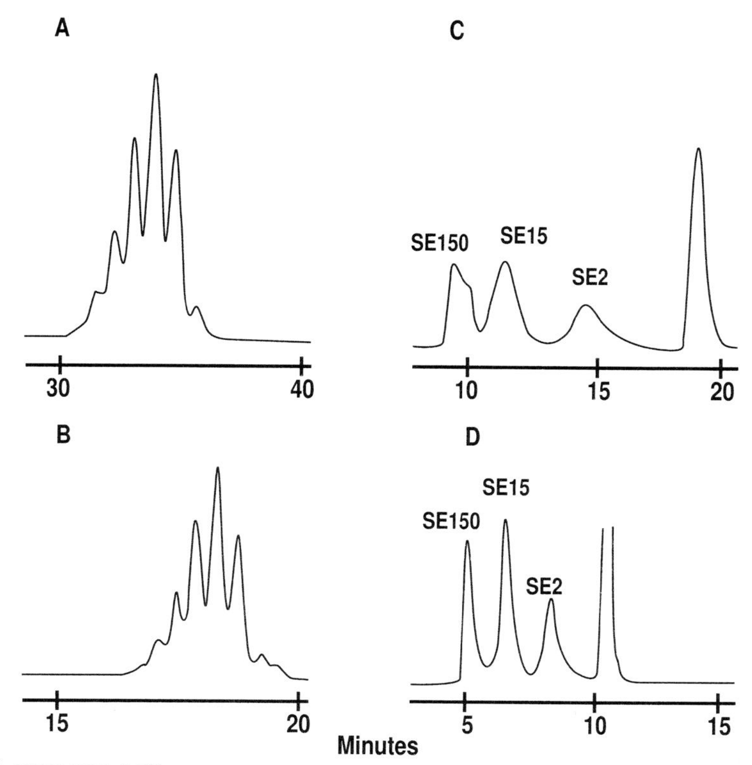

FIGURE 4.12 Higher and faster resolution with TSK-GEL PW_{XL} columns. Column: (A) TSK-GEL G2500PW, two 10-μm, 7.5 mm × 60 cm columns in series; (B) TSK-GEL G2500PW_{XL}, two 6-μm, 7.8 mm × 30 cm columns in series; (C) TSK-GEL G4000PW, 17 μm, 7.5 mm × 60 cm; and (D) TSK-GEL G4000PW_{XL}, 10 μm, 7.8 mm × 30 cm. Sample: (A and B) polyethylene glycol 200 and (C and D) polyethylene oxide standards: SE-150, SE-15, and SE-2 in 100 μl. Elution: (A and B) distilled water and (C and D) 0.1 *M* NaCl. Flow rate: 1.0 ml/min. Temperature: (A and B) 25°C and (C and D) 50°C. Detection: RI.

resolution of chitooligosaccharides obtained by using the smaller (6-μm) particle size packing in TSK-GEL G-Oligo-PW columns as compared to the resolution obtained with a TSK-GEL G2000PW column. The pore sizes in both TSK-GEL G-Oligo-PW and TSK-GEL G2000PW columns are about 125 Å and both resins bear approximately 0.2 meq/ml of cationic groups. Due to the presence of cationic groups, neither column is recommended for separating anionic materials.

The elution volumes for typical amino acids on TSK-GEL G-Oligo-PW and TSK-GEL G2500PW_{XL} columns are compared in Table 4.7. Differences in elution behavior are attributed to the cationic and anionic groups on the resin surfaces of the TSK-GEL G-Oligo-PW and TSK-GEL G2500PW_{XL} packings, respectively. Although most amino acids behave similarly to glycine on both columns, acidic amino acids are retarded on TSK-GEL G-Oligo-PW and are excluded from TSK-GEL G2500PW_{XL}. The elution of all amino acids is pH independent on TSK-GEL G2500PW_{XL} whereas most amino acids are retarded

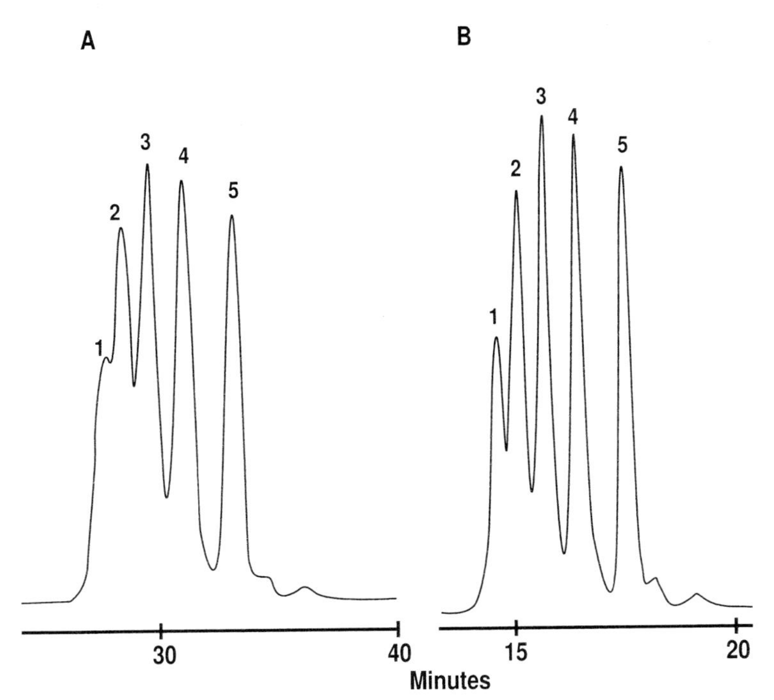

FIGURE 4.13 Faster and higher resolution of chitooligosaccharides on a TSK-GEL G-Oligo-PW column. Column: (A) TSK-GEL G2000PW, two 10-μm, 7.5 mm $\times$ 60 cm columns in series and (B) TSK-GEL G-Oligo-PW, two 6-μm, 7.8 mm $\times$ 30 cm columns in series. Sample: (1) chitohexaose, (2) chitopentaose, (3) chitotetraose, (4) chitotriose, and (5) chitobiose. Elution: Distilled water. Flow rate: 1.0 ml/min. Detection: RI.

on TSK-GEL G-Oligo-PW at high pH. Due to hydrophobic interaction effects, aromatic amino acids are retarded on both columns.

TSK-GEL G2500PW$_{XL}$ columns are recommended over TSK-GEL G-Oligo-PW columns for anionic compounds such as oligonucleotides and peptides. However, for nonionic oligomers, TSK-GEL G-Oligo-PW columns provide higher resolution than TSK-GEL G2500PW$_{XL}$ columns, as shown in Fig. 4.14. The calibration curve for hydrolyzed β-cyclodextrin on two TSK-GEL G-Oligo-PW columns in series is shown in Fig. 4.15A. The polyethylene glycol and oxide calibration curves for both TSK-GEL G-Oligo-PW (not shown) and TSK-GEL G2500 PW$_{XL}$ (Fig. 4.11) columns are nearly identical.

The TSK-GEL G-DNA-PW column is dedicated to the separation of large polynucleotides, such as DNA and RNA fragments of 500 to 5,000 bp. The calibration curve for double-stranded DNA fragments on a TSK-GEL G-DNA-PW column is shown in Fig. 4.15B. The packing has very large pores (~4000 Å) and small particle size, ensuring good resolution (13). Although small nucleic acid fragments can be analyzed with TSK-GEL SW columns, large fragments require large-pore TSK-GEL PW columns for proper chromatography.

For the separation of large DNA fragments greater than 1000 bp, a four-column system is typically required. The baseline resolution of DNA fragments

TABLE 4.7 Comparison of Amino Acid Elution Behavior

Sample/remark	Water 7.0	Elution volumes in buffered solutions (pH) 5.5	6.8	8.5
TSK-GEL G2500PW$_{XL}$				
Glycine Normal elution, pH independent	8.11	7.91	7.91	8.03
Glutamic acid Excluded elution, pH independent	4.96	5.95	5.81	5.44
Tyrosine Retarded elution, pH independent	12.35	12.54	12.50	11.97
Phenylalanine Retarded elution, pH independent	12.54	12.77	12.70	12.18
Tryptophan Retarded elution, pH independent	27.66	27.50	27.86	26.75
TSK-GEL G-Oligo-PW				
Glycine Normal elution, pH dependent	9.94	9.26	9.26	9.99
Glutamic acid Retarded elution, pH dependent	12.24	11.73	12.92	22.47
Tyrosine Retarded elution, pH dependent	24.92	14.78	15.24	32.07
Phenylalanine Retarded elution, pH dependent	22.37	12.82	13.21	26.63
Tryptophan Retarded elution, pH dependent	53.43	27.86	29.01	58.86

up to approximately 7000 bp can be achieved, provided there is a twofold difference in the chain length of the fragments. Figure 4.16 shows the elution of double-stranded DNA fragments obtained from pBR322 DNA cleaved by both *Eco*RI and *Bst*NI on four TSK-GEL G-DNA-PW columns in series. The eluted peaks were collected and subjected to polyacrylamide gel electrophoresis, which showed almost complete separation of the 1060-, 1857-, and 4362-bp fragments. Although lower flow rates typically yield better separations of most fragments, resolution of the 1857- and 4362-bp fragments was slightly greater at the higher flow rate, as shown in Fig. 4.16B.

2. Mobile Phase

In an ideal SEC separation, the mechanism is purely sieving, with no chemical interaction between the column matrix and the sample molecules. In practice, however, a small number of weakly charged groups on the surface of all TSK-GEL PW type packings can cause changes in elution order from that of an ideal system. Fortunately, the eluent composition can be varied greatly with TSK-GEL PW columns to be compatible with a wide range of neutral, polar, anionic, and cationic samples. Table 4.8 lists appropriate eluents for GFC of all polymer types on TSK-GEL PW type columns (11).

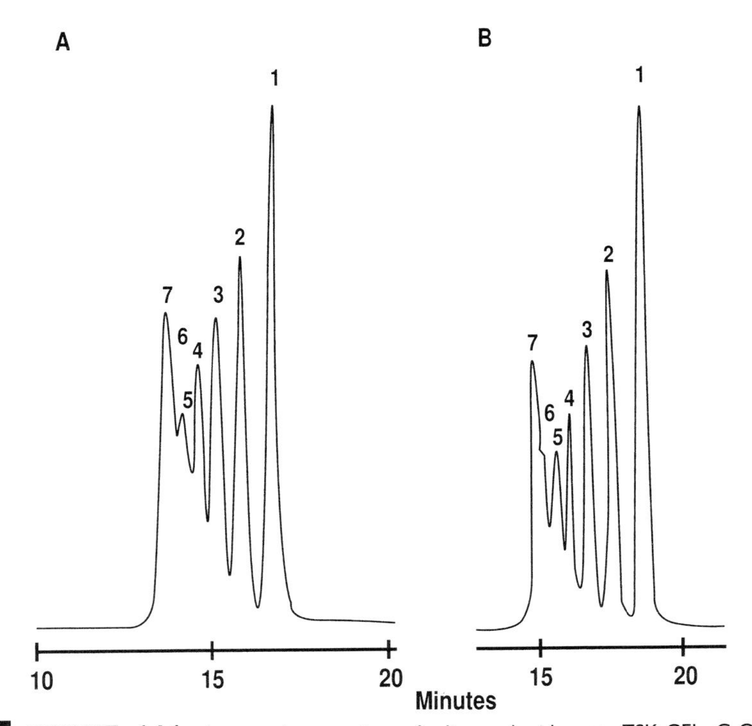

FIGURE 4.14 Improved separation of oligosaccharides on TSK-GEL G-Oligo-PW. Column: (A) TSK-GEL G2500PW$_{XL}$, two 6-μm, 7.8 mm × 30 cm columns in series and (B) TSK-GEL G-Oligo-PW, two 6-μm, 7.8 mm × 30 cm columns in series. Sample: Hydrolyzed β-cyclodextrin. Elution: Distilled water. Flow rate: 1.0 ml/min. Temperature: 60°C. Detection: RI.

For some nonionic, nonpolar polymers, such as polyethylene glycols, normal chromatograms can be obtained by using distilled water. Some more polar nonionic polymers exhibit abnormal peak shapes or minor peaks near the void volume when eluted with distilled water due to ionic interactions between the sample and the charged groups on the resin surface. To eliminate ionic interactions, a neutral salt, such as sodium nitrate or sodium sulfate, is added to the aqueous eluent. Generally, a salt concentration of 0.1–0.5 *M* is sufficient to overcome undesired ionic interactions.

Ionic repulsion between anionic samples and the resin causes poor resolution. As shown in Fig. 4.17, the addition of only 0.01 *M* $NaNO_3$ results in normal elution and peak shape for an anionic polymer, sodium polyacrylate.

Cationic samples can be adsorbed on the resin by electrostatic interaction. If the polymer is strongly cationic, a fairly high salt concentration is required to prevent ionic interactions. Figure 4.18 demonstrates the effect of increasing sodium nitrate concentration on peak shapes for a cationic polymer, DEAE-dextran. A mobile phase of 0.5 *M* acetic acid with 0.3 *M* Na_2SO_4 can also be used.

3. Sample Load

Figure 4.19 demonstrates the effect of sample concentration on the separation of polyethylene oxide (PEO). At a concentration of 1.6 mg/ml of each

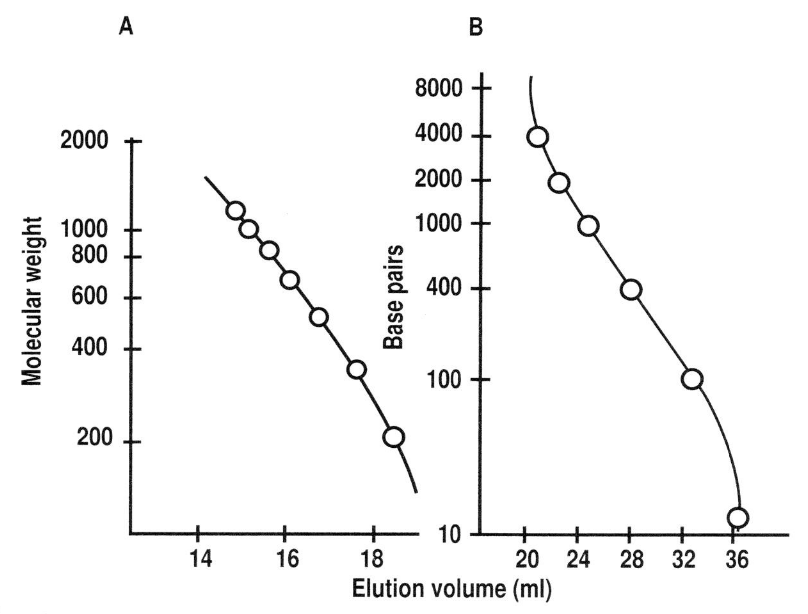

FIGURE 4.15 Calibration curves for oligosaccharides and double-stranded DNA. Column: (A) TSK-GEL G-Oligo-PW, two 6-μm, 7.8 mm $\times$ 30 cm columns in series and (B) TSK-GEL G-DNA-PW, four 10-μm, 7.8 mm $\times$ 30 cm columns in series. Sample: (A) hydrolyzed β-cyclodextrin and (B) *EcoRI- and Bst*NI-cleaved pBR322 DNA, void volume determined with λ-DNA. Elution: (A) distilled water and (B) 0.3 *M* NaCl in 0.1 *M* Tris–HCl, pH 7.5, plus 1 m*M* EDTA. Flow rate: (A) 1.0 ml/min and (B) 0.15 ml/min. Detection: UV at 260 nm.

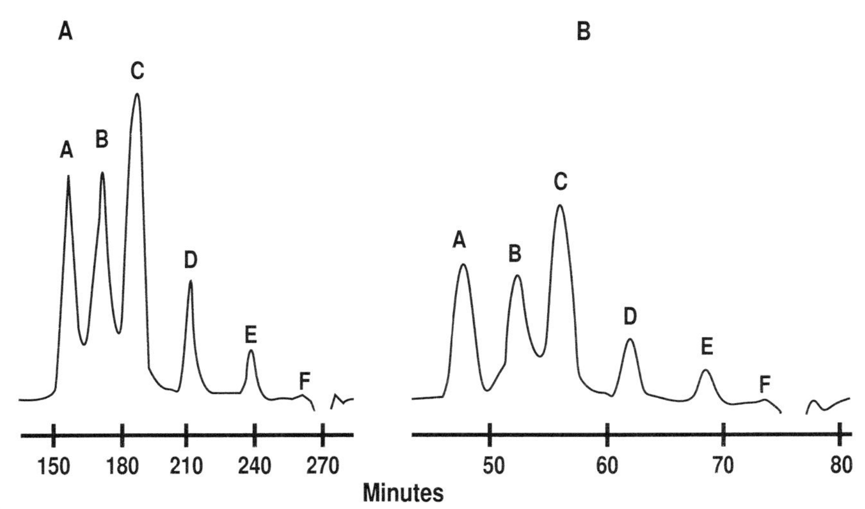

FIGURE 4.16 Separation of large DNA fragments on a TSK-GEL G-DNA-PW column. Column: TSK-GEL G-DNA-PW, four 10-μm, 7.8 mm $\times$ 30 cm columns in series. Sample: 60 μl of *Eco*-RI- and *Bst*NI-cleaved pBR322 DNA. Base pairs: (A) 4362, (B) 1857, (C) 1060 and 928, (D) 383, (E) 121, and (F) 13. Elution: 0.3 *M* NaCl in 0.1 *M* Tris-HCl, pH 7.5, plus 1 m*M* EDTA. Flow rate: (A) 0.15 ml/min and (B) 0.5 ml/min. Detection: UV at 260 nm.

TABLE 4.8 Recommended Eluents for GPC of Water-Soluble Polymers on TSK-GEL PW Type Columns

Type of polymer	Typical sample	Suitable eluent
Nonionic hydrophilic	Polyethylene glycol Soluble starch, methyl cellulose, pullulan Dextran Above samples plus hydroxyethyl cellulose, polyvinyl alcohol, polyacrylamide	Distilled water 0.01 *N* NaOH DMSO Buffer or salt solution (e.g., 0.1–0.5 *M* $NaNO_3$)
Nonionic hydrophobic	Polyvinylpyrrolidone	Buffer or salt solution with organic solvent (e.g., 20% CH_3CN in 0.1 *M* $NaNO_3$)
Anionic hydrophilic	Sodium chondroitinsulfate, sodium alginate, carboxymethyl cellulose, sodium polyacrylate, sodium hyaluronate	Buffer or salt solution (e.g., 0.1 *M* $NaNO_3$)
Anionic hydrophobic	Sulfonated lignin sodium salt, sodium polystyrenesulfonate	Buffer or salt solution with organic solvent (e.g., 20% CH_3CN in 0.1 *M* $NaNO_3$)
Cationic hydrophilic	Glycol chitosan, DEAE-dextran, poly(ethyleneimine), poly(trimethylaminoethyl methacrylate) iodide salt	0.5 *M* acetic acid with 0.3 *M* Na_2SO_4, or 0.8 *M* $NaNO_3$
Cationic hydrophobic	Poly(4-vinylbenzyltrimethylammonium chloride), poly(*N*-methyl-2-vinylpyridinium) iodide salt	0.5 *M* acetic acid with 0.3 *M* Na_2SO_4
Amphoteric hydrophilic	Peptides, proteins, poly and oligosaccharides, DNA, RNA	Buffer or salt solution (e.g., 0.1 *M* $NaNO_3$)
Amphoteric hydrophobic	Blue dextran, collagen, gelatin, hydrophobic proteins Hydrophobic peptides	Buffer or salt solution with organic solvent (e.g., 20% CH_3CN in 0.1 *M* $NaNO_3$) 35–45% CH_3CN in 0.1% TFA

PEO, resolution of the three PEOs is lost. In addition, sample volumes greater than 100 μl will result in increased HETPs and reduced resolution.

4. Flow Rate

Flow rate has a tremendous effect on resolution with aqueous GFC as shown in Fig. 4.20. Lower flow rates will decrease HETPs (up to a point) and increase resolution. In general, a flow rate range of 0.5–0.8 ml/minute is recommended for TSK-GEL PW analytical columns.

5. Temperature

Figure 4.21 demonstrates the effect of temperature on the resolution of PEOs on a TSK-GEL G6000PW_{XL} and G3000PW_{XL} in series. Increased temperature will decrease mobile phase viscosity and improve diffusion, which will improve resolution.

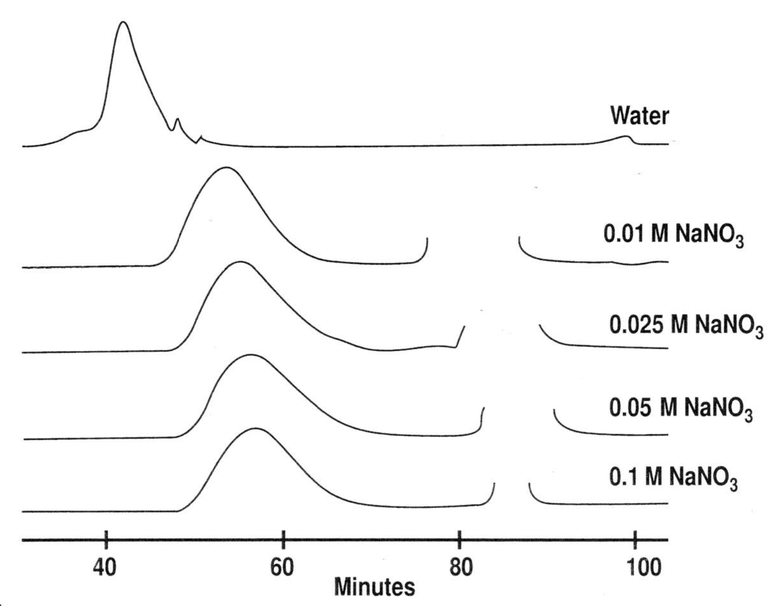

FIGURE 4.17 Effect of ionic strength on the elution of anionic polymers. Column: TSK-GEL GMPW, two 17 μm, 7.5 mm $\times$ 60 cm columns in series. Sample: 0.5 ml of 0.05-0.1% of the sodium salt of polyacrylic acid, an anionic polymer. Elution: Water 0.01, 0.025, 0.05, or 0.1 *M* $NaNO_3$ in water. Flow rate: 0.5 ml/min. Detection: RI.

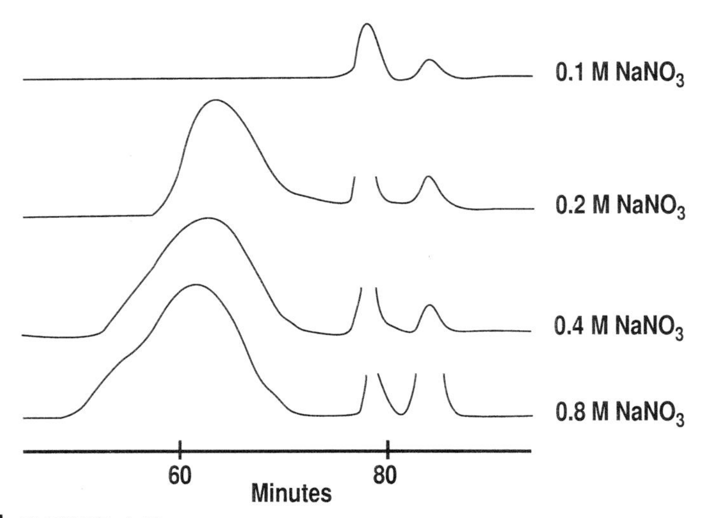

FIGURE 4.18 Use of sodium nitrate of elute cationic polymers. Column: TSK-GEL GMPW, two 17-μm, 7.5 mm $\times$ 60 cm columns in series. Sample: 0.5 ml of 0.05–0.1% of the cationic polymer DEAE-dextran. Elution: 0.1, 0.2, 0.4, or 0.8 *M* $NaNO_3$ in water. Flow rate: 0.5 ml/min. Detection: RI.

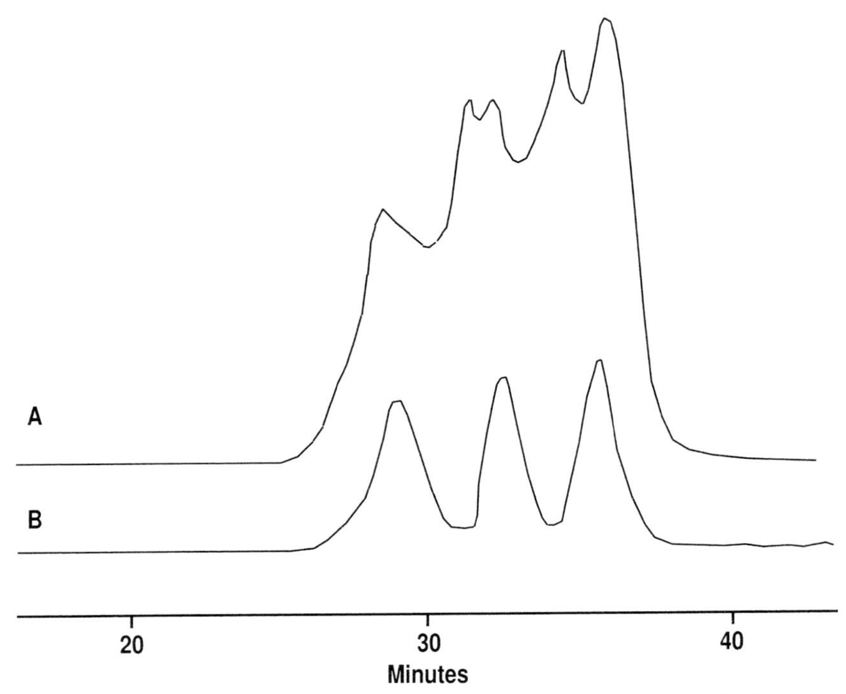

FIGURE 4.19 Effect of concentration of sample solution in the separation of polyethylene oxide. Column: G6000PW$_{XL}$ + G3000PW$_{XL}$ (two 7.8 mm × 30 cm). Sample: Polyethylene oxide; SE150, SE30, SE8, 100 μl; (A) 1.6 mg/ml each and (B) 0.4 mg/ml each. Elution: 0.1 *M* NaCl. Flow rate: 0.5 ml/min. Temperature: 50°C. Detection: RI.

6. Hydrophobic Sample Considerations

TSK-GEL PW type resins are more hydrophobic than polysaccharide gels such as cross-linked dextran. The hydrophobic interaction increases as the salt concentration of the eluent increases, while it can be reduced by the addition of an organic solvent modifier such as acetonitrile. Water-soluble organic solvents are frequently used as modifiers to suppress hydrophobic interactions between the sample and the resin surface. Modifiers are used for the proper elution of both charged and neutral hydrophobic polymers. Typical examples for a variety of sample types are given in Table 4.8. All TSK-GEL PW type column packings are compatible with up to 20% aqueous solutions of methanol, ethanol, propanol, acetonitrile, formic acid, acetic acid, dimethylformamide, dimethyl sulfoxide, or 50% acetone. Solvent exchange must be carried out slowly. As shown in Fig. 4.22, successful elution of the hydrophobic amphoteric polymers collagen and gelatin requires the addition of 20% CH_3CN to 0.1 *M* $NaNO_3$. Peak areas are reduced and elution is not reproducible when the organic solvent is omitted from the elution buffer for hydrophobic samples.

Small peptides may be difficult to chromatograph by aqueous GFC due to complex nonsize effects such as ionic and hydrophobic interactions. Elution

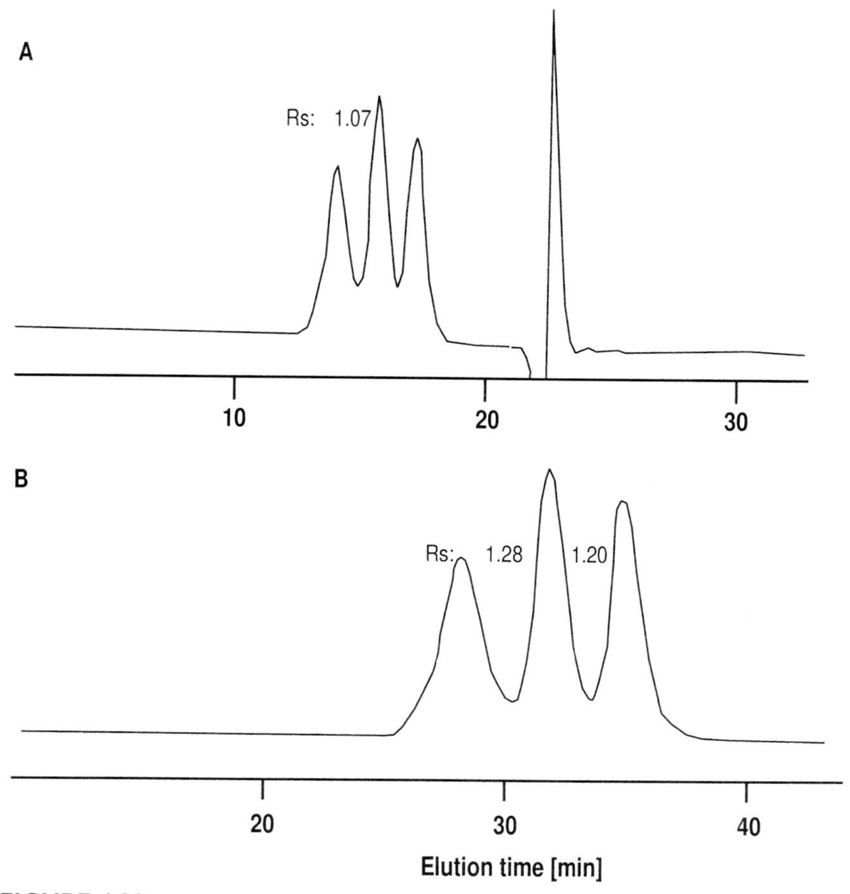

FIGURE 4.20 Effect of flow rate in the separation of polyethylene oxide. Column: G6000PW$_{XL}$ + G3000PW$_{XL}$ (two 7.8 mm × 30 cm). Sample: polyethylene oxide, SE150, SE30, SE8, 100 μl. Elution: 0.1 *M* NaCl. Flow rate: (A) 1.0 ml/min and (B) 0.5 ml/min. Temperature: 50°C. Detection: RI.

with organic solvents and buffered salt solutions overcomes these effects (12). For example, the separation of a peptide mixture, using a high acetonitrile concentration in 0.1% TFA, is shown in Fig. 4.23 (page 120).

In most situations the eluent composition is chosen to minimize the effects of hydrophobic interaction, but these secondary effects can be used to advantage. By careful selection of a salt and its concentration, specific selectivities for analytes can be achieved without the use of organic solvents. Therefore, many separations usually run by solvent gradient reversed-phase methods can be completed with a purely aqueous isocratic eluent (13,14).

In Fig. 4.24 (page 121), a mixture of acidic, basic, and neutral aromatic compounds is separated in a single analysis with good efficiency and peak shape by using an aqueous mobile phase and a TSK-GEL G2500WP$_{XL}$ column. Column temperature changes of only 5°C, however, can strongly influence retention due to changes in mobile-phase viscosity.

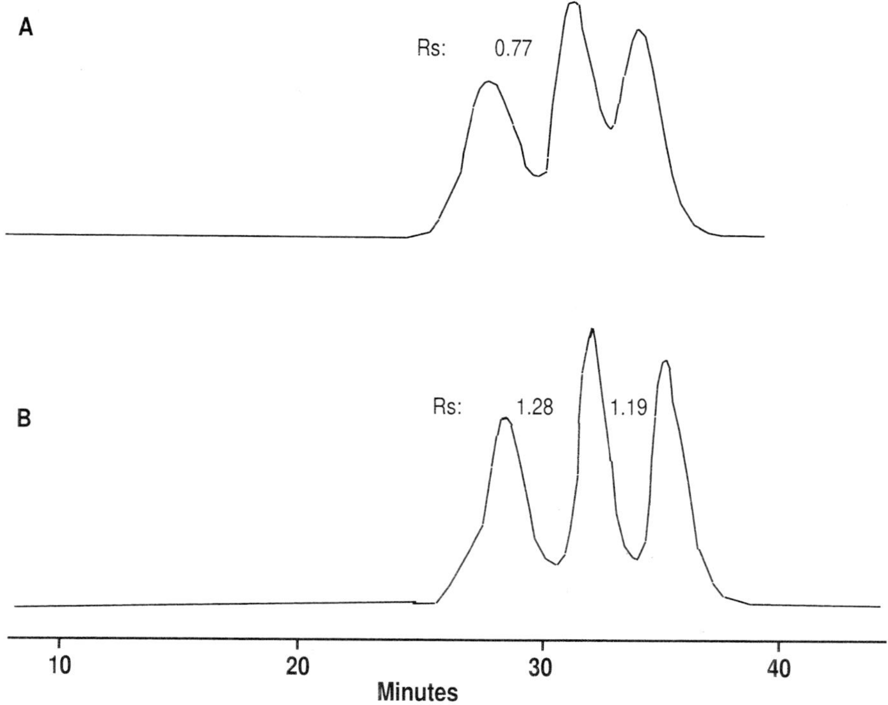

FIGURE 4.21 Effect of temperature in the separation of polyethylene oxide. Column: G6000PW$_{XL}$ + G3000PWX$_L$ (two 7.8 mm × 30 cm). Sample: Polyethylene oxide, SE150, SE30, SE8, 40 μg in 100 μl. Elution: 0.1 *M* NaCl. Flow rate: 0.5 ml/min. Temperature: (A) 25°C and (B) 50°C. Detection: RI.

C. Applications

1. Polymers

Nonionic polysaccharides are one of the most simple substances to analyze by size exclusion chromatography because they seldom exhibit nonsize exclusion effects. Due to their wide molecular weight distribution, TSK-GEL PW columns are recommended for their analysis.

Figure 4.25 (page 122) shows results obtained on TSK-GEL SW and TSK-GEL PW columns for low molecular weight polyethylene glycol (PEG) oligomers and high molecular weight dextrans. The TSK-GEL G2000PW column successfully resolved components of PEG 200, whereas the TSK-GEL G2000SW column did not (Fig. 4.25A). Therefore, the TSK-GEL G2000PW column would be preferable for this analysis.

For the profile of higher molecular weight dextrans, TSK-GEL G5000PW columns and TSK-GEL G4000SW columns can be used. This is because the TSK-GEL G4000SW column has the packing with the largest pores available in the TSK-GEL SW columns, and the TSK-GEL G5000PW column has a comparable molecular weight range. In the analysis of dextran standard T2000, with an average molecular size of 2,000,000 Da, the TSK-GEL SW

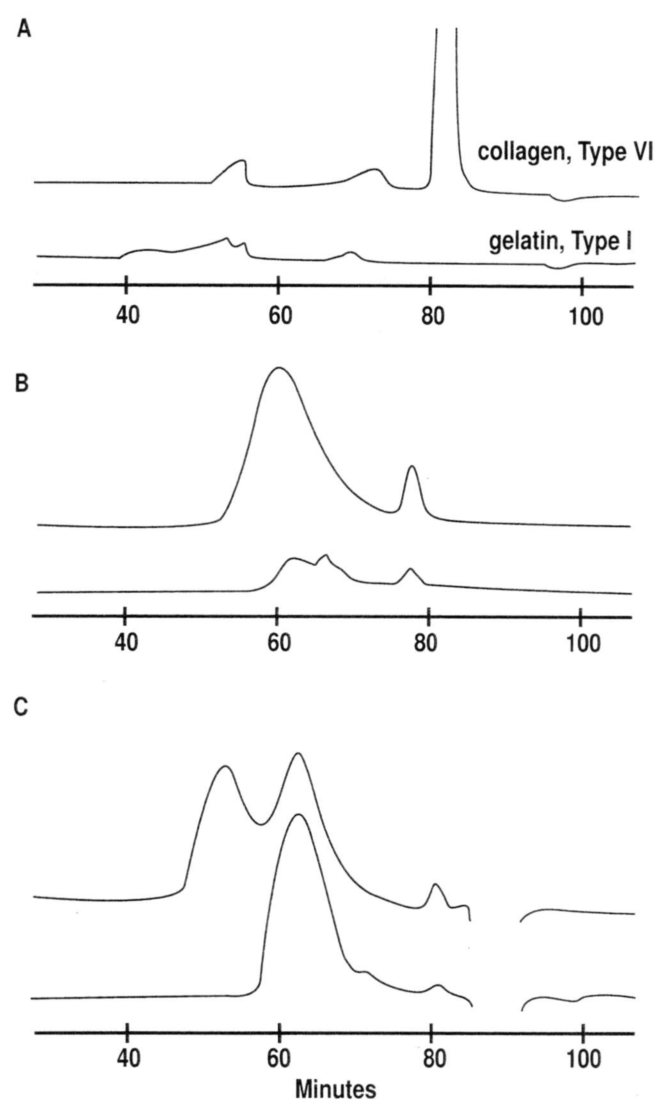

FIGURE 4.22 Use of mobile-phase additives to improve the elution of profile of amphoteric polymers. Column: TSK-GEL GMPW, two 17-μm, 7.5 mm × 60 cm columns in series. Sample: 0.5 ml of 0.05–0.1% collagen, Type VI, or gelatin, Type I. Elution: (A) distilled water, (B) 0.1 *M* $NaNO_3$ in water, and (C) 20% CH_3CN in 0.1 *M* $NaNO_3$. Flow rate: 0.5 ml/min. Detection: RI.

column almost completely excluded the sample. In contrast, the sample was effectively chromatographed on the TSK-GEL PW type column (Fig. 4.25B). Characterization of pullulan, lily amylose, poly(vinylalcohol), and poly(4-vinylbenzyl trimethylammonium chloride) are other examples of the use of TSK-GEL PW columns in series for the analysis of natural and synthetic polydisperse samples.

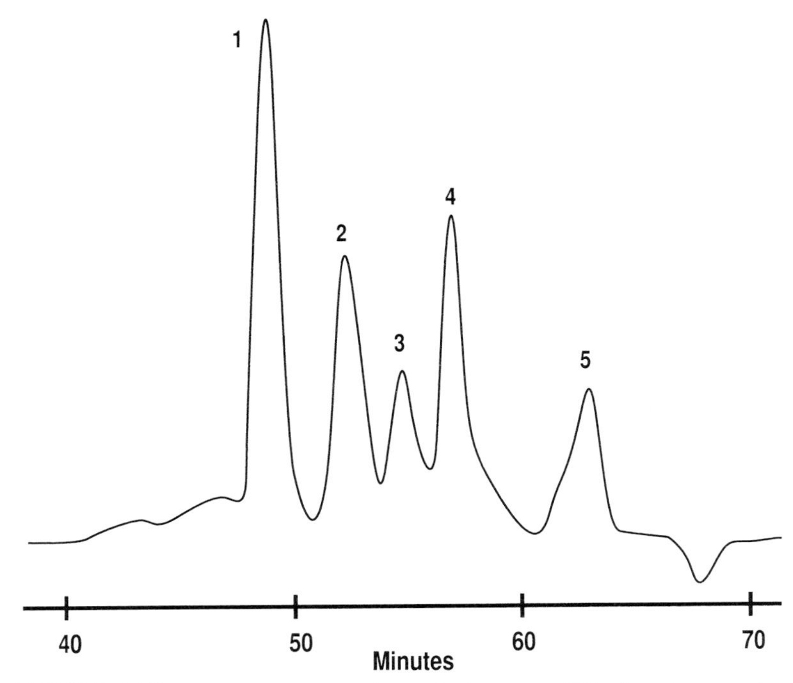

FIGURE 4.23 Use of an organic solvent eliminates non-SEC effects in separation of a peptide mixture. Column: TSK-GEL G3000PW$_{XL}$, two 6-μm, 7.8 mm × 30 cm columns in series. Sample: (1) aprotinin, (2) insulin B chain, (3) α-MSH, (4) bradykinin potentiator C, and (5) glutathione. Elution: 45% CH_3CN in 0.1% TFA. Flow rate: 0.3 ml/min. Detection: UV at 220 nm.

If the sample molecular weight range is broad or is unknown, TSK-GEL GMPW and TSK-GEL GMPW$_{XL}$ mixed-bed columns are effective. These mixed-bed columns provide linear calibration curves for polyethylene glycols and oxides over the range of 500–8,000,000 Da. They can replace multicolumn systems and are recommended for performing preliminary analysis of molecular weight distribution for unknown samples.

2. Proteins and Peptides

As Fig. 4.26 shows, the calibration curves for TSK-GEL G25000PW$_{XL}$ and TSK-GEL G2000SW$_{XL}$ are very similar and are almost identical for samples below 3000 Da. The curves were calculated using 17 samples ranging in size from myoglobin (17,800 Da) to glycine (75 Da). Although the curves are similar in shape through this range of sample sizes, each sample molecule behaved differently on the two columns, indicating a further role of sample–column packing interactions. For example, although the organic solvent was used to reduce hydrophobic effects, the hydrophobic peptide leu-enkephalin was retarded on the TSK-GEL G2500PW$_{XL}$ column.

Small peptides are among the most difficult compounds to analyze by aqueous SEC, although conditions for good separations have been developed for both TSK-GEL SW and TSK-GEL PW column types (15,16). Figure 4.27 (page 124) compares the results of separating two mixtures of peptides on

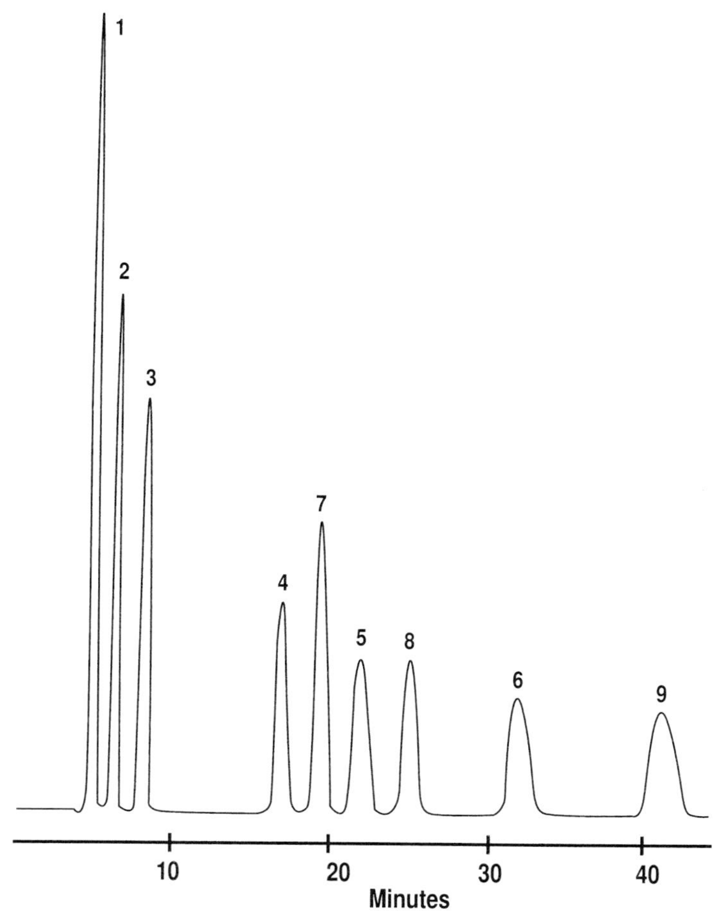

FIGURE 4.24 Adsorption chromatography of small molecules with a TSK-GEL G2500PW$_{XL}$ column. Column: TSK-GEL G2500PW$_{XL}$, 6 μm, 7.8 mm × 30 cm. Sample: (1) phenylacetic acid, (2) 3-phenylpropionic acid, (3) 4-phenylbutyric acid, (4) benzylamine, (5) 2-phenylethylamine, (6) 3-phenylpropylamine, (7) benzyl alcohol, (8) 2-phenylethanol, and (9) 3-phenyl-1-propanol. Elution: 0.1 *M* $NaClO_4$ in water. Flow rate: 2.0 ml/min. Temperature: 65°C. Detection: UV at 215 nm.

both TSK-GEL G2000SW$_{XL}$ and TSK-GEL G2500SW$_{XL}$ columns to demonstrate which might be superior for a particular type of peptide. The first group of peptides had molecular masses ranging from 6500 to 555 Da. In the second group, the range was extended from 17,800 to 75 Da. The chromatograms confirm that TSK-GEL G2000SW$_{XL}$ columns give higher resolution for most peptide mixtures, but do not perform as well as TSK-GEL G2500SW$_{XL}$ columns at peptide molecular weights lower than 1000. For very small peptides, the TSK-GEL PW column type is preferable. [See Table 4.9 (page 125), which lists the elution volumes for several peptides of differing molecular weights on TSK-GEL G2000PW$_{XL}$ and TSK-GEL G2000SW$_{XL}$ columns.]

For very large proteins, such as low-density lipoproteins or gelatin, TSK-GEL PW columns with a large pore size have been shown to be very effective

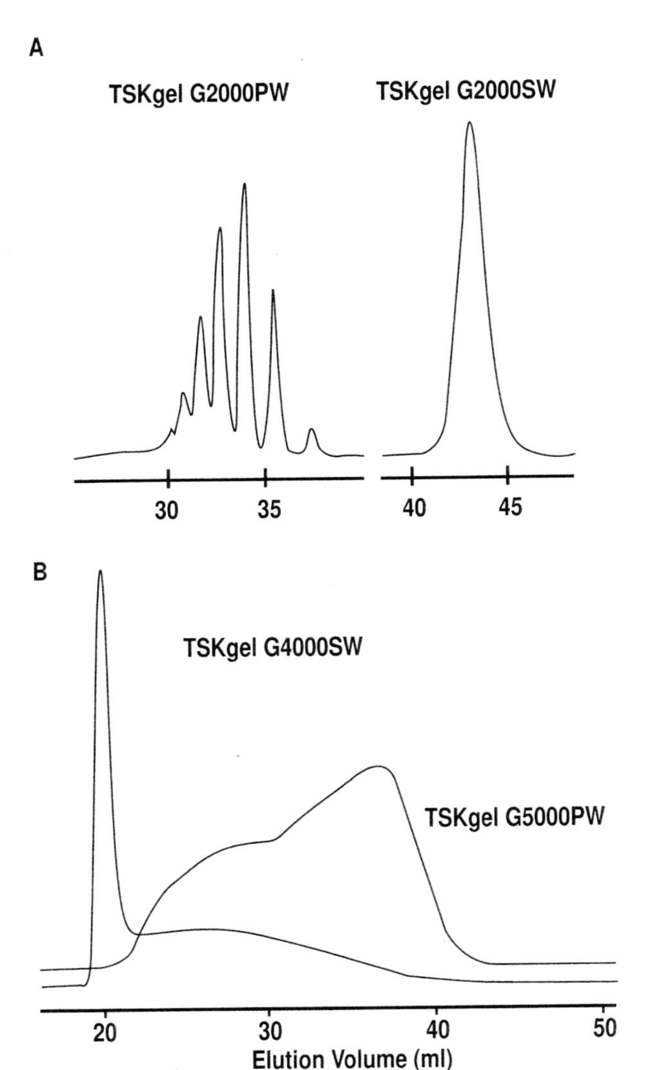

FIGURE 4.25 Elution of oligomers and polymers on TSK-GEL PW and TSK-GEL SW columns. Column: TSK-GEL G2000PW, two 10-μm, 7.5 mm × 60 cm columns in series; TSK-GEL G2000SW, two 10-μm, 7.5 mm × 60 cm columns in series; TSK-GEL G5000PW, two 17-μm, 7.5 mm × 60 cml columns in series; TSK-GEL G4000SW, two 13-mm, 7.5 mm × 60 cm columns in series. Sample: (A) Polyethylene glycol 200, 200 average MW and (B) dextran standard T2000, 2,000,000 average MW. Elution: Distilled water. Detection: RI.

(17–21). Analytical methods that utilize combinations of TSK-GEL SW and TSK-GEL PW columns in series for isolating chylomicron, high-density lipoprotein (HDL), and other large lipoproteins have been published by Hara *et al.* (17–21), along with methods for cholesterol, phospholipids, and triglycerides analysis using on-line postcolumn reaction procedures. Figure 4.28 (page 126) shows a calibration curve for proteins on TSK-GEL PW_{XL} columns. As seen in

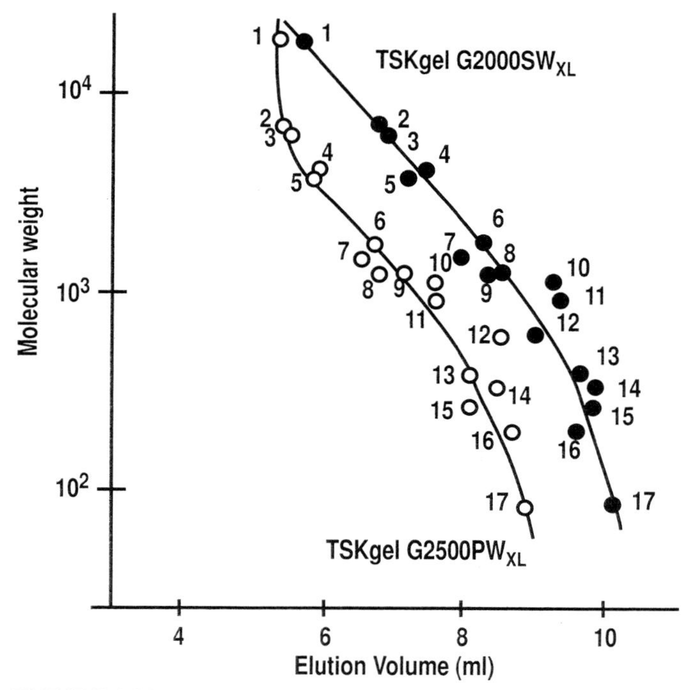

FIGURE 4.26 Calibration curves for peptides on TSK-GEL G2500PW$_{XL}$ and TSK-GEL G2000SW$_{XL}$ columns. Column: TSK-GEL G2500PW$_{XL}$, 6 μm, 7.8 mm $\times$ 30 cm; TSK-GEL G2000SW$_{XL}$, 5 μm, 7.8 mm $\times$ 30 cm. Sample: (1) myoglobin (17,800 MW), (2) aprotinin (6500 MW), (3) insulin (5807 MW), (4) big gastrin (3849 MW), (5) glucagon (3482 MW), (6) bombesin (1619 MW), (7) substance P (1347 MW), (8) bradykinin-potentiator B (1182 MW), (9) LHRH (1182 MW), (10) oxytocin (1007 MW), (11) DSIP (848 MW), (12) leu-enkephalin (555 MW), (13) TRH (362 MW), (14) glutathione (307 MW), (15) tetraglycine (246 MW), (16) alanylvaline (188 MW), and (17) glycine (75 MW). Elution: 45% CH_3CN in 0.1% TFA (pH 3). Flow rate: 1.0 ml/min. Detection: UV at 215 nm.

Fig. 4.29 (page 126), very large proteins, such as low-density lipoproteins (LDL and VLDL), gelatin, and sea worm chlorocruorin, which are excluded even by G4000SW columns, can be covered by PW columns of large pore size such as the G5000PW$_{XL}$ and G6000PW$_{XL}$ columns.

3. Nucleic Acids

a. DNA Fragments

TSK-GEL G2000SW, G3000SW, G4000SW, G5000PW columns were evaluated for their effectiveness for separating double-stranded DNA fragments and ribosomal and transfer RNAs (22). The choice of column is dependent on sample molecular weight. Small nucleic acids can be adequately analyzed using TSK-GEL SW columns. Larger nucleic acids should be analyzed with TSK-GEL PW columns of larger pore size, such as the TSK-GEL G-DNA-PW and TSK-GEL G5000PW columns. Calibration curves for double-stranded DNA fragments on TSK-GEL SW type columns and a TSK-GEL G5000PW column are shown in Fig. 4.30 (page 127); the calibration curve for DNA fragments on a TSK-GEL

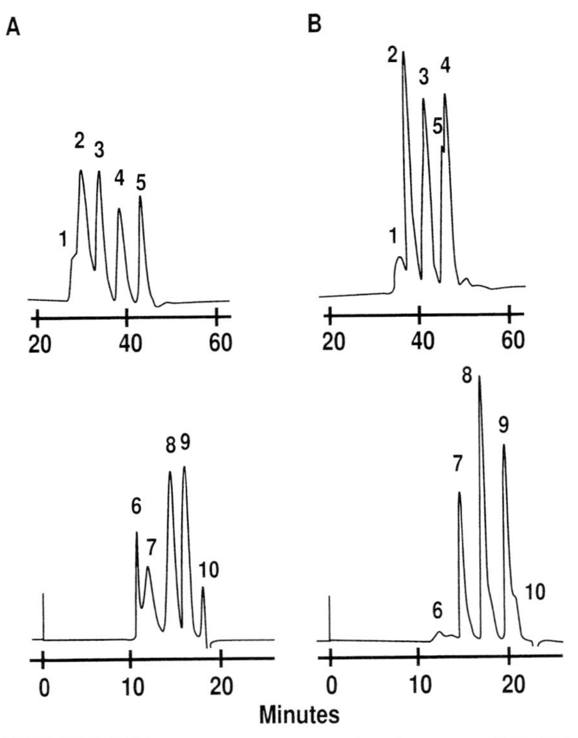

FIGURE 4.27 Separations of peptide mixtures on TSK-GEL PW_{XL} and TSK-GEL SW_{XL} columns. Column: (A) TSK-GEL G2500PW_{XL}, 6 μm, 7.8 mm × 30 cm and (B) TSK-GEL G2000SW_{XL}, 5 μm, 7.8 mm × 30 cm. Sample: approximately 2.5 μg/40 μl each of (1) aprotinin (6500 MW), (2) big gastrin (3849 MW), (3) bombesin (1619 MW), (4) oxytocin (1007 MW), (5) leu-enkephalin (555 MW), (6) myoglobin (17,800 MW), (7) glucagon (3482 MW), (8) LHRH (1182 MW), (9) TRH (362 MW), and (10) glycine (75 MW). Elution: 45% CH_3CN in 0.1% TFA. Flow rate: 0.2 ml/min (top pair); 0.5 ml/min (bottom pair). Detection: UV at 215 nm.

G-DNA-PW column is shown in Fig. 4.15. The exclusion limits for double-stranded DNA fragments are lower than those for rRNAs, indicating that double-stranded DNA fragments have a larger effective molecular weight in solution than rRNAs of the same molecular weight. Often, two or more columns can be used in series to improve resolution. Table 4.10 lists the recommended TSK-GEL SW and TSK-GEL PW columns for separating double-stranded DNA and RNA fragments.

Three experiments illustrate the differences between the performances of TSK-GEL SW and TSK-GEL PW columns for nucleic acids of differing molecular weight: separation of *Hae*III-cleaved pBR322 DNA fragments. *Eco*RII-cleaved pBR322 DNA fragments, and of total *Escherichia coli* tRNA. Results are shown in Fig. 4.31 through 4.33 (pages 128 through 130). To calculate the approximate molecular weight of the sample components, the number of base pairs should be multiplied by 650 (22). The *Hae*III reaction produces 22 fragments, which were not resolved successfully with any single column. When using two columns in series, however, resolution improved dramatically. Although neither the TSK-GEL SW, or the TSK-GEL PW column resolved all 22 fragments to baseline,

TABLE 4.9 Comparison of Peptide Elution Volumes on TSK-GEL G2500PW$_{XL}$ and TSK-GEL G2000SW$_{XL}$ Columns

Peptide	Da	Elution volume	
		G2500PW$_{XL}$	G2000SW$_{XL}$
Glycine	75	8.8	10.0
Alanyl-valine	188	8.6	9.5
Tetraglycine	246	8.0	9.7
Glutathione	307	8.4	9.8
TRH	362	8.0	9.6
Leu-enkephalin	555	8.5	9.0
DSIP	848	7.6	9.3
Oxytocin	1007	7.6	9.2
LHRH	1182	7.1	8.3
Bradykinin potentiator B	1182	6.7	8.4
Substance P	1347	6.5	7.9
Bombesin	1619	6.7	8.2
Glucagon	3482	5.9	7.2
Big gastrin	3849	5.9	7.4
Insulin	5807	5.5	6.9
Aprotinin	6500	5.4	6.7
Myoglobin	17800	(5.35)	5.6

the TSK-GEL G3000SW column clearly provided a higher level of resolution than the TSK-GEL G5000PW column (Fig. 4.31).

The pBR322 DNA is cleaved by *Eco*RII into six fragments of 13 to 1857 bp. Again, two columns used in series gave the best resolution. As shown in Fig. 4.32, a TSK-GEL G5000PW column was able to isolate the 1857-bp fragment from the 928 and 1060 fragments, indicating its superiority in the molecular weight range above 1×10^6. (To compare differences in resolution for this sample at different flow rates, see Fig. 4.16.) Separation of the four *E. coli* RNAs, shown in Fig. 4.33, confirms the better performance of TSK-GEL SW columns for samples with a wide molecular weight range. The sample consists of 4S tRNA (25,000 Da), 5S rRNA (39,000 Da), 16S rRNA (560,000 Da), and 23 S rRNA (1,100,000 Da); all four molecules are within the molecular weight range recommended for TSK-GEL SW type columns. The two chromatograms demonstrate a superior separation with the TSK-GEL G4000SW column.

b. Plasmids

The large pore structure of the TSK-GEL G6000PW allows it to separate large molecules such as pBR322 plasmid from contaminating RNAs and proteins in a much shorter time frame than other methods (23). A two column system of G6000PW (7.5 mm i.d. $\times$ 60 cm) was used to separate the cleared lysate and phenol extract of the plasmid as shown in Fig. 4.34 (page 130). The plasmid

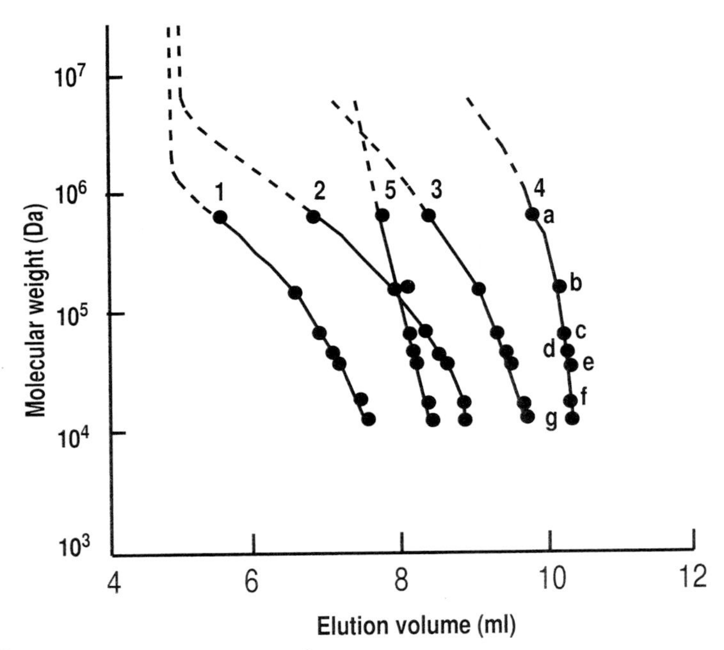

FIGURE 4.28 Calibration curves of TSK-GEL PW_{XL} columns for proteins. Column: (1) G3000PW_{XL}, (2) G4000PW_{XL}, (3) G5000PW_{XL}, (4) G6000PW_{XL}, and (5) GMPW_{XL}. Sample: (a) thyroglobulin, (b) γ-globulin, (c) albumin, (d) ovalbumin, (e) β-lactoglobulin, (f) myoglobin, (g) cytochrome c. Elution: 0.2 M phosphate buffer (pH 6.8). Flow rate: 1.0 ml/min

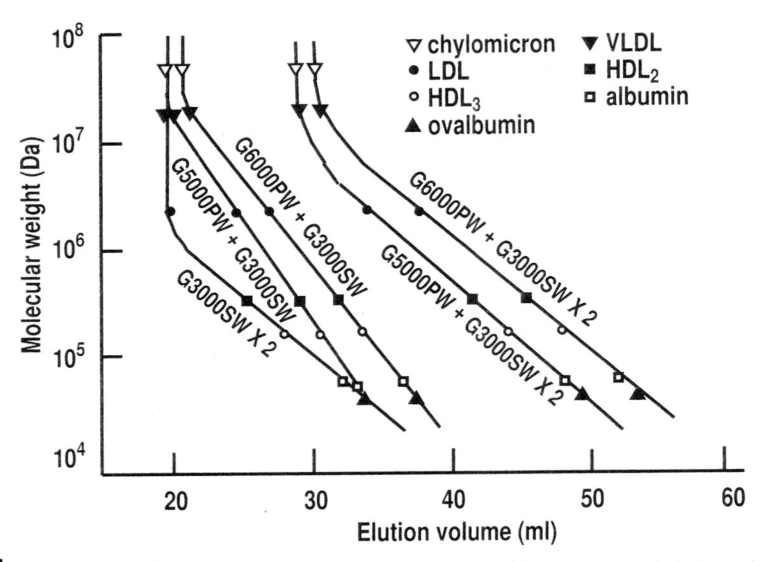

FIGURE 4.29 Relation between molecular weight of lipoproteins and elution volume for combination GFC columns. Column: 7.5 mm i.d. × 60 cm. Sample: Chylomicron, VLDL, LDL, HDL_2, HDL_3, albumin, and ovalbumin. Elution: 0.1 *M* Tris–HCl buffer (pH 7.4). Flow rate: 1.0 ml/min.

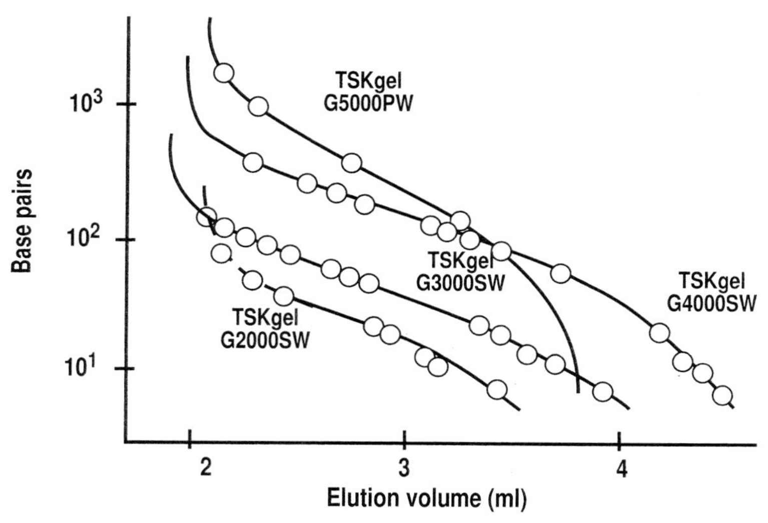

FIGURE 4.30 Calibration curves for double-stranded DNA fragments on TSK-GEL SW and TSK-GEL PW columns. Column: TSK-GEL SW or TSK-GEL PW, two 7.5 mm × 60 cm columns in series. Sample: 22 fragments from *Hae*III-cleaved pBR322 DNA and 6 fragments from *Eco*RI-cleaved pBR322 DNA. Elution: 0.1 *M* NaCl in 0.1 *M* phosphate buffer, pH 7.0, plus 1 m*M* EDTA. Flow rate: 1.0 ml/min. Detection: UV at 260 nm.

peaks are under the horizontal bar in Fig. 4.34. Plasmid DNA was eluted from TSK-GEL G5000PW columns in under 2 min with greater than 98% recovery (24).

4. Viruses

The TSK-GEL G6000PW column has been reported to provide superior separations of viruses when compared to Sephacryl S-1000 (25). Figure 4.35 (page

TABLE 4.10 Recommended TSK-GEL SW and TSK-GEL PW Columns for Separating Double-Stranded DNA and RNA Fragments

	Recommended TSK-GEL column
Base pairs of DNA	
<55	G2000 or G3000SW_{XL} or SW
55–110	G3000SW_{XL} or SW
110–375	G4000SW_{XL} or SW
375–1500	G5000PW_{XL}
1000–7000	G-DNA-PW
RNA daltons	
<60,000	G2000 or G3000SW_{XL} or SW
60,000–120,000	G3000SW_{XL} or SW
120,000–1,200,000	G4000SW_{XL} or SW
1,200,000–10,000,000	G5000PW_{XL}

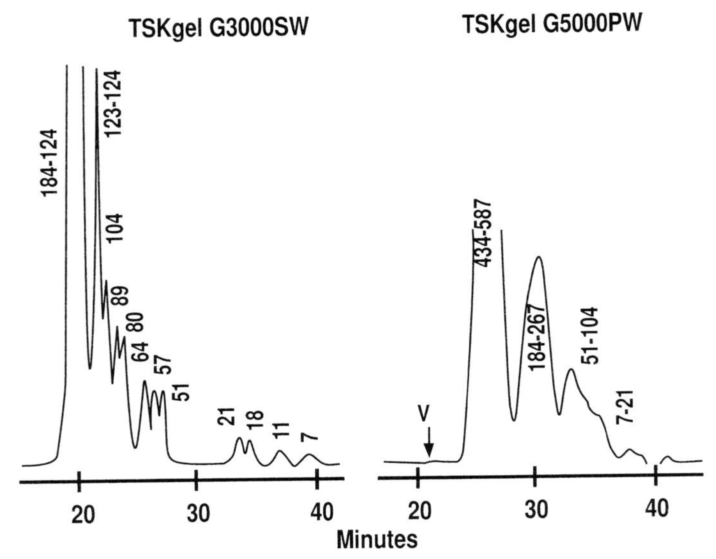

FIGURE 4.31 Improved resolution of small DNA fragments on TSK-GEL SW columns compared to TSK-GEL PW columns. Column: TSK-GEL G3000SW, two 10-μm, 7.5 mm × 30 cm columns in series; TSK-GEL G5000PW, two 17-μm, 7.5 mm × 30 cm columns in series. Sample: 0.1 ml of *Hae*III-cleaved pBR322 DNA, 0.015% (w/v). Elution: 0.1 *M* NaCl in 0.1 *M* phosphate buffer, pH 7.0, and 1 mM EDTA. Flow rate: 1.0 ml/min. Detection: UV at 260 nm.

131) shows the elution profiles of adenovirus and vesicular stomatitus virus on a TSK-GEL G6000PW and G5000PW in series. Himmel and Squire (26) studied the elution parameters of viruses on a preparative TSK-GEL G5000PW column.

5. Liposomes

Figure 4.36 shows a comparison of elution profiles of dipalmitoylphosphatidylcholine (DPPC) vesicles on a TSK-GEL G6000PW column and a Sephacryl S-1000 column (25).

III. CHOOSING TSK-GEL SW TYPE OR TSK-GEL PW TYPE COLUMNS FOR GFC

The main criterion to use in choosing between the TSK-GEL SW and TSK-GEL PW size exclusion columns is the molecular weight of the sample material. The fact that TSK-GEL SW columns are based on silica whereas TSK-GEL PW columns are derived from a hydrophilic polymer network has relatively little impact on the separation. The chemical differences between the two column types play a minor role as compared to particle and pore size differences. In general, TSK-GEL SW columns are suitable for the separation of monodisperse biopolymers such as proteins and nucleic acids due to a higher resolving power. TSK-GEL PW columns should be chosen for samples of particularly small or large effective molecular mass, namely less than 1000 or greater than 100,000 Da. While a TSK-GEL SW column is typically the first column to try for biopolymers, TSK-GEL PW columns have demonstrated good results for

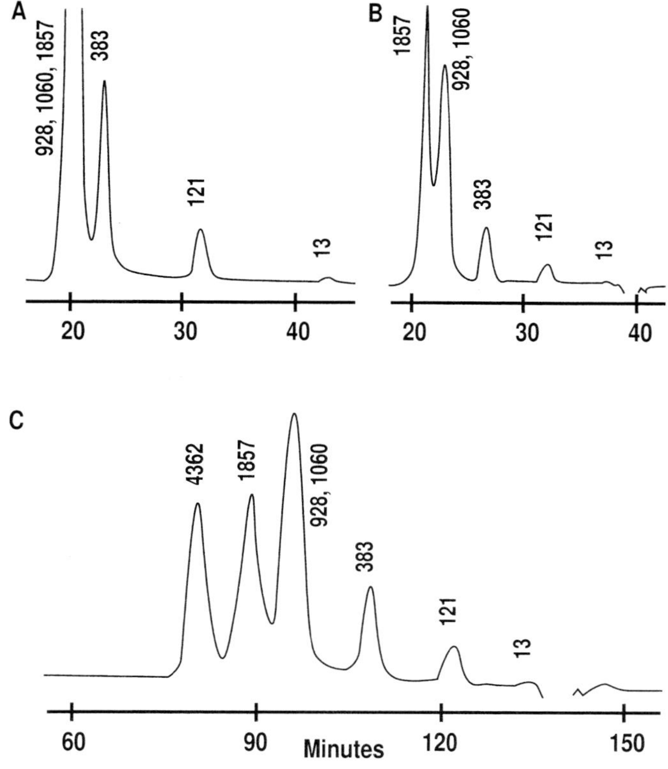

FIGURE 4.32 Separate larger DNA fragments on TSK-GEL G5000PW or TSK-GEL G-DNA-PW. Column: (A) TSK-GEL G4000SW, two 13-μm, 7.5 mm $\times$ 30 cm columns in series; (B) TSK-GEL G5000PW, two 17-μm, 7.5 mm $\times$ 30 cm columns in series; and (C) TSK-GEL G-DNA-PW, four 10-μm, 7.8 mm $\times$ 30 cm columns in series. Sample: (A and B) 0.1 ml of *Eco*RI-cleaved pBR322 DNA, 0.01–0.1% (w/v); (C) 1.7 μg of *Eco*RI-cleaved pBR322 DNA and 8 μg of *Bst*NI-cleaved pBR322 DNA. Elution: (A and B) 0.1 *M* NaCl in 0.1 *M* phosphate buffer, pH 7.0, and 1 m*M* EDTA; (C) 0.3 *M* NaCl in 0.1 *M* Tris-HCl, pH 7.5, and 1 m*M* EDTA. Flow rate: (A and B) 1.0 ml/min, (C) 0.3 ml/min. Detection: UV at 260 nm.

large proteins and DNA fragments as well as for smaller peptides. TSK-GEL PW columns are recommended for separating polydisperse compounds, such as polysaccharides and synthetic polymers. [See Table 4.11 (page 132) for suggestions on how to choose between TSK-GEL SW and TSK-GEL PW column types].

Comparisons were also made between TSK-GEL G3000SW and TSK-GEL G3000PW columns for separating larger sample molecules that could benefit from the larger pore size in these packings. Figure 4.37 (page 133) shows chromatograms obtained for a mix of thyroglobulin, IgG, bovine serum albumin, α-chymotrypsinogen A, and myoglobin. The TSK-GEL G3000SW column provided superior resolution for the higher molecular weight proteins, but was unable to resolve α-chymotrypsinogen A from myglobin. The TSK-GEL G3000PW column did not provide baseline separation for any of the five proteins, but it was more successful in splitting the two lower molecular weight proteins. The most suitable column or system will depend on the particular components that need to be measured.

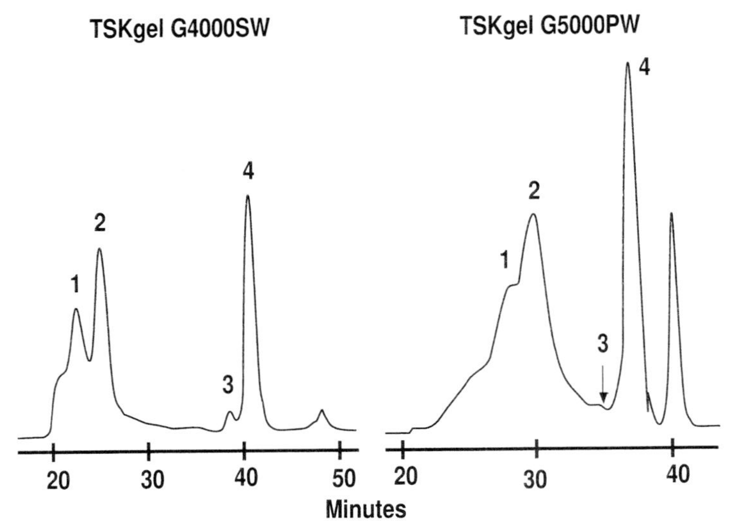

FIGURE 4.33 Separation of total *E. coli* RNA on TSK-GEL SW and TSK-GEL PW columns. Column: TSK-GEL G4000SW, two 13-μm, 7.5 mm $\times$ 30 cm columns in series; TSK-GEL G5000PW, two 17-μm, 7.5 mm $\times$ 30 cm columns in series. Sample: 0.1 ml of 1 : 10 diluted solution of total *E. coli* RNA: (1) 23S rRNA (1,100,000 MW), (2) 16S rRNA (560,000 MW), (3) 5S rRNA (39,000 MW), and (4) 4S tRNA (25,000 MW). Elution: 0.1 *M* NaCl in 0.1 *M* phosphate buffer, pH 7.0, plus 1 m*M* EDTA. Flow rate: 1.0 ml/min. Detection: UV at 260 nm.

Although sample size will most greatly influence the choice of column, the chemical and physical differences between TSK-GEL SW columns and TSK-GEL PW columns will also affect choice. For example:

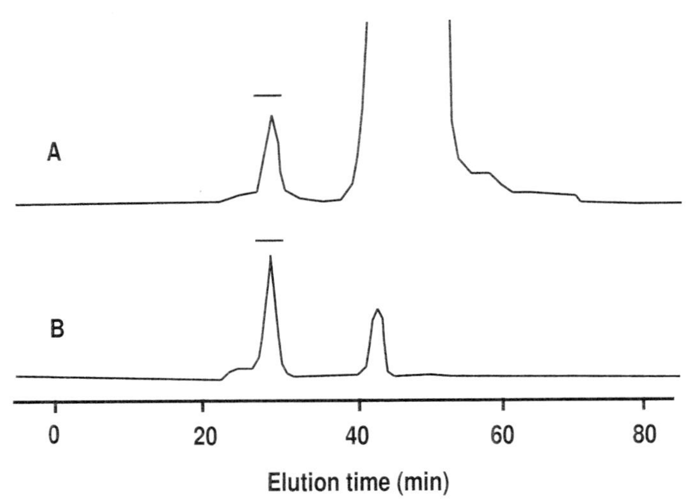

FIGURE 4.34 Separation of plasmid pBR322 on TSK-GEL G6000PW. Column: TSK-GEL G6000PW, two 7.5 mm $\times$ 60 cm in series. Sample: (A) 80 μl cleared lysate and (B) 20 μl phenol extract. Elution: 0.3 *M* NaCl with 1 m*M* EDTA in 0.1 *M* Tris–HCl, pH 7.5. Flow rate: 1 ml/min. Detection: UV at 260 nm.

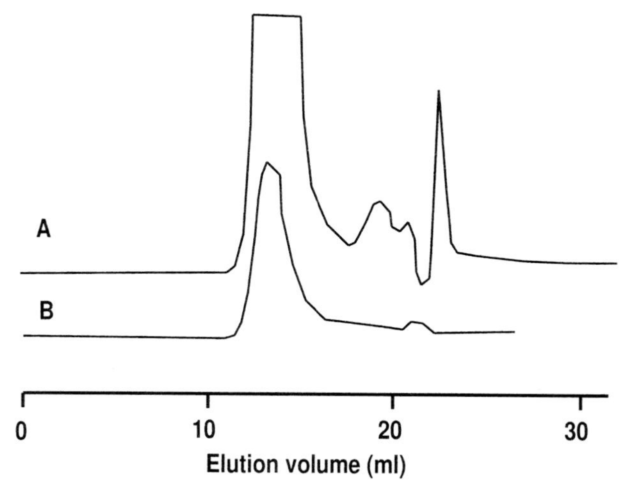

FIGURE 4.35 Elution profiles of adenovirus and vesicular stomatitus virus on TSK-GEL G6000PW and G50000PW. Column: TSK-GEL G6000PW + G5000PW in series. Sample: (A) adenovirus and (B) vesicular stomatitus virus. Elution: 145 m*M* NaCl in 10 m*M* Na-HEPES buffer, pH 7.4. Flow rate: 1 ml/min. Detection: UV at 254 nm.

1. Certain degrees of hydrophobic and ionic interaction can occur between the sample and the TSK-GEL PW column packing, which results in greater resolving power than seen by the pure size exclusion mechanism of the TSK-GEL SW columns. The hydrophobic properties can be manipulated by changing the salt concentration of the mobile phase to, in effect, combine hydrophobic interaction chromatography with size exclusion chromatography.

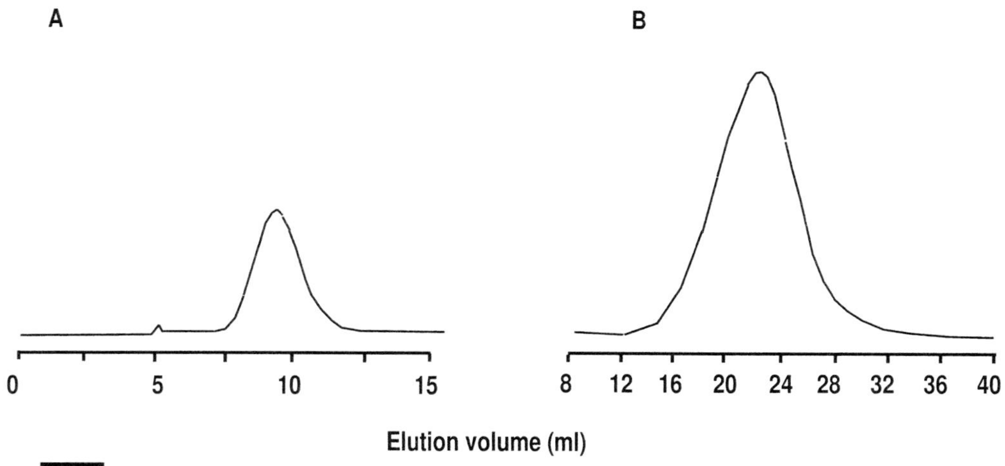

FIGURE 4.36 Elution profiles of liposome vesicles on TSK-GEL G6000PW and Sephacryl S-1000. Column: (A) TSK-GEL G6000PW, 7.5 mm × 30 cm and (B) Sephacryl S-1000, 10 mm × 45 cm. Sample: Dipalmitoylphosphatidylcholine. Elution: 145 m*M* NaCl in 10 m*M* Na–HEPES buffer, pH 7.4. Flow rate: 1 ml/min. Detection: UV at 254 nm.

TABLE 4.11 Recommended Column Selection Guide for High-Performance Gel-Filtration Chromatography

Sample	Column selection: First choice	Column selection: Alternative	Selection criteria
Carbohydrates			
Polysaccharides	TSK-GEL $GMPW_{XL}$	TSK-GEL $G5000PW_{XL}$ + TSK-GEL $G3000PW_{XL}$	Large pore size, linear calibration curve, small particles, high resolving power
Oligosaccharides	TSK-GEL G-Oligo-PW, TSK-GEL G2000PW	TSK-GEL $G2500PW_{XL}$	Small particles, high resolving power
Nucleic acids			
DNA fragments			
Large	TSK-GEL G-DNA-PW, TSK-GEL $G5000PW_{XL}$	—	Large pore size, small particles, high resolving power
Medium and small	TSK-GEL $G4000SW_{XL}$ or SW, TSK-GEL $G3000SW_{XL}$ or SW	—	Suitable pore sizes
RNA	TSK-GEL $G4000SW_{XL}$ or SW, TSK-GEL $G3000SW_{XL}$ or SW	—	Suitable pore sizes
Oligonucleotides	TSK-GEL $G2500PW_{XL}$	—	Small pore size, ionic interaction
Proteins			
Normal size proteins	TSK-GEL Super SW3000, TSK-GEL $G3000SW_{XL}$ or SW, TSK-GEL $G4000SW_{XL}$ or SW, TSK-GEL Super SW2000	TSK-GEL $G3000PW_{XL}$ TSK-GEL $G4000PW_{XL}$	Small particles, small to medium range pore sizes
Large proteins			
LDL	TSK-GEL $G6000PW_{XL}$, TSK-GEL $G5000PW_{XL}$	—	Large pore size
Gelatin	TSK-GEL $GMPW_{XL}$	TSK-GEL $G5000PW_{XL}$ + $G3000PW_{XL}$	Large pore size, linear calibration curve
Peptides			
Large	TSK-GEL Super SW3000, TSK-GEL $G3000SW_{XL}$ or SW, TSK-GEL $G2000SW_{XL}$ or SW	TSK-GEL $G3000PW_{XL}$	Small to medium range pore size, versatile
Small	TSK-GEL $G2500PW_{XL}$	TSK-GEL Super SW2000, TSK-GEL G2000SWXL or SW	Linear calibration curve, high resolving power
Virus	TSK-GEL $G6000PW_{XL}$, TSK-GEL $G5000PW_{XL}$	—	Large pore size, high resolving power
Synthetic polymers	TSK-GEL $GMPW_{XL}$	TSK-GEL $G5000PW_{XL}$ + $G3000PW_{XL}$	Large pore size, low adsorption, linear calibration curve
Synthetic oligomers			
Nonionic and cationic	TSK-GEL G-Oligo-PW	TSK-GEL $G2500PW_{XL}$	Small pore size, high resolving power
Anionic	TSK-GEL $G2500PW_{XL}$	—	Small pore size, ionic interaction

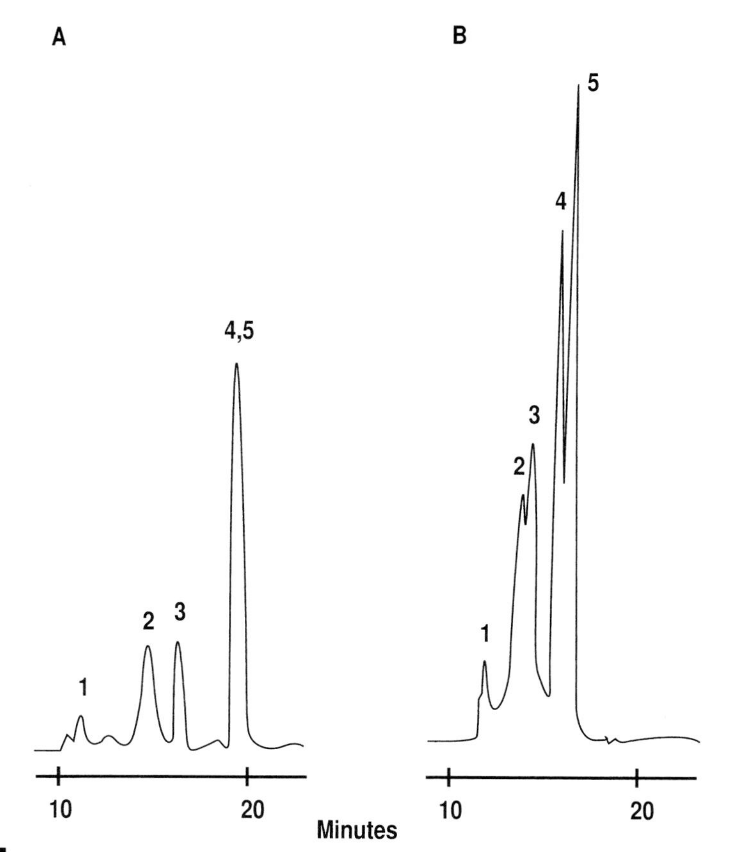

FIGURE 4.37 Better resolution of larger proteins on silica-based packing than on polymer-based packing. Column: (A) TSK-GEL G3000SW$_{XL}$ 5 μm, 7.8 mm $\times$ 30 cm and (B) TSK-GEL G3000PW$_{XL}$, 6 μm, 7.8 mm $\times$ 30 cm. Sample: (1) thyroglobulin (660, 000 MW), (2) IgG (156,000 MW), (3) bovine serum albumin (67,000 MW), (4) α-chymotrysinogen (25,700 MW), and (5) myoglobin (17,600 MW). Elution: 0.2 *M* phosphate buffer, pH 6.8. Flow rate: 0.5 ml/min. Detection: UV at 215 nm.

2. TSK-GEL PW columns exhibit lower adsorptive properties for synthetic water-soluble polymers than TSK-GEL SW columns. This is believed to be due to the linear structure of many such polymers, which might enable the molecule to more easily penetrate the pores and interact with the silanol groups of the TSK-GEL SW packings. In contrast to the residual silanol groups of the TSK-GEL SW packings, which reside in the pores, the carboxylic acid groups characteristic of most TSK-GEL PW packings are on the resin surface. Therefore, depending on the type of analyte, interaction with surface charges on TSK-GEL PW type packings can be neutralized at eluent ionic strengths from 0.01 to 0.1 *M*, whereas neutralization of charges on TSK-GEL SW type packings require higher eluent ionic strengths of 0.3 to 0.5 *M*.

3. Resin-based TSK-GEL PW columns are durable to alkaline conditions, which cannot be used with silica-based columns. Such durability facilitates separations with basic eluents and allows the use of alkaline cleaning reagents.

TSK-GEL SW columns allow use of elution buffers comprised completely of water-soluble organic solvents, whereas the TSK-GEL PW packings limit organic cosolvent use to a maximum of 50%.

4. Physically, TSK-GEL SW columns can withstand greater back pressures.

5. Operating temperatures are only from 10 to 30°C for TSK-GEL SW columns, but are from 10 to 80°C for TSK-GEL PW columns. The TSK-GEL SW and TSK-GEL PW sections of this chapter discuss the individual column properties and usage in more detail.

IV. CLEANING AND REHYDRATION OF TSK-GEL COLUMNS

A. Cleaning Procedures

1. Establishing Baseline Parameters

Occasionally, samples are run that adsorb onto the packing material. Generally, if one of the performance characteristics of the column changes by 10% or more, it is prudent to clean the column. These performance characteristics are (1) asymmetry factor, retention time, resolution, and theoretical plates.

A standard test probe is not absolutely necessary to monitor the column. Any well-resolved peak in the sample may be used. To use a sample component, baseline data must be established when the column is new and performing well. After establishing that the column is performing properly using the manufacturer's standard test procedure, calculate the assymetry factor, theoretical plates, and resolution of one or more of the sample components. Also note the retention time. This will become the "baseline test mix," which will be used for later comparison.

2. Cleaning Guidelines

There are several basic guidelines to follow when cleaning a column. These rules apply regardless of the type of TSK-GEL column that is being cleaned.

1. Clean the column in the reverse flow direction. Most of the contamination will reside on the top portion of the column, which will prevent the contamination from traversing the entire length of the column.

2. Do not connect the column to the detector. This precaution will prevent any contaminants precipitating in the flow cell that could damage it.

3. Operate the column at half of the maximum recommended flow rate, taking special care to monitor the pressure, because the cleaning solution may be of different viscosity than the normal mobile phase.

4. If the cleaning solution has a high or low pH, ensure that the rest of the chromatographic system is compatible.

3. Recommended Cleaning Solutions

a. TSK-GEL SW/SW$_{XL}$ Columns

1. If the column is contaminated with basic compounds, clean it with a concentrated salt solution at pH 3, e.g., 0.5–1.0 *M* K_2SO_4. Avoid the use of halides, as they will corrode stainless steel over time.

2. If the column is contaminated with hydrophobic compounds, clean it with water-miscible organic solvents such as MeOH, CH_3CN, or EtOH. Concentrations of 10–20% organic in water should be sufficient for cleaning.

3. For precipitated protein, buffered solutions containing chaotropic reagents such as 0.1% SDS, 8 *M* urea, or 6 *M* guanidine or proteolytic enzymes such as pepsin may be used. However, an extended washing with buffer is required to remove SDS and guanidine. Unexpected elution behavior can occur if these reagents are not removed completely.

b. TSK-GEL PW/PW_{XL} Columns

1. If the column is contaminated with basic compounds, clean it with a concentrated salt in the normal mobile phase, e.g., 0.5–1.0 *M* K_2SO_4. Avoid the use of halides, as they will corrode stainless steel over time. Buffered solutions at low pH (2–3) or high pH (11–12) can also be used.

2. If the column is contaminated with hydrophobic compounds, clean it with water-miscible organic solvents such as MeOH, CH_3CN, or EtOH. Concentrations of 10–20% organic in water should be sufficient for cleaning.

3. For precipitated protein, buffered solutions containing chaotropic reagents such as 0.1% SDS, 8 *M* urea, or 6 *M* guanidine or proteolytic enzymes such as pepsin may be used. However, an extended washing with buffer is required to remove SDS and guanidine. Unexpected elution behavior can occur if these reagents are not removed completely.

B. Rehydration Procedure

Dehydration of TSK-GEL columns can result from improper use or during long-term storage. This condition can be remedied by the following procedure.

1. Connect the column to the high-performance liquid chromatography (HPLC) system in the reverse flow direction.
2. Do not connect the column to the detector.
3. Pump a filtered mobile phase of 20% methanol in distilled, deonized water over the column at half of the recommended maximum flow rate.
4. Continue this procedure for at least 10 column volumes. Air bubbles may be detectable in the effluent.
5. Connect the column to the HPLC system in the normal flow direction.
6. Equilibrate with the running mobile phase.
7. Perform the manufacturer's recommended QC test to ensure that the column is running properly.

V. TSK-GEL H Type Columns

A. Introduction

TSK-GEL H type columns are for gel-permeation chromatography (GPC) in organic solvents. They are packed with porous poly(styrene–divinylbenzene) resins that have a high degree of cross-linking.

TABLE 4.12 TSK-GEL H_6 Columns

Description	Particle size (μm)	Minimum theoretical plates/30 cm	Exclusion limit (polystyrene molecular weight)	Standard flow rate (ml/min)	Maximum flow rate (ml/min)	Maximum pressure drop/30 cm (bar)
G1000H6	13	6000	1,000	1.0–1.8	2.4	30
G2000H6	13	6000	10,000	1.0–1.8	3.0	30
G2500H6	13	6000	20,000	1.0–1.8	3.0	30
G3000H6	13	6000	60,000	1.0–1.8	3.0	20
G4000H6	13	6000	400,000	1.0–1.8	3.0	20
G5000H6	13	6000	4,000,000	1.0–1.8	3.0	20
G6000H6	13	6000	40,000,000	1.0–1.8	3.0	20
G7000H6	13	6000	400,000,000	1.0–1.8	3.0	20
GMH6	13	6000	400,000,000	1.0–1.8	3.0	20
GMH6-HT	13	4500	400,000,000	0.8–1.5	2.0	20

H type columns were commercialized in 1972 by Tosoh as the first high-performance GPC columns in the world (29,30). Since then, they have been used for analyses of a wide range of polymers, oligomers, polymer additives, and other small molecules that are soluble in organic solvents. The first products introduced in 1972 were 10, 13, and 17 μm in resin particle diameter and had theoretical plates of 4000–8000 plates/30 cm, which were 5–10 times higher than those of columns available commercially in those days. Smaller particles of 5 and 3 μm were also introduced early 1980s and 1990s, respectively. As a result, very high theoretical plates of more than 32,000 plates/30 cm can now be attained on H type columns.

B. Product Offerings

Five series of H type columns are now available, each packed with a different particle size grade of spherical resin, as shown in Tables 4.12–4.16. H_6, H_8,

TABLE 4.13 TSK-GEL H_8 Columns

Description	Particle size (μm)	Minimum theoretical plates/30 cm	Exclusion limit (polystyrene molecular weight)	Standard flow rate (ml/min)	Maximum flow rate (ml/min)	Maximum pressure drop/30 cm (bar)
G1000H8	10	8000	1,000	0.8–1.2	1.6	30
G2000H6	10	8000	10,000	0.8–1.2	2.0	30
G2500H8	10	8000	20,000	0.8–1.2	2.0	30
G3000H8	10	8000	60,000	0.8–1.2	2.0	20
G4000H8	10	8000	400,000	0.8–1.2	2.0	20

TABLE 4.14 TSK-GEL H_{XL} Columns

Description	Particle size (μm)	Minimum theoretical plates/30 cm	Exclusion limit (polystyrene molecular weight)	Standard flow rate (ml/min)	Maximum flow rate (ml/min)	Maximum pressure drop/30 cm (bar)
G1000HXL	5	16,000	1,000	0.5–1.0	1.0	50
G2000HXL	5	16,000	10,000	0.5–1.0	1.2	50
G2500HXL	5	16,000	20,000	0.5–1.0	1.2	50
G3000HXL	6	16,000	60,000	0.5–1.0	1.2	35
G4000HXL	6	16,000	400,000	0.5–1.0	1.2	35
G5000HXL	9	14,000	4,000,000	0.5–1.0	1.2	35
G6000HXL	9	14,000	40,000,000	0.5–1.0	1.2	15
G7000HXL	9	14,000	400,000,000	0.5–1.0	1.2	15
GMHXL	9	14,000	400,000,000	0.5–1.0	1.2	15
GMHXL-HT	13	5,500	400,000,000	0.5–1.0	1.2	15

and H_{HR} columns are packed with 13-, 10-, and 5-μm particles, respectively. H_{XL} columns contain 5-, 6-, 9-, or 13-μm particles. SuperH columns contain 3-, 4-, or 5-μm particles. The small particle size of the resins in SuperH columns provides a high resolution of samples and an efficiency over 16,000 plates per 15-cm column.

TABLE 4.15 TSK-GEL H_{HR} Columns

Description	Particle size (μm)	Minimum theoretical plates/30 cm	Exclusion limit (polystyrene molecular weight)	Standard flow rate (ml/min)	Maximum flow rate (ml/min)	Maximum pressure drop/30 cm (bar)
G1000HHR	5	16,000	1,000	0.5–1.0	2.0	50
G2000HHR	5	16,000	10,000	0.5–1.0	2.0	50
G2500HHR	5	16,000	20,000	0.5–1.0	2.0	50
G3000HHR	5	16,000	60,000	0.5–1.0	2.0	50
G4000HHR	5	16,000	400,000	0.5–1.0	2.0	50
G5000HHR	5	16,000	4,000,000	0.5–1.0	2.0	50
G6000HHR	5	16,000	40,000,000	0.5–1.0	2.0	50
G7000HHR	5	16,000	400,000,000	0.5–1.0	2.0	50
GMHHR-L	5	16,000	4,000,000	0.5–1.0	2.0	50
GMHHR-N	5	16,000	400,000	0.5–1.0	2.0	50
GMHHR-M	5	16,000	4,000,000	0.5–1.0	2.0	50
GMHHR-H	5	16,000	400,000,000	0.5–1.0	2.0	50

TABLE 4.16 TSK-GEL SuperH Columns

Description	Particle size (μm)	Minimum theoretical plates/30 cm	Exclusion limit (polystyrene molecular weight)	Standard flow rate (ml/min)	Maximum flow rate (ml/min)	Maximum pressure drop/30 cm (bar)
SuperH1000	3	16,000	1,000	0.3–0.6	0.8	70
SuperH2000	3	16,000	10,000	0.3–0.6	0.8	60
SuperH2500	3	16,000	20,000	0.3–0.6	0.8	60
SuperH3000	3	16,000	60,000	0.3–0.6	0.8	40
SuperH4000	3	16,000	400,000	0.3–0.6	0.8	40
SuperH5000	3	16,000	4,000,000	0.3–0.6	0.8	40
SuperH6000	4	10,000	40,000,000	0.3–0.6	0.8	30
SuperH7000	5	10,000	400,000,000	0.3–0.6	0.8	30
SuperHM-L	3	16,000	4,000,000	0.3–0.6	0.8	40
SuperHM-N	3	16,000	400,000	0.3–0.6	0.8	40
SuperHM-M	3	16,000	4,000,000	0.3–0.6	0.8	40
SuperHM-H	4	16,000	400,000,000	0.3–0.6	0.8	40

H type resins are available in different pore sizes. Examples of calibration curves for polystyrene standards are shown in Figs. 4.38 and 4.39. Other series of H type columns have similar calibration curves. Exclusion limits are listed in Tables 4.12–4.16.

H type columns have a greatly expanded molecular weight separation range of samples that can be analyzed. Resins of several pore sizes are provided in the mixed-bed columns, GMH, for the analyses of polymer samples with a broad molecular weight range. GMH columns have linear calibration curves and are convenient for converting chromatograms to molecular weight distribution curves.

All H type columns are stainless steel. H_6 and H_8 columns are 7.5 mm i.d. × 30 cm or 7.5 mm i.d. × 60 cm. H_{XL} and H_{HR} columns are 7.8 mm i.d. × 30 cm. SuperH columns are 6 mm i.d. × 15 cm. The small size of SuperH columns has a merit of low consumption of solvent. This is advantageous not only economically but also for minimizing environmental problem. Selected H type columns are available in 21.5 mm i.d. × 30 cm or 21.5 mm i.d. × 60 cm semipreparative columns and 55 mm i.d. × 60 cm preparative columns. Connection fittings are 1/16 in. regular Swaige-lock type, and H type columns can be connected with most commercial HPLC instruments.

C. Limitation of Operation

1. Elution Solvent

All H type analytical columns are supplied containing tetrahydrofuran, with the exception of GMH-HT columns, which are only shipped in *o*-dichlorobenzene. Semipreparative and preparative columns contain chloroform

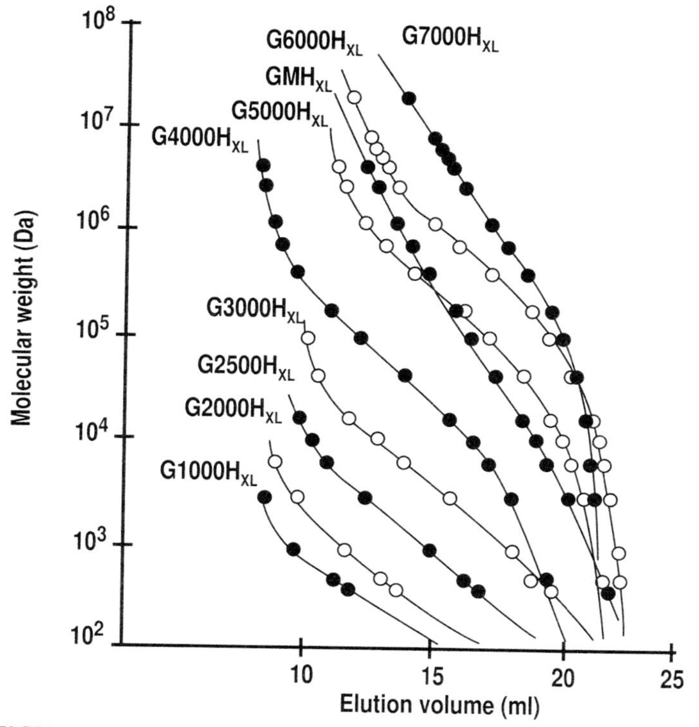

FIGURE 4.38 Calibration curves for TSK-GEL H_{XL} columns with polystyrene standards. Column: TSK-GEL H_{XL} series, two 7.8 mm × 30 cm columns in series. Sample: Polystyrene standards. Elution: Tetrahydrofuran. Flow rate: 1.0 ml/min. Detection: RI.

because it can be removed easily after fractionation. In addition, in tetrahydrofuran, analytical columns are also available packed in acetone, chloroform, dimethylformamide, and *o*-dichlorobenzene. The shipping solvent of a column determines the elution solvents that can be used. All H_6, H_8, and H_{XL} columns supplied containing tetrahydrofuran can be replaced by benzene, chloroform, toluene, xylene, dichloromethane, and dichloroethane. For applications requiring the use of different solvents, columns shipped in other solvents must be used. Table 4.17 lists the solvents that may be used to replace the original shipping solvent. The solvent can be changed only once. It is important that the substitution be made by a linear gradient with a 2% per minute rate of change with a flow rate of less than half the standard flow rate, which avoids degradation of column performance by creating space at the inlet of a column. However, H_{HR} and SuperH columns tolerate solvent exchange over a wide range of solvents due to the excellent swelling properties of resins. Table 4.18 shows an example of the minimal swelling ratio on the usage of different solvents.

Column performance is maintained during solvent exchange. H_{HR} and SuperH columns are compatible with the following solvents: acetone, benzene, carbon tetrachloride, chloroform, 1-chloronaphthalene, *o*-chlorophenol,

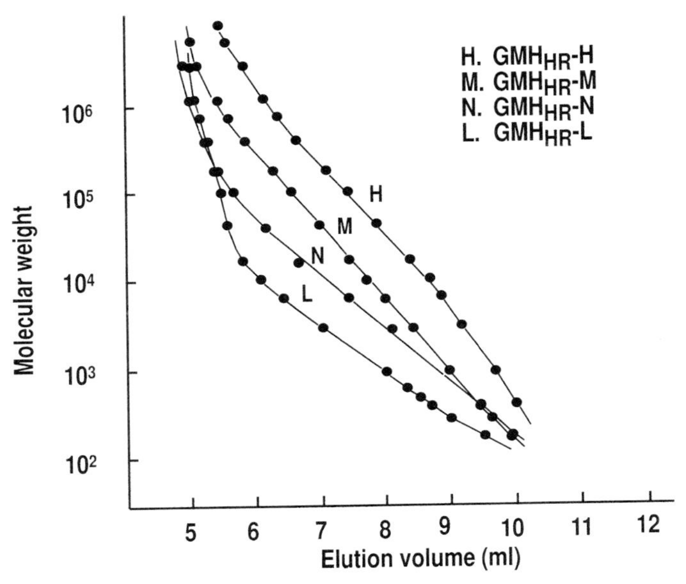

FIGURE 4.39 Calibration curves of TSK-GEL H_{HR} columns with standard polystyrene. Column: TSK-GEL H_{HR}, 7.8 mm i.d. × 30 cm. Elution: THF. Flow rate: 1.0 ml/min. Detection: UV at 254 nm. Temperature: 25°C.

o-chlorophenol/chloroform, *m*-cresol/chloroform, decahydronaphthalene, *o*-dichlorobenzene, dichloromethane, 1,4-dioxane, dimethylacetoamide, dimethylformamide, dimethyl sulfoxide, ethanol, ethylacetate, *n*-hexane, hexafluoroisopropanol, hexafluoroisopropanol/chloroform, methylethylketone, *N*-methylpyrrolidone, pyridine, quinoline, tetrahydrofuran, toluene, and trichlorobenzene.

TABLE 4.17 Solvent Compatibility of H_6, H_8, and H_{XL} Columns

Shipping solvent	Can be replaced by
Tetrahydrofuran	Benzene, chloroform, dichloroethane, dichloromethane, toluene, xylene
Acetone	Carbon tetrachloride, chloroform/*o*-chlorophenol, chloroform/*m*-cresol, chloroform/hexafluoroisopropanol, chloroform/methanol (up to 60%), *o*-dichlorobenzene, dimethylformamide, dimethyl sulfoxide, dioxane, ethylacetate, FC-113, haxane, methylethylketone, *N*-methylpyrrolidone, pyridine, quinoline, cyclohexane, dodecane
Chloroform	Chloroform/*m*-cresol, chloroform/hexafluoroisopropanol, chloroform/methanol (up to 20%), dimethyl sulfoxide, dioxane
o-Dichlorobenzene	l-Chloronaphthalene, trichlorobenzene
Dimethylformamide	Dimethyl sulfoxide, dioxane, tetrahydrofuran, toluene

TABLE 4.18 Swelling Property of H_{HR} Resin (G2000HHR)

Solvent	Volume in solvent/volume in tetrahydrofuran
Toluene	1.01
Benzene	1.00
Tetrahydrofuran	1.00
Acetone	0.99
Dimethylformamide	0.99
Methanol	0.98

2. Temperature

It is recommended that G1000H through G3000H be used at a temperature up to 60°C and G4000H through GMH up to 80°C. For applications requiring the operation at higher temperature, special high-temperature columns, GMH-HT, must be used. GMH-HT columns can be used up to 140°C.

3. Flow Rate and Pressure Drop

H type columns must be used at a flow rate and pressure drop below maximum values listed in Tables 4.12–4.16. Standard flow rates are also listed in these tables. They are flow rate range recommendable for long-term usage in tetrahydrofuran at 25°C and vary with temperature. H type columns can be operated at a higher flow rate at elevated temperatures. They also vary with solvent depending on the viscosity. They are approximately inversely proportional to the solvent viscosity. The maximum pressure drop listed in the tables is for one column. When some columns are used in series, the total maximum pressure drop is a summation of values of all columns.

D. Separation Conditions

1. Pore Size of Resin

It is very important to select the resin pore size that will best match the molecular weight of the sample to attain satisfactory results because each resin has the limited separation range. Small molecules are primarily analyzed on G1000H. An example is shown in Fig. 4.40. Phthalate esters and some other small molecules are separated close to baseline in about 10 min on $G1000H_{XL}$, even though the molecular weights of components differ by less than 50 Da. Oligomers are analyzed using G2000H, G2500H, or G3000H, depending on the molecular weights of sample components. Polymers are analyzed on G3000H through G7000H or GMH. When the range of molecular weight of sample components is unknown, GMH columns are recommended for the first trial because they have a broad separation range. The molecular weight range in the sample can be estimated from the calibration curve of the column. Best match columns can then be selected.

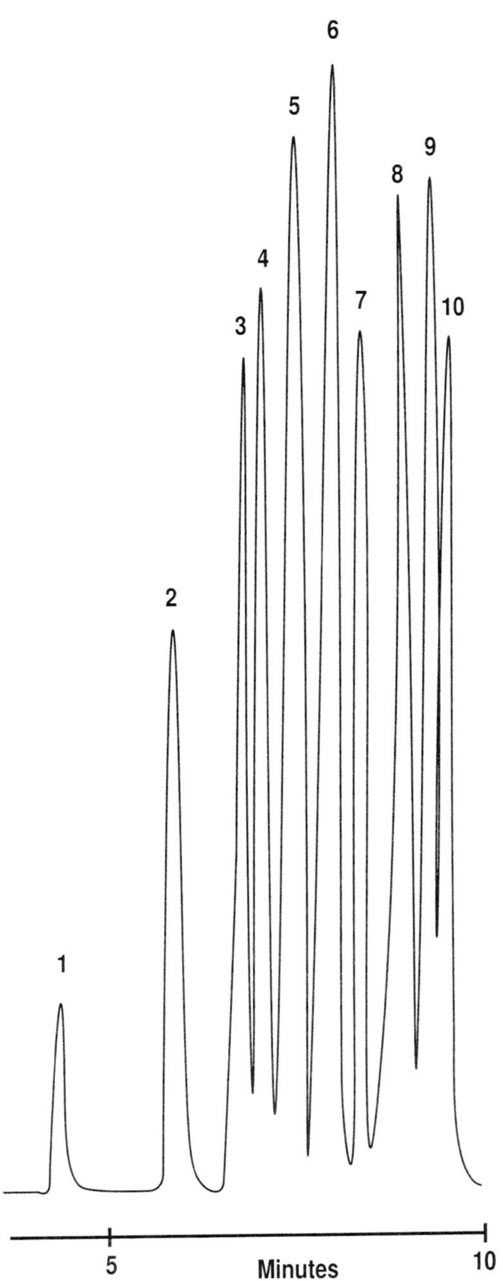

FIGURE 4.40 High resolution of phthalate esters on TSK-GEL G1000H_{XL}. Column: TSK-GEL G1000H_{XL}, 7.8 mm × 30 cm. Sample (1) polystyrene (10,200 MW), (2) dioctylphthalate (391 MW), (3) dibutylphthalate (278 MW), (4) dipropylphthalate (250 MW), (5) diethylphthalate (222 MW), (6) dimethylphthalate (194 MW), (7) *n*-propylbenzene (120 MW), (8) ethylbenzene (116 MW), (9) toluene (92 MW), and (10) benzene (78 MW). Elution: Tetrahydrofuran. Flow rate: 1.0 ml/min. Detection: UV at 254 nm.

2. Particle Size of Resin

Small particle size resins provide higher resolution, as demonstrated in Fig. 4.41. Low molecular weight polystyrene standards are better separated on a G2000H_{XL} column packed with 5 μm resin than a G2000H_8 column packed with 10 μm resin when compared in the same analysis time. Therefore, smaller particle size resins generally attain a better required resolution in a shorter time. In this context, SuperH columns are best, and H_{HR} and H_{XL} columns are second best. Most analyses have been carried out on these three series of H type columns. However, the performance of columns packed with smaller particle size resins is susceptible to some experimental conditions such as the sample concentration of solution, injection volume, and detector cell volume. They must be kept as low as possible to obtain the maximum resolution. Chain scissions of polymer molecules are also easier to occur in columns packed with smaller particle size resins. The flow rate should be kept low in order to prevent this problem, particularly in the analyses of high molecular weight polymers.

3. Column Length

Column length has an effect on resolution and analysis time. Longer columns provide higher resolution at the expense of longer analysis time, as shown in Fig. 4.41. The 60-cm-long G2000H_8 column has nearly identical resolution

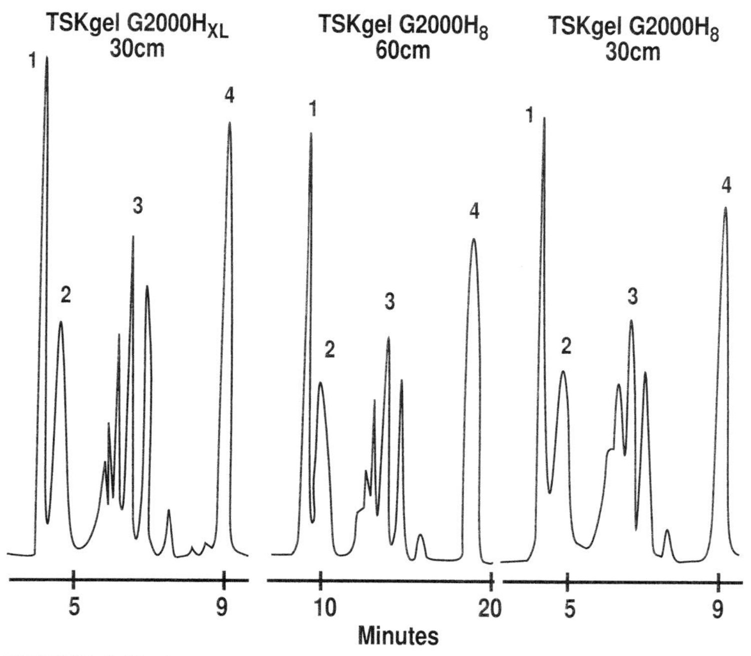

FIGURE 4.41 Low MW polystyrene standards separated on TSK-GEL H8 and TSK-GEL H_{XL} columns. Column: TSK-GEL G2000H_{XL}, 6 μm, 7.8 mm × 30 cm; TSK-GEL G2000H8, 10 μm, 7.5 mm × 60 cm; TSK-GEL G2000H8, 10-μm, 7.5 mm × 30 cm. Sample: Polystyrene standards: (1) 42,800 MW, (2) 2,800 MW, (3) 370 MW, and (4) benzene. Elution: Tetrahydrofuran. Flow rate: 1.0 ml/min. Detection: UV at 254 nm.

as the 30-cm-long G2000H_{XL} column, although analysis time is doubled. Therefore, column length should be selected to an optimum according to required resolution.

4. Elution Solvent

It is generally recommended that elution solvents are good solvents of samples having similar polarities or solubility parameters as sample molecules. When poor solvents are used, adsorption or partition of samples on resin sometimes happens and samples are not separated in accordance with molecular size. This means molecular weights or molecular weight distributions cannot be determined correctly. However, poor solvents sometimes have the advantage if adsorption or partition of samples does not happen because the concentration effect on elution volume of polymer becomes minimum (31,32). For example, elution volume is independent of sample concentration in the analyses of polychloroprene in methylethylketone, which is a theta solvent for polychloroprene at 25°C. When polychloroprene is analyzed in a good solvent such as tetrahydrofuran, the elution of sample becomes earlier as sample concentration increases, which causes an error in molecular weight estimation. Sample loading capacity also becomes higher when poor solvents are used. This is advantageous in the preparative fractionation of polymers.

It is well known that anionic samples tend to adsorb on poly(styrene–divinylbenzene) resins. However, cationic samples tend to be repelled from the resins. The mechanism seems to be an ionic interaction, although the poly(styrene–divinylbenzene) resin should be neutral. The reason is not well clarified. Therefore, it is recommended to add some salt in the elution solvent when adsorption or repulsion is observed in the analyses of polar samples. For example, polysulfone can be analyzed successfully using dimethylformamide containing 10 m*M* lithium bromide as an elution solvent, as shown in Fig. 4.42.

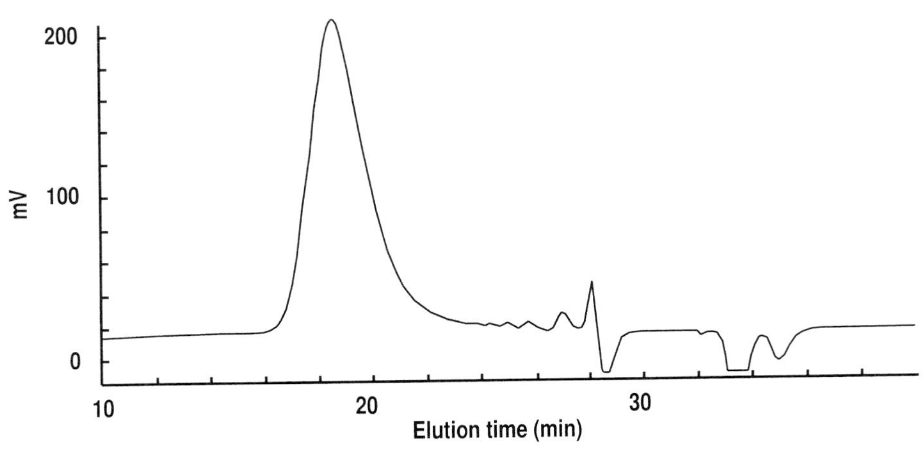

FIGURE 4.42 Chromatogram of polysulfone on TSK-GEL GMH_{HR}-M columns. Column: TSK-GEL GMH_{HR}-M, 7.8 mm i.d. × 30 cm. Elution: 10 m*M* lithium bromide in DMF. Flow rate: 1.0 ml/min. Pressure: 60 bar. Detection: RI. Temperature: 40°C.

If lithium bromide is not included in dimethylformamide, polysulfone adsorbs on the resin. Some other examples are shown in the application list (Table 4.19).

5. Temperature

The temperature must be raised when there is no solvent that can dissolve samples at ambient temperature. For example, polyolefines such as polyethylene and polypropylene are usually analyzed at 130–140°C because no solvent can dissolve these polyolefines at lower temperatures. It is also preferable to perform analyses at elevated temperatures when the viscosity of the elution solvent is considerably higher at ambient temperature. However, a temperature around 25–40°C is recommended when good solvents having low viscosity are available at such a temperature. It is much more convenient to operate a GPC instrument at 25–40°C than to operate at higher temperatures.

6. Injection Volume

The injection volume should be kept as small as possible to attain maximum resolution in analyses. This is particularly important in analyses on columns packed with small particle size resins such as SuperH. Injection volumes of 0.1% or less of the total column volume are recommended on SuperH columns. A few times larger injection volumes may be applied to other series of H type columns.

Larger injection volumes, e.g., 2% of the total column volume, are sometimes advantageous in the preparative fractionation of polymers (33). More samples can be injected using larger injection volumes with a slight decrease in resolution. When the same amount of sample is injected with a smaller injection volume and a higher sample concentration, the resolution decreases more significantly.

7. Sample Concentration

The sample concentration also should be kept as low as possible, particularly in analyses of polymers on columns packed with small particle size resins. The maximum sample concentration to achieve maximum resolution decreases as the sample molecular weight becomes higher and the resin particle size becomes smaller. It is usually in the range of 0.05–5 mg/ml, depending on the sample molecular weight and resin particle size.

E. Applications

A wide range of application for small molecules through very high molecular weight polymers on H type columns has been reported since the commercialization in 1972 (34,35). Some examples of applications are summarized in Table 4.19. Table 4.19 contains the type of sample, column, and elution solvent. More applications can be found in technical literatures published by TosoHaas or Tosoh.

VI. TOYOPEARL RESINS FOR SIZE EXCLUSION CHROMATOGRAPHY

Toyopearl HW size exclusion chromatography resins are macroporous packings for bioprocessing chromatography. They are applicable for process-scale

TABLE 4.19 Applications on TSK-GEL H Type Columns

Sample	Example of column	Elution solvent[a]
Acrylonitrile-butadiene	SuperHM-H-styrene copolymer	THF
Acrylonitrile-styrene	SuperHM-H copolymer	THF
Acrylonitrile-vinylidene	SuperHM-H chloride copolymer	THF
Alkyd resin	G3000HHR	THF, DMF containing LiBr
Arylalcohol-styrene	SuperHM-H copolymer	THF
Butadiene-styrene copolymer	SuperHM-H	Toluene
Butylene-ethylene-styrene	SuperHM-H copolymer	THF
Butylmethacrylate-styrene	SuperHM-H copolymer	THF
Cellulose acetate	SuperHM-H	THF
Cellulose acetate/butyrate	SuperHM-H	THF
Cellulose nitrate	G2000HXL + GMHXL	THF
Cellulose propionate	SuperHM-H	THF
Cresol resin	G2000HHR + G3000HHR + G4000HHR	DMF containing LiBr
Cyanoacrylate resin	SuperHM-H	Toluene
Epoxy resin	G3000HXL	THF
Ethylcellulose	G2000HXL + G3000HXL + G4000HXL + G5000HXL	THF
Ethylene glycol oligomer	SuperH2000	THF, HFIP containing TFA-Na
Ethylene oligomer	G2000HXL + G4000HXL	Xylene
Ethylene-vinylacetate	SuperHM-M copolymer	THF
Fluoro rubber	GMHXL	THF
Isoprene-styrene copolymer	SuperHM-H	THF
Maleic anhydride-styrene	G2000HXL + G4000HXL copolymer	THF
Melanine-modified urea resin	SuperHM-M	DMF containing LiBr
Melanine resin	GMHHR-L	DMSO containing LiBr
Nylon 6	SuperHM-M	HFIP containing TFA-Na
Nylon 12	SuperHM-M	HFIP containing TFA-Na
Nylon 66	SuperHM-M	HFIP containing TFA-Na
Phenol resin	G3000HHR	THF, DMF, DMF containing LiBr
Polyacrylonitrile	GMH8	DMF containing LiBr
Polyaminobismaleimide	G2000HXL + GMHXL	DMF containing LiBr
Polubutadiene	SuperHM-H	THF
Polybutylene terephthalate	GMHHR-M	Chloroform
Polybutylmethacrylate	SuperHM-H	THF
Polycaprolactone	SuperHM-H	THF
Polycarbonate	SuperHM-H	THF
Polyester polyol	G2000H XL	THF
Polyetherimide	G4000HHR + G5000HHR	HFIP containing TFA-Na
Polyethersulfone	G2000HHR + G3000HHR + G4000HHR	DMF containing LiBr

TABLE 4.19 *(Continued)*

Sample	Example of column	Elution solvent[a]
Polyethylene (high density)	GMHHR-H	ODCB
Polyethylene (low density)	GMHHR-H	ODCB
Polyethylene (linear low density)	GMHHR-H	ODCB
Polyethylene terephthalate	SuperHM-M	HFIP containing TFA-Na
Polyethylmethacrylate	SuperHM-H	THF
Polyimide	GMHXL	DMF containing LiBr
Polyisoprene	GMHXL	THF
Polymethylmethacrylate	SuperHM-H	THF, HFIP containing TFA-Na
Poly(α-methylstyrene)	SuperHM-H	THF
Polyoxymethylene	SuperHM-H	HFIP containing TFA-Na
Polypropylene	GMHHR-H	ODCB
Polypropylene (isotactic)	SuperHM-H	THF
Polystyrene	SuperHM-H	THF
Polysulfone	SuperHM-H	THF
Polytetraethoxysilane	G2000H8 + G3000H8	THF
Polyurethane	G4000HXL + G5000HXL	THF
Polyvinylacetate	SuperHM-H	THF
Polyvinylbutyral	SuperHM-H	THF, HFIP containing TFA-Na
Polyvinylchloride	SuperHM-H	THF
Polyvinylfluoride	G2000HXL + GMHXL	DMF containing LiBr
Polyvinylformal	SuperHM-H	THF
Polyvinylidene fluoride	GMHXL	DMF containing LiBr
Silicone oil	SuperH2000	Toluene
Silicone resin	G1000HXL + G2000HXL + G3000HXL	THF
Unsaturated polyester	GMHXL	Chloroform/HFIP, THF
Urea resin	G1000HXL + G2000HXL + G2500HXL	DMF
Vinylacetate-vinylalcohol	SuperHM-H-vinylchloride copolymer	THF
Vinylacetate-vinylchloride	SuperHM-H copolymer	THF
Vinylacetate	SuperHM-H-N-vinylpyrrolidone copolymer	THF
Vinylchloride	SuperHM-H-vinylidenechloride copolymer	THF

[a]DMF, dimethylformamide; DMSO, dimethyl sulfoxide; HFIP, hexafluoroisopropanol; LiBr, lithium bromide; ODCB, *o*-dichlorobenzene; TFA-Na, sodium trifluoroacetate; THF, tetrahydrofuran.

purifications of globular proteins and molecular weight estimations of unknown biomolecules. These resins are semirigid, spherical beads synthesized by a copolymerization of ethylene glycol and methacrylate type polymers. The resins have hydrophilic surfaces due to the presence of ether and hydroxyl groups.

The numerous surface hydroxyl groups provide attachment points for other functional groups and ligands.

Toyopearl HW resins are available in a series of five types, HW-40, HW-50, HW-55, HW-65, and HW-75, each of which is defined by a molecular size separation range, as shown in Table 4.20. The resins are available in three particle diameter ranges: high-performance S or superfine grade (20–40 μm), F or fine grade (30–60 μm), and C or coarse grade (50–100 μm). The Toyopearl HW resin range spans peptide and protein molecular masses from 100 to 50,000,000 Da. Each Toyopearl HW resin has a characteristic calibration curve and exclusion limit for globular proteins as shown in Fig. 4.43. The properties of Toyopearl HW resins and the recommended separation range for each resin for proteins and other types of samples are listed in Table 4.20.

A. Advantages of Toyopearl HW Resins

Toyopearl HW resins overcome several disadvantages of conventional gels, which do not function well at higher flow rates or pressures, are unstable to extremes of pH, salt, and organic solvent concentrations, and can leach saccharide derivatives into the process material. For process-scale SEC, Toyopearl HW resins provide the following advantages:

- high resolution
- improved purification of polysaccharides and glycoproteins
- high mass and activity recovery for proteins
- reproducible scaleup from analytical to production levels
- high physical and chemical stability
- greater speed and reproducibility in column packing

TABLE 4.20 Properties and Molecular Weight Separation Ranges for Toyopearl HW Resins

			Molecular weight of sample		
Toyopearl resin	Particle size (μm)	Pore size (Å)	Polyethylene glycols and oxides	Dextrans	Globular proteins
HW-40S	20–40	50	100–3000	100–7000	100–10,000
HW-40F	30–60				
HW-40C	50–100				
HW-50S	20–40	125	100–18,000	500–20,000	500–80,000
HW-50F	30–60				
HW-55S	20–40	500	100–150,000	1000–200,000	1000–700,000
HW-55F	30–60				
HW-65S	20–40	1000	500–1,000,000	10,000–1,000,000	40,000–5,000,000
HW-65F	30–60				
HW-75F	30–60	>1000	4000–5,000,000	100,000–10,000,000	500,000–50,000,000

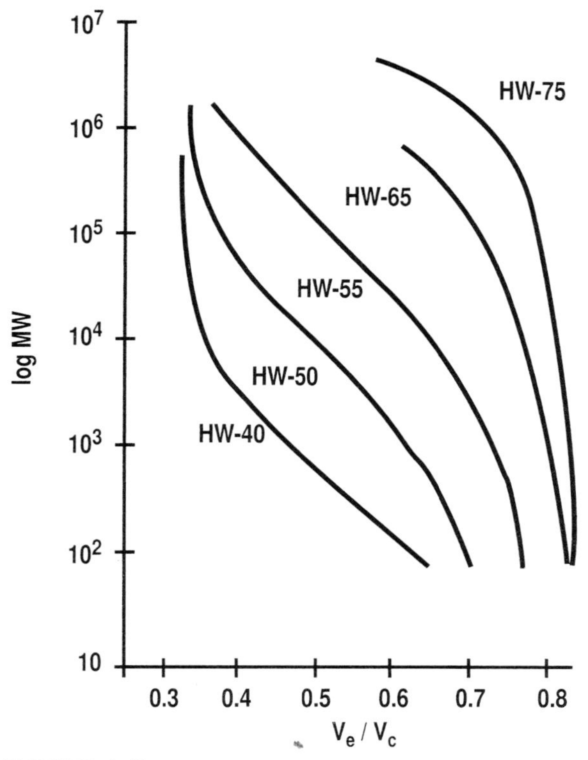

FIGURE 4.43 Calibration curves for globular proteins on toyopearl resins. Column: 22 mm × 30 cm. Sample: Protein standards. Elution: 0.06 *M* phosphate buffer, pH 7, in 0.06 *M* KCl. Legend: V_e, elution volume; V_c, column volume.

1. High Resolution

The pore characteristics and narrow particle size distribution ranges of Toyopearl HW resins provide high chromatographic resolution with sharp peaks and minimal sample dilution, in contrast to the resolution obtained with conventional soft gels. The smaller particle sizes and greater physical stability of Toyopearl resins also enable the high resolution of proteins at faster flow rates than conventional dextran gels (36). Although Toyopearl HW resins are designed primarily for aqueous GFC, they can also perform partition–adsorption chromatography when used with organic solvents. This ability adds a further dimension to the separation and improves resolution (37).

2. Improved Purifications

Like other GFC matrices, including TSK-GEL SW and TSK-GEL PW packings, and dextran and agarose gels, Toyopearl HW resins exhibit some ionic and hydrophobic interaction with samples. The hydrophobic properties of Toyopearl HW resins, however, can be utilized more effectively for improved protein purifications because, unlike numerous other GFC packing materials, Toyopearl HW resins can be used with high levels of organic solvent (38).

Purifications are made simpler with Toyopearl HW media because there is no risk of leached polysaccharides to contaminating eluted fractions. Saccharide derivatives have been known to leach from conventional low-pressure column packings, such as dextran or agarose gels.

3. High Mass and Activity Recovery for Proteins

For many proteins, especially glycoproteins, the physical characteristics, particularly the hydrophilic nature of Toyopearl HW resins, improve mass and activity recovery rates. Toyopearl HW media do not adsorb proteins, as conventional gels can, and thus do not interfere with sample recovery (39).

4. Reproducible Scaleup

The hydrophilic surface characteristics and the chemical nature of the polymer backbone in Toyopearl HW resins are the same as for packings in TSK-GEL PW HPLC columns. Consequently, Toyopearl HW packings are ideal scaleup resins for analytical separation methods developed with TSK-GEL HPLC columns. Figure 4.44 shows a protein mixture first analyzed on TSK-GEL G3000 SW_{XL} and TSK-GEL G3000 PW_{XL} columns, then purified with the same mobile-phase conditions in a preparative Toyopearl HW-55 column. The elution profile and resolution remained similar from the analytical separation on the TSK-GEL G3000 PW_{XL} column to the process-scale Toyopearl column. Scaleup from TSK-GEL PW columns can be direct and more predictable with Toyopearl HW resins.

5. Physical and Chemical Stability

The methacrylic backbone structure makes the spherical Toyopearl particles rigid, which in turn allows linear pressure flow curves up to nearly 120 psi (<10 bar), as seen in Fig. 4.45. Toyopearl HW resins are highly resistant to chemical and microbial attack and are stable over a wide pH range (pH 2–12 for operation, and from pH 1 to 13 for routine cleaning and sanitization). Toyopearl HW resins are compatible with solvents such as methanol, ethanol, acetone, isopropanol, *n*-propanol, and chloroform. Toyopearl HW media have been used with harsh denaturants such as guanidine chloride, sodium dodecyl sulfate, and urea with no loss of efficiency or resolution (40). Studies in which Toyopearl HW media were exposed to 50% trifluoroacetic acid at 40°C for 4 weeks revealed no change in the retention of various proteins. Similarly, the repeated exposure of Toyopearl HW-55S to 0.1 *N* NaOH did not change retention times or efficiencies for marker compounds (41).

6. Packing a Toyopearl HW Column

With soft gels, column packing has often been plagued with such problems as inferior reproducibility and excessive time requirements. These problems are alleviated with physically stable Toyopearl HW media. However, an improperly packed column can have significantly reduced efficiency. The two key variables for the successful packing of Toyopearl HW media, packing velocity and column size, have been evaluated to determine the optimal packing conditions.

Figure 4.46 shows the relationship between packing velocity and sample resolution using Toyopearl HW-55F and Toyopearl HW-55S columns and

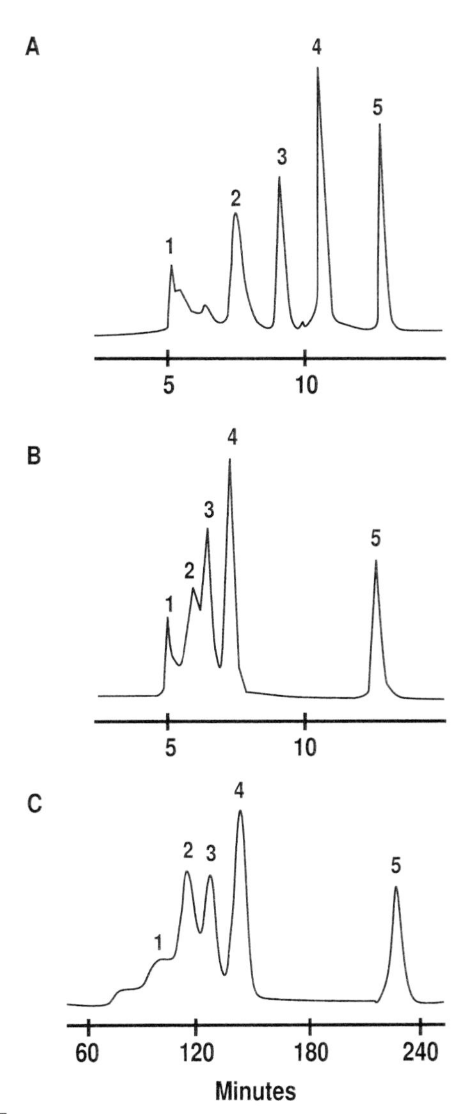

FIGURE 4.44 Scaleup of protein separation on TSK-GEL columns and Toyopearl HW-55F resin. Column: (A) TSK-GEL G3000SW$_{XL}$, 5 μm, 7.8 mm × 30 cm; (B) TSK-GEL G3000PW$_{XL}$, 6 μm, 7.8 mm × 30 cm; and (C) Toyopearl HW-55F, 30–60 μm, 22 mm × 60 cm. Sample: (1) thyroglobulin, (2) IgG, (3) ovalbumin, (4) ribonuclease A, and (5) *p*-aminobenzoic acid. Elution: 0.05 *M* phosphate buffer, pH 7, in 0.3 *M* NaCl. Flow rate: 1.0 ml/min. Detection: UV at 280 nm.

bovine serum albumin (67,000 Da) and myoglobin (18,000 Da) as sample proteins. Resolution is constant at packing velocities above the critical values shown in Fig. 4.46 and decreases with packing velocities below these critical values. Therefore, the packing velocity should generally be high. Extremely high packing velocities are unnecessary, however, and can result in high pressure drops (42,43).

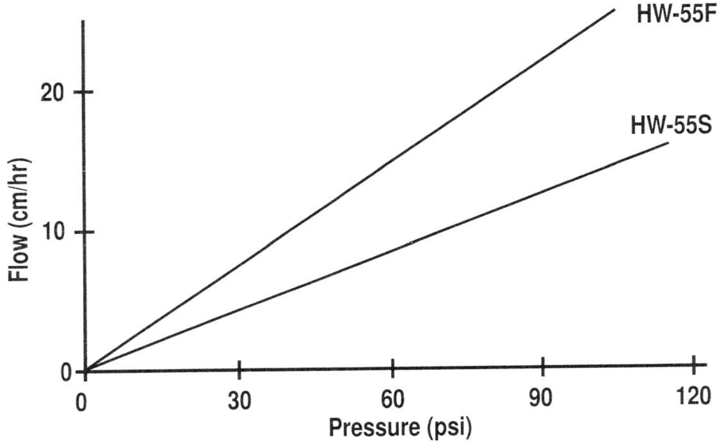

FIGURE 4.45 Linear pressure/flow curves for Toyopearl HW-55 resins. Column: Toyopearl HW-55F, 30–60 μm, 22 mm × 90 cm; Toyopearl HW-55S, 20–40 μm. 22 mm × 90 cm. Elution: Distilled water.

Column length has a large impact on the optimal packing velocity. By optimizing packing velocity, efficient long and short columns can be packed. Optimal packing velocity is related inversely to column length. Optimal packing velocity increases with decreasing column inner diameter, especially with col-

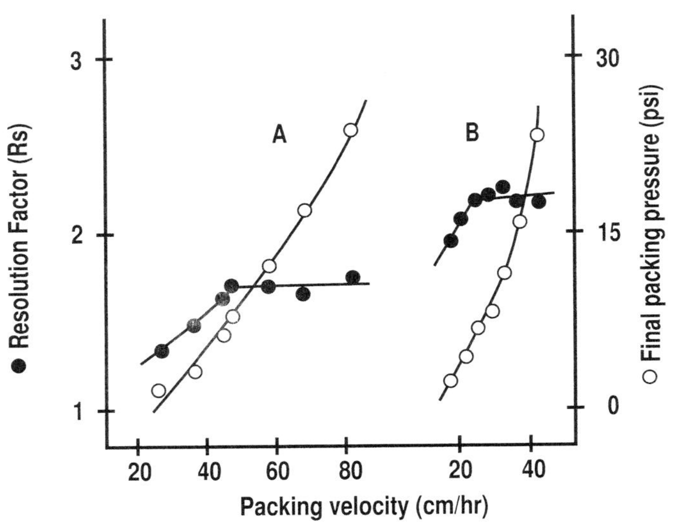

FIGURE 4.46 Influence of packing velocity and particle size on resolution factor. Column: (A) Toyopearl HW-55F, 30–60 μm, 22 mm × 60 cm and (B) Toyopearl HW-55S, 20–40 μm, 22 mm × 60 cm. Sample: 100 μl of 0.25% solution of bovine serum albumin and myoglobin. Elution: 0.1 *M* phosphate buffer, pH 7, in 0.1 *M* KCl. Flow rate: 13 cm/hr. Detection: UV at 220 nm.

TABLE 4.21 Optimal Packing Velocities for Toyopearl HW Resins

Toyopearl resin	Column size	Linear velocity (cm/hr)	Flow rate (ml/min)
HW-40F	16 mm × 60 cm	70–90	2.4–3.0
HW-50F	22 mm × 60 cm	60–80	4.0–5.0
HW-55F	16 mm × 60 cm	60–85	2.0–3.0
HW-65F	22 mm × 60 cm	50–70	3.0–5.0
HW-75F	22 mm × 60 cm	40–150	2.5–10.0

umns having a small (<16 mm) inner diameter. Table 4.21 shows the optimal velocities to use for packing Toyopearl HW media in columns of different sizes (42,43).

B. Optimizing Separation Conditions

Many operating variables, such as sample volume, flow rate, column length, and temperature, must be considered when performing any separation. The relative importance of these variables for Toyopearl HW-55F resin columns has been specifically evaluated. For example, Fig. 4.47 shows the relationship between column efficiency, or height equivalent of a theoretical plate (HETP),

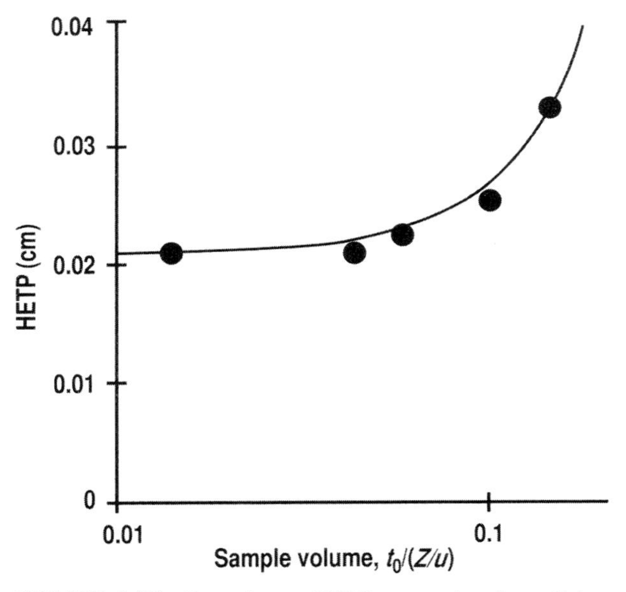

FIGURE 4.47 Dependence of HETP on sample volume. Column: Toyopearl HW-55F, 22 mm × 30 cm. Sample: 0.1% myoglobin. Elution: 14 m*M* Tris–HCl, pH 7.9, in 0.3 *M* NaCl. Flow rate: 52 cm/hr. Detection: UV at 220 nm. Legend: t_0, sample injection time; *Z*, column length; *u*, linear velocity.

and sample volume using myoglobin as the sample. HETP remains constant up to a certain sample volume. Above that volume, HETP increases rapidly and, accordingly, column efficiency decreases. The plots of efficiency against sample volume generally follow the same behavior for all general protein samples, but differ in the point at which the curve increases. Each protein has a unique critical sample concentration, above which HETP will rise rapidly (44,45).

The effect of flow rate on resolution by Toyopearl HW-55F and Toyopearl HW-55S columns has been studied using a bovine serum albumin sample. For both columns, resolution decreased with increasing flow rate (46). Resolution is increased, however, with decreasing particle size (47). Resolution is proportional to the square root of the column length, as theoretically expected, and indicates that longer columns can be packed as well as shorter columns. Therefore, for samples difficult to resolve, the solution may be to increase the column length.

Finally, temperature can have a profound effect on column performance. Tanaka *et al.* (48) studied the effects of temperature on the separation of hydrolyzed β-cyclodextrin. In their studies, resolution increased with temperature on a Toyopearl HW-40S column.

C. Applications

1. Separation of Low Molecular Weight Samples Using Toyopearl HW-40 Columns

Separation of low molecular weight nucleic acids by ion exchangers involves troublesome desalting of the eluent. Because Toyopearl HW-40S media can be used with a wider range of mobile phase eluents, a volatile solvent, which is easier to remove, could be chosen. Figure 4.48 gives an example of separating low molecular weight nucleosides and nucleotides using 0.1 *M* ammonium acetate as the elution buffer (49).

Toyopearl HW-40 media are ideal for the rapid desalting of proteins and other large molecules. This is useful for conditioning a crude sample prior to loading onto an ion exchange column or for the removal of high concentrations of salt after ion exchange or hydrophobic interaction chromatography.

The chemical stability of Toyopearl HW-40 resin to organic eluents allows this material to be used for a variety of applications, including the purification of synthetic functionalized surfactants, polyphenolics, and phenolic glycosides (50,51).

2. Separation of Medium Molecular Weight Samples Using Toyopearl HW-50 and Toyopearl HW-55 Columns

Toyopearl HW-50S resin has been used to help isolate the ubiquitin–histone conjugate *u*H2A from the unicellular ciliated protozoan *Tetrahymena pyriformis*. Figure 4.49 shows the separation of *u*H2A from the histone, H2A. The sole difference between these two components is a small polypeptide, ubiquitin (approximately 8500 Da). The *u*H2A fraction was then further purified by HPLC on a Tosoh ODS-silica column (52). One of the many benefits

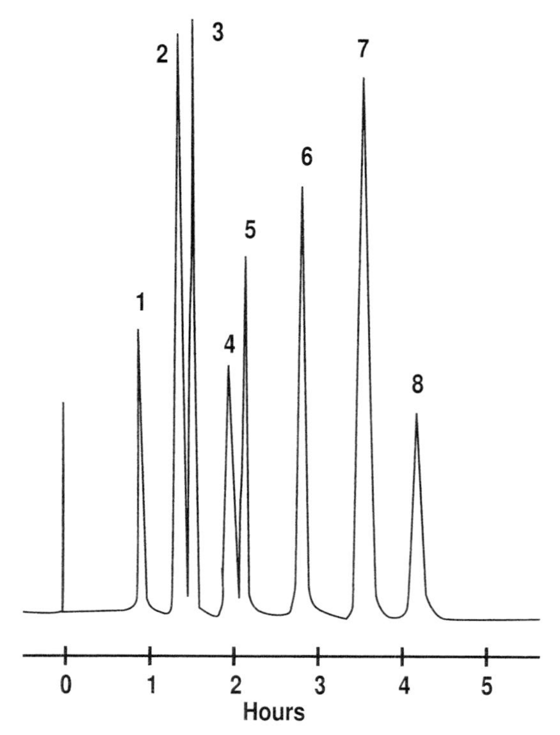

FIGURE 4.48 Separation of nucleosides and nucleotides on Toyopearl HW-40S. Column: Toyopearl HW-40S, 26 mm × 29 cm. Sample: (1) blue dextran, (2) ATP, (3) ADP, (4) 2′-AMP, (5) 3′-AMP, (6) 3′,5′-cAMP, (7) adenosine, and (8) adenine. Elution: 0.1 *M* ammonium acetate, pH 4.6. Flow rate: 80 ml/hr. Detector: UV at 280 nm.

offered by Toyopearl HW media is high recovery rates from proteins following separation.

3. Separation of Large Molecular Weight Samples Using Toyopearl HW-65 and Toyopearl HW-75 Columns

In general, Toyopearl HW media offer an advantage over commonly used polyacrylamide or agarose gels in high-speed separations of biopolymers, especially when rigorous denaturing conditions are necessary. Figure 4.50, for example, shows the purification of mRNA (approximately 300,000 Da) from *B. mori* silkworm on Toyopearl HW-65F resin using 0.5% sodium dodecyl sulfate in 6 *M* urea as the eluent. Figure 4.50 also shows the high recovery of total mRNA activity and mRNA activity coding for the major plasma proteins (53).

Toyopearl HW-75 resin, with pores larger than 1000 Å, have been used in place of ultracentrifugation steps for the purification of plasmid DNA. Ultracentrifugation is a time-consuming process and requires expensive chemicals, such as cesium chloride. Toyopearl HW-75 resin provides superior separation performance for plasmid DNA and also provides high yields (54).

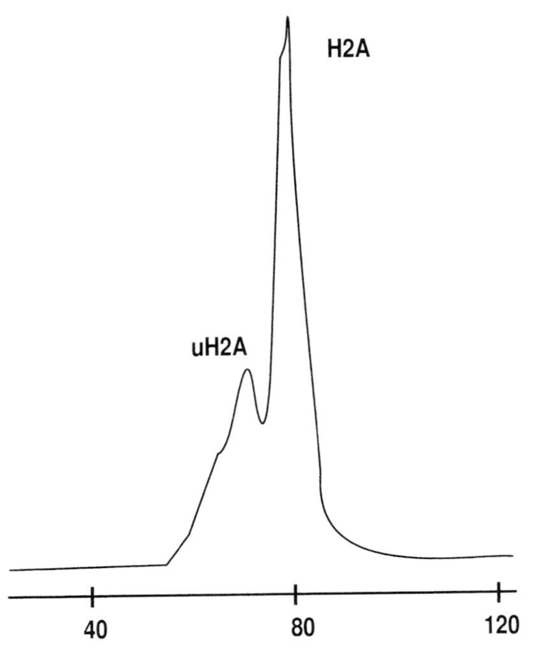

FIGURE 4.49 Isolation of a complex protein conjugate on Toyopearl HW-50S. Column: 22 mm × 83 cm. Sample: Fraction from crude *Tetrahymena* H2A containing the ubiquitin–histone conjugate uH2A. Elution: 10 m*M* HCl. Flow rate: 0.1 ml/min. Detection: UV at 230 nm.

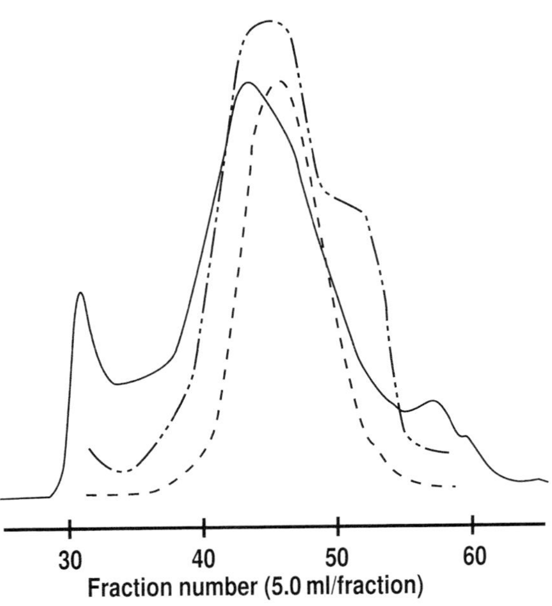

FIGURE 4.50 Purification of mRNA on Toyopearl HW-65F. Column: Toyopearl HW-65F, 25 mm × 90 cm. Sample: 40 mg of poly(A) RNA/5 ml of formamide, sample from silkworm. Elution: 20 m*M* sodium citrate, 5 m*M* EDTA, 0.5% SDS, 6 *M* urea, pH 3.5. Flow rate: 24 ml/hr. Detection: UV at 254, total mRNA activity (---), mRNA activity for major plasma proteins (– - - –).

REFERENCES

1. Watanabe, H., Umino, M., and Sasagawa, T. (1984). Scientific Report of Toyo Soda, **28**, 1–20.
2. Rokushika, S., Ohkawas, T., and Hatano, H. (1979). *J. Chromatogr.* **176**, 456–461.
3. Snyder, L. R., and Kirkland, J. J. (1979). *Introduction to Modern Liquid Chromatography*, 2nd edition, pp. 34–36, John Wiley & Sons, Inc.: New York.
4. Kato, Y., Yamasaki, Y., Moriyama, H., Tokunaga, K., and Hashimoto, T. (1987). *J. Chromatogr.* **404**, 333–339.
5. Chen, M., Wyatt Technology Corporation. (1996). "Application Notes, Volume Two", Company Publication.
6. Yamasaki, Y., and Kato, Y. (1989). *J. Chromatogr.* **467**, 436–440.
7. Kast, E., Pathmanbhan, N., Wong, J., O'Connor, B., and Klein, M. (1996). Poster presented at Protein Society 1996, "Preparative Isolation of Immunoglobulin Heavy and Light Chains by Size-Exclusion Chromatography".
8. Kato, Y., Sasaki, M., Hashimoto, T., Murotsu, T., Fukushige, S., and Matsubara, K. (1984). *J. Biochem.* **95**, 83–86.
9. Sasaki, H., Matsuda, T., Ishikawa, O., Takamatsu, T., Tanaka, K., Kato, Y., and Hashimoto, T. (1985). Scientific Report of Tosoh Corporation, **29**, 37–54.
10. TosoHaas Technical Report No. 12: "Separation of Water-Soluble Polymers by High Performance Gel Permeation Chromatography on TSK-GEL PWXL".
11. Kato, Y., Matsuda, T., and Hashimoto, T. (1985). *J. Chromatogr.* **332**, 39–46.
12. TosoHaas Technical Report No. 10: "Separation of Water-Soluble Oligomers by High Performance Gel Filtration Chromatography on TSKgel PWXL".
13. Kato, Y., Yamasaki, Y., and Hashimoto, T. (1985). *J. Chromatogr.* **320**, 440–444.
14. Muller, E., and Jork, H. (1991). Proceedings of the 6th European Conference on Food Chemistry, **2**, 557–562.
15. Alsop, R. M., and Vlachogiannis, G. J. (1982). *J. Chromatogr.* **246**, 227–240.
16. Hara, I., Okazaki, M., and Ohno, Y. (1980). *J. Biochem.* **87**, 1863–1868.
17. Okazaki, M., Ohno, Y., and Hara, I. (1980). *J. Chromatogr.* **221**, 257–264.
18. Hara, I., Shiraishi, K., and Okazaki, M. (1982). *J. Chromatogr.* **239**, 549–557.
19. Okazaki, M., Hagiwara, N., and Hara, I. (1982). *J. Biochem.* **91**, 1381–1389.
20. Okazaki, M., Shiraishi, K., Ohno, Y., and Hara, I. (1981). *J. Chromatogr.* **223**, 285–293.
21. Hiyoy, Y., Yoshida, H., and Nakajima, T. (1982). *J. Chromatogr.* **240**, 341–348.
22. Swergold, G. D., and Rubin, C. S. (1983). *Anal. Biochem.* **131**, 295–300.
23. Yamasaki, Y., Kato, Y., Murotsu, T., Fukushige, S., and Matsubara, K. (1987). *J. High Res. Chromatogr. and Chromatogr. Comm.* **10**, 45–46.
24. Himmel, M., Perna, P., and McDonell, M. (1982). *J. Chromatogr.* **240**, 155–163.
25. Ollivon, M., Walter, A., and Blumenthal, R. (1986). *Anal. Biochem.* **152**, 262–274.
26. Himmel, M., and Squire, P. (1981). *J. Chromatogr.* **210**, 443–452.
27. Alfredson, T. V., Wher, C. T., Tallman, L., and Klink, F. E. (1982). *J. Liquid Chromatogr.* **5**, 489–524.
28. Kato, Y., Sasaki, M., Hashimoto, T., Murotsu, T., Fudushige, S., and Matsubara, K. (1983). *J. Chromatogr.* **266**, 341–349.
29. Kato, Y., Kido, S., and T. Hashimoto (1973). *J. Polymer Sci. Polymer Phys. Ed.* **11**, 2329–2337.
30. Kato, Y., Kido, S., Yamamoto, M., and Hashimoto, T. (1974). *J. Polymer Sci. Polymer Phys. Ed.* **12**, 1339–1345.
31. Kato, Y., Kido, S., and Hashimoto, T. (1974). *J. Appl. Polymer Sci.* **18**, 1239–1241.
32. Kato, Y., Kido, S., and Hashimoto, T. (1974). *J. Polymer Sci. Polymer Phys. Ed.* **12**, 813–814.
33. Kato, Y., Kametani, T., Furukawa, K., and Hashimoto, T. (1975). *J. Polymer Sci. Polymer Phys. Ed.* **13**, 1695–1703.
34. Kato, Y., Kido, S., Watanabe, H., Yamamoto, M., and Hashimoto, T. (1975). *J. Appl. Polymer Sci.* **19**, 629–631.
35. Kato, Y., Kametani, T., and Hashimoto, T. (1976). *J. Polymer Sci. Polymer Phys. Ed.* **14**, 2105–2108.
36. Toyopearl News No. 7: "Peak Sharpness in Chromatography with Toyopearl Gels."
37. Sasaki, H., Asada, R., Kato, Y., and Hashimoto, T. (1983). *Toyo Soda Kenkyuhohoku*, **27**, 49–65.

38. Kato, Y., Komiya, K., and Hashimoto, T. (1982). *J. Chromatogr.* **247**, 184–186.
39. Taguchi, H., Matsuzawa, M., and Ohta, T. (1980). Data presented at the annual meeting of the Agricultural Chemical Society of Japan, (April 1980).
40. Okumura, K. and Hori, R. (1980). Data presented at the 100th Meeting of the Pharmaceutical Society of Japan, *Toyopearl News* 10.
41. Unpublished results.
42. Kato, Y., Komiya, K., Iwaeda, T., Sasaki, H., and Hashimoto, T. (1981). *J. Chromatogr.* **205**, 185–188.
43. Kato, Y., Komiya, K., Iwaeda, T., Sasaki, H., and Hashimoto, T. (1981). *J. Chromatogr.* **206**, 135–138.
44. Yamamoto, S., Nomura, M., and Sano, Y. (1986). *J. Chemical Engineering of Japan,* **19**, 227–231.
45. Yamamoto, S., Nomura, M., and Sano, Y. (1987). *J. Chromatogr.* **394**, 363–367.
46. Kato, Y., Komiya, K., Iwaeda, T., Sasaki, H., and Hashimoto, T. (1981). *J. HRC & CC* **4**, 135.
47. Gurkin, M., and Patel, V. (1982). *American Laboratory,* **14**, 64–73.
48. Tanaka, K., Kitamura, T., Matsuda, T., Yamasaki, H., and Sasaki, H. (1981). *Toyo Soda Kenkyohohoku,* **25**, 21–30.
49. Aoyama, S., Hirayanagi, K., Yoshimura, T., and Ishikawa, T. (1982). *J. Chromatogr.* **253**, 133–137.
50. Murakami, Y., and Nakano, T. *Toyopearl News* 21.
51. Ozawa, T. (1982). *Agric. Biol. Chem.* **46**, 1079–1081.
52. Fusauchi, Y., and Iwai, K. (1985). *J. Biochem.* **97**, 1467–1476.
53. Izumi, S., and Tomino, S. *Toyopearl News* 15.
54. Yoshinaga, K., and Suzuki, Y. (1983). *Agric. Biol. Chem.* **47**, 919–920.

5

AMERICAN POLYMER STANDARDS COLUMNS FOR SIZE EXCLUSION CHROMATOGRAGHY

JOHN E. ARMONAS
BRIAN H. PEABODY

American Polymer Standards Corporation, Mentor, Ohio 44060

I. INTRODUCTION

During the period of 1951 through 1964 much work was done in the development of macroporous styrene–divinylbenzene (ST-DVB) copolymers (1–9). These gels were successful in separating synthetic polymers that are soluble in organic solvents. The large particle size (70 μm) and low efficiency (300 plates per foot) resulted in poor resolution and long analysis times. Since 1964 the particle sizes of gel-permeation/size exclusion chromatography (GPC/SEC) gels have been reduced and efficiency has increased, resulting in resolution improvements with a corresponding reduction in run times. In recent years, GPC/SEC columns with 5-, 10-, and 15-μm particle sizes have become commonplace.

By 1997, GPC/SEC packings, for use with organic solvents, reached a plateau in terms of new product development. Columns are generally available in individual pore sizes ranging from 50 to 10^6 Å, as measured by the size of the polystyrene molecule that can be resolved, as well as by linear columns that contain a mixture of gels with various pore sizes. Use of particle sizes below 5 μm further increases resolution; however, column back pressure and the shear rate also increase, thus rendering any 3-μm gels useless for higher molecular weight polymers.

With large companies such as Waters Corp., TOSOH Corp., Polymer Laboratories Ltd., Shodex Corp., and Phenomenex Inc. and smaller companies such as American Polymer Standards Corp., Polymer Standards Service, and Jordi Associates, all manufacturing ST-DVB GPC/SEC gel, the market has become saturated with analogous columns. All the manufacturers just men-

Column Handbook for Size Exclusion Chromatography

tioned produce columns with very similar back pressure, resolution, and run times. No matter which company produced the column, the results will be equivalent. Therefore, the major difference in columns becomes one of economics. How much each producer spends on advertising, packaging, and so on is reflected directly in the cost of their respective columns. For the consumer, price is the only real difference in the GPC/SEC columns available commercially today, as GPC/SEC columns have become a commodity.

II. AMERICAN POLYMER GEL-PERMEATION/SIZE EXCLUSION CHROMATOGRAPHY (GPC/SEC) COLUMN REPACKING

Many researchers choose to buy expensive GPC/SEC columns from one of the major producers because that producer's columns had been used in the past or because of a successful marketing campaign by one particular producer. It should be noted that repacked columns can be obtained for a fraction of the cost of new columns. American Polymer Standards repacked columns are guaranteed to perform just as well as new columns from any company. When a column is repacked the only parts reused are the stainless-steel tube and end caps. This hardware is then repacked using new frits and new ST-DVB gel. Each column is individually tested in a quality control laboratory and shipped in the customer's choice of solvent. American Polymer Standards offers a column repacking service because it is a practical, inexpensive way for customers to acquire state of the art GPC/SEC columns.

Without sacrificing performance, researchers can save money with repacked GPC/SEC columns in several ways. The most important savings comes from the reuse of the expensive, durable stainless-steel hardware. As long as the hardware has a zero dead volume union it will not contribute to or detract from column performance. Valco, Parker, Upchurch, Swagelok, Waters, and others all produce columns with zero dead volume unions. Column performance will be determined by the gel and frits inside the hardware. Both the gel and the frits are replaced as part of any sound column repacking procedure. Columns repacked by American Polymer Standards are tested for theoretical plates, asymmetry, and back pressure. This assures the end user of equal performance compared to any new column. American GPC/SEC columns are compatible with a variety of organic solvents, including dimethylformamide (DMF), NMP, dimethy sulfoxide (DMSO), hexafluoroisopropanol (HFIP), acetone, toluene, trichlorobenzene (TCB), *o*-dichlorobenzene (ODCB), chloroform, tetrahydrofuran (THF), and others. Figures 5.1–5.4 show examples of work done on American Polymer Standards repacked GPC/SEC columns in DMSO, THF, toluene, and NMP, respectively.

The point to consider with repacked GPC/SEC columns is the warranty. Because the columns are fragile, warranties are very important. American Polymer Standards offers a 90-day unconditional warranty on every repacked column. This type of warranty is especially useful for beginners, who would typically make mistakes within the first 3 months of GPC/SEC work.

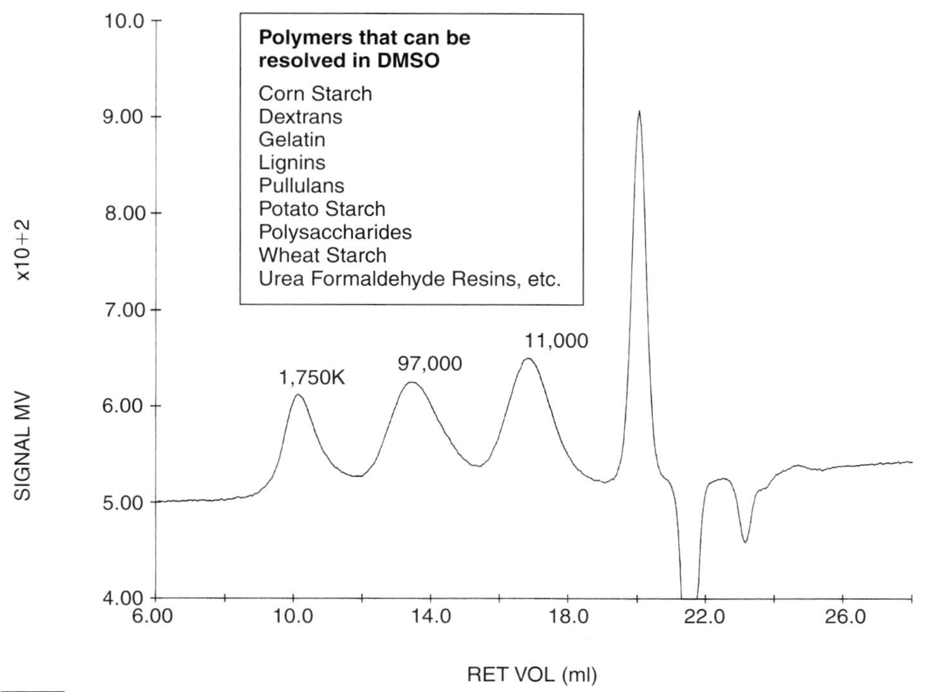

FIGURE 5.1 Analysis of American Polymer Standards dextran standards, two columns AMGEL Linear 300 × 7.8 mm, eluant DMSO, flow rate 1 ml/min, temperature 50°C, detector (DRI).

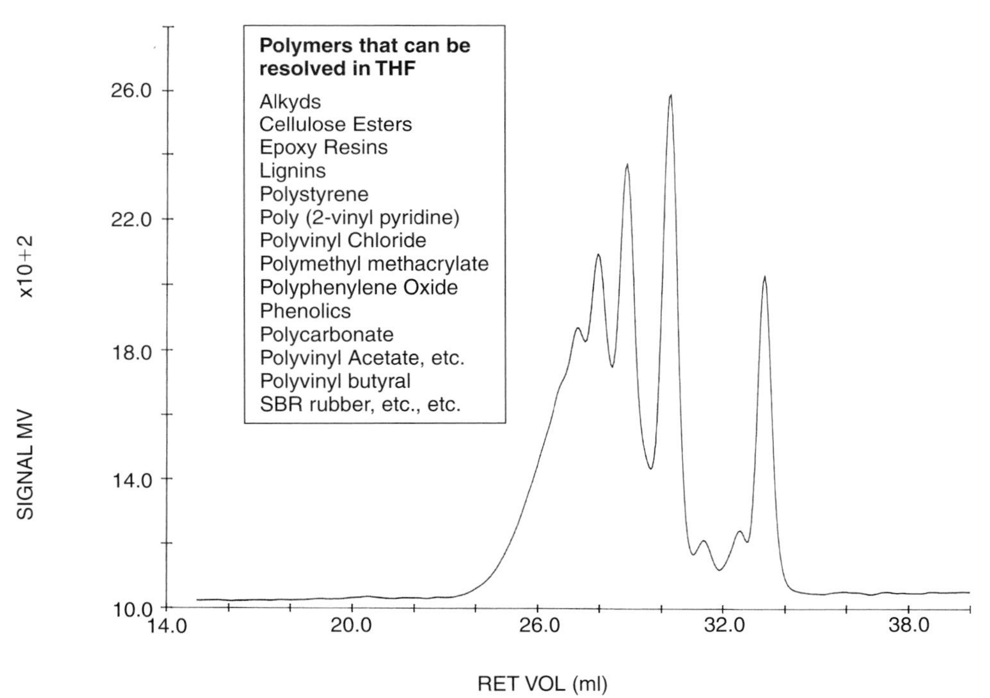

FIGURE 5.2 Analysis of epoxy resin, three columns AMGEL 10^4 Å, AMGEL 10^3 Å, AMGEL 500Å 300 × 7.8 mm, eluant THF, flow rate 1 ml/min, temperature 30°C, detector (DRI).

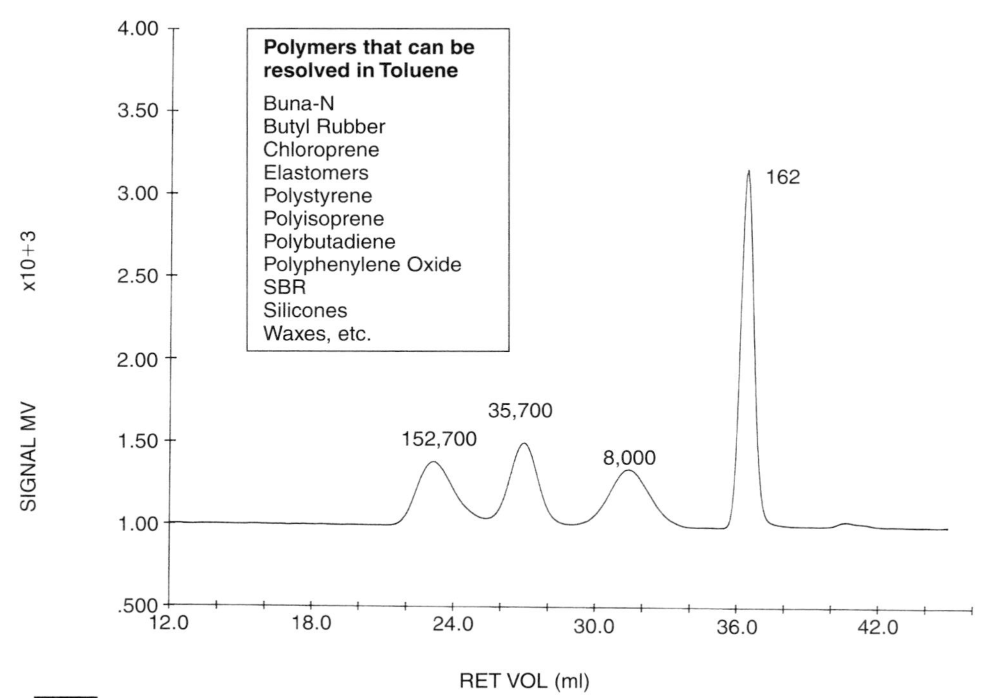

FIGURE 5.3 Analysis of American Polymer Standards polydimethylsiloxane standards, three columns two AMGEL Linear, one AMGEL 500Å 300 × 7.8 mm, eluant toluene, flow rate 1 ml/min, temperature 50°C, detector (DRI).

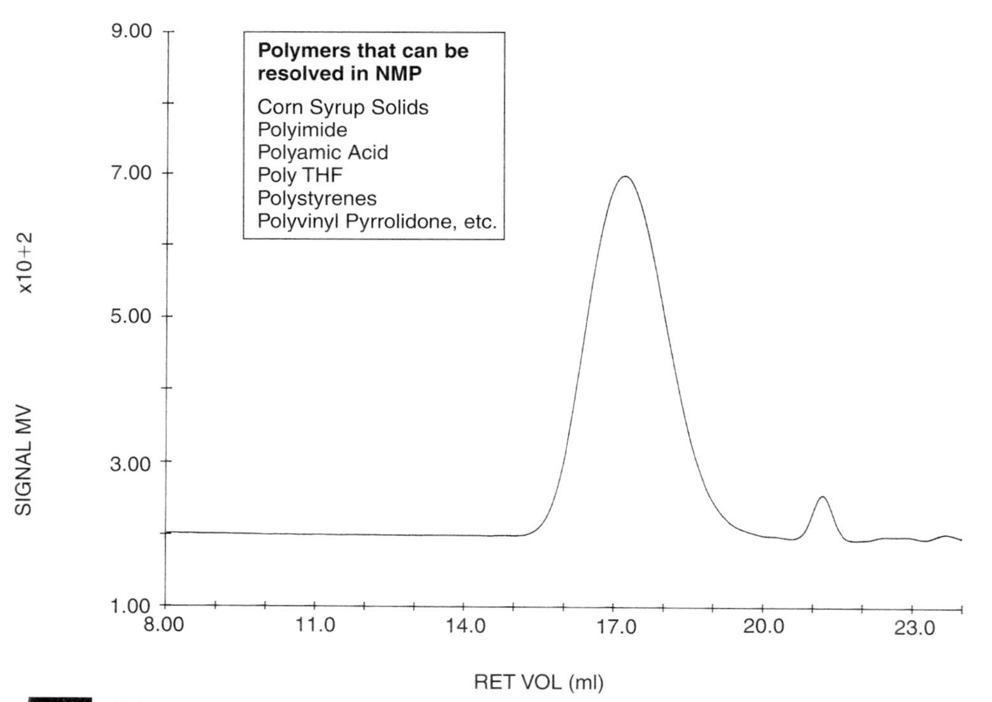

FIGURE 5.4 Analysis of polyamic acid, two columns AMGEL Linear 300 × 7.8 mm, eluant NMP, flow rate 1 ml/min, temperature 50°C, detector (DRI).

III. GPC/SEC STYRENE–DIVINYLBENZENE GEL

The literature contains numerous articles on how to prepare macroporous styrene–divinylbenzene copolymers that can be used as packings for the separation of nonionic substances by molecular size (1–9). Pepper in 1951 (1) and Pepper and Reichenberg in 1953 (2) both described the method for suspension polymerization of ST-DVB into porous copolymer spheres. In 1953, Wheaton, Bauman, and Ann (3) were the first to report the separation of nonionic substances by molecular size using ST-DVB spheres. This work was done with low molecular weight materials. In 1956, Lathe and Ruthven (4) reported the separation of high molecular weight materials using ST-DVB spheres. Mikes (5) reported the preparation of three-dimensional polymer networks using styrene and divinylbenzene monomers in 1958. Meitzuer and Oline (6) described the use of a nonsolvent as diluent to prepare gels with good mechanical stability and permeability in a 1959 patent application. In 1960 Millar (7) described the large-scale preparation of spherical porous resins from ST-DVB. Vaughan (8) reported that samples dissolved in organic solvents were separated by molecular weight differences.

In 1964 Moore (9) coined the term gel-permeation chromatography in his landmark article. For the first time Moore described GPC and the use of ST-DVB spheres as packings for GPC columns. Three patents were issued to the Dow Chemical Company covering the manufacture and use of GPC gels (10–12).

IV. MANUFACTURE OF GPC/SEC GELS

This section deals with the procedure used by American Polymer Standards Corp. in the manufacture of GPC/SEC gels. The reaction is performed in a three-neck flask equipped with a reflux condenser, a mechanical stirrer, and a thermometer. First, prepare the water phase and then the organic phase. After mixing the organic phase into the water phase, stir at 300 to 400 rpm for 2 hr at 40°C. Heat to 70°C and continue mixing at 150 rpm for 10 hr. Cool to room temperature, (RT), dilute with water, and filter. Wash the gel with water, acetone, toluene, and again with acetone. Dry at 70°C for 12 hr, classify the gel, and package.

A. Discussion

A key factor in doing a successful suspension polymerization is the composition of the aqueous phase or stabilizer. Too much stabilizer results in emulsion polymerization, which produces small particles (less than 1 μm). Too little stabilizer results in bulk polymerization. For the production of GPC gels, the ratio of aqueous phase to organic phase should be about 2 : 1.

Individual components in the formulation of the aqueous phase all contribute to the successful production of a GPC/SEC gel. The stabilizer acts as a protective coating to prevent the agglomeration of the monomer droplets. Polyvinyl alcohol, gelatin, polyacrylic acids, methylcellulose, and hydroxypropyl methylcellulose are among the polymers commonly used as stabilizers. The hot water allows the stabilizer to disperse without forming clumps. NH_4OH helps build the viscosity of the stabilizer solution. The vacuum helps remove trapped air from the viscous solution, thereby preventing the formation of hollow particles.

One important factor to consider in the preparation of the organic phase is the presence of inhibitors in the monomers. Some formulae call for the removal of inhibitors, primarily TCB, from the monomers. The TCB inhibitor forms highly colored complexes with metallic salts rendering the final product colored. Styrene has about 50 ppm of TCB. DVB, being more reactive, contains about 1000 ppm of TCB. There are several options for the removal of inhibitors. Columns packed with DOWEX MSA-1 or DOWEX 11 ion-exchange resins (Dow Chemical Company) can be used. White drierite or activated alumina also works well.

If the inhibitor is not removed, dissolved oxygen reacting with the inhibitor TCB will prevent polymerization. In the absence of dissolved oxygen, polymerization will proceed at essentially the same rate as that of the uninhibited monomer. A nitrogen blanket can be used to remove the oxygen from the mixture prior to polymerization.

Several components of the organic phase contribute greatly to the character of the final product. The pore size of the gel is chiefly determined by the amount and type of the nonsolvent used. Dodecane, dodecanol, isoamyl alcohol, and odorless paint thinner have all been used successfully as nonsolvents for the polymerization of a GPC/SEC gel. Surfactants are also very important because they balance the surface tension and interfacial tension of the monomer droplets. They allow the initiator molecules to diffuse in and out of the droplets. For this reason a small amount of surfactant is crucial. Normally the amount of surfactant in the formula should be from 0.1 to 1.0 weight percent of the monomers, as large amounts tend to emulsify and produce particles less than 1 μm in size.

During the actual preparation of the GPC/SEC gel, there are several noteworthy items in the procedure. When combining aqueous and organic phases, always pour the organic phase into the water phase as the reverse procedure produces very large particles. This mixture must be held at 40°C to prevent the initiator from starting the reaction before the right size particles are formed. Rotor speed determines the particle size of the spheres: the faster the speed the smaller the particles. Constant torque mixers produce the best results with more narrow particle-size distributions. The initial mixture should be stirred at 300–400 rpm to ensure a particle-size distribution from 2 to 20 μm.

Once the actual reaction begins, a defoamer can be used, but it is simpler not to overfill the reaction flask. Pressure vessels allow the use of higher temperatures and result in much shorter reaction times, simultaneously overcoming the foaming problems. After the reaction is complete and the GPC/SEC gel has been cleaned it is important not to dry the gel above 85°C as self-ignition can occur.

B. Experimental

Formulae for the various pore size gels all have the same basic composition. The aqueous phase consists of water, methocel, and NH_4OH (28%). The organic phase consists of DVB, toluene, nonsolvent, surfacant, and initiator. Note in Table 5.1 that the only difference between a small pore gel and a large pore gel is the replacement of some toluene with a nonsolvent.

To prepare the aqueous phase, one should begin by taking one-third of the water and heating it to 90°C. With good mixing at 120 rpm, slowly introduce the methocel into the hot water. Once the methocel is well dispersed, add the remaining two-thirds of cold water and cool the mixture to 25°C. Mix at 25°C until the solution becomes dear. Add the NH_4OH to the solution and mix for at least 30 min. Heat the water phase to 40°C and apply a vacuum for at least 30 min.

The preparation of the organic phase is done in a second flask by mixing the toluene, nonsolvent (dodecane, etc.), DVB, styrene or any other monomer, surfactant, and azobis. Mix at 120 rpm under a nitrogen blanket and bring the solution to 40°C.

The actual polymerization should be performed in a clean glass flask with at least three necks. The flask should be equipped with a reflux condenser, a thermometer, a stirring motor capable of 400 rpm with a banana-type blade, and a heating source that can keep constant temperature at 40° and 70°C. Begin by pouring the water phase into the reaction flask and adjust the temperature to 40°C. Slowly pour the organic phase into the reaction flask. Start the nitrogen purge, keeping the temperature at 40°C, and begin mixing at 300–400 rpm. At this point it is vital to hold at 40°C and continue mixing at 300–400 rpm for 2 hr.

The actual reaction begins when the mixing is slowed to 150 rpm and the temperature is increased to 70°C. Some foaming may occur once the reaction mixture reaches 70°C. Hold at 70°C for at least 10 hr to complete the reaction. At this point the nitrogen bleed can be discontinued.

After the reaction is completed, the mixture should be cooled to RT and diluted with water. The gel should separate by floating up. If this does not happen the organic solvents need to be steam distilled off first. Filter off the floating gel and wash it with water, acetone, toluene, and again with acetone.

TABLE 5.1 Formulae for the Production of GPC/SEC Gels

Phase	Small pore gel	Large pore gel
Aqueous		
Water	5000 ml	5000 ml
Methocel[a]	100 g	100 g
NH_4OH	10 ml	10 ml
Organic		
DVB (80%)	750 ml	750 ml
Toluene	1500 ml	750 ml
Dodecane	0	750 ml
Surfactant (Tween 80)[b]	10 ml	10 ml
Azobis[c]	25 g	25 g

[a]Dow Chemical Company.
[b]Aldrich Chemical Company.
[c]2,2-Azobis(2-methylpropionitrile).

After the final acetone wash, the gel can be dried at 70°C, using a vacuum if possible.

C. Optimizing the Formulation

Because all the variables that influence the properties of the final product are known, one can use a statistical design (known as a one-half factorial) to optimize the properties of the GPC/SEC gels. Factorial experiments are described in detail by Hafner (10). For example, four variables at two levels can be examined in eight observations. From these observations the significance of each variable as related to the performance of the gel can be determined. An example of a one-half factorial experiment applied to the production of GPC/SEC gel is set up in Table 5.2. The four variables are the type of DVB, amount of dodecane, type of methocel, and rate of stirring.

Because 100% DVB is not available commercially, one has to choose either 56% DVB or 80% DVB for the preparation of GPC/SEC gels. The potential components of commercial divinylbenzene are shown in Fig. 5.5. The composition of commercial 56% DVB consists of 56% DVB isomers and 44% ethylstyrene and diethylbenzene isomers. The diethylbenzene is nonreactive, but the ethylstyrene acts as "styrene" in the formulation and tends to produce softer, more swellable GPC gels. Therefore, it is not necessary to actually add styrene to the formulation to produce a ST-DVB gel. Because 80% DVB has only 20% ethylstyrene and diethylbenzene isomers the GPC/SEC gel produced using it would be more rigid and less swellable.

Other variables in the factorial experiment also have an impact on the character of the final product. The amount of nonsolvent is a very important variable to examine as the pore size of the gel depends on the amount of it present in the formulation. The stabilizer acts as a suspending agent and influences the particle size of the GPC/SEC gel. Lower viscosity suspending agents

TABLE 5.2 One-Half Factorial Experiment for Optimization of GPC/SEC Gels

Experiment	Variable 1[a]	Variable 2[b]	Variable 3[c]	Variable 4[d]
1	−	−	−	−
2	+	−	−	+
3	−	+	−	+
4	+	+	−	−
5	−	−	+	+
6	+	−	+	−
7	−	+	+	−
8	+	+	+	+

[a]Type of DVB (+ value is 80%,− value is 56%).
[b]Amount of nonsolvent (+ value is 0, − value is replace 50% of toluene).
[c]Type of methocel (+ value is K-100, − value is F-50).
[d]Rate of stirring (+ value is 150 rpm, − value is 300 rpm).

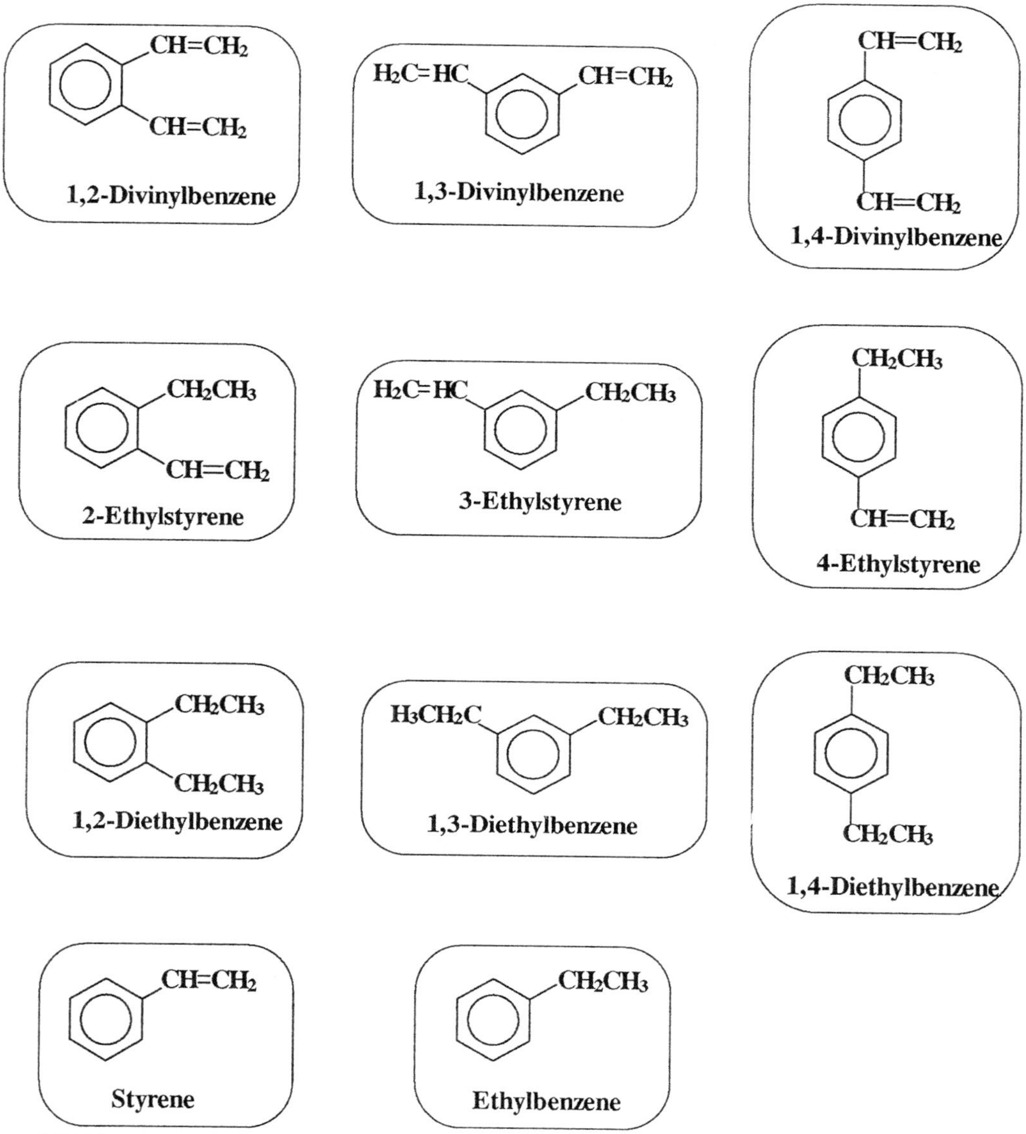

1,2-Divinylbenzene

1,3-Divinylbenzene

1,4-Divinylbenzene

2-Ethylstyrene

3-Ethylstyrene

4-Ethylstyrene

1,2-Diethylbenzene

1,3-Diethylbenzene

1,4-Diethylbenzene

Styrene

Ethylbenzene

FIGURE 5.5 Potential components of commerical divinylbenzene, DVB isomers, ethylstyrene isomers, diethylbenzene isomers, styrene, and ethylbenzene.

tend to produce smaller particles. However, the rate of stirring during the droplet formation stage has a more dramatic effect on particle size than the suspending agent. Particle size and rotor speed have an inverse relationship. The use of the factorial experiment allows the researcher to determine the optimal formulation for a given pore size gel or to develop different pore size gels.

D. Classification of the GPC/SEC Gel

Once the GPC/SEC gel has been prepared, cleaned, and dried it needs to be classified to narrow the particle size distribution. Most GPC/SEC gel preparations have a particle size distribution ranging from 1 to 50 μm. At present, most practical columns have one of three particle sizes. Five-micron particle columns are generally used for lower molecular weight samples (500,000 MW and lower). These columns can also be used for polymers with higher molecular weights, but one should be aware of the potential for sheering of the polymer molecules. Ten-micron columns are excellent for most polymeric samples at room and high temperatures with both fluid and viscous solvents. Fifteen-micron columns are normally used for very high molecular weight samples (above 1,000,000 MW) and for high temperature work. The GPC/SEC gel particle classification is done most easily by air classification using a Donaldson Acucut Classifier or a comparable unit.

E. Packing and Testing GPC/SEC Columns

Once the gel is classified, it can be packed into GPC/SEC column hardware to produce the desired particle size distribution. Packing of GPC/SEC columns with the desired efficiency can be difficult at times. Over the years, researchers have tried a number of techniques with various degrees of success. The balanced slurry method allows the GPC/SEC gel to be suspended in the solvent mix so that settling due to gravity does not influence the packing of the GPC/SEC column. A mixture of 3 volume parts acetone and 2.5 volume parts perchloroethylene will produce a balanced slurry as the density of most GPC/SEC gels is 1.0 to 1.1 g/cc. A mixture of 52.7 volume parts toluene with 47.3 volume parts chloroform also produces a balanced slurry. Although the balanced slurry method has been used with good results, most GPC/SEC gels can be packed with a nonbalanced slurry. Solvents for nonbalanced slurries include MEK, toluene/isopropanol, methylene/*n*-heptane, and so on. A good starting point for making the slurry is 10 g of dry gel in 200 ml of solvent. This amount is sufficient to pack a single 7.8 mm × 30 cm column.

Once a column has been packed successfully it needs to be evaluated for performance. The most common way of evaluating the performance of a GPC/SEC column is to calculate the theoretical plates. Most manufacturers use the formula

$$N = [5.54(D/W)]^2,$$

where N is the number of plates per 30 cm, D is the distance from injection to pea elution, and W is the peak width at half height. The marker for injection should be a 3% solution of ODCB for 100 to 10^5-Å columns. A 5% solution of dicyclohexyl phthalate should be used for 10^6-Å and linear columns. The dicyclohexyl phthalate is necessary for the linear columns due to the high percentage of the 10^6-Å gel in the linear gel. Columns should also be checked for back pressure. The back pressure for any one GPC/SEC column should not exceed 500 psi.

After packing and testing the column, one may find that the plate count or the peak symmetry is unsatisfactory. In this case some adjustments need to be made when packing the next column. If one observes tailing on the inlet side of the peak on testing, the amount of gel should be increased for the next packing. One might also try packing for a longer time or decreasing the flow rate by 5–20%. If tailing is observed on the outlet side of the peak, one would generally make the opposite adjustments. These adjustments in the packing procedure would include decreasing the amount of gel, packing for a shorter time, and increasing the flow rate by 5–20%.

V. CONCLUSION

With many large manufacturers producing equivalent, high-efficiency GPC/SEC columns, the market has become saturated. Individual researchers consuming these columns are overwhelmed with each manufactures' marketing claims and campaigns. The bottom line for the consumer should be price alone as there is no difference in performance among the commercially available columns. American Polymer Standards repacked columns offer the consumer new column performance at a fraction of the cost of a new column. This is accomplished by using new GPC/SEC gels in the regeneration process. Another way researchers can use state of the art columns at a fraction of the cost of new columns is to produce their own gel. With the formulations and instructions offered here one could reproduce all the individual porosities and particle sizes quite easily.

When purchasing GPC/SEC columns it is imperative to realize that no manufacturer has a technical edge on another. The products are analogous, with the primary difference being price. When one buys a GPC/SEC column, it is the service of gel preparation and column packing that is really purchased. We have provided the formulae and directions for the preparation of GPC gels because we do not want our customers to pay for a service that they themselves can do for less.

REFERENCES

1. Pepper, J. (1951). *J. Appl. Chem.* **1**, 124.
2. Pepper, J. and Reichenberg, Z. (1953). *Elektrochem* **57**, 183.
3. Wheaton, R. M., Bauman, W. C., and Ann, N. V. (1956). *Acad. Sci.* 57, 159.
4. Lathe, G. M., and Ruthven, C. R. (1956). *Biochem. T.* **62**, 665.
5. Mikes, H. (1958). *J. Polym. Sci.* **30**, 615.
6. Meitzner, E. F., and Online, T. A. Union S. Africa (Patent Application 59/2393 May 19, 1959), to Rohm and Haas.
7. Millar, J. (1960). *Polym. Sci.* 1311.
8. Vaughan, H. F. (1960). *Nature* 55.
9. Moore, J. C. (1964). *J. Polym. Sci. A* **2**, 835.
10. U.S. Patent 3,322,695, May 30, 1967.
11. U.S. Patent 3,326,875, June 20, 1967.
12. U.S. Patent 3,649,200, March 4, 1972.

6

Shodex COLUMNS FOR SIZE EXCLUSION CHROMATOGRAPHY

HIROSHI SUZUKI

Shodex Group, Showa Denko K.K. 13-9, Minato-ku, Tokyo 105-8518, Japan

SADAO MORI

Faculty of Engineering, Department of Industrial Chemistry, Mie University, Tsu Mie 514, Japan

I. INTRODUCTION

Showa Denko K.K. started the Shodex HPLC business in 1973 by developing columns to determine the molecular weight distributions of polymers produced at its petrochemical plant. Since then, more than 600 items of columns have been developed to achieve various kinds of analyses. Among them are several series of columns that can be used for size exclusion chromatography. The abundant variety of columns is one of the important characteristics of Shodex. Any kind of analytical requirements can be satisfied by choosing the appropriate column supplied by Showa Denko.

II. GEL TECHNOLOGY OF Shodex COLUMNS

Four series of Shodex columns are suited for size exclusion chromatography.

Series name	Packing material
GPC KF (K, KD, HFIP, UT, HT, AT)	Styrene–divinylbenzene copolymer
OHpak SB-HQ	Polyhydroxymethacrylate
Asahipak GS/GF -HQ	Polyvinyl alcohol
PROTEIN KW	Polyhydroxylated silica

The hydrophobicity of the gel surface differs depending on the type of gel. In polymer-based series, the GPC KF series has the most hydrophobic gel surface whereas the SB-HQ series has the most hydrophilic gel surface. In order to obtain an ideal GFC separation mode, a suitable column should be chosen whose gel surface has hydrophobicity similar to that of the sample.

III. Shodex ORGANIC GPC SERIES

A. Column Characteristics

Shodex has a wide variety of columns for organic GPC using organic solvents. The columns are packed with porous styrene–divinylbenzene copolymer gels especially developed for GPC use. Five types of standard-size GPC columns packed with different solvents are available. Downsized GPC columns are also available.

1. Standard-Size GPC Columns

Figure 6.1 shows the calibration curves of the KF-800 series using standard polystyrene. Appropriate columns can be chosen according to the molecular weight range of samples. KF-806M is packed with mixed gel and can be used for samples having a wide molecular weight distribution. In order to ensure good resolution, columns are available that are packed with different types of solvents such as quinoline, *o*-dichlorobenzene, ethyl acetate, tetrachloroethane, and dimethylacetamide (Table 6.1).

2. Linear-Type Columns

Linear type columns are especially designed to have wider linear molecular weight ranges. These linear-type columns are highly recommended for correcting nonlinear sections of molecular weight calibration curves (Table 6.2).

Figure 6.2 shows the calibration curves of the linear series using standard polystyrene. The high linearity of the calibration curves helps in calculating the molecular weight of polymers with a wide molecular weight distribution. The KF-800L linear series has improved the efficiency of oligomer domain separation.

Figure 6.3 shows a comparison of elution patterns of standard polystyrene between a linear-type column and a standard-type column. Because of the high linearity of its calibration curve, the linear series has improved the efficiency of oligomer domain separation.

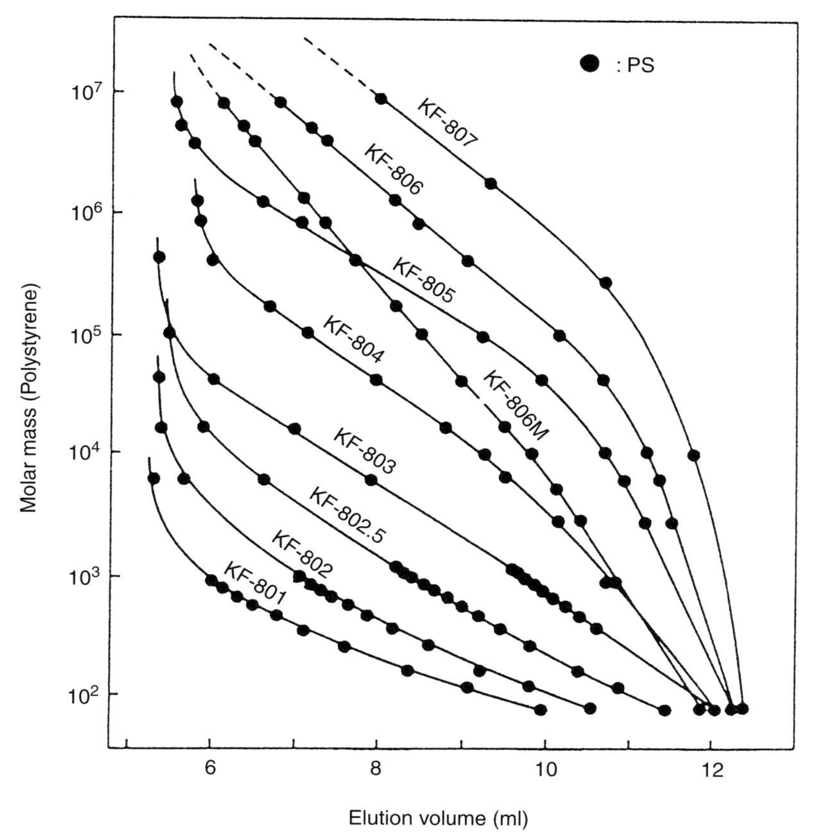

FIGURE 6.1 Calibration curves of Shodex GPC KF-800 series. Column: Shodex GPC KF-800 series 8 mm i.d. × 300 mm. Eluent: THF. Sample: Polystyrene standards.

3. High-Temperature GPC Columns

Shodex AT-800, HT-800, and UT-800 series are designed for use at high temperature of up to 140, 140, and 210°C, respectively. These columns are suited for the analysis of polymers such as polyolefins, which are insoluble at room temperature. The UT-800 series is recommended if the sample contains ultrahigh molecular weight components (Table 6.3).

Figure 6.4 (page 178) shows the chromatograms of high density polyethylene A using a multiangle laser light-scattering detector (MALLS) and refractive index (RI) detector. Polymers having a molecular weight range of up to about 5×10^6 can be detected using a conventional-type column. However, it has been revealed that an ultrahigh molecular weight component with a range of about 2×10^7 is contained in the polymer using UT-806M. This is because intrinsic

TABLE 6.1 Specifications of Shodex GPC KF, K, KD-800 Series

In-column solvents			Column size (mm)	Theoretical plate number	Exclusion limit Polystyrene PEG[a]	Particle size (μm)	Pore size (Å)	Flow Rate (ml/min)		Maximum pressure (kgf/cm^2)	Maximum temperature (°C)
THF	Chloroform	DMF						Range	Max		
KF-801	K-801	KD-801	8 × 300	>16,000	1,500[b]	6	20	0.5–1.0	2.0	35	60
KF-802	K-802	KD-802	8 × 300	>16,000	5,000	6	60	0.5–1.0	2.0	35	60
KF-802.5	K-802.5	KD-802.5	8 × 300	>16,000	20,000	6	80	0.5–1.0	2.0	35	60
KF-803	K-803	KD-803	8 × 300	>16,000	70,000	6	100	0.5–1.0	2.0	35	60
KF-804	K-804	KD-804	8 × 300	>16,000	400,000	7	200	0.5–1.0	2.0	35	60
KF-805	K-805	KD-805	8 × 300	>10,000	4,000,000	10	500	0.5–1.0	2.0	35	60
KF-806	K-806	KD-806	8 × 300	>10,000	(20,000,000)[c]	10	1000	0.5–1.0	2.0	35	60
KF-807	K-807	KD-807	8 × 300	>5,000	(200,000,000)[c]	18	>1000	0.5–1.0	2.0	35	60
KF-G	K-G	KD-G	4.6 × 10		Guard column	8		0.5–1.0	2.0	35	60

[a] For KF-800 and K-800, polystyrene is used with THF. For KD-800, polyethylene glycol is used with DMF.
[b] In the case of KD-801, the exclution limit is 2500.
[c] Figure in parentheses is estimated value.

TABLE 6.2 Specifications of Shodex GPC Linear Columns

In-column solvents						Flow rate (ml/min)			
THF	Chloroform	Column size (mm)	Theoretical plate number	Exclusion limit Polystyrene[a]	Particle size (μm)	Range	Max	Maximum pressure (kgf/cm^2)	Maximum temperature (°C)
KF-803L	K-803L	8 × 300	>16,000	70,000	6	0.5–1.0	2.0	35	60
KF-804L	K-804L	8 × 300	>16,000	400,000	7	0.5–1.0	2.0	35	60
KF-805L	K-805L	8 × 300	>10,000	4,000,000	10	0.5–1.0	2.0	35	60
KF-806L	K-806L	8 × 300	>10,000	(20,000,000)[b]	10	0.5–1.0	2.0	35	60
KF-806M	K-806M	8 × 300	>12,000	(20,000,000)[b]	10	0.5–1.0	2.0	35	60
KF-807L	K-807L	8 × 300	>5,000	(200,000,000)[b]	18	0.5–1.0	2.0	35	60

[a]Polystyrene is used with THF.
[b]Figure in parentheses is estimated value.

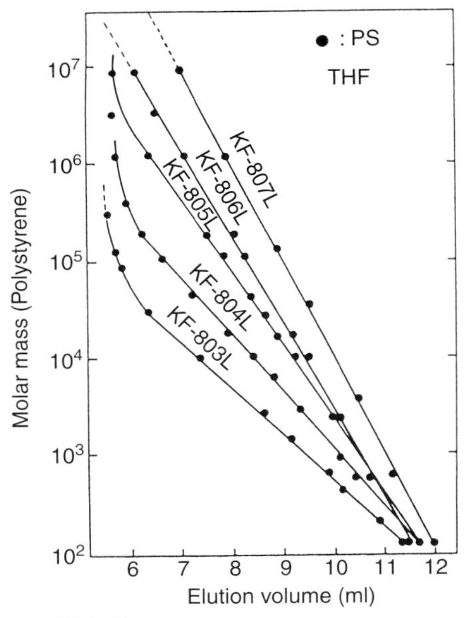

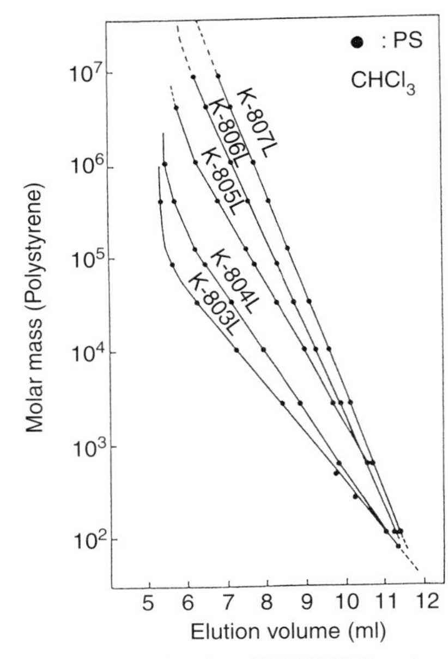

FIGURE 6.2 Calibration curves of GPC linear series. Column: *Left:* Shodex GPC KF-800L series, 8 mm i.d. × 300 mm. *Right:* Shodex GPC K-800L series, 8 mm i.d. × 300 mm. Eluent: *Left:* THF. *Right:* $CHCl_3$. Sample: Polystyrene standards.

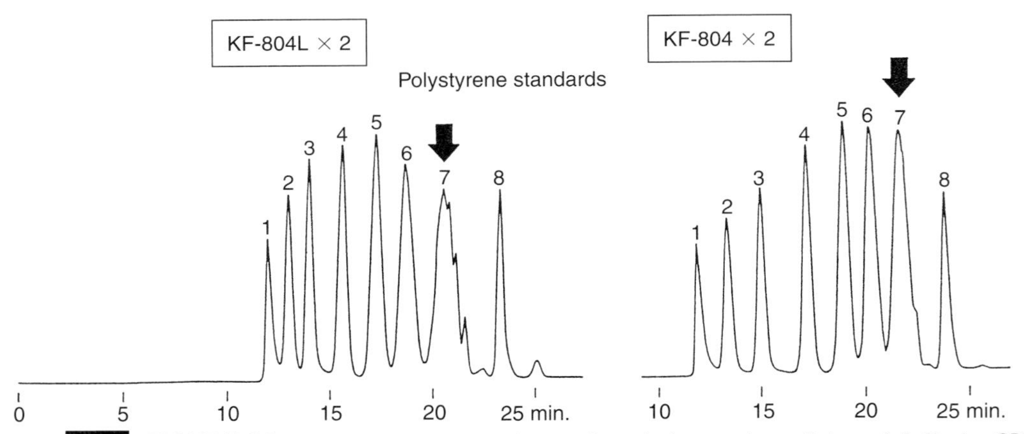

FIGURE 6.3 Comparison of linear column and standard-type column. Column: *Left:* Shodex GPC KF-804L × 2, 8 mm i.d. × 300 mm × 2. *Right:* Shodex GPC KF-804 × 2, 8 mm i.d. × 300 mm × 2. Eluent: THF. Flow rate: 1.0 mL/min. Detector: Shodex UV (254 nm). Column temp.: Ambient. Sample: 100 μL: (1) Polystyrene, 501,000; (2) Polystyrene, 156,000; (3) Polystyrene, 66,000; (4) Polystyrene, 21,900; (5) Polystyrene, 7,000; (6) Polystyrene, 2,400; (7) Polystyrene, 580; (8) Ethylbenzene, 106.

TABLE 6.3 Specifications of Shodex GPC HT, UT-800 Series

In-column solvent							Flow Rate (ml/min)		
Toluene	Usable temperature	Column size (mm)	Theoretical plate number	Exclusion limit Polystyrene[a]	Particle size (μm)	Pore size (Å)	Range	Max	Maximum pressure (kgf/cm^2)
HT-803	100–140	8 × 300	>6000	70,000	13	100	0.5–1.0	1.5	15
HT-804	100–140	8 × 300	>6000	400,000	13	200	0.5–1.0	1.5	15
HT-805	100–140	8 × 300	>6000	4,000,000	13	500	0.5–1.0	1.5	15
HT-806	100–140	8 × 300	>6000	(20,000,000)[b]	13	1000	0.5–1.0	1.5	15
HT-806M	100–140	8 × 300	>6000	(20,000,000)[b]	13	1000	0.5–1.0	1.5	15
HT-G	100–140	8 × 50		Guard column	13	1000	0.5–1.0	1.5	
UT-802.5	100–210	8 × 300	>4000	20,000	30	80	0.3–1.0	1.5	10
UT-806M	100–210	8 × 300	>4000	(20,000,000)[b]	30	1000	0.3–1.0	1.5	10
UT-807	100–210	8 × 300	>3000	(200,000,000)[b]	30	>1000	0.3–1.0	1.5	10
UT-G	100–210	8 × 50		Guard column	30	1000	0.3–1.0	1.5	
AT-806MS	R.T.–140	8 × 250	>5000	(20,000,000)	12	1000	0.5–1.0	1.5	20
AT-G	R.T.–140	8 × 50		Guard column	20	100	0.5–1.0	1.5	

[a]Polystyrene is used with THF.
[b]Figure in parenthese is estimated value.

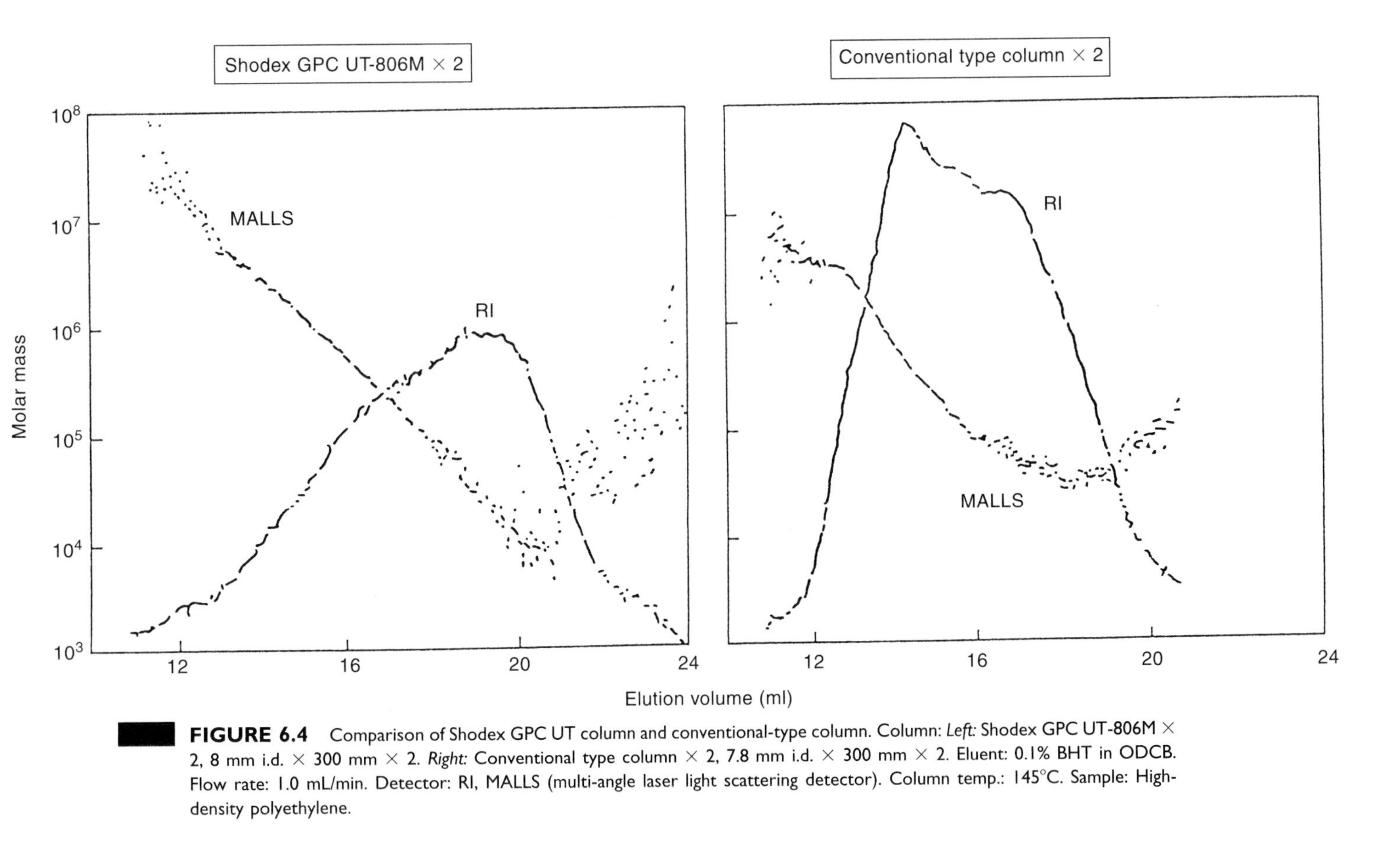

FIGURE 6.4 Comparison of Shodex GPC UT column and conventional-type column. Column: *Left:* Shodex GPC UT-806M × 2, 8 mm i.d. × 300 mm × 2. *Right:* Conventional type column × 2, 7.8 mm i.d. × 300 mm × 2. Eluent: 0.1% BHT in ODCB. Flow rate: 1.0 mL/min. Detector: RI, MALLS (multi-angle laser light scattering detector). Column temp.: 145°C. Sample: High-density polyethylene.

TABLE 6.4 Specifications of Shodex GPC HFIP-800 Series

In-column solvent						Flow Rate (ml/min)			
HFIP	Column size (mm)	Theoretical plate number	Exclusion limit PMMA[a]	Particle size (μm)	Pore size (Å)	Range	Max	Maximum pressure (kgf/cm^2)	Maximum temperature (°C)
HFIP-803	8 × 300	>10,000	50,000	10	100	0.5–1.0	1.5	40	50
HFIP-804	8 × 300	>10,000	150,000	7	200	0.5–1.0	1.5	40	50
HFIP-805	8 × 300	> 8,000	400,000	10	500	0.5–1.0	1.5	40	50
HFIP-806	8 × 300	> 8,000	(5,000,000)[b]	10	1000	0.5–1.0	1.5	40	50
HFIP-806M	8 × 300	> 8,000	(5,000,000)[b]	10		0.5–1.0	1.5	40	50
HFIP-807	8 × 300	> 3,000		18	>1000	0.5–1.0	1.5	40	50
HFIP-LG	8 × 50		Guard column	10		0.5–1.0	1.5	40	50

[a]Polymethylmethacrylate is used with HFIP.
[b]Figure in parenthese is estimated value.

TABLE 6.5 Specifications of Shodex GPC KF-600 Series

In-column solvent						Flow Rate (ml/min)			
THF	Column size (mm)	Theoretical plate number	Exclusion limit Polystyrene	Particle size (μm)	Pore size (Å)	Range	Max	Maximum pressure (kgf/cm^2)	Maximum temperature (°C)
KF-601	6 × 150	>17,000	1,500	3	20	0.4–0.6	0.8	70	45
KF-602	6 × 150	>17,000	5,000	3	60	0.4–0.6	0.8	70	45
KF-602.5	6 × 150	>17,000	20,000	3	80	0.4–0.6	0.8	70	45
KF-603	6 × 150	>17,000	70,000	3	100	0.4–0.6	0.8	70	45
KF-604	6 × 150	>16,000	400,000	3	200	0.4–0.6	0.8	70	45
KF-605	6 × 150	> 7,000	4,000,000	10	500	0.4–0.6	0.8	6	45
KF-606	6 × 150	> 7,000	(20,000,000)[a]	10	1000	0.4–0.6	0.8	6	45
KF-606M	6 × 150	> 8,000	(20,000,000)[a]	10		0.4–0.6	0.8	6	45
KF-607	6 × 150	> 5,000	(200,000,000)[a]	18	>1000	0.4–0.6	0.8	6	45
KF-G	4.6 × 10		Guard column	8		0.4–0.6	0.8		

[a]Figure in parenthese is estimated value.

molecular weight distribution can be determined using UT-806M, whereas a shear occurs in a conventional column.

4. HFIP-Packed Columns

The Shodex GPC HFIP series is packed with a hexafluoroisopropanol (HFIP) solvent. Engineered plastics, such as polyamides (nylon) and polyethylene terephthalate, were analyzed previously at a high temperature of about 140°C. Using HFIP as an eluent, such engineered plastics can be analyzed at ordinary temperatures (Table 6.4).

5. Downsized GPC Columns

The Shodex GPC KF-600 series is packed with 3-μm styrene–divinylbenzene copolymer gels in a column having a volume of about one-third compared to standard-types of columns, which are best suited for reducing the organic solvents consumption, shortening the analysis time, and lowering the detection limit (Table 6.5).

Figure 6.5 shows a chromatogram of Epikote 1004 using the KF-600 series and SYSTEM-24H (a newly developed GPC system used with the KF-600 series for full utilization of the efficiency of the downsized columns). Compared with conventional-sized columns such as the KF-800 series, the time of analysis is shortened to one-half and the chromatographic resolution is improved.

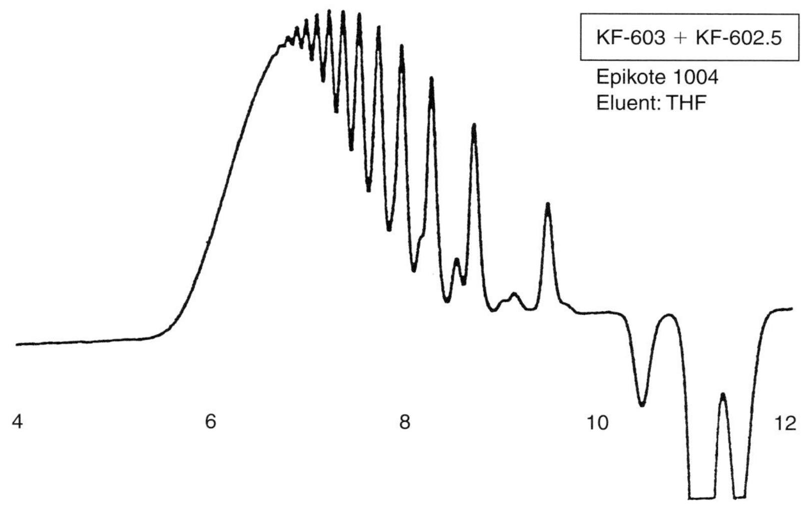

FIGURE 6.5 Analysis of epoxy resin by downsized column. Instrument: Shodex GPC SYSTEM-24. Column: Shodex GPC KF-603 + KF-602.5, 6 mm i.d. × 150 mm × 2. Eluent: THF. Flow rate: 0.6 mL/min. Detector: Shodex RI-74. Column temp.: 40°C. Sample: 0.1%, 100 μL, Epikote 1004.

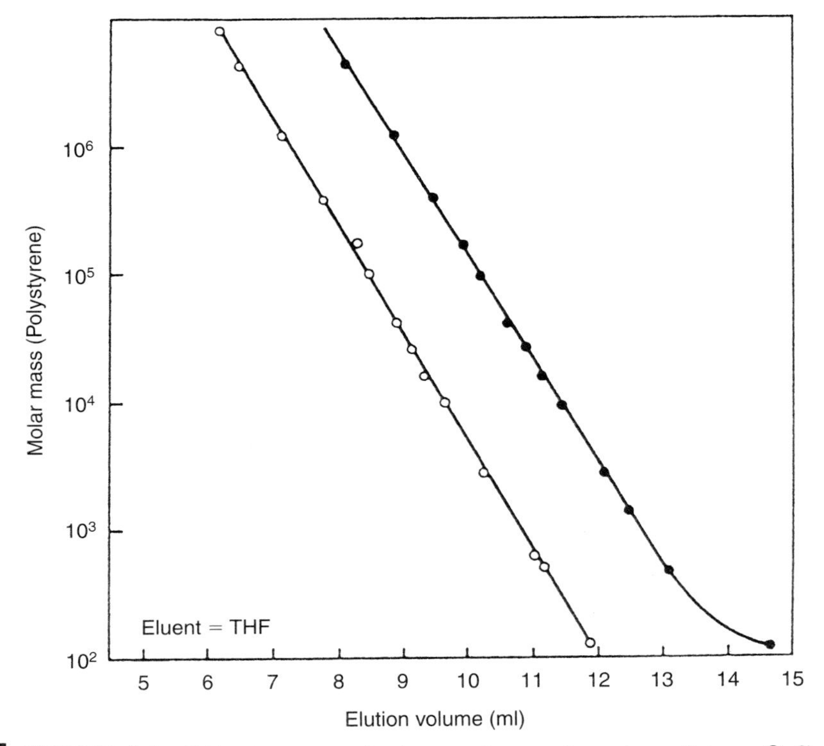

FIGURE 6.6 Calibration curves of solvent–peak separation column. Column: ●, Shodex GPC KF-806L + KF-800D, 8 mm i.d. × 300 mm + 8 mm i.d. × 100 mm. ○, Shodex GPC KF-806L, 8 mm i.d. × 300 mm. Eluent: THF. Sample: Polystyrene.

6. Solvent Peak Separation Columns

These columns make the elution of low molecular weight components slower and can be used to separate a solvent peak or other troublesome peaks to realize accurate measurements of the molecular weight distribution. The use of this column with 805L, 806L, 806M, and 807L columns is recommended.

Figures 6.6 and 6.7 show the effect of a solvent separation column. In the case of Fig. 6.7, the upper part of the figure shows the chromatogram of polyvinyl chrolide, which contains dioctyl phthalate (DOP), using KF-806L. In this case, DOP is not separated from a solvent peak. However, DOP can be separated from the solvent peak using KF-800D in conjunction with KF-806L (Table 6.6).

7. Preparative Columns

Table 6.7 shows the available preparative columns and corresponding analytical columns. Preparative columns are also available in larger diameters of up to 50 mm.

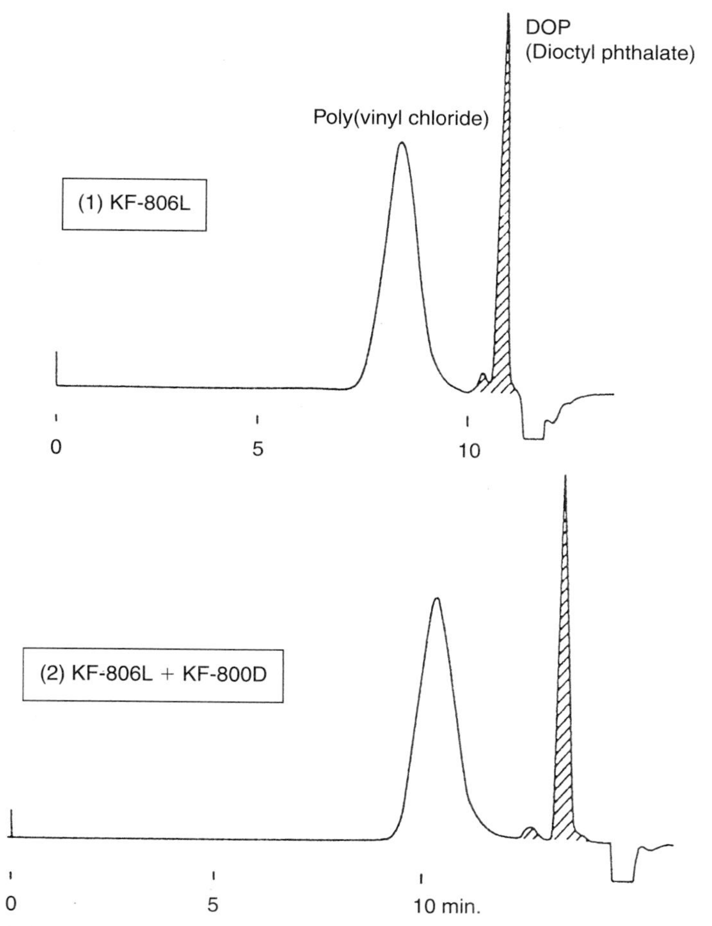

FIGURE 6.7 Effect of solvent-peak separation column. Column: (1) Shodex GPC KF-806L, 8 mm i.d. × 300 mm. (2) Shodex GPC KF-806L + KF-800D, 8 mm i.d. × 300 mm + 8 mm i.d. × 100 mm. Eluent: THF. Flow rate: 1.0 mL/min. Detector: Shodex RI. Sample: Poly(vinyl chloride).

TABLE 6.6 Specifications of Shodex GPC KF-800D and K-800D Columns

In-column solvents		Column size (mm)	Particle size (μm)	Pore size (Å)	Maximum temperature (°C)
THF	Chloroform				
KF-800D	K-800D	8 × 100	10	20	60

TABLE 6.7 Specifications of Shodex GPC KF, K-2000 Series

In-column solvents										
THF	Chloroform	Column size (mm)	Theoretical plate number	Exclusion limit Polystyrene[a]	Particle size (μm)	Pore size (Å)	Maximum flow rate (ml/min)	Maximum pressure (kgf/cm^2)	Maximum temperature (°C)	Analytical column
KF-2001	K-2001	20 × 300	>16,000	1,500	6	20	4.5	20	60	KF(k)-801
KF-2002	K-2002	20 × 300	>16,000	5,000	6	60	4.5	20	60	KF(k)-802
KF-2002.5	K-2002.5	20 × 300	>16,000	20,000	6	80	4.5	20	60	KF(k)-802.5
KF-2003	K-2003	20 × 300	>16,000	70,000	6	100	4.5	20	60	KF(k)-803
KF-2004	K-2004	20 × 300	>12,000	400,000	7	200	4.5	20	60	KF(k)-804
KF-2005	K-2005	20 × 300	>8,000	4,000,000	10	500	4.5	20	60	KF(k)-805
KF-2006	K-2006	20 × 300	>8,000	(20,000,000)[b]	10	1000	4.5	20	60	KF(k)-806
KF-2006M	K-2006M	20 × 300	>8,000	(20,000,000)[b]	10	1000	4.5	20	60	KF(k)-806M
KF-LG	K-LG	8 × 50		Guard column	15	200	4.5		60	KF(k)-G

[a]Polystyrene is used with THF.
[b]Figure in parentheses is estimated value.

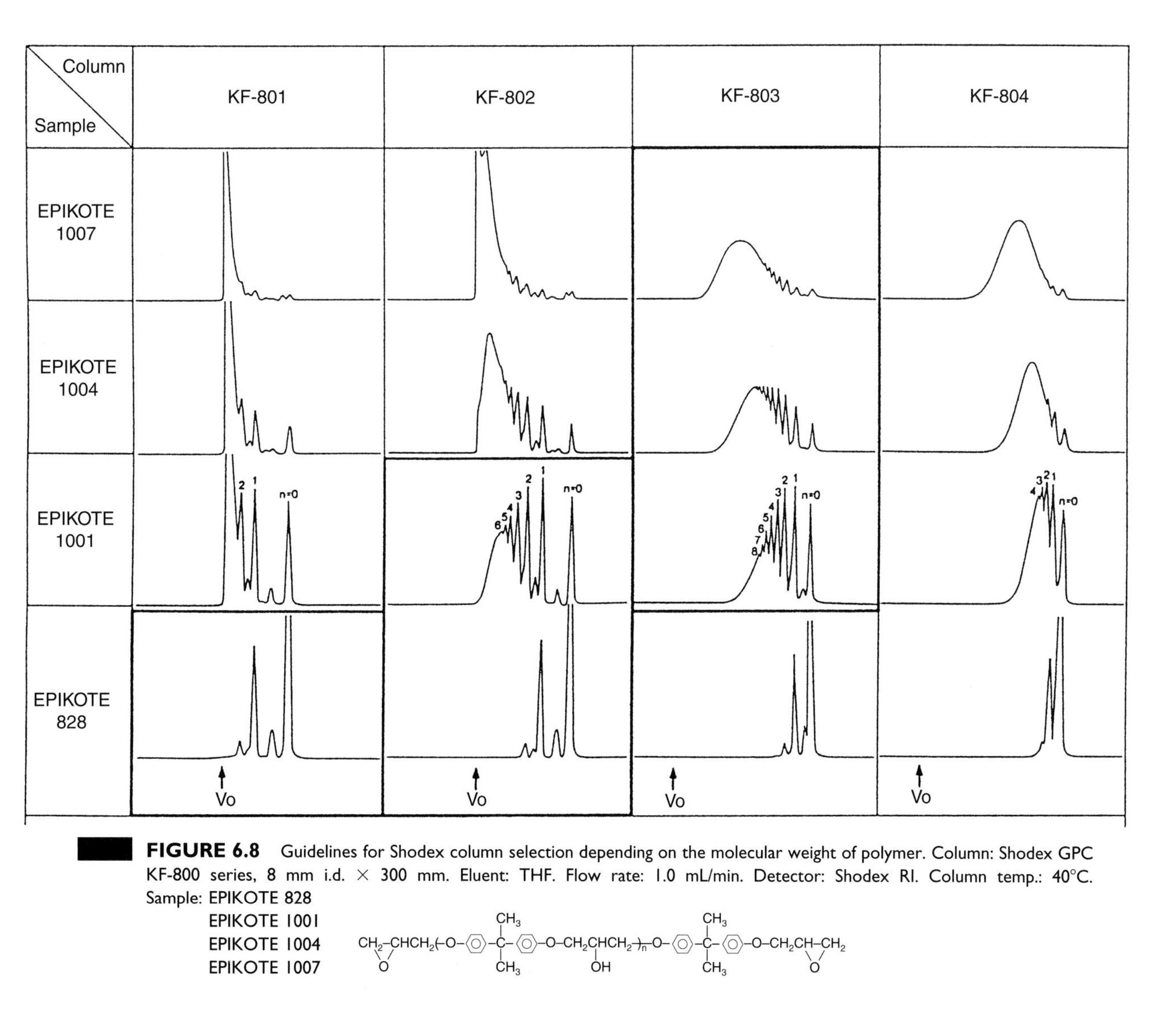

FIGURE 6.8 Guidelines for Shodex column selection depending on the molecular weight of polymer. Column: Shodex GPC KF-800 series, 8 mm i.d. × 300 mm. Eluent: THF. Flow rate: 1.0 mL/min. Detector: Shodex RI. Column temp.: 40°C.

Sample: EPIKOTE 828
EPIKOTE 1001
EPIKOTE 1004
EPIKOTE 1007

$$CH_2\overset{O}{-}CHCH_2(-O-\langle\bigcirc\rangle-\underset{CH_3}{\overset{CH_3}{C}}-\langle\bigcirc\rangle-O-CH_2\underset{OH}{C}HCH_2-)_n-O-\langle\bigcirc\rangle-\underset{CH_3}{\overset{CH_3}{C}}-\langle\bigcirc\rangle-O-CH_2CH\overset{O}{-}CH_2$$

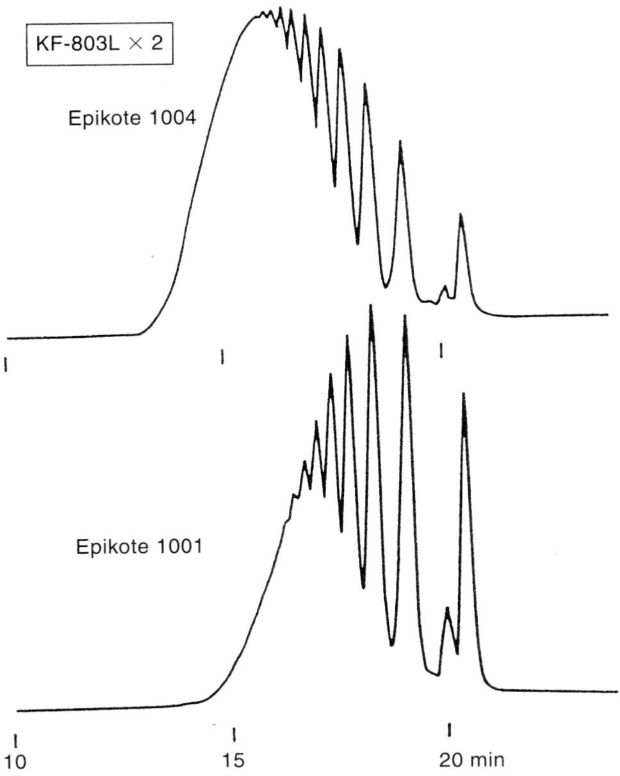

FIGURE 6.9 Epoxy resin. Column: Shodex GPC KF-803L × 2, 8 mm i.d. × 300 mm × 2. Eluent: THF. Flow rate: 1.0 mL/min. Detector: Shodex UV (254 nm). Column temp.: Ambient. Sample: 100 μL each, Epikote 1004, Epikote 1001.

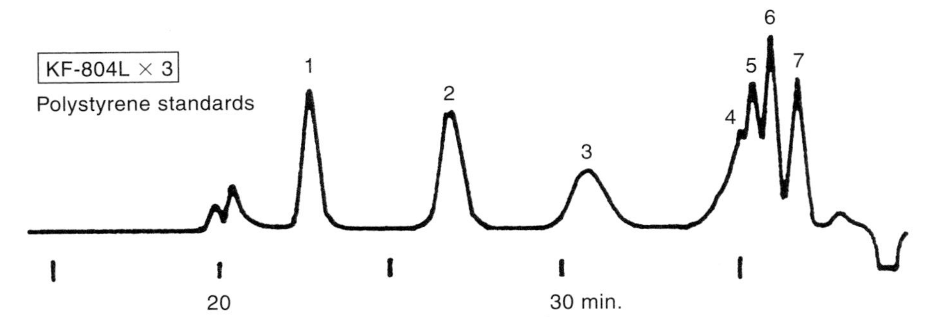

FIGURE 6.10 Polystyrene standards. Column: Shodex GPC KF-804L × 3, 8 mm i.d. × 300 mm × 3. Eluent: THF. Flow rate: 1.0 mL/min. Detector: Shodex RI. Column temp.: Ambient. Sample: 75 μL: (1) Polystyrene—107,000, 0.01%; (2) Polystyrene—19,000, 0.01%; (3) Polystyrene—3,600, 0.01%; (4) Polystyrene—578, 0.01%; (5) Polystyrene—474, 0.02%; (6) Polystyrene—370, 0.02%; (7) Polystyrene—266, 0.02%.

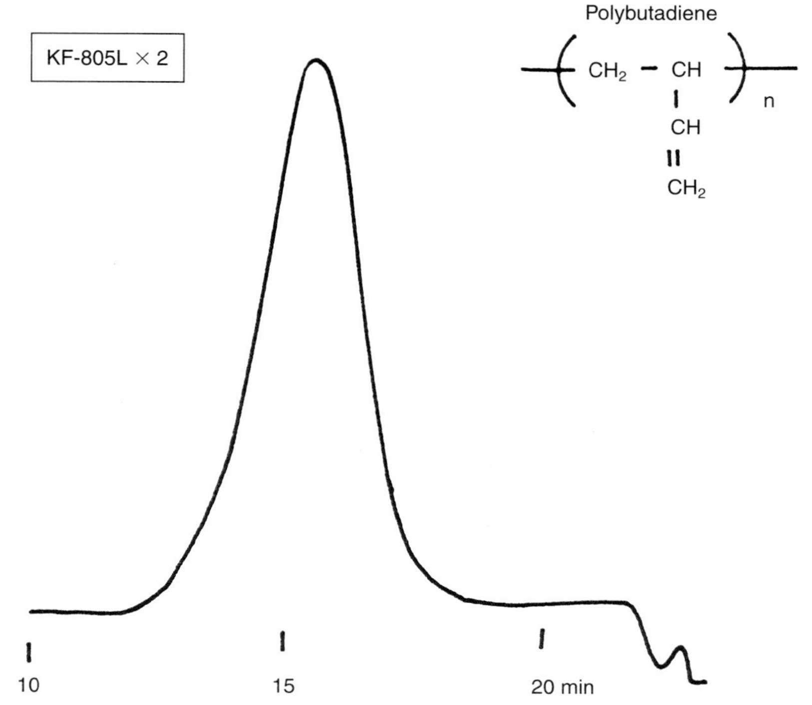

FIGURE 6.11 1,2-Polybutadiene. Column: Shodex GPC KF-805L × 2, 8 mm i.d. × 300 mm × 2. Eluent: THF. Flow rate: 1.0 mL/min. Detector: Shodex RI. Column temp.: 40°C. Sample: 0.1%, 100 μL, 1,2-Polybutadiene.

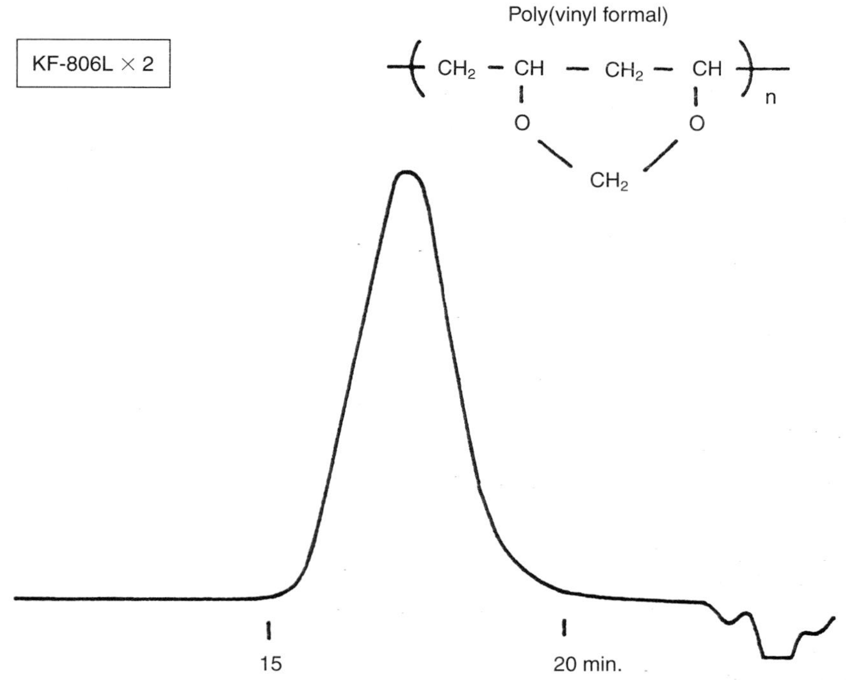

FIGURE 6.12 Poly(vinyl formal). Column: Shodex GPC KF-806L × 2, 8 mm i.d. × 300 mm × 2. Eluent: THF. Flow rate: 1.0 mL/min. Detector: Shodex RI. Column temp.: 40°C. Sample: 0.1%, 100 μL, Poly(vinyl formal).

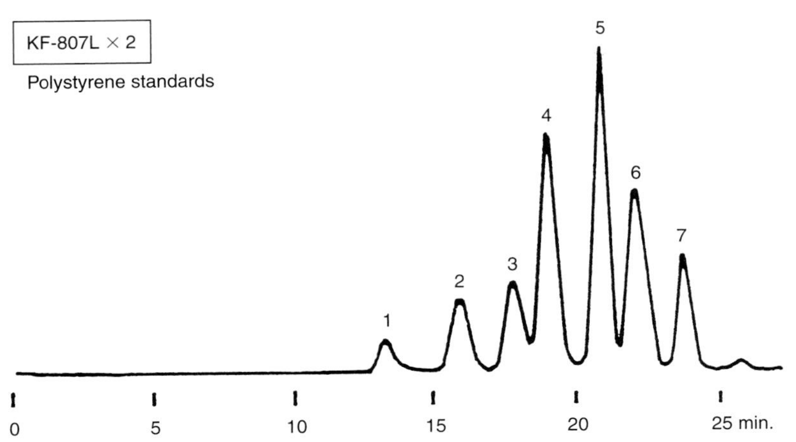

FIGURE 6.13 Polystyrene standards. Column: Shodex GPC KF-807L × 2, 8 mm i.d. × 300 mm × 2. Eluent: THF. Flow rate: 1.0 mL/min. Detector: Shodex UV (254 nm). Column temp.: 40°C. Sample: 100 μL: (1) Polystyrene—20,700,000, 0.005%; (2) Polystyrene—1,100,000, 0.01%; (3) Polystyrene—128,000, 0.01%; (4) Polystyrene—33,000, 0.02%; (5) Polystyrene—3,560, 0.02%; (6) Polystyrene—580, 0.01%; (7) Ethylbenzene—106, 0.01%.

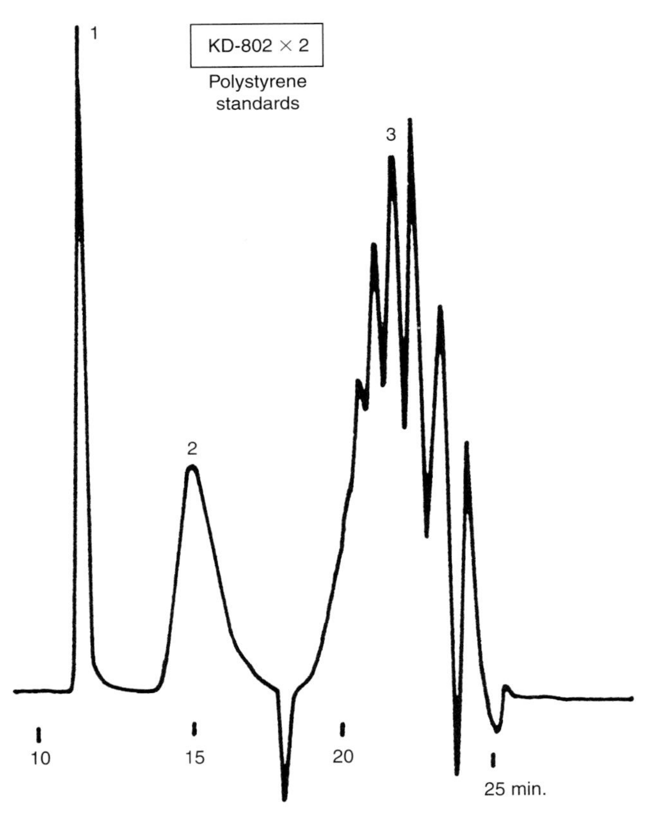

FIGURE 6.14 Polystyrene standards. Column: Shodex GPC KD-802 × 2, 8 mm i.d. × 300 mm × 2. Eluent: 10 m*M* LiBr/DMF. Flow rate: 1.0 mL/min. Detector: Shodex RI. Column temp.: 50°C. Sample: (1) Polystyrene—37,000, 0.1%; (2) Polystyrene—4,000, 0.02%; (3) Polystyrene—580, 1.0%

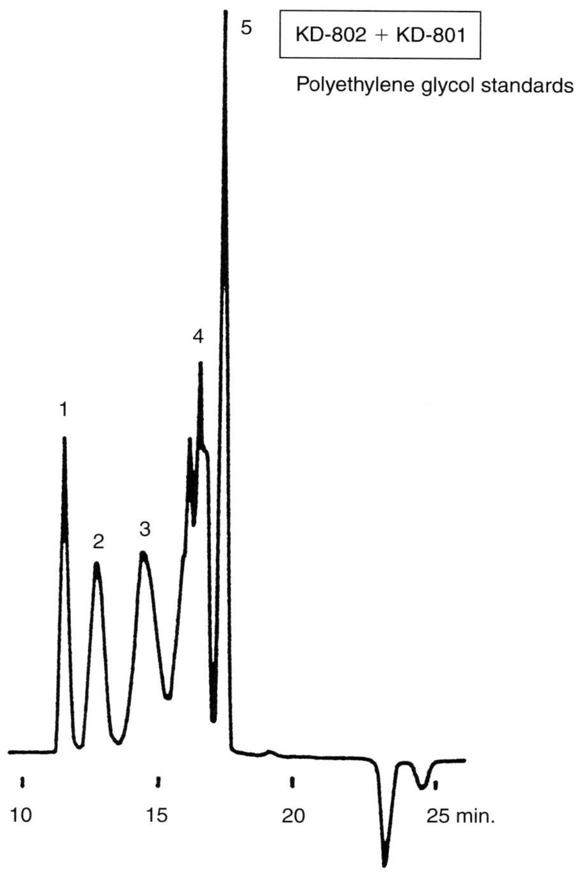

FIGURE 6.15 Polyethylene glycol standards. Column: Shodex GPC KD-802 + KD-801, 8 mm i.d. × 300 mm × 2. Eluent: 10 m*M* LiBr/DMF. Flow rate: 1.0 mL/min. Detector: Shodex RI. Column temp.: 50°C. Sample: 100 μL: (1) PEG—20,000, 0.15%; (2) PEG—2,000, 0.21%; (3) PEG—600, 0.41%; (4) PEG—200, 0.85%; (5) EG—62, 1.33%

B. Application Examples

1. General-Purpose GPC

Figure 6.8 (page 185) shows chromatograms of four kinds of epoxyde resins having different molecular weight distributions. Based on the molecular weight distribution of each sample, the most suitable column with the proper pore size for each sample should be chosen. KF-801 or KF-802 is recommended for the analysis of Epikote 828, which has the lowest molecular weight distribution. KF-802 or KP-803 is recommended for Epikote 1001, and KF-803 is recommended for Epikote 1004 and 1007, which have rather high molecular weight distributions. KF-804 is suited to the analysis of samples having higher molecular weight distributions.

Figures 6.9–6.13 (pages 186 through 188) show typical chromatograms of general-purpose GPC.

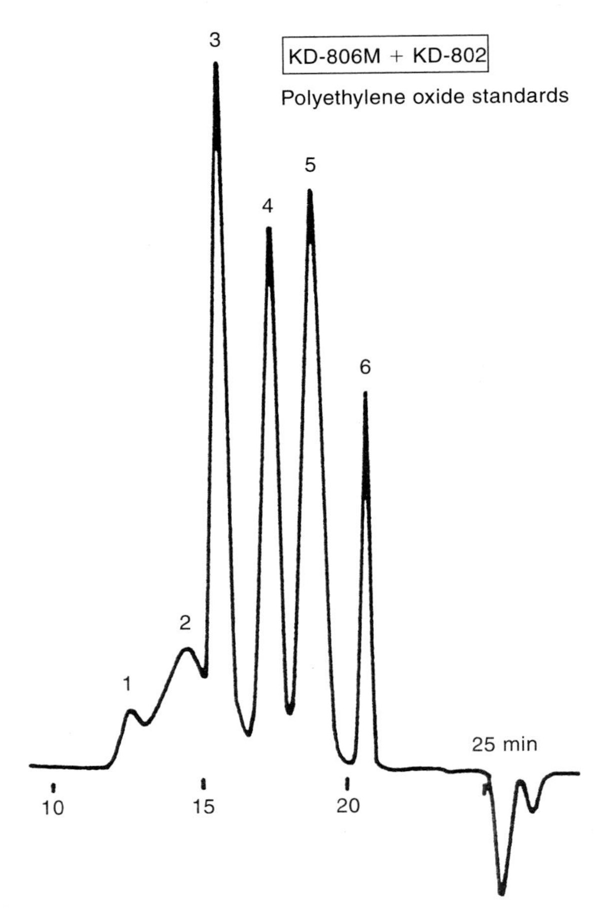

FIGURE 6.16 Polyethylene oxide standards. Column: Shodex GPC KD-806M + KD-802, 8 mm i.d. × 300 mm × 2. Eluent: 10 m*M* LiBr/DMF. Flow rate: 1.0 mL/min. Detector: Shodex RI. Column temp.: 50°C. Sample: 100 μL: (1) PEO—1,200,000, 0.18%; (2) PEO—100,000, 0.3%; (3) PEO—10,000, 0.4%; (4) PEO—2,000, 0.4%; (5) PEO—600, 0.55%; (6) EG—62, 0.67%.

2. Polar Compounds

Figures 6.14–6.16 show the chromatograms of polystyrene, polyethylene glycol, and polyethylene oxide standards using dimethylformamide (DMF) as an eluent.

If DMF is used in the KD-800 series as eluent, the retention volume of polystyrene increases slightly because of the adsorption of polystyrene to the column (Fig. 6.17).

3. Engineered Plastics

Figures 6.18–6.20 show the chromatograms of engineered plastics such as polyamide (nylon) and polyethylene terephthalate at ordinary temperature.

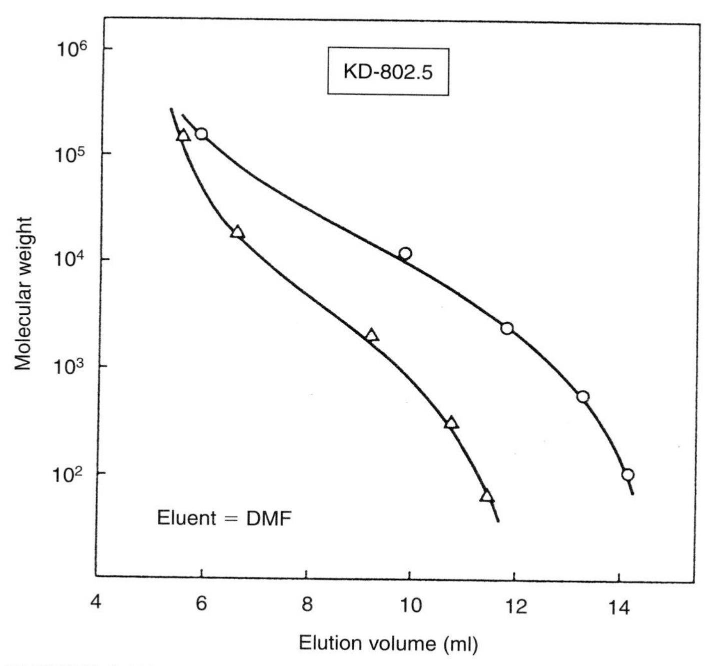

FIGURE 6.17 Comparison of calibration curves using various standard samples. Column: Shodex GPC KD-802.5, 8 mm i.d. × 300 mm. Eluent: DMF. Sample: ○, Polystyrene. △, PEG.

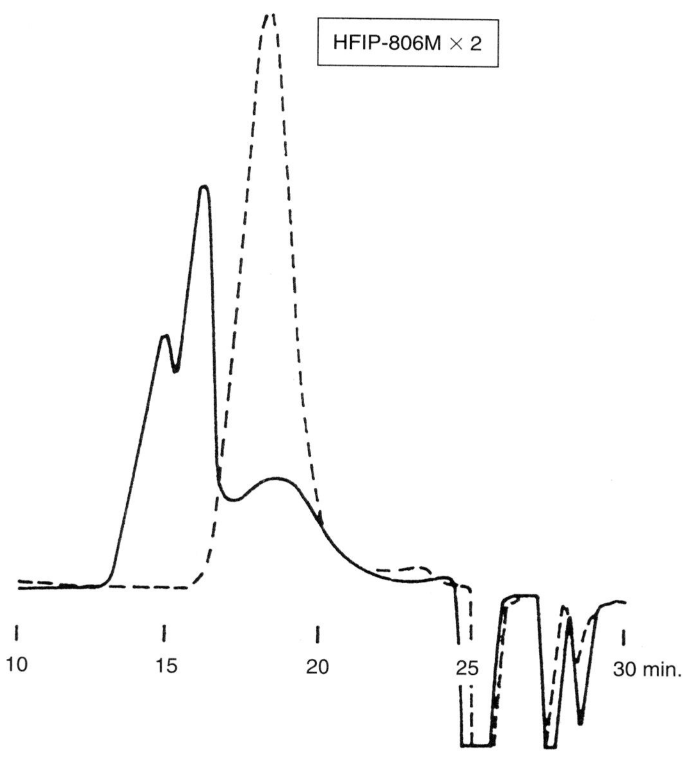

FIGURE 6.18 Polyamide. Column: Shodex GPC HFIP-806M × 2, 8 mm i.d. × 300 mm × 2. Eluent: ——, HFIP. --------, 5 m*M* CF_3COONa/HFIP. Flow rate: 1.0 mL/min. Detector: Shodex RI. Column temp.: 40°C. Sample: Polycaprolactum (Nylon 6).

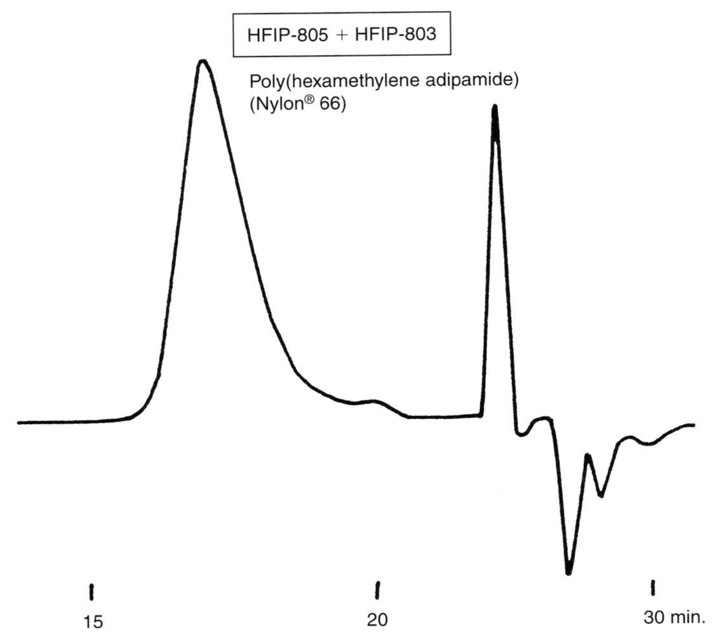

FIGURE 6.19 Polyamide. Column: Shodex GPC HFIP-805 + HFIP-803, 8 mm i.d. × 300 mm × 2. Eluent: 5 m*M* CF_3COONa/HFIP. Flow rate: 1.0 mL/min. Detector: Shodex RI. Column temp.: 40°C. Sample: 0.05%, 500 μL, Poly(hexamethylene adipamide) (Nylon 66).

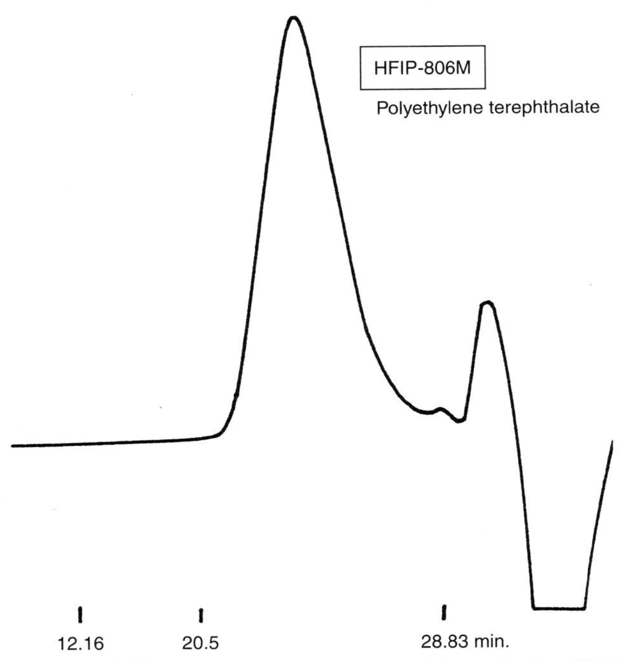

FIGURE 6.20 Polyethylene terephthalate. Column: Shodex GPC HFIP-806M, 8 mm i.d. × 300 mm. Eluent: 5 m*M* CF_3COONa/HFIP. Flow rate: 0.5 mL/min. Detector: Shodex RI. Column temp.: 40°C. Sample: 0.05%, 500 μL, PET.

In the case of functional polymers such as nylon, the molecular size can increase because ionic functional groups contained in the molecule repel each other. To decrease ionic repulsion, sodium trifluoroacetate should be added to HFIP.

The polystyrene–terephthalate copolymer is used as a raw material for transparent bottles of refreshing drinks. It becomes clear that the polymer contains a small amount of oligomer.

IV. Shodex OHpak SB-800HQ SERIES

A. Column Characteristics

The packed columns of Shodex OHpak SB-800HQ series are packed with polyhydroxymethacrylate gels and are designed for use with high-resolution, high-speed aqueous size exclusion chromatography. The packed columns are best suited for the analysis of water-soluble polymers and proteins (Table 6.8).

Figure 6.21 shows the calibration curves of the SB-800 HQ series using standard pullulan. Because a high molecular weight standard sample is not available, the calibration curves of 805 and 806 are partly estimates (dotted lines). The difference in the conformation between polyethylene oxide (PEO) and pullulan in the solvent causes a shift of the calibration curves of pullulan slightly higher than those of PEO. The OHpak SB-800HQ series is better suited for the analysis of hydrophilic samples than the Asahipak GS/GF series.

As shown in Table 6.9, the SB-800HQ series can be used by adding various solvents to the eluent.

B. Application Examples

1. Ionic and Nonionic Hydrophilic Polymers

Deionized water can be used as an eluent for the analysis of nonionic polymers such as pullulan and polyethylene glycol. However, in most cases, salt solutions or buffer solutions are used to decrease ionic or other interactions between samples and the stationary phase or to prevent sample association (Figs. 6.22 and 6.23, pages 196 and 197).

2. Nonionic and Ionic Hydrophobic Polymers

The addition of polar organic solvents to the eluent is recommended with the goal of decreasing the hydrophobic adsorption. The addition of salts to the eluent is also recommended when the sample is ionic (Figs. 6.24–6.26, pages 198 and 199).

Figure 6.24 shows the chromatogram of polyvinylpyrrolidone. Although PVP is slightly hydrophobic, GPC elution can be achieved by adding acetonitrile in the eluent.

3. Proteins

The Shodex OHpak SB-800 HQ series is usually not the best choice for the analysis of proteins. However, in some cases, such as when an alkaline

TABLE 6.8 Specifications of Shodex OHpak SB-800 HQ Series

Column type	Column size (mm)	Theoretical plate number	Particle size (μm)	Pore size (Å)	Maximum flow rate (ml/min)	Maximum pressure (kgf/cm^2)	Maximum temperature (°C)	Exclusion limit	
								Pullulan	PEG
SB-802HQ	8 × 300	>10,000	8	20	1.2	60	70	4,000	2,000
SB-802.5HQ	8 × 300	>15,000	6	60	1.2	60	70	10,000	8,500
SB-803HQ	8 × 300	>15,000	6	80	1.2	60	70	100,000	55,000
SB-804HQ	8 × 300	>15,000	10	100	1.2	30	70	1,000,000	300,000
SB-805HQ	8 × 300	>10,000	13	200	1.2	30	70	4,000,000	2,000,000
SB-806HQ	8 × 300	>10,000	13	500	1.2	30	70	(20,000,000)[a]	(10,000,000)[a]
SB-806MHQ	8 × 300	>10,000	13		1.2	30	70	(20,000,000)[a]	(10,000,000)[a]
SB-G	6 × 50		10		1.2			Guard column	Guard column

[a]Figure in parentheses is estimated value.

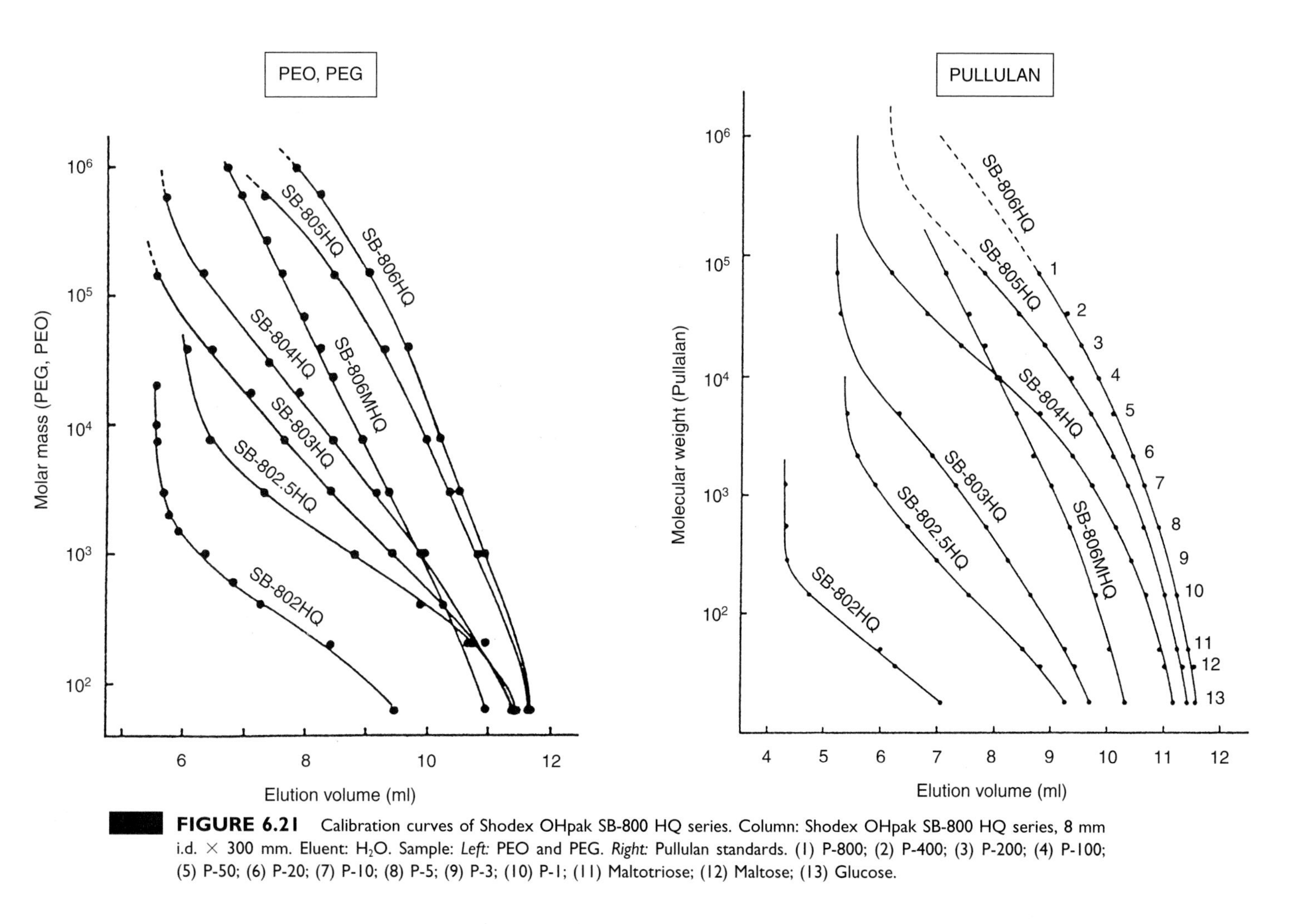

FIGURE 6.21 Calibration curves of Shodex OHpak SB-800 HQ series. Column: Shodex OHpak SB-800 HQ series, 8 mm i.d. × 300 mm. Eluent: H_2O. Sample: *Left:* PEO and PEG. *Right:* Pullulan standards. (1) P-800; (2) P-400; (3) P-200; (4) P-100; (5) P-50; (6) P-20; (7) P-10; (8) P-5; (9) P-3; (10) P-1; (11) Maltotriose; (12) Maltose; (13) Glucose.

TABLE 6.9 Usable Organic Solvent for Shodex OHpak SB-800 HQ Series

Column type	Usable pH range	Usable organic solvents (%)		
		MeOH	MeCN	DMF
SB-802HQ	3–10	0	0	0
SB-802.5HQ	3–10	<100	<75	<100
SB-803HQ	3–10	<100	<75	<100
SB-804HQ	3–10	<75	<75	<100
SB-805HQ	3–10	<75	<75	<100
SB-806HQ	3–10	<75	<75	<100
SB-806MHQ	3–10	<75	<75	<100
SB-G	3–10			

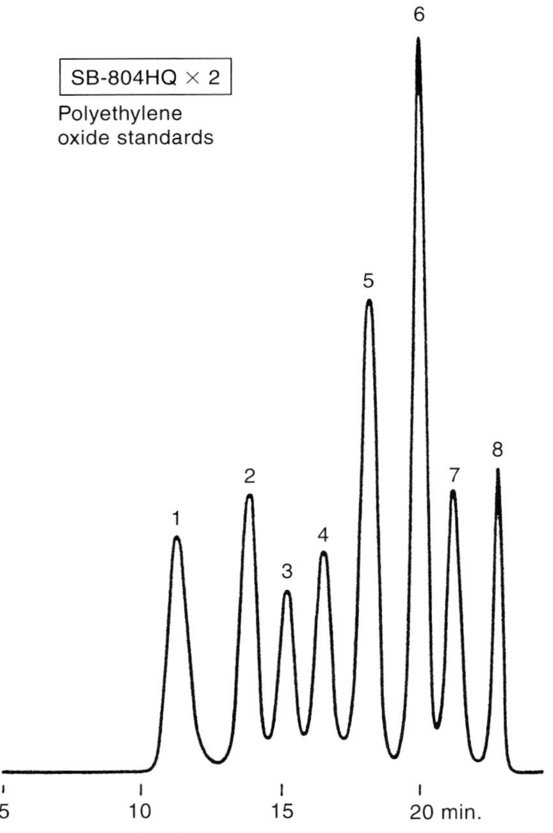

FIGURE 6.22 PEO standards. Column: Shodex OHpak SB-804 HQ × 2, 8 mm i.d. × 300 mm × 2. Eluent: 0.02% NaN_3. Flow rate: 1.0 mL/min. Detector: Shodex RI. Column temp.: 30°C. Sample: (1) PEO, 594,000; (2) PEO, 145,000; (3) PEO, 73,000; (4) PEO, 40,000; (5) PEO, 18,000; (6) PEO, 3,000; (7) PEO, 600; (8) EG, 62.

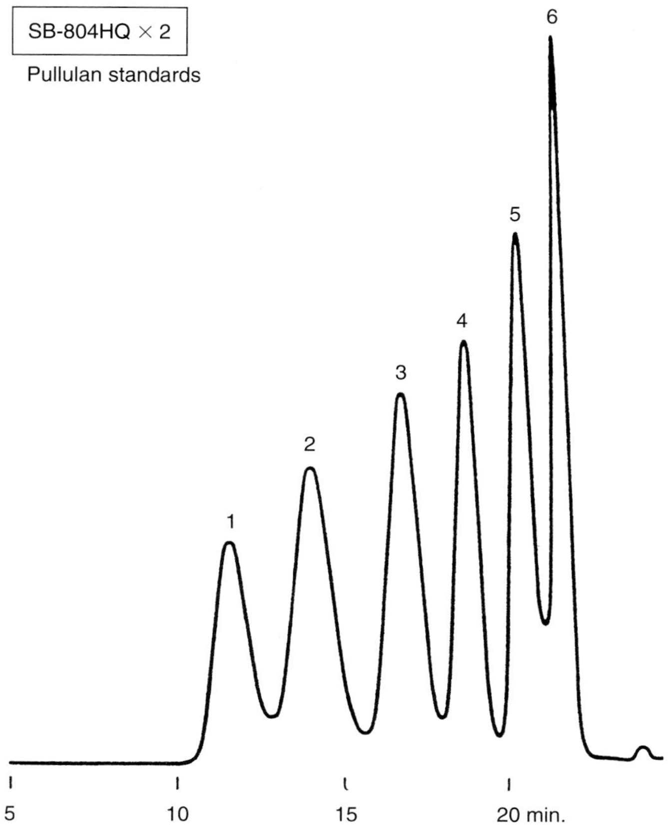

FIGURE 6.23 Pullulan standards. Column: Shodex OHpak SB-804 HQ × 2, 8 mm i.d. × 300 mm × 2. Eluent: 0.02% NaN_3. Flow rate: 1.0 mL/min. Detector: Shodex RI. Column temp.: 30°C. Sample: (1) P-800, 758,000; (2) P-200, 194,000; (3) P-50, 46,700; (4) P-10, 12,200; (5) P-1, 1,300; (6) Glucose, 180.

condition is acquired to keep the biological activity of the sample, the SB-800 HQ series should be used because of its chemical stability. Protein modifiers, such as urea and guanidine–HCl, can also be added to the eluent. Surfactants, such as sodium dodecyl sulfate or Brij-35, can be added to the eluent to increase the solubility of proteins, including membrane proteins.

C. Preparative and Semipreparative Columns

Shodex preparative and semipreparative columns are supplied in Table 6.10 (page 201).

V. Shodex Asahipak GS/GF SERIES

A. Column Characteristics

The packed columns of Shodex Asahipak GF/GS HQ series are made of an especially modified polyvinyl alcohol based resin (Table 6.11, page 202).

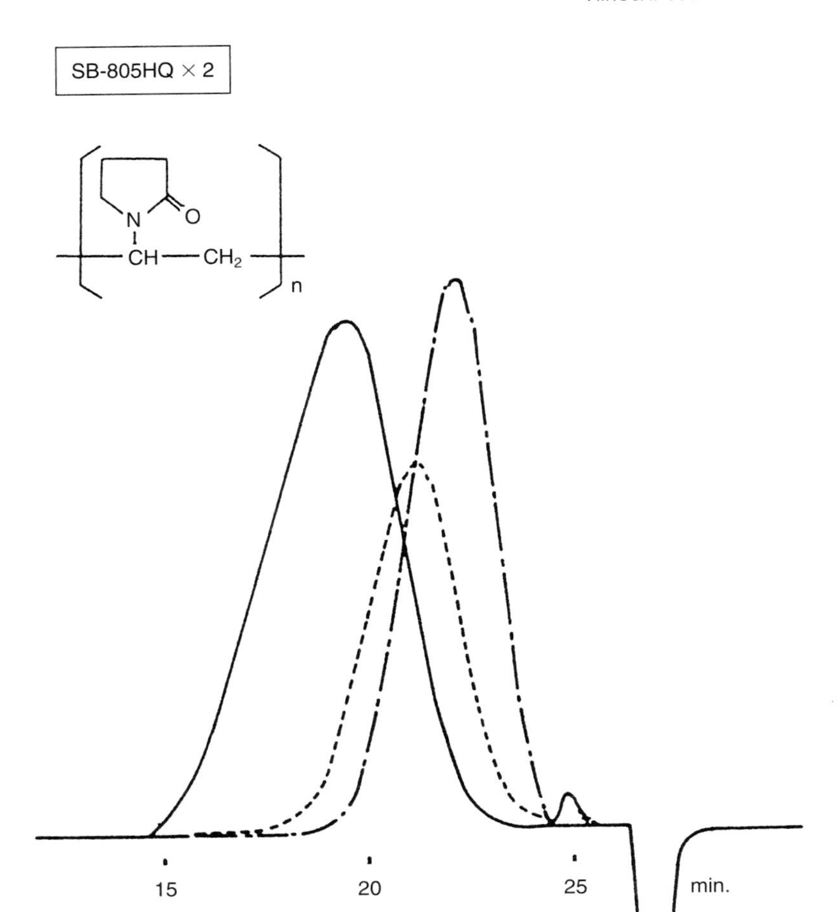

FIGURE 6.24 Polyvinylpyrrolidone. Column: Shodex OHpak SB-805 HQ × 2, 8 mm i.d. × 300 mm × 2. Eluent: 100 m*M* $NaNO_3/CH_3CH$ = 90/10. Flow rate: 1.0 mL/min. Detector: Shodex RI. Column temp.: 30°C. Sample: Polyvinylpyrrolidone: ———: K-60 Mw 160,000; --------: K-30 Mw 40,000; — - —: K-15 Mw 10,000.

The Shodex Asahipak GF-HQ series can be used with various types of solvents as an eluent (Table 6.12, page 203).

B. Application Examples

1. Ionic and Nonionic Hydrophilic Polymers

Deionized water can be used as an eluent for the analysis of nonionic polymers such as pullulan and polyethylene glycol. However, in most cases, salt solutions or buffer solutions are used to decrease ionic or other interaction between samples and the stationary phase or to prevent sample association (Figs. 6.27 and 6.28, pages 203 and 204).

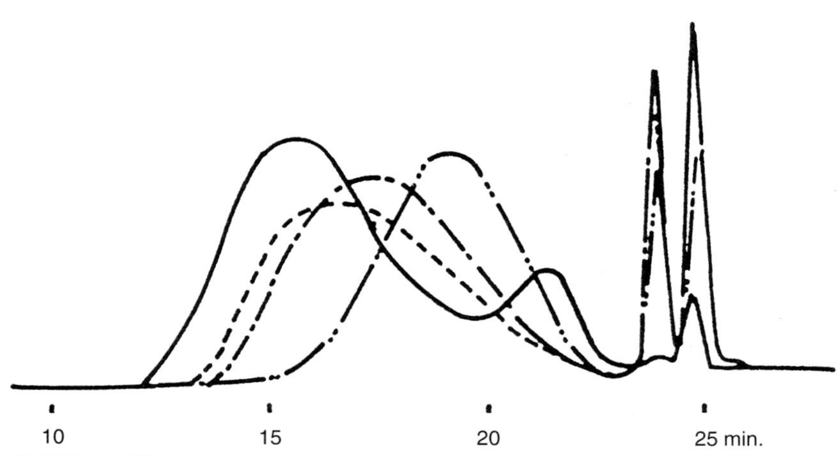

FIGURE 6.25 Hydroxymethyl cellulose. Column: Shodex OHpak SB-805 HQ × 2, 8 mm i.d. × 300 mm × 2. Eluent: 100 m*M* $NaNO_3$. Flow rate: 1.0 mL/min. Detector: Shodex RI. Column temp.: 30°C. Sample: Hydroxyethyl cellulose: ——— : 4,500 ~ 6,500 cps.; -------- : 800 ~ 1,500 cps.; — - —: 200 ~ 300 cps.; — - - —: 20 ~ 30 cps.

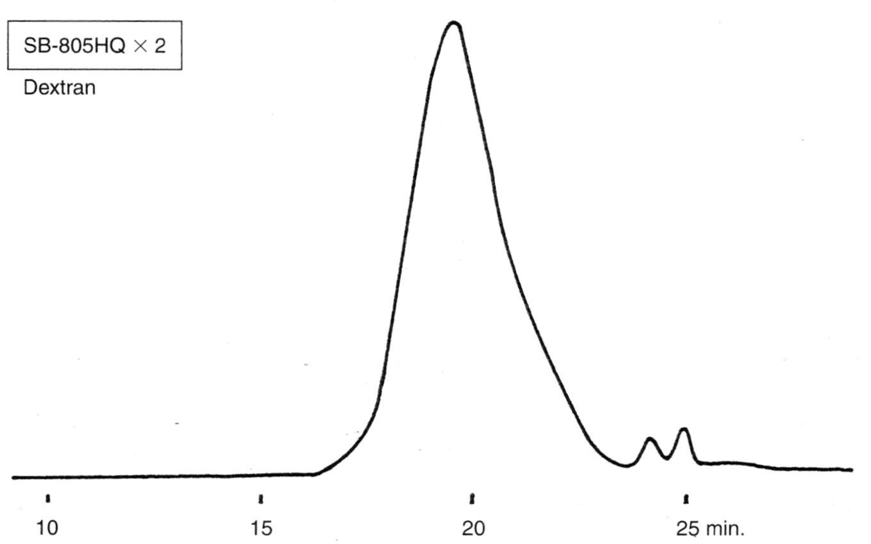

FIGURE 6.26 Dextran. Column: Shodex OHpak SB-805 HQ × 2, 8 mm i.d. × 300 mm × 2. Eluent: 100 m*M* $NaNO_3$. Flow rate: 1.0 mL/min. Detector: Shodex RI. Column temp.: 30°C. Sample: Dextran (Industrial grade).

Figure 6.29 (page 205) shows the chromatogram of poly(allylamine hydrochloride) using the Asahipak GF-HQ series. The peak shape is affected by changing the ionic strength. It was confirmed that the best peak shape is obtained when the concentration of LiCl in the eluent is 250 m*M*.

2. Nonionic and Ionic Hydrophobic Polymers

The Shodex Asahipak GF-HQ series is unique in that both organic and aqueous size exclusion chromatography can be performed with one column.

As packing material, the series employs highly cross-linked polyvinyl alcohol gels that have an affinity for a variety of solvents and show very little shrinkage or swelling when the in-column solvent is replaced.

A multisolvent system can be adapted to achieve the GPC elution mode. Figures 6.30 and 6.31 (page 206) show how to achieve GPC elution by changing the mixing ratio of chloroform and methanol in the analysis of Epikote 1001.

Figures 6.32 to 6.35 (pages 207 and 208) show the chromatograms of polystyrene sulfonates using the Asahipak GF-HQ series. Because polystyrene sulfonates have benzene rings and show hydrophobic nature, acetonitrile is added to the eluent by up to 40% in order to reduce the hydrophobic interaction between gel and sample. As a result, normal GPC elution behavior is obtained. When GF-510HQ or GF-310HQ is used, whose molecular weight limit (ML) is 300,000 and 40,000, respectively, samples having a molecular weight higher than ML of each column are eluted exclusively. When samples having a lower molecular weight are applied, better resolution is obtained using those columns.

Figures 6.36 to 6.38 (pages 209 and 210) show the chromatograms of polymethacrylates using the Asahipak GF-HQ series. As polymethacrylates also have hydrophobic structure, acetonitrile is added to the eluent by up to 10%. As a result, normal GPC elution behavior is obtained.

In the case of Figure 6.39 (page 211), similar chromatograms were apparently obtained with each column. However, when the elution patterns were examined with a MALLS, it was revealed that a kind of adsorption occurred in the case of GF-710 HQ, and GPC elution was not conducted properly.

It is very difficult to answer the question of which column should be chosen: OHpak or Asahipak. For the best results, the best column should be chosen and the optimum chromatographic conditions should be determined in view of the physical and chemical nature of each polymer. In general, Asahipak is recommended for analyses of ionic polymer, and OHpak is recommended for nonionic polymers. However, this does not necessarily apply to all cases. In order to find a definitive answer to the question, further studies will be conducted.

C. Preparative and Semipreparative Columns

Shodex preparative and semipreparative columns are supplied in Table 6.13 on page 212.

TABLE 6.10 Specifications of Shodex OHpak SB-2000 Series

Column type	Column size (mm)	Theoretical plate number	Particle size (μm)	Pore size (Å)	Maximum flow rate (ml/min)	Maximum pressure (kgf/cm^2)	Maximum temperature (°C)	Exclusion limit		Analytical column
								Pullulan	PEG	
SB-2002	20 × 300	>8,000	15	20	5	25	50	4,000	2,000	SB-802HQ
SB-2002.5	20 × 300	>10,000	15	60	5	25	50	10,000	8,500	SB-802.5HQ
SB-2003	20 × 300	>10,000	15	80	5	25	50	100,000	55,000	SB-803HQ
SB-2004	20 × 300	>10,000	20	100	5	25	50	1,000,000	300,000	SB-804HQ
SB-2005	20 × 300	>10,000	20	200	5	25	50	4,000,000	2,000,000	SB-805HQ
SB-2006	20 × 300	>10,000	20	500	5	25	50	(20,000,000)[a]	(10,000,000)[a]	SB-806HQ
SB-2006M	20 × 300	>10,000	13	1000	5	25	50	(20,000,000)[a]	(10,000,000)[a]	SB-806MHQ
SB-LG	8 × 50		20		5	25	50	Guard column	Guard column	SB-G

[a]Figure in parentheses is estimated value.

TABLE 6.11 Specifications of Shodex Asahipak GF, GS-HQ Series

Column type	Column size (mm)	Theoretical plate number	Flow rate (ml/min)		Exclusion limit Pullulan	Usable pH range	Particle size (μm)	Maximum pressure (kgf/cm^2)	Maximum temperature (°C)
			Range	Max					
GS-220HQ	7.6 × 300	>16,000	0.4–0.6	1.0	3,000	2–9	6	60	60
GS-320HQ	7.6 × 300	>16,000	0.4–0.6	1.0	40,000	2–12	6	50	60
GS-520HQ	7.6 × 300	>15,000	0.4–0.6	1.0	300,000	2–12	7	30	60
GS-620HQ	7.6 × 300	>15,000	0.4–0.6	1.0	1,000,000	2–9	7	35	60
GS-2G7B	7.6 × 50		0.4–0.6	1.0	Guard column	2–12	9		60
GF-310HQ	7.6 × 300	>16,000	0.4–0.6	1.0	40,000	2–9	5	70	60
GF-510HQ	7.6 × 300	>16,000	0.4–0.6	1.0	300,000	2–9	5	65	60
GF-710HQ	7.6 × 300	>10,000	0.4–0.6	1.0	10,000,000	2–9	9	25	60
GF-7MHQ	7.6 × 300	>11,000	0.4–0.6	1.0	10,000,000	2–9	7	45	60
GF-1G7B	7.6 × 50		0.4–0.6	1.0	Guard column	2–9	9		60

TABLE 6.12 Usable Organic Solvents for Shodex Asahipak GF-HQ Series

Usable range (%)	Organic solvent
0–100	H_2O, MeOH, EtOH, MeCN, THF, DMF, acetone, chloroform, ethylacetate
0–50	Dimethyl sulfoxide

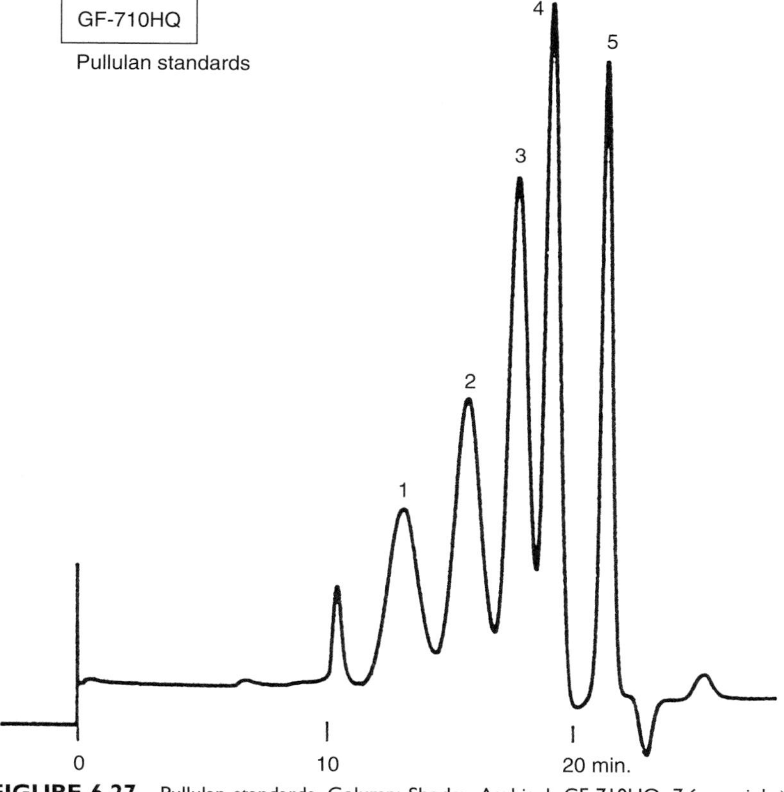

FIGURE 6.27 Pullulan standards. Column: Shodex Asahipak GF-710HQ, 7.6 mm i.d. × 300 mm. Eluent: H_2O. Flow rate: 0.5 mL/min. Detector: Shodex RI. Column temp.: 50°C. Sample: 0.1% each, 50 μL; (1) P-800; (2) P-200; (3) P-50; (4) P-10; (5) EG.

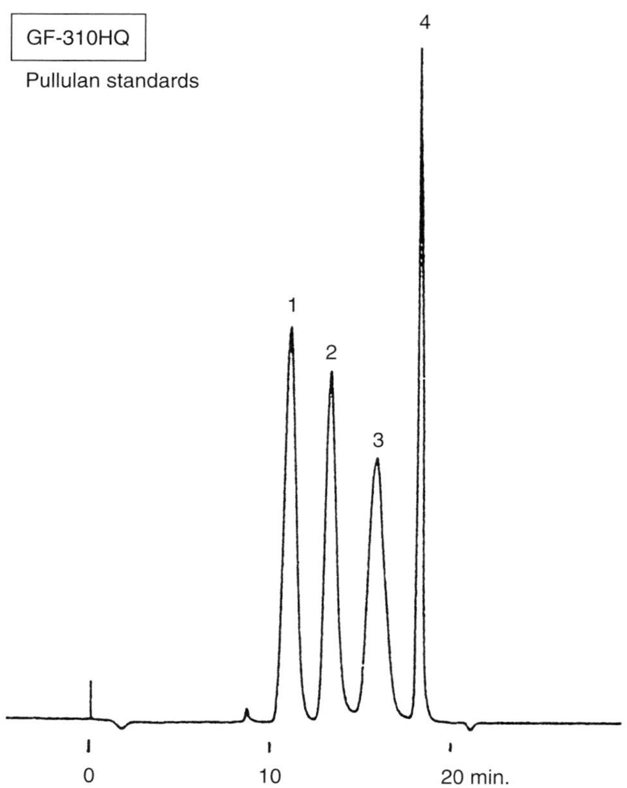

FIGURE 6.28 Pullulan standards. Column: Shodex Asahipak GF-310HQ, 7.6 mm i.d. × 300 mm. Eluent: H_2O. Flow rate: 0.5 mL/min. Detector: Shodex RI. Column temp.: 30°C. Sample: 1% each, 50 μL; (1) Pullulan P-20, 23,700; (2) PEG, 3,000; (3) PEG, 1,000; (4) EG, 62.

VI. Shodex PROTEIN KW SERIES

A. Column Characteristics

The packed columns of Shodex PROTEIN KW-800 series are packed with hydrophilic silica-based gels and are best suited for analyses of proteins and water-soluble polymers (Table 6.14, page 213).

Figure 6.40 (page 214) shows the calibration curves of the Shodex PROTEIN KW-800 series obtained using protein, pullulan, and polyethylene glycol (PEG) and polyethylene oxide (PEO) standards as a sample. KW-802.5 is best suited for the analysis of samples containing portions with a molecular weight of less than 100,000 Da. KW-803 and KW-804 are best suited for the analysis of samples containing higher molecular weight portions (<300,000 and >300,000, respectively). Three calibration curves obtained using protein,

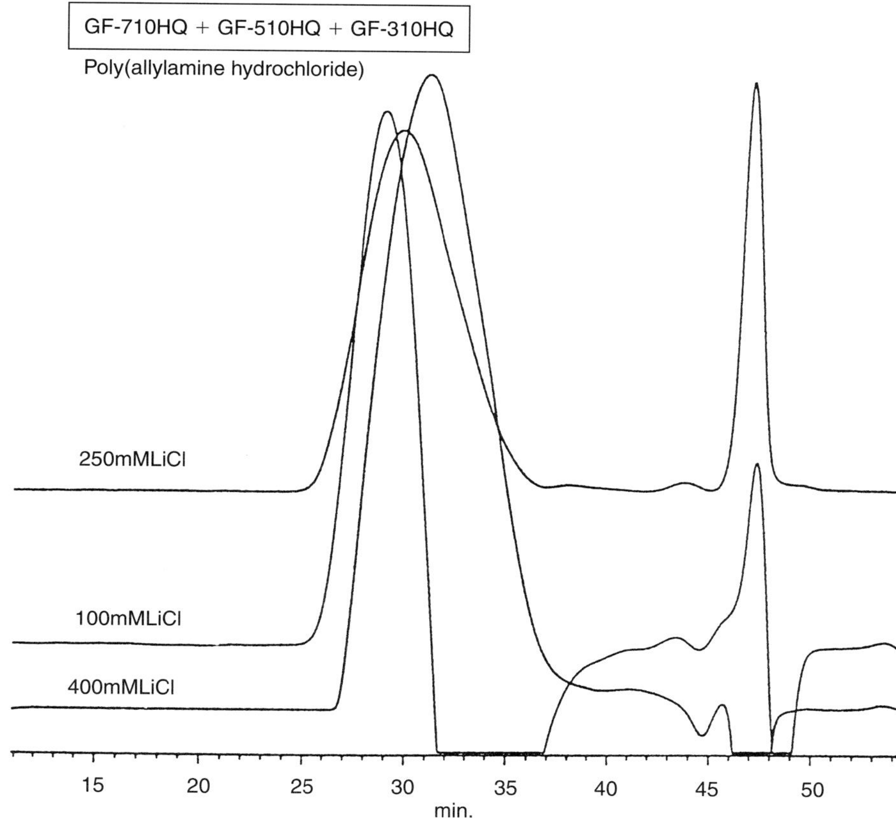

FIGURE 6.29 Poly(allylamine hydrochloride). Column: Shodex Asahipak GF-710 HQ + GF-510 HQ + GF-310 HQ, 7.6 mm i.d. × 300 mm × 3. Eluent: LiCl aqueous solution. Flow rate: 0.6 mL/min. Detector: Shodex RI. Column temp.: 50°C. Sample: 1%, 50 μL: Poly(allylamine hydrochloride).

pullulan, and PEG and PEO do not coincide with each other because the shape and density of each molecule in an eluent are different from each other.

B. Application Examples

Three different types of columns packed with gels of different pore sizes are available. Columns should be selected that are suitable for the molecular weight range of specific samples, as each type has a different exclusion limit (Fig. 6.41, page 215). Bovine serum albumin (BSA), myoglobin, and lysozyme show good peak shapes using only 100 m*M* of sodium phosphate buffer as an eluent. There is no need to add any salt to the eluent to reduce the ionic interaction between protein and gel.

Shodex PROTEIN KW-803 shows good efficiency of up to 0.5 mg of BSA

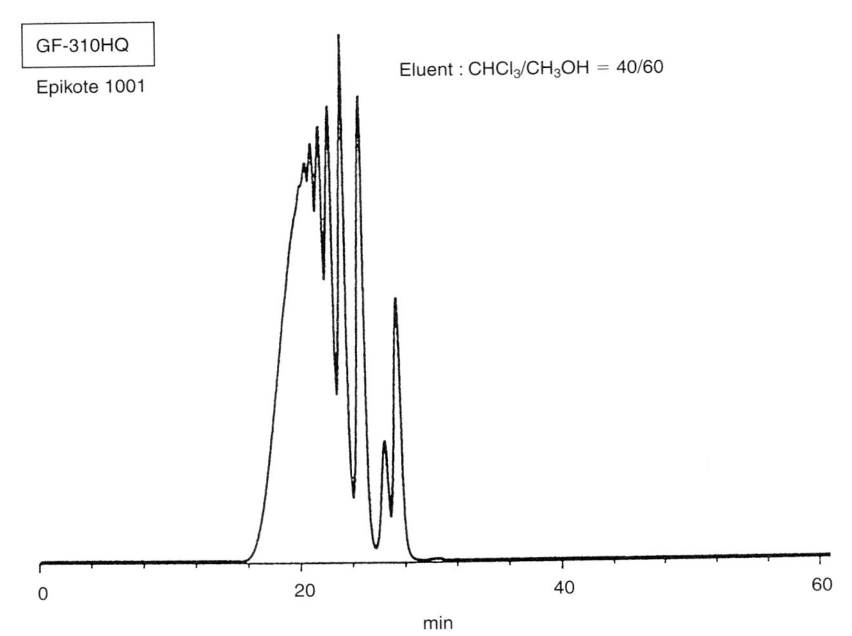

FIGURE 6.30 Epoxy resin. Column: Shodex Asahipak GF-310 HQ, 7.6 mm i.d. × 300 mm. Eluent: $CHCl_3/CH_3OH$ = 40/60. Flow rate: 0.5 mL/min. Detector: Shodex UV (254 nm). Column temp.: 30°C. Sample: 0.5%, 100 μL, Epikote 1001.

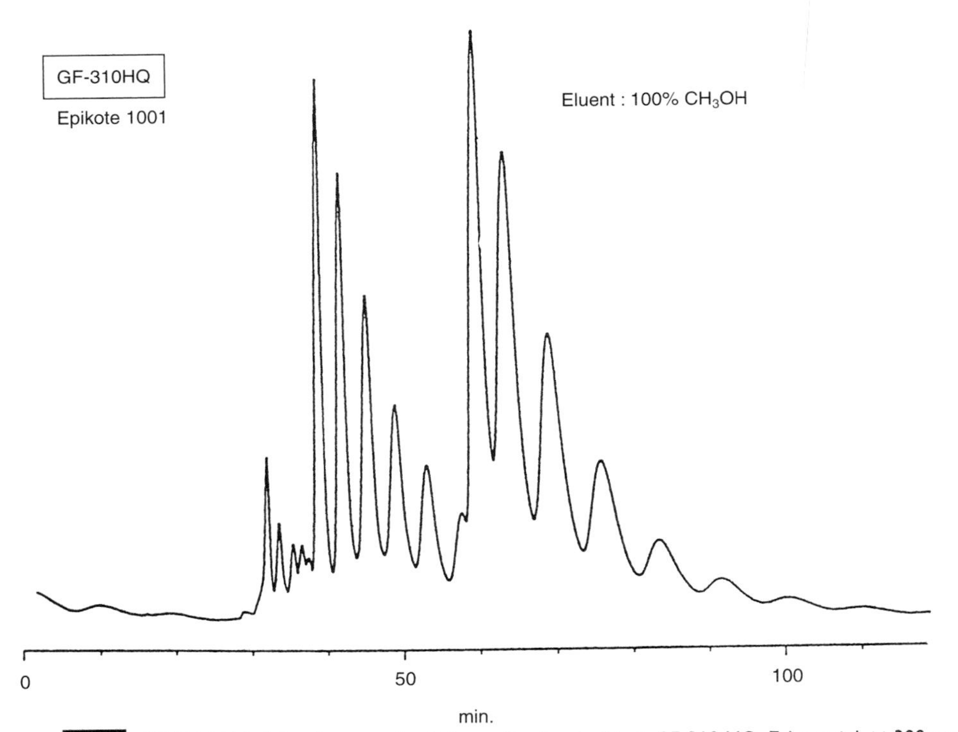

FIGURE 6.31 Epoxy resin. Column: Shodex Asahipak GF-310 HQ, 7.6 mm i.d. × 300 mm. Eluent: CH_3OH. Flow rate: 0.5 mL/min. Detector: Shodex UV (254 nm). Column temp.: 30°C. Sample: 0.5%, 100 μL, Epikote 1001.

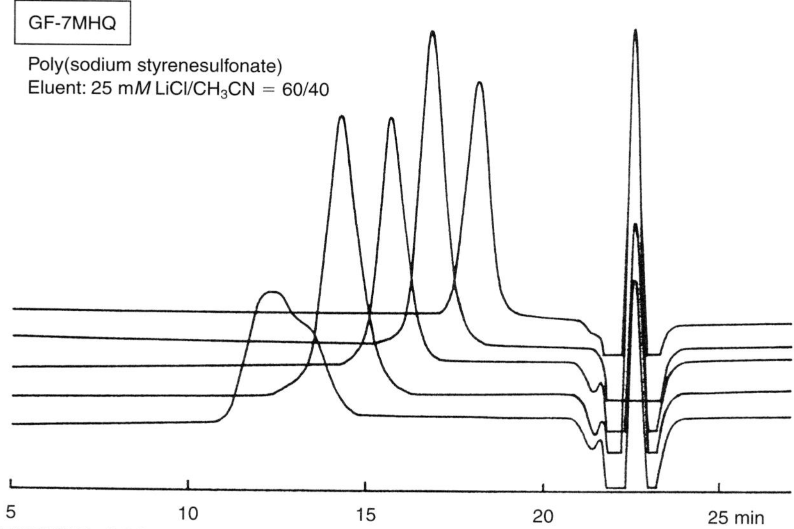

FIGURE 6.32 Poly(sodium styrenesulfonate). Column: Shodex Asahipak GF-7M HQ, 7.6 mm i.d. × 300 mm. Eluent: 25 mM LiCl/CH_3CN = 60/40. Flow rate: 0.4 mL/min. Detector: Shodex RI. Column temp.: 30°C. Sample: 200 μL each: (1) Poly(sodium styrenesulfonate)—990,000, 0.1%; (2) Poly(sodium styrenesulfonate)—149,000, 0.1%; (3) Poly(sodium styrenesulfonate)—48,600, 0.1%; (4) Poly(sodium styrenesulfonate)—16,800, 0.1%; (5) Poly(sodium styrenesulfonate)—4,300, 0.1%.

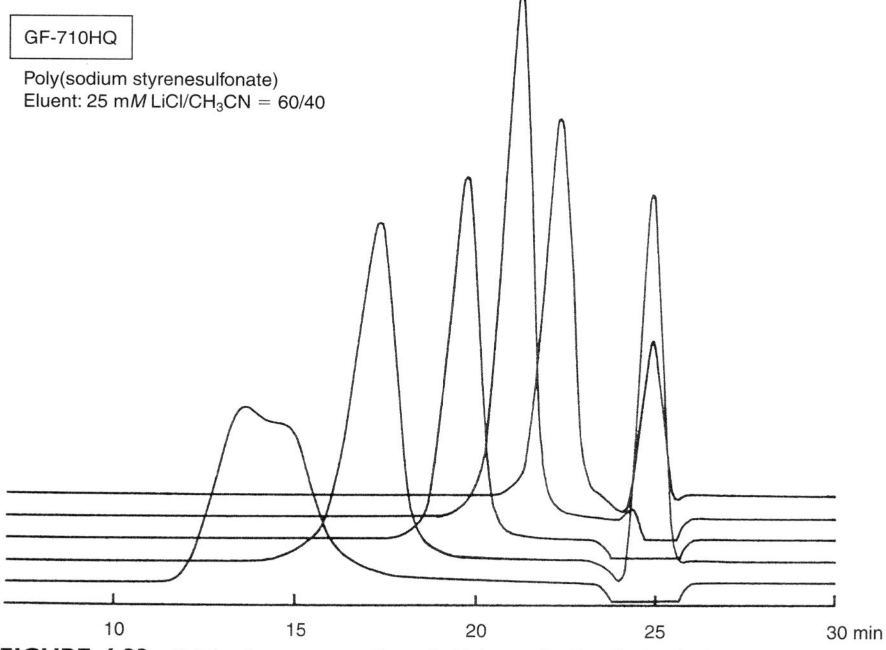

FIGURE 6.33 Poly(sodium styrenesulfonate). Column: Shodex Asahipak GF-710 HQ, 7.6 mm i.d. × 300 mm. Eluent: 25 mM LiCl/CH_3CN = 60/40. Flow rate: 0.4 mL/min. Detector: Shodex RI. Column temp.: 30°C. Sample: 200 μL each: (1) Poly(sodium styrenesulfonate)—990,000, 0.1%; (2) Poly(sodium styrenesulfonate)—149,000, 0.1%; (3) Poly(sodium styrenesulfonate)—48,600, 0.1%; (4) Poly(sodium styrenesulfonate)—16,800, 0.1%; (5) Poly(sodium styrenesulfonate)—4,300, 0.1%.

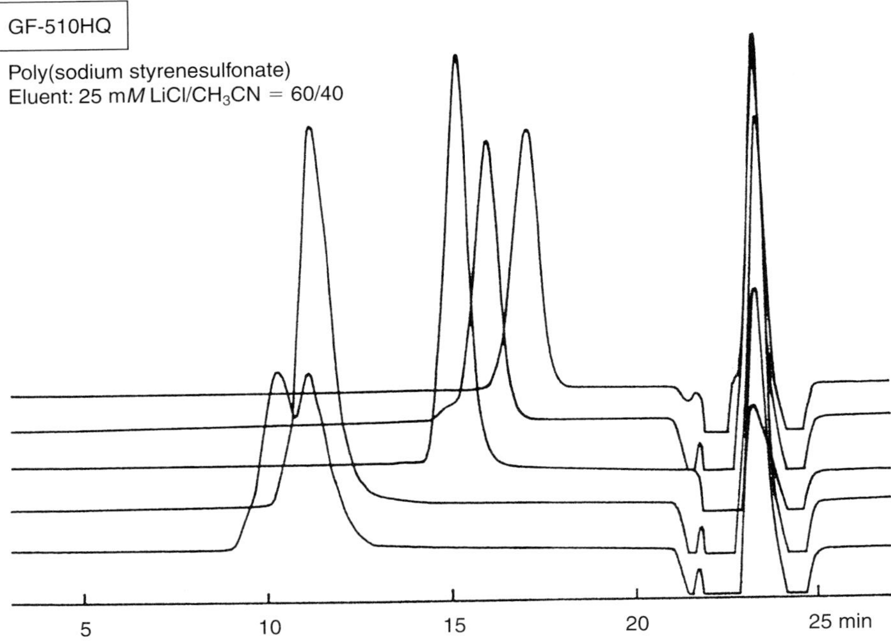

FIGURE 6.34 Poly(sodium styrenesulfonate). Column: Shodex Asahipak GF-510 HQ, 7.6 mm i.d. × 300 mm. Eluent: 25 m*M* LiCl/CH_3CN = 60/40. Flow rate: 0.4 mL/min. Detector: Shodex RI. Column temp.: 30°C. Sample: 200 μL each: (1) Poly(sodium styrenesulfonate)—990,000, 0.1%; (2) Poly(sodium styrenesulfonate)—149,000, 0.1%; (3) Poly(sodium styrenesulfonate)—48,600, 0.1%; (4) Poly(sodium styrenesulfonate)—16,800, 0.1%; (5) Poly(sodium styrenesulfonate)—4,300, 0.1%.

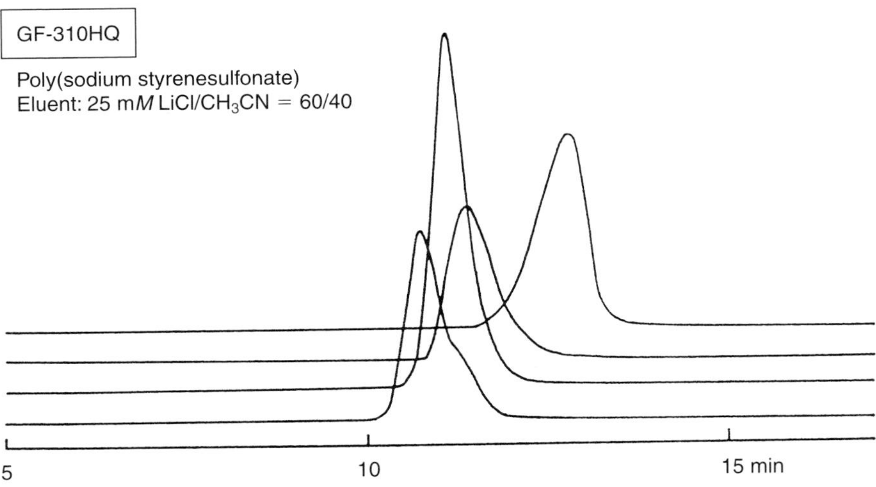

FIGURE 6.35 Poly(sodium styrenesulfonate). Column: Shodex Asahipak GF-310 HQ, 7.6 mm i.d. × 300 mm. Eluent: 25 m*M* LiCl/CH_3CN = 60/40. Flow rate: 0.4 mL/min. Detector: Shodex RI. Column temp.: 30°C. Sample: 200 μL each: (1) Poly(sodium styrenesulfonate)—149,000, 0.1%; (2) Poly(sodium styrenesulfonate)—48,600, 0.1%; (3) Poly(sodium styrenesulfonate)—16,800, 0.1%; (4) Poly(sodium styrenesulfonate)—4,300, 0.1%.

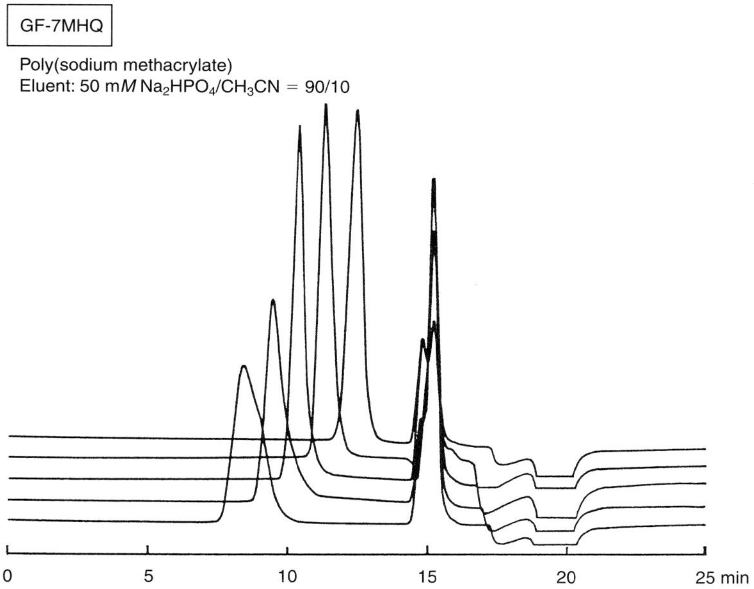

GF-7MHQ

Poly(sodium methacrylate)
Eluent: 50 m*M* Na_2HPO_4/CH_3CN = 90/10

0 5 10 15 20 25 min

FIGURE 6.36 Poly(sodium methacrylate). Column: Shodex Asahipak GF-7M HQ, 7.6 mm i.d. × 300 mm. Eluent: 50 m*M* Na_2HPO_4/CH_3CN = 90/10. Flow rate: 0.6 mL/min. Detector: Shodex RI. Column temp.: 50°C. Sample: 100 μL each: (1) Poly(sodium methacrylate)—700,000, 0.1%; (2) Poly(sodium methacrylate)—210,000, 0.1%; (3) Poly(sodium methacrylate)—86,600, 0.1%; (4) Poly(sodium methacrylate)—25,900, 0.1%; (5) Poly(sodium methacrylate)—6,900, 0.1%.

GF-710HQ

Poly(sodium methacrylate)
Eluent: 50 m*M* Na_2HPO_4/CH_3CN = 90/10

5 10 15 20 25 min

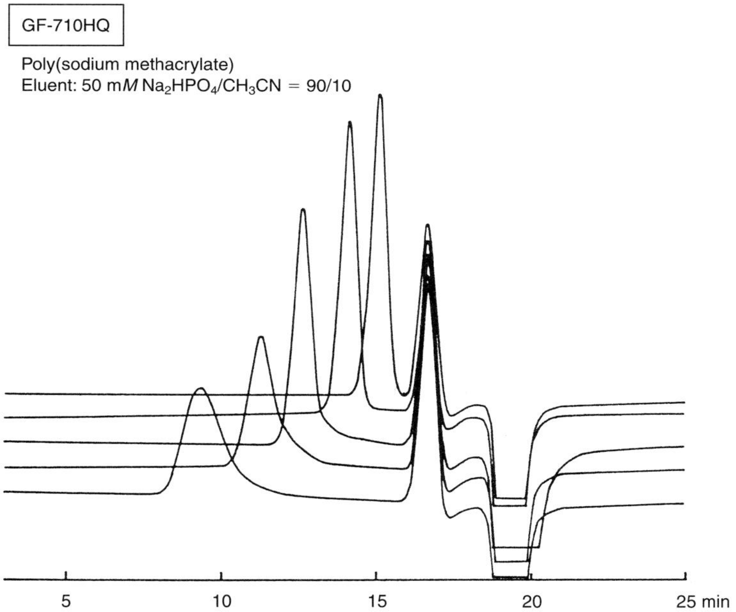

FIGURE 6.37 Poly(sodium methacrylate). Column: Shodex Asahipak GF-710 HQ, 7.6 mm i.d. × 300 mm. Eluent: 50 m*M* Na_2HPO_4/CH_3CN = 90/10. Flow rate: 0.6 mL/min. Detector: Shodex RI. Column temp.: 50°C. Sample: 100 μL each: (1) Poly(sodium methacrylate)—700,000, 0.1%; (2) Poly(sodium methacrylate)—210,000, 0.1%; (3) Poly(sodium methacrylate)—86,600, 0.1%; (4) Poly(sodium methacrylate)—25,900, 0.1%; (5) Poly(sodium methacrylate)—6,900, 0.1%.

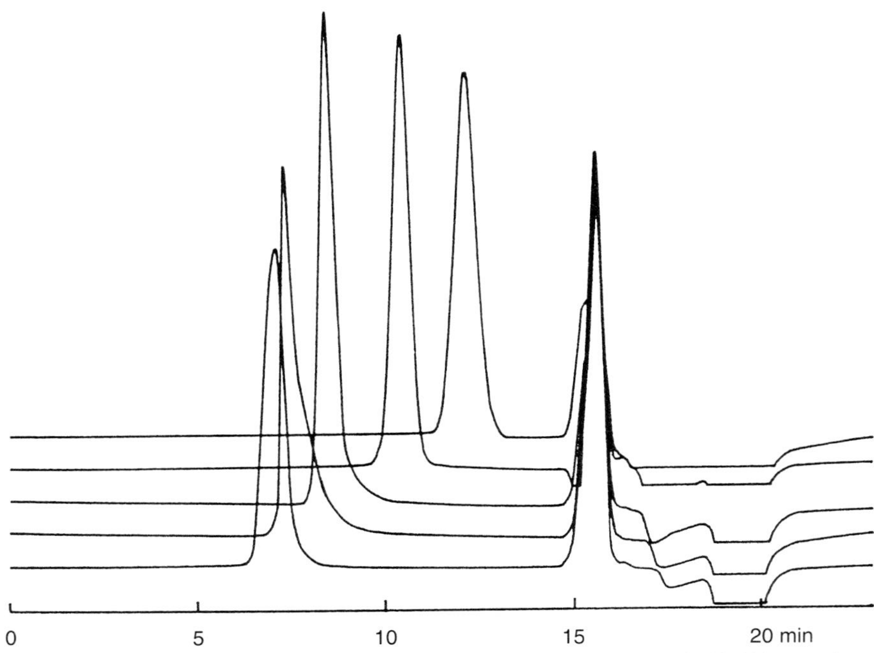

FIGURE 6.38 Poly(sodium methacrylate). Column: Shodex Asahipak GF-510 HQ, 7.6 mm i.d. × 300 mm. Eluent: 50 m*M* Na_2HPO_4/CH_3CN = 90/10. Flow rate: 0.6 mL/min. Detector: Shodex RI. Column temp.: 50°C. Sample: 100 μL each: (1) Poly(sodium methacrylate)—700,000, 0.1%; (2) Poly(sodium methacrylate)—210,000, 0.1%; (3) Poly(sodium methacrylate)—86,600, 0.1%; (4) Poly(sodium methacrylate)—25,900, 0.1%; (5) Poly(sodium methacrylate)—6,900, 0.1%.

loading (Fig. 6.42, page 215). If the sample injection to KW-803 exceeds 1 mg, height equivalent to a theoretical plate (HETP) becomes gradually larger and the peak shape of BSA becomes broader.

C. Preparative and Semipreparative Columns

Shodex preparative and semipreparative columns are supplied in Table 6.15 on page 216.

VII. Conclusion

As a result of continuous technical improvements, new versions of Shodex columns with higher performance are being offered. Some of the conventional Shodex columns are now classified as old versions as seen in Table 6.16 on page 217.

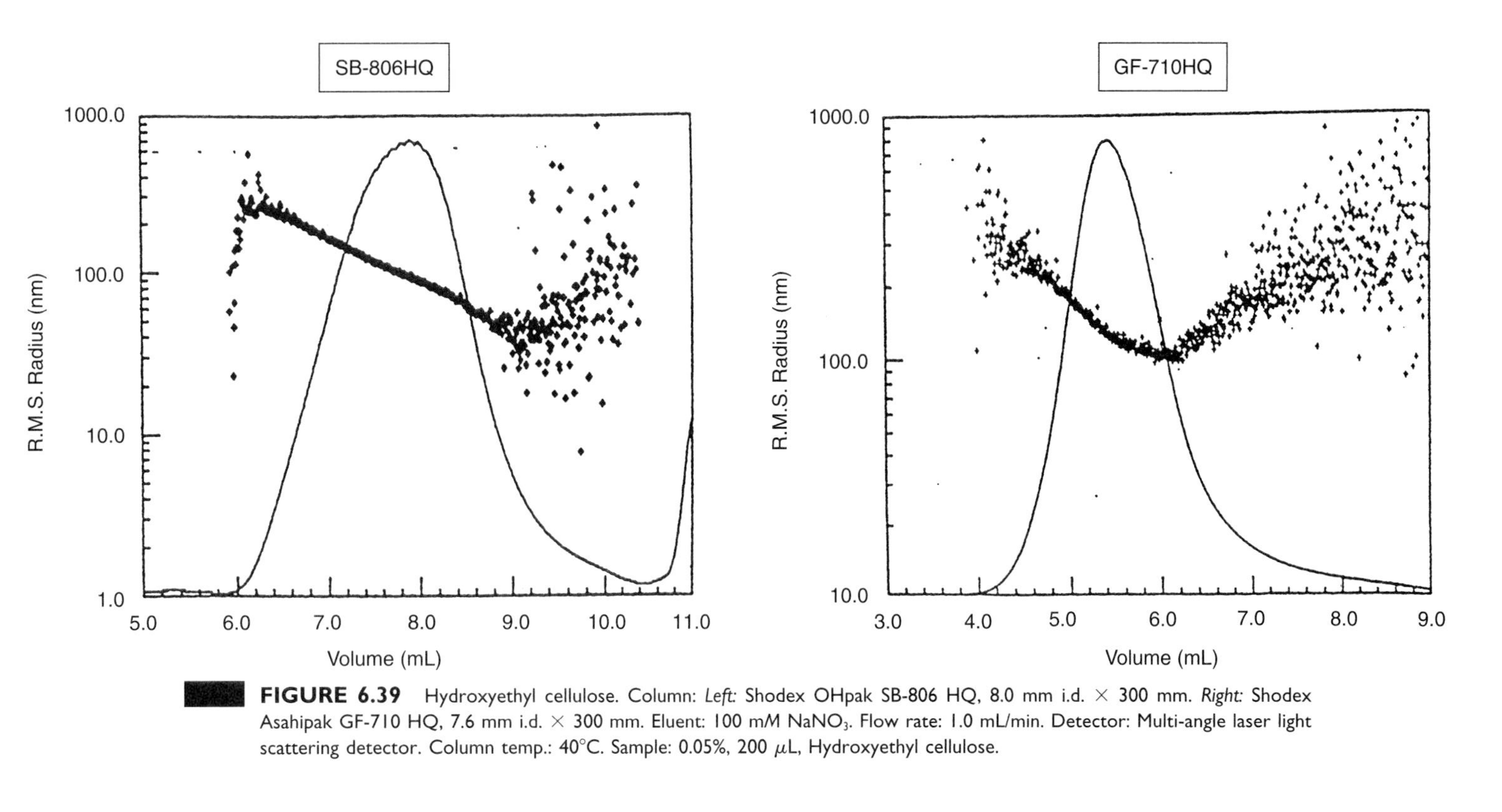

FIGURE 6.39 Hydroxyethyl cellulose. Column: *Left:* Shodex OHpak SB-806 HQ, 8.0 mm i.d. × 300 mm. *Right:* Shodex Asahipak GF-710 HQ, 7.6 mm i.d. × 300 mm. Eluent: 100 m*M* $NaNO_3$. Flow rate: 1.0 mL/min. Detector: Multi-angle laser light scattering detector. Column temp.: 40°C. Sample: 0.05%, 200 μL, Hydroxyethyl cellulose.

TABLE 6.13 Specifications of Shodex Asahipak GS Preparative Columns

Column type	Column size (mm)	Theoretical plate number	Flow rate (ml/min)		Analytical column
			Range	Max	
GS-220 21F	21.5 × 300	>7,000	4–6	8	GS-220HQ
GS-220 21G	21.5 × 500	>12,000	4–6	8	GS-220HQ
GS-620 21F	21.5 × 300	>7,000	4–6	8	GS-620HQ
GS-620 21G	21.5 × 500	>12,000	4–6	8	GS-620HQ
GS-710 21F	21.5 × 300	>7,000	4–6	8	GF-710HQ
GS-710 21G	21.5 × 500	>12,000	5–8	12	GF-710HQ
GS-10G 7B	7.6 × 50		5–8	12	Guard column
GS-320 21F	21.5 × 300	>7,000	2–3	4	GS-320HQ
GS-320 21G	21.5 × 500	>12,000	2–3	4	GS-320HQ
GS-520 21F	21.5 × 300	>7,000	4–6	8	GS-520HQ
GS-520 21G	21.5 × 500	>12,000	4–6	8	GS-520HQ
GS-20G 7B	7.6 × 50				Guard column
GS-310 21F	21.5 × 300	>7,000	5–8	12	GF-310HQ
GS-310 21G	21.5 × 500	>12,000	5–8	12	GF-310HQ
GS-510 21F	21.5 × 300	>7,000	5–8	12	GF-510HQ
GS-510 21G	21.5 × 500	>12,000	5–8	12	GF-510HQ
GSM-700 21F	21.5 × 300	>7,000	2–3	4	GF-7MHQ
GS-20G 7B	7.6 × 50				Guard column

TABLE 6.14 Specifications of Shodex PROTEIN KW-800 Series

Column type	Column size (mm)	Particle size (μm)	Theoretical plate number	Separation range (kDa)		Pore size (Å)	Flow rate (ml/min)		Maximum pressure (kgf/cm^2)	Maximum temperature	pH range	Eluent ionic strength (*M*)
				Pullulan	Protein		Range	Max				
KW-802.5	8 × 300	5	>20,000	1–60	5–150	150	0.5–1.0	1.5	70	45	3–7.5	0.1–0.3
KW-803	8 × 300	5	>20,000	2–170	5–700	300	0.5–1.0	1.5	70	45	3–7.5	0.1–0.3
KW-804	8 × 300	7	>10,000	4–500	10–1000	500	0.5–1.0	1.5	70	45	3–7.5	0.1–0.3

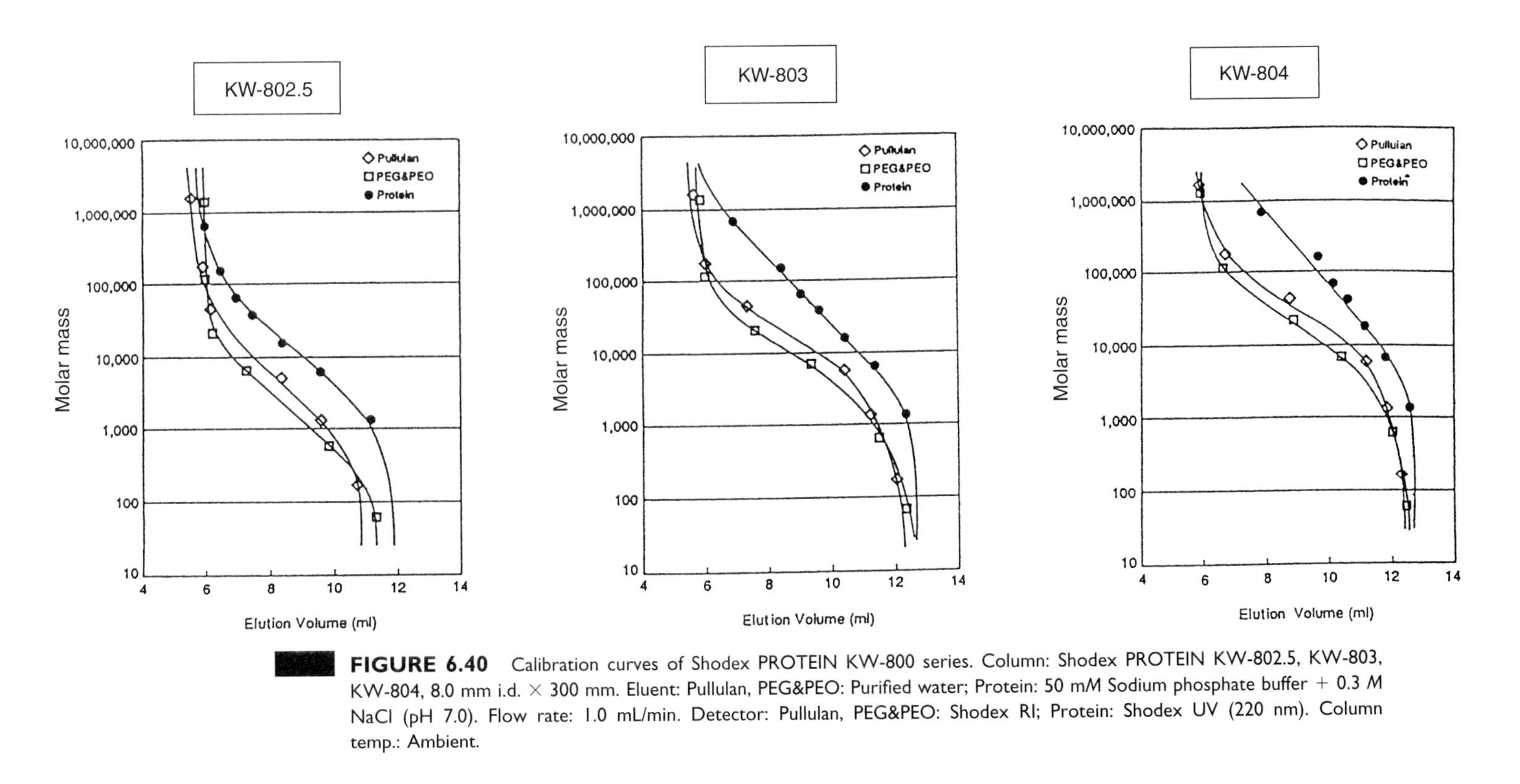

FIGURE 6.40 Calibration curves of Shodex PROTEIN KW-800 series. Column: Shodex PROTEIN KW-802.5, KW-803, KW-804, 8.0 mm i.d. × 300 mm. Eluent: Pullulan, PEG&PEO: Purified water; Protein: 50 m*M* Sodium phosphate buffer + 0.3 *M* NaCl (pH 7.0). Flow rate: 1.0 mL/min. Detector: Pullulan, PEG&PEO: Shodex RI; Protein: Shodex UV (220 nm). Column temp.: Ambient.

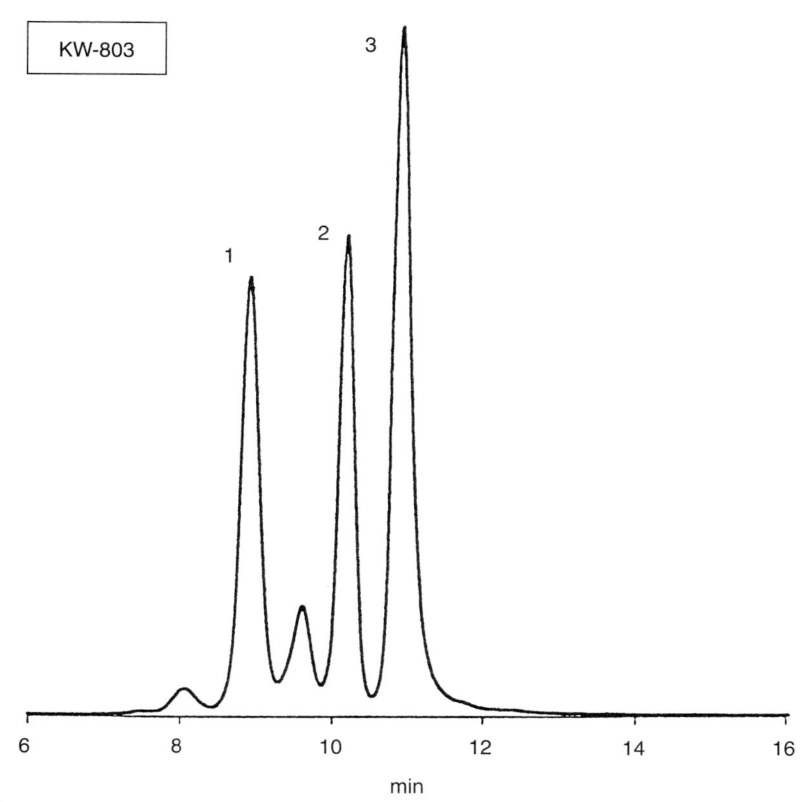

FIGURE 6.41 Proteins. Column: Shodex PROTEIN KW-803, 8.0 mm i.d. × 300 mm. Eluent: 100 m*M* Sodium phosphate buffer (pH 7.0). Flow rate: 1.0 mL/min. Detector: Shodex UV (220 nm). Column temp.: Ambient. Sample: 20 μL: (1) BSA, 0.5 mg/mL; (2) Myoglobin, 0.5 mg/mL; (3) Lysozyme, 0.5mg/mL.

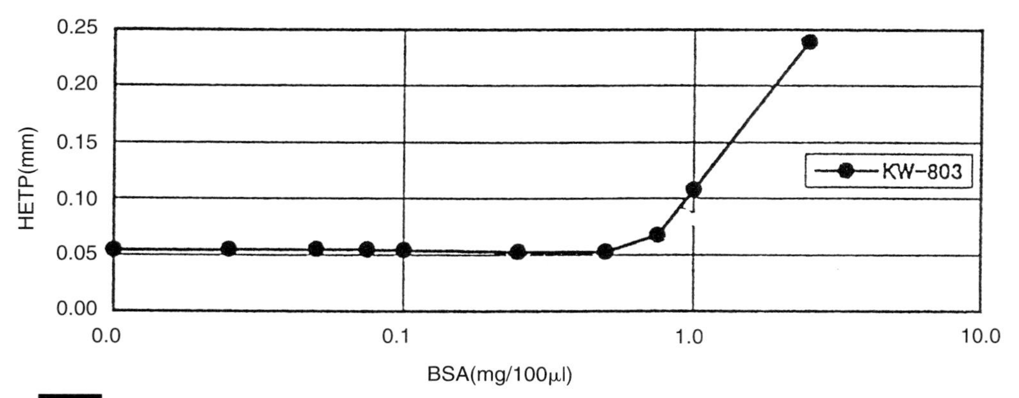

FIGURE 6.42 Effect of sample loading on efficiency. Column: Shodex PROTEIN KW-803, 8.0 mm i.d. × 300 mm. Eluent: 50 m*M* Sodium phosphate buffer + 0.3 *M* NaCl (pH 7.0). Flow rate: 1.0 mL/min. Detector: Shodex UV (280 nm). Column temp.: Ambient. Sample: 0.5 mg/mL, 100 μL, BSA.

TABLE 6.15 Specifications of Shodex PROTEIN KW-2000 Series

Column type	Column size (mm)	Particle size (μm)	Theoretical plate number	Separation range (kDa)		Pore size (Å)	Flow rate (ml/min)		Maximum pressure (kgf/cm^2)	Maximum temperature	pH range	Eluent ionic strength (*M*)	Analytical column
				Pullulan	Protein		Range	Max					
KW-2002.5	20 × 300	5	>16,000	1–60	5–150	150	2–4	6	70	45	3–7.5	0.1–0.3	KW-802.5
KW-2003	20 × 300	5	>16,000	2–170	10–700	300	2–4	6	70	45	3–7.5	0.1–0.3	KW-803
KW-2004	20 × 300	7	>10,000	4–500	10–2000	500	2–4	6	70	45	3–7.5	0.1–0.3	KW-804

TABLE 6.16 Correspondence of New Series with Old Series

Old series	New series
GPC A-800	GPC KF-800
GPC AC-800	GPC K-800
GPC AD-800	GPC KD-800
GPC AT-800	GPC HT-800, UT-800
OHpak B-800	OHpak SB-800HQ
OHpak KB-800	OHpak SB-800HQ
Asahipak GS-()10	Asahipak GF-()10HQ
Asahipak GS-()20	Asahipak GS-()20HQ
PROTEIN WS-800	PROTEIN KW-800

For further information, see the Shodex home pages (http://www.sdk.co.jp/shodex), comprising more than 2500 pages of technical data on high-performance liquid chromatography (HPLC). Anyone can obtain relevant information from the home pages whenever they have questions about HPLC.

7

SIZE EXCLUSION CHROMATOGRAPHY ON FRACTOGEL EMD BioSEC

LOTHAR R. JACOB
LOTHAR BRITSCH

Merck KGaA, 64271 Darmstadt, Germany

I. INTRODUCTION TO SIZE EXCLUSION CHROMATOGRAPHY (SEC)

Size exclusion chromatography is the predominant chromatographic technique for rapidly determining the molecular weight distribution of macromolecules. The method was developed by Porath and Flodin in 1959 (1), although similar effects had already been utilized earlier for the separation of molecules. The separation mode is based on the differential penetration of molecules of different size and shape into the pores of the stationary phase. Thus, in gels made for size exclusion chromatography (SEC), the structure of the pores must have a controlled range of size. SEC, also referred to as gel-filtration or molecular-sieving chromatography (the term gel-permeation chromatography is used when the macromolecules are soluble in organic solvents), is the only method in which no interaction between the chromatographic support and the molecules is supposed to take place. Therefore, this technique is unique among the chromatographic methods. In the very beginning the method was mainly used for desalting protein preparations. A detailed description of the principles and the methodology is reviewed in Hagel (2).

High-performance size exclusion chromatography is used for the characterization of copolymers, as well as for biopolymers (3). The packings for analyses of water-soluble polymers mainly consist of 5- to 10-μm particles derived from deactivated silica or hydrophilic polymeric supports. For the investigation of organosoluble polymers, cross-linked polystyrene beads are still the column packing of choice.

Column Handbook for Size Exclusion Chromatography

In addition to process and quality control the preparative separation of peptides and proteins is an important application area of SEC. One main advantage of SEC is that very mild separation conditions for biological macromolecules can be utilized, such as the presence of essential ions or cofactors, high or low ionic strength, or other buffer additives (detergents, urea, glycerol) and any kind of buffer system and pH. Because the separation mechanism itself is based on the size and shape of analytes, the elution behavior of globular molecules can sometimes be predicted. Because no adsorptive effects occur, high yields of biologically active molecules are often obtained. However, one disadvantage of the technique itself is the limited sample volume.

Different polymeric stationary phases for SEC have been described, including cellulose beads, cross-linked polyvinyl alcohol, and cross-linked glucomannan. However, the most suited and widely distributed packings consist of glycidyl methacrylate copolymers, agarose, or composite gels containing dextran. The methacrylate-based copolymer Fractogel EMD BioSEC for the large-scale purification of biological macromolecules has been introduced to overcome the drawback derived from the poor pressure stability of soft gels. One main advantage of the new Fractogel EMD BioSEC is its wide application range for proteins and large peptides over a molecular weight range from 5 up to 1000 kDa. Compared to conventional soft gels for SEC, the rigid polymer Fractogel EMD BioSEC is easier to pack into large columns, gives very stable gel packings, and, at the same time, allows relatively high linear flow rates. High concentrations of proteins can be loaded onto Fractogel EMD BioSEC columns with no loss of recovery. The high selectivity of Fractogel EMD BioSEC allows efficient purification steps, yielding good purification factors. Because the rigid sorbent is very stable to alkali treatment, the requirements of production-scale chromatography are completely fulfilled.

II. SYNTHESIS OF FRACTOGEL EMD BioSEC

A large range of media for SEC are available commercially for low and medium pressure applications. They differ with respect to the chemical structure of the support, the pore size distribution, and the particle size. At a minimum, several properties must be fulfilled for high-performance chromatographic phases to be used in biological applications. The matrix has to be biocompatible (nontoxic) and hydrophilic. Chemical stability against acid, alkali, and organic solvents is also of importance. Additionally, the gels have to be resistant to biological degradations. Good mechanical stability is necessary for packing the chromatographic phase into big columns with large bed heights.

In contrast to the first gels for SEC that were mainly produced from carbohydrate-derived materials, Fractogel EMD BioSEC is based on the synthetic copolymer Fractogel, a well-established chromatographic matrix that has been used since the mid-1980s. The Fractogel matrix is a synthetic, cross-linked polymethacrylate copolymer. Fractogel EMD BioSEC is the product of a graft polymerization of 2-methoxyethyl acrylamide to the underivatized Fractogel matrix. The synthesis is achieved by a cerium(IV)-induced polymerization reaction that takes advantage of the capability of the cerium ions to produce radicals from aliphatic hydroxy functions on the surface of the matrix polymer. Therefore, the polymerization reaction can only start on the Fractogel matrix, preventing the formation of homopolymer in solution as well as from significant cross-linking of the grafted polymer (4). Due to the possibility of adjusting the correct polymerization conditions, the ligand density can be chosen in a way that the chromatographic support provides an excellent separation of proteins according to their size and shape.

III. PROPERTIES OF FRACTOGEL EMD BioSEC

Fractogel EMD BioSEC is designed for all chromatographic applications up to the production scale. The selectivity curve for a particular Fractogel, obtained by plotting the distribution coefficient versus the logarithm of the molecular weight of different proteins, is shown in Fig. 7.1. The distribution coefficient is independent of the column dimension and can be calculated from the elution volume. The plot yields a sigmoidal curve. The steeper the slope of the curve in the fractionation range, the better the resolution of the support. A standard separation is shown in Fig. 7.2A. As an internal quality control test to achieve the release of the material to the inventory, the dimer of bovine serum albumin (BSA) must be visible in the chromatogram as a distinct peak. The slope of the selectivity curve is affected by the polymerization reaction. During the synthesis of Fractogel EMD BioSEC, the practical working range is adjusted to molecular weights from 5000 up to 1,000,000 Da. The influence of the tentacle modification is illustrated in Fig. 7.2B where the nonmodified base matrix Fractogel HW 65 is compared to Fractogel EMD BioSEC. Only the latter shows the resolving power needed for high-performance SEC.

One striking property of the methacrylate polymer network is the high stability of the methacrylate ester groups against alkaline attack. Therefore, even harsh cleaning procedures utilizing strong alkaline solutions can be applied. As for all tentacle materials, the grafted linear tentacle polymer prevents proteins from coming into contact with the matrix, resulting in very high recoveries for proteins.

The mechanical stability of the material allows higher flow rates compared to conventional soft gels. An example for a pressure drop curve is shown in Fig. 7.3 (page 224). The pressure stability is very helpful during regeneration and equilibration of the columns as higher flow rates can be used. Furthermore, the pressure stability facilitates the packing of larger columns used for the preparative polishing of proteins. The high stability against alkali treatment enables users to set up production-scale separations utilizing Fractogel EMD BioSEC, including alkaline cleaning-in-place (CIP) procedures with 0.1–1 *M* sodium hydroxide. The long-term pH stability of the sorbent is in the range of pH 1–12. The narrow particle size distribution of 20–40 μm provides good efficiency. Fractogel EMD BioSEC is resistant to detergents and organic solvents. Figure 7.4 (page 225) demonstrates the long-term stability after treatment with acetonitrile. Fractogel EMD BioSEC also has constant gel bed volumes at different salt concentrations without significant shrinking or swelling.

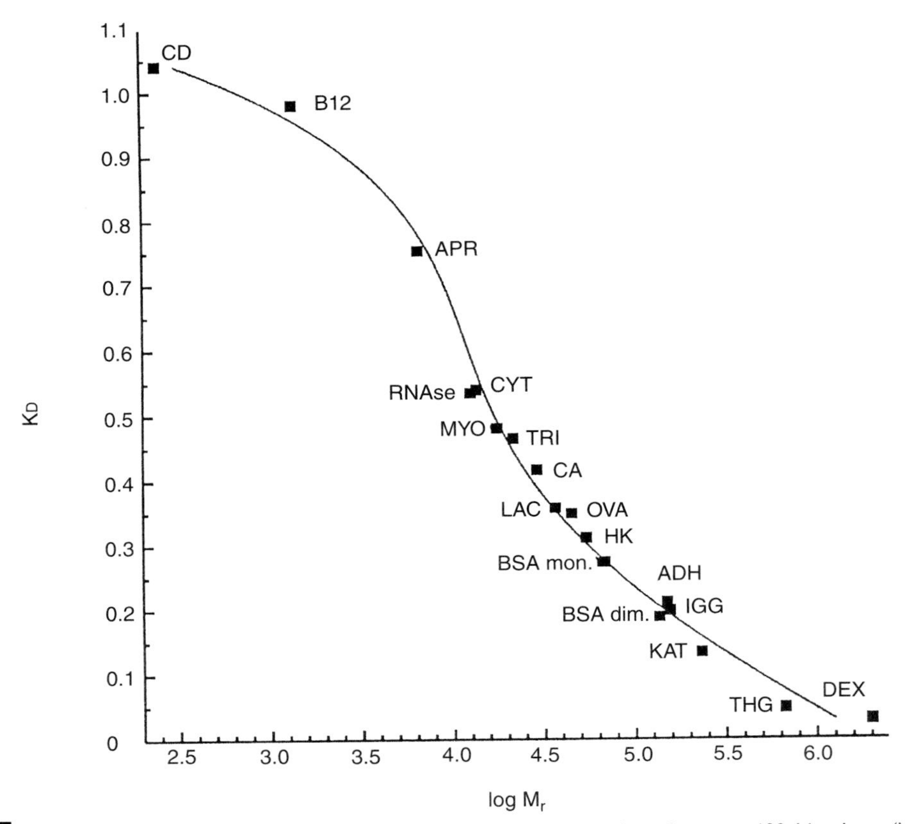

FIGURE 7.1 Selectivity curve of Fractogel EMD BioSEC (S). A Superformance 600-16 column (V_t: 121.6 ml; V_0: 41.1 ml) was used with a linear flow rate of 30 cm/hr utilizing 20 m*M* sodium phosphate containing 0.3 *M* sodium chloride (pH 7.2) as eluent. Molecular mass standards used were as follows (M_r): DEX, Dextran Blue 2000 (2,000,000); THG, thyroglobulin (669,000); KAT, catalase (bovine) (232,000) IGG, immunoglobulin G (156,000); ADH, alcohol dehydrogenase (*S. cerevisiae*) (150,000); BSA dim., dimer of bovine serum albumin (136,000); BSA mon., monomer of BSA (68,000); HK, hexokinase (54,000); OVA, ovalbumin, (45,000); LAC, β-lactoglobulin A (36,600); CA, carbonic anhydrase (29,000); TRI, trypsin inhibitor (soybean) (21,500); MYO, myoglobin (17,500); RNase, ribonuclease A (13,500); CYT, cytochrome c (12,500); APR, aprotinin (6500); B12, vitamin B_{12} (1355); and CD, cytidine (243).

Because the separation is nearly independent of the total amount of protein loaded onto the column, no limitation with respect to the protein concentration exists for concentrations at least up to 60 mg. The high mechanical stability enables the injection of even highly viscous samples with high concentrations of protein.

IV. ADEQUATE BUFFERS

In theory, SEC of proteins depends only on their molecular size. Sometimes the size of a protein varies with the ionic strength of the buffer (5,6). The concentration of salt not only affects the conformation of the protein, but can also influence the chromatographic separation itself. Additional retention

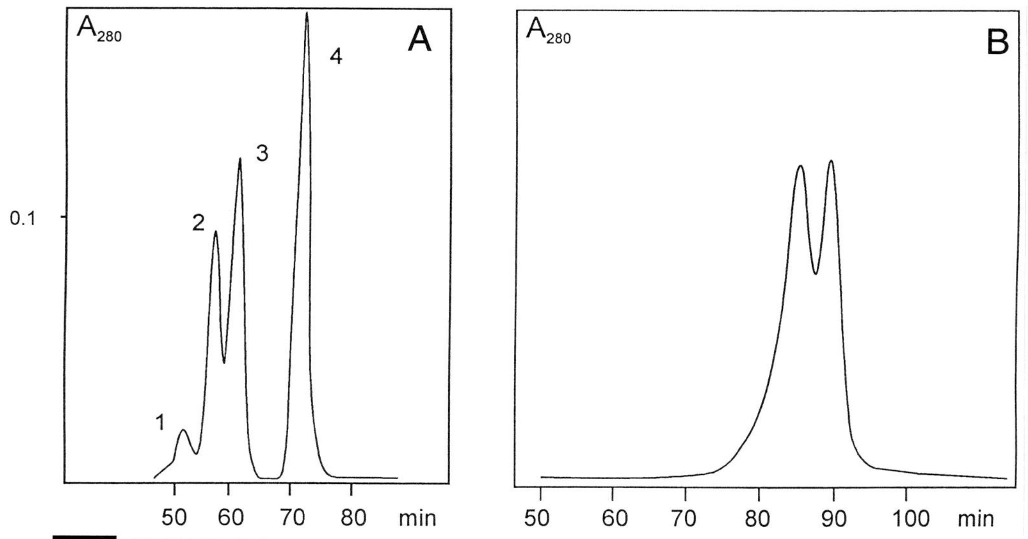

FIGURE 7.2 (A) Separation of a standard protein mixture. A test mixture consisting of BSA (5 mg), dimer (peak 1) and monomer (peak 2), ovalbumin (5 mg) (peak 3), and cytochrome c (3 mg) (peak 4) was loaded onto a Fractogel EMD BioSEC column (600 × 16 mm) with a bed height of 600 mm. PBS (pH 7.2) was used as the eluent at a flow rate of 1 ml/min; the sample volume was 0.5 ml. (B) The same protein sample as in A was injected onto a column of identical dimensions packed with unmodified Fractogel HW 65. Without the tentacle modification the base matrix displays only a poor resolution of the test mixture.

effects may be due to electrostatic and hydrophobic interactions. For example, the retention time of cytochrome c is affected significantly at low ionic strength (Fig. 7.5, page 226). Cytochrome c is positively charged at pH 7.2 and may thus be retained by ionic interactions with carboxy groups of the methacrylate-based matrix. However, at concentrations above 50 m*M* NaCl (up to 300 m*M*) no interaction is observed. Because the negative charges on ovalbumin are sparse compared to the positive charges on cytochrome c, ovalbumin is not repulsed from the gel under the same conditions. Probably, not only weak ionic interactions can occur in SEC. Sometimes, weak nonspecific interactions also cause a delayed elution of certain compounds. However, the modification of the matrix with hydrophilic tentacles leads to a reduction of contacts between the proteins and the gel. Despite this fact, the buffer should contain 50 to 300 m*M* NaCl to suppress the ionic interactions between the proteins and the matrix. The user should also take care that no precipitation of protein by salt occurs, which could clog the column. Very high concentrations of salt are not suitable because a precipitation of proteins can occur (salting out), and lower but elevated concentrations can result in hydrophobic interactions (which will cause later elution). As a result, poor resolution or incorrect values with respect to the estimated molecular mass of unknown proteins can be observed. A pure sizing effect is difficult to obtain when a complex protein mixture is applied. Thus, the best buffer composition for a given separation problem has to be worked out experimentally.

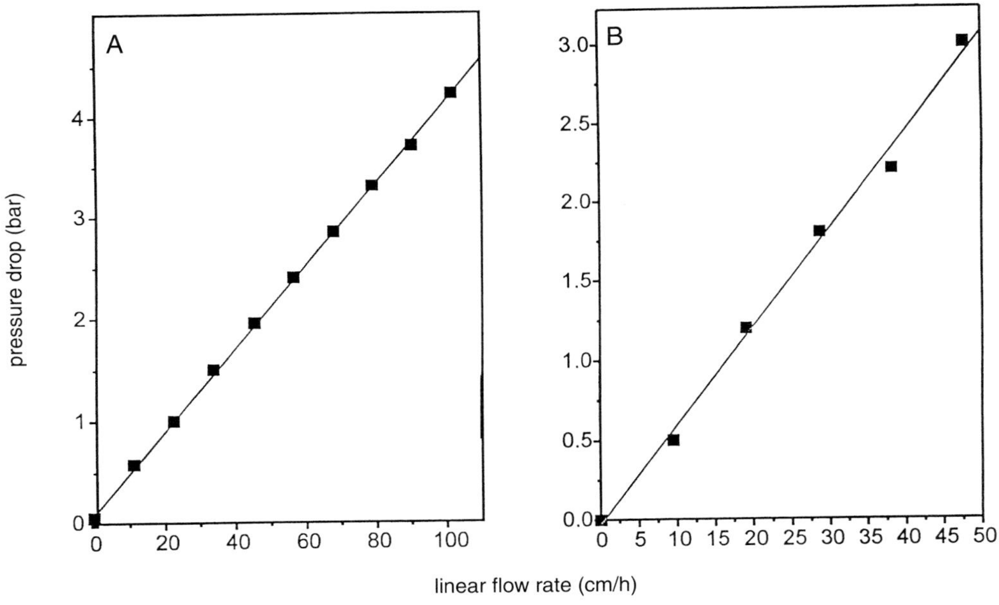

FIGURE 7.3 (A) To demonstrate the dependence of pressure drop from the linear flow rate of Fractogel EMD BioSEC (S) a Superformance column 600-26 packed with Fractogel EMD BioSEC (S) (V_t 318.5 ml; 600 mm bed height) was connected to an external pressure gauge (0.1–6.0 bar), and a HPLC pump. The linear flow rate was increased successively from 10 up to 100 cm/hr and the pressure drop was measured. A pressure drop of 1 bar was reached at a linear flow rate of approximately 25 cm/hr. (B) The same experiment was performed using a Superformance 1000-200 preparative column (200 mm i.d.) with a bed volume of 20 liter (63 cm bed height) and a maximum flow rate of 50 cm/hr. A pressure drop of 1 bar was reached at a linear flow rate of approximately 17 cm/hr.

V. RECOMMENDED COLUMNS FOR SEC

It should always be kept in mind that the column length will influence the resolution of any size exclusion column. For laboratory purposes the column dimensions should range from 600 × 16 mm to 1000 × 50 mm. Using smaller columns (e.g., 1 cm i.d.), sometimes a loss of resolution can be observed due to wall effects or inhomogeneous column packing. However, longer columns of 16 mm inner diameter can be packed more easily with Fractogel EMD BioSEC. Figure 7.6 (page 227) shows chromatograms obtained from two different SEC columns with increasing length (16 mm i.d.).

To obtain high resolutions the columns should be of high quality. Usually, columns made of glass or transparent plastic tubes are preferred to ensure the control of the packing and operating steps by eye. As important as the absence of dead volumes is the even distribution of sample in the column adapter. Depending on the quality of the column, linear flow rates of 100 cm/hr and higher can be applied. Such high flow rates are used mostly for the equilibration or regeneration of the column. Columns other than Superformance may be used with Fractogel EMD BioSEC (S). However, the column manufacturers' instructions must be followed with respect to the pressure limits. Certain col-

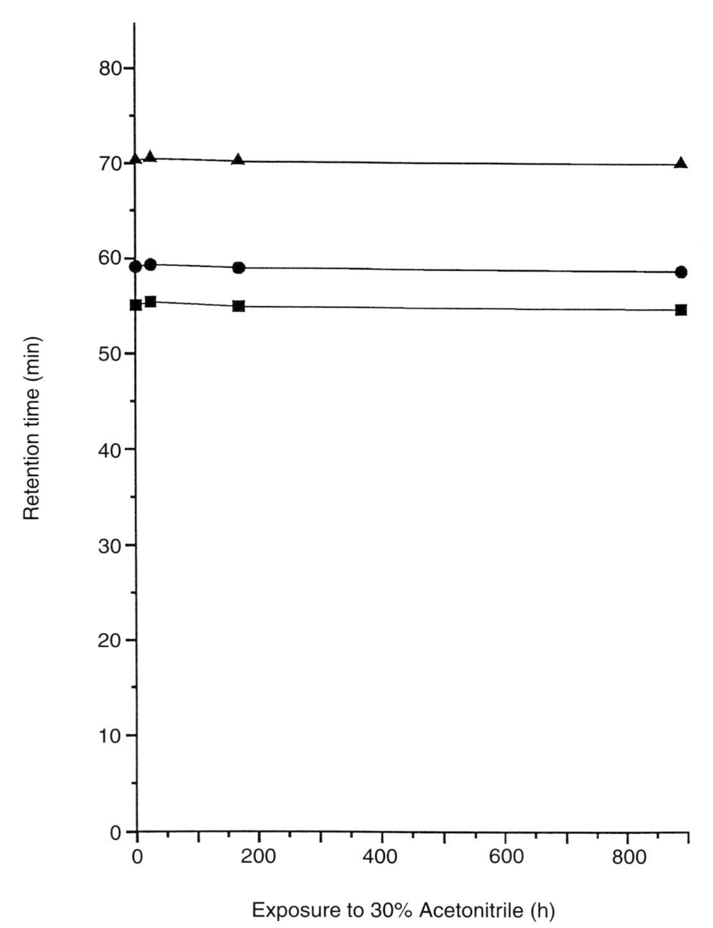

FIGURE 7.4 Separation of a standard protein mixture on a Fractogel EMD BioSEC-column (600-16 mm) after incubation with 30% acetonitrile. The sample contained BSA (■), ovalbumin (●), and cytochrome c (▲) (sample volume: 500 μl; flow rate: 1.0 ml/min). No significant shifts of the retention times and no loss of the resolution were observed even after 900 hr of exposure.

umns will not allow the high linear flow rates during packing or equilibration, which are generally applicable with Superformance columns. In this case the actual flow rates have to be reduced appropriately.

For production-scale separations, column diameters up to 30 cm are recommended. Usually the length of the column is in the range of 600–1200 mm for smaller column diameters (less than 50 mm). Columns with larger diameters can be packed up to 900 mm.

Although the viscosity of the sample solution may affect the resolution, for practical reasons highly concentrated protein samples will give the best separations in the case of SEC with respect to the process economy. Although the actual loading capacity depends on the separation problem and on the

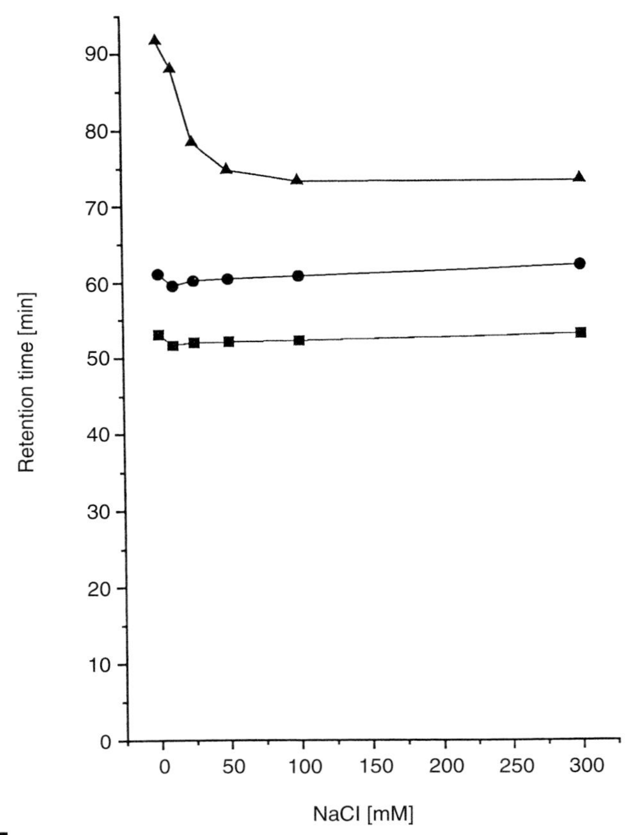

FIGURE 7.5 Dependence of retention time of various proteins from ionic strength on Fractogel EMD BioSEC (S). A sample consisting of 0.5 ml of BSA (■), ovalbumin (●), and cytochrome c (▲) (5/5/3 mg/ml) was chromatographed at various concentrations of sodium chloride (20 m*M* sodium phosphate buffer, pH 7.2). The retention times are obtained from a Fractogel EMD BioSEC (S) 600-16 column (600 mm bed height) at a flow rate of 1 ml/min corresponding to a linear flow rate of 30 cm/hr.

sample composition, the sample volumes given in Table 7.1 can be used for the first attempt. Further experiments must be carried out to obtain the best experimental conditions.

VI. PACKING THE COLUMN

In general, for size exclusion chromatography, packing of the column must be carried out very thoroughly to obtain optimal results. With soft gels, column packing has often been an obstacle and much time was required. Although no such problems occur with physically stable methacrylate polymers, an improperly packed column has to be avoided. The packing of Fractogel and the dependence of column performance on packing velocity have been described (7). In addition, the optimal packing velocity and its relationship to the column dimensions have been investigated (8). High column efficiency was obtained

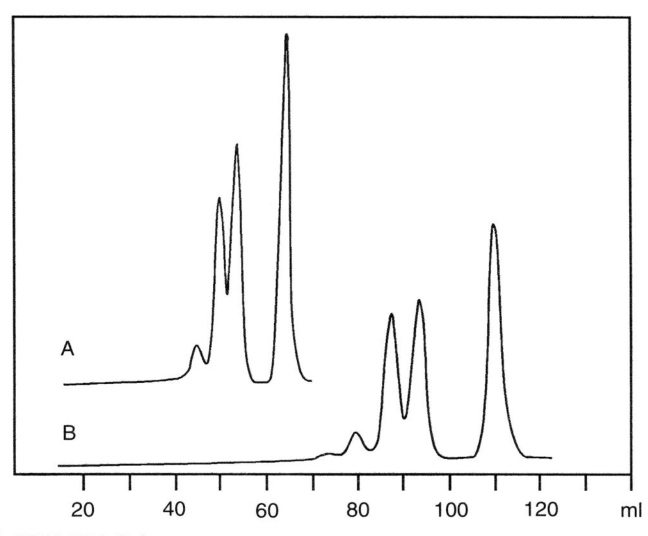

FIGURE 7.6 Effect of column length on the separation efficiency. Two different Fractogel EMD BioSEC columns (A: 600 × 16 mm, B: 1000 × 16 mm) were tested using BSA, ovalbumin, and cytochrome c (5/5/3 mg/ml) as sample (20 m*M* sodium dihydrogen phosphate, 300 m*M* NaCl, pH 7.2; 0.5 ml/min). Better resolution can be achieved using longer columns.

at high packing velocities. As a result of the high velocity, a dense and even packing can be achieved. For small-scale applications, columns with inner diameters between 16 and 50 mm should be used. However, it is also possible to use smaller columns for SEC. With decreasing column inner diameters, the optimal packing velocity increases. A further important issue is the pressure stability of the chromatographic column. For example, 10 ml/min is the recommended starting flow rate for Superformance columns (Merck KGaA) with a 16-mm-inner diameter and a maximum pressure stability of 80 bar. If the

TABLE 7.1 Recommended Maximum Sample Volumes for Different SEC Column Dimensions

Column dimension	Bed height (cm)	Maximum sample volume (ml)
600-16	60	6
1200-16	120	12
600-26	60	16
1200-26	120	32
1000-50	60	57
1000-50	95	90
1000-100	60	234
1000-100	95	370
1000-200	56	879
1000-200	95	1491

column is less pressure stable (maximum pressure rate 10 bar), the starting flow rate may be 8 ml/min. The flow rate should be reduced to 5 ml/min if a pressure of 8 bar is reached to avoid cracking of the glass tube.

Because temperature shifts may also influence the packing quality, the temperature should not be changed during the chromatographic step and the packing of the column should be done at the operation temperature. To prevent the denaturation of sensitive proteins, the chromatography is carried out in a cold chamber (or cabinet). For this purpose the column packing has to be performed at the same ambient temperature (store the gel before use at the same temperature!).

For packing and operation the column can be connected to any high-performance liquid chromatography (HPLC) system or medium pressure equipment if the pump can work precisely with constant flow rates. Using a peristaltic pump the volumetric flow rate should be calibrated with equilibration buffer. In general, all tubing should be as short as possible to avoid dilution of the sample either prior to reaching the gel bed or after leaving the bottom adapter. However, the inner diameter of the tubing has to be large enough to avoid increased back pressure of the system. Thus, HPLC tubing with inner diameters less than 0.5 mm are not recommended for laboratory-scale columns. Because the buffers used for this technique contain salt, inert pump systems are preferred. Also, the column lifetime may be prolonged if connected to inert systems. Stainless-steel components may corrode if buffers are used, giving rise to the formation of metal oxides that can be deposited in the chromatographic support. This may result in a loss of resolution and recovery and will probably damage the column packing after a while.

A. Preparing the Gel

The gel is supplied as a suspension in 20% aqueous ethanol. Before packing the storage solution should be removed from the gel.

Place the appropriate amount of gel suspension in a sintered-glass funnel, remove the supernatant through suction, and resuspend the gel three times in a threefold excess of water, followed by suction (see also Section IX).

B. Packing a Superformance 600-16 Column

1. Mount the column, including the lower adaptor and the filling tube, according to the column instruction manual.

2. Prepare a homogeneous gel slurry by mixing 160 g of gel from the sintered glass funnel with 160 ml of a 1 *M* sodium chloride solution.

3. Pour the slurry into the vertically mounted column with the column outlet closed and add water on top of the slurry so that the level just reaches the upper end of the filling tube.

4. Mount the upper end piece of the filling tube and immediately connect it to the pump, open the column outlet, and start the pump at a flow rate of 10 ml/min (delivering water or any desired buffer system of low ionic strength). Optionally, the column outlet can be additionally connected to a water jet pump, which has to be operated during the first 2 min of packing. The water

jet vacuum may be disconnected after 2 min while the column runs under a high flow rate.

5. Allow the gel to pack completely within 10–15 min and then stop the pump and close the column outlet.

6. Decant or aspirate the supernatant from the gel bed and disconnect the filling tube.

7. Transfer excess gel from the filling tube into a bottle supplied with storage medium.

8. Rinse the column end piece with water and remove excess gel until the gel surface has a distance of 5–10 mm to the end of the glass tube.

9. Utilize the filter tool to insert the polypropylene filter plate on top of the gel. Get rid of any gel particles above the filter by thoroughly rinsing with water.

10. Mount the upper adaptor, thereby compressing the gel bed, until the filter plate reaches the height of the column end piece. The lower adaptor should be adjusted before starting the packing procedure in order to reach 600 ± 10 mm total bed height.

11. Equilibrate the packed gel with at least 3 bed volumes at 1 ml/min of the desired buffer system. It is recommended to include 50–300 m*M* of salt in the buffer.

C. Packing of Preparative Columns

The packing material of Fractogel EMD BioSEC columns is rather rigid compared to soft gels such as agaroses. As with small columns, the packing velocity should also be high for preparative columns. However, the optimal packing velocities are inversely proportional to the column length. The optimal packing velocity decreases with increasing column inner diameter. For the packing of larger columns it is also necessary to know that the Fractogel EMD BioSEC material can be compressed to a certain extent. Figure 7.7 shows the compressibility of the sorbent. The results were obtained from 25 ml of settled Fractogel EMD BioSEC after packing into a Superformance 150-16 column (16 mm i.d.) at 9 ml/min using different metal cylinders of increasing weight to compress the gel bed. The compressed bed height was measured each time after reaching a constant level. The gel bed was allowed to relax for 1 min after each compression treatment. Compression forces were calculated, including the weight of the insertion tool, e.g., compression of a completely settled gel using a pressure of 10 N/cm^2 (roughly equivalent to the pressure caused by a mass of 1 kg, which is allowed to press down an area of 1 cm^2 by its weight) will result in a reduction of bed volume to approximately 93% of the initial volume.

As a practical result, the amount of gel to be prepared for a preparative column must exceed the nominal volume of the final column by 10%. For the packing of production-scale columns the maximum pressure rate of the column has to be considered. The large columns consist mostly of borosilicate glass tubes with similar pressure stabilities. For example, a Superformance column with dimensions of 1000 mm in length and 50 mm in width is pressure stable up to 14 bar. Therefore, Fractogel EMD BioSEC should be packed with a

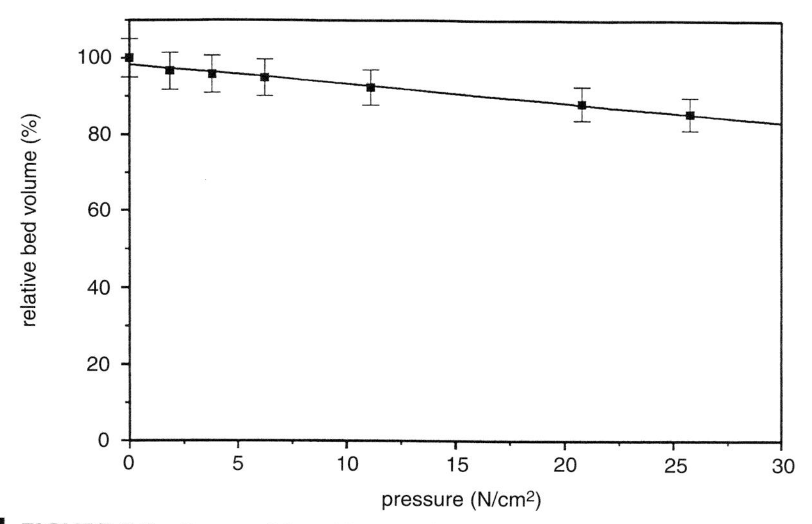

FIGURE 7.7 Compressibility of Fractogel EMD BioSEC (S) (for details, see text; Section VI,C).

starting flow rate of 50 ml/min. If a packing device is used, the filling tube should contain a solution of 2 *M* NaCl. The pressure is about 9 bar and must be monitored during the packing procedure. If the pressure increases, the flow rate has to be lowered from 50 to 25 ml/min in several steps. After the packing is completed the adapter has to be adjusted to the upper gel bed surface. The column is then equilibrated with the corresponding working buffer. Figure 7.8 summarizes the packing protocol for a column with dimensions of 1000 × 100 mm. The column was packed to a bed height of approximately 79 cm at a packing pressure as indicated. The flow rate was lowered from a starting velocity of 107 to 32 cm/hr in subsequent steps not to exceed the maximum pressure rate of the column. The packing of a 200-mm-diameter column can be performed in the same way; however, one should start at a flow rate of 30–40 liter/hr. The final velocity for a column with a bed height of about 60 cm is in the range of 30–55 cm/hr. Table 7.2 lists the starting packing velocities for different column dimensions.

D. Checking the Quality of the Packed Gel Bed

The performance and basic characteristics of the packed column can be checked by running the test chromatograms as follows:

1. Equilibrate the column with 3 bed volumes of 20 m*M* sodium phosphate buffer, 300 m*M* sodium chloride, pH 7.2.
2. Run the column with a flow rate of 1 ml/min (30 cm/hr) and inject 0.5 ml of 0.3% acetone in water; monitor the eluate at 280 nm.
3. Run the column with a flow rate of 1 ml/min (30 cm/hr) and inject 0.5 ml of a protein solution (5 mg/ml bovine serum albumin, 5 mg/ml ovalbumin, 3 mg/ml cytochrome c) in the same buffer as before; monitor the eluate at 280 nm.

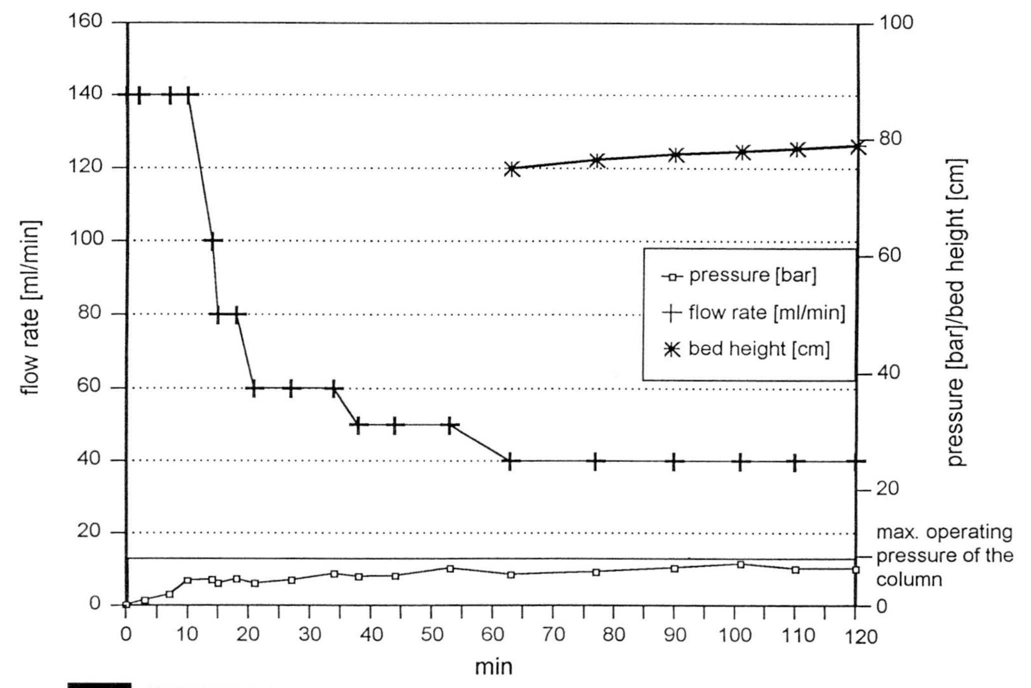

FIGURE 7.8 Illustration of the packing procedure developed for a Superformance 1000-100 column (100 mm i.d.; 950 mm bed height).

4. Run the column with a flow rate of 1 ml/min (30 cm/hr) and inject a thoroughly prepared mixture of 25 μl of high molecular weight calf thymus DNA (Boehringer, Mannheim, article No. 104 167; stock solution with 2–3 mg/ml; avoid vigorous shaking during preparation of the mixture) and 0.475 ml of a solution of 2.5 g sodium nitrate in equilibration buffer; monitor the eluate at 280 nm.

TABLE 7.2 Recommended Packing Flow Rates for SEC Columns

Column	Maximum pressure rate (bar)	Bed height (cm)	Recommended starting flow rate		Final packing and operating flow rate	
			ml/min	cm/hr	ml/min	cm/hr
Superformance 600-16	80	60	10	298	1	29.4
Pharmacia XK 16 × 600	10	60	8	240	1	29.4
Superformance 600-26	60	60	27	302	2.6	29.1
Superformance 1000-50	14	95	50	158	10	31.6
Superformance 1000-100	8	79	140	107	42	32.3
Superformance 1000-200	4	56	500	95	157	30.0

5. Run the column at a flow rate of 1 ml/min (30 cm/hr) and inject 0.5 ml of a solution of Dextran Blue 2000 (1 mg/ml in buffer, centrifuged and filtered through a 0.22-μm filter membrane); monitor the eluate at 280 nm.

The elution volumes of Dextran Blue or DNA are equivalent to the void volume (V_0) of the column. Usually, a small difference appears between the two values due to differences in molecular shape and slight nonspecific interaction with the matrix.

The elution volume of sodium nitrate [V_e ($NaNO_3$)] allows one to calculate the volume of solid matrix (V_m) in the column according to

$$V_m = V_t - V_e\,(NaNO_3),$$

where V_t is the total bed volume in the column. In turn, the intraporous volume (V_p) is given by

$$V_p = V_t - V_0 - V_m.$$

The intraporous volumes and solid matrix volumes can be expressed as a percentage of the total bed volumes and should be equivalent to 50–55 and 15–18%, respectively. The solid matrix volume depends on the packing density and can be taken as a qualitative number for the control of the reproducibility of repeated packing procedures.

The chromatogram of the protein mixture should show the partial separation of serum albumin and ovalbumin with a trough of at least 30% of height between their peak signals and baseline separation between ovalbumin and cytochrome c. If present in the sample, the dimeric form of serum albumin should also appear as an individual peak signal before elution of the monomeric form.

The position and shape of the peak signal in a chromatogram of acetone can be used to calculate the number of theoretical plates per meter of bed height (N) according to

$$N = 5.54 \times \left(\frac{V_e}{W_{1/2}}\right)^2 \times \frac{100}{BH},$$

where V_e is the elution volume of acetone, $W_{1/2}$ the peak width at half-height of the peak signal (given in volume units to abolish the dependence on recorder speed), and BH is the bed height in centimeters.

VII. RECOMMENDED FLOW RATES FOR SEC COLUMNS

The flow rate in SEC significantly affects the resolution. Depending on the selectivity wanted, linear flow rates have to be adapted to the column dimensions. In general, running the column at a low flow rate results in higher resolution, but diffusion may produce diminishing resolution when the flow rate is too low. The flow rates recommended for a particular column diameter should not be increased. In the case of Superformance columns, the best results can be obtained by applying linear flow rates of about 30–80 cm/hr. Of course, linear flow rates below 30 cm/hr can contribute to further increased resolution.

However, sometimes only a little loss of resolution is obtained at flow rates up to 100 cm/hr or more depending on the difference in size of the analytes. All equilibration and regeneration steps may be performed at high linear flow rates (for aqueous buffers or solutions up to 150 cm/hr) because of the high mechanical stability of the Fractogel EMD matrix. The high flow rate properties of Fractogel EMD BioSEC media enable the user to reduce separation times to maximize productivity. The influence of the flow rate on the resolution is demonstrated in Fig. 7.9. The rate of decrease in resolution with the increasing linear flow rate is relatively low at flow rates above 60 cm/hr.

VIII. LOADING CAPACITY

The actual loading capacity always depends on the sample composition and the separation problem. As a rule the volume of the loaded sample should not exceed 5% of the column volume. However, this recommendation is valid only for preparative runs. For analytical applications when a high resolution is needed, the volume of the injected sample should be about 1% of the total column volume or even less. For a preparative run on a 1000 × 200-mm column (bed height 60 cm), two different sample volumes were injected. If the sample volume is 0.3% of the total bed volume, the separation is more efficient

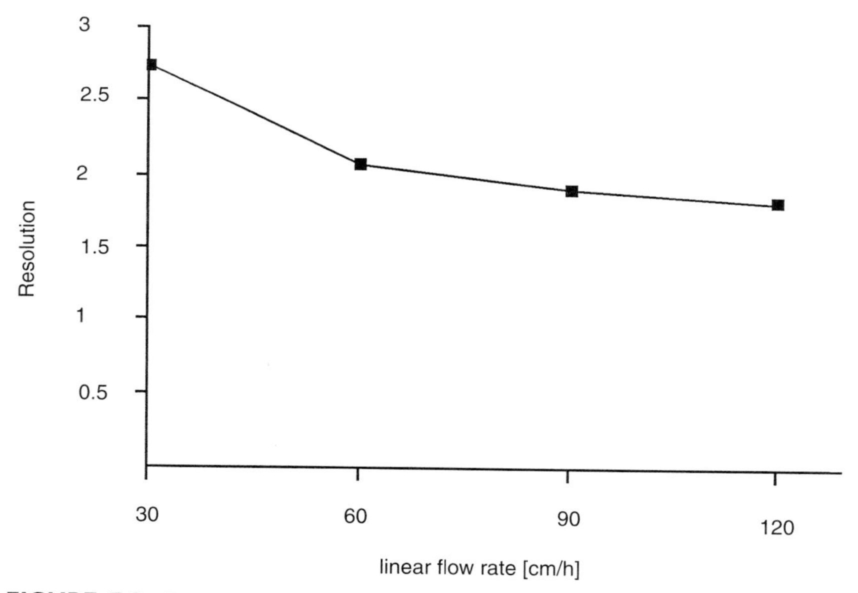

FIGURE 7.9 Dependence of linear flow rate on the resolution of commercial ovalbumin and cytochrome c on Fractogel EMD BioSEC. A column with dimensions of 600 × 16 mm was used at different flow rates (1, 2, 3, and 4 ml/min) (all other chromatographic conditions as in Fig. 7.2A). The resolution (R_s) was calculated according to the equation $R_s = 2(V_{e2} - V_{e1})/(W_{h2} - W_{h1})$, where V_{e1} and V_{e2} are the elution volumes for peaks 1 and 2, respectively, and W_{h1} and W_{h2} are the peak widths at half height.

compared to a sample volume of 1.5% (Fig. 7.10). The total amount of protein loaded onto the column does not influence the resolution significantly (Fig. 7.11). In this case, the effect of the amount of protein is minimal at sample concentrations up to 65 mg/ml when 0.5 ml corresponding to 0.4% of the total bed volume is injected. It must be mentioned that the viscosity of the sample often limits the use of soft gels for SEC and a dilution is necessary before chromatography. For the more rigid methacrylate polymers the viscosity of the sample is not that critical and highly concentrated protein samples can be applied directly onto the column. The productivity of size exclusion chromatography is best optimized by choosing the highest possible linear flow

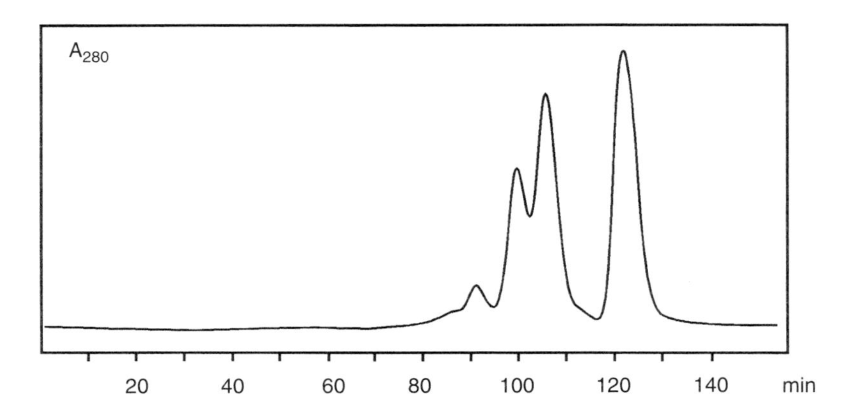

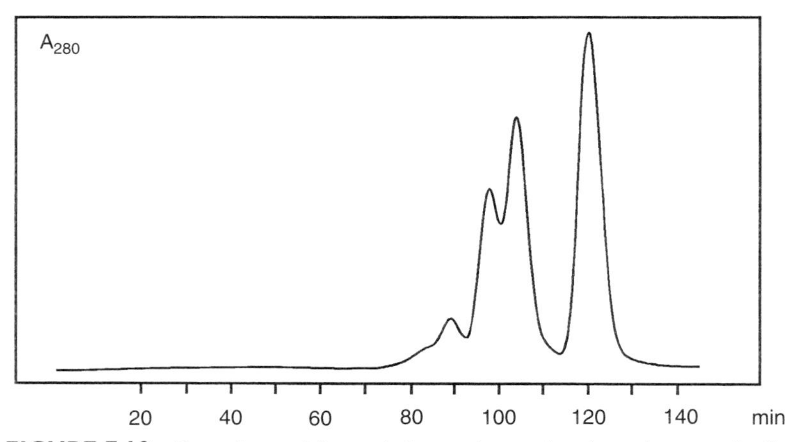

FIGURE 7.10 Dependence of the resolution on the sample volume. A preparative Superformance column 1000-200 (bed volume: 20 liters) packed with Fractogel EMD BioSEC (S) (bed height: 63 cm) was loaded with 60 ml (top) and 300 ml of a mixture of bovine serum albumin (5 mg/ml), ovalbumin (5 mg/ml), and cytochrome c (3 mg/ml) (bottom) (20 m*M* sodium phosphate buffer, 0.3 *M* NaCl, pH 7.2; flow rate: 100 ml/min corresponding to 19 cm/hr). When the sample volume is 300 ml the separation efficiency for BSA and ovalbumin is similar. Thus the column can be loaded with larger sample volumes, resulting in reasonable separations.

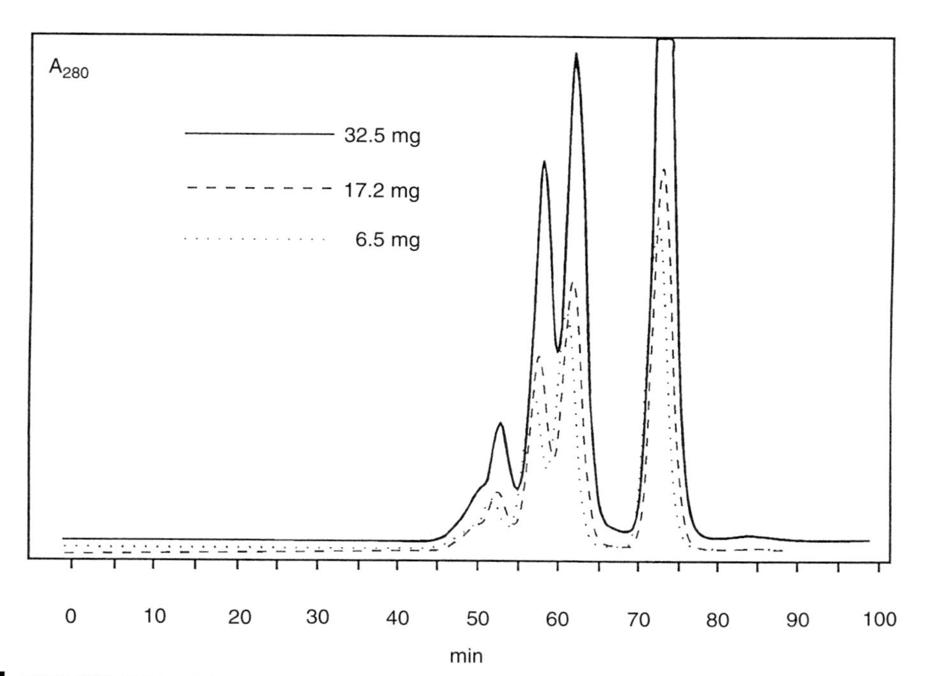

FIGURE 7.11 Influence of the protein concentration on the resolution. The standard separation from Fig. 7.2A was investigated with an increasing protein concentration (6.5, 17.2, and 32.5 mg protein/run). Nearly the same resolution was obtained independent of the sample concentration.

rate, a protein concentration as high as possible, and reducing the sample volume to obtain the required purity.

IX. DETERMINATION OF MOLECULAR MASS OF PROTEINS

Because the migration distance correlates with the size of the molecule, this technique can also be used for the determination of the molecular mass of a protein. However, both the size and the shape of the protein are important for the retention time (9). Therefore, the term "apparent" molecular mass should be used for results obtained by SEC. The molecular mass of proteins can be assessed after a column is calibrated using standard proteins of known molecular mass (calibration proteins are listed in Table 7.3). The void volume should be determined by injecting a molecule large enough to be totally excluded from the gel matrix. The most commonly used substance is Dextran Blue with a molecular weight of approximately 2×10^6 and a hydrodynamic radius of 350 Å. A filtered solution containing 5 mg Dextran Blue/ml is suitable. The pore volume can be determined using sodium nitrate (see Section VI,D). Figure 7.12 shows an example of the separation of four calibration proteins obtained from a Fractogel EMD BioSEC (S) column with a length of 60 cm and an inner diameter of 1.6 cm (121 ml bed volume). The K_D values of the individual proteins can be plotted against the logarithm of the molecular mass. The

TABLE 7.3 Selection of Calibration Proteins Commonly Used for the Calibration of Fractogel EMD BioSEC Columns[a]

Protein	Molecular mass (Da)	Recommended concentration (mg/ml)
Thyroglobulin	669,000	2
Ferritin	440,000	0.6
Catalase	232,000	2.5
Immunoglobulin G	156,000	5
Bovine serum albumin (BSA), dimer	134,000	5
BSA, monomer	67,000	
Hexokinase	54,000	5
Ovalbumin	43,000	10
β-Lactoglobulin A	36,600	5
Carbonic anhydrase	29,000	8
Trypsin inhibitor	21,500	5
Myoglobin	17,500	3
Ribonuclease A	13,700	10
Aprotinin	6,500	5

[a]The void volume can be estimated with a solution of Dextran Blue (5 mg/ml). The BSA preparation contains the monomeric as well as the dimeric form of the protein.

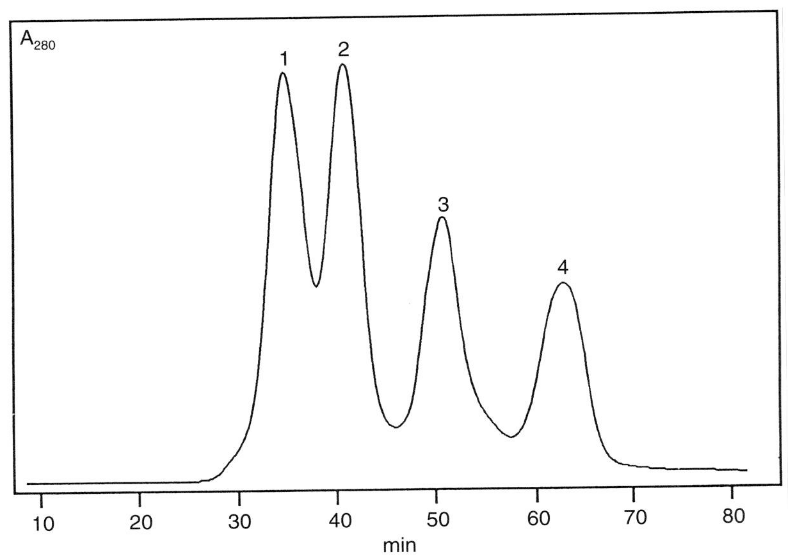

FIGURE 7.12 Standard separation of a calibration mixture on Fractogel EMD BioSEC (S) with a column dimension of 600-16 mm. The calibration mixture contained thyroglobulin (1), immunoglobulin G (2), β-lactoglobulin (3), and ribonuclease (4). Four hundred microliters of the mixture was injected onto the column at a flow rate of 1 ml/min (eluent: 20 m*M* sodium phosphate buffer, 0.1 *M* NaCl; pH 7.2). The K_D values of the individual proteins can be determined by plotting the corresponding elution volumes against the logarithms of the molecular weight.

resulting curve can be used to determine the molecular mass of unknown proteins by their retention times only. Preparative columns can be calibrated in the same way (Fig. 7.13).

X. COLUMN AND GEL MAINTENANCE, CLEANING, REGENERATION, AND STORAGE METHODS

One question of chief interest concerns the number of runs that can be run with one individual SEC column. The lifetime of the sorbent itself must be tested as well as the maximum run number for the packed column. Because column packing procedures for SEC columns are rather time-consuming and all SEC columns have to be checked very carefully with respect to performance, very frequent repacking of the column is unreasonable. Therefore, CIP protocols are generally necessary. The CIP protocol should be developed as part of the process validation program.

The particular regeneration procedure and its frequency depend on the nature of the samples and their size. A typical regeneration procedure consists of washing the column with 1 or 2 bed volumes of an alkaline solution (0.1–0.5 *M* NaOH) and reequilibrating the column with the initial buffer solution (or storage buffer). After repeated cycles, contaminating substances that cannot be removed easily by this regeneration procedure tend to accumulate on the column. For the removal of lipids and very hydrophobic compounds, organic solvents are suitable. A linear flow rate of 60 cm/hr should not be exceeded

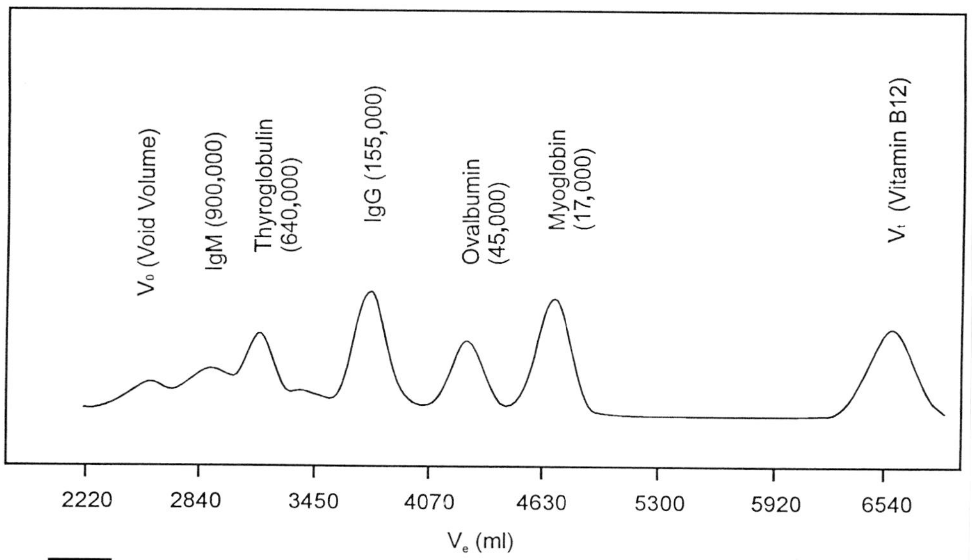

FIGURE 7.13 Preparative separation of various proteins on Fractogel EMD BioSEC (S). The length of the column was 1000 mm and the inner diameter 100 mm. The flow rate was 6.2 ml/min with 20 m*M* sodium phosphate buffer (pH 7.2) containing 0.3 *M* NaCl as the eluent. The injected standard proteins can be used to create a calibration curve.

in the case of methacrylate-based gels for this procedure. In the presence of organic solvents, an upper limit of column pressure drop of 1 bar should not be exceeded. Alternatively, a 2% solution of sodium lauroylsarcosinate can be used. Sometimes treatment with urea (6 *M*), pure water, organic solvents (ethanol, 2-propanol), or detergents (0.1–0.5%) is suitable.

To demonstrate the reproducibility of results obtained from Fractogel EMD BioSEC (S), a separation of a protein mixture was repeated 100 times. After each run with standard proteins the column was regenerated with 1 *M* sodium hydroxide for 60 min at a flow rate of 2 ml/min (60 cm/hr) before reequilibration with running buffer. The total exposure time of the column to 1 *M* NaOH in this experiment was at least 100 hr. In Fig. 7.14, results are given indicating that Fractogel EMD BioSEC could be used for many chromatographic cycles while maintaining original function. The plot of retention times proves that the proteins are separated with exactly the same selectivity.

For production purposes it is convenient to remove or inactivate microorganisms, including vegetative cells, using sodium hydroxide solutions as this chemical is a very inexpensive and efficient cleaning agent. Normally, the concentration of NaOH is in the range of 0.1 to 1 *M*. The result of the column hygiene procedure is also affected by the incubation time. Usually, incubation

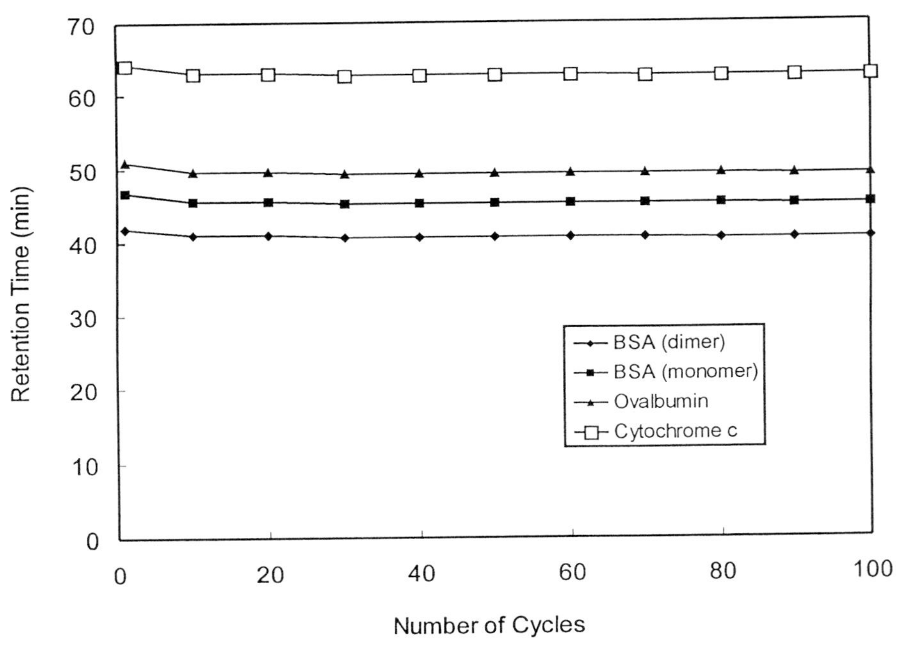

FIGURE 7.14 A Fractogel EMD BioSEC Superformance column (600-16) was loaded with 500 μl of BSA, ovalbumin, and cytochrome c (5/5/3 mg/ml) at 1 ml/min. The test covered 100 individual runs with the standard proteins as samples. The buffer system used was 20 m*M* sodium dihydrogen phosphate, 300 m*M* NaCl, pH 7.2. After each individual run the column was rinsed with 1 *M* NaOH (60 min with 1 *M* NaOH at 2 ml/min). No significant change in retention times and resolution was observed after 100 cycles.

times of 30 min to several hours at room temperature are effective. Thus, the stability of the sorbent against alkaline exposure has to be demonstrated, as the robustness of the chromatographic sorbents used in the production of proteins is very important, not only for the economics of the manufacturing process but also for product safety.

The stability of Fractogel EMD BioSEC was investigated under harsh alkaline conditions. Fractogel EMD BioSEC (S) was stored in the presence of 1 *M* NaOH at room temperature up to 700 hr. After time intervals from 20 up to 700 hr and before each test the column was washed with water, neutralized, and equilibrated with elution buffer until the pH value was constant, and the chromatography of standard proteins was carried out as before. After a 700-hr treatment with 1 *M* NaOH, no significant change in the retention times was detected (Fig. 7.15). However, it can be observed that Fractogel may gradually become more acidic after harsh treatment with sodium hydroxide. Fractogel and its derivatives with neutral functional groups can act as weak cation exchangers due to the presence of a certain amount of carboxylate groups on the methacrylate polymer surface. This property can only be monitored using equilibration buffers of low ionic strength. Lysozyme-binding capacity is used

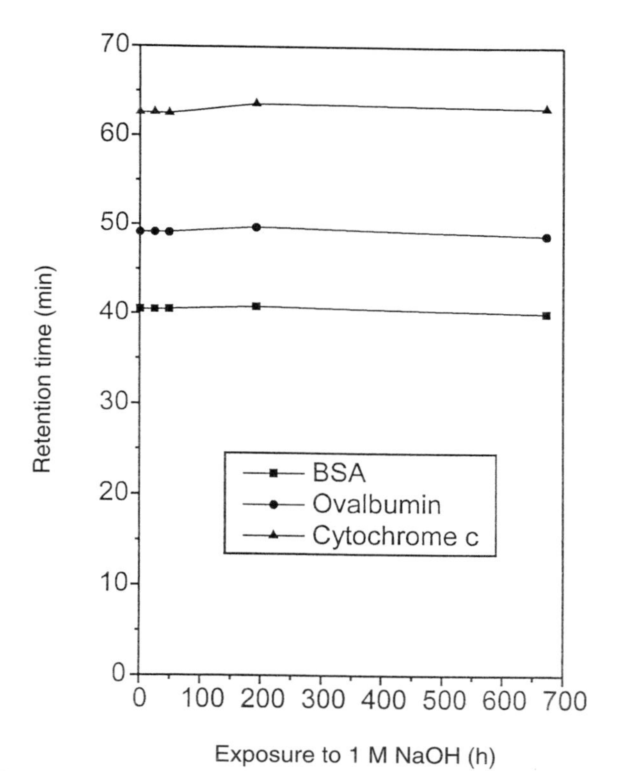

FIGURE 7.15 Long-term stability of Fractogel EMD BioSEC (S) after long-term treatment with sodium hydroxide. The chromatography of standard proteins (for conditions, see Fig. 7.2A) was carried out after various times of exposure to 1 *M* sodium hydroxide solution and reequilibration of the column with the buffer.

routinely as a means to determine the residual acidity of Fractogel-derived chromatography supports. Fractogel EMD BioSEC has a higher residual acidity compared to the starting material Fractogel HW65 due to harsh alkaline washing procedures during the manufacturing process. However, the further increase of acidity during many CIP cycles is slower with the BioSEC material because the grafted substituted acrylamide polymer provides significant shielding of the matrix surface. Generally, the ion exchanger functionality of the Fractogel BioSEC gel is suppressed completely at ionic strenghts of 50 to 100 m*M* of salt in the buffer. Because the gel is very stable, the recommended CIP concentration for Fractogel EMD BioSEC is 0.5 *M* NaOH. Higher concentrations of NaOH can be used, if necessary.

Fractogel EMD BioSEC (S) can also be sterilized repeatedly using an autoclave (120°C, 30 min) without any loss of resolution or changes in retention times. No significant changes of the intraporous volume, the solid matrix volume, and the retention times of the standard protein mixture can be detected after six autoclaving cycles. However, gel to be used again should be submitted to autoclaving only after a thorough CIP procedure has been completed.

After the chromatographic run the column can be stored in 20% ethanol at 3–8°C. Ethanol has to be removed before the next run by rinsing with several volumes (at least five) of buffer. The removal of ethanol during the washing step is demonstrated in Fig. 7.16.

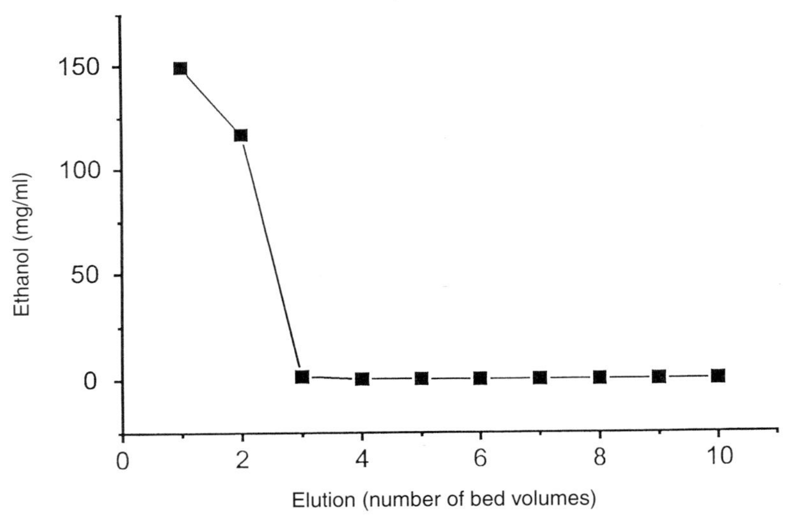

FIGURE 7.16 Removal of ethanol during the column washing procedure. A Superformance 50-10 column packed with Fractogel EMD BioSEC (S) was washed after storage in 20% ethanol/150 m*M* NaCl with 20 m*M* sodium phosphate (pH 6.0). Fractions of 1 bed volume in size with a flow rate of 1 ml/min were collected. The individual fractions were analyzed for ethanol content by gas chromatography. The detection limit was 0.01 mg ethanol per milliliter. After washing with 4 bed volumes of buffer, the ethanol concentration had dropped to 0.26 mg/ml.

XI. APPLICATION AREAS

SEC provides relatively poor resolution due to the fact that none of the protein is retained during chromatography. As observed by Stellwagen (10), fewer than 10 proteins can be separated from one another by baseline. In Fig. 7.17, a complex mixture of seven compounds containing ferritin, IgG, transferrin, ovalbumin, myoglobin, aprotinin, and vitamin B_{12} is separated using a semi-preparative column. The chromatogram represents a fractionation range for proteins from 440,000 down to 17,500 Da. Additionally, the low molecular mass compounds (vitamin B_{12} and aprotinin) can be isolated.

Despite the relative low resolution power, many protein purification protocols contain at least one sizing step. SEC is generally performed late in the procedure when the number of proteins is small. The term "polishing" describes the final step of a protein purification where the objective is to remove possible impurities, such as structurally very similar or closely related forms of the product (aggregates, deamidated or oxidized product, isoforms, etc.), host proteins, reagents and buffer substances, leachables from chromatographic supports, endotoxins, nucleic acids (DNA, RNA), or viruses. This is usually the step where SEC is used. It can separate high and low molecular mass impurities from the target product in one step while simultaneously exchanging the buffer. However, if a target protein is extremely large or small, SEC can be utilized earlier in the purification procedure.

SEC is very easy to perform up to the production scale and the results are very reproducible. Because no gradient elution has to be applied, no programmable gradient mixing system is necessary and only comparatively simple equipment is needed for the operation. Additionally, the method can be integrated easily in purification schemes and most of the operational steps can be

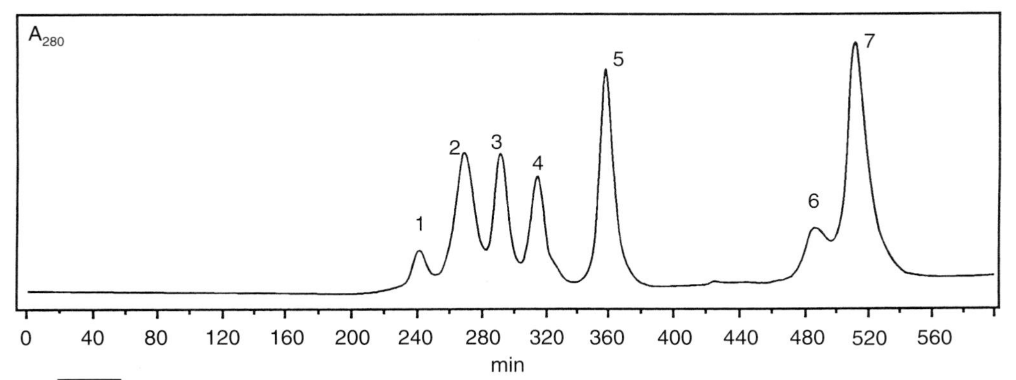

FIGURE 7.17 Separation of a complex mixture on Fractogel EMD BioSEC (S) with a column dimension of 1000 × 50 mm (Superformance glass column). The sample contained ferritin (1), immunoglobulin G (2), transferrin (3), ovalbumin (4), myoglobin (5), aprotinin (6), and vitamin B_{12} (7). Five milliliters of the mixture was injected onto the column at a flow rate of 3 ml/min (eluent: 20 m*M* sodium phosphate buffer, 0.1 *M* NaCl, pH 7.2).

automated. Because the regeneration of Fractogel EMD BioSEC is easy and the stability of the support is high enough to run many chromatographic cycles, the material provides a very economic way to perform small-scale as well as production-scale steps.

The principal applications of SEC on Fractogel EMD BioSEC are purifications of cytosolic proteins as well as recombinant proteins and, simultaneously, the determination of their molecular mass. Membrane proteins can also be isolated with this technique. SEC can also be helpful in plasma fractionation (11). With the introduction of recombinant proteins there has been an emphasis on the determination of aggregates or other by-products copurified with the target protein. Size exclusion chromatography is also very useful for the purification of monoclonal and polyclonal antibodies. Albumins with a molecular weight of about 60 kDa elute earlier than IgG or IgM molecules (Fig. 7.18). Figure 18 demonstrates that Fractogel EMD BioSEC may be helpful in the removal of unwanted (dimeric) forms of IgG.

Another application shows the preparative purification and "polishing" of a therapeutic fusion protein with a humanized recombinant IgG protein. The fusion protein was expressed by the fermentation of baby hamster kidney cells. The filtered culture supernatant (155 liters) contained 2.2 g of IgG and 75.5 g of total protein. After the immunoglobulins were isolated by expanded bed adsorption and rebuffering, the IgG fraction was bound to Fractogel EMD SO_3^- (M). This column achieved baseline separation of complete antibodies (fusion protein) from small amounts of antibodies lacking the fusion part. The resulting highly purified IgG fraction (110 ml) was diluted to 150 ml and

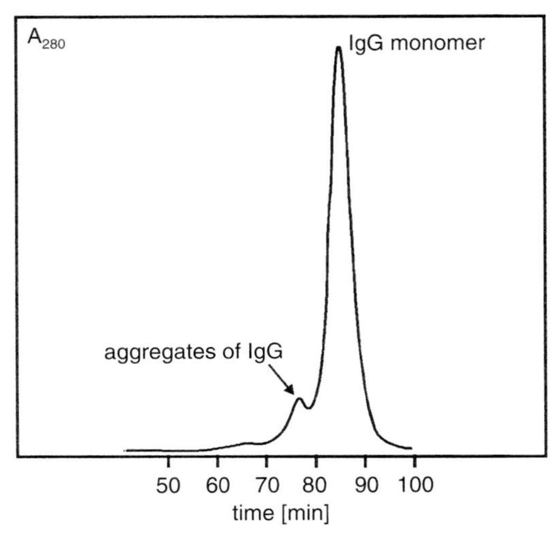

FIGURE 7.18 Preparative separation of aggregates from monomer of human IgG on a Fractogel EMD BioSEC (S) column (1000 × 50 mm; column volume: 1.88 liter). Five milliliters of a sample consisting of IgG (10 mg/ml) was applied at a flow rate of 10 ml/min (21.4 cm/hr) to remove aggregates (eluent: 20 m*M* sodium phosphate, 0.1 *M* sodium chloride, pH 7.2).

applied to a Superformance 1000-100 column (95 cm bed height) packed with 7.5 liters of Fractogel EMD BioSEC (S). Apparently homogeneous antibodies (0.91 g) suitable for clinical studies were recovered from this column in the 260 ml of the eluted protein fraction. The yield in this step was up to 80%; the overall yield was 41% (Fig. 7.19). Misfolded protein, aggregates, and degradation products were completely absent in the main IgG fraction. Endotoxins eluted earlier than the bulk of antibody from the size exclusion column. The endotoxin

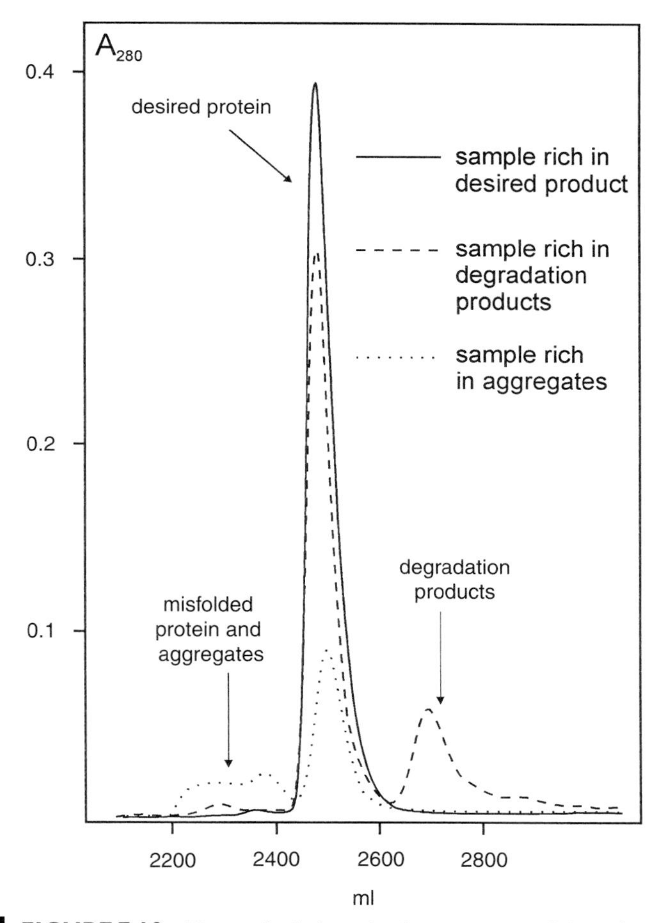

FIGURE 7.19 Three individual runs for the preparative polishing of recombinant humanized immunoglobulin fusion protein by size exclusion chromatography using a Superformance 1000-100 column packed with 7.5 liters of Fractogel EMD BioSEC (S) (eluent: PBS, pH 7.4; flow rate 30 cm/hr). The sample consists of 150 ml containing up to 0.9 g of prepurified protein. The sample composition for the three runs was different. The first injection represents a sample containing only a small amount of aggregated proteins. During the second run (dashed line), aggregates and degradation products were separated. For the last run (dotted line) a sample was injected containing large amounts of aggregates and misfolded protein.

TABLE 7.4 Troubleshooting for SEC Columns

Observation	Cause	What to do
Increased back pressure	Column is clogged due to the presence of protein aggregates and/or insoluble residuals from biological samples	Replace filters and regenerate the column Improve sample preparation
	Column is clogged due to precipitation of proteins	Regenerate column and lower the concentration of salt
	Microbial growth in the column	Regenerate column with NaOH; store column/media in 20% ethanol or 0.05% sodium azide
Sudden increase of back pressure	Tubing is blocked by dust particles	Disconnect and rinse or replace all tubing
Change of color of the media	Microbial growth; incomplete regeneration; contamination with (Lipo)proteins or dyes	Perform appropriate column hygiene steps; use cleaning in place protocols; if necessary, repack the column
Loss of resolution after a few steps	Microbial growth in the column Accumulation of contaminating compounds	Perform regeneration procedures more often
Elution profiles are not reproducible	Degradation of the sample during storage Microbial growth in the column; column is contaminated	Use fresh sample Perform regeneration procedures after each run
K_D values above 1	Nonspecific interaction with the matrix Ionic interactions	Add more salt to avoid ionic interactions (avoid salting out!)
	Hydrophobic interactions	Add a suitable detergent; reduce the salt concentration to avoid hydrophobic interactions or switch from structure-forming salts (phosphate; sulfate) to chlorides or acetates

K_D values below 0	Column packing is damaged	Check column performance (acetone peak)
Unexpected elution position	Molecular shape of the protein is not globular; protein aggregates	—
Poor resolution	Sample volume is too high	Apply smaller sample volumes
	Sample has been applied improperly	Check column performance (acetone peak)
	Dead volume of the system too large	Check mixing spaces, detector flow cell, and diameter of tubing
	Flow rate too high	Adapt flow rate
	Column is packed poorly	Check the column packing quality
Low recovery of biological activity	Biological assay is influenced by the buffer used for SEC	Buffer exchange before testing
Poor recovery of biological activity with good recovery of protein	Target protein is not stable in the chosen buffer system for SEC	Change the buffer system
	Separation of enzyme and cofactor(s)	Test by pooling fractions to detect cofactor
Poor mass recovery	Target protein may have been proteolyzed	Add protease inhibitors
	Nonspecific interaction	Add ethylene glycol or organic solvents; increase the concentration of salt
Recovery of biological activity too high (compared to the amount loaded)	Separation of inhibitors	Test by pooling
Very rounded peaks in the chromatogram	Column is packed poorly	Check the column packing quality
	Overloaded column	Decrease the sample volume

level in the final preparation was determined by the limulus amebocyte lysate (LAL) test to be less than 1 IU per milligram of protein.

XII. TROUBLESHOOTING

In general, by improving the packing quality, reducing the flow rate, and using longer columns (or coupling of two columns), the resolution can also be improved. A comparison of three different column lengths is given in Fig. 7.6. A summary of various problems that may occur during SEC is listed in Table 7.4. This list may help in troubleshooting during the actual separation using SEC columns. However, not all basic requirements for biochromatography are described in detail here.

Note that if K_D values are above 1, there must be a nonspecific interaction between the target molecule and the matrix. In the case of K_D values below 0, the column packing is probably damaged.

In the case of poor recovery the following points should be considered. The matrix itself is hydrophilic enough for the application of proteins. Due to the tentacle modification, a further reduction of nonspecific interactions with very hydrophobic compounds can be detected. The shielding effect of the linear tentacle chains prevents proteins from contacting the matrix. However, the addition of detergents or organic solvents may sometimes result in higher mass recoveries in the case of very hydrophobic molecules. To avoid poor recoveries due to ionic interactions, 50–100 mM of salt is always recommended. Sometimes this concentration can be increased further to optimize the yield. If the salt concentration is too high, however, the recovery can be influenced negatively due to hydrophobic interactions.

Although the mechanism of SEC is in principle nonadsorptive, effects derived from different matrices can be observed. As a result, quite different chromatographic behaviors of one particular protein can be obtained depending on the type of sorbent used.

Sometimes elution profiles may also depend on the number of cycles. This can cause problems in the reproducibility of SEC separations. It should be mentioned that no regeneration procedure will give a completely new SEC support.

During the separation itself some trouble can occur concerning the back pressure. An increasing back pressure indicates contamination of the column and thus should be monitored. If this happens or if a visible contamination of the sorbent is noticed, a regeneration of the column is necessary. However, if the back pressure rises very rapidly the column may be clogged by denatured proteins. As a first attempt, the frits should be replaced by new ones, trying the top adapter first.

If unidentified peaks are detected the stability of the protein under the chromatographic conditions should be checked. In all analytical investigations of proteins on SEC columns it is desirable to be able to monitor the eluted peaks at a very high sensitivity of the ultraviolet detector. Therefore, very pure (analytical grade) salts and buffers should be used.

REFERENCES

1. Porath, J., and Flodin, P. (1959). *Nature* **183,** 1657–1659.
2. L. Hagel, *in* "Protein purification: principles, high resolution methods and applications" (J. C. Janson and L. Ryden, eds.), pp. 63–106. VCH Publishers, New York, 1989.
3. Magiera, D. J., and Krull, I. S. (1992). *J. Chromatogr.* **606,** 264–271.
4. Müller, W. (1990). *J. Chromatogr.* **510,** 133–140.
5. Westerlund, J., and Yao, Z. (1995). *J. Chromatogr.* **718,** 59–66.
6. Cai, C., Romano, V. A., and Dubin, P. L. (1995). *J. Chromatogr.* **693,** 251–261.
7. Kato, Y., Komiya, K., Iwaeda, T., Sasaki, H., and Hashimoto, T. (1981). *J. Chromatogr.* **205,** 185–188.
8. Kato, Y., Komiya, K., Iwaeda, T., Sasaki, H., and Hashimoto, T. (1981). *J. Chromatogr.* **206,** 135–138.
9. Dubin, P. L., Edwards, S. L., Mehta, M. S., and Tomalia, D. *J. Chromatogr.* **635,** 51–60.
10. Stellwagen, E. (1990). *In* "Methods of Enzymology" (M. P. Deutscher, ed.), Vol. 182, pp. 317–328. Academic Press, New York.
11. Josic, D., Horn, H., Schulz, P., Schwinn, H., and Britsch, L. (1998). *J. Chromatogr.* **A796,** 289–298.

8 SIZE EXCLUSION HIGH-PERFORMANCE LIQUID CHROMATOGRAPHY OF SMALL SOLUTES

ANDREW J. ALPERT

PolyLC Inc., Columbia, Maryland 21045

I. OVERVIEW OF SIZE EXCLUSION CHROMATOGRAPHY (SEC) OF SMALL SOLUTES

Size-exclusion chromatography was introduced in 1959, with the invention of Sephadex. Some of the first applications involved attempts to fractionate small peptides and amino acids by size (1,2). These efforts were not very successful. The reason is that the fractionation range in SEC is determined by the pore diameter of the stationary phase. In polymer-based materials such as Sephadex, this is controlled by the degree of cross-linking; the more highly cross-linked the material, the narrower the pores. The agents used for cross-linking interact with some solutes, especially aromatic ones. Thus, phenylalanine, tyrosine, and tryptophan (Phe, Tyr, and Trp, respectively) elute later than V_t on Sephadex G-10, G-15, and G-25, the most highly cross-linked grades (2,3). Phe and oligophenylalanines elute in order of smallest to largest, the opposite of the sequence expected in SEC (4). The same adsorption phenomenon has been noted with Bio-Gel P2, a polyacrylamide-based medium (5,6). Silica and other inorganic materials can be made with very narrow pores, but difficulty in diffusion into and out of micropores (<20 Å) leads to poor efficiency (7). Thus, with most commercial SEC columns, the lower limit for V_t is around 1000 Da, which would correspond to a peptide of about eight to nine amino acid residues.

Column Handbook for Size Exclusion Chromatography

II. INTRODUCTION OF PolyHYDROXYETHYL ASPARTAMIDE AND THE EFFECT OF CHAOTROPES ON THE FRACTIONATION RANGE

A SEC material should be hydrophilic if it is to be used for biological applications. One such material, introduced by PolyLC in 1990 (8), is silica with a covalently attached coating of poly(2-hydroxyethyl aspartamide); the trade name is PolyHYDROXYETHYL Aspartamide (PolyHEA). This material was evaluated for SEC of polypeptides by P.C. Andrews (University of Michigan) and worked well for the purpose (Fig. 8.1). Because formic acid is a good solvent for polypeptides, Dr. Andrews tried a mobile phase of 50 m*M* formic acid. The result was a dramatic shift to a lower fractionation range for both V_0 and V_t (Fig. 8.2) to the point that V_t was defined by the elution position of water,

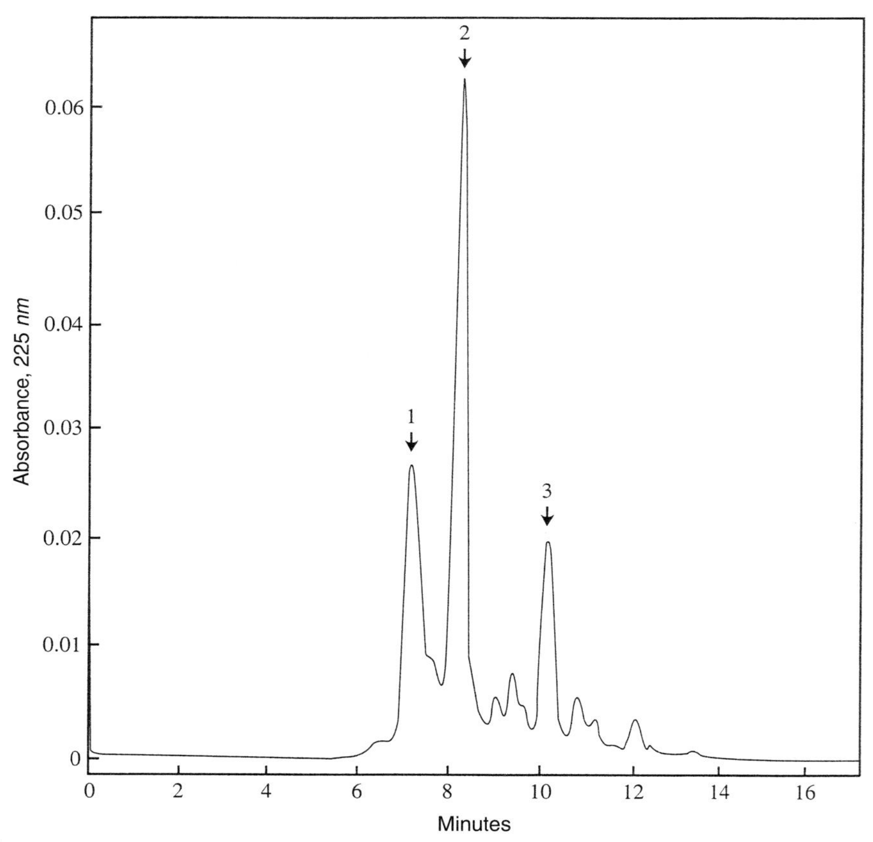

FIGURE 8.1 SEC of an acid/ethanol extract of lamprey pancreas. When performing SEC of peptides, one generally obtains the best correlation of retention times and molecular weights if the mobile phase is acidic and contains some organic solvent. Column: PolyHEA, 200 × 9.4 mm; 5 μm, 200 Å. Flow rate: 2.0 ml/min. Mobile phase: 5 m*M* sodium phosphate + 200 m*M* sodium sulfate, pH 3.0, with 25% (v/v) acetonitrile. Peaks: (1) Plasma lipid-binding protein (11,500 Da); (2) Insulin (6241 Da), glucagon (3900 Da), and Somatostatin-37 (4052 Da); and (3) somatostatin-14 (1623 Da). (Courtesy of P. C. Andrews, University of Michigan.)

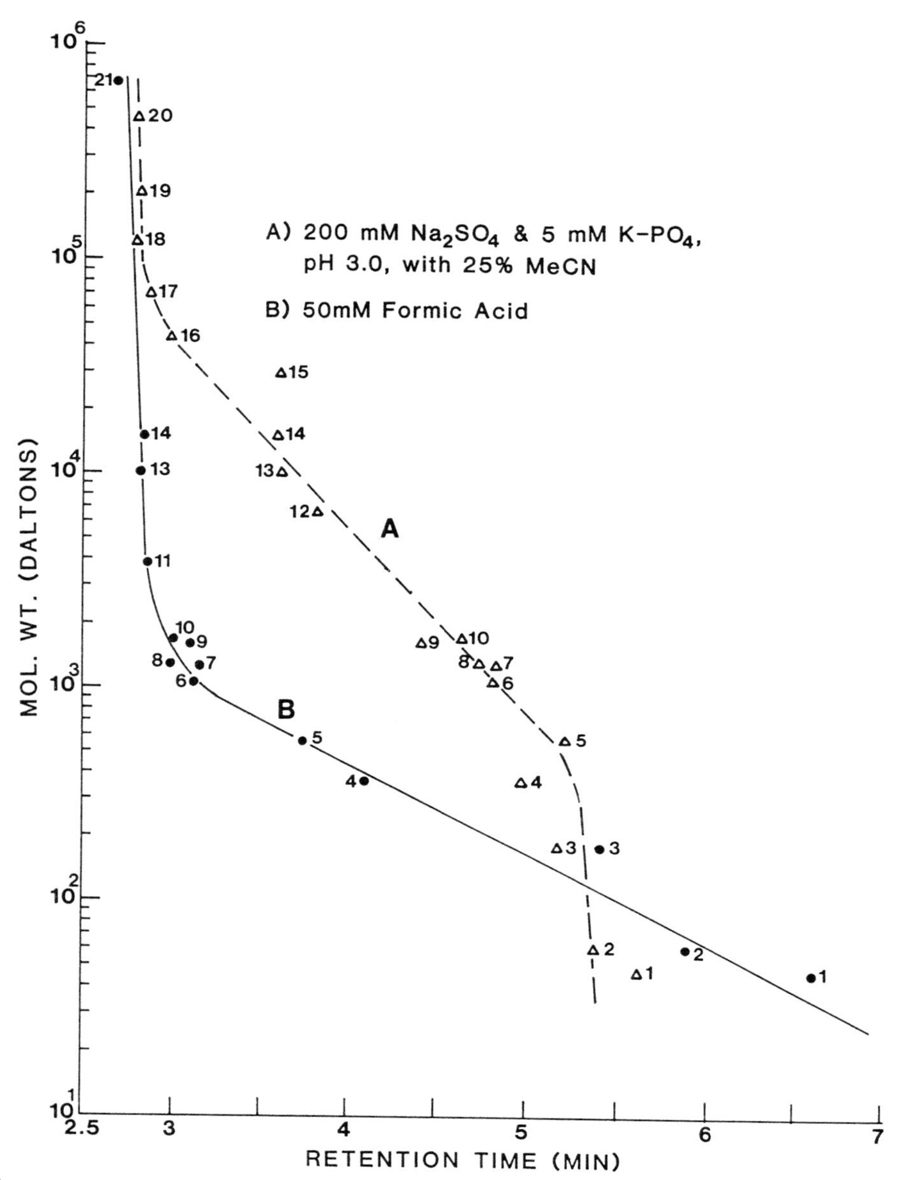

FIGURE 8.2 SEC ranges on a PolyHEA column (200 Å) with (A) a nondenaturing and (B) a denaturing mobile phase. Column and flow rate: Same as Fig. 8.1. Sample Key: (1) Formic acid (46 Da); (2) acetic acid (60 Da); (3) *N*-chloroacetyl-Tris base (177 Da); (4) thyrotropin-releasing hormone (362 Da); (5) [Leu]^5-enkephalinamide (554 Da); (6) angiotensin II (1046 Da); (7) luteinizing hormone-releasing hormone (1236 Da); (8) angiotensin I (1296 Da); (9) bombesin (1592 Da); (10) α-melanocyte-stimulating hormone (α-MSH) (1665 Da); (11) poly-L-lysine (3800 Da); (12) insulin (bovine) (6500 Da); (13) ubiquitin (10,000 Da); (14) ribonuclease (15,000 Da); (15) carbonic anhydrase (29,000 Da); 16) ovalbumin (chicken) (43,000 Da); (17) bovine serum albumin (66,000 Da); (18) IgG (150,000 Da); (19) β-amylase (200,000 Da); (20) apoferritin (443,000 Da); and (21) thyroglobulin (669,000 Da). (Data courtesy of P. C. Andrews, University of Michigan.)

and acetic acid, amino acids, and the smallest peptides were included in the range (Fig. 8.3). Despite the high absorbancy of the mobile phase (1.0 AU at 220 nm), it was possible to subtract the elevated baseline and monitor peptides at low wavelengths because elution was isocratic.

Shifts in the SEC fractionation range are not new. It has been known for decades that adding chaotropes to mobile phases causes proteins to elute as if they were much larger molecules. Sodium dodecyl sulfate (SDS) (9) and guanidinium hydrochloride (Gd.HCl) (9–12) have been used for this purpose. It has not been clearly determined in every case if these shifts reflect effects of the chaotropes on the solutes or on the stationary phase. Proteins are denatured by chaotropes; the loss of tertiary structure increases their hydrodynamic radius. However, a similar shift in elution times has been observed with SEC of peptides in 0.1% trifluoroacetic acid (TFA) (13–15) or 0.1 *M* formic acid (16), even if they were too small to have significant tertiary structure. Speculation as to the cause involved solvation effects that decreased the effective pore size of the

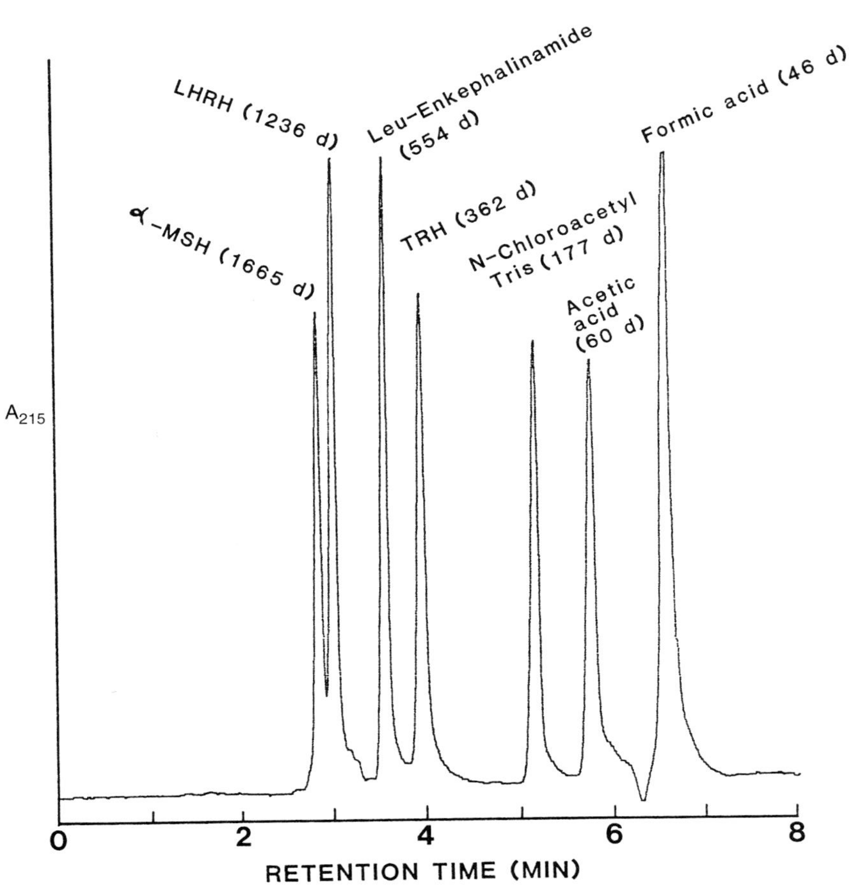

FIGURE 8.3 SEC in 50 m*M* formic acid. Column and flow rate: Same as Fig. 8.1. (Courtesy of P. C. Andrews, University of Michigan.)

column. Such a simplistic explanation does not account for the ability of a SEC material with a pore diameter of 200 Å to separate solutes as small as amino acids (Figs. 8.1 and 8.2).

With a *nondenaturing* mobile phase (i.e., one that does not contain a strong chaotrope), the fractionation range of PolyHEA with a 200-Å-pore diameter is comparable to that of other columns with pore diameters of 50 or 60 Å. This suggests that the PolyHEA coating is unusually thick, about 75 Å or so, compared with 15 Å with the typical coating on silica. When the mobile phase does contain a chaotrope, the resulting fractionation range resembles one that might be obtained with a material with a pore diameter around 15 Å (similar to a hypothetical Sephadex G-7 material). It is plausible that the coating of this material is normally rendered impermeable by numerous hydrogen bonds between adjacent polypeptide chains; PolyHEA has a high concentration of amide groups, and amide–amide hydrogen bonds are about twice as strong as amide–hydroxyl hydrogen bonds (17). The coating is also probably hydrated by a highly ordered layer of water. Similar statements have been made about the hydration of Sephadex rendering 20% of the potential pore volume inaccessible in SEC (2,3). Chaotropes are weakly hydrated (18) and form hydrogen bonds to stationary phases in preference to water. Thus, similar to aromatic amino acids, they are retained on highly cross-linked SEC materials past V_t. When included in the mobile phase, the chaotrope outcompetes adjacent polymer chains for forming hydrogen bonds. The consequence is a great increase in their steric radius, to the point that they occlude a 200-Å pore (Fig. 8.4). At the same time, the chaotrope would disrupt the hydration layer. This has two important consequences: (1) The available pore volume increases. This is consistent with observations made with PolyHYDROXYETHYL A (Figs. 8.2 and 8.11 vs. Fig. 8.12). (2) The space between the polymer chains becomes permeable; this distance between chains is effec-

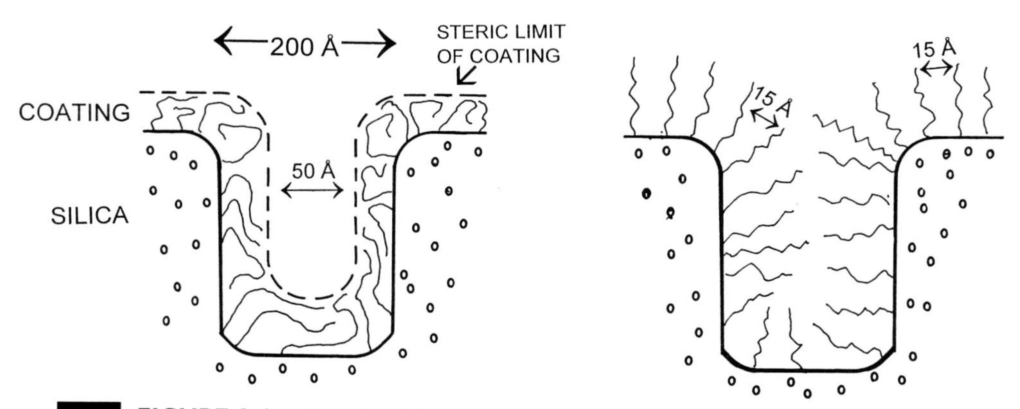

FIGURE 8.4 Effect of mobile phase on PolyHEA coating. (Left) Nondenaturing mobile phase (e.g., Fig. 8.1). (Right) Denaturing mobile phase (e.g., Fig. 8.3). Material with a 200-Å-pore diameter is shown. The coating swells in the presence of a chaotrope, becoming permeable and occluding the pores. The effective pore diameter now becomes the distance between polymeric chains in the permeable coating: about 15 Å.

tively the new pore diameter. Judging from the fractionation range, the chains are spaced about 15 Å apart. Evidently it is easy to diffuse into and out of such a permeable network of "soft" pores, and solutes elute in peaks much sharper and more symmetrical than would be afforded by a SEC material with "hard" pores 15 Å in diameter.

A review of the literature discloses similar effects with other SEC materials. Richter and Schwandt (16) eluted a TSK 2000P-SW column with 0.1 *M* formic acid and were able to desalt peptides as small as 794 Da. With 0.01 *M* formic acid, the range shifted so that peptides <3500 Da were not desalted. Irvine and Shaw (15) eluted a TSK G-2000-SW column with 0.1% TFA and obtained a range of approximately 500–50,000 Da. With the addition of 0.25 *M* NaCl (a structure-forming salt), the range shifted to 5000–200,000 Da. Both Montelaro *et al.* (9) and Kato *et al.* (10) observed shifts of V_0 and V_t to lower values with a TSK G2000-SW column upon addition of 6 *M* Gd.HCl to the mobile phase; the latter group reported a range of 1000–25,000 Da. Swergold and Rubin (14) eluted a TSK G3000-PW column with 0.1% TFA containing 40% acetonitrile (ACN) and obtained a range of approximately 300–130,000 Da. They stated that 0.3% phosphoric acid could be substituted for the TFA. The ACN was essential to eliminate hydrophobic interactions with this polymeric stationary phase. Bennett *et al.* (13) used an I-125 column from Waters; with 0.1% TFA + 40% ACN, the range shifts from 2000–80,000 to approximately 500–60,000 Da.

These observations suggest that chaotropes render the coatings of a number of SEC materials permeable, albeit to a lesser extent than with PolyHEA. For example, TSK G2000-SW is reported by the manufacturer to have a pore diameter of 125 Å. With a denaturing mobile phase, the separation of the smallest solutes is about one-fourth as good as with PolyHEA with 200-Å pores. The implication is that the TSK coating is considerably thinner or more tightly bonded to the surface than that of PolyHEA. A similar comparison may be made between the separation of thyrotropin-releasing hormone (TRH) and acetic acid on PolyHEA (Fig. 8.3) and the Waters I-125 column (19).

III. VERIFICATION OF SEC MECHANISM FOR SMALL SOLUTES

Amino acids are a convenient set of standards for elucidating the forces involved in SEC of small solutes. One can resolve to baseline as many as 5 of the 20 natural amino acids within the range of a 200-Å PolyHEA column (and 7 within the range of a 60-Å column). In general, the order of elution is most to least polar. This order appears at first to reflect hydrophobic interaction between the solute and the stationary phase. With other SEC media, the late elution of aromatic amino acids has been attributed to electron donor–acceptor interactions (20). Were this the mechanism, then oligopeptides of the most hydrophobic amino acids should elute later than the amino acids themselves, as overall retention in adsorption chromatography reflects the additive contribution of each subunit of a solute. Such is the case with Phe and oligophenylalanines on Sephadex (4) but is not the case with PolyHEA. $(Phe)_2$ elutes before

Phe and $(Trp)_2$ elutes before Trp (Fig. 8.5), the sequence to be expected in SEC. Also, none of the amino acids elutes from PolyHEA after V_t. Finally, the adsorption effects are generally attributed to the cross-linking agents used in high concentration to prepare the small-pore SEC media. With the chaotropic mobile phase, however, the network of "soft" pores is generated without cross-linking agents. It would seem that the most hydrophobic amino acids *look* the smallest, particularly the aromatic ones. Before making such a claim, however, one must rule out effects from some of the forces potentially responsible for the adsorption of amino acids to PolyHEA: *hydrophobic interaction, hydrophilic interaction,* and *electrostatic effects.*

Controlling for these forces requires variation in the amount of salt, organic solvent, and the pH of the mobile phase. It is impractical to perform such experiments with 50 m*M* formic acid; an alternative additive must be used that maintains its chaotropic properties independent of salt content or pH. Fortunately, mobile phases containing 50 m*M* hexafluoro-2-propanol (HFIP) afford a fractionation range comparable to that of the formic acid (Fig. 8.6), permitting the effects of these variables to be studied systematically.

A. Effect of pH

At pH 3.0, Lys has a (+) charge, whereas Asp is protonated (and neutral). Varying the pH of the mobile phase leads to a decrease in the retention of Lys and and an increase in that of Asp [as it gains (−) charge]. The plot of this effect (Fig. 8.7) is, in effect, a titration curve of PolyHEA. The coating has a

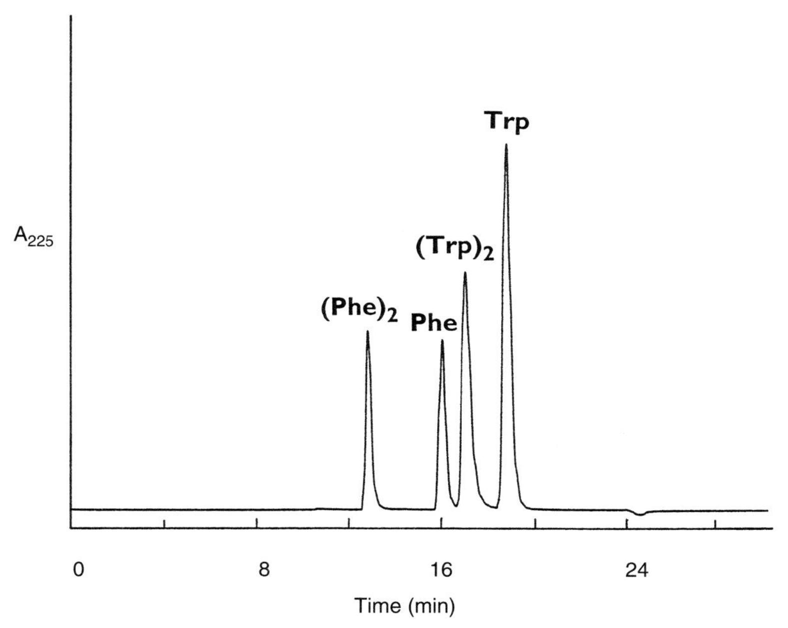

FIGURE 8.5 SEC of aromatic amino acids and dipeptides. Column: Same as Fig. 8.1. Flow rate: 0.6 ml/min. Mobile phase: 50 m*M* formic acid. Detection: A_{225} = 0.5 AUFS.

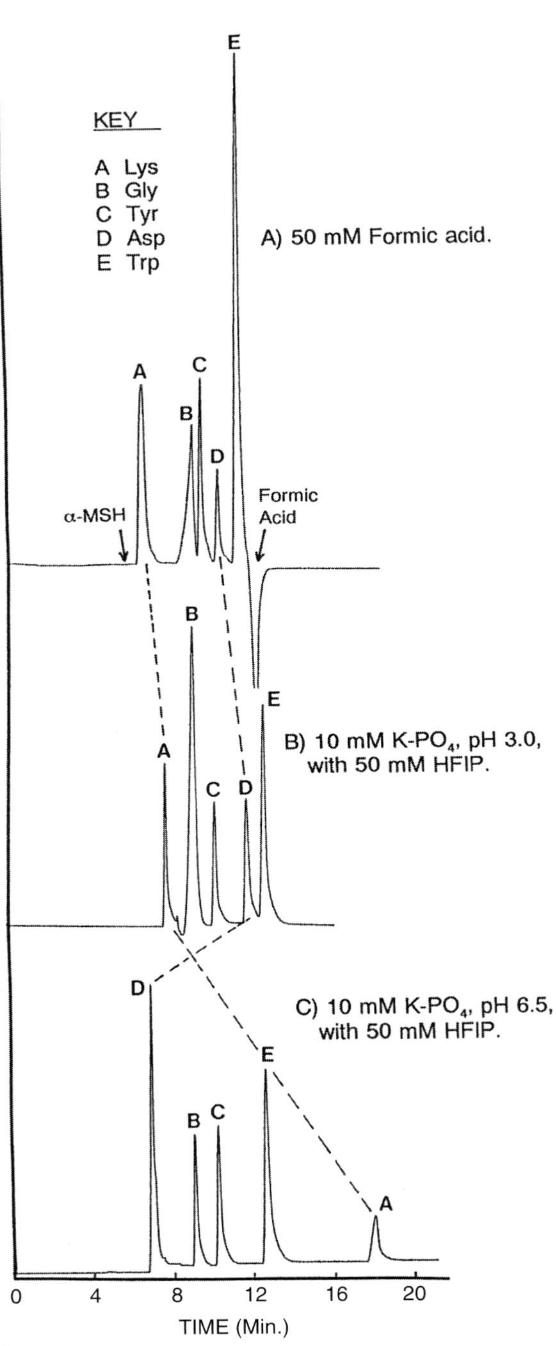

FIGURE 8.6 Comparison of hexafluoro-2-propanol (HFIP) with formic acid as a denaturing agent in SEC. Elution positions of neutral amino acids were similar with both agents. The elution positions of Lys and Asp shifted dramatically in C, as shown by the tie lines, but this was an effect of pH (see Fig. 8.7). The elution positions of α-MSH and formic acid are shown to demonstrate that the amino acids eluted within V_0 and V_t. Column: Same as Fig. 8.1. Flow rate: 1.0 ml/min. Mobile phase: As noted. Detection: A_{215} = 0.1 AUFS.

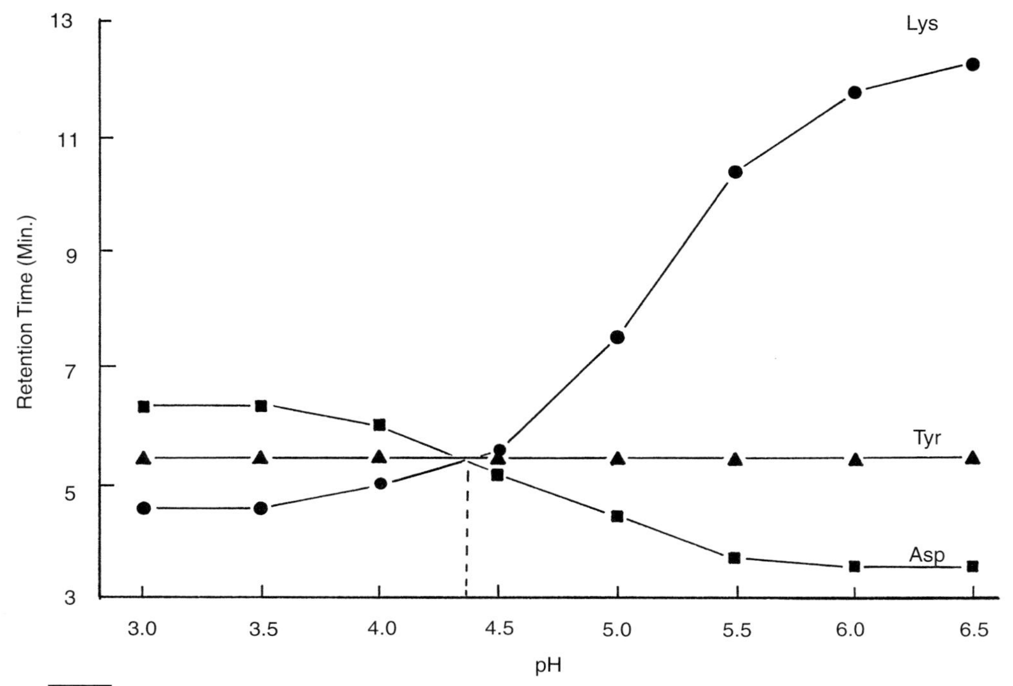

FIGURE 8.7 Effect of pH on retention of amino acids. Column and flow rate: Same as Fig. 8.1. Mobile phase: 10 m*M* potassium phosphate with 50 m*M* HFIP; pH as indicated (adjusted prior to the addition of HFIP).

small positive charge below pH 4.4 and a small negative charge above pH 4.4. At pH 4.4 (where the peaks of Lys and Asp coincide), the charges are in balance; the coating is zwitterionic. Presumably these charged groups are the termini of the polypeptide coating. The retention of neutral amino acids was not affected by pH, indicating that electrostatic effects did not determine their elution order.

B. Effect of Salt

Electrostatic effects have long been recognized in commercial HPLC columns for SEC of proteins (15,21,22). The usual remedy is to add 100 m*M* salt to the mobile phase. This works here too; the Lys and Asp peaks collapse into the Gly peak with 100 m*M* salt (Fig. 8.8). High concentrations of sodium sulfate were added to determine the role played in SEC by hydrophobic interactions (sodium sulfate, a structure-forming salt, strengthens such interactions). Sodium sulfate increased the retention only of the most hydrophobic amino acids to any extent, and then only when the concentration approached 1 *M*. Clearly, hydrophobic interaction cannot account for the elution order of amino acids on PolyHEA.

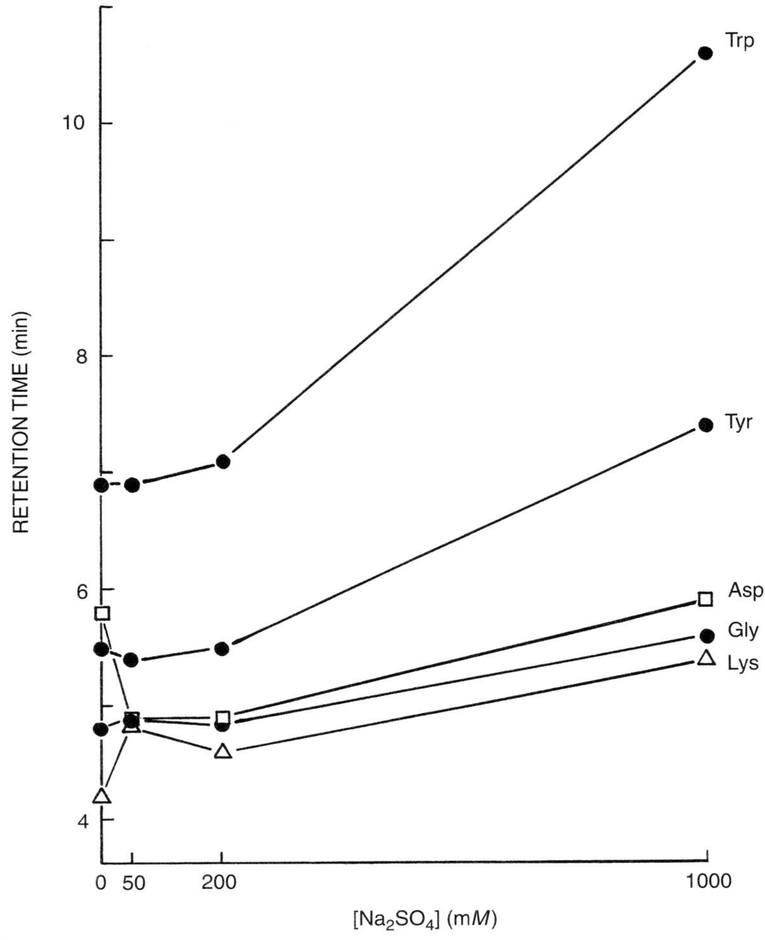

FIGURE 8.8 Effect of salt on retention of amino acids. Column and flow rate: Same as Fig. 8.1. Mobile phase: 10 m*M* potassium phosphate + sodium sulfate (as noted), pH 3.0, with 50 m*M* HFIP.

C. Effect of Organic Solvents

As a final check on these results, ACN was added to the mobile phase; 25% generally suffices to eliminate hydrophobic interactions with reasonably hydrophilic stationary phases (23). At the same time, a sufficiently high concentration of organic solvent induces hydrophilic interaction, a situation where the stationary phase is more polar than the mobile phase (8). Adsorption of solutes then occurs through a partitioning mechanism between the dynamic mobile phase and the stagnant hydration layer. A "normal phase" elution order is obtained: least to most polar. This is clearly the case with 60% ACN with PolyHEA (Fig. 8.9) as well as with SynChropak GPC (24). With 25% ACN, the SEC elution order is not affected significantly except for Lys-, which elutes significantly later (basic amino acids are the most hydrophilic of all). It is clear that

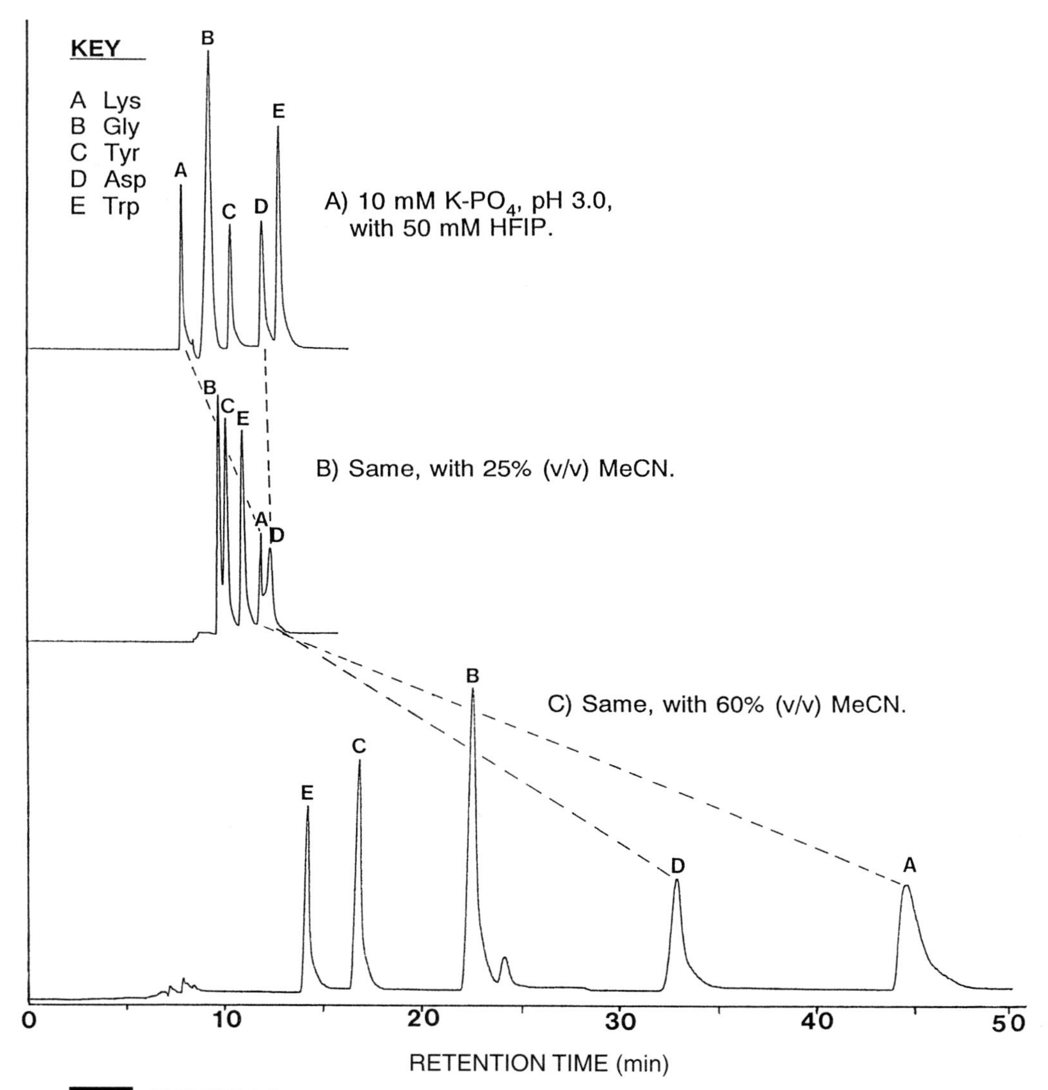

FIGURE 8.9 Effect of organic solvent on retention of amino acids. Small amounts of organic solvent condense the fractionation range somewhat for small solutes (B vs A), presumably by decreasing the swelling of the PolyHEA coating. Higher concentrations of organic solvent induce hydrophilic interactions, occasioning an inversion of the elution order. Both Asp and Lys are well retained (shown with the tie lines), distinguishing this effect from electrostatic effects. Column and flow rate: Same as Fig. 8.6. Mobile phase: As noted. Detection: A_{215} = 0.1 AUFS.

hydrophilic interactions do not account for the elution order of amino acids in SEC, in the absence of appreciable levels of organic solvent.

D. How Big Are Amino Acids?

These controlled experiments eliminate adsorption as an explanation for the elution order of neutral amino acids from PolyHEA. Perhaps this order does reflect their actual sizes, which seem to range from approximately 8 to 3 Å (in the case of Gly vs. Trp). The perceived size of a solute in SEC is really the size of its sphere of hydration. This can be one to three water molecules thick (for chaotropes and structure-forming ions, respectively) (18). In general, homopeptides of polar amino acids are more highly hydrated than those of nonpolar amino acids (25). Actually, it may be more accurate here to speak of spheres of solvation. The separations are taking place in the presence of a chaotrope in the mobile phase. This may disrupt the hydration layer around the solutes as it does with the stationary phase [and which would presumably eliminate the electron donor–acceptor interactions postulated by Porath (20)]. It is unclear to what extent these results reflect the relative size of amino acids in more conventional solvents. In any case, they do establish that the aromatic amino acids look smaller than the other amino acids. Their specific adsorption to some highly cross-linked SEC matrices is coincidence; no such adsorption occurs with PolyHEA.

Caution: The order of elution of solutes smaller than a tetrapeptide may be in order of decreasing polarity rather than decreasing molecular weight. This reflects the relative size of their spheres of hydration.

E. Superdex Peptide

Pharmacia has introduced an SEC material named Superdex (26). This has a shell of agarose to confer rigidity, with size exclusion performed by an interior network of dextran. The Superdex Peptide column represents the low end of the molecular weight scale, with a fractionation range of 100–7000 Da. The company literature (27) shows reasonably good separations of small solutes such as Gly, $(Gly)_3$, and $(Gly)_6$. However, peptides with several aromatic residues elute after V_t (28), indicating that this matrix has the same problem with the adsorption of aromatic amino acids as do other highly cross-linked SEC media. Also, graphs of the fractionation range (27) show no change in the presence of a number of chaotropes or 70% ACN (which is high enough to induce significant hydrophilic interaction with so polar a material). Either the solvation properties of this material differ dramatically from those of all other materials used for SEC of small solutes or else they should be evaluated more carefully with controlled experiments.

IV. FRACTIONATION RANGE WITH PolyHEA AS A FUNCTION OF PORE DIAMETER

When a column of PolyHEA with a 300-Å-pore diameter is eluted with 50 m*M* formic acid, a plot of the fractionation range resembles two lines of different slopes joined together (Fig. 8.10). The line at the lower end covers the same molecular weight range as the 200-Å material: 20–1600 Da. An interpretation of the data is that the swollen, permeable coating does not completely fill a pore with a diameter of 300 Å. The lower end of the fractionation range reflects the pore volume within the coating, whereas the upper end of the range reflects the pore volume within the pores but above the coating. Thus, one can fractionate both small and large solutes with the same column. Figure 8.11

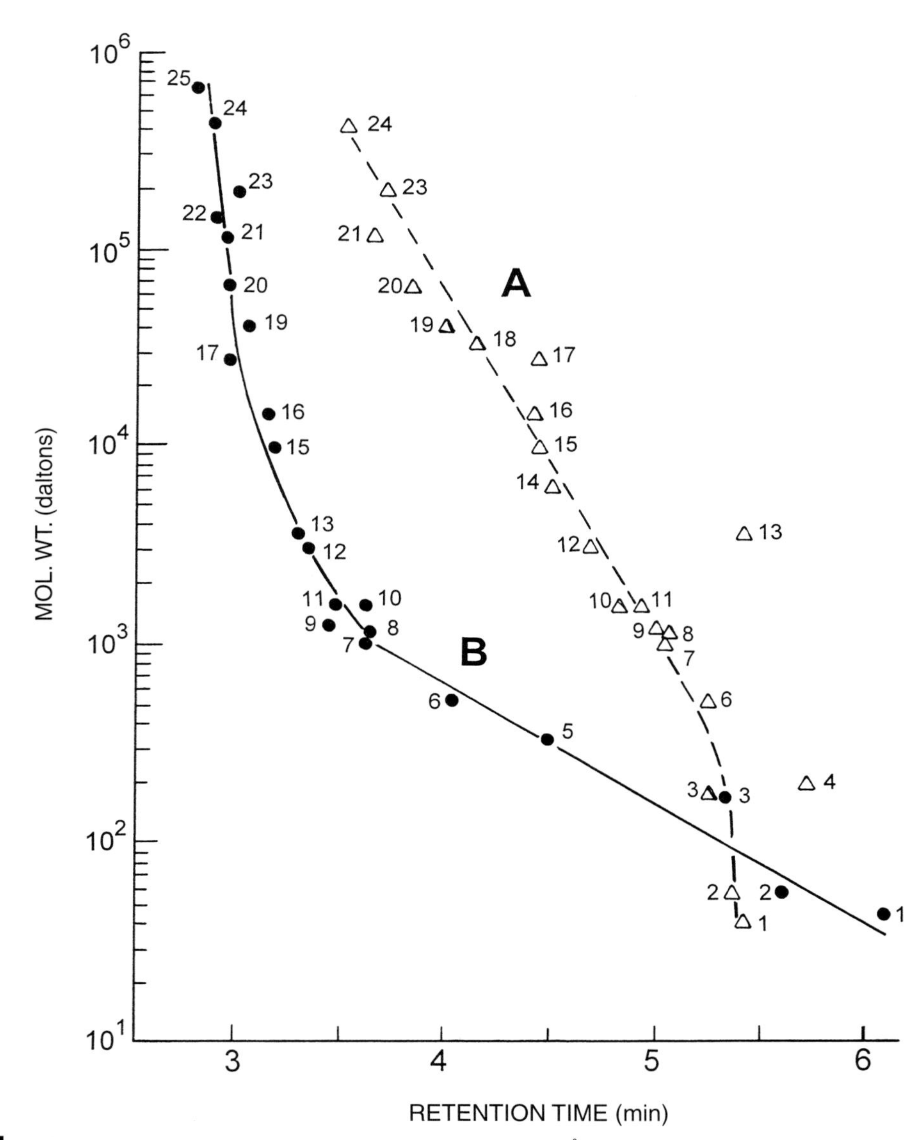

FIGURE 8.10 SEC ranges on a PolyHEA column with 300-Å pores. Column, flow rate, and mobile phases: As in Fig. 8.2, but the pore diameter is 300 Å. Sample key: (1) Formic acid (46 Da); (2) acetic acid (60 Da); (3) *N*-chloroacetyl-Tris base (177 Da); (4) citric acid (192 Da); (5) thyrotropin-releasing hormone (362 Da); (6) [Leu[5]]-enkephalinamide (554 Da); (7) angiotensin II (1046 Da); (8) luteinizing hormone-releasing hormone (1236 Da); (9) angiotensin I (1296 Da); (10) bombesin (1592 Da); (11) α-melanocyte-stimulating hormone (1665 Da); (12) somatostatin-28 (3149 Da); (13) poly-L-lysine (3800 Da); (14) insulin (bovine) (6500 Da); (15) ubiquitin (10,000 Da); (16) ribonuclease (15,000 Da); (17) carbonic anhydrase (29,000 Da); (18) β-lactoglobulin (29,200 Da); (19) ovalbumin (chicken) (43,000 Da); (20) bovine serum albumin (66,000 Da); (21) IgG (150,000 Da); (22) alcohol dehydrogenase (150,000 Da); (23) β-amylase (200,000 Da); (24) apoferritin (443,000 Da); and (25) thyroglobulin (669,000 Da).

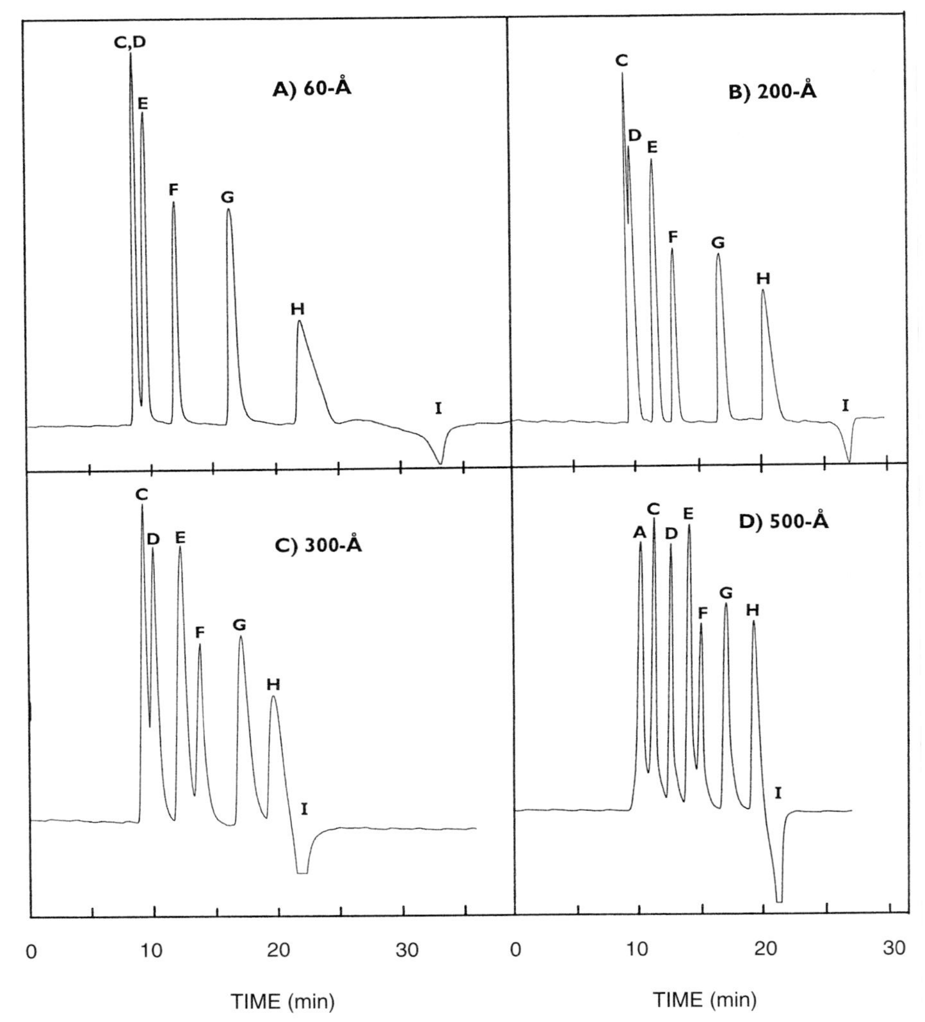

FIGURE 8.11 Effect of pore diameter on SEC of standards (denaturing mobile phase). Columns, flow rate and mobile phase: As in Fig. 8.5 except for pore diameters as noted. Sample Key: (A) IgG (150,000 Da); (C) cytochrome c (12,384 Da); (D) α-melanocyte-stimulating hormone (1665 Da); (E) [Met]^5-enkephalinamide (573 Da); (F) aspartame (294 Da); (G) phenylalanine (165 Da); (H) acetic acid (60 Da); and (I) water (18 Da).

shows the fractionation of some standard solutes with columns of various pore diameters. All of the materials include the region 20–1600 Da in their fractionation range, but this region accounts for a decreasing percentage of the total range as the pore diameter increases. Wider pore materials are of limited utility with denaturing mobile phases. Presumably they would be employed with solutes as large as proteins, and denaturation would be a significant

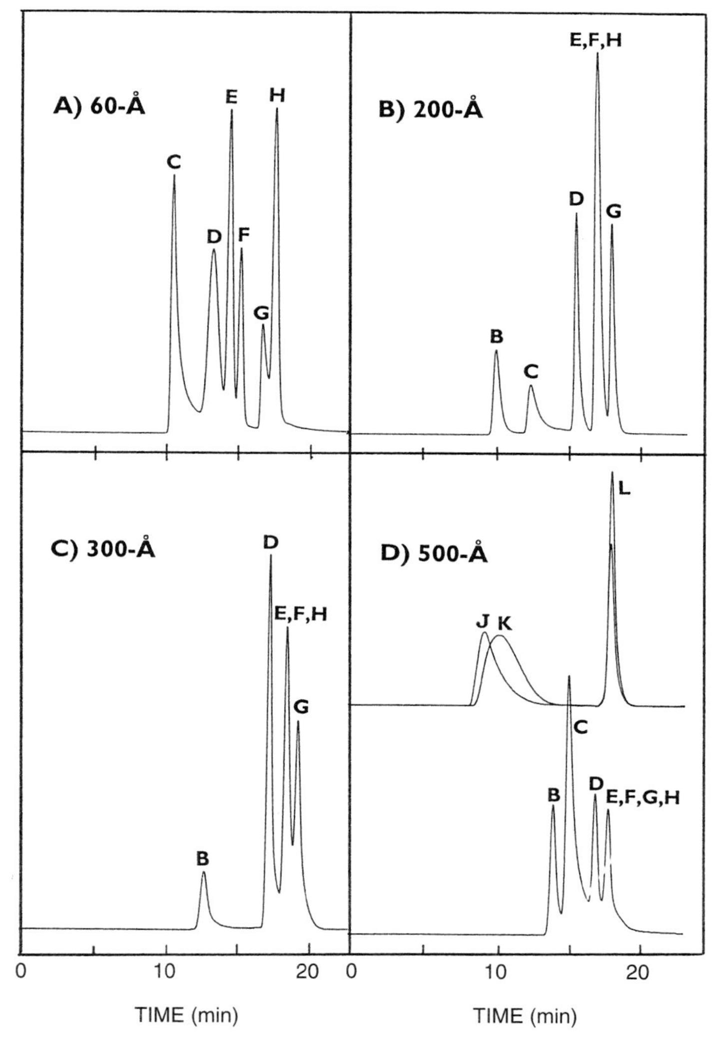

FIGURE 8.12 Effect of pore diameter on SEC of standards (nondenaturing mobile phase). "Nondenaturing" refers to the effect on the stationary phase. Most large proteins were in fact denatured by this mobile phase (which was optimized for use with peptides, not proteins). Accordingly, it was necessary to use polyacrylamide to demonstrate the approximate range and position of V_0 under these conditions. The polyacrylamide standards both eluted at V_o with the 300-Å column (not shown). Columns and flow rate: Same as in Fig. 8.11. Mobile phase: Same as in Fig. 8.1. Sample key: (B) Ovalbumin (43,000 Da); (J) polyacrylamide (1,000,000 Da); (K) polyacrylamide (400,000 Da); (L) low molecular weight impurity in the polyacrylamide standards. Other samples as in Fig. 8.11.

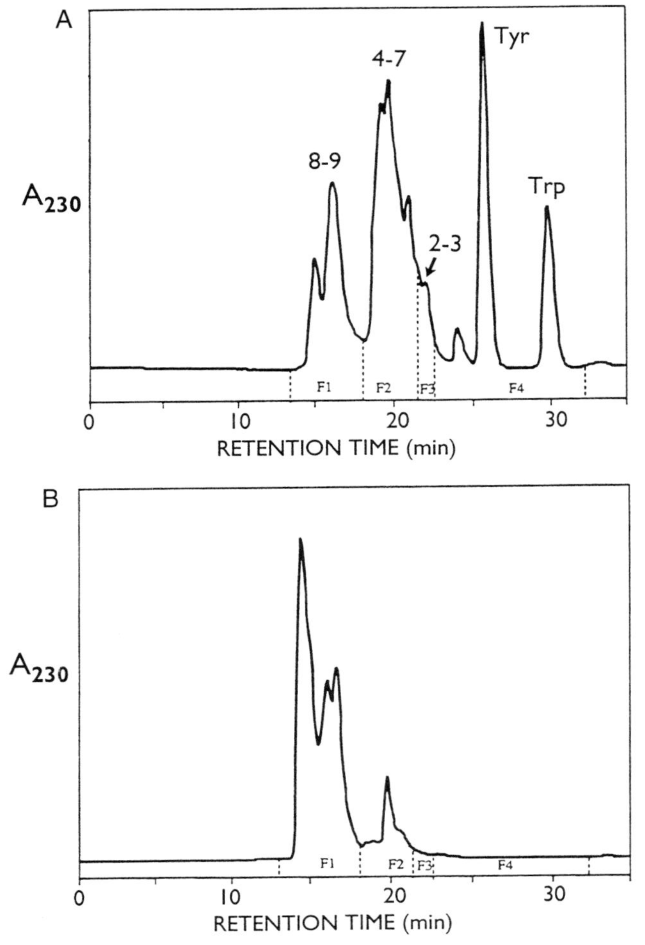

FIGURE 8.13 SEC of casein hydrolyzates. Numbers above the peaks refer to the number of amino acid residues in the typical peptide in the indicated fraction. Column: PolyHEA, 200 × 9.4 mm; 5 μm, 200 Å. Flow rate: 0.5 ml/min. Mobile phase: 50 m*M* Formic acid. Detection: A_{230}. Samples: (A) Pancreatin hydrolyzate and (B) tryptic hydrolyzate. (Adapted from Ref. 29 with permission from Silvestre *et al.* Copyright 1994, American Chemical Society.)

concern with such solutes. Also, if the fractionation of small solutes is of no interest in such analyses, then the pore volume occupied by the swollen coating is "wasted," as it decreases the portion available for fractionation in the size range of interest. In general, nondenaturing mobile phases will afford better resolutions in the size ranges of interest with large solutes such as proteins. Figure 8.12 demonstrates this point with the same columns and standards as Fig. 8.11. It should be noted that *nondenaturing* refers here to the effect on the stationary phase; the organic solvent and low pH used in Figs. 8.1 and 8.12 may well denature larger proteins. These conditions were selected as suitable for SEC of peptides. In general, SEC of proteins is usually best performed with neutral buffers and an absence of organic solvent.

V. APPLICATIONS

A. SEC of Peptides and Amino Acids

The proteins in food supplements are often hydrolyzed to short peptides to make them easier to absorb. A high content of amino acids is deleterious, however. Thus, there is ongoing interest in determining the size distribution of peptides in protein hydrolyzates. Silvestre *et al.* (29,30) used a PolyHEA column to compare casein hydrolyzates prepared through various methods. They were able to assess the content of the smallest peptides, as well as amino acids (Fig. 8.13).

B. SEC of Small Solutes Other Than Peptides

In an attempt to isolate a factor responsible for stimulating hepatocyte growth, Nelson *et al.* (31) used a PolyHEA column to fractionate an extract of liver by size. The active fraction eluted at a position corresponding to approximately 200 Da; the actual molecular weight (electrospray mass spectrometry; ES-MS) was 215 Da. The compound of interest proved to be glycerophosphorylethanolamine.

C. Mass Spectrometry: On-Line Desalting

Steven Carr (SmithKline Beecham) has used microbore columns to desalt proteins prior to ES-MS (32). The pore diameter of PolyHEA used (usually 200 Å) was selected so that all proteins of interest would elute at V_0 with 50 m*M* formic acid. Only the V_0 peak was allowed to flow into the ES-MS nebularizer; the rest of the SEC effluent (including the salts) was diverted to waste by opening a microdumper valve between the column and the nebularizer. The properties of the mobile phase were quite compatible with ES-MS analysis.

ACKNOWLEDGMENTS

I am grateful to Drs. P. C. Andrews and Marialice Silvestre for permission to include their data in this chapter. Some of these data were presented previously as poster 110, 10th ISPPP (Wiesbaden, October 1990). Bio-Gel is a trademark of Bio-Rad Corp. Sephadex and Superdex are trademarks of Pharmacia Biotech. TSK-GEL is a trademark of Tosoh Corp. PolyHYDROXYETHYL Aspartamide and PolyHYDROXYETHYL A are trademarks of PolyLC Inc.

REFERENCES

1. Porath, J. (1960). *Biochim. Biophys. Acta* **39**, 193.
2. Gelotte, B. (1960). *J. Chromatogr.* **3**, 330.
3. Janson, J.-C. (1967). *J. Chromatogr.* **28**, 12.
4. Bretthauer, R. K., and Golichowski, A. M. (1967). *Biochim. Biophys. Acta* **155**, 549.
5. Schwartz, A. N., and Zabin, B. A. (1966). *Anal. Biochem.* **14**, 321.
6. Thornhill, D. P. (1972). *Biochim. Biophys. Acta* **279**, 1.
7. Chappell, I., Baines, P., and Carpenter, P. K. (1992). *J. Chromatogr.* **603**, 49.

8. Alpert, A. J. (1990). *J. Chromatogr.* **499,** 177.
9. Montelaro, R. C., West, M., and Issel, C. J. (1981). *Anal. Biochem.* **114,** 398.
10. Kato, Y., Komiya, K., Sasaki, H., and Hashimoto, T. (1980). *J. Chromatogr.* **193,** 458.
11. Lazure, C., Dennis, M., Rochemont, J., Seidah, N. G., and Chrétien, M. (1982). *Anal. Biochem.* **125,** 406.
12. Richter, W. O., Jacob, B., and Schwandt, P. (1983). *Anal. Biochem.* **133,** 288.
13. Bennett, H. P. J., Browne, C. A., and Solomon, S. (1983). *Anal. Biochem.* **128,** 121.
14. Swergold, G. D., and Rubin, C. S. (1983). *Anal. Biochem.* **131,** 295.
15. Irvine, G .B., and Shaw, C. (1986) *Anal. Biochem.* **155,** 141.
16. Richter, W. O., and Schwandt, P. (1984). *J. Chromatogr.* **288,** 212.
17. Habermann, S. M., and Murphy, K. P. (1996). *Protein Sci.* **5,** 1229.
18. Collins, K. D., and Washabaugh, M. W. (1985). *Quart. Rev. Biophys.* **18,** 323.
19. Rivier, J. E. (1980). *J. Chromatogr.* **202,** 211.
20. Porath, J. (1997). *J. Protein Chem.* **16,** 463.
21. Kopaciewicz, W., and Regnier, F. E. (1982). *Anal. Biochem.* **126,** 8.
22. Golovchenko, N. P., Kataeva, I. A., and Akimenko, V. K. (1992). *J. Chromatogr.* **591,** 121.
23. Zhu, B. Y., Mant, C. T., and Hodges, R. S. (1991). *J. Chromatogr.* **548,** 13.
24. Mant, C. T., Parker, J. M. R., and Hodges, R. S. (1987). *J. Chromatogr.* **397,** 99.
25. Kuntz, I. D. (1971). *J. Am. Chem. Soc.* **93,** 514.
26. Hagel, L. (1993). *J. Chromatogr.* **648,** 19.
27. Pharmacia Biotech "Data File" bulletin 18-1119-46 (10/96).
28. Hedlund, H., Kärf, L., Lindberg, U., Nyhammar, T., and Winter, A., Poster P-222, ISPPP '97 Conference (October 26–29, 1997), Rockville, MD.
29. Silvestre, M. P. C., Hamon, M., and Yvon, M. (1994). *J. Agric. Food Chem.* **42,** 2778.
30. Silvestre, M. P. C., Hamon, M., and Yvon, M. (1994). *J. Agric. Food Chem.* **42,** 2783.
31. Nelson, C., Moffat, B., Jacobsen, N., Henzel, W. J., Stults, J. T., King, K. L., McMurtrey, A., Vandlen, R., and Spencer, S. A. (1996). *Exp. Cell Res.* **229,** 20.
32. Shushan, B., *in Views, Fall 1994* (Perkin-Elmer newsletter), pp. 16–17.

9

DESIGN, PROPERTIES, AND TESTING OF POLYMER STANDARDS SERVICE SIZE EXCLUSION CHROMATOGRAPHY (SEC) COLUMNS AND OPTIMIZATION OF SEC SEPARATIONS

P. KILZ

PSS Polymer Standards Service, D-55023 Mainz, Germany

I. INTRODUCTION

Size exclusion chromatography (SEC, also known as GPC and GFC) has become a very well accepted separation method since its introduction in the late-1950s by works of Porath and Flodin (1) and Moore (2). Polymers Standards Service (PSS) packings for SEC/SEC columns share this long-standing tradition as universal and stable sorbents for all types of polymer applications. In general, PSS SEC columns are filled with spherical, macroporous cross-linked, pressure-stable, and pH-resistant polymeric gels.

PSS SEC/SEC columns cover the full range of applications for polymer characterization. They comprise columns for all types of separations, eluents, and tasks.

1. PSS columns for organic eluents: PSS SDV columns are based on proven styrene–divinylbenzene type sorbents with improved sorbent characteristics and column technology.
2. PSS columns for aqueous separations: PSS HEMA Bio and the further improved PSS SUPREMA columns are based on methacrylic ester copolymer technology and exhibit very good stability while maintaining high efficiency.
3. PSS columns for medium polar or mixed solvents: PSS HEMA and PSS SUPREMA Basic were designed to allow SEC separations in polar media such

Column Handbook for Size Exclusion Chromatography

as pure alcohols or mixed solvents, e.g., aqueous buffers with high tetrahydrofuran (THF) concentrations. These columns can also be used as "universal" columns, as they allow separations in aqueous and organic solvents.

4. PSS columns for fluorinated eluents: PSS PFG columns were developed by PSS because users worldwide were unsatisfied with the stability of conventional organic SEC columns when running solvents such as hexafluroisopropanol (HFIP). Because polymeric gels tend to be unstable in fluorinated media, PSS modified silica to achieve better stability while maintaining perfect chromatographic performance.

All PSS columns combine the following advantages:

excellent separation properties
very good stability (physically, chemically, and biologically)
very high solvent compatibility
expert advice and custom support for applications by PSS developers and manufacturer
comprehensive service (e.g., column refill services for all columns of all manufacturers)

PSS SEC/SEC columns can be employed for all

Applications: polymer characterization, preparative-scale fractionations, sample preparation, ultraquick separations, separations with highest efficiency
Eluents: all organic and aqueous systems (alcohols, fluorinated hydrocarbons (FCHC), etc.), solvent mixtures
Samples: ultrahigh to very low molecular weight; synthetic and bio polymers
Dimensions: micro-SEC (2 × 100 mm) up to preparative-scale SEC (40 × 1000 mm)

Details and applications are given in Section VII.

II. BASICS OF SEC COLUMN DESIGN

All SEC columns have to be designed and synthesized by the polymer chemist to meet the specific requirements of the separation mechanism (3). With regard to efficient SEC separations, there are a number of important aspects to consider:

- inertness of packing
- good accessibility of pore structure
- high pore volume and porosity
- diffusion control (fast mass transfer)

Unfortunately, most column and sorbent manufacturers do not develop column packing materials mainly for SEC work, but for the bigger high-performance liquid chromatography (HPLC) market. However, there are many important differences to consider when designing packings for different modes of chromatography. Table 9.1 illustrates some of the design differences between SEC and HPLC packings.

Moreover, polymer applications determine a number of important column (hardware) design properties to get reproducible results and the most efficient separations. Most of them are related to the polymer conformation in the injection band moving through the column.

The column end fittings must have optimized flow characteristics to distribute the sometimes very viscous injection solution evenly across the column cross section. Wall effects have to be minimized in the column to achieve the best results and the column hardware itself has to be truly inert. This is especially important when running aggressive eluents such as highly corrosive solvent mixtures or buffers. Column hardware (optimized frits) and the packing material itself have to be designed for low shear. This is most important when investigating high molar mass samples, which otherwise will degrade during elution from the column.

A. Major SEC Column Design Characteristics

The chemical nature of the packing has the largest influence on the retention of molecules and a big impact on the efficiency of the separation itself. The chemical and physical properties of the sorbent are determined by the choice of the comonomers for the copolymerization. The type of the copolymerization process employed by the synthetic chemist introduces the macroporous structure into the sorbent and determines the surface topology (accessibility, resolution) and the surface chemistry of the packing (4).

Additional factors influencing column performance are the type and quality of the packing process, which mainly determines the theoretical plate count (N) of the column. In contrast to HPLC columns the efficiency of the separation itself is determined predominantly by the quality of the sorbent alone (pore

TABLE 9.1 Different Requirements for Column Packings for HPLC and SEC Applications

	Requirements for	
Property of sorbent	**HPLC columns**	**SEC columns**
Pore size distribution	Narrow	Medium
Pore size	Less important	Small to large
Pore volume	Less important	High
Particle size distribution	Narrow	Narrow
Surface area	Large	Not important
Surface chemistry	Homogeneous	Inert
Pore architecture	High surface area	High accessibility
Mass transfer	High	High
Axial dispersion	Low	Low

access). The optimum design and proper synthesis of the sorbent will mainly influence polymer resolution (R_{sp}).

PSS did a market survey of SEC users to find out which features SEC columns should have. Table 9.2 lists the most frequently mentioned features (5). It is obvious from Table 9.2 that the requirements for SEC columns are difficult to meet and that the polymer chemists and manufacturers have to find the optimum property combinations. This is a very challenging and thrilling area of research for everyone involved in this type of work. Further improved products are expected in the near future.

III. REVIEW: SEC SORBENTS AND REQUIREMENTS

For a long time there have been discussions about which type of sorbent is the best for SEC separations in various mobile phases. In principle, organic (copolymer) and inorganic packings can be used. Each type of packing has its benefits and drawbacks. Table 9.3 summarizes major sorbent properties and reveals some interesting aspects of SEC separations and its requirements on packings.

1. Inorganic packings (silica, alumina, etc.) are very stable (yet brittle) and show very high pore volumes (i.e., efficiency). However, their chemical stability is very limited and the surface is very active (this is also true for reversed-phase columns), allowing their use in special applications only.
2. Polymer packings for organic eluents overall show good properties, but not outstanding performance in a special feature. This makes them ideal for most SEC work.
3. Column sorbents for aqueous media show just average properties. This is due to the different copolymerization process, which does not allow easy formation of macroporous beads with proper pore topology. This fact also reflects many experiences of SEC users, who have to input much more effort to get good aqueous SEC work accomplished.

TABLE 9.2 User's Response to Desired SEC Column Features

Important column feature	Main influence by
High resolution	Sorbent topology, packing process
"Universal" column	Monomer and additives choice
Interaction-free surface	Monomer and additives choice
Good physical stability	Copolymerization process and pore
Good chemical stability	Sorbent technology
Good biological stability	Copolymerization process
Low shear	Particle size distribution, packing process
Low wear	Polymerization chemistry, packing process
Easy column combinations	Pore structure control

TABLE 9.3 Comparison of Major SEC Sorbent Properties

Property	Polymer packing for eluents: Organic	Polymer packing for eluents: Aqueous	Inorganic packing[a]
Physical stability	+	0	++
Chemical stability	+	+	−−
Biological stability	+	0 PSS Hema: +	0
Pore volume	+	0 PSS Suprema: +	++
Eluent compatibility	0	0 PSS Hema: ++	+

[a]Symbols indicate property strengths and weaknesses: ++, very good; +, good; 0, appropriate; −, poor; and −−, very poor.

4. Table 9.3 also shows that there have been some improvements in SEC sorbents for aqueous applications since the early 1990s with the availability of PSS HEMA and PSS SUPREMA columns. Table 9.4 gives an overview of available packings and their major use (this table is not complete and is intended for showing the major features and uses in SEC applications).

IV. SEC COLUMN SELECTION CRITERIA

The major parameter for column selection is the intended application. A balance of mobile-phase polarity in comparison with the polarity of the stationary

TABLE 9.4 Synopsis of Major Packings for SEC Applications

	Polymer packings			Inorganic packings		
Polarity	Non	Medium	Polar	Non	Medium	Polar
Chemistry	St-DVB	Acrylic	OH-acrylic (ionic)	SiO_2-C_{18}	SiO_2-diol	SiO_2
Solvents	Nonmedium	All	Aqueous		Selective only	
Producers	Few	Two	Few		Many	
Examples	PSS SDV PL Gel Styragel TSK-H Shodex A,K	PSS HEMA Shodex — — —	PSS HEMA Bio PSS Suprema PL Aquagel UltraHydrogel TSK-PW Shodex OH (ionic: SO_3H, amides)		Very many	
Use	Lipophilic samples	Universal	OH: universal SO_3H: saccharides	Limited	Some proteins	Very limited

phase and sample polarity is important for pure SEC separations (6) (Fig. 9.1). In general, users will select their columns according to the mobile phase (see Table 9.5 for details) they need to use.

Another important parameter for column selection is the proper choice of sorbent porosity. The pore size of the sorbent determines the fractionation range of the column. The best way of doing this is by looking at the calibration curves of the columns, which are normally documented by the column vendor (cf. Fig. 9.3 for PSS SDV column calibration curves and PSS SDV fractionation ranges) (7).

A. Chromatographic Modes of Column Separation

All SEC separations require an interaction-free diffusion of the sample molecule in solution into and out of the pore structure of the column packing material (8–11). In general, this goal is easier to achieve in organic media than in aqueous eluents. The reason being that aqueous mobile phases have many more parameters to adjust correctly (e.g., type of salt, salt concentration, pH, addition of organic modifier, concentration of cosolvent). Additionally, water-soluble macromolecules have more ways to interfere with the stationary phase due to charged functional groups, hydrophobic and/or hydrophilic regions in the molecule, and so on. These types of parameters have to be balanced in a proper SEC experiment. In order to obtain a "pure" SEC separation, the polarity of the stationary phase, the polarity of the eluent, and the polarity of

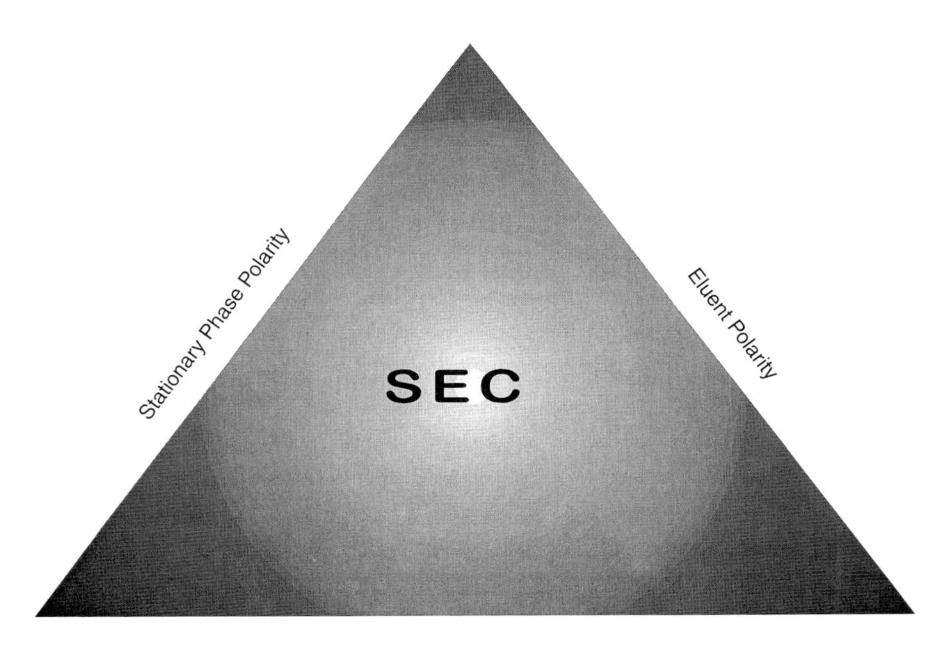

FIGURE 9.1 Balancing polarities of phase system in SEC applications for interaction-free separations.

TABLE 9.5 Properties of PSS SEC Columns

Property	PSS SDV	PSS PFG	PSS HEMA	PSS Suprema
Type of eluent	Organic	Polar organic	Organic, aqueous, and mixtures	Aqueous
Particle sizes (μm)	3, 5, 10, 20	7	10	5, 10, 20
Porosities [Å]	50, 100, 500, 10^3, 10^4, 10^5, 10^6, 10^7, linear	100, 300, 1000, 4000; linear	40, 100, 300, 1000; linear	30, 100, 300, 1000, 3000, 10000, 30000; linear
Molecular mass range (Da)	<100 to >30 million	100 to >30 million	100 to 3 million	<100 to >100 million
Pore volume	High	Very high	Medium high	High
Inertness	Very good	Good	Good	Good
Chemical stability	pH 1–13	pH 2–8	pH 2–12	pH 2–12
Physical stability	150 bar, 150°C	300 bar, 120°C	200 bar, 120°C	100 bar, 120°C
Biological stability	Very good	Very good	Very good	Very good
Mass transfer	Good	Very good	Average	Good
Efficiency	N = 30–100,000/m	N = 20–40,000/m	N = 20–40,000/m	N = 20–70,000/m
Resolution	Good to excellent	Very good	Average	Good to very good
Solvent compatibility	Good	Very good	Very good	Good

the sample have to be matched, as otherwise specific interactions will occur, which will overlay with the normal SEC elution behavior.

The basic principle of chromatography separations can be described by thermodynamics using the distribution coefficient K (12):

$$K = a_s/a_m = \exp(-\Delta G/RT),$$

where a is the activity (concentration) of the molecule in the stationary phase (index s) and the mobile phase (index m) and ΔG is the change in free energy of the species between the stationary phase and the mobile phase.

As is known, SEC separations require interaction-free conditions. Therefore, the enthalpic contribution to the free energy term vanishes when no enthalpic interaction is postulated between analyte and sorbent:

$$K_{SEC} = \exp(\Delta S/RT), \qquad 0 < K_{SEC} \leq 1, \Delta H = 0,$$

where ΔS is the entropy loss when a molecule enters the pore of the stationary phase.

In the case of no steric exclusion of the molecule from parts of the stationary phase, the retention can be described by the enthalpic term alone:

$$K_{HPLC} = \exp(-\Delta H/RT), \qquad K_{HPLC} \geq 1\ \Delta S = 0,$$

where ΔH is the enthalpy change when a molecule is adsorbed by the stationary phase.

So far there are two modes (SEC and HPLC) where there is no change in entropy or enthalpy, respectively. Another mode of chromatographic behavior exists when enthalpic and entropic contributions balance out, i.e., when the change in free energy disappears ($\Delta G = 0$) (13–17). This mode is called liquid chromatography at the critical point of adsorption (LC-CAP). The polymeric nature of the sample (i.e., the repeating units) does not contribute to the retention of the species. Only defects (end groups, comonomers, branching points) contribute to the separation of the molecule. Figure 9.2 illustrates this behavior and shows the dependence of retention volume on the molar mass for the different modes of chromatography.

Using this basic theory of separation, the experimental conditions can be modified to shift the separation into the chromatographic mode one would like to operate in. In many cases this can be done without buying new columns, but by just adjusting the polarity of the mobile phase.

Let us consider the separation of polymethylmethacrylate (PMMA) on a nonmodified silica column as an example. In THF (medium polar eluent) the PMMA eludes in size exclusion mode because the dipoles of the methylmethacrylate (MMA) are masked by the dipoles of the THF. Using the nonpolar toluene as the eluent on the same column, the separation is governed by adsorption because the dipoles of the carbonyl group in the PMMA will interact with the dipoles on the surface of the stationary phase. The separation of PMMA in the critical mode of adsorption can be achieved by selecting an appropriate THF/toluene mixture as the eluent. In this case all PMMA samples

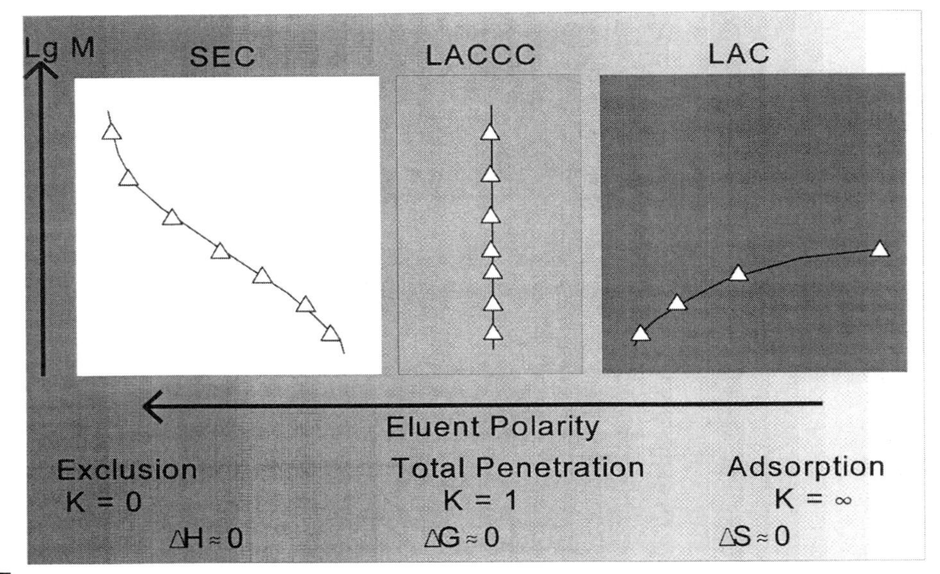

FIGURE 9.2 Different modes of chromatography as seen in the elution order of samples with different molar masses.

will elute at the same time regardless of their different molar masses. PMMA samples with different end groups will be separated with high selectivity with no overlapping of size separation effects (4).

B. Selection of Pore Size and Separation Range

The molar mass range of the samples to be investigated determines the column porosity to choose for most efficient separations. The larger the pores in the column packing the higher molar mass samples can be characterized. Unfortunately, there is no general nomenclature that will allow the easy selection of column pore sizes. Each manufacturer has its own system for pore size designation. The most easy method to find out which columns will be useful for a selected task uses the calibration curve that every manufacturer shows in their literature. Figure 9.3 shows the calibration curves (on the top) and the recommended molar mass separation ranges (on the bottom) for PSS SDV columns (18). The highest efficiency for a separation is determined by the lowest slope of the calibration curve.

Figure 9.4 illustrates a simple way of selecting the best column for SEC work based on the calibration curves of two sorbents with different pore sizes (4).

1. Column Combinations

Column combinations are used often in SEC to obtain the necessary chromatographic resolution. In general, a single SEC column is not efficient enough to allow the determination of small differences in samples. Column combinations will influence the molar mass fractionation range and the separation efficiency of the column set (7). This means that a combination of two identical columns will improve the separation, whereas the molar mass fractionation range remains unchanged. The higher efficiency of this column set is paid for by longer retention times and larger eluent volumes needed.

If two columns with different porosity are used in a column combination, the efficiency of the separation will not change much, but the fractionation range will be increased immensely. In this case, longer chromatography times allow a better separation of broader samples (or samples with high and low molar mass components).

2. Advantages and Disadvantages of Mixed-Bed Columns

Discussing pore size selection of columns leads directly to the issue of using single porosity columns or so-called linear or mixed-bed columns, which contain mixtures of different pore sizes in a single column (3,19). Both types of columns have advantages and disadvantages, as shown in Table 9.6.

An example may show how the different concepts come into effect in a real-life laboratory environment. This example is based on column selections that many laboratories use for ordinary, general-purpose work.

A. Conventional column set consisting of a series of columns with single pore sizes.

1. "Universal" column set: PSS SDV 5 μm 10^6, 10^5, 1000 Å. Separation range: 10^7 to 200 Da. Resolution: $R_{sp} \approx 15$. Plate count: $N_{th} \approx 50{,}000$/m.

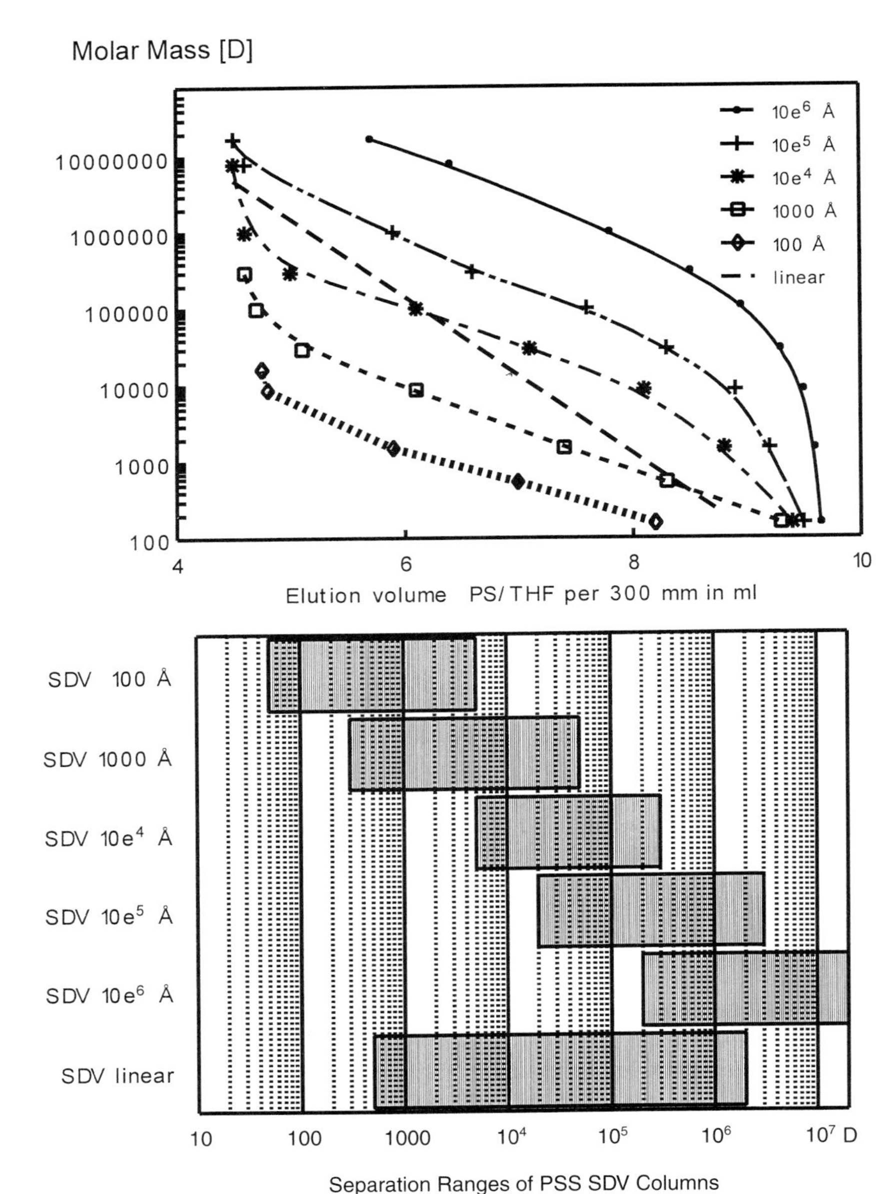

FIGURE 9.3 Separation characteristics of PSS SDV columns for organic eluents.

2. Optimized oligomer column set: PSS SDV 3 μm 1000, 100, 50 Å. Separation range: 30 K to 50 Da. Resolution: $R_{sp} \approx 25$. Plate count: $N_{th} \approx$ 85,000/m.

B. Mixed-bed column combination with PSS SDV 5 μm linear (3×). Separation range: 3×10^6 to 500. Resolution: $R_{sp} \approx 15$. Plate count: $N_{th} \approx$ 50,000/m.

The performances of column set A1 and B are similar, although column set B is more expensive. Using a specially selected column set for oligomer

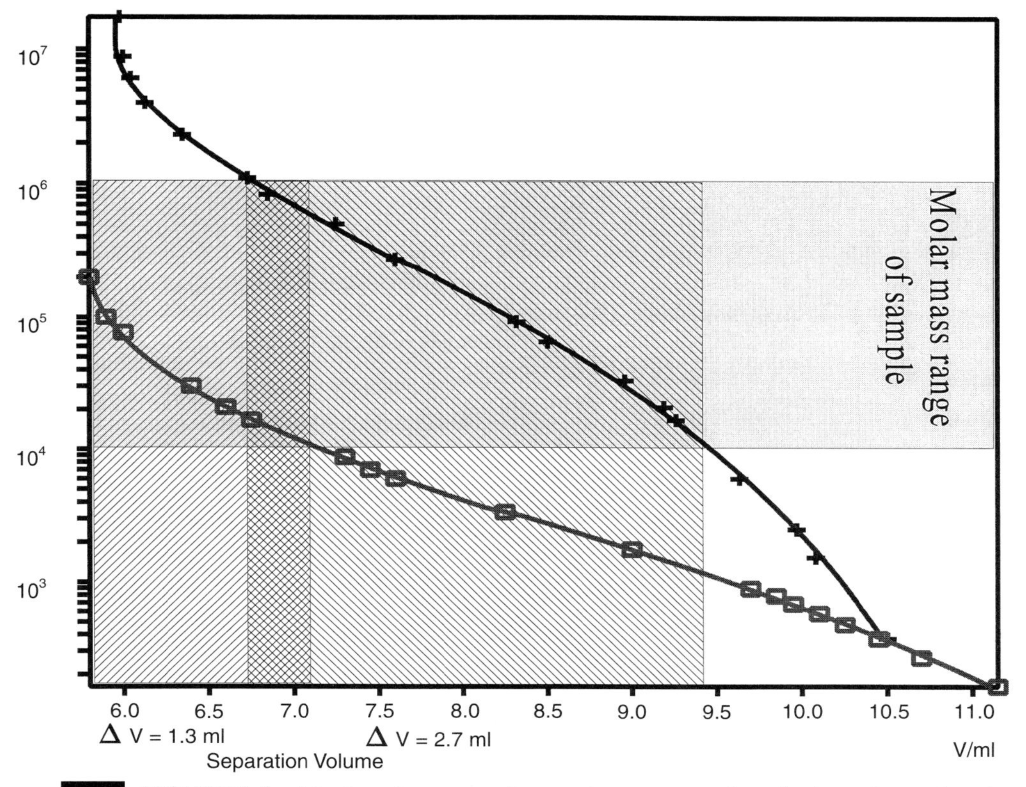

FIGURE 9.4 Selecting columns according to molar mass range of sample: the small pore size column has a separation volume of 1.3 ml (and partial exclusion above 200 kDa) and the other column is well selected with a separation volume of 2.7 ml.

separation (as in column set A2), the user gets a much improved resolution and a separation range that corresponds directly to the samples to be investigated. In the case of mixed-bed columns, the separation range is determined by the

TABLE 9.6 Comparison of Single Porosity Type and Linear Mixed-Bed SEC Columns

Pore type	Advantage	Disadvantage
Single porosity	Efficient Optimized Flexible Low cost Good for QC	Viscous fingering
Mixed bed or linear	Fast (screening work) "Universal" Inject band dilution	Low efficiency Column combination

column, not by the application or the user, therefore sacrificing flexibility. This can be seen easily in Fig. 9.5 (4).

The linear column (PSS SDV 5 μm linear) has a wider molar mass fractionation range while keeping the analysis time roughly the same. Therefore the slope of the calibration curve is much steeper and the resolution will be poorer in this case. The second column with a single pore size (PSS SDV 5 μm 1000 Å) separates only below 50,000 Da, but does this very efficiently in the same time.

C. Influence of Particle Size on Separation Efficiency

SEC columns have become much more efficient since they were introduced in the late 1950s. The major factor for this has been the ability of synthetic polymer chemists to produce smaller particle sizes of column packing materials. The first sorbents were several 100 μm wide in diameter (20), whereas modern columns are filled with particles in the range between 3 and 20 μm, which caused an immense improvement in separation power. The major drawback

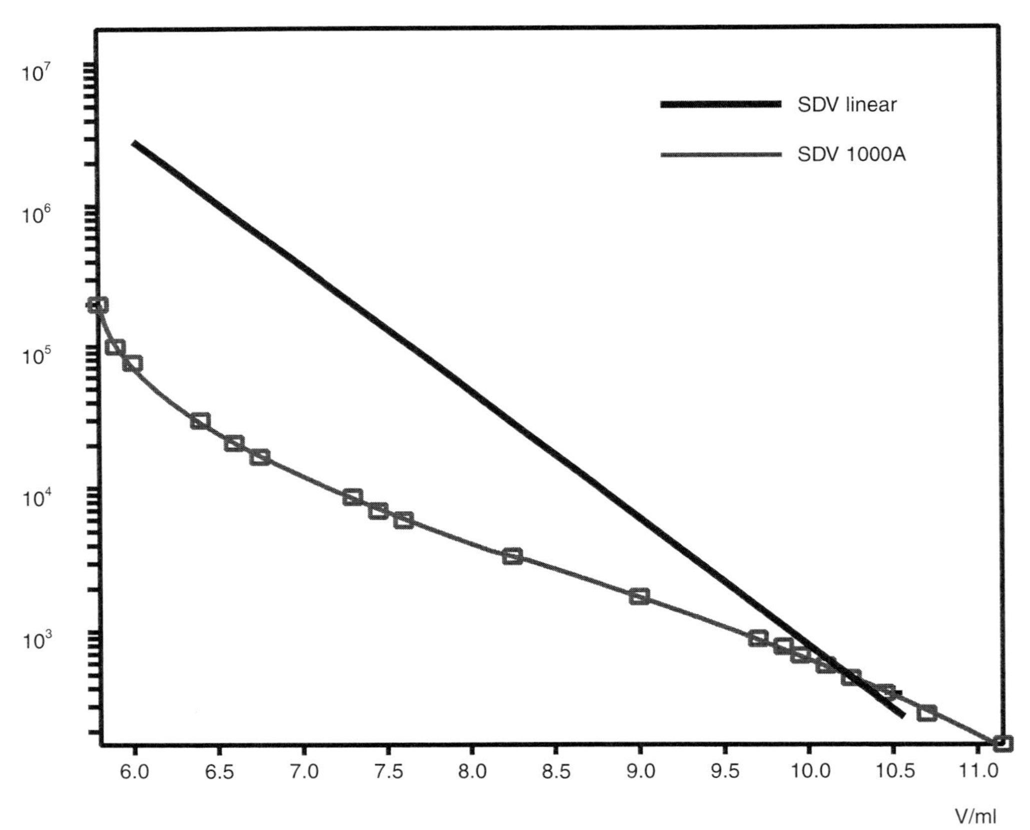

FIGURE 9.5 Comparison of molar mass separation range and separation efficiency of linear mixed-bed (upper curve) and single pore size column (lower curve).

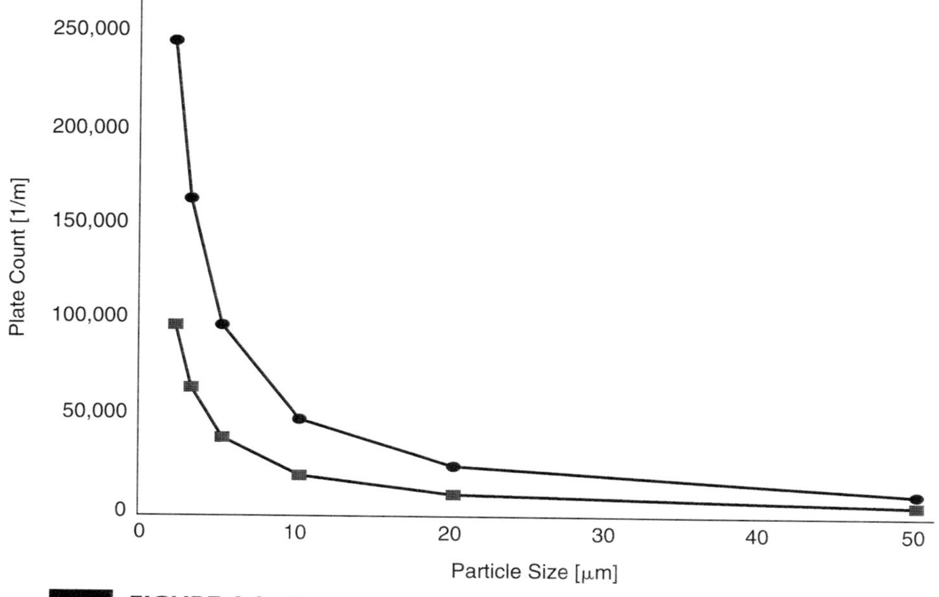

FIGURE 9.6 Theoretical (●) and normally found (■) column efficiencies in SEC columns with packings of different particle size.

of such small particle columns is their increased backpressure and shear forces in the column. This is especially important for high molar mass samples or samples that contain microgels, which will clog those columns more easily than the older ones with larger particle size packings. Column manufacturers have taken note of these requirements and have been developing columns with less shear and optimized flow characteristics while keeping efficiency high (Fig. 9.6).

Table 9.7 summarizes the comparison of SEC column performance with regard to particle size of the packing material. The author tried to create this table using the test results of different manufacturers of styrene–divinylbenzene columns.

TABLE 9.7 Dependence of Theoretical Plate Count and Resolution on Sorbent Particle Size[a]

Particle size (μm)	3	5	10	20
Plate count (1/m)	>80,000	>45,000	>30,000	>15,000
Resolution, R_{sp}	6–9	4–6	2–4	1–2

[a]Average values obtained experimentally from SEC columns of different major vendors.

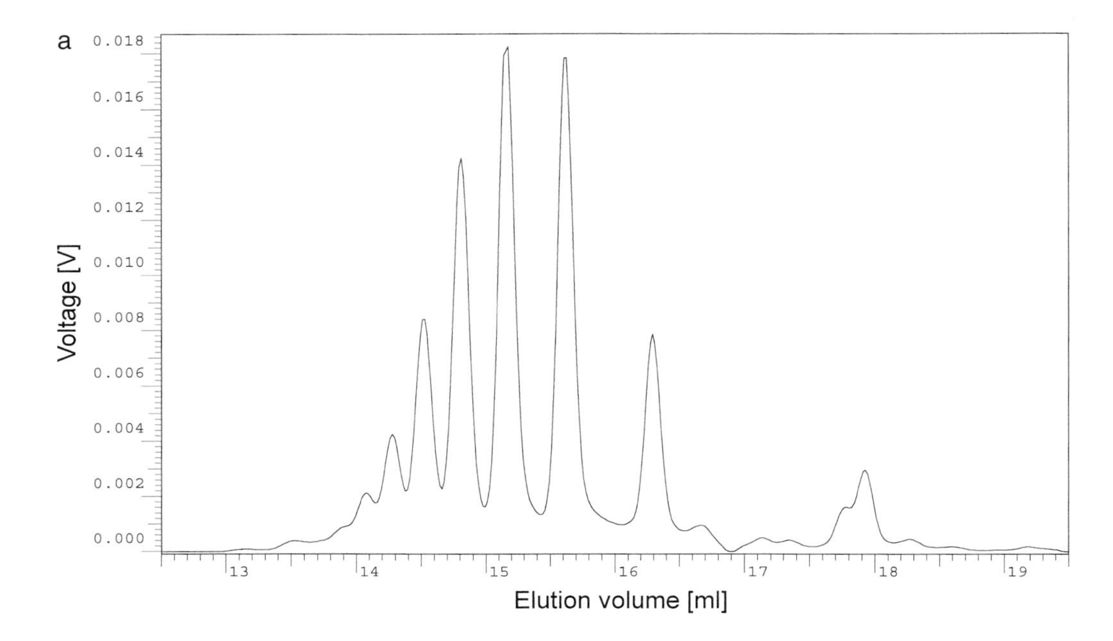

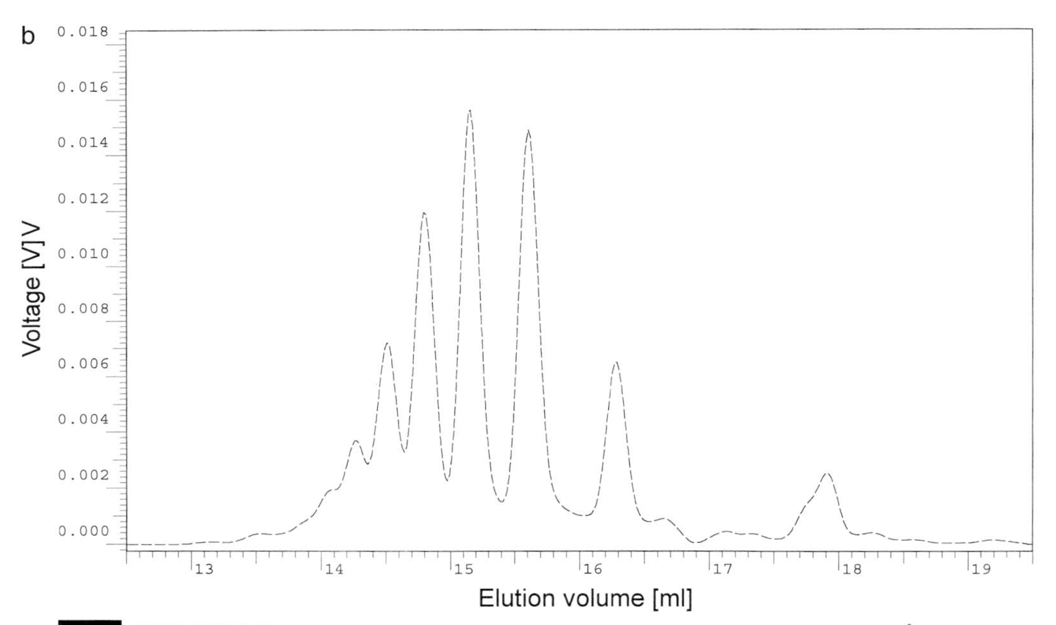

FIGURE 9.7 (a) Separation of PSS Oligostyrene Standard 400 D on PSS SDV 3 μm 100 Å (dimensions: 8 × 600 mm). (b) Separation of PSS Oligostyrene Standard 400 D on PSS SDV 5 μm 100 Å (dimensions: 8 × 600 mm); please note the decrease in resolution. (c) Separation of PSS Oligostyrene Standard 400 D on PSS SDV 10 μm 100 Å (dimensions: 8 × 600 mm); please note the obvious loss off efficiency as compared to a 3-μm column with otherwise identical characteristics.

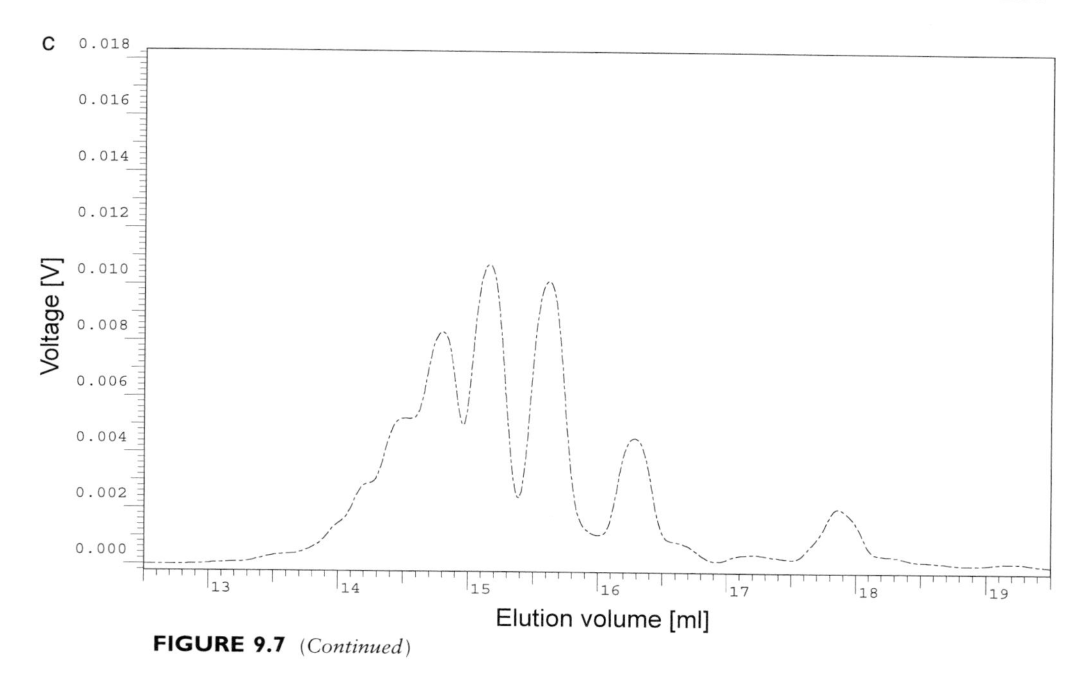

FIGURE 9.7 (*Continued*)

Figure 9.7 shows separations under identical conditions using PSS SDV columns with 3 (Fig. 9.7a)-, 5 (Fig. 9.7b)- and 10 (Fig. 9.7c)-μm particle size columns. A polystyrene oligomer standard was injected and all analyses were performed in THF as the eluent. The much higher efficiencies of small particle size columns are obvious, which is important in the SEC separation of low molecular weight compounds such as additives, by-products, and resins. The reader should note that all chromatograms are area normalized and have the same *Y* axis to show the differences in peak width and height.

Because of high shear forces in 3-μm columns, their use is limited to samples below about 100,000 Da. This is the reason why PSS only offers PSS SDV 3-μm columns up to 1000-Å pore size.

D. Influence of Column Dimensions

Column dimensions mainly determine the quantity of sample to be separated. However, because the SEC process is driven by size separation and is diffusion controlled, special care has to be taken to keep optimized separation conditions, especially when going to smaller internal diameter columns. Overloading and excessive linear flow rates can be observed quite often in these typese of columns. For this reason, standard 8-mm i.d. columns are commonly used, as they are rugged and have a good tolerance toward separation conditions.

PSS SEC column dimensions were chosen to allow easy scaling of chromatography conditions without the need to optimize separations for each column dimension separately. The volume flow rate and the sample load can be calculated easily when changing from a standard analytical column to a microbore or a prep-scale column. The separation efficiency will not change and the elution times will also remain constant.

The volume flow rate between different column dimensions can be calculated easily (assuming constant linear flow velocity) according to

$$F_{\text{new}} = F_{\text{old}} \left(\frac{\text{i.d.}_{\text{new}}}{\text{i.d.}_{\text{old}}} \right)^2.$$

The sample load can be determined by

$$m_{\text{new}} = m_{\text{old}} \left(\frac{\text{i.d.}_{\text{new}}}{\text{i.d.}_{\text{old}}} \right)^2 \left(\frac{\text{L}_{\text{new}}}{\text{L}_{\text{old}}} \right).$$

As an example, a separation on a standard PSS SEC column (8 × 300 mm dimension), which is done at a flow rate of 1.0 ml/min with a 100-μl injection of a 1% sample solution (sample load 1 mg), can be reproduced exactly on a 4 × 250 mm PSS SEC column when using a flow rate of 0.25 ml/min and injecting 20 μl of the same 1% sample solution. This corresponds to 0.2 mg of injected mass.

Table 9.8 shows examples of preparative separation conditions that allow a simple transfer of one method to a different column dimension (6).

E. Influence of Operating Parameters

In order to achieve the best efficiency the SEC column should be operated at optimized operating parameters. The most important ones are flow rate [cf. van Deemter equation for band-broadening effects (21)], sample viscosity (depends on molar mass and concentration of the sample), and injection volume (7).

Ideally, the sample should be injected onto the column as an infinitely thin disc, which covers the total cross section of the column. Because this is impossible, PSS has injected finite volumes onto the columns. In theory, these injection volumes should be as low as possible. In order to be able to detect the sample with significance, a certain (high) concentration of the sample has to be injected. This concept works well for low molar mass compounds, which do not generate much sample viscosity. However, when working with samples

TABLE 9.8 Scaling of Experimental Conditions for Different Column Dimensions

Column dimension i.d. × L (mm)	Typical flow rate (ml/min)	Sample load (mg)	Instrument requirements
2 × 250	0.17	0.05–0.5	Microbore pump, injector, and detector
4 × 250	0.25	0.2–2	Normal analytical instrumentation
8 × 300	1.0	1–10	Normal analytical instrumentation
20 × 300	6.25	6–60	Normal analytical instrumentation
40 × 300	25.0	25–250	Prep pump and detector cell

TABLE 9.9 Experimental Parameters for Running Polymer Standards in SEC

Molar mass (example)	Flow rate (example)	Injection volume (example)	Concentration (example)
High (>1000 kDa)	Small (<0.5 ml/min)	High (250 μl)	Low (<0.1g/liter)
Medium (10–1000 kDa)	Medium (1 ml/min)	Medium (100 μl)	Medium (<2 g/liter)
Low (<10 kDa)	Small–medium (0.5–1 ml/min)	Small (<20 μl)	High (<20 g/liter)

of higher molar mass, an increased viscosity of the injection solution will cause so-called viscous fingering in the column. This influences sample retention (normally they will elute later) and peak shape (peak broadening). For high molar mass samples it is therefore more important to keep solution viscosity low. This means the injection volume has to be increased for similar detectability.

Similar considerations apply to best volume flow rates for samples of different molar mass. For high molar mass samples, flow rates should be reduced to avoid shearing the macromolecule in the column. Moreover, a reduced flow rate is necessary because the diffusion coefficients of large molecules will get pretty small. This means that the macromolecule will pass by a pore in the packing material without having the time to enter it, if the linear flow rate is too high.

Table 9.9 (6) gives some guidelines for proper SEC separation conditions when analyzing polymer standards with narrow molar mass distribution on a single 30-cm column. The conditions have to be adjusted when running industrial polymers (which are normally much wider in molar mass distribution). Depending on the width of the MMD, concentrations can be increased by a factor of 3 to 10 for such samples. As a general rule, it is advisable to keep the concentration of the injected solution lower than $c \cdot [\eta] < 0.2$.

V. TESTING AND EVALUATION OF SEC COLUMNS

In order to maintain good column performance the separation efficiency of SEC columns should be checked regularly. Because some column manufacturers do not test columns individually prior to shipping, a new column should always be tested before first use. All PSS SEC columns are tested individually before they are shipped. PSS delivers all columns in the solvent the user wants to run it in and tests the column using these conditions. This guarantees maximum certainty for the user to receive exactly what they pay for. Additionally, the risk of reconditioning columns from one to another solvent is taken over by PSS as the manufacturer.

All PSS columns are tested prior to shipping on standard SEC equipment (not on optimized test instruments, which will give ideal results that can hardly be met in a laboratory) for

- column identity
- plate count
- resolution
- asymmetry
- pore volume
- back pressure

Each column is shipped with a PSS column quality certificate, a column connector, and the test kit, which was used at the PSS quality inspection laboratory to test the column resolution. This allows the user to reproduce the column tests without problems. It should be noted, however, that column test results not only depend on the performance of the column alone, but also on the instrument employed for the test. Figure 9.8 shows an example of a PSS column quality certificate.

PSS uses the following formulae (22,23) to calculate plate count, asymmetry, and resolution. PSS uses test conditions that conform to the ISO/EN 13885 and DIN 55672 requirements for SEC sample testing.

A. Calculation of Plate Count

There are different ways to calculate the theoretical plate count of a column. PSS SEC column plate counts are measured using the so-called half-height method. Figure 9.9 shows a graphic representation of this test.

The following formula is used to calculate the plate count:

$$N = \left(\frac{V_p}{\sigma}\right)^2 = 5.54\left(\frac{V_p}{w_{0.5}}\right)^2,$$

where V_p is the elution volume at peak maximum, σ is the peak variance ($w_{0.66} \equiv 2\,\sigma$ for Gaussian peaks), and $w_{0.5}$ is the peak width at half-height. PSS column plate counts are always normalized to 1-m column length for the easy comparison of different columns.

Plate count should always be tested with a monodisperse sample of low molecular weight. Polymers can also be used, but they show much lower plate counts because their diffusion coefficients are much smaller than those of low molecular weight compounds.

The reader should note that the value of the theoretical plate count depends on the sample chosen for testing. PSS always specifies on the PSS column quality certificate which sample has been used for measuring plate count and the exact test conditions.

Using other methods for the calculation of plate count can result in different numbers, depending on peak shape. It should also be kept in mind that many other operational parameters, such as eluent viscosity, column temperature, flow rate, and injection volume, will influence the results of the plate count determination.

column:	PSS SDV 10^4 Å
serial no.:	7121702
dimension [mm]:	8 x 300 mm
particle size [μm]:	5
shipped in:	THF

test conditions

eluent:	THF
flow rate [ml/min]:	1.0
temperature [°C]:	20
pressure [bar]:	25
detection:	UV 254 nm
sample:	BHT, polystyrene mixture
inject volume [μl]:	20

test chromatogram:

BHT:

theoretical plates / m: 60000
asymmetry: 1.09

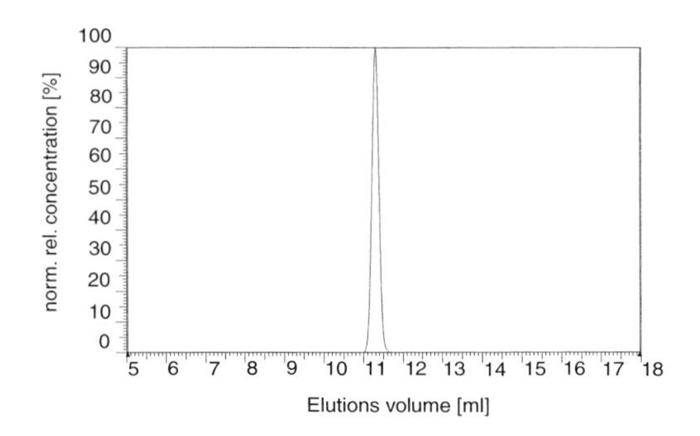

PS-MIX:

M_P = 390000
M_P = 133000
M_P = 46000
M_P = 18100
M_P = 5600
M_P = 1900

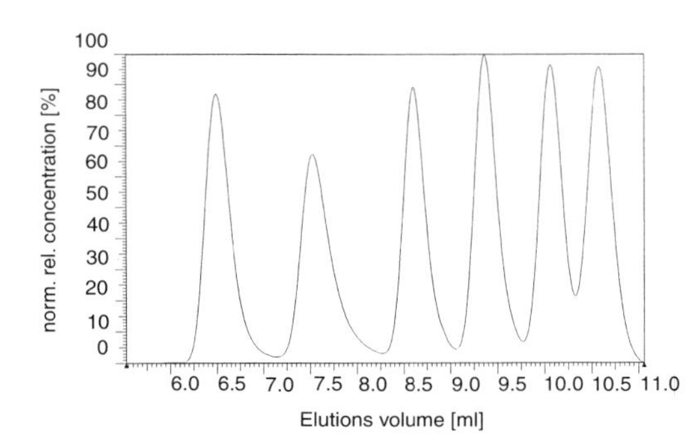

FIGURE 9.8 Information provided on a PSS column quality certificate, which is shipped with every single column.

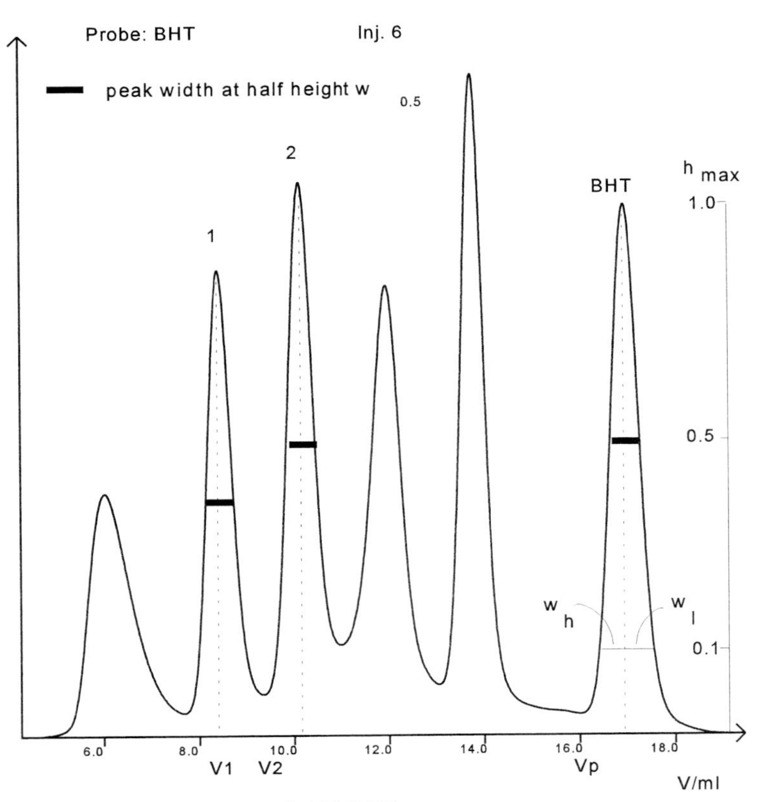

FIGURE 9.9 Definition of column test parameters using a SEC chromatogram of a polymer standard mixture (PSS Polystyrene ReadyCal) and a low molar mass compound (BHT).

B. Determination of Peak Asymmetry

Peak asymmetry at PSS is measured at 10% of the height of the peak. The asymmetry calculation is done according to ISO/EN (23) and DIN (22) requirements:

$$A = w_h/w_l.$$

Figure 9.9 defines these parameters.

The reader should be aware that these definitions are different from those normally used in chromatography, where peak tailing leads to values larger than unity and peak fronting to values smaller than unity. The ISO/EN/DIN definitions (22,23) are just reciprocal.

C. Calculation of Resolution

For SEC separations of polymers, column efficiency is better characterized by specific resolution, R_{sp}, and efficiency, T, than by theoretical plate count. Peak resolution, R_s, is calculated according to (7):

$$R_s = \frac{V_2 - V_1}{2(\sigma_1 + \sigma_2)} = \frac{\log (M_1/M_2)}{2\ D\ (\sigma_1 + \sigma_2)},$$

where V is the elution volume at peak maximum, σ is the peak variance, M is the molar mass of the sample, and D is the slope of the calibration curve.

Unfortunately, this value is sample and system dependent: just by using polymer standards with different molar masses, different values of peak resolution can be generated.

A sample independent parameter is the so-called specific resolution, R_{sp}, which can be obtained from the peak resolution by normalizing the values according to the molar masses of the samples used for testing (24):

$$R_{sp} = \frac{R_s}{\log(M_1/M_2)}.$$

D. Calculation of Efficiency

This parameter was introduced in the ISO/EN and DIN standard quoted earlier. In order to have good polymer separation efficiency, the following criterion has to be met:

$$T = \frac{V_{e,\,M_x} - V_{e\,(10\times M_x)}}{\pi r^2} > 6$$

where $V_{e,\,M_x}$ is the elution volume of molar mass M_x, $V_{e,\,(10\times M_x)}$ is the elution volume for 10 times the molar mass, and r is the internal radius of the column.

VI. PSS COLUMN QUALITY CONTROL SYSTEM

All processes in the production of PSS columns are controlled by an efficient multistep quality control (QC) system (25). This QC system requires complete tests and documentation for all materials used in all production stages. All QC work has to be performed by specially trained and highly skilled polymer chemists.

A. Raw Materials

All raw materials and chemicals are checked for meeting the specifications set in the manufacturing procedures by standardized methods. PSS uses extremely high quality steel (V4A quality), which has proven to be especially resistant to corrosive compounds and/or conditions.

TABLE 9.10 Results of PSS Inverse SEC Tests on PSS SEC Packings[a]

Pore volume (%)	Surface area (m^2/ml)	Mean pore size (nm) (dispersity)	Mass range (kDa) (optimum)	Size range (nm) (optimum)	SEC selectivity (%)
59	72	$D_n = 28$, $D_v = 28$ (2.7)	4–1200 (100)	4–90 (22)	58

[a]SEC porosimetry results for PSS SDV 5 μm 10^5 Å (lot: A52059).

B. Sorbent Manufacturing

The synthesis of the packing materials is done by experienced polymer chemists using standardized equipment and procedures. PSS takes special care in cleaning the sorbents after polymerization to achieve constant quality and surface chemistry characteristics. Each production step is checked separately for quality control conformity.

C. Testing of Sorbents

All packing materials produced at PSS are tested for all relevant properties. This includes physical tests (e.g., pressure stability, temperature stability, permeability, particle size distribution, porosity) as well as chromatographic tests using packed columns (plate count, resolution, peak symmetry, calibration curves). PSS uses inverse SEC methodology (26,27) to determine chromatographic-active sorbent properties such as surface area, pore volume, average pore size, and pore size distribution. Table 9.10 shows details on inverse SEC tests on PSS SDV sorbent as an example. Fig. 9.10 shows the dependence

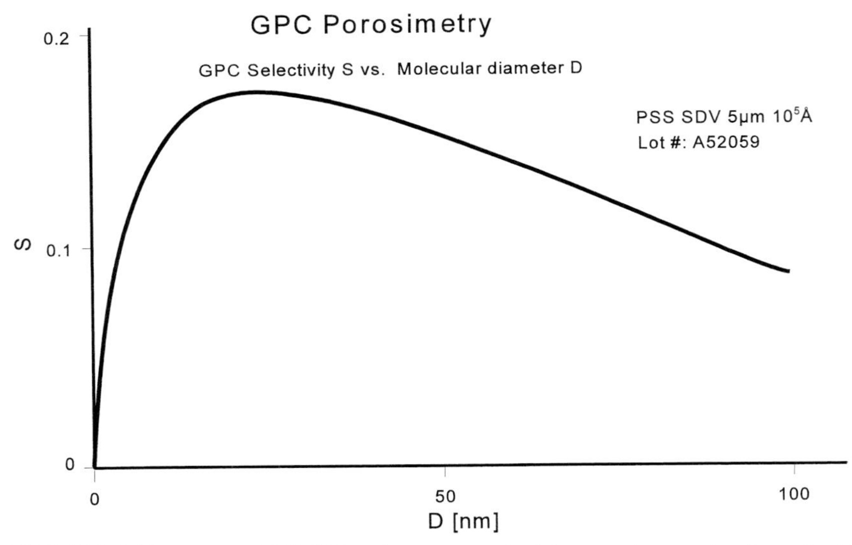

FIGURE 9.10 Result of PSS SEC sorbent tests: selectivity vs molecular size determined by inverse SEC.

of SEC selectivity S *vs* molecular size for different sized macromolecules as determined by inverse SEC.

D. Column Packing Process

PSS has developed proprietary packing procedures for its sorbents, that allow a homogeneous filling of the column hardware with no change in particle properties. This thoroughness in the packing procedure is reflected in the superior performance of PSS SEC columns and their long life, even in difficult conditions.

The complete packing process is computer monitored and computer controlled. Potential problems in column packing can be seen directly and the affected column is removed from the production cycle. Fig. 9.11 shows a graphic representation of a column packing process by monitoring the packing pressure with time.

After column packing, each column is tested for theoretical plate count, peak symmetry, resolution, pore volume, and back pressure. If one of these tests fails the column is removed from the production cycle. If a PSS SEC column is kept in storage for a longer time, it is retested for theoretical plate count, peak symmetry, resolution, pore volume, and back pressure prior to shipping to the customer to prove up-to-date column performance.

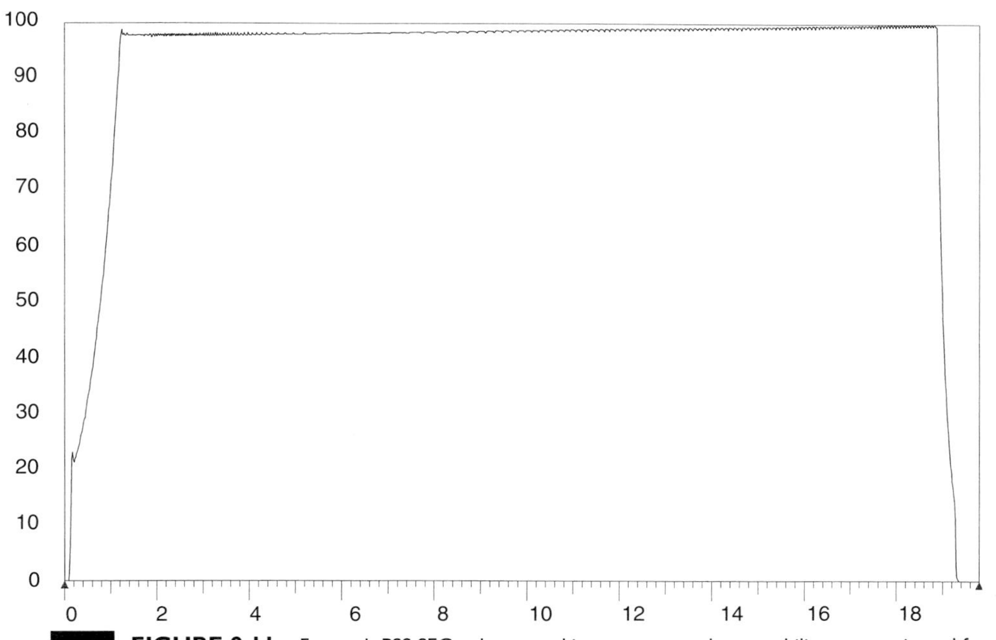

FIGURE 9.11 For each PSS SEC column, packing pressure and permeability are monitored for the best reproducibility of column performance.

PSS is constantly working to maintain its high column quality by continuously optimizing manufacturing processes and quality control procedures and incorporating the latest technology advances. Figure 9.12 shows the results of plate count testing of PSS SDV columns over a 10-year period.

In more than 10 years, PSS SDV columns showed relative standard deviations with regard to plate count measurements between 2.5 and 4.5%, depending on the particle size of the column packing. PSS columns consistently exceeded the specified plate counts: for PSS SDV 10-μm columns, a long-term average of 49,000 plates/m was measured (specified minimum value: 35,000/m), for PSS SDV 5-μm columns the average was 68,000/m (45,000/m), and for PSS SDV 3-μm packings, an average of 93,500/m (85,000/m) was calculated.

The PSS column certification procedure requires that all columns are rigorously tested individually as described earlier. After final testing, individual column quality certificates are issued in a three-step process.

1. The person who has tested the column collects all necessary columns parameters and prepares the PSS column quality certificate.
2. The PSS columns quality manager inspects each column and certificate for formal and physical correctness.
3. Columns and underlying quality data are entered into the PSS computer system. These entries are cross checked and verified by the PSS quality manager.

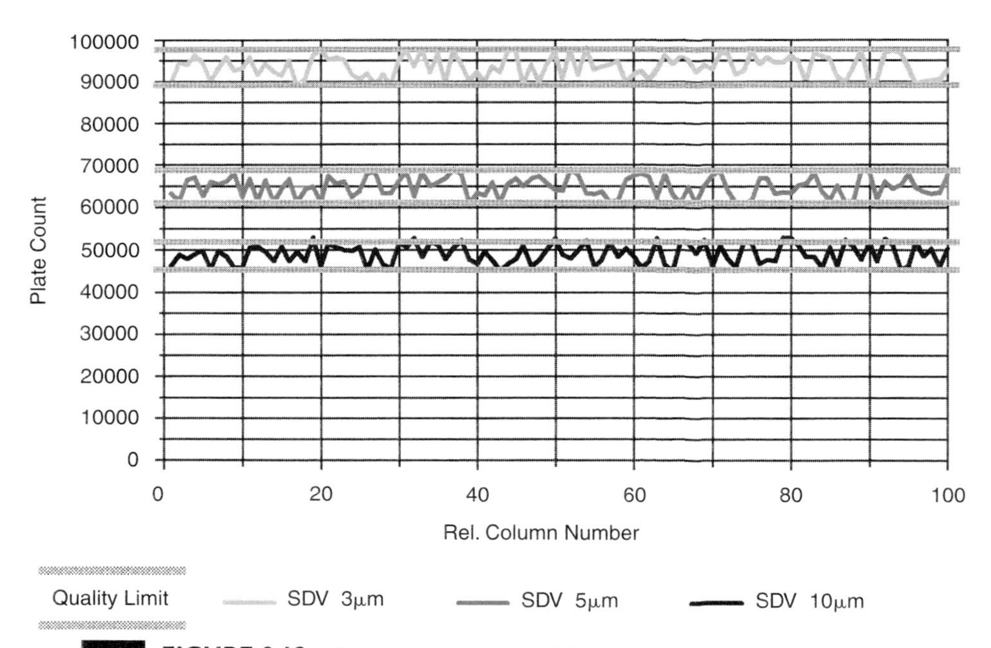

FIGURE 9.12 Quality control chart of PSS SDV columns plate count (per m) measurements over a 10-year period; ±5% limits are also shown.

VII. APPLICATIONS OF PSS SEC COLUMNS

This section briefly describes some interesting applications of PSS SEC that have not been published before. For general application literature or application questions, please contact your local PSS representative or PSS Germany directly.

A. Organic SEC Applications

PSS SDV columns can be used for all applications requiring organic eluents. The exception to the rule is the exclusion of lower aliphatic alcohols (e.g. methanol) from the otherwise complete list (28). For fluorinated solvents such as TFE and HFIP, PSS recommends its specially designed PFG columns (cf. Section VII,C), which have a much longer life in this kind of demanding eluents. Figures 9.13 through 9.19 show some unusual applications that illustrate the variety of solvents and the feasibility of the columns.

B. Aqueous Applications

Advances in aqueous column packings are reflected in this application section. PSS HEMA and HEMA Bio columns (cf. Fig. 9.21) have proven to be extremely stable columns in the middle molecular weight range. PSS HEMA columns

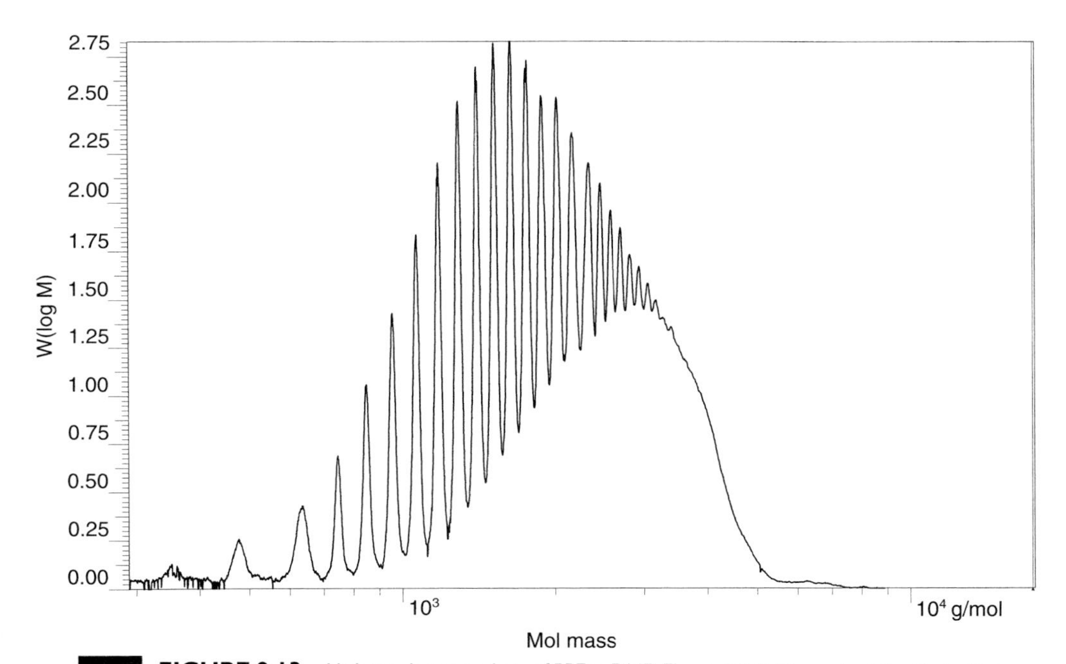

FIGURE 9.13 High resolution analysis of PPE in DMF. Eluent: DMF. Flow rate: 0.3 ml/min. Columns: PSS SDV 5 μm 2× 1000 Å + 3× 100 Å (8 × 300 mm each). Oven temp: 80°C. Detection: UV at 285 and RI. Standards: PSS PMMA oligomer calibration kit 8 × 1 g.

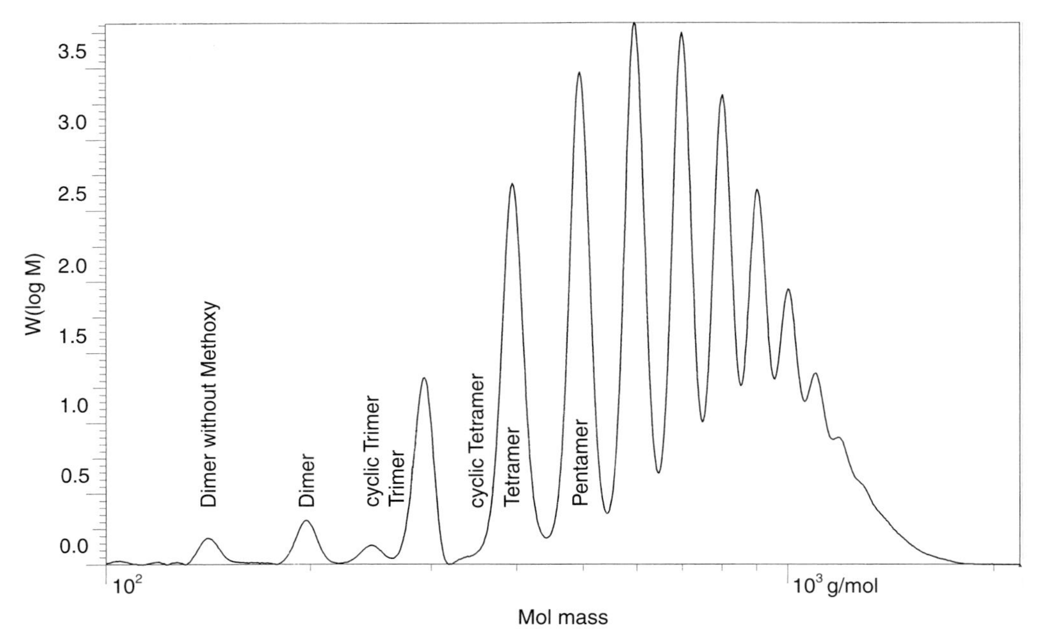

FIGURE 9.14 PMMA Oligomers with cyclic products. Eluent: THF. Flow rate: 1.0 ml/min. Columns: PSS SDV 5 μm 1000Å + 500 Å + 100 Å + 50 Å (8 × 300 mm each). Oven temp: 25°C. Detection: UV at 230 nm and RI. Standards: PSS PMMA oligomer calibration kit 8 × 1 g.

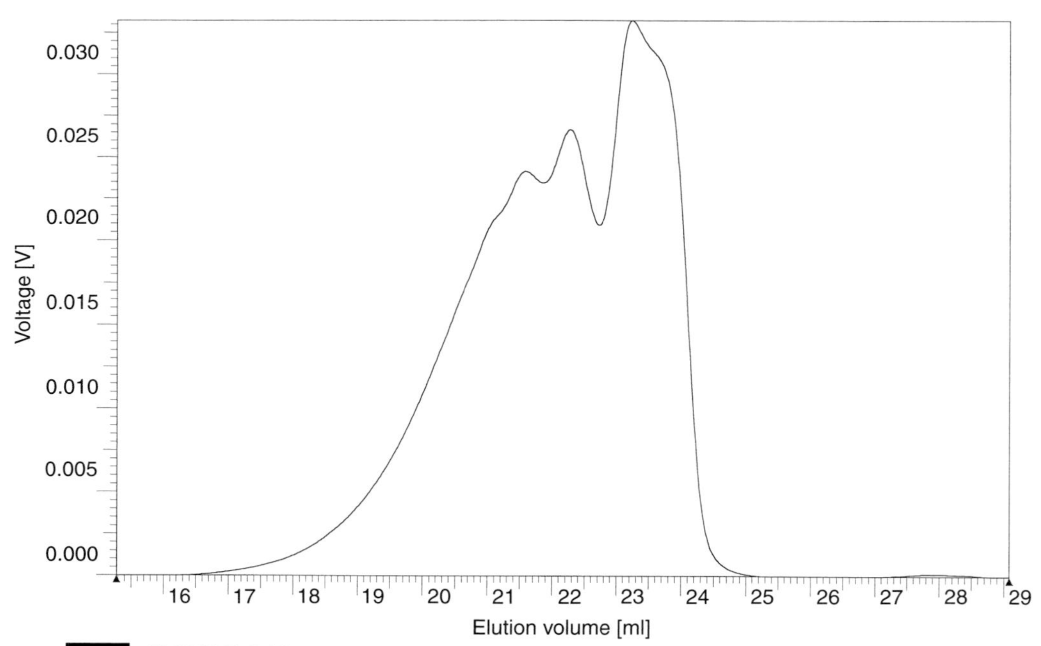

FIGURE 9.15 Coal tar pitch analysis. Eluent: TCB. Flow rate: 1.0 ml/min. Columns: PSS SDV 10 μm 1000 Å + 500 Å + 100 Å (8 × 300 mm each). Oven temp: 135°C. Detection: RI. Standards: PSS polystyrene ReadyCal standards.

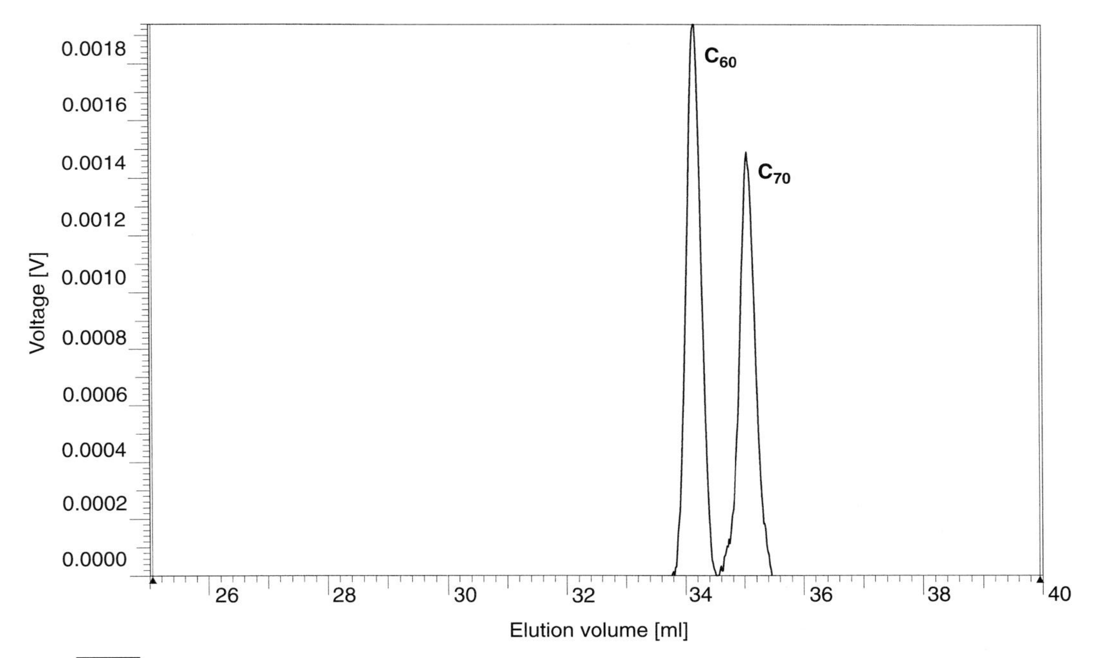

FIGURE 9.16 Preparative separation of fullerenes. Eluent: Toluene. Flow rate: 8.0 ml/min. Injection volume: 500 μl. Columns: PSS SDV 5 μm, 100 Å, 20 × 300 mm. Oven temp: 30°C. Detection: UV at 370 nm. Standards: PSS polystyrene oligomer calibration kit 10 × 1 g.

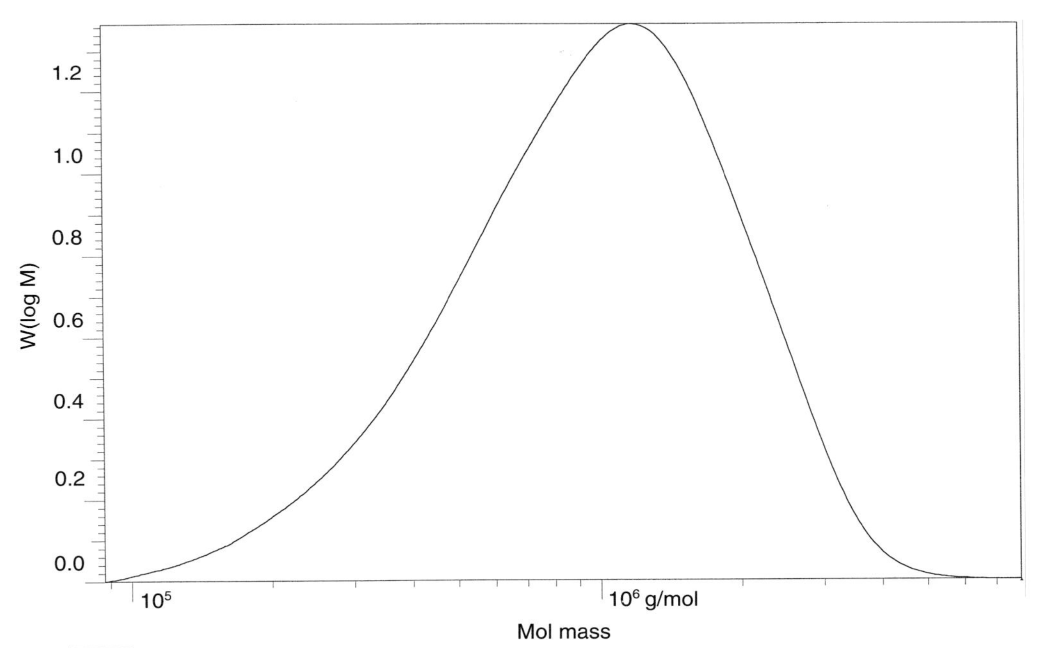

FIGURE 9.17 Characterization of high molar mass PIB. Eluent: THF. Flow rate: 0.5 ml/min. Columns: PSS SDV 20 μm 10^7 Å and 10^5 Å, (8 × 300 mm each). Oven temp: 30°C. Detection: RI. Standards: PSS PIB calibration kit 10 × 0.5 g.

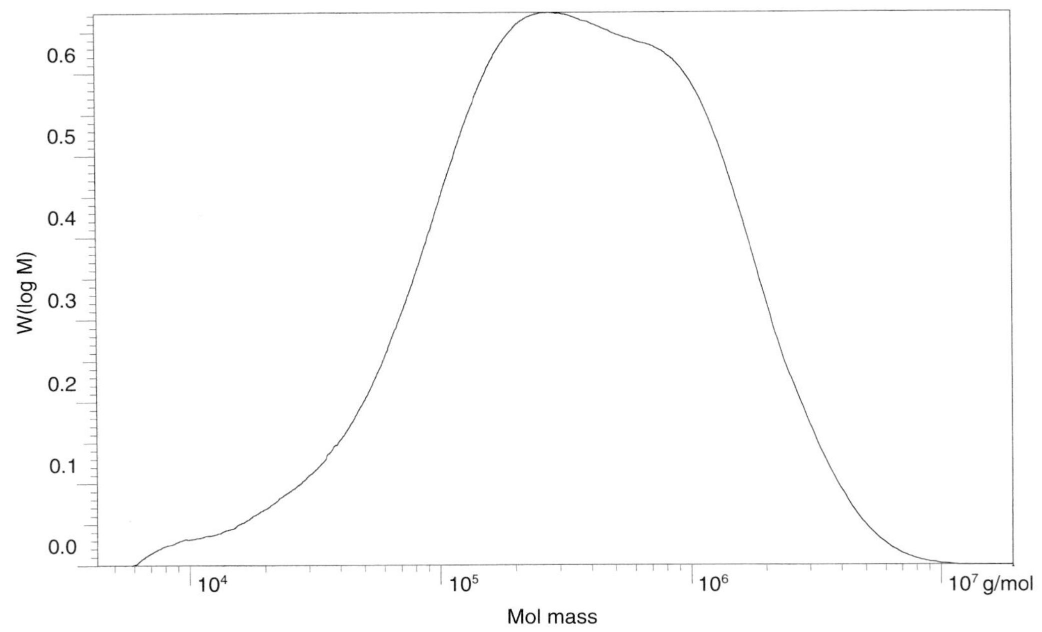

FIGURE 9.18 Analysis of PVP in DMAc. Eluent: DMAc + 0.1% LiBr. Flow rate: 1.0 ml/min. Columns: PSS SDV 10 μm 10^6 Å + 10^6 Å + 1000 Å (8 × 300 mm each). Oven temp: 70°C. Detection: RI. Standards: PSS PMMA calibration kit 12 × 1 g.

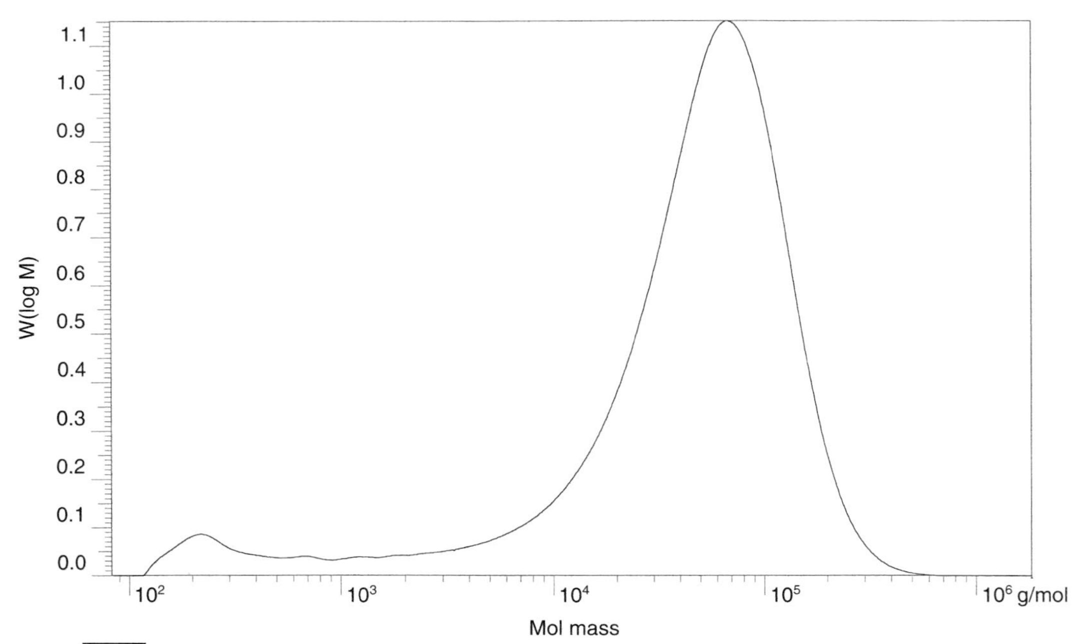

FIGURE 9.19 Analysis of poly(sulfone) in DMAc. Eluent: DMAc + 0.1% LiBr. Flow rate: 1.0 ml/min. Columns: PSS SDV 10 μm 10^6 Å + 10^6 Å + 1000 Å (8 × 300 mm each). Oven temp: 70°C. Detection: RI. Standards: PSS PMMA calibration kit 12 × 1 g.

have been used for many years with high fractions of organic modifiers (up to 100%) to allow for the separation of copolymers with hydrophobic segments. The PSS Suprema columns series was designed to overcome the limitations of HEMA and HEMA Bio columns in the separation of very small and very large molecules. The successful separation of extremely high molar mass polymers such as polyethylene oxide (PEO) ($M_r > 30$ million Da), polyacrylamides ($M_r > 45$ million Da), and polysaccharides (M_r about 10 million Da) is now possible with PSS Suprema columns (cf. Figs. 9.20 to 9.25 for application details).

C. Applications of PSS Polar Fluoro Gel Columns

This PSS column packing material has been especially designed for

- use of fluorinated solvents
- analysis of semicrystalline polymers such as aromatic polyesters (PET, PBT, etc.), aliphatic polyesters (polylactide, etc.), polyamides (PA-6,6, PA-6, etc.), polyacetals (POM), natural and synthetic silks, and cellulose and lignins
- better performance and longer lifetimes than conventional polymeric packings

PSS Polar Fluoro Gel (PSS PFG) is available in four individual porosities (100, 300, 1000, and 4000 Å) and in a linear mixed-bed column covering all

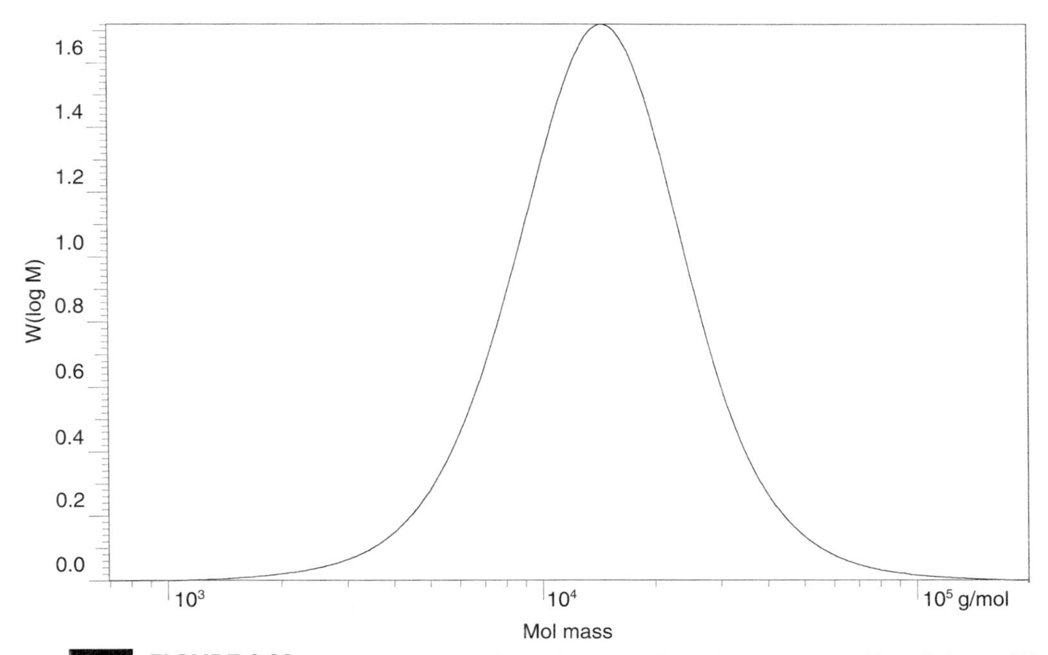

FIGURE 9.20 Analysis of heparin. Eluent: 0.1 *M* Na_2SO_4, pH 5. Flow rate: 1 ml/min. Columns: PSS Suprema 10 μm, 100 + 1000, 8 × 300 mm. Oven temp: 30°C. Detection: UV at 234 and RI. Standards: PSS heparin standard. Calibration: According to European Pharmacopeia with PSS WINGPC Heparin Software Module.

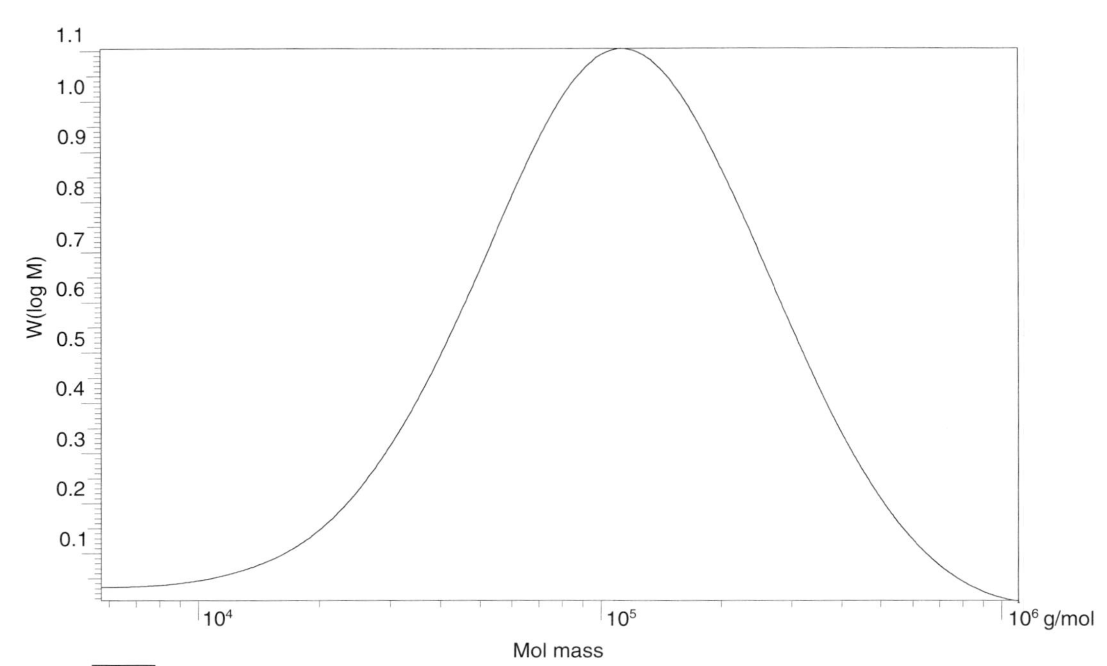

FIGURE 9.21 Analysis of polyamine polymer. Eluent: 0.15% formic acid. Flow rate: 1.0 ml/min. Columns: PSS HEMA 10 μm, 40 + 100 + 1000 + 2000, 8 × 300 mm. Oven temp: 30°C. Detection: RI. Standards: PSS polyvinylpyridinium standards.

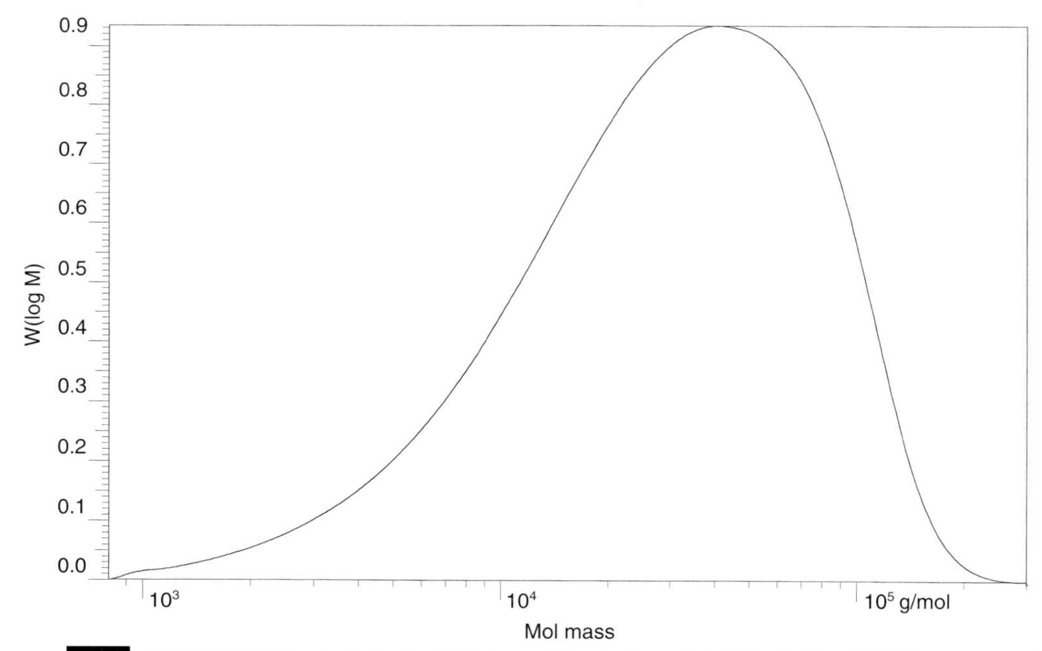

FIGURE 9.22 Analysis of poly(vinyl pyrrolidone). Eluent: 0.1 *M* Tris buffer, pH 7. Flow rate: 1 ml/min. Columns: PSS Suprema 10 μm, 100 + 1000, 8 × 300 mm. Oven temp: 30°C. Detection: RI. Standards: PSS polyvinylpyridin standards.

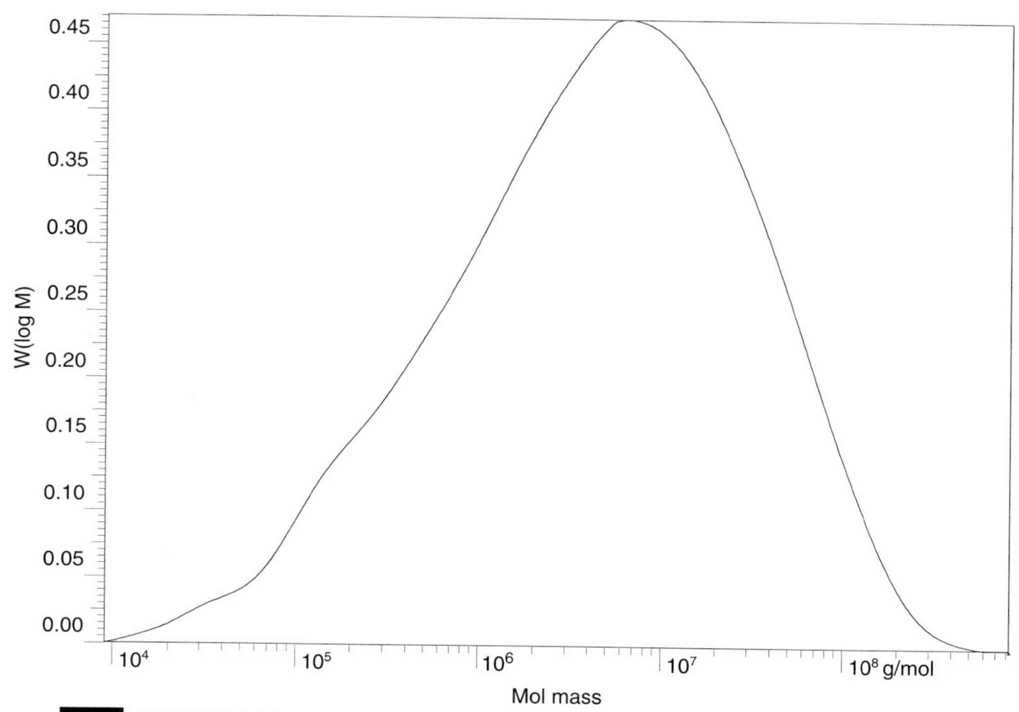

FIGURE 9.23 Analysis of ultrahigh poly(acrylamide). MW 48 million by analytical ultracentrifugation. Eluent: 0.1 *M* Na_2SO_4. Flow rate: 0.3 ml/min. Columns: PSS Suprema 30000, 20 μm, 8 × 300 mm. Oven temp: 30°C. Detector: RI. Standards: PSS polyacrylamide standards.

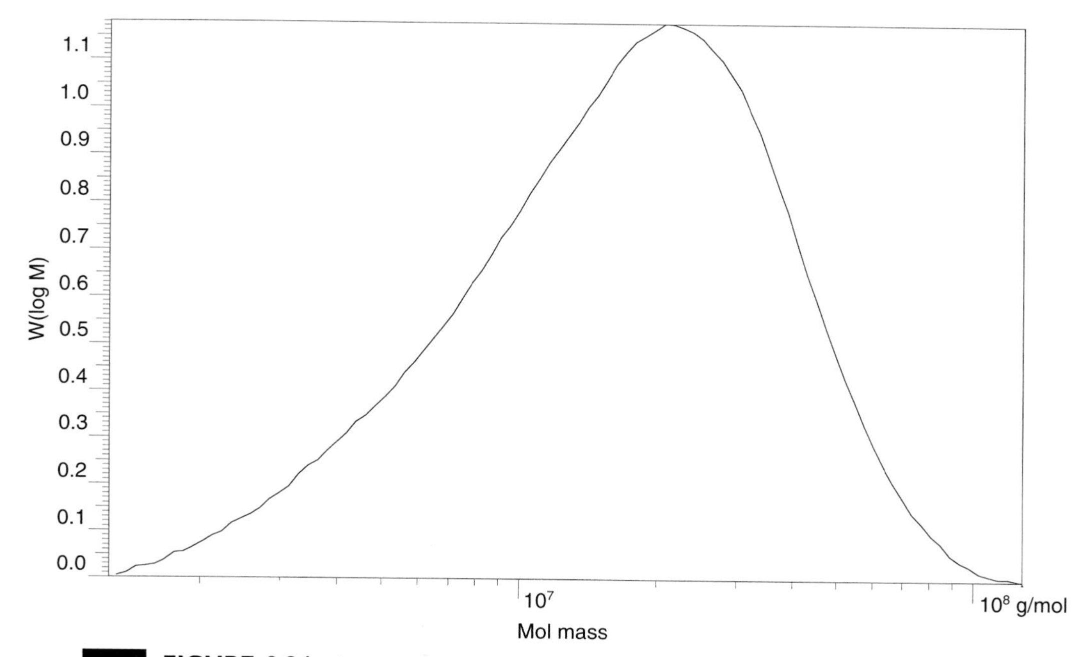

FIGURE 9.24 Analysis of ultrahigh poly(ethylene oxide), MW about 37 million. Eluent: 0.05 *M* $NaNO_3$. Flow rate: 0.3 ml/min. Columns: PSS Suprema 20 μm, 30000, 8 × 300 mm. Oven temp: 30°C. Detector: RI. Standards: PSS PEO standards.

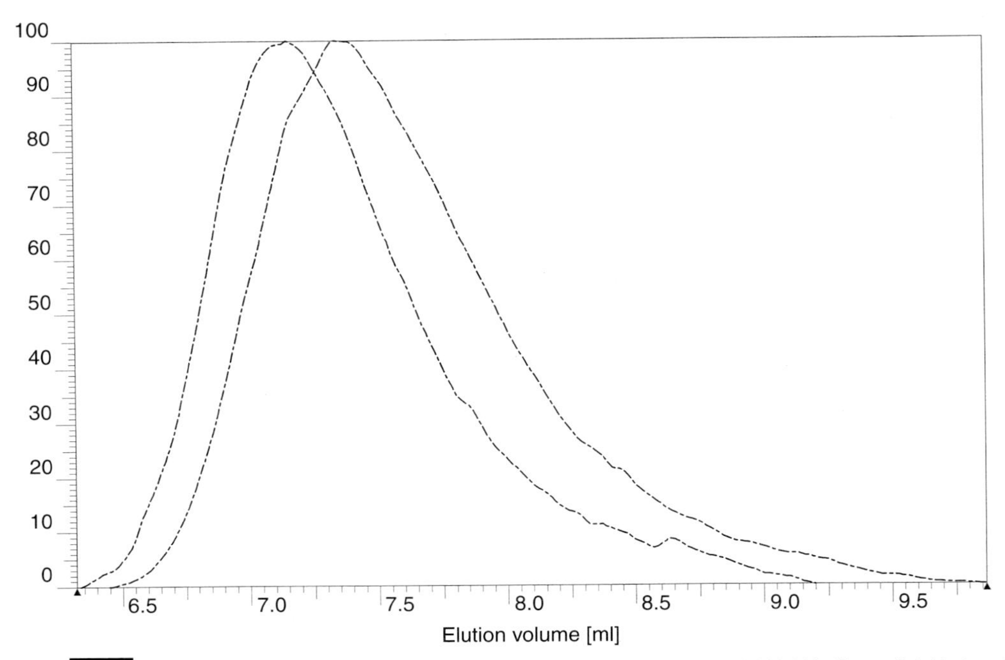

FIGURE 9.25 Analysis of xanthanes: MW 13,000,000 and MW 9,000,000. Eluent: 0.1 *M* phosphate buffer, pH 7.1. Flow rate: 0.5 ml/min. Columns: PSS Suprema 10 μm, 10000, 8 $\times$ 300 mm. Oven temp: 30°C. Detector: RI. Standards: PSS dextran standards.

molar masses of polycondensation polymers and other semicrystalline macromolecules, which can be dissolved in polar eluents (Figs. 9.26–9.33).

VIII. CONCLUSION

This chapter illustrates the improvements in SEC column technology and modern applications of SEC separations. The better understanding of SEC column design and separation parameters described in the theoretical sections of this chapter will help the reader fine-tune his or her own work. The same is true for column performance tests, which should be applied regularly, especially after a column purchase. In order to obtain reproducible results, it is recommended to choose column manufacturers who can assure constant quality and performance and to invest in knowledgeable, well-trained support personnel and experienced application chemists.

More and more interesting and specialized SEC columns will appear on the market in the near feature, enabling us to do our work more reliably and efficiently. However, the author does not expect "quantum jumps" in SEC column technology in the next decade, which have been seen since the late 1970s.

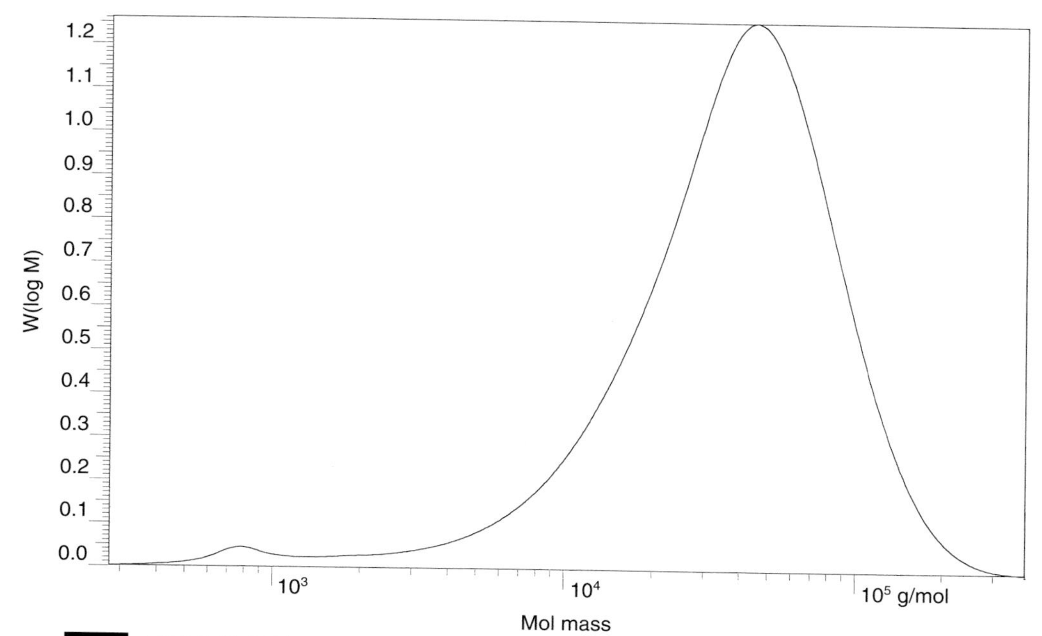

FIGURE 9.26 Room temperature analysis of polyethylene terephthalate. Columns: PSS PFG 100 + 1000. Eluent: HFIP + 0.1 *M* NatFat. Temp: 25°C. Detection: UV 254 nm, RI. Calibration: PSS PET standards (broad).

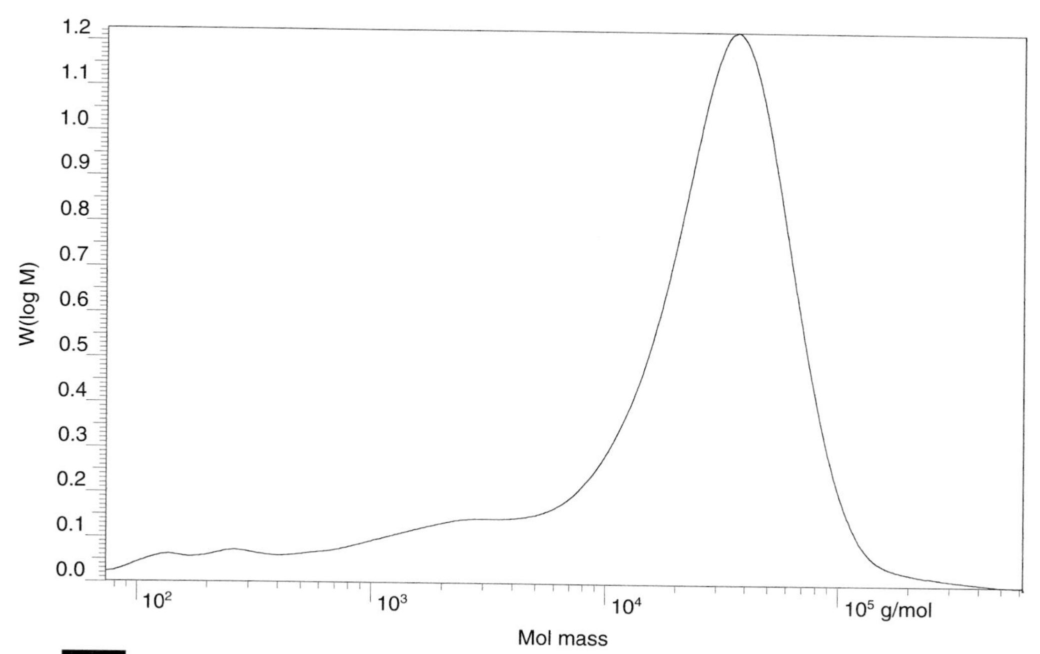

FIGURE 9.27 Analysis of recycled polybutylene terephthalate. Columns: PSS PFG 100 + 1000. Eluent: HFIP. Temp: 25°C. Detection: UV 254 nm, RI. Calibration: PSS PBT standards (broad).

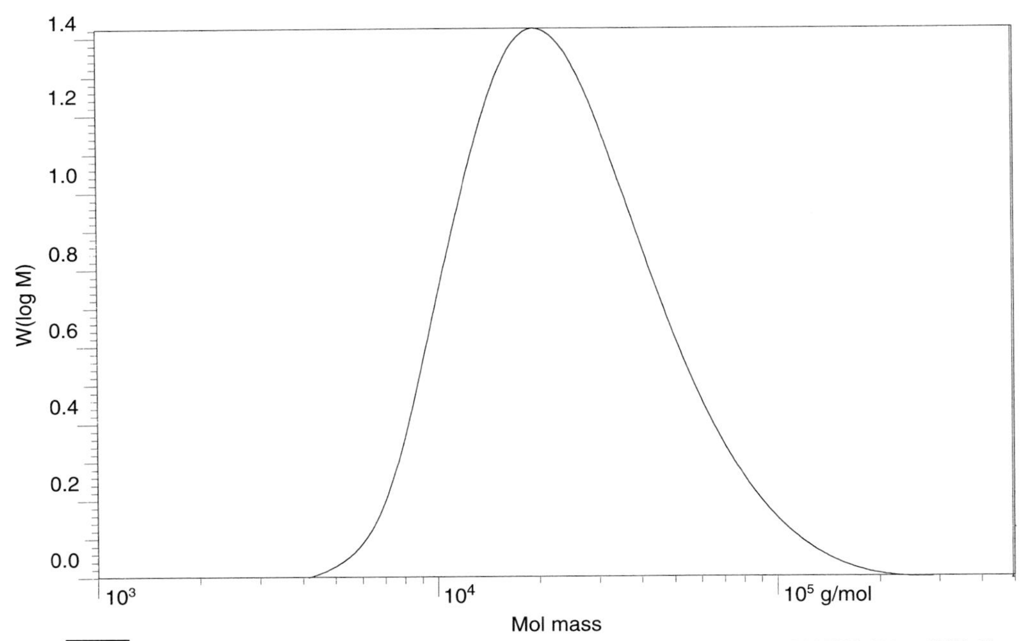

FIGURE 9.28 Room temperature analysis of poly(amide-6). Columns: PSS PFG 100 + 1000. Eluent: TFE + 0.1 *M* NatFat. Temp: 25°C. Detection: RI. Calibration: PSS PA-6 standards (broad).

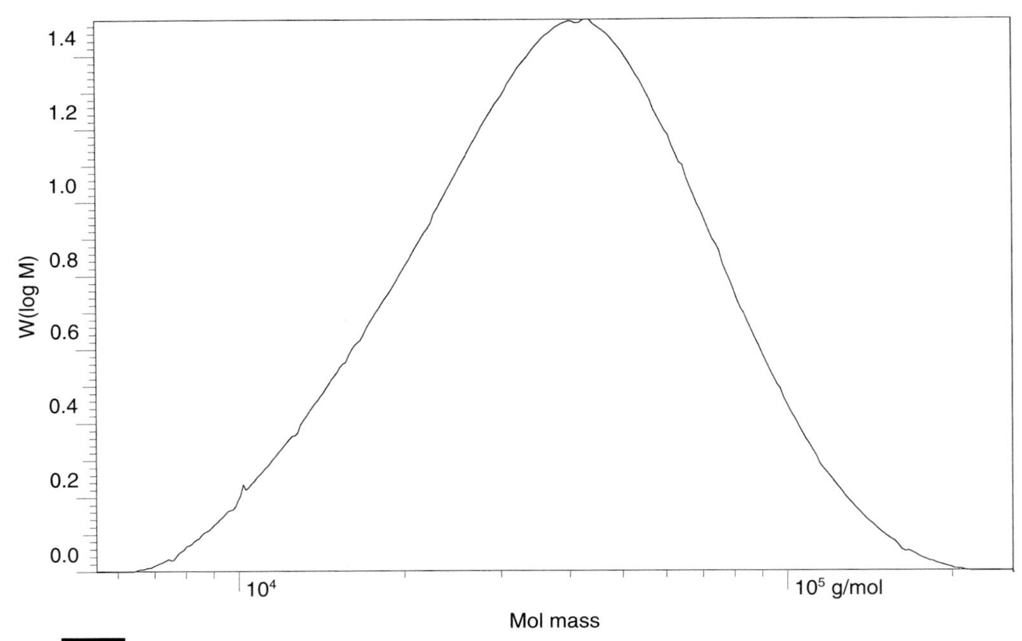

FIGURE 9.29 Analysis of natural spider silk (spider web). Columns: PSS PFG 100 + 1000. Eluent: HFIP + 0.1 *M* KtFat. Temp: 25°C. Detection: UV 254 nm, RI. Calibration: PSS PMMA ReadyCal kit.

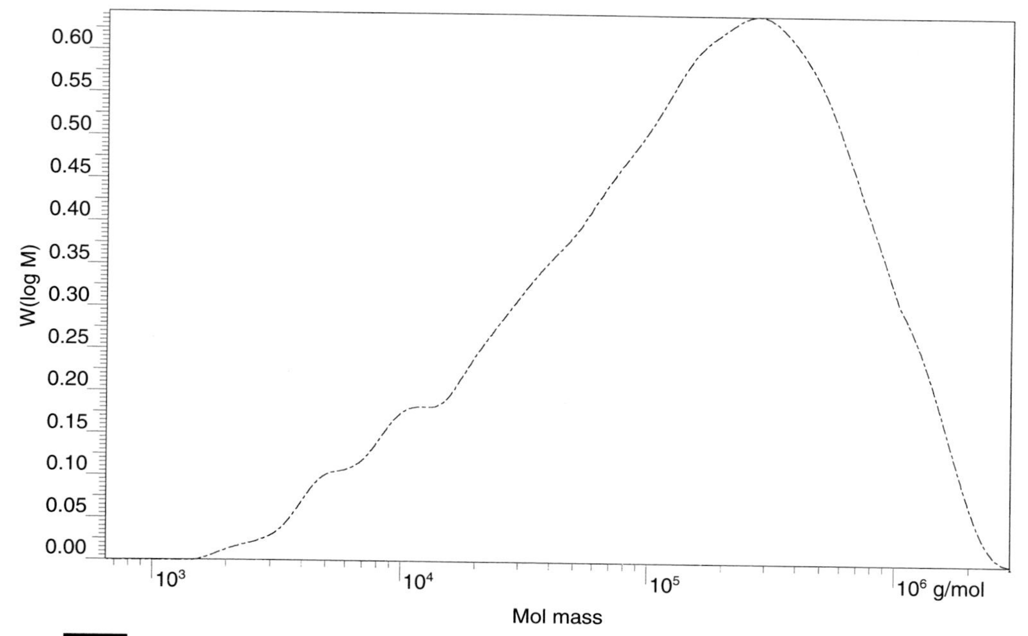

FIGURE 9.30 Analysis of methylcellulose. Columns: PSS PFG 100 + 1000. Eluent: HFIP + 0.1 *M* KtFat. Temp: 25°C. Detection: RI. Calibration: PSS PMMA ReadyCal kit.

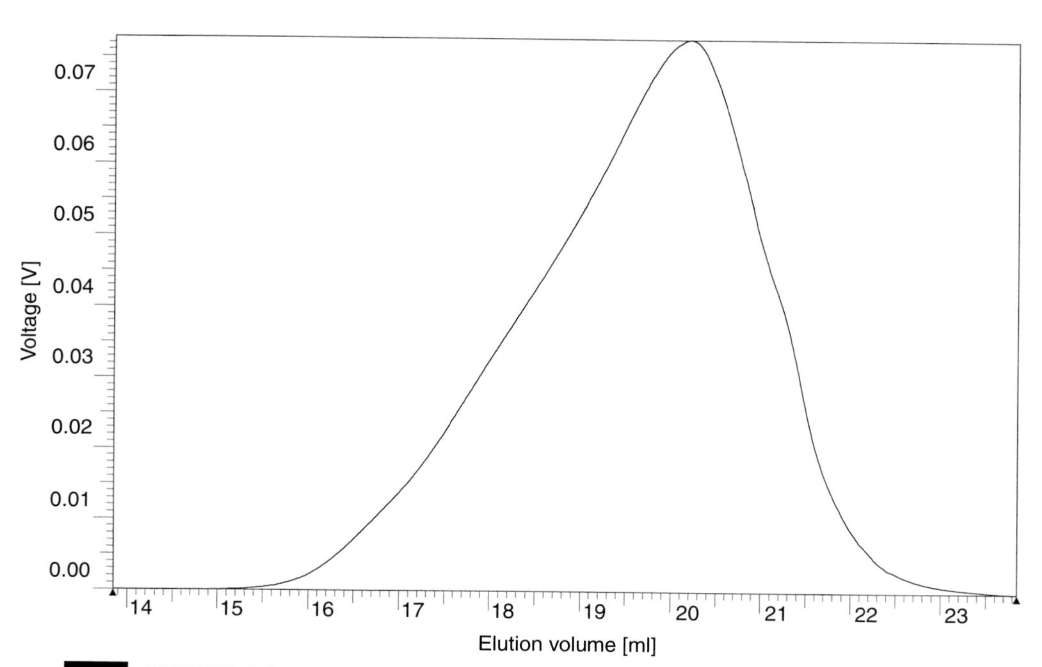

FIGURE 9.31 Analysis of lignin. Columns: PSS PFG 100 + 1000. Eluent: HFIP + 0.1 *M* KtFat. Temp: 25°C. Detection: RI. Calibration: PSS PMMA ReadyCal kit.

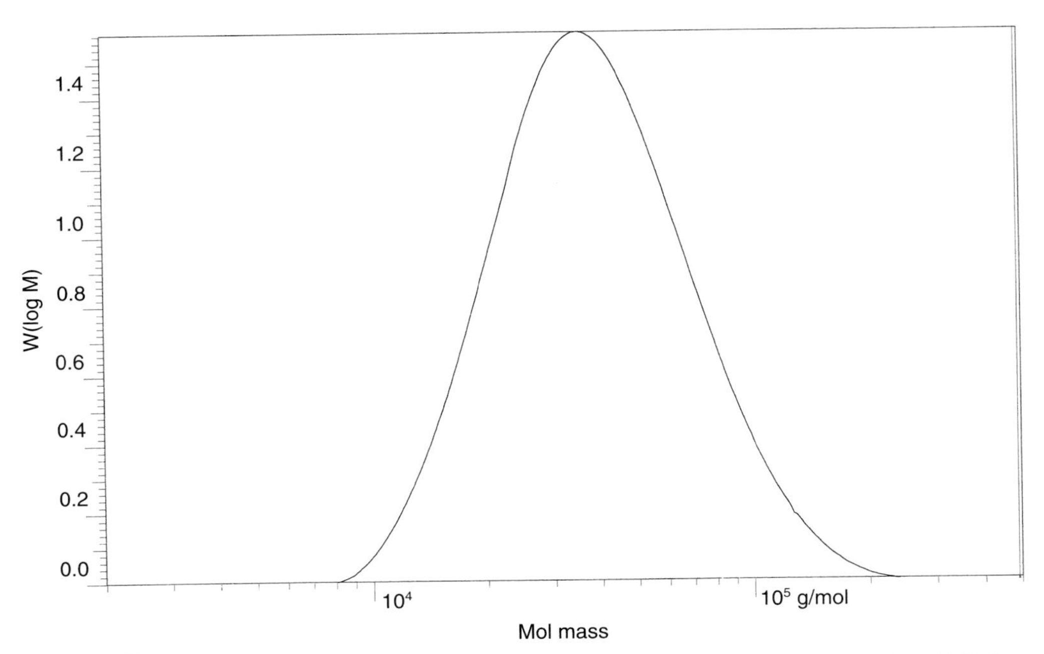

FIGURE 9.32 Analysis of biodegradable poly(lactic acid). Columns: PSS PFG 100 + 1000. Eluent: TFE + 0.1 *M* NatFat. Temp: 25°C. Detection: UV 230 nm, RI. Calibration: PSS PMMA ReadyCal kit.

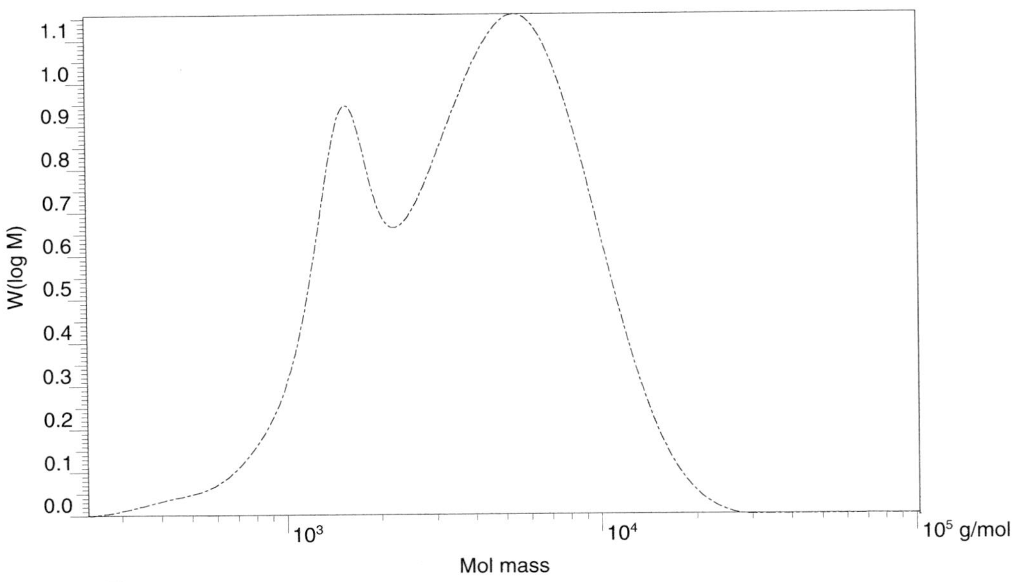

FIGURE 9.33 Analysis of cationic poly(amino siloxane). Columns: PSS PFG 100 + 1000. Eluent: HFIP + 0.1 *M* KtFat. Temp: 25°C. Detection: UV 230 nm, RI. Calibration: PSS PDMS siloxane standards.

ACKNOWLEDGMENTS

The author thanks the editor for his continuous support and patience and all colleagues at PSS who contributed their work to made this chapter as interesting as the author hopes readers will find it.

REFERENCES

1. Porath, J., and Flodin, P. (1959). *Nature* **183,** 1657.
2. Moore, J. C. (1964). *J. Polym. Sci.* **A2,** 835.
3. Kilz, P. (1995). *Proc. East. Anal. Symp.* p. 136, Sommerset.
4. Kilz, P. (1996). *Proc. InCom Symp.* p. 251, Düsseldorf.
5. PSS SEC Market Study, Mainz, 1994.
6. PSS GPC Training School Handbook, Mainz, 1991.
7. Yau, W. W., Kirkland, J. J., and Bly, D. D. (1979). "Modern Size-Exclusion Liquid Chromatography." Wiley, New York.
8. Flodin, P. (1962). Dissertation, Uppsala.
9. Altgelt, K. H., and Moore, J. C. (1996). *In* "Polymer Fractionation" (M. R. J. Cantow, ed.). New York.
10. Grubisic, Z., Rempp, R., and Benoit, H. (1967). *J. Polym. Sci.* **B5,** 753.
11. Cantow, M. R. J., Proter, R. S., and Johnson, J. F. (1967). *J. Polym. Sci. A-1* **5,** 987.
12. Yau, W. W., Malone, C. P., and Suchan, H. L. (1970). *Sep. Sci.* **5,** 259.
13. Belenkii, B. G., and Gankina, E. S. (1977). *J. Chromatogr.* **141,** 13.
14. Entelis, S. G., Evreinov, V. V., and Gorshkov, A. V. (1986). *Adv. Polym. Sci.* **76,** 129.
15. Schulz, G., Much, H., Krüger, H., and Wehrsted, G. (1990). *J. Liq. Chromatogr.* **13,** 1745.
16. Gorshkov, A. V., Much, H., Becker, H., Pasch, H., Evreinov, V. V., and Enteilis, S. G. (1990). *J. Chromatogr.* **523,** 91.
17. Pasch, H., Much, H., Schulz, G., and Gorshkov, A. V. (1992). *LC GC Intl.* **5,** 38.
18. PSS Catalogue, Mainz, 1996.
19. Neue, U. D. (1995). *Proc. Internat. GPC Symp.* p. 779.
20. Determann, H. (1967). "Gelchromatographie." Springer, Berlin.
21. Glöckner, G. (1982). "Liquid Chromatography of Polymers." Hüthig, Heidelberg; and literature quoted therein.
22. German Institute of Standardization, DIN 55672 Teil 1, Beuth Verlag, Berlin, 1996.
23. ISO, ISO/EN 13885, yellow print, Bruxelles, 1997.
24. Yau, W. W., Kirkland, J. J., Bly, D. D., and Stoklosa, H. J. (1976). *J. Chromatogr.* **125,** 219.
25. PSS Quality Management Handbook, Mainz, 1995.
26. Cassasa, E. F., and Tagami, Y. (1969). *Macromolecules* **2,** 14.
27. Gorbunov, A. A., and Skvortsov, A. M. (1991). *Polymer* **32,** 3001.
28. PSS SEC Columns User's Manual, Mainz, 1996.

10

SynChropak SIZE EXCLUSION COLUMNS

KAREN M. GOODING

Formerly of SynChrom, Inc., Lafayette, Indiana 47901 and MICRA Scientific, Northbrook, Illinois 60062

I. INTRODUCTION

SynChropak GPC supports were introduced in 1978 as the first commercial columns for high-performance liquid chromatography of proteins. SynChropak GPC columns were based on research developed by Fred Regnier and coworkers in 1976 (1,2). The first columns were only available in 10-μm particles with a 100-Å pore diameter, but as silica technology advanced, the range of available pore diameters increased and 5-μm particle diameters became available. SynChropak GPC and CATSEC occasionally were prepared on larger particles on a custom basis, but generally these products have been intended for analytical applications.

Cationic polymers, such as polyvinylpyridines, cannot be analyzed by standard high-performance size exclusion (HPSEC) columns such as SynChropak GPC because there is always a slight residual negative charge on both silica-based and polymer-based supports. SynChropak CATSEC was developed to overcome those problems (3). SynChropak CATSEC columns have a bonded phase of polymerized polyamine that totally masks all silica sites and carries a slight residual positive charge so that cationic polymers can be effectively analyzed.

II. SUPPORT TECHNOLOGY

A. Physical Properties

SynChropak size exclusion supports are composed of spherical uniformly porous silica that has been derivatized with a suitable layer. SynChropak GPC supports are available in six pore diameters ranging from 50 to 4000 Å and particle diameters from 5 to 10 μm. SynChropak CATSEC supports are available in four pore diameters. Table 10.1 details the physical characteristics of the product lines.

B. Chemical Properties

SynChropak GPC supports are bonded with γ-glycidoxypropylsilane by a proprietary process that results in a thin, neutral hydrophilic layer that totally covers the silanol sites of the silica. The silica backbone prevents the supports from swelling.

SynChropak CATSEC supports are bonded by polymerizing polyamines by a proprietary process, resulting in a thin, slightly cationic layer that effectively shields cationic polymers from any negative charges of the silica particles (3). Figure 10.1 illustrates the analysis of a mixture of polyvinylpyridines of various molecular weights on a CATSEC 1000 column.

C. Calibration

The neutral hydrophilic surface and the wide range of pore diameters available for SynChropak GPC allow many compounds from small peptides to nucleic acids and other polymers to be analyzed. Table 10.2 lists the approximate exclusion limits for both linear and globular solutes. Although this information

TABLE 10.1 Physical Characteristics of SynChropak SEC Supports

	Diameter	
SynChropak	Particle (μm)	Pore (Å)
GPC PEPTIDE	5	50
GPC 100	5	100
GPC 300	5	300
GPC 500	7	500
GPC 1000	7	1000
GPC 4000	10	4000
CATSEC 100	5	100
CATSEC 300	5	300
CATSEC 1000	7	1000
CATSEC 4000	10	4000

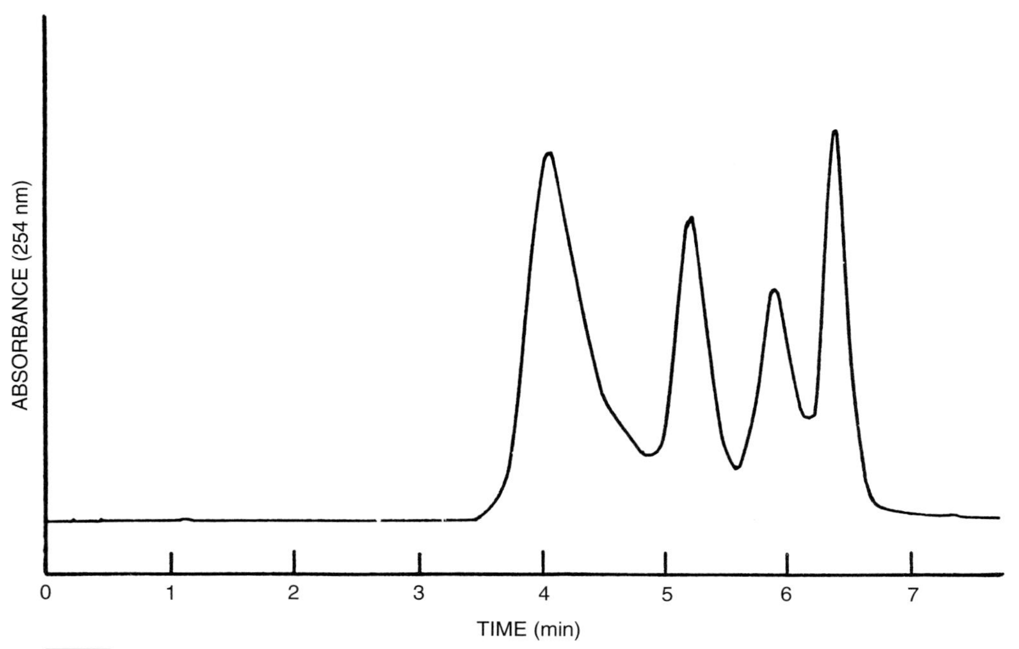

FIGURE 10.1 Analysis of polyvinylpyridines on SynChropak CATSEC 1000. Mobile phase: 0.2 *M* sodium chloride in 0.1% trifluoroacetic acid. Column: 250 × 4.6 mm i.d. Flow rate: 0.5 ml/min. Sample: (1) MW 600,000; (2) MW 100,000; (3) MW 20,000; (4) MW 3000. (From MICRA Scientific, Inc., with permission.)

TABLE 10.2 Molecular Weight Range for SynChropak SEC Supports[a]

	Proteins		Polymers	
SynChropak[b]	Included	Excluded	Included	Excluded
GPC PEPTIDE	5,000	35,000	500	10,000
GPC 100	5,000	160,000	500	25,000
GPC 300	10,000	1,000,000	2,000	100,000
GPC 500	40,000	1,000,000	10,000	350,000
GPC 1000	400,000	10,000,000	40,000	1,000,000
GPC 4000			70,000	10,000,000
CATSEC 100			500	20,000
CATSEC 300			2,000	50,000
CATSEC 1000			10,000	600,000
CATSEC 4000			20,000	1,000,000

[a]Reprinted from MICRA Scientific with permission.

[b]GPC: DNA and glycyltyrosine (cytidine, GPC PEPTIDE). 0.1 *M* potassium phosphate. pH 7. CATSEC: PVP 600,000 and cytidine: 0.1% trifluoroacetic acid with 0.2 *M* NaCl.

is useful, it is best to choose a column based on the calibration curve rather than this kind of table because a calibration curve is never linear at the high and low molecular weight extremes. Figures 10.2–10.4 show the calibration curves associated with the columns and standards listed in Table 10.2.

For proteins, the most useful columns are those with pores of 100–500 Å, as seen in Fig. 10.2, because most proteins elute on the linear portions of the calibration curves. Figure 10.5 illustrates an analysis of a protein mixture on SynChropak GPC 100. Small peptides can be analyzed on the 50-Å SynChropak GPC Peptide column with appropriate mobile-phase modifications. Many peptides have poor solubility in mobile phases standardly used for protein analysis, as discussed later in this chapter.

Linear polymers yield different calibration curves than globular proteins because their volumes in solution (radius of gyration) are so much larger than those of globular proteins (4). Figure 10.6 (page 312) shows the analysis of anionic sulfonated polystyrenes on the same column used for proteins in Fig. 10.5. It is obvious that the molecular weight ranges separated by this column for the two kinds of molecules are vastly different. The calibration curves in Fig. 10.3 show that solutes with molecular weights ranging from 500 to 10,000,000 can be analyzed on at least one of these columns. When samples contain a very wide range of molecular weights, two or three different columns in series or a mixed bed column (GPC Linear) may be able to give resolution of all the peaks. The calibration of SynChropak GPC Linear compared to that of GPC 100 and GPC 1000 is shown in Fig. 10.7 (page 313). A sample chromatogram on this mixed bed column is seen in Fig. 10.8 (page 313).

When cationic polymers are run on SynChropak CATSEC columns, the calibration curves, as shown in Fig. 10.4, are not identical to those produced

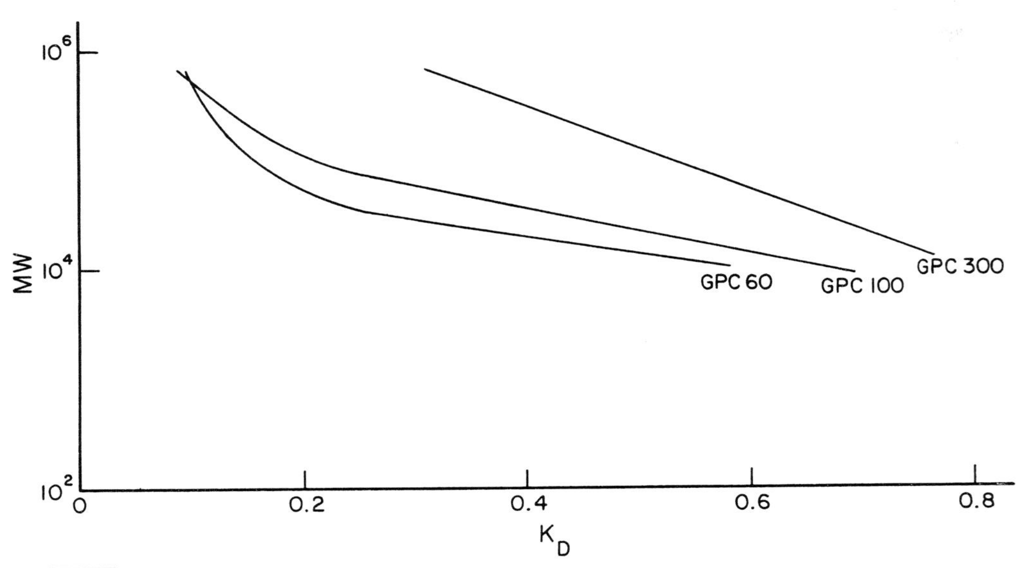

FIGURE 10.2 Calibration curves for proteins on SynChropak GPC columns. Mobile phase: 0.1 *M* potassium phosphate, pH 7. (From MICRA Scientific, Inc., with permission.)

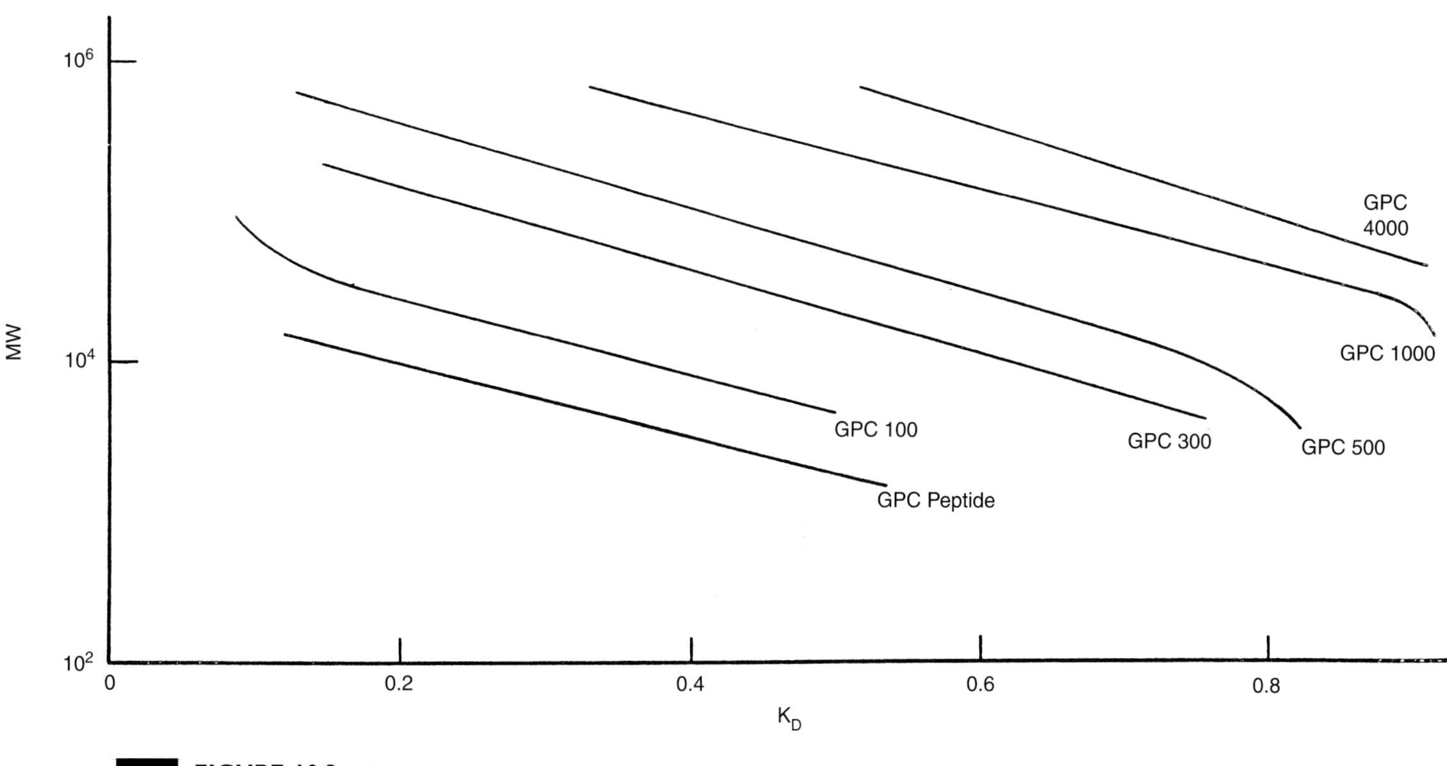

FIGURE 10.3 Calibration curves for sulfonated polystyrenes on SynChropak GPC columns. Mobile phase: 0.1 *M* sodium sulfate. (From MICRA Scientific, Inc., with permission.)

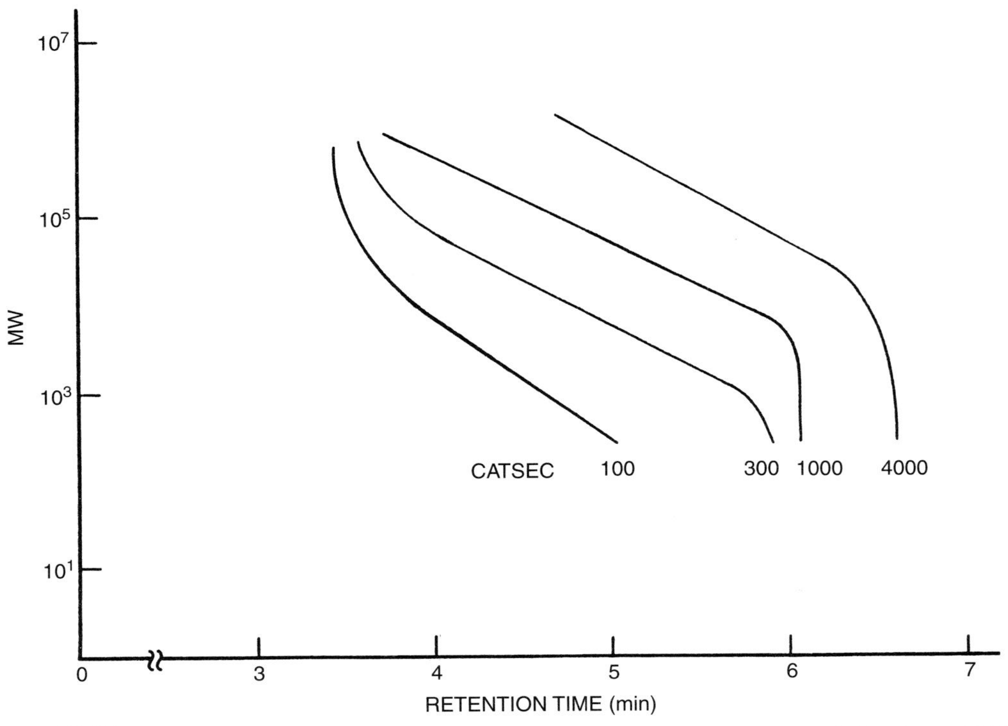

FIGURE 10.4 Calibration curves for polyvinylpyridines on SynChropak CATSEC columns. Mobile phase: 0.2 *M* sodium chloride in 0.1% trifluoroacetic acid. (From MICRA Scientific, Inc., with permission.)

by anionic polymers on SynChropak GPC (Fig. 10.3). This effect is due to a combination of distinct bonded phase layers on the supports and different molar volumes of the anionic vs cationic polymeric solutes (5).

D. Quality Assurance

Quality assurance for size exclusion supports is based primarily on the reproducibility of molecular weight calibrations. Although the reproducibility of the exclusion and inclusion limits is important, the distribution coefficients (K_D) of included standards are a better indication of duplication. Table 10.3 (page 314) shows such data for the SynChropak GPC and CATSEC supports.

III. COLUMN TECHNOLOGY

A. Hardware

SynChropak GPC and CATSEC columns are packed in either stainless steel or PEEK columns with standard inverted fittings so that they readily connect to most instruments. Stainless-steel columns are available in 2.1, 4.6, 7.8, 10,

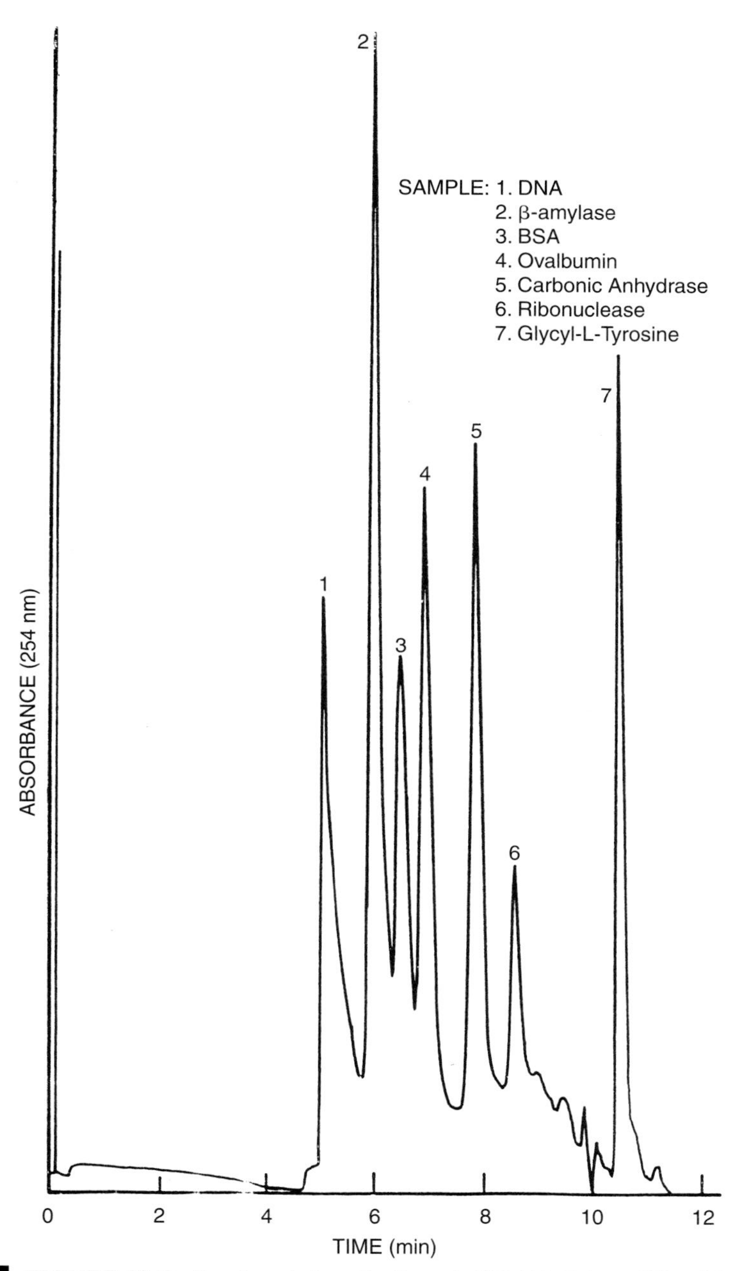

FIGURE 10.5 Protein analysis on SynChropak GPC 100. Column: 300 × 7.8 mm i.d. Mobile phase: 0.1 *M* potassium phosphate, pH 7. Flow rate: 1 ml/min. (From MICRA Scientific, Inc., with permission.)

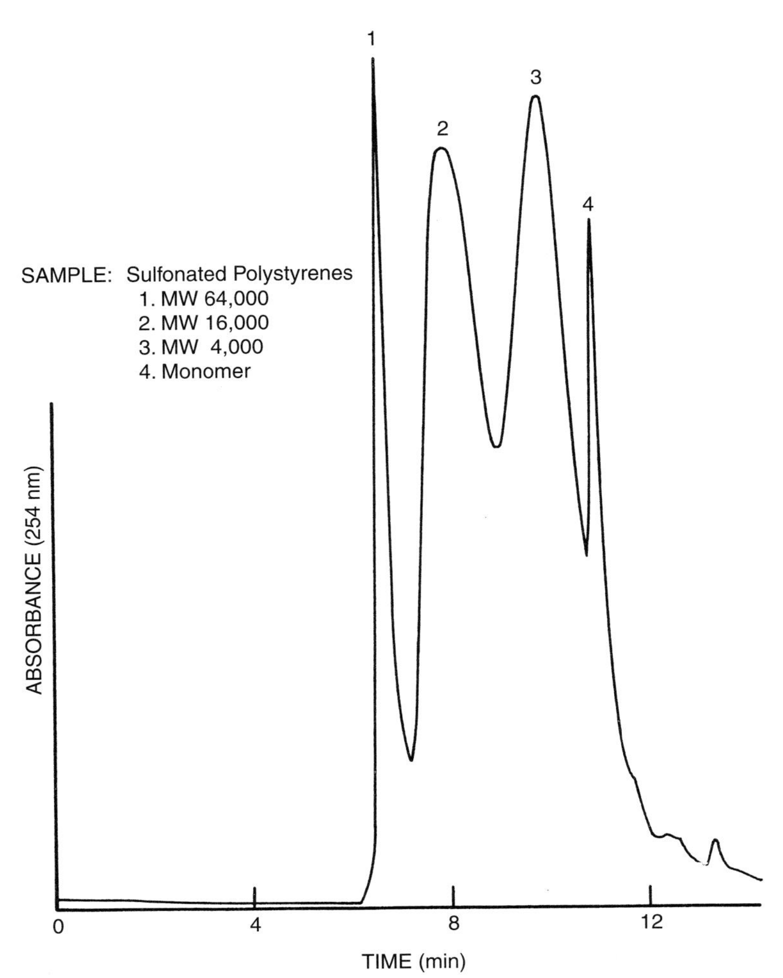

FIGURE 10.6 Analysis of sulfonated polystyrenes on SynChropak GPC 100. Column: 250 × 4.6 mm i.d. Mobile phase: 0.1 *M* sodium sulfate. Flow rate: 0.25 ml/min. (From MICRA Scientific, Inc., with permission.)

and 21 mm i.d., all of which have 2-μm frits with filter areas corresponding to the internal cross-sectional areas of the columns. PEEK columns are available in 4.6 and 7.8 mm i.d. All columns have standard lengths of 25 cm except the 7.8-mm i.d. columns, which are 30 cm long. The small pore SynChropak GPC Peptide is also available as a desalting column in a 10-cm length.

B. Packing Techniques

SynChropak GPC supports can be packed by slurrying in methanol and packing upward with methanol to a pressure of 4000 psi. SynChropak CATSEC supports can be packed by slurrying in a mixture of isopropanol and methanol and packing upward with methanol to a pressure of 4000 psi. The high pore

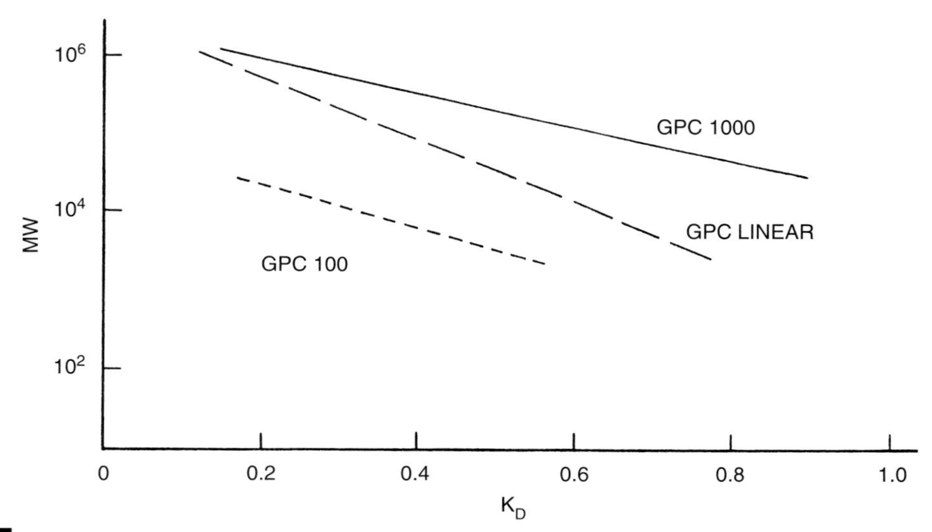

FIGURE 10.7 Calibration curves for sulfonated polystyrenes on SynChropak GPC linear, GPC 100 and GPC 1000. Conditions as in Fig. 10.3. (From MICRA Scientific, Inc., with permission.)

SAMPLE:	RETENTION TIME (min)
1. DNA	3.33
2. SPS 354,000	3.99
3. SPS 31,000	4.85
4. SPS 6,500	5.39
5. Cytidine	6.44

DETECTION: 254 nm

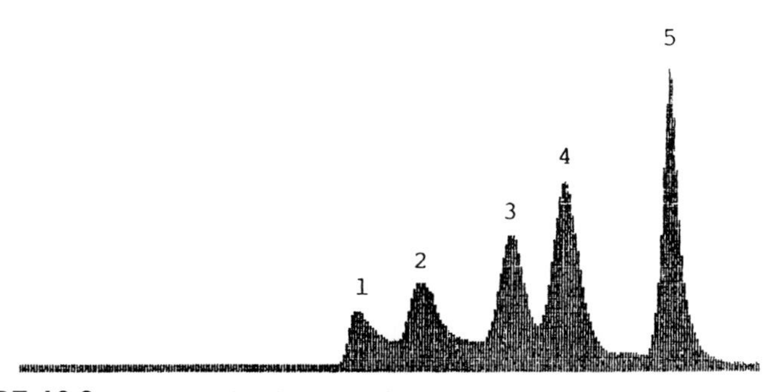

FIGURE 10.8 Analysis of sulfonated polystyrenes on SynChropak GPC Linear. Column: 250 × 4.6 mm i.d. Flow rate: 0.5 ml/min. Mobile phase: 0.1 *M* sodium sulfate, pH 7. (From MICRA Scientific, Inc., with permission.)

TABLE 10.3 QA and Standard Deviation for SynChropak SEC Columns[a]

SynChropak[b]	Protein	K_D	STD	Polymer MW	Polymer K_D	Polymer STD
				Sulfonated polystyrenes		
GPC peptide	Cytochrome c	0.41	0.03	4000	0.32	0.05
GPC 100	β-Lactoglobulin	0.39	0.03	4000	0.48	0.05
GPC 300	Ovalbumin	0.62	0.04	31,000	0.44	0.04
GPC 500	Thyroglobulin	0.51	0.03	65,000	0.55	0.05
GPC 1000				195,000	0.53	0.02
GPC 4000				690,000	0.55	0.03
				Polyvinylpyridines		
CATSEC 100				3000	0.41	0.03
CATSEC 300				6000	0.55	0.02
CATSEC 1000				51,000	0.52	0.05

[a]Reprinted from MICRA Scientific with permission.
[b]GPC: DNA and glycyltyrosine (cytidine, GPC PEPTIDE); 0.1 *M* potassium phosphate, pH 7, 0.5 ml/min. CATSEC: PVP 600,000 and cytidine; 0.1% trifluoroacetic acid with 0.2 *M* NaCl; 0.5 ml/min.

volumes of size exclusion supports render them more fragile than low pore volume materials and necessitate lower packing pressures than those used for more robust materials, such as reversed-phase supports, which can be packed at pressures exceeding 10,000 psi.

C. Quality Control

Each SynChropak column is tested chromatographically to assure that it has been packed according to specifications. For SynChropak GPC columns, a mixture of a high molecular weight DNA and glycyltyrosine, a dipeptide, is used to evaluate internal volume and efficiency. The mobile phase used for the test is 0.1 *M* potassium phosphate, pH 7, and the flow rate is 0.5 ml/min for 4.6-mm i.d. columns. Minimum plate count values and operational flow rates are listed in Table 10.4 for 4.6-mm i.d. columns of all supports and the various diameters of the SynChropak GPC 100 columns.

SynChropak CATSEC columns are evaluated similarly using a polyvinylpyridine standard of molecular weight 600,000 and cytidine. The mobile phase is 0.1% trifluoroacetic (TFA) acid containing 0.2 *M* sodium chloride. Minimum plate counts are listed in Table 10.4.

IV. OPERATION

A. Mobile Phase Selection

In size exclusion chromatography, the mobile phase must be selected to totally solubilize the sample and eliminate all interactions of the solutes with the

TABLE 10.4 Minimum Plate Counts[a,b]

SynChropak	Minimum plates 250 × 4. 6 mm i.d. 0.5 ml/min
	Glycyltyrosine
GPC peptide	9,800
GPC 100	13,600
GPC 300	14,300
GPC 500	13,000
GPC 1000	12,300
GPC 4000	8,300
	Cytidine
CATSEC 100	15000
CATSEC 300	14900
CATSEC 1000	16100
CATSEC 4000	8900

Column (min)	GPC 100 minimum plates Glycyltyrosine	Flow rate (ml/min)
250 × 2.1	8,500	0.1
250 × 4.6	13,600	0.5
300 × 7.8	23,000	1.0
250 × 10	20,700	1.0
250 × 21.2	20,700	6.0

[a]From MICRA Scientific with permission.
[b]Mobile phases as in Table 10.2.

support and bonded phase. Because most samples in size exclusion chromatography include macromolecules, mobile-phase selection is not trivial. It has the additional function of defining the shape and thus the radius of gyration and volume of the molecule. This, rather than molecular weight, is the true dimension measured during size exclusion chromatography (4).

1. Proteins

For many proteins, a simple buffer such as 0.1 *M* phosphate, pH 7, produces excellent separations on SynChropak GPC columns. Generally, minimal interaction is achieved when the ionic strength is 0.05–0.2 *M*. To prevent denaturation or deactivation of proteins, the pH is generally kept near neutrality. For denatured proteins, 0.1% sodium dodecyl sulfate (SDS) in 0.1 *M* sodium phosphate, pH 7, is recommended.

2. Peptides

Peptides are much more difficult to run by size exclusion chromatography than proteins because their solubilities differ greatly. Their shapes also vary from linear to semidefined. For a three peptide standard, Mant and Hodges

(6) found that a mobile phase of 0.005 *M* potassium phosphate with 0.2 *M* potassium chloride, pH 6.5, yielded no interaction with the support whereas lower concentrations of potassium chloride caused inverted elution of the peaks due to support–solute interaction. In an unpublished study (7), it was observed that for hydrophobic peptides such as certain angiotensins, a standard phosphate mobile phase produced interaction of some peptides with the column (Fig. 10.9a), whereas the addition of 35% methanol yielded elution by size, as seen in Fig. 10.9b. These were not adequate conditions for angiotensin III, however, which still had slight hydrophobic interaction with the support. This study and others have evaluated 0.1% TFA with isopropanol as an eluent for size exclusion chromatography of peptides due to its success as an eluent for reversed-phase chromatography. Although this system worked successfully for some peptides, others had interaction and most proteins were partially denatured or exhibited ion exclusion. A mobile phase containing 0.1% SDS was the most universal mobile phase for peptides (Fig. 10.10); however, it denatured the proteins, yielded broader peaks than many other solvents, and increased the molar volume of the solutes.

3. Neutral and Anionic Polymers

Anionic and neutral polymers are usually analyzed successfully on SynChropak GPC columns because they have minimal interaction with the appropriate mobile-phase selection; however, cationic polymers adsorb to these columns, often irreversibly. Mobile-phase selection for hydrophilic polymers is similar to that for proteins but the solubilities are of primary importance. Organic solvents can be added to the mobile phase to increase solubility. In polymer analysis, ionic strength and pH can change the shape of the solute from mostly linear to globular; therefore, it is very important to use the same conditions during calibration and analysis of unknowns (8). Many mobile phases have been used, but 0.05–0.2 *M* sodium sulfate or sodium nitrate is common.

4. Cationic Polymers

Mobile-phase selection for cationic polymers is similar to that for the other polymers in that ionic strength and pH can change the shape of the solute from linear to globular (9). Mobile phases are often low pH; e.g., 0.1% trifluoroacetic acid, including 0.2 *M* sodium chloride, has been used successfully for polyvinylpyridines. Sodium nitrate can be substituted for the chloride to avoid corrosive effects. Some salt must be included so that ion exclusion does not occur (3).

B. Flow Rate

Although SynChropak GPC supports have excellent efficiencies for small molecules at various flow rates, macromolecules, because of their low diffusion constants, exhibit band spreading when linear velocities are increased. This effect increases with molecular weight, as seen in Fig. 10.11 (4). It should be noted that proteins are usually homogeneous in size and thus yield better efficiencies than polymers, which are usually heterogeneous. For preliminary analy-

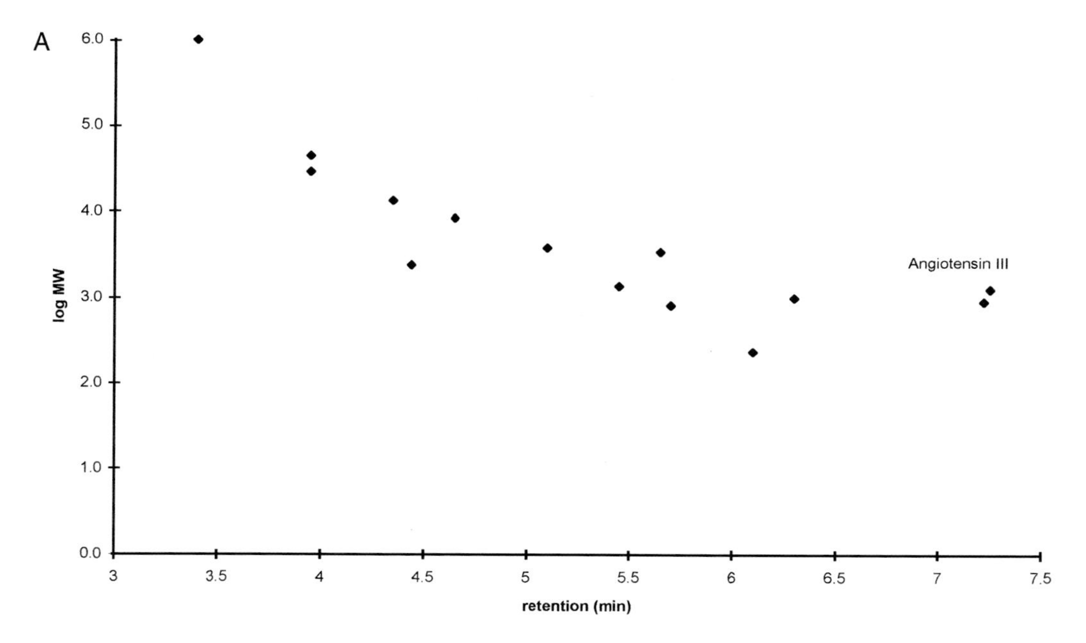

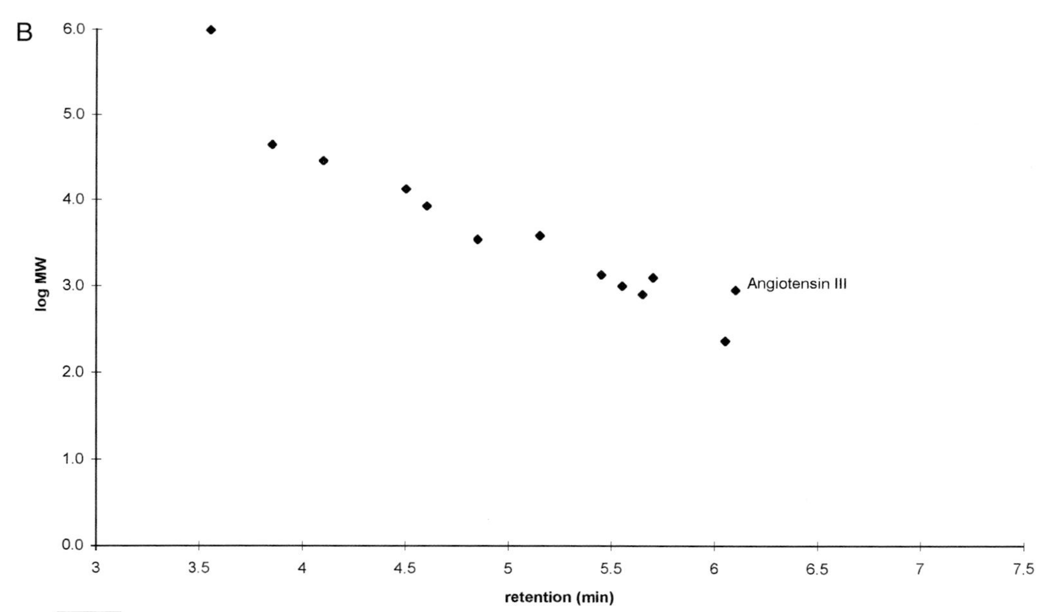

FIGURE 10.9 Analysis of peptides on SynChopak GPC peptide. Mobile phase: (a) 0.1 *M* potassium phosphate, pH 7; and (b) 0.1 *M* potassium phosphate, pH 7, 35% methanol. (From MICRA Scientific, Inc., with permission.)

ses, a flow rate of 0.5 ml/min (0.6 mm/sec) on a 4.6-mm i.d. column usually yields adequate efficiency, but for maximum resolution, flow rates of 0.25 ml/min, or even lower, may be necessary. Equivalent flow rates for other dimensions of col-

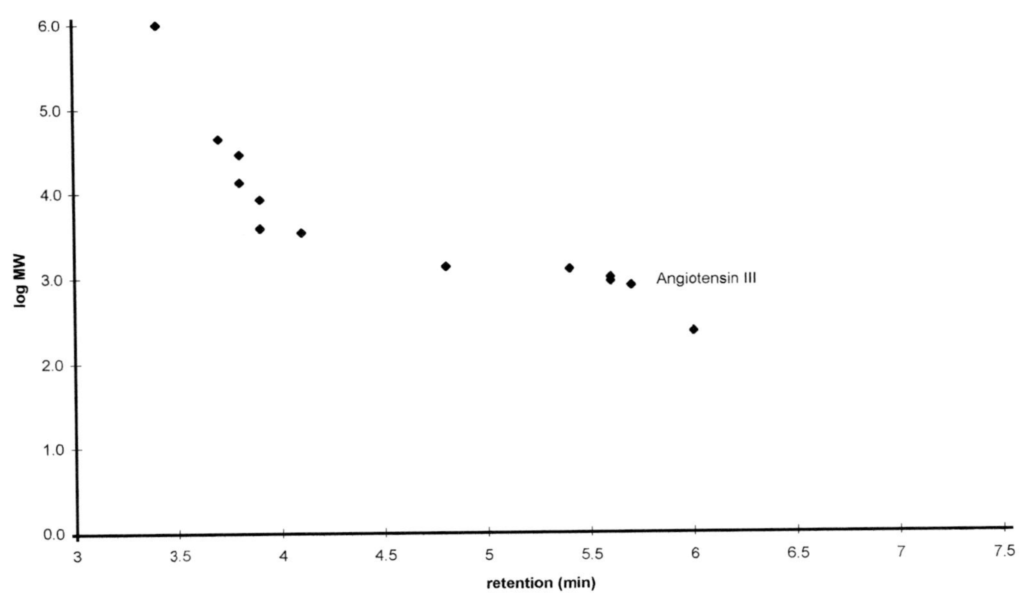

FIGURE 10.10 Analysis of peptides on SynChopak GPC peptide. Mobile phase: 0.1 *M* sodium phosphate, pH 7, 0.1% sodium dodecyl sulfate. (From MICRA Scientific, Inc., with permission.)

umns can be calculated by multiplying the flow rate on the 4.6-mm i.d. column by $d_c^2/(4.6)^2$, where d_c is the internal diameter of the column of interest.

C. Loading

Loading capacities in size exclusion chromatography are very low because all separation occurs within the liquid volume of the column. The small diffusion coefficients of macromolecules also contribute to bandspreading when loads are increased. The mass loading capacities for ovalbumin (MW 45,000) on various sizes of columns can be seen in Table 10.5. The maximum volume that can be injected in size exclusion chromatography before bandspreading occurs is about 2% of the liquid column volume. The maximum injection volumes for columns of different dimensions can also be seen in Table 10.5.

D. Restrictions

Supports that are optimum for size exclusion chromatography have high pore volumes because all separation occurs within the pore. One consequence of high pore volume, however, is increased fragility to pressure. SynChropak size exclusion supports have a maximum pressure limit of 4000 psi, which is significantly lower than that of other supports, such as those used for reversed-phase chromatography. If pressures are exceeded, the support will crush, creating a void at the front of the column and resultant band broadening. Another consequence of high pressure is that in fracturing some of the silica, silanol groups, which can interact with molecules containing cationic or silanophilic groups, can be exposed.

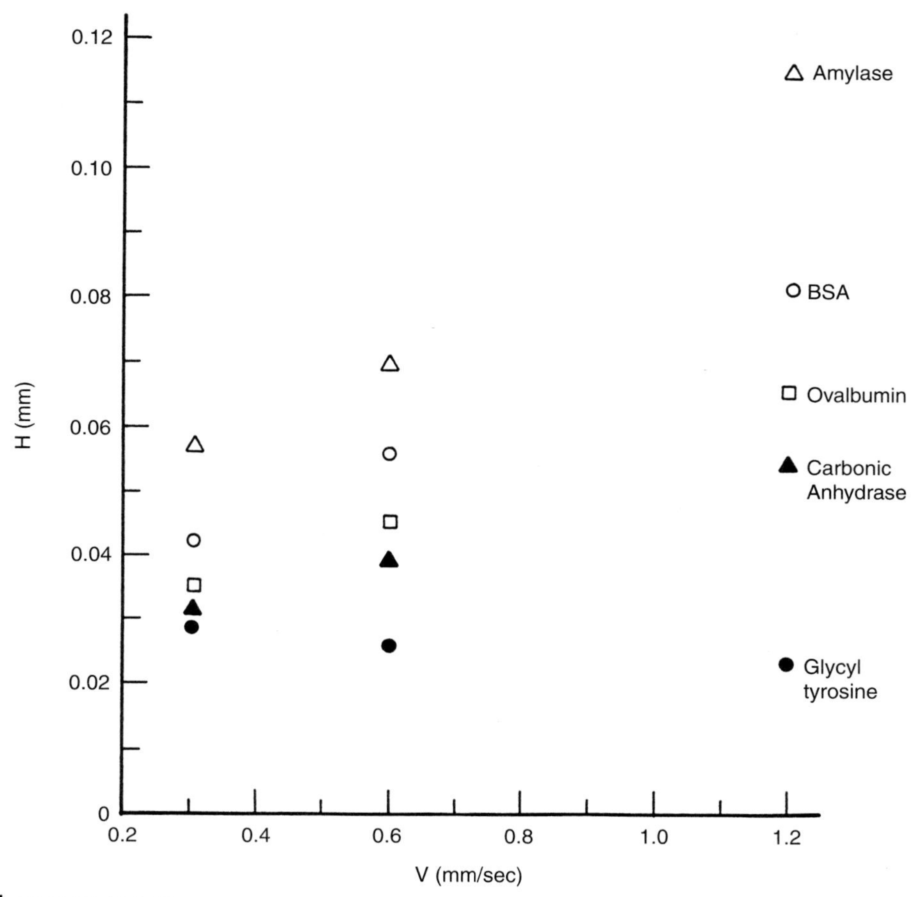

FIGURE 10.11 Effect of linear velocity on plate height for proteins. Column: SynChropak GPC 500, 250 × 4.6 mm i.d. Mobile phase: 0.1 *M* potassium phosphate, pH 7. (Reprinted from Ref. 4 by courtesy of Marcel Dekker, Inc.)

TABLE 10.5 Loading Capacities of SynChropak SEC Columns[a,b]

i.d. (mm)	Flow rate (ml/min)	Mass (mg)	Volume (μl)
2.1	0.1	0.05–0.10	2
4.6	0.5	0.2–0.4	8
7.8	1.5	1.0–2.0	30
10.0	2.0	2.0–4.0	50
21.2	8.0	8.0–16.0	200

[a]From MICRA Scientific with permission.
[b]GPC Mobile phases as in Table 10.2; ovalbumin solute.

TABLE 10.6 Applications on SynChropak SEC Columns

Solute	Column	Reference
Acrylamide-based polyelectrolytes	CATSEC	*Intern. J. Environ. Anal. Chem.* **29,** 1–14 (1987)
Albumin, human serum	GPC	*Anal. Chem.* **65,** 601–605 (1993)
ATP synthase	GPC	*Biochemistry* **27,** 6704 (1988)
Cationic polymers	CATSEC	*J. Liq. Chromatogr.* **12,** 1431–1449 (1989)
Cellulosics	GPC	*J. Chromatogr.* **192,** 275–293 (1980)
Complex polysaccharides	GPC	*J. Agric. Food Chem.* **41,** 1274–1281 (1991)
Corn starch	GPC	*Anal. Chem.* **53,** 736–737 (1981)
Cytochrome c	GPC	*J. Chromatogr.* **53,** 379–389 (1987)
Fulvic acids	GPC	*Anal. Chem.* **57,** 279–283 (1985)
Guar gum	GPC	*J. Chromatogr.* **206,** 410–415 (1981)
Hydrolyzed plant proteins	GPC	*Anal. Biochem.* **124,** 191–200 (1982)
Immune complexes, immunoglobulins	GPC	*J. Chromatogr.* **421,** 434–436 (1987)
Leucine zipper mutants	GPC	*Science* **262,** 1401–1406 (1993)
Lignite coal polymer	GPC	*Resources, Conservation and Recycling* (1990)
Lignite coal polymer	GPC	*Appl. Biochem. Biotech.* **24/25,** 889 (1990)
mRNA	GPC	*BioTechniques* 114–118 (1985)
Macrophage-derived growth factor	GPC	*Br. J. Rheumatol.* **24,** 197–202 (1985)
Microcolumn	GPC	*J. Chromatogr.* **448,** 73–86 (1988)
Microcolumns	GPC	*Anal. Chem.* **60,** 1826–1829 (1988)
Nonideal	GPC	*Anal. Biochem.* **126,** 8–16 (1982)
Pectin	GPC	*Carbohydr. Res.* **215,** 91–104 (1991)
Pectin	GPC	*J. Food Sci.* **58,** 680–687 (1993)
Pectin	GPC	*Arch. Biochem. Biophys.* **274,** 179–191 (1989)
Pectin Subunits	GPC	*Arch. Biochem. Biophys.* **294,** 253–260 (1992)
Pectins	GPC	*J. Agric. Food Chem.* **37,** 584–591 (1989)
Pectins	GPC	*Carbohydr. Polym.* **15,** 89–104 (1991)
Pectins	GPC	*J. Liq. Chromatogr.* **3,** 1481–1496 (1980)
Peptide standards	GPC	*J. Chromatogr.* **397**, 99–112 (1987)
Peptides, methods	GPC	*HPLC of Peptides and Proteins* 125–134 (1991)
Peptides, methods	GPC	*HPLC of Peptides and Proteins* 11–22 (1991)
Peptides, methods	GPC	*J. Chromatogr.* **458,** 147–167 (1988)
Poly(vinylalcohol)	CATSEC	*Am. Lab.* 47J-47V (1995)
Polyacrylamide	GPC	*Anal. Chem.* **57,** 2098–2101 (1985)
Polyelectrolytes, wastewater	CATSEC	*J. Environ. Eng.* **116,** 343–360 (1990)
Polymers	GPC	*Adv. Carbohydr. Chem. Biochem.* (1984)
Polymers	GPC	*J. Chromatogr. Sci.* **18,** 430–441 (1980)
Polymers	GPC	*J. Chromatogr. Sci.* **18,** 409–429 (1980)
Polynucleotides	GPC	*J. Chromatogr.* **436,** 299–307 (1988)
Polysaccharides	GPC	*Chromatography of Polymers, Char. by SEC and FFF Ch.* 22

(*continues*)

TABLE 10.6 *(continued)*

Solute	Column	Reference
Polyvinylpyridines	CATSEC	*J. Liq. Chromatogr.* **5**, 2259–2270 (1982)
Protein aggregates	GPC	*J. Biol. Chem.* **264**, 15863–15868 (1989)
Proteins	GPC	*Anal. Biochem.* **97**, 176–183 (1979)
Proteins, methods	GPC	*HPLC of Peptides and Proteins* 135–144 (1991)
Proteins, methods	GPC	*J. Chromatogr.* **554**, 125–135 (1991)
Serum proteins	GPC	*Anal. Chem.* **65**, 2972–2976 (1993)
Starches	GPC	*Carbohydr. Polym.* **23**, 175–183 (1994)
Surface proteins	GPC	*J. Agric. Food Chem.* **40**, 1613–1616 (1992)
Theory and practice	GPC	*HPLC of Biological Macromolecules* 47–75 (1990)
Viscosity	GPC	*J. Chromatogr.* **550**, 705–719 (1991)

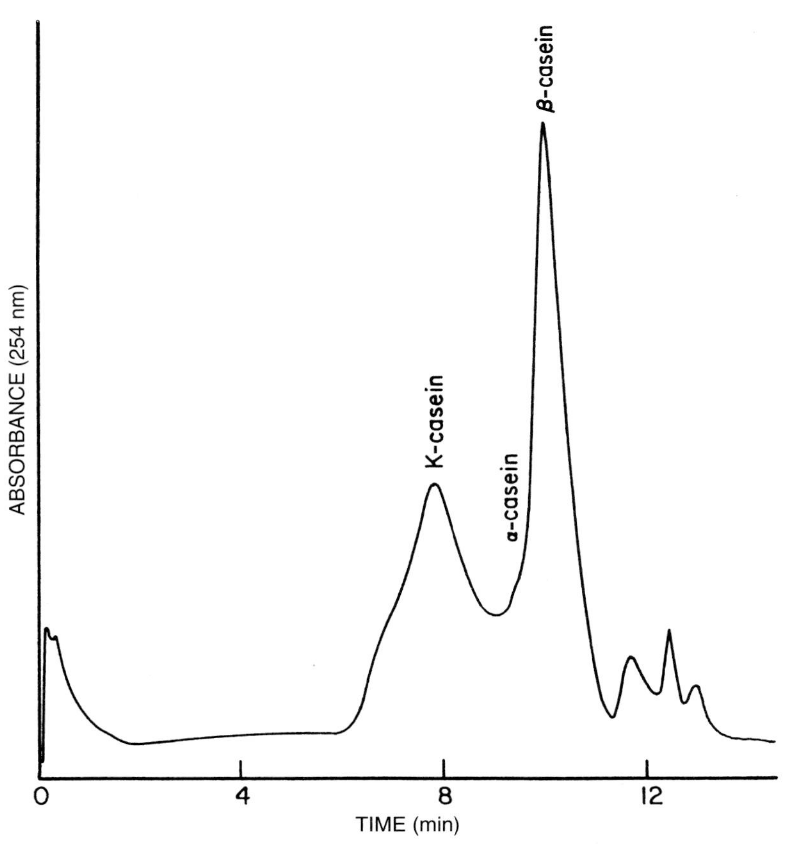

FIGURE 10.12 Analysis of caseins on SynChropak GPC 500. Column: 250 × 4.6 mm i.d. Mobile phase: 0.1 *M* potassium phosphate, pH 7. Flow rate: 0.25 ml/min. (From MICRA Scientific, Inc., with permission.)

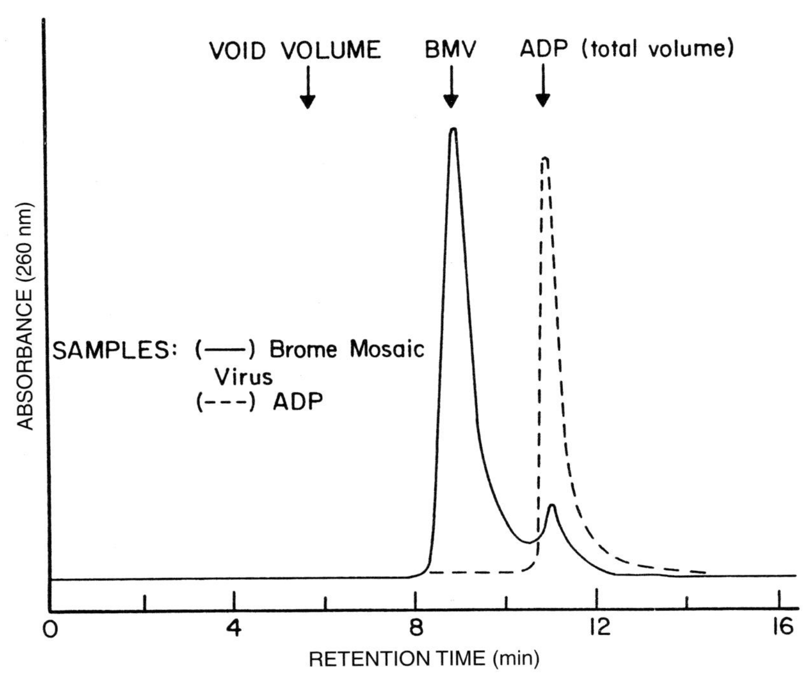

FIGURE 10.13 Analysis of brome mosaic virus BMV on SynChropak GPC 500. Column: 250 × 4.6 mm i.d. Mobile phase: 0.05 *M* Bis–Tris, 0.5 *M* sodium acetate, pH 5.9. Flow rate: 0.3 ml/min. (Reprinted from Jerson Silva and MICRA Scientific, Inc., with permission.)

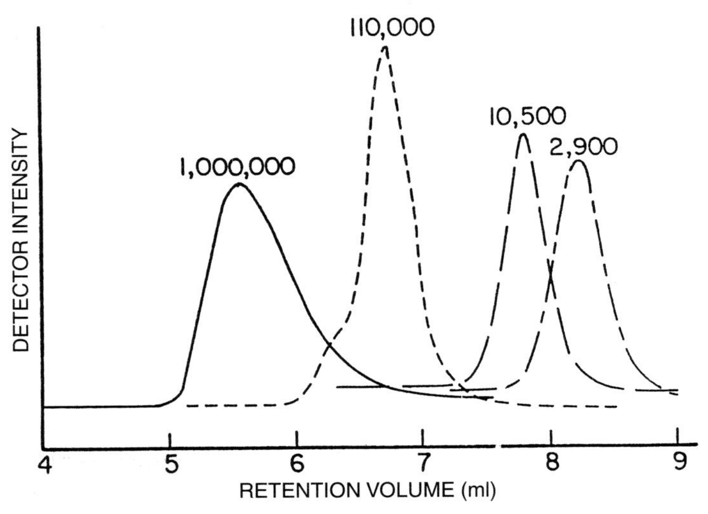

FIGURE 10.14 Analysis of polyvinylpyridines on SynChropak CATSEC 100, 300, and 1000 columns in series (250 × 4.6 mm i.d.). Flow rate: 0.37 ml/min. Mobile phase: 0.1% trifluoroacetic acid in 0.2 *N* sodium nitrate. Detection by differential viscometry. (Reprinted from Ref. 9 with permission.)

The other restriction of SynChropak size exclusion columns is a general one for silica-based supports, that of pH. The most harmful pH is that above 7.5 due to silica dissolution. The bonded phase of SynChropak GPC has some polymeric properties; therefore, it is not removed rapidly from the silica at pH 2–3. The bonded phase of SynChropak CATSEC is polymeric and stable at pH 2–7.5.

Elevated temperatures may be used with SynChropak GPC or CATSEC supports if necessary for solubilization, speed, or reduction of band spreading. Such conditions are not recommended for routine analyses, however, because column degradation is enhanced as temperatures are raised.

V. APPLICATIONS

Many proteins and polymers have been analyzed on SynChropak GPC and CATSEC columns. Table 10.6 lists some of the published applications. The use of a surfactant to analyze the caseins in milk is illustrated in Fig. 10.12. Viruses have also been analyzed on SynChropak GPC columns, as seen in the chromatogram from Dr. Jerson Silva of the University of Illinois (Fig. 10.13). Dr. Nagy and Mr. Terwilliger analyzed cationic polymers on a series of CATSEC columns using differential viscometry as detection (Fig. 10.14) (9).

REFERENCES

1. Chang, S. H., Gooding, K. M., and Regnier, F. E. (1976). *J. Chromatogr.* **125**, 103–114.
2. Regnier, F. E., and Noel, R. (1976). *J. Chromatogr. Sci.* **14**, 316–320.
3. Gooding, D. L., Schmuck, M. N., and Gooding, K. M. (1982) *J. Liq. Chromatogr.* **5**, 2259–2270.
4. Gooding, K. M., and Regnier, F. E. (1990). *In* "HPLC of Biological Macromolecules: Methods and Applications" (K. M. Gooding and F. E. Regnier, eds.), pp. 47–475. Dekker, New York.
5. "SynChronotes" (1991). Vol. 6, No. 2. SynChrom, Inc., Lafayette, IN (now MICRA Scientific, Inc., Northbrook, IL).
6. Mant, C. T., and Hodges, R. S. (1991). *In* "HPLC of Peptides and Proteins" (C. T. Mant and R. S. Hodges, eds.), pp. 125–134, CRC Press, Boca Raton, FL.
7. Freiser, H., Schmuck, M. N., and Gooding, K. M. Presented at HPLC '92.
8. Barth, H. G. *J. Chromatogr. Sci.* **18**, 409–429.
9. Nagy, D. J. (1995). *Am. Lab.* Feb. 1995, 47J–47V.

11 WATERS COLUMNS FOR SIZE EXCLUSION CHROMATOGRAPHY

UWE DIETER NEUE

Waters Corporation, Milford, Massachusetts 01757

I. INTRODUCTION

Size exclusion chromatography (SEC) is by far the simplest form of chromatography. It is used primarily to separate molecules of different sizes: large molecules elute first, smaller molecules elute last. The most important application areas are the characterization of polymers in general and the separation of biopolymers such as proteins in particular. When used for the characterization of industrial polymers, the technique is often also called gel-permeation chromatography (GPC). When used for the separation of proteins, it is referred to as gel-filtration chromatography (GFC). Both older designations have fallen in disfavor.

The ideal packing for size exclusion chromatography exhibits no adsorptive interaction with the analytes. The driving mechanism of SEC is therefore the steric exclusion of analytes from the pores of the packing (1). Because the range of the separation is limited by the pore volume in a column, the specific pore volume of a packing (together with its packing density) determines the performance of a packing. The selectivity of different packings is determined by the average pore size and the pore size distribution of a packing.

In size exclusion chromatography, large molecular weight compounds are excluded from some or all of the pore volume and elute *earlier* than compounds with a lower molecular weight. Also, the elution of a polymer increases with increasing pore size of the packing. Packings with several different pore sizes are offered for all types of size exclusion packings. More information is available in the section dealing with the specific types of packings available for size exclusion chromatography.

For the classical form of size exclusion chromatography in organic solvents, packings based on highly cross-linked styrene–divinylbenzene are used. For SEC of polar polymers using polar or aqueous solvents, packings based on a polar methacrylate polymer are used. Diol-derivatized silica is used for the separation of proteins and other polar polymers. The different packings will be discussed in sections dedicated to their different application areas.

The initial discussions in this chapter center on the packing types used in different applications, the decision of whether to use a mixed-bed or a single pore size packing, and the selection of the particle size and the column dimensions. The chapter then discusses Waters products for size exclusion chromatography: the Styragel line for general SEC applications using organic solvents, the Ultrahydrogel line for water-soluble polymers using aqueous mobile phases, and finally the silica-based Protein-Pak line of columns for proteins and related compounds. Every section covers product selection, a few application examples, and some aspects of the care and maintenance of the products. Additional information is available from Waters Corporation.

II. COLUMN SELECTION PRINCIPLES

A user of size exclusion chromatography is first faced with the decision which columns and packings should be used for the types of problems that may be encountered. The appropriate packing and the best set of columns need to be selected. In many cases, it is best to choose column banks for specific applications and to always use them for these applications only. However, in smaller laboratories it might be best to select the column banks based on the most typical and frequent applications problems. The following sections give the user an overview of the column selection process by (1) discussing the packing type, i.e., Styragel packings versus Ultrahydrogel packings versus Protein-Pak packings; (2) considering the issues of the selection of the best pore sizes of the packing and whether the choice of a single pore size or a mixed pore size is more appropriate; (3) reviewing the particle size choices; and (4) exploring the use of smaller diameter columns, which are designed for a reduction in solvent consumption.

A. Packing Type

The basic choice of packing material requires the user to make a decision as to which type of packing is the most suitable for the application or applications that will be used. Waters offers three types of packing materials: Styragel packings based on styrene–divinylbenzene copolymers, Ultrahydrogel packings based on methacrylate polymers, and Protein-Pak packings based on diol-derivatized silica.

The Styragel family of packings represents the classical packing of size exclusion chromatography (2). It is based on cross-linked styrene–divinylbenzene particles. Pore sizes range from around 20 Å for the Styragel

HR 0.5 column to above 1 μm for the Styragel HMW 7 column, a broad range covering every possible size necessary for SEC applications. Exclusion ranges for polystyrene standards stretch from about 1000 Da for the Styragel HR 0.5 column to a size exceeding 10^8 Da for the Styragel HMW 7 column. Smaller pore sizes, below Styragel HR 2, are only obtained after the packing swells in the solvent in which the column is used. The pore sizes of Styragel HR 3 and higher are permanent and are fairly unaffected by the choice of mobile phase. The subject of the best choice of pore size for a particular application is covered in Section II,B.

Styragel packings are the most versatile packings in size exclusion chromatography. Interaction of polymers with the divinylbenzene matrix is rare and can be suppressed with the correct solvent (3), which is the primary reason for the popularity of styrene–divinylbenzene-based packings. Styragel-based columns are packed in toluene, tetrahydrofuran, and dimethylformamide for a broad range of applications. For example, many applications can be run using toluene at room temperature or at an elevated temperature, trichlorobenzene at 140°C, tetrahydrofuran at 30–45°C, or dimethylformamide at 85–145°C. In addition, 4.6-mm i.d. columns are available for use with hexafluoroisopropanol. A complete guide to the applications of Styragel packings and appropriate solvent choices is given in Section III,A.

The user should be aware though of the presence of other functions in the matrix of the packing; the initiator used, azobisisobutyronitrile, leaves a small amount of CN functions in the packings. Also, small amounts of alcohol functions cannot be excluded, especially as the packing ages. For most applications, these small amounts of residual side functions are of little concern.

The styrene–divinylbenzene matrix of the Styragel packings is chemically very inert, which makes this family of packings useful for a broad range of applications. The chromatographic conditions for the analysis of many polymers have been worked out in detail. A more specific discussion of the solvents recommended for the different polymer types is included in Section III,A,4.

The specific pore volume of all packings for size exclusion chromatography is important. All Styragel packings have a porosity of approximately two-thirds of the particle volume. This gives a large peak capacity while maintaining the structural integrity of the packing.

Ultrahydrogel size exclusion columns are designed for the analysis of very polar, mostly water-soluble macromolecules. The packing is based on a highly hydroxylated methacrylate polymer, with pore size exclusion ranges for polyethyleneoxide standards from a molecular mass of about 1000 Da to 10^6 Da. Ultrahydrogel columns are generally used for neutral water-soluble polymers such as polysaccharides and cationic, anionic, and amphoteric polymers. Most applications use aqueous mobile phases, but a certain amount of organic solvents (up to 50%) are quite compatible with these packings. Detailed solvent compositions are discussed in Section III,B.

Generally, the interaction of polar analytes with the packing is rather weak due to the hydrophilic polyhydroxy functions on the surface of the packing. However, small amounts of acidic functional groups are present on the surface of the packing. The influence of these functional groups can be suppressed easily with the use of salts in the mobile phase.

The compatibility of the matrix with acids and bases is quite good. Ultrahydrogel columns can be used with mobile phases from pH 2 to pH 12 for extended periods of time.

Protein-Pak packings are designed for the size exclusion chromatography of proteins and related compounds. They are based on silica, which is deactivated with glycidylpropylsilane. The diol function prevents the interaction of the target analytes with the silica surface. However, because coverage of the silica surface is always incomplete, residual acidic silanols can interact with the analytes. For this reason, most applications are carried out with a salt concentration above 0.2 mol/liter, which eliminates the interaction of analytes with surface silanols. Protein-Pak packings are stable from pH 2 to pH 8.

Due to the fact that the application area for Protein-Pak columns is more limited than the application area of other SEC packings, only a few pore sizes are needed for standard protein separations. A more detailed discussion of this product line follows in Section III,C.

B. Pore Size Selection

For any size exclusion separation, the selection of the right pore size columns is of primary importance. All the different packings for size exclusion chromatography are available in a wide range of pore sizes, and the user needs to decide which pore sizes fit best to the requirements of the application. For this purpose, manufacturers publish either calibration curves obtained with standard polymers (e.g., polystyrenes for the Styragel line) or molecular weight ranges for standard polymers that can be separated using a given column type. Because the hydrodynamic size of a polymer in a particular solvent depends on the chain length, the chain length can be used as a primary factor in converting the standard polymer to the polymer of interest. For example, for Styragel type packings, polystyrenes are commonly used as standards. If one wants to know the exclusion limit of a Styragel type packing for polyethylene, one simply converts the exclusion limit for polystyrene to an exclusion limit for polyethylene. If, for example, the exclusion limit for polystyrene (MW of styrene is 104) is 100,000 Da, the corresponding exclusion limit for polyethylene (MW of ethylene is 28) is $100{,}000 \times 28/104 = 27{,}000$ Da. Similar procedures can be used for Ultrahydrogel type packings that are calibrated using polyethylene glycol.

Tables 11.1 through 11.4 give useful ranges for the various Styragel and Ultrahydrogel packings. It should be pointed out that Styragel packings with the designation E at the end of the name are mixed-bed columns with an "extended" range of the calibration curve. To view the full range of each calibration curve, go to the sections that discuss the individual column types.

To select a column for a particular analytical problem, the first step is to make a choice about the pore size(s) to be used for the separation. In general, one cannot expect that a single pore size will fulfill the needs of a separation. In size exclusion chromatography, it is more common that columns of different types are combined with each other to deliver the separation range needed for a particular analysis. Therefore, column banks with different pore sizes are frequently combined with each other to maximize the separation power for

TABLE 11.1 Effective Molecular Weight Ranges for Styragel HR Packings

Column	MW range (polystyrene)
Styragel HR 0.5	−1000
Styragel HR 1	100–5000
Styragel HR 2	500–20,000
Styragel HR 3	500–30,000
Styragel HR 4	5000–600,000
Styragel HR 5	50,000–4,000,000
Styragel HR 6	200,000–10,000,000
Styragel HR 4E	50–100,000
Styragel HR 5E	2000–4,000,000

TABLE 11.2 Effective Molecular Weight Ranges for Styragel HT Packings

Column	MW range (polystyrene)
Styragel HT 2	100–10,000
Styragel HT 3	500–30,000
Styragel HT 4	5000–600,000
Styragel HT 5	50,000–4,000,000
Styragel HT 6	200,000–10,000,000
Styragel HT 6E	5000–10,000,000

TABLE 11.3 Effective Molecular Weight Ranges for Styragel HMW Packings

Column	MW range (polystyrene)
Styragel HMW 2	100–10,000
Styragel HMW 7	500,000–100,000,000
Styragel HMW 6E	5000–100,000,000

TABLE 11.4 Effective Molecular Weight Ranges for Ultrahydrogel Packings

Column	MW range (polyethyleneoxide)
Ultrahydrogel 120	−5000
Ultrahydrogel 250	20–80,000
Ultrahydrogel 500	500–200,000
Ultrahydrogel 1000	500–1,000,000
Ultrahydrogel 2000	10,000–20,000,000
Ultrahydrogel Linear	20–20,000,000
Ultrahydrogel DP	5000

the molecular weight range of the polymers to be analyzed. It should be pointed out though that columns that do not contribute to the separation are actually detrimental to the separation. This is due to the fact that columns that are not relevant for the separation still add to the spreading of the peaks, but not to their resolution. This can best be examined by looking at the equation for the selectivity of a size exclusion separation S_{SEC}:

$$S_{SEC} = \frac{\varepsilon_p \, \Delta K_0}{\log(M_2) - \log(M_1)}, \qquad (1)$$

where e_p is the porosity of the column and ΔK_0 is the difference in the partition coefficients between the molecular weights M_1 and M_2. This is nothing but the inverse slope of the calibration curve. The flatter the calibration curve, the larger the difference in the elution volume between two species, and the selectivity increases. If a column does not contribute to a separation, the difference in the partition coefficients is zero and therefore the selectivity of the separation vanishes. However, because the column that does not contribute to the separation still contributes to the total bandspreading, an overall inferior separation results. It is important to consider this phenomenon for maximizing the separation obtainable with a column bank. This is one of the reasons why many users dedicate a set of columns to a particular group of analytes and use different sets of columns for different types of analyses or analytes.

If several columns appear to be useful for the separation, it is generally advantageous to choose the columns with the flattest calibration curves for the separation. As shown by Eq. (1), this requires a column with either a large porosity (ε_p) or a large difference in the partition coefficients. For the latter reason, it is usually beneficial to select the column with the smallest pore size for the separation. Furthermore, mixed pore size columns have a smaller difference in the partition coefficients than single pore size columns. Therefore, single pore size columns generally have a resolution advantage over mixed pore size columns. This is shown in Fig. 11.1, where the resolution of a single pore size column is compared to the resolution obtained with a mixed bed column. As a consequence, single pore size columns are used preferentially to focus on the desired molecular weight information.

To provide the user with the information necessary to choose the correct column, the selectivity for a broad range of Styragel size exclusion columns is plotted in Fig. 11.2. As in Fig. 11.1, the resolution is plotted for a range of Styragel columns versus the molecular weight of the polystyrene polymers. One can easily see the ranges of the different types of packings and the advantages of the different columns. For example, Styragel HR 3 columns have an excellent resolution in the lower molecular weight range, well below 10^5 Da. Styragel HR 4 columns cover the intermediate molecular weight range very well. For larger molecular weights, the Styragel HR 5 column can be used.

While it is preferred to select the best column set for a particular separation, Styragel HR 4E and 5E columns may have advantages if a broader molecular weight range needs to be covered. The resolution is not as good as individual pore size columns in a given molecular weight range, but they cover a much broader molecular weight range than individual columns. This can be advantageous in situations where a wider range of analyses need to be covered by a

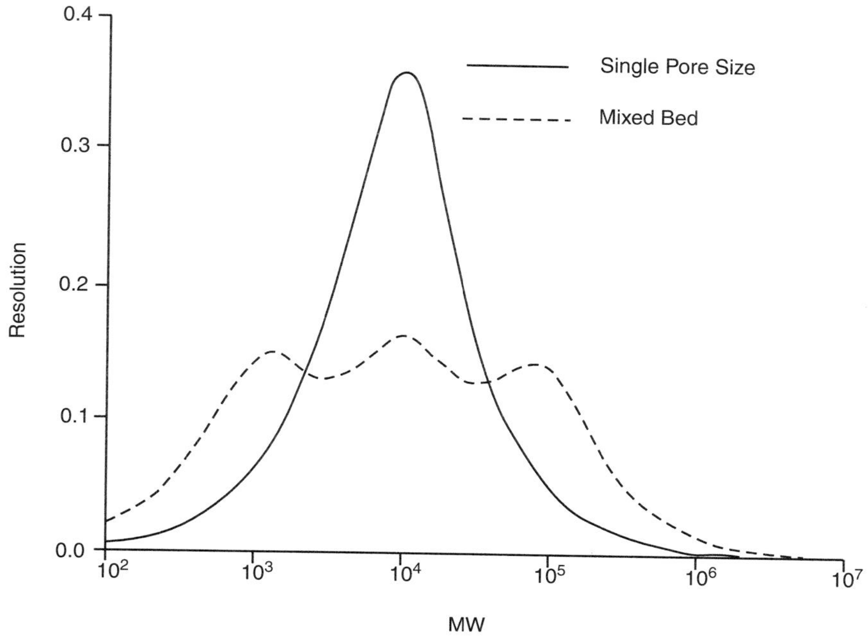

FIGURE 11.1 Resolution of a single pore size column and a mixed-bed column. Individual pore size columns deliver a significantly higher resolution than mixed-bed columns. The theoretical resolution of a single pore size column is compared to the resolution obtainable with a mixed-bed column. (Courtesy of Waters Corp.)

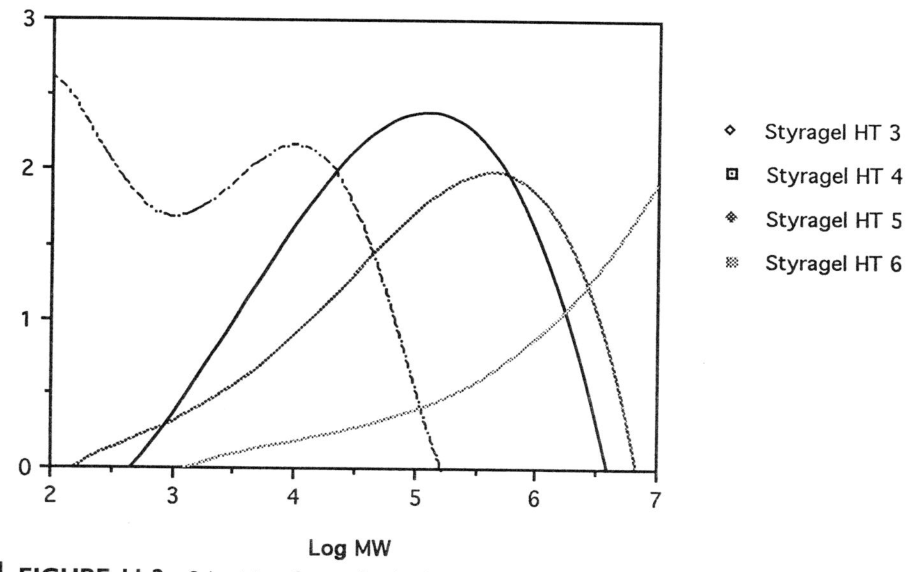

FIGURE 11.2 Selectivity of several individual pore size columns. This graph shows the resolution capability of Styragel HT 3, HT 4, HT 5, and HT 6 columns. The selectivity [Eq. (1)] of these columns vs the molecular weight of polystyrene is plotted. (Courtesy of Waters Corp.)

single set of columns in a single mobile phase. For a more detailed discussion of the compromises in the column choice, see Moore (2) and Neue (4).

C. Particle Size Selection

Several compromises are involved in the selection of the correct particle size. On one hand, one desires the highest possible resolution in the shortest amount of time. Therefore, the smallest particle size should be chosen that still gives resolution of the polymer without causing excessive column back pressure. On the other hand, there are constraints on both the strength of the particle and the strength of the polymer. This section discusses the selection of the best particle size.

The band spreading of peaks in SEC is described by a form of the van Deemter equation (5):

$$h = \frac{2 \cdot D_M \cdot (1 + k_e)}{u_i} + 2 \cdot d_p + 0.6 \cdot \frac{k_e}{(1 + k_e)^2} \cdot \frac{d_p^2}{D_M} \cdot u_i \qquad (2)$$

where h is the theoretical plate height, D_M is the diffusion coefficient in the mobile phase, u_i is the linear velocity, d_p is the particle size, and k_e is the partition coefficient for size exclusion chromatography. Because we want to minimize h in this equation, one needs to look at the parameters that can be changed by the user. The partition coefficient of the polymer is given by its nature and the choice of the pore size of the column. The diffusion coefficient of the sample varies over several orders of magnitude in most size exclusion separations. It is of course of concern to minimize its influence, but since the diffusion coefficient appears in both the denominator and the numerator of Eq. (2), the range of appropriate conditions is very broad. The same is true for the linear velocity. The dependence of h on the linear velocity and therefore the optimal linear velocity varies with the molecular weight of the polymer. Because the goal is trying to separate the polymer by molecular weight, it is best to preselect standard running conditions independently from the molecular weight of the polymer.

For most polymer analyses, choosing 5-μm particles is the best choice for the particle size. This particle size still gives low back pressure at reasonable run times for most polymer analyses, and column stability is not an issue. However, as the molecular weight of the analyte increases, the polymer becomes sensitive to shear and can degrade during transport through the column. This is especially an issue for ultrahigh molecular weight polymers such as ultrahigh molecular weight polyethylene. For this reason, packings based on larger particle sizes are preferred for this type of analysis. As a consequence, packings with a particle size of 20 μm are offered for this type of analysis (Styragel HMW columns).

Similarly, for the analysis of polymers using high-temperature solvents, the important concern is column stability and durability. For this reason, 10-μm particles are the best column choice. Waters Styragel HT columns are designed for this kind of application. Similarly, these columns are also more tolerant to eluent changes. Therefore, these columns are also recommended for room temperature applications, where the user needs to change eluents frequently. Hence, the decision of the particle size for a particular analysis is a trade-off between maximum resolution as offered by the Styragel HR line, column ruggedness as offered by the Styragel HT line, and avoidance of polymer degradation as offered by the Styragel HMW line of packings.

D. Narrow-Bore Columns

Generally, size exclusion chromatography is carried out using columns with an internal diameter of 7.8 mm. However, some SEC applications require the use of expensive solvents. For this purpose, size exclusion columns with a smaller internal diameter (4.6 mm) have been developed. Of course one should use proportionally lower flow rates with narrow-bore columns. If the standard column size uses a flow rate of 1 ml/min, then the smaller 4.6-mm columns should be used at a flow rate of 0.35 ml/min. This provides the same linear velocity as 1 ml/min on 7.8-mm columns. The decreased flow rate reduces solvent consumption and solvent disposal cost. The performance of the smaller diameter columns is not compromised if properly optimized instrumentation is used.

Smaller diameter columns are especially useful when expensive solvents are used. Figure 11.3 shows the analysis of poly(1,4-butylene terephthalate) using a Waters Alliance narrow-bore GPC system, quantitated against narrow polymethylmethacrylate standards. In this case, the solvent used is hexafluoro-2-isopropanol with 0.05 *M* sodium trifluoroacetic acid at a flow rate of

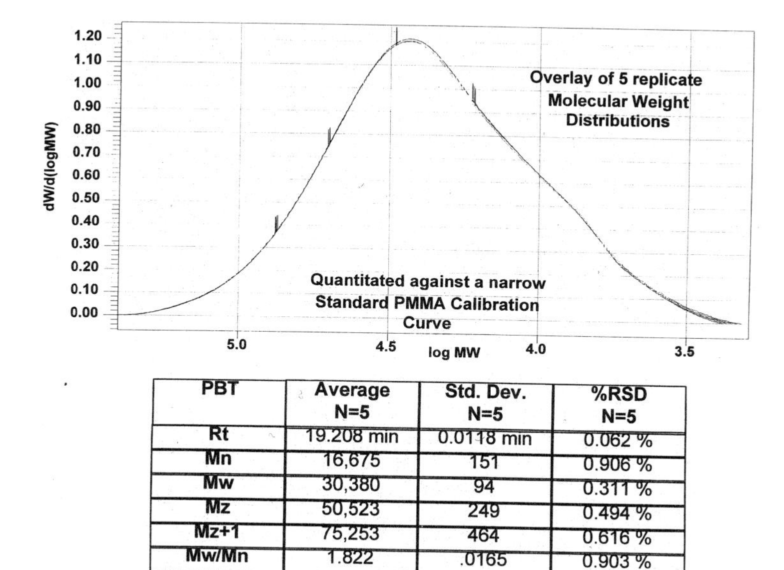

PBT	Average N=5	Std. Dev. N=5	%RSD N=5
Rt	19.208 min	0.0118 min	0.062 %
Mn	16,675	151	0.906 %
Mw	30,380	94	0.311 %
Mz	50,523	249	0.494 %
Mz+1	75,253	464	0.616 %
Mw/Mn	1.822	.0165	0.903 %

FIGURE 11.3 Analysis of poly(1,4-butylene terephthalate) using a Waters Alliance narrow-bore GPC system. Columns: 4.6 × 300 mm Styragel HR 2, HR 3, and HR 4. Mobile phase: hexafluoroisopropanol, 0.35 ml/min at 30°C. (Chromatogram courtesy of Peter Alden, Waters Corp.)

0.35 ml/min and at a temperature of 30°C. The columns used were narrow-bore Styragel HR 2, HR 3, and HR 4 columns that were specially packed for use with hexafluoro-2-isopropanol. An overlay of five replicate molecular weight distributions is shown. The relative standard deviation of the MW distribution is 0.31%, whereas RSDs for the number-average distribution M_n and z-averaged distribution M_z were 0.91 and 0.49, respectively.

Narrow-bore columns are most useful for the analysis of polymers that are difficult to analyze in inexpensive solvents. However, if the appropriate equipment is available, good results can be obtained for a broad range of standard analyses. A comparison of an analysis of standards between an equivalent bank of conventional 7.8-mm and solvent efficient 4.6-mm columns is shown in Fig. 11.4. The columns used were Styragel HR 0.5, 1, 2, and 3 columns at 35°C with tetrahydrofuran (THF) as the solvent. The flow rate was 1 ml/min for the conventional columns (Fig. 11.4A) and 0.35 ml/min for the solvent-efficient 4.6-mm columns (Fig. 11.4B). If the correct equipment is available, the reduced solvent consumption of these solvent-efficient Styragel columns is of value to the environmentally conscious user.

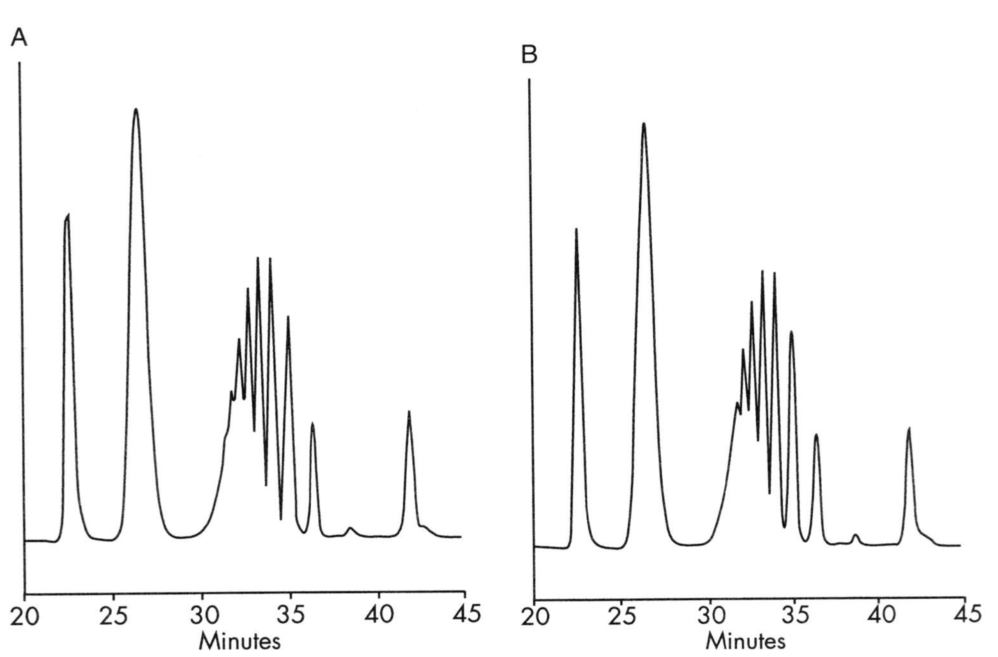

FIGURE 11.4 Comparison of chromatograms obtained on conventional (A) and solvent-efficient Styragel columns (B). In each case the column bank was a bank of Styragel HR 0.5, HR 1, HR 2, and HR 3 columns at 35°C with THF as the solvent. The sample is a mixture of polystyrene standards. With proper care and optimized instrumentation, good resolution can be obtained with solvent-efficient Styragel columns. (Courtesy of Waters Corp.)

III. WATERS PRODUCTS FOR SIZE EXCLUSION CHROMATOGRAPHY

This section discusses in detail the column types that are available for the size exclusion chromatography of both polar and nonpolar analytes. It first discusses the various columns available for standard nonaqueous size exclusion chromatography. It then reviews the columns available for general size exclusion chromatography using aqueous mobile phases. Finally, it examines the columns designed for size exclusion chromatography of proteins and peptides.

A. The Styragel Line of Size Exclusion Columns

The classical packing for size exclusion chromatography is Styragel. It is offered in a wide range of pore sizes, including mixed pore size packings. This packing is used for the broadest scope of size exclusion applications, extending from the analysis of additives to the size exclusion chromatography of ultrahigh molecular weight polyethylene.

There are basically three ranges of analytical problems that Styragel columns are capable of solving. For low molecular weight analysis and general high-resolution analysis, the Styragel HR line of columns is recommended. For high-temperature applications, ruggedness, and excellent column lifetime, Styragel HT columns are the best choice. For the analysis of ultrahigh molecular weight compounds that are susceptible to shearing, Styragel HMW columns are most useful. The range of packings offered is demonstrated in the column selection guide shown in Fig. 11.5, which also serves as a guide to the different types of Styragel packings.

Styragel columns are available not only in a broad range of pore sizes, but they are also shipped in three different solvents. See Section III,A,5 for more details.

1. Styragel HR Columns for High-Resolution SEC

Styragel HR columns are the workhorse columns for the analysis of low molecular weight samples. They are designed for the high-resolution analysis of oligomers, epoxies, or polymer additives where resolution is critical. This family of packings contains the largest group of different pore size options, especially for the analysis of smaller molecules. The calibration curves are shown in Fig. 11.6. There are seven individual pore size columns in this family of packings. In addition, two mixed-bed columns are available. The separation range for these 5-μm columns extends from oligomers to a styrene molecular weight of approximately 10^7 Da. These columns are recommended for the small to midrange molecular weight region.

The Styragel HR family contains a large number of packings with a small pore size. They are aimed at the resolution of low molecular weight components such as oligomers, additives, or epoxy resins. An example of a chromatogram of an amine-based epoxy resin is shown in Fig. 11.7. The chromatogram was obtained using a bank of Styragel HR 0.5, HR 1, HR 2, and HR 3 columns. As shown, individual peaks are obtained for different molecular weights con-

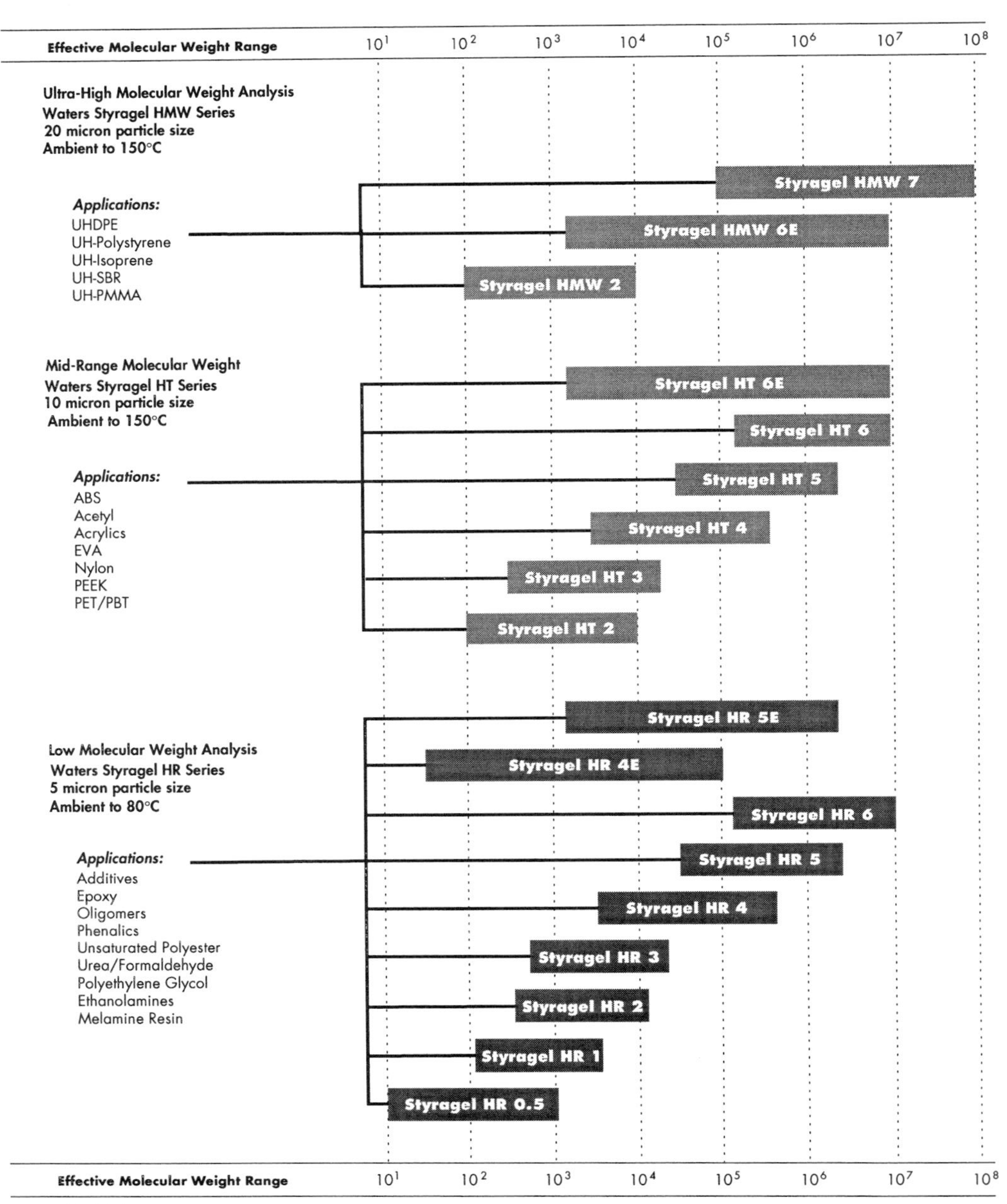

FIGURE 11.5 Column selection guide for Styragel columns. Columns are grouped into three categories: low molecular weight analysis, midrange molecular weight analysis, and ultrahigh molecular weight analysis. Typical applications are listed for each type of columns. (Courtesy of Waters Corp.)

tained in the raw material. This is important for the quality control of the composition of the epoxy resin. In general, Styragel HR columns are designed for this kind of high-resolution analysis.

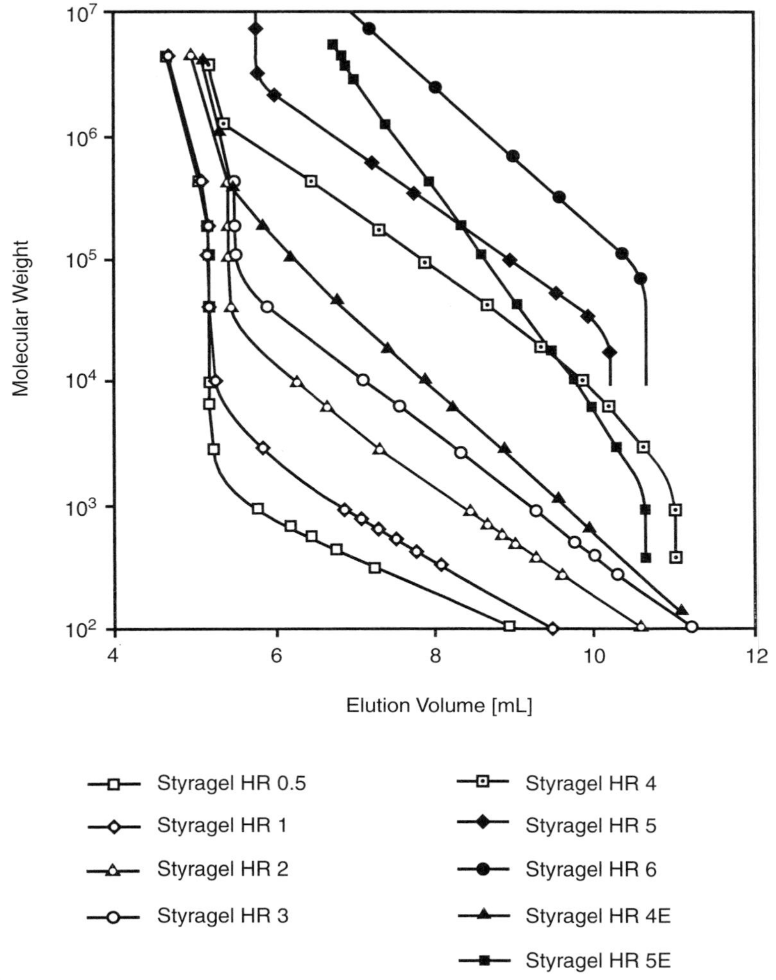

FIGURE 11.6 Calibration curves of Waters Styragel HR columns. (Courtesy of Waters Corp.)

2. Styragel HT Columns for High-Temperature SEC

Styragel HT columns are the workhorse columns for size exclusion chromatography. They can be used at ambient and high temperature and with aggressive solvents. They are designed for use in the mid-to-high molecular weight range, especially for applications requiring high-temperature solvents. Calibration curves for Waters Styragel HT columns are shown in Fig. 11.8. The focus of this group of products is the general-purpose size exclusion chromatography of a broad range of polymers. Therefore, five different types of packings are available, with pore size ranges from low to very high molecular weight. In addition, a mixed bed reaching from a molecular weight of 5000 to a molecular weight of 10,000,000 is available as a general purpose workhorse column. A common column choice for many analyses is a combination of one Styragel

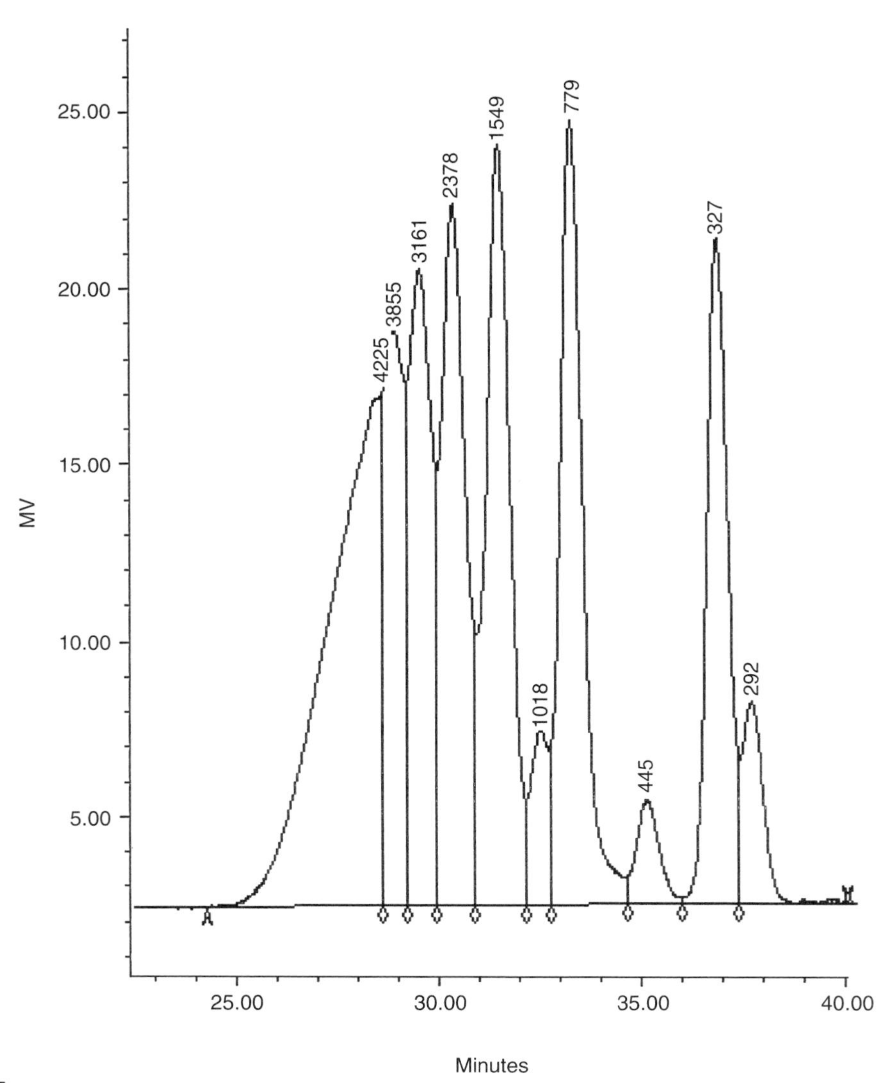

FIGURE 11.7 Chromatogram of an amine-based epoxy resin. The analysis of an amine-based epoxy resin was obtained on a column bank of 7.8 x 300 mm Styragel HR 0.5, HR 1, HR 2, and HR 3 columns. (Courtesy of Waters Corp.)

HT 2 column with two Styragel HT 6E columns. While such a combination does not provide the highest resolution analysis, it is the best scouting tool for unknown samples. The best column combination can then be chosen for the routine analysis of the polymer.

It is especially important to note that Styragel HT columns can be purchased in three different solvents. Conversion procedures for these columns to most of the other commonly used SEC solvents are available in the care and use manual of the Styragel HT column line. Although the same conversion procedures are suitable for other Styragel columns as well, the column stability of

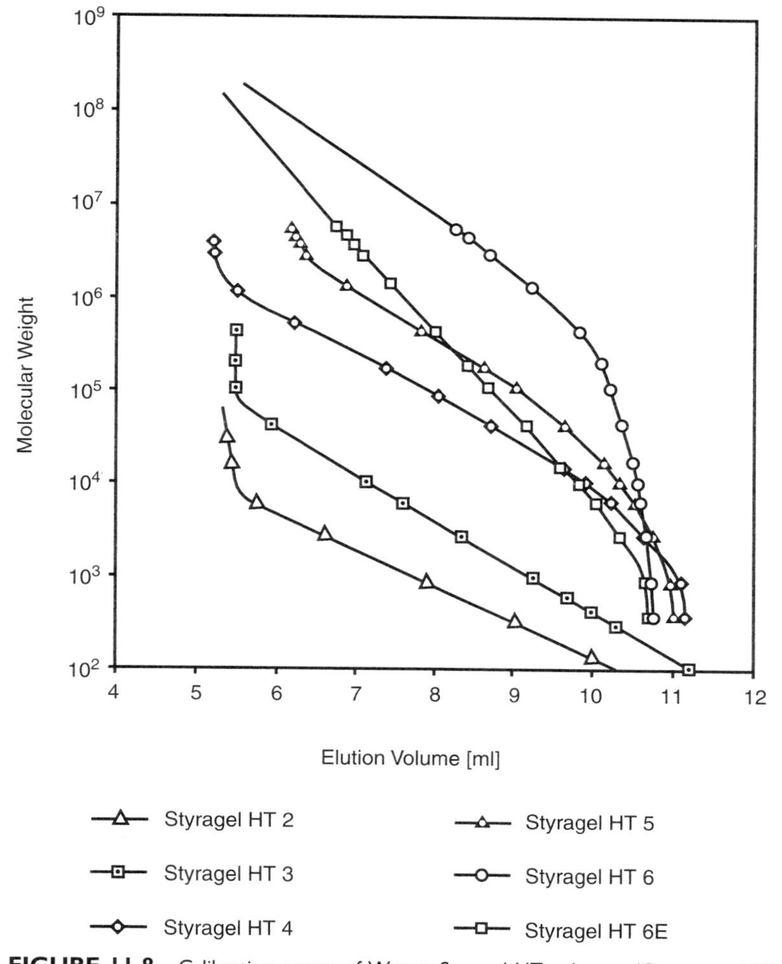

FIGURE 11.8 Calibration curves of Waters Styragel HT columns. (Courtesy of Waters Corp.)

Styragel HT columns has been optimized to tolerate a conversion to all commonly used SEC solvents.

3. Styragel HMW Column for the SEC of Ultrahigh Molecular Weight Polymers

The size exclusion chromatography of ultrahigh molecular weight polymers requires the use of a larger particle size than standard size exclusion chromatography. Ultrahigh molecular weight polymers can be susceptible to shearing with standard columns. The large particle size of Styragel HMW columns (20 μm), together with a large, high-porosity 10-μm frit, eliminates the shear effect encountered with smaller particle columns and makes it possible to analyze molecular weight distributions up to a styrene molecular mass of over 10^8 Da. The calibration curves of Styragel HMW columns are shown in Fig.

11.9. It should be pointed out though that the analysis of polymers of that large a molecular weight requires special precautions in dissolution and handling.

4. Styragel Guard Column

A general-purpose Styragel guard column is available for the protection of Styragel columns. To minimize the influence of the guard column on the size exclusion analysis, the column dimensions are 4.6 × 30 mm. The same guard column can be used with any Styragel SEC column. This standardization of the guard column to a single type as well as the choice of the smaller diameter eliminates the hassle of the selection of the guard column type.

5. Column and Eluent Selection Guides

Styragel columns can be used in a wide range of organic solvents. Elution protocols have been worked out for all polymers that are soluble in organic

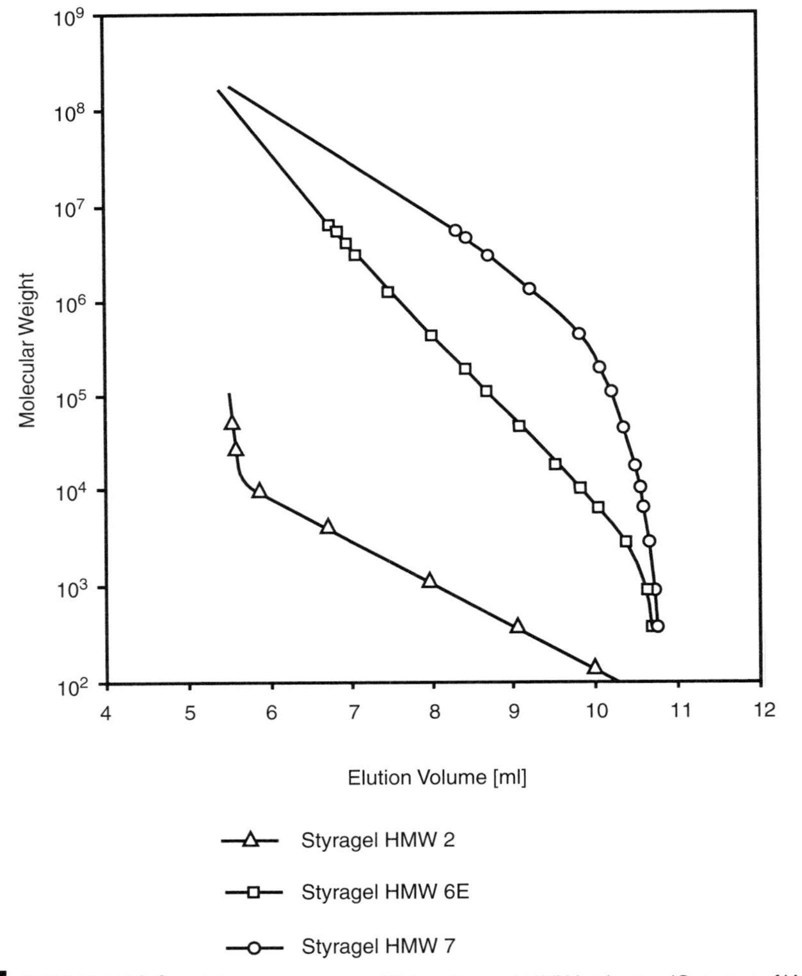

FIGURE 11.9 Calibration curves of Waters Styragel HMW columns. (Courtesy of Waters Corp.)

solvents. Columns are shipped in toluene, tetrahydrofuran, or dimethylformamide. Many applications are run in these solvents. A few applications require special solvents. An example is the high-temperature applications in trichlorobenzene at 140°C for polyolefins and related polymers or phenol/trichlorobenzene 1:1 at 145°C for PEEK, but the conversion of columns to these solvents is very straightforward. Alternatively, columns shipped in dimethylformamide can be changed to dimethylacetamide for the analysis of poly(furan-formaldehyde) resins. Detailed protocols have been worked out that are available in the care/use manuals for Styragel columns. Also, a solvent selection and conversion chart has been developed for the use of Styragel columns with a broad range of polymers. Detailed information is contained in Fig. 11.10, the solvent selection and conversion chart, which contains the column shipping solvent as well as the SEC solvent for a broad range of polymers.

6. Care and Maintenance

Styragel columns are compatible with most solvents commonly used in size exclusion chromatography. Exceptions are found on both sides of the polarity scale; the use of standard general-purpose Styragel columns with aliphatic hydrocarbons or with alcohols (except hexafluoroisopropanol) and water is generally not recommended. However, it is possible to pack columns in special solvents for special-purpose applications. The interested user should contact Waters for additional information.

Styragel columns can be used with a broad range of solvents and at elevated temperature. They can be converted to different solvents and temperatures following the general guidelines given in this section. Detailed conversion protocols are available in the care and use manual.

Generally, conversion from one solvent to another is carried out at low flow rates. The commonly used flow rate for this conversion is 0.2 ml/min for standard columns and 0.1 ml/min for solvent-efficient columns. This minimizes any swelling/shrinking stress put on the column. The temperature of a solvent conversion is chosen to minimize any pressure stress on the column bank. As a general rule, the pressure per column should never exceed 3.5 MPa (500 psi) during solvent conversion. For example, the conversion of a column bank from toluene to trichlorobenzene (TCB) or o-dichlorobenzene (ODCB) is commonly carried out at 90°C. This minimizes the stress on the column due to the higher viscosity of the target solvents.

All Styragel columns can be used for high-temperature SEC, but the Styragel HT line of columns has been optimized for this type of application. After a conversion to the typical solvents used for this type of SEC, Styragel SEC columns can be used for extended periods of time at these conditions without detriment. Commonly used conditions are at temperatures of around 140 to 150°C using solvents such as TCB or ODCB, but higher temperatures have also been used without difficulties. Figure 11.11 shows an example. Low-density polyethylene was analyzed at 140°C using a set of three Styragel HT 6E columns. In addition to the molecular weight distribution, information on the branching of the polymer was obtained by viscometry detection.

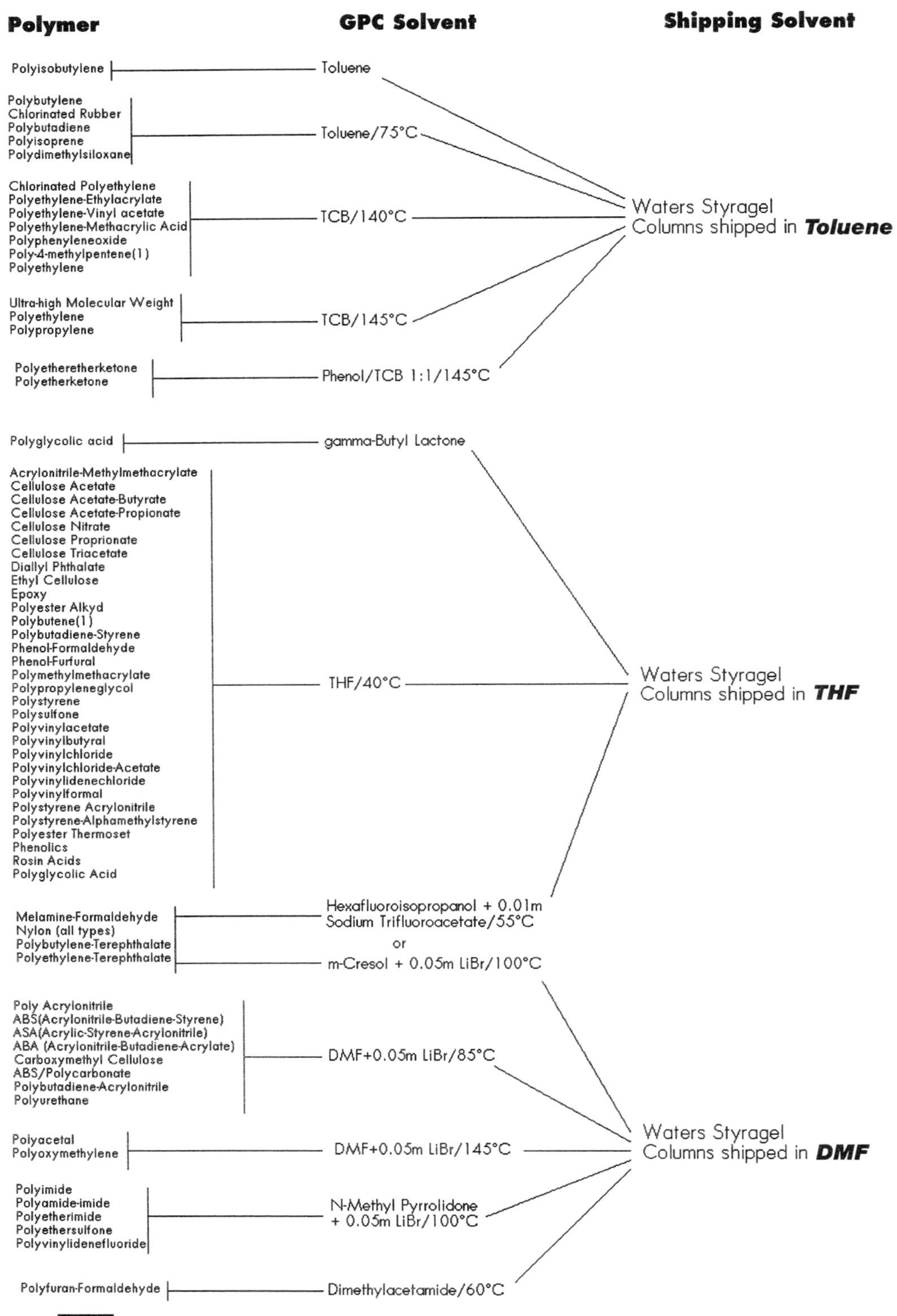

FIGURE 11.10 Solvent selection and conversion chart for Waters Styragel columns. (Courtesy of Waters Corp.)

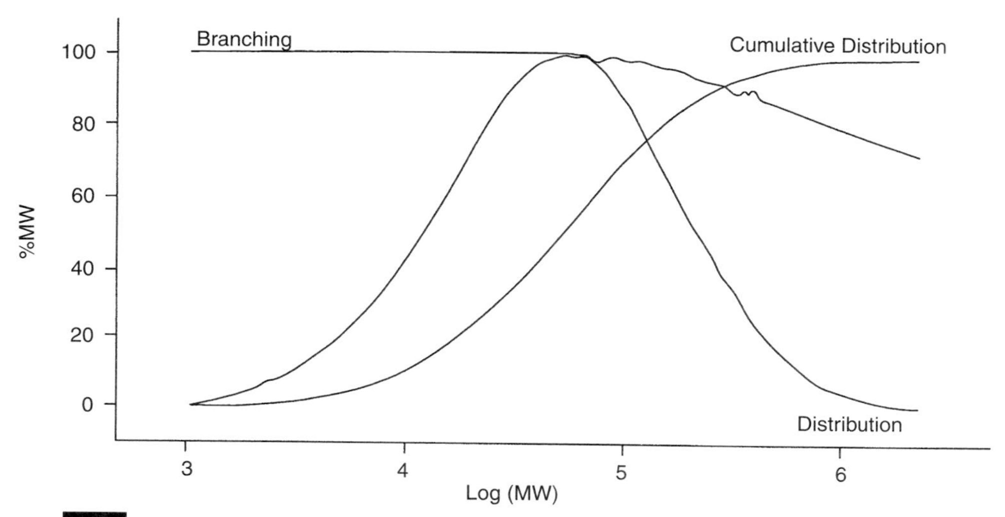

FIGURE 11.11 High-temperature size exclusion chromatography of low-density polyethylene. This SEC profile of low-density polyethylene was obtained at 140°C using a bank of three Styragel HT 6E columns with a mobile phase of TCB. Polymer branching information was obtained by viscometry detection. (Courtesy of Waters Corp.)

The dissolution of high molecular weight analytes increases the viscosity of the samples. Samples with a high viscosity can exhibit viscous fingering on the column top. Because this phenomenon can impair resolution, it is recommended to keep the sample concentration below certain limits. These limits are shown in Table 11.5.

B. The Ultrahydrogel Line of Size Exclusion Columns

Ultrahydrogel columns are size exclusion columns that are designed to be used with aqueous eluents. The packing is a highly hydroxylated polymethacrylate that is compatible with high concentrations of organic solvents (up to 50%).

TABLE 11.5 Maximum Allowable Sample Concentrations as a Function of Molecular Weight Range

Molecular weight range	Sample concentration (%)
0–25,000	<0.25
25,000–200,000	<0.1
200,000–2,000,000	<0.05
>2,000,000	<0.02

I. Column and Eluent Selection Guides

Due to the fact that the primary structure of the Ultrahydrogel packing is a hydroxylated methacrylate, the interaction of many polar polymers with the packing is minimized easily. The presence of small amounts of anionic functions on the surface of the polymer usually requires the addition of salt to the mobile phase. A common mobile phase for many applications is 0.1 M $NaNO_3$. Detailed eluent selection guidelines are given in Table 11.6.

The polymer types analyzed by aqueous size exclusion cover anionic and cationic polyelectrolytes as well as nonionic polymers with a limited level of hydrophobicity. Ionic interactions between the polymer and the packing material as well as between the polymer molecules themselves can be suppressed by increasing the ionic strength of the mobile phase. Generally, 0.1 M sodium nitrate is a good solvent for many polymers. However, higher concentrations of salt are recommended for some ionic polymers. Hydrophobic interactions can be suppressed by the addition of small amounts of an organic solvent. In many cases, 20% acetonitrile is sufficient to suppress the hydrophobic interaction. In a few cases, the suppression of hydrophobic interaction and the suppression of ionic effects are simultaneously necessary. In such a case, a high ionic

TABLE 11.6 Eluent Selection Chart for Ultrahydrogel Columns

Class	Polymer	Eluent
Nonionic	Polyethylene oxide	0.1 M $NaNO_3$
	Polyethylene glycol	
	Polysaccharides	
	Pullulans	
	Dextrans	
	Cellulosics	
	Polyvinyl alcohol	
	Polyacrylamide	
Nonionic hydrophobic	Polyvinyl pyrrolidone	80/20 0.1 M $NaNO_3/CH_3CN$
Anionic	Polyacrylic acid (Na)	0.1 M $NaNO_3$
	Polyalginic acid (Na)	
	Hyaluronic acid	
	Carrageenan	
Anionic hydrophobic	Polystyrene sulfonate (Na)	80/20 0.1 M $NaNO_3/CH_3CN$
	Lignin, sulfonated	
Cationic	DEAE-dextran	0.8 M $NaNO_3$
	Polyvinylamine	
	Polyepiamine	0.1 M TEA
	N-Acetylglucosamine	0.1 M TEA/1% AcOH
Cationic hydrophobic	Polyethyleneimine	0.5 M NaOAc/ 0.5 M AcOH
	Poly(*N*-methyl-2-vinylpyridinium) salt	
	Chitosan	0.5 M AcOH/0.3 M Na_2SO_4
	Peptides	0.1% TFA/40% CH_3CN
Amphoteric	Collagen/gelatin	80/20 0.1 M $NaNO_3/CH_3CN$

strength is combined with the addition of an organic solvent such as acetonitrile. Alternatively, acetic acid or acetate buffer can be used.

Calibration curves for the Ultrahydrogel column family using using polyethylene oxide standards and water as the mobile phase are shown in Fig. 11.12.

2. Care and Maintenance

Generally, Ultrahydrogel columns are compatible with aqueous mobile phases from pH 2 to pH 12. The addition of organic modifiers such as acetonitrile is recommended up to 50%, but most common applications do not require such a high concentration of organic modifier. These columns have been tested in applications from 10 to 80°C using aqueous mobile phases.

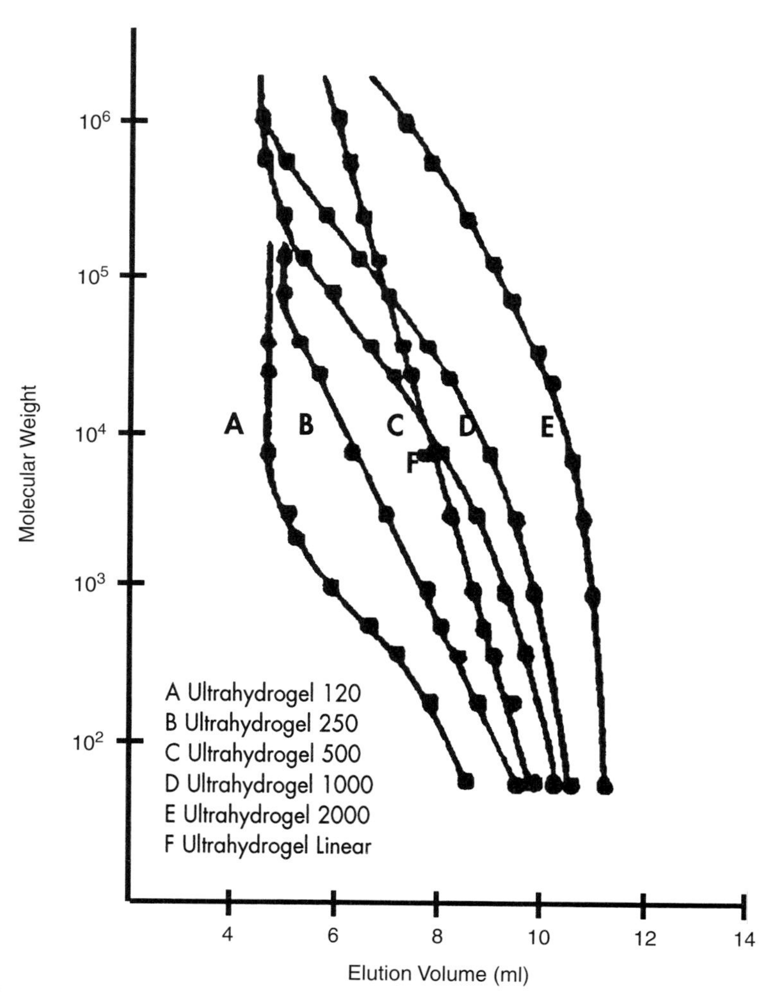

FIGURE 11.12 Calibration curves of the Ultrahydrogel family of columns. (Courtesy of Waters Corp.)

It should be pointed out up front that guard columns are available for this column series. Although the use of guard columns is not a common practice in size exclusion chromatography, it is highly recommended as many inadvertent contaminants can be removed easily using guard columns. Ultrahydrogel guard columns can be regenerated with high pH or low pH buffer, organic solvents, urea, and other mobile phases.

Nonionic samples can generally be analyzed without an adjustment of the pH or the salt concentration of the mobile phase. However, many typical samples are ionic or ionogenic. Under these circumstances, the addition of salt to the mobile phase is often required to prevent exclusion effects that are not related to the size of the analyte molecule. Ultrahydrogel columns are compatible with a broad range of salt and buffer solutions. Recommended compositions can be found in Table 11.6, but a broader range of buffers can be used.

Solvent conversion of columns designed for aqueous size exclusion chromatography is rarely a problem. However, it should always be carried out at slow flow rates. For Ultrahydrogel columns, the recommended flow rate for a solvent conversion is below 0.3 ml/min. One should typically use 0.1 ml/min for most solvent conversion procedures.

For long-term column storage, it is recommended to replace the mobile phase with distilled water with 0.05% sodium azide. The columns can then be stored at room temperature. For shorter storage times, e.g., a long weekend, the column should be stored in the mobile phase after making sure that the end plugs of the columns have been tightened well.

C. The Protein-Pak Line of Size Exclusion Columns

Several silica-based columns are available for the size exclusion chromatography of proteins: Protein-Pak 60, Protein-Pak 125, and Protein-Pak 300SW. Packings are based on different pore size silicas that have been derivatized with glycidyl functions. These functions shield the interaction of protein molecules with the silica surface. Therefore, these packings can be used for the size exclusion separation of proteins with aqueous mobile phases containing 0.1 *M* salt solutions. Good recovery of active protein is the consequence, and many practical applications of these packings are preparative.

1. Description of Columns

An important difference between Protein-Pak columns and other size exclusion columns is the silica backbone of the Protein-Pak columns. Because the silica structure is unaffected by the solvent, these columns do not swell or shrink as a function of the solvent. This is a general advantage compared to other size exclusion columns. However, silica-based columns can only be used up to pH 8, which limits their applicability. Also, surface silanols are accessible for interaction with the analytes, but this phenomenon has been minimized by proper derivatization techniques. Generally, a small amount of salt in the mobile phase eliminates interaction with silanols.

2. Special Considerations for the Size Exclusion Chromatography of Proteins

For the size exclusion chromatography of proteins on silica-based diol packings, it is generally recommended to use fully aqueous mobile phases with a salt concentration between 0.1 and 0.3 *M*. In general, a phosphate buffer around pH 7 is used as the mobile phase. Under these circumstances, the tertiary structure of most proteins is preserved without difficulty and the interaction of proteins with each other is minimized. However, other inorganic buffers or combinations of buffers with organic solvents can be used without difficulties for special applications.

3. Care and Maintenance

Generally, the care and maintenance of these columns are easier compared to polymer-based packings due to the fact that there is no swelling and shrinking associated with silica-based phases. The compatibility with solvents is very broad. A clear exception is the stability at high pH values. Alkaline pH values beyond pH 8 should be avoided. Otherwise, Protein-Pak columns are compatible with many salt and buffer solutions such as phosphate, acetate, citrate, formate, or biobuffers. It should be pointed out that the storage of the columns over extended periods of time in highly concentrated salt solutions is not recommended. Also, the formation of microbial contamination can be avoided by storage in an aqueous solution containing 0.05% sodium azide.

Columns can be washed with solvents and solvent combinations suitable to remove adsorbed contaminants. When considering the adsorption of analytes, think not only of the diol functionality, but also of the adsorption to residual silanols. Often, the injection of small amounts (500 μl) of dimethyl sulfoxide removes contamination that has accumulated on the column. Aqueous solutions of sodium dodecyl sulfate, guanidine hydrochloride, or urea are compatible with Protein-Pak columns.

The pressure capability of silica-based columns extends beyond the range of organic polymers. Protein-Pak columns are stable to above 25 MPa (4000 psi), the smaller pore sizes to above 40 MPa (6000 psi).

IV. SUMMARY

This chapter gives a summary of the size exclusion columns currently available from Waters Corporation. It includes a description of the principles used to select the column type as well as a discussion of the properties and use of different columns. A few application examples are included as well. This review is designed to give the user a good overview over the properties and capabilities provided by Styragel, Ultrahydrogel, and Protein-Pak columns.

TRADEMARKS

Waters, Styragel, and Ultrahydrogel are registered trademarks of Waters Corporation. Protein-Pak is a trademark of Waters Corporation.

REFERENCES

1. Malawer, E. G. (1995). Introduction to size-exclusion chromatography. *In* "Handbook of Size-Exclusion Chromatography (Chi-san Wu, Ed.), Dekker, New York.
2. Moore, J. C. (1964). *J. Polym. Sci. A* **2**, 835.
3. Neue, U. D. (1997). "HPLC Columns: Theory, Technology and Practice." Wiley, New York.
4. Neue, U. D. (1994). Waters Column V.1, 21–24.
5. Groh, R., and Halász, I. (1981). *Anal. Chem.* **53**, 1325–1335.

12 SIZE EXCLUSION CHROMATOGRAPHY COLUMNS FROM POLYMER LABORATORIES

ELIZABETH MEEHAN

Polymer Laboratories, Shrophshire SY6 6AX, United Kingdom

I. INTRODUCTION

Polymer Laboratories was founded in Shropshire, England in 1976 to produce columns and polymer standards for size exclusion chromatography (SEC). From a small operation situated on the site of the Rubber and Plastics Research Association (now RAPRA Technology) the company expanded quickly, making it necessary to set up its own United Kingdom facility on the current site. The product range of the company has diversified over the years, taking advantage of expertise in other areas of polymer particle technology, such as high-performance liquid chromatography (HPLC) packings and bulk media, polymeric supports for peptide synthesis, and latex particles. However, SEC has remained the major line of business for Polymer Laboratories, reinforced by significant investment in SEC instrument development, which has seen the release of SEC systems and detectors since the early 1990s. The company has offices in the United States (Amherst, MA), the Netherlands, and Germany. In 1998 Polymer Laboratories was accredited with the ISO 9001 quality standard for the production and development of all its polymer-based products.

For organic SEC separations the use of polystyrene/divinylbenzene (PS/DVB) particles is almost universal throughout the industry. Polymer Laboratories PS/DVB material, PLgel, which is produced in a series of individual pore sizes, formed the basis for the original product line of SEC columns. Developments in the refinement of particle sizing introduced the benefits of smaller particle size and more efficient columns, which significantly reduced SEC analysis time through a reduction in the number of columns required for a high-resolution separation. The concept of mixed gel columns, although not new as Yau had already introduced the idea for silica based packings (1), was commercialized for PS/DVB packings in 1984. As the benefits of this approach to column technology became increasingly apparent, Polymer Laboratories introduced a complete range of PLgel mixed gel columns in 1990.

Traditional methods of preparation of PS/DVB particles generally involved a suspension polymerization to produce macroporous particles with a relatively narrow range of porosity but with a very wide range of particle sizes that must be refined to produce distributions narrow enough to successfully pack into SEC columns. More recently there have been several alternative preparation techniques described in the literature that produce PS/DVB beads with quite different physical characteristics. Polymer Laboratories have seized on these new particle technologies for specific applications where the resultant particles can offer advantages to the chromatographer and one example of this, PL HFIPgel, is described later in this chapter.

The field of aqueous SEC is of growing interest and also represents an area of active development by several companies, with Polymer Laboratories included. In this area, contrary to the organic SEC scenario, a variety of polymer chemistries are employed in the preparation of the column packing materials. This fact alone makes the practice of aqueous SEC more difficult than its organic counterpart and requires the suppliers of such columns to offer a good level of technical support to the column user. This chapter outlines the characteristics and applications of PL aquagel-OH aqueous SEC columns.

II. PLgel ORGANIC SIZE EXCLUSION CHROMATOGRAPHY (SEC) COLUMNS

PLgel consists of macroporous, spherical PS/DVB particles produced by suspension polymerization (2). The porosity of the beads is controlled by the use of various diluent systems and the final product exhibits a very high degree of cross-linking. The physical and chemical stability of the beads rely heavily on the fact that microporous structure is avoided. Subsequent to the suspension polymerization process, PLgel materials are washed thoroughly to remove any residual surfactants that could potentially render the surface of the beads more polar and induce nonsize exclusion effects for certain types of analysis or cause column stability problems, particularly in polar solvents. This factor can sometimes account for chromatographic variability observed between PLgel columns and PS/DVB columns supplied by other vendors.

A. PLgel Individual Pore Size Columns

A range of individual pore size PLgel packing materials is produced and their pore size distribution is conveniently represented by a SEC calibration curve as illustrated in Fig. 12.1. It should be pointed out that the descriptors used for the different pore sizes, 50 Å, 100 Å, and so on, are not the actual pore sizes of the beads but relate to the size of a polystyrene molecule just excluded from the packing material. This nomenclature comes from the original work carried out by Moore (3) and should only be viewed in the context of differentiating

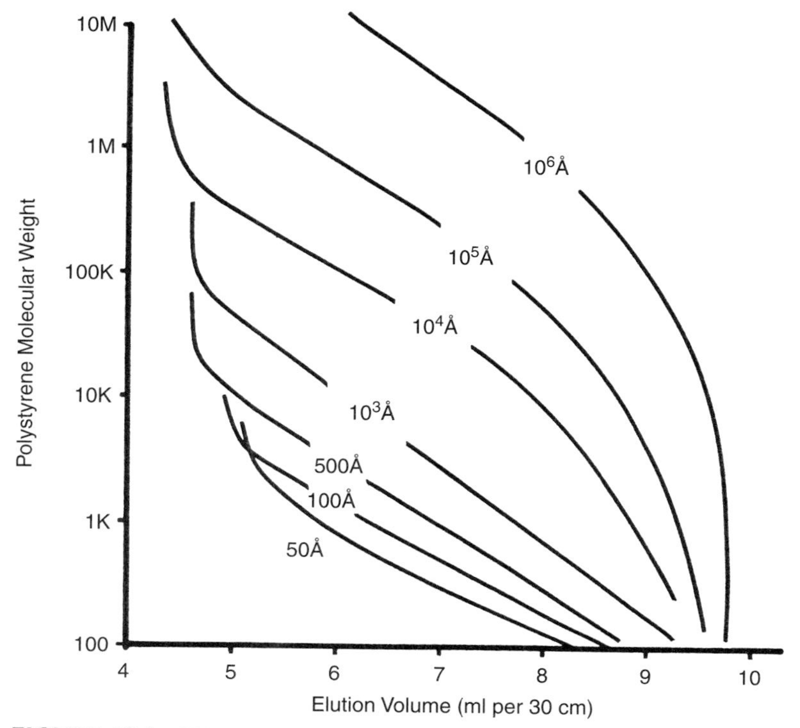

FIGURE 12.1 SEC calibration curves for PLgel individual pore size columns (300 × 7.5 mm), eluent THF at 1.0 ml/min, polystyrene calibrants.

between the individual pore sizes in the family of products. The SEC calibration curve, and the information derived from it, is, however, a useful reference for the performance criteria of the column. In each case the parameters of interest are (1) the exclusion limit, in terms of polystyrene molecular weight, which indicates the upper molecular weight limit of resolution; (2) the pore volume, V_p, which can be calculated as the difference between the total permeation volume (V_t) and the exclusion volume (V_o); and (3) the effective operating range of the column where the curve is relatively linear and has a shallow slope.

Exclusion limits and practical operating ranges specified for PLgel individual pore size columns are summarized in Table 12.1. Table 12.1 illustrates the versatility of this family of columns. Pore volume can also be expressed as

$$V_p = \frac{(V_t - V_o)}{V_t} \times 100\%,$$

and for PLgel columns the actual pore volume varies from one pore size to another between 40 and 55%, being lower at the extremes of the range and maximum in the midrange of the pore sizes. This characteristic is inherent in the manufacturing process and reflects a compromise between optimum resolution and enhanced column durability. For consistent column performance these just-mentioned parameters must be controlled carefully and, in practice,

TABLE 12.1 Operating Ranges for PLgel Individual Pore Size Columns

Column type	Exclusion limit (polystyrene equivalent)	Operating range (polystyrene equivalent)
PLgel 50Å	2,000	0–2,000
PLgel 100 Å	4,000	0–4,000
PLgel 500 Å	30,000	500–30,000
PLgel 10E3 Å	60,000	500–60,000
PLgel 10E4 Å	600,000	10,000–600,000
PLgel 10E5 Å	2,000,000	60,000–2,000,000
PLgel 10E6 Å	10,000,000	600,000–10,000,000

each individual pore size PLgel has a specification and tolerance assigned to them. For example, if the pore volume were too low then resolution would be reduced, whereas if pore volume were too high the beads may suffer from poor mechanical stability and columns may be prone to void.

Each of the PLgel individual pore sizes is produced by suspension polymerization, which yields a fairly diverse range of particle sizes. For optimum performance in a chromatographic column the particle size distribution of the beads should be narrow; this is achieved by air classification after the cross-linked beads have been washed and dried thoroughly. Similarly, for consistent column performance, the particle size distribution is critical and is another quality control aspect where both the median particle size and the width of the distribution are specified. The efficiency of the packed column is extremely sensitive to the median particle size, as predicted by the van Deemter equation (4), whereas the width of the particle size distribution can affect column operating pressure and packed bed stability.

The excellent chemical and physical stability of PLgel columns arises from the high degree of polymer cross-linking in the beads and, to some extent, from the column packing technique. The degree of cross-linking can be assessed very simply by measuring the volume change of the beads when exposed to solvents of varying polarity. All polymers swell to some extent, even when cross-linked, and this swelling will be at a maximum in a solvent with a solubility parameter similar to that of the polymer. The key to optimizing column performance is to minimize the change in swell as a function of solvent polarity; this aspect of column design has been addressed with PLgel beads and is again a quality-controlled parameter.

PLgel columns are produced using a slurry packing technique (5) with packing pressures of the order of 2200–3000 psi, depending on the media in question. In SEC operation it is vital that the packing pressure of the column is not exceeded as this will result in voiding of the column. Therefore it is important to measure the physical stability of a packed column in order to ensure that it will operate successfully within the guidelines issued to the user. The packing techniques employed for PLgel ensure a high packed bed density and a guaranteed maximum operating pressure of 2200 psi (150 bar), which is a quality-controlled parameter.

B. PLgel Mixed Gel Columns

Any individual pore size SEC column has a finite molecular weight operating range where the resolution is maximized because the calibration curve is relatively linear and at its shallowest slope. The range of most individual pore size columns is limited to 1–2 decades of molecular weight, which for some SEC applications is adequate. However, SEC is commonly used for the characterization of polydisperse materials, which require resolution over several decades of molecular weight; in these cases the traditional approach has been to combine individual pore size columns together in a column bank to increase the overall resolving range. There are two major drawbacks (6) to this approach.

1. Individual pore size columns have variable pore volume, and because the column dimensions are fixed, the combination of different columns must result in variable slope of the overall calibration curve and hence variable degrees of resolution as a function of molecular weight.
2. The user must have a good understanding of the individual calibration curves for each individual pore size column in order to avoid mismatching them, which can result in artifacts in polymer distributions.

In order to circumvent these aspects of column selection, as well as make life easier for the user, the concept of mixed gel columns was introduced. Here a homogeneous mixture of varying quantities of individual pore size beads is blended together and packed so as to produce a column that exhibits a linear calibration curve over a specified range of molecular weight and in which there is no mismatch of pore sizes, so avoiding spurious chromatographic peaks for polydisperse samples. The resultant mixed gel column has a similar pore volume to an individual pore size column, around 45–50%, but operates over a much extended range of molecular weight. By selecting specific mixtures of different pore size gels, Polymer Laboratories has developed a whole family of PLgel mixed gel columns, designated A to E, whose calibration curves are shown in Fig. 12.2. The exclusion limits and operating ranges have been designed for specific application areas and are summarized in Table 12.2. It should also be noted that the particle size of the beads used varies according to the type of mixed gel column. It has been shown (7) that when macromolecules pass through SEC columns packed with small particle size beads (5 μm and less) they can undergo mechanical shear degradation. Therefore, in PLgel mixed gel columns designed for higher molecular weight applications, larger particle size beads are employed to ensure that the polymer is not unduly affected by passage through the chromatographic system.

A linear calibration curve achieved using PLgel mixed gel columns infers several key advantages in SEC analyses.

1. Column selection is greatly simplified and the column is specifically designed for particular application areas.
2. The resolving capability is the same over the full operating range of the column. This offers an ideal situation for the analysis of very polydisperse polymers that may also contain lower molecular weight additives.
3. Overall resolution can be improved by simply adding another column of the same type, thus decreasing the slope of the calibration curve and increasing the efficiency of the system.

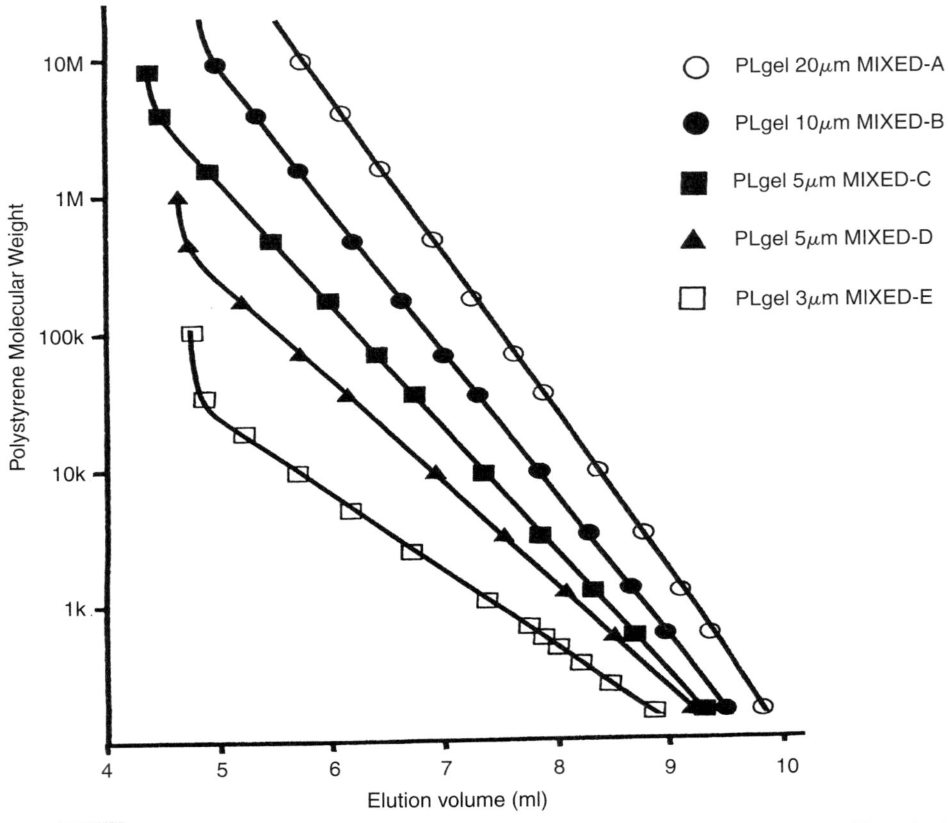

FIGURE 12.2 SEC calibration curves for PLgel mixed gel columns (300 × 7.5 mm), eluent THF at 1.0 ml/min, polystyrene calibrants.

4. Replacement inventory is reduced.

5. For SEC employing light scattering or viscosity detection it is often necessary to extrapolate the log M versus retention time data derived from the integration to account for the lack of signal sensitivity at one or another extreme of the distribution. If the characteristics of the column are known to be linear

TABLE 12.2 Operating Ranges for PLgel Mixed Gel Columns

Column type	Exclusion limit (polystyrene equivalent)	Linear operating range (polystyrene equivalent)
PLgel 20 μm MIXED-A	40,000,000	2,000–40,000,000
PLgel 10 μm MIXED-B	10,000,000	500–10,000,000
PLgel 5 μm MIXED-C	2,000,000	200–2,000,000
PLgel 5 μm MIXED-D	400,000	200–400,000
PLgel 3 μm MIXED-E	30,000	Up to 30,000

in this respect, the user can have greater confidence in making a linear extrapolation.

The use of mixed gel technology has become widespread since the late 1980s. If the column is well designed it should not be necessary to supplement resolution in certain areas of the operating range (particularly the extremes) by the addition of individual pore size columns. This practice is not necessary or recommended for PLgel mixed gel columns, although it is a common practice for other commercial products.

C. PLgel Column Performance Testing

PLgel beads are made in a batch process, and the characteristics of each batch of material are assessed and quality controlled as described previously. As PS/DVB is highly cross-linked and extremely stable, it therefore has a good shelf life and repeated assessment of material performance is not really an issue. Although the material properties remain consistent, by its very nature the column packing process can be susceptible to a number of variables which will ultimately affect the performance of the finished column. Therefore, every column that is packed must be tested individually to ensure a consistent product. The column performance, i.e., how well the column has been packed, is assessed by measuring the number of theoretical plates and the symmetry for a test probe eluting at total permeation (8). It is important to remember that this type of measurement will itself be affected by a number of chromatographic factors: flow rate, temperature, choice of test probe, and system dispersion. Therefore, it is vital to establish a protocol for this test that is rigidly adhered to in order to ensure consistent column performance.

PLgel columns are normally tested using tetrahydrofuran (THF) as the eluent at a flow rate of 1.0 ml/min and toluene as a test probe. The detector can be either a differential refractometer or a UV spectrophotometer at 254 nm, but the injection loop volume never exceeds 20 μl and all connections throughout the system comprise short lengths of 0.010-inch i.d. tubing. By definition the chromatographic system has very low dead volume and therefore low dispersion so that the measurement made is a true reflection of the column performance independent of the system. Table 12.3 summarizes the guaranteed minimum efficiency specifications for PLgel columns and in all cases, the column symmetry is specified to 0.90–1.35. Conventional PLgel columns are supplied in toluene as this is a very stable, less volatile solvent for shipping. However, the columns can be supplied in alternative solvents on request.

D. Column Selection and Operation

The main criterium for column selection is pore size distribution as it is desirable to have maximum pore volume for separation in the molecular weight range of interest. Having determined the upper molecular weight limit required, a column with a suitable exclusion limit should be selected. In the case of individual pore size columns, it is then a question of selecting other columns with complementary calibration curves to comprise a column set covering the re-

TABLE 12.3 Guaranteed Minimum Column Efficiencies for PLgel Columns

Column type	Guaranteed minimum efficiency (plates/meter)
PLgel 5 μm 50 Å	60,000
PLgel 5 μm 100 Å	60,000
PLgel 5 μm 500 Å	60,000
PLgel 5 μm 10E3 Å	50,000
PLgel 5 μm 10E4 Å	50,000
PLgel 5 μm 10E5 Å	50,000
PLgel 10 μm 50 Å	35,000
PLgel 10 μm 100 Å	35,000
PLgel 10 μm 500 Å	35,000
PLgel 10 μm 10E3 Å	35,000
PLgel 10 μm 10E4 Å	35,000
PLgel 10 μm 10E5 Å	35,000
PLgel 10 μm 10E6 Å	35,000
PLgel 20 μm MIXED-A	17,000
PLgel 10 μm MIXED-B	35,000
PLgel 5 μm MIXED-C	50,000
PLgel 5 μm MIXED-D	50,000
PLgel 3 μm MIXED-E	80,000

quired molecular weight range. With the advent of PLgel mixed gel columns, this selection process is very much simplified and the user needs only to consider what upper molecular weight limit is required and to choose a PLgel mixed gel column with an appropriate exclusion limit.

The particle size of the packing material will determine the efficiency of the column, and for improved resolution the user should select a smaller particle size column. Indeed, with the introduction of PLgel 3-μm MIXED-E columns, SEC is now able to compete with other HPLC techniques for the high-resolution separation of small molecules as illustrated in Fig. 12.3. However, as mentioned earlier, for the SEC analysis of high molecular weight polymers, columns packed with smaller beads should be avoided. Obviously columns packed with larger particle size beads have lower efficiency and it is therefore necessary to utilize more columns together in series in order to ensure the same degree of resolution.

Resolution in SEC can be described by a specific relationship

$$R_{sp} = \frac{0.25}{D\sigma},$$

where R_{sp} is specific resolution, D is the slope of the calibration curve ($d\log M/dt$), and σ is the peak variance (related to the peak width) of a single narrow standard eluting in the linear range of the calibration curve. R_{sp} values have

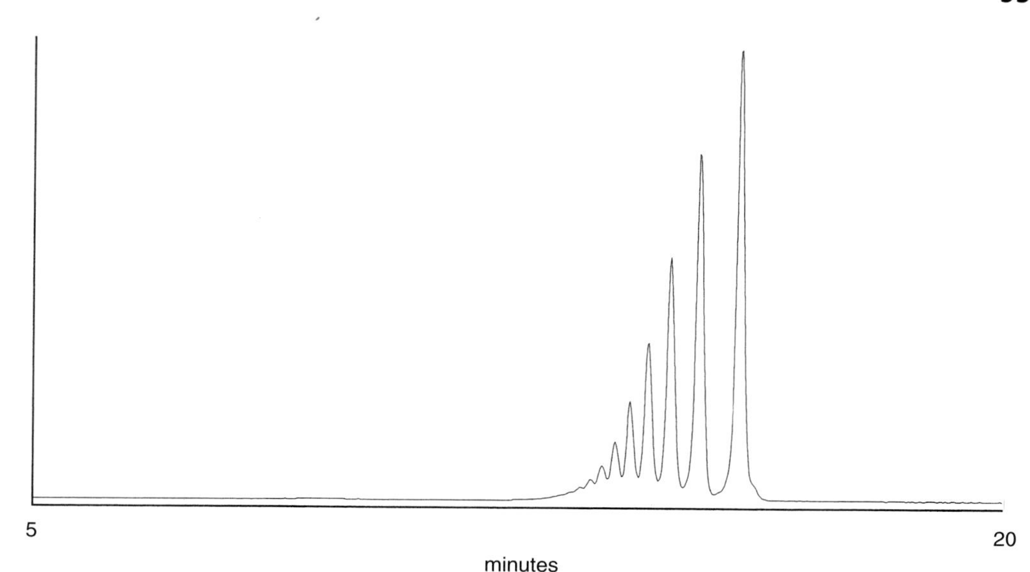

FIGURE 12.3 Separation of polyester oligomers, 2 × PLgel 3 μm MIXED-E 300 × 7.5 mm, eluent THF at 1.0 ml/min, evaporative light scattering detector

been determined for the PLgel mixed gel column series using a polystyrene narrow standard calibration curve and selecting a single polystyrene standard ($M_p = 9200$) as a test probe (9). The results, summarized in Table 12.4, suggest that column sets should comprise a minimum of one 3-μm column, two 5-μm columns, three 10-μm columns, and four 20-μm columns for equivalent resolution. Of course the specific resolution equation also indicates that more columns should be used in a column set for further improvement in resolution.

The optimum flow rate for most SEC separations using conventional PLgel column dimensions (internal diameter 7.5 mm) is 1.0 ml/min. It may be of some benefit to work with lower flow rates, particularly for the analysis of higher molecular weight polymers where the reduced flow rate improves resolution through enhanced mass transfer and further reduces the risk of shear

TABLE 12.4 Specific Resolution (R_{sp}) Calculations for PLgel Mixed Gel Columns

Column type	Peak variance (σ)	Calibration slope (D)	Specific resolution (R_{sp})
PLgel 20 μm MIXED-A	0.204	1.155	1.061
PLgel 10 μm MIXED-B	0.166	1.138	1.730
PLgel 5 μm MIXED-C	0.115	0.915	2.376
PLgel 5 μm MIXED-D	0.115	0.737	2.949
PLgel 3 μm MIXED-E	0.106	0.605	3.898

degradation. Increased flow rate is not common, but is permissible as long as the maximum allowable pressure on the PLgel column (150 bar) is not exceeded.

The choice of eluent for an SEC separation is determined primarily by the solubility of the sample under investigation, although secondary considerations are detector compatibility, hazard, cost, and so on. The choice of SEC eluent is becoming increasingly more diverse as the chemistry of new polymers develops, and SEC columns must be able to tolerate the demands of these more novel solvents. PLgel columns are compatible with solvents covering a very wide range of solvent polarity because the beads have a very high degree of cross-linking. Furthermore, the excellent column stability permits regular exchanges between solvents with very different polarity with no deterioration in column performance. Table 12.5 shows the diversity of eluents used in SEC applications with PLgel columns. Many of the more polar solvents exhibit relatively high viscosity and for these an elevated temperature operation is highly recommended in order to reduce the solvent viscosity, which will result in improved chromatography and reduced column operating pressure. In other applications the use of elevated temperature is required in order to achieve and maintain solubility of the sample. In this category the most typical examples are polyolefin polymers, but others exist generally in the field of engineering polymers which require even more aggressive solvents and ever increasing temperatures. For example, PLgel columns have been used routinely in the

TABLE 12.5 Range of Solvents Used Routinely with PLgel Columns

Solvent	Solubility parameter
Hexane	7.3
Cyclohexane	8.2
Toluene	8.9
Ethyl acetate	9.1
Tetrahydrofuran	9.1
Chloroform	9.3
Methyl ethyl ketone	9.3
Dichloromethane	9.7
Acetone	9.9
o-Dichlorobenzene	10.0
1,2,4-Trichlorobenzene	10.0
m-Cresol	10.2
o-Chlorophenol	10.2
Dimethyl acetamide	10.8
n-Methyl pyrolidone	11.3
Dimethyl sulfoxide	12.0
Dimethylformamide	12.1
Benzyl alcohol	12.1

analysis of poly(phenylene sulfide) using *o*-chloronaphthalene as the eluent at 210°C as illustrated in Fig. 12.4.

There is increasing interest in copolymer systems, which, due to their chemical heterogeneity, may require very complex eluent systems in order to dissolve the sample and ensure that the separation ensues by a pure size exclusion mechanism. In these examples, the PLgel is also compatible with eluent systems containing mixed solvents of different polarity (including water as a cosolvent up to 10% by volume) and in organic solvents modified with acids or bases (e.g., acetic or formic acid, triethanolamine) as it is stable in the pH range of 1–14.

III. PL HFIPgel COLUMNS

The use of hexafluoroisopropanol (HFIP) as an SEC eluent has become popular for the analysis of polyesters and polyamides. Conventional PS/DVB-based SEC columns have been widely used for HFIP applications, although the relatively high polarity of HFIP has led to some practical difficulties: (1) the SEC calibration curve can exhibit excessive curvature, (2) polydisperse samples can exhibit dislocations or shoulders on the peaks, and (3) low molecular weight resolution can be lost, causing additive/system peaks to coelute with the low molecular weight tail of the polymer distribution

These problems can be associated with the fact that various pore size SEC packings, either in a bank of individual pore size columns or in a mixed gel column set, respond differently in this extremely polar solvent. In addition the

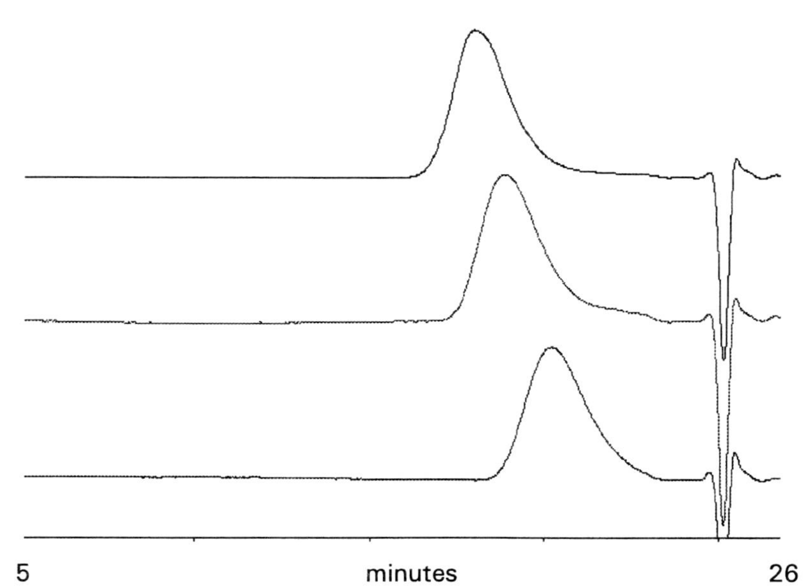

FIGURE 12.4 Analysis of three samples of poly(phenylene sulfide), 3 × PLgel 10 μm MIXED-B 300 × 7.5 mm, eluent *o*-chloronaphthalene, temperature 210°C, DRI detector.

relatively high viscosity of HFIP can give problems with high column pressure for small particle size, higher efficiency columns. It was because of these observations that a new particle technology was applied to produce columns more suitable for use with the HFIP eluent, namely PL HFIPgel columns.

PL HFIPgel is produced by a two stage polymerization process (10). First, polystyrene template particles are produced by dispersion polymerization. These uncross-linked particles are then brought into an aqueous emulsion of PS/DVB and initiator. In a diffusion controlled process in the presence of water-soluble and sterically hindered stabilizers, the template particles are swollen and cross-linked, resulting in beads that exhibit a uniform particle size but polydisperse porosity (11). Table 12.6 compares the properties of SEC columns packed from conventional PS/DVB beads and these novel PL HFIPgel beads. The narrow particle size distribution, compared with a conventional air-classified PLgel material, helps reduce column operating pressure while maintaining a high column efficiency. The resultant column has an extended operating range of molecular weight with a high pore volume as illustrated in the SEC calibration curve in the HFIP eluent shown in Fig. 12.5. As this material is derived from a single reaction, there is no combination of different pore sizes involved. Subsequently polymer distributions obtained using this type of column, as illustrated in Fig. 12.6, show high resolution over a wide range of molecular weight without any evidence of shoulders and with relatively low operating pressures. PL HFIPgel columns are shipped ready to use in HFIP solvent.

IV. PL aquagel-OH AQUEOUS SEC COLUMNS

Whereas for organic SEC column technology a particular type of bead (PS/DVB) is used almost universally, in the field of aqueous SEC there have been a variety of approaches to derive polymeric beads suitable for the application. For this reason there is more secrecy about the chemical composition of the packing materials and columns produced by different manufacturers.

The basic requirements for an aqueous SEC column are (1) the beads must exhibit an extremely hydrophilic surface chemistry, (2) the beads should exhibit

TABLE 12.6 Comparison of Packing Material Properties

Property	Conventional PS/DVB	PL HFIPgel
Particle size distribution	Polydisperse, requires refinement to give narrower fraction before use in column packing	Monodisperse as produced in the reactor
Pore size distribution	Useable range limited to 1–2 decades of molecular weight	Extended range covering 4 decades of molecular weight
Pore volume	Varies according to pore size, 40–55%	High pore volume, 50–55%
Application to polymer analysis	Combination of individual pore sizes (columns or gels) to cover required molecular weight range	Molecular weight range covered by single pore size

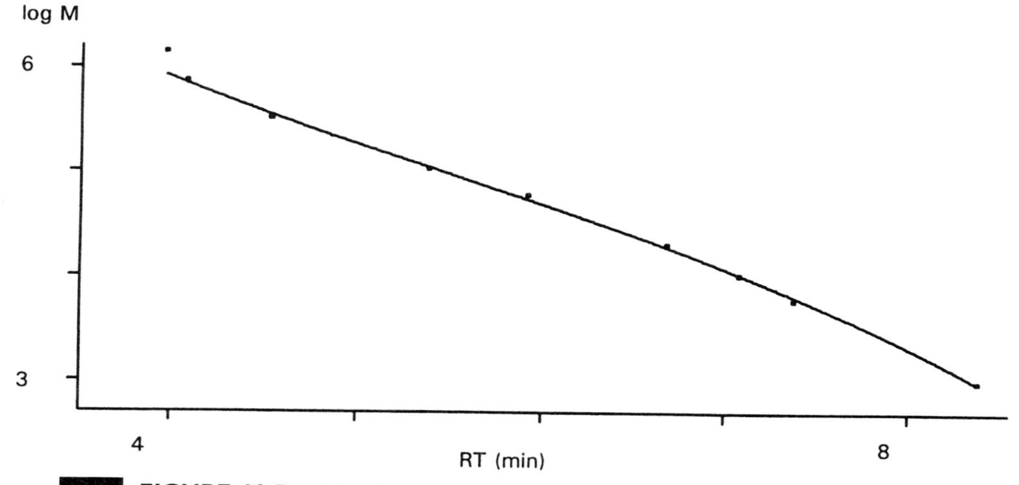

FIGURE 12.5 SEC calibration curve for PL HFIPgel 300 × 7.5 mm, eluent HFIP at 1.0 ml/min, temperature 40°C, polymethylmethacrylate calibrants.

minimal ionic or hydrophobic sites, (3) the column must have good chemical and mechanical stability, and (4) the column must have high efficiency.

PL aquagel-OH consists of macroporous, hydrophilic particles that exhibit a polyhydroxyl functionality. The columns exhibit extremely good stability during operation; they will tolerate an operating pressure up to 140 bar and an eluent composition containing organic modifier up to 50% by volume and eluent pH in the range of 2–10.

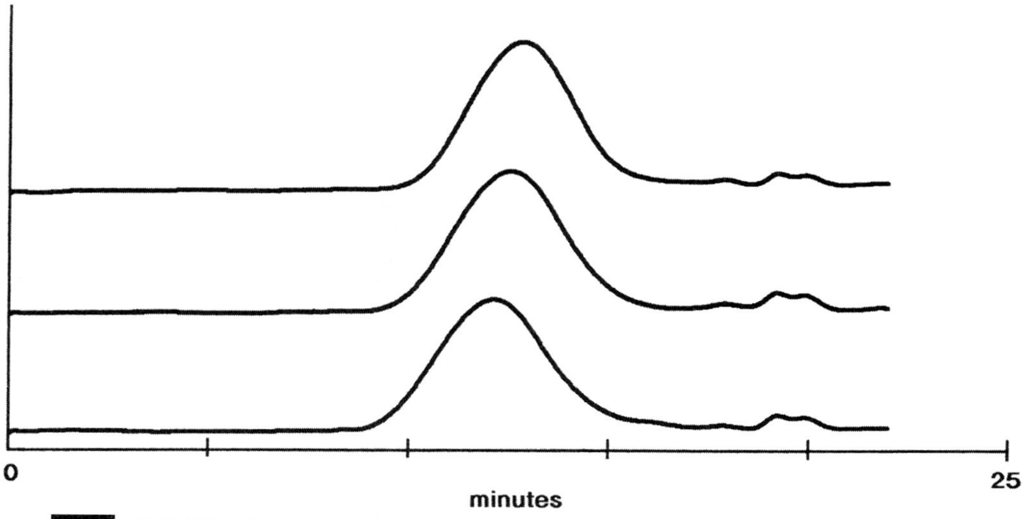

FIGURE 12.6 Analysis of three samples of PET, 2 × PL HFIPgel 300 × 7.5 mm, eluent HFIP at 1.0 ml/min, temperature 40°C, DRI detector.

PL aquagel-OH columns are available as individual pore sizes or in a mixed gel format as illustrated in the calibration curves shown in Fig. 12.7. Exclusion limits and operating ranges for this range of columns are expressed in terms of polyethylene oxide/glycol (PEO/PEG) standards. In choosing aqueous SEC columns it is important to remember that the calibration curves may be expressed in terms of PEO/PEG, polysaccharide, or protein molecular weight. As these reference polymers exhibit very different hydrodynamic volume the exclusion limits and operating ranges will vary significantly in molecular weight for the different types of calibrant.

PL aquagel-OH is available in two particle sizes: 8 μm is used in higher performance columns and 15 μm in columns designed for the analysis of very high molecular weight water soluble polymers where shear degradation becomes a concern. PL aquagel-OH columns are normally tested using water as the eluent at a flow rate of 1.0 ml/min and glycerol as a test probe. The detector used is a differential refractometer, the injection loop volume never exceeds 20 μl, and all connections throughout the system comprise short lengths of 0.010-inch i.d. tubing. By definition the chromatographic system has very low dead volume and therefore low dispersion so that the measurement made

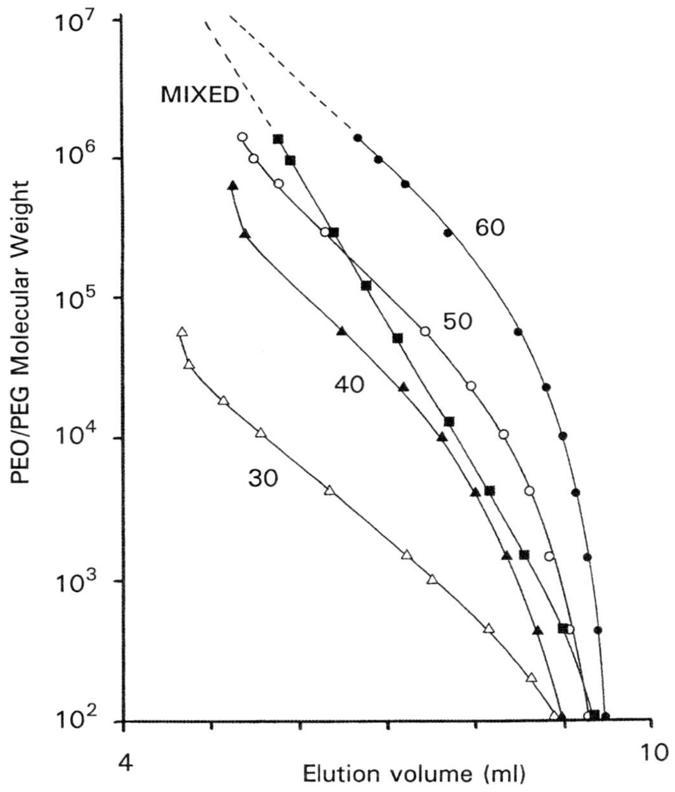

FIGURE 12.7 SEC calibration curves for PL aquagel-OH columns (300 × 7.5 mm), eluent water at 1.0 ml/min, polyethylene oxide/glycol calibrants.

is a true reflection of the column performance independent of the system. Table 12.7 summarizes the exclusion limits and operating ranges for the range of columns as well as the guaranteed minimum efficiencies. Many of the column selection criteria have already been covered in previous sections and apply equally to aqueous SEC separations as they do to organic SEC. However, a major topic for method development in aqueous SEC is eluent selection.

Water-soluble polymers can exhibit a very diverse range of properties, both in terms of their molecular weight distribution and, more importantly for SEC analysis, their chemistry. Both natural and synthetic water-soluble polymers can be either nonionic (neutral) or ionic (polyelectrolyte) and in turn may be either hydrophilic or relatively hydrophobic. These properties of the polymers, together with imperfections in the surface chemistry of the aqueous SEC packings, mean that in practice pure water is not an ideal eluent for SEC work. Modifications to both the ionic strength and the polarity of the eluent are used to stabilize the polymer solution as well as to suppress any sample to column interaction (12). Generally, ionic interactions are stronger and can be addressed by the addition of salts/buffers and the adjustment of pH in the eluent. Cationic polymers are the most difficult to analyze, requiring fairly extreme conditions to counteract the sample–sample and sample–column ionic effects. Weaker but still significant, hydrophobic interactions can be suppressed by adding organic modifier to the eluent, and in the case of PL aquagel-OH columns, methanol can be used to this effect up to 50% by volume in the eluent. Although it is impossible to state a suitable eluent for every polymer type, general rules can be applied to generic classes of polymers as summarized in Table 12.8.

V. COLUMN HARDWARE

Polymer Laboratories employs their own individual design of column hardware, which was developed to give low dispersion, high-performance SEC columns. The components used in the design are specified to very low tolerances dimen-

TABLE 12.7 PL aquagel-OH Column Specification

Column type	Operating range (PEO/PEG equivalent)	Guaranteed minimum efficiency (plates/meter)
PL aquagel-OH 30 8 μm	100–30,000	35,000
PL aquagel-OH 40 8 μm	10,000–200,000	35,000
PL aquagel-OH 50 8 μm	50,000–1,000,000	35,000
PL aquagel-OH 60 8 μm	200,000–10,000,000	35,000
PL aquagel-OH MIXED 8 μm	100–10,000,000	35,000
PL aquagel-OH 40 15 μm	10,000–200,000	15,000
PL aquagel-OH 50 15 μm	50,000–1,000,000	15,000
PL aquagel-OH 60 15 μm	200,000–10,000,000	15,000

TABLE 12.8 Typical Eluent Systems for Generic Types of Water-Soluble Polymers

Polymer type	Typical examples	Suitable eluent for use with PL aquagel-OH columns
Nonionic, hydrophilic	Polyethylene oxide, polyethylene glycol	Pure water
	Polyviny alcohol, hydroxyethyl cellulose, polyacrylamide	0.1–0.2 *M* salt/buffer, pH 7
Nonionic, hydrophobic	Polyvinylpyrrolidone	0.1–3 *M* salt/buffer with 20% methanol
Anionic, hydrophilic	Sodium polyacrylate, sodium hyaluronate, carboxymethyl cellulose	0.1–0.3 *M* salt/buffer, pH 7–9
Anionic hydrophobic	Sodium polystyrene sulfonate	0.1–0.3 *M* salt/buffer, pH 7–9 with 20% methanol
Cationic, hydrophilic, and hydrophobic	Chitosan, poly-2-vinyl pyridine	0.3–1.0 *M* salt/buffer, pH 2–7 with the addition of methanol for more hydrophobic polymers

sionally as well as being made from the highest quality materials. The porosity of the frit used in the column depends entirely on the particle size of the beads packed in the column, the larger the particles the greater the porosity of the frit. This ensures that the frit holds back the particles without imparting additional shear on the polymer solution as it passes through the column, a very important consideration for SEC analysis of very high molecular weight polymers.

A. Narrow-Bore SEC Columns

Conventionally, analytical SEC columns have been produced with an internal diameter of 7.5 mm and column lengths of 300 and 600 mm. In recent years environmental and safety issues have led to concerns over the reduction of organic solvent consumption, which has resulted in the development of columns for organic SEC that are more solvent efficient (13). By reducing the internal diameter of the column, the volumetric flow rate must be reduced in order to maintain the same linear velocity through the column. This reduction is carried out in the ratio of the cross sectional areas (or internal diameters) of the two columns. For example, if a 7.5-mm i.d. column operates at 1.0 ml/min, then in order to maintain the same linear velocity through a 4.6-mm i.d. column the flow rate would be

$$1.0 \text{ ml/min} \times \frac{(4.6^2)}{(7.5^2)} = 0.38 \text{ ml/min}.$$

PLgel mixed gel packings are available in 250 × 4.6-mm columns, referred to as PLgel MiniMIX columns, and are designed to reduce organic solvent consumption. Because of the reduced internal diameter and the slightly reduced column length, a PLgel MiniMIX 250 × 4.6-mm column operated at 0.3 ml/min will give the same performance as a conventional PLgel mixed 300 × 7.5-mm column operated at 1.0 ml/min.

Several issues need to be considered when operating narrow-bore SEC columns.

1. The solvent delivery system must be capable of pumping solvent at a constant, pulse-free flow rate of 0.3 ml/min. Traditionally, SEC pumps have been optimized to operate at 1.0 ml/min.

2. The system dead volume must be reduced to an absolute minimum, particularly when using very efficient narrow-bore SEC columns. Extra column dispersion becomes a greater consideration as the column volume is reduced, and dead volume should be minimized in all parts of the system, including injection valves, connecting tubing, and detectors, if the column performance is to be realized.

3. Sample loading must be reduced in accordance with the column inside diameter. Polymers exhibit high solution viscosity, and in order to avoid band broadening due to viscous streaming the sample concentration must be reduced for narrow-bore columns. Overloading effects become noticeable at much lower concentrations using 4.6-mm columns compared to 7.5-mm columns because of the effective sample concentration in a smaller volume column.

B. Preparative SEC Columns

Preparative SEC provides a simple and relatively rapid method for the fractionation and isolation of components in a sample. Although a certain level of resolution is obviously required, the aim then is to increase the loading on the column in order to maximize the yield of each fraction with each pass through the column. For this reason the particle size of the beads used to manufacture preparative SEC columns tends to be large as this permits the use of higher loadings. In addition, the dimensions of the column are also increased as this permits an increase in sample loading. As preparative SEC columns are relatively expensive and quite large volumes of solvent are consumed in their operation, method development can be carried out conveniently using 7.5-mm i.d. columns packed with the same beads and then scaled up in terms of sample loading and eluent flow rate.

Preparative SEC columns from Polymer Laboratories are packed in 25-mm internal diameter tubing, 300 or 600 mm long, using 10-μm beads. Compared to the 7.5-mm i.d. column, the sample loading can be increased in terms of both injection volume and sample concentration, although the extent of the latter parameter depends largely on the molecular weight of the compounds of interest. Typically the injection volume can be increased by 10 to 20 times (2000-μl loops are commonly used) and the concentration by as much as 10 times, depending on the sample molecular weight. As the column inside diameter is increased, the volumetric flow rate through the column must be increased in order to maintain the same linear velocity as a 7.5-mm i.d. column, a similar but opposite case to that of narrow-bore columns. Therefore, preparative SEC separations are carried out at flow rates of around 10.0 ml/min to equate to the 7.5-mm column separation performed using a flow rate of 1.0 ml/min.

VI. SUMMARY

Although the design of modern size exclusion chromatography columns still retains some of the basic polymer science founded in the 1960s, there have been some significant developments in particle technology that have expanded the scope of SEC for both organic and aqueous applications. Increased activity in the field of polymer research places greater demands on SEC columns as the technique continues to provide the polymer scientist with a very useful tool for characterization. Polymer Laboratories has a very well established, successful product range but is continuing to explore new areas of column development where they may be useful in solving some of the more complex SEC applications problems.

REFERENCES

1. Yau, W. W., Ginnard, C. R., and Kirkland, J. J. (1978). *J. Chromatogr.* **149**, 465–487.
2. Seidl, J., Malinsky, J., Dusek, K., and Heitz, W. (1967). *Adv. Polym. Sc.* **5**, 113.
3. Moore, J. C. (1964). *J. Polym. Sci. A* **2**, 835.
4. Yau, W. W., Kirkland, J. J., and Bly, D. D. (1979). *In* "Modern Size-Exclusion Liquid Chromatography," p. 63. Wiley, New York.
5. Ravindranath, B. (1989). *In* "Principles and Practice of Chromatography," p. 317. Ellis Horwood, Chichester.
6. Warner, F. P., Dryzek, Z., and Lloyd, L. L. (1986). "New Criteria Influencing the Selection of High Performance GPC Columns for Polymer Analysis." Presented at Antec, Boston, MA.
7. Meehan, E., and O'Donohue, S. J. (1992). "The Role of Column and Media Design in the SEC Characterisation of High Molecular Weight Polymers." Presented at ISPAC 5, Inuyama, Japan.
8. Bristow, P. A. (1976). *In* "Liquid Chromatography in Practice," p. 16. Hept, UK.
9. Meehan, E., McConville, J. A., Oakley, S. A., and Warner, F. P. (1991). "Performance Criteria for Mixed Gel GPC Columns." Presented at the International GPC Symposium, San Francisco, CA.
10. Wang, Q. C., Svec, F., and Frechet, J. M. (1992). *Polym. Bull.* **28**, 569–576.
11. Meehan, E., Oakley, S. A., and Warner, F. P. (1996). "The Application of a Novel Particle Technology for GPC Using HFIP as Eluent." Presented at the International GPC Symposium, San Diego, CA.
12. Barth, H. G. (1980). *J. Chromatogr. Sci.* **18**.
13. Meehan, E., Oakley, S. A., Warner, F. P., (1992). "Narrow Bore Columns for Size Exclusion Chromatography." Presented at Pittcon '92, New Orleans.

13 JORDI GEL COLUMNS FOR SIZE EXCLUSION CHROMATOGRAPHY

HOWARD JORDI

Jordi Associates, Inc., Bellingham, Massachusetts 02019

I. INTRODUCTION

The technique of size exclusion chromatography (SEC) has developed into a highly polished series of techniques for analyzing polymers of many types. Most of the initial gels used were based on polystyrene/polydivinylbenzene (PDVB) copolymers for nonaqueous applications. Even today most nonaqueous gels are advertised as copolymers of styrene and divinylbenzene, although other types are available as well as polymethylmethacrylate gels and gels based on polyvinylalcohol, for example. The early history of SEC is given in Cantow (1) where the basic process for gel manufacture is described in some detail. Because the gels described are copolymers of styrene and divinylbenzene, all the gels produced will tend to swell and shrink with changes in mobile-phase polarity. It seemed logical that a series of rigid polydivinylbenzene gels made with as pure a divinylbenzene as possible might largely solve the problems of softer, less cross-linked gels. Figure 13.1 shows the results of running a PDVB column and a polystyrene/polydivinylbenzene copolymer column in tetrahydrofuran before and after exposure to methanol. The copolymer bed collapsed while the PDVB bed was unaffected. In fact, PDVB gels can be run in polar solvents such as methanol, acetonitrile, 2-propanol, and acetone without harmful effects.

Underivatized PDVB gels are generally unsuitable, however, for work in aqueous mobile-phase systems because (1) the gels cannot be wetted to pack in water and (2) organic materials will absorb irreversibly to the gels in aqueous systems. Work on this problem is continuing by first trying to modify the

Column Handbook for Size Exclusion Chromatography

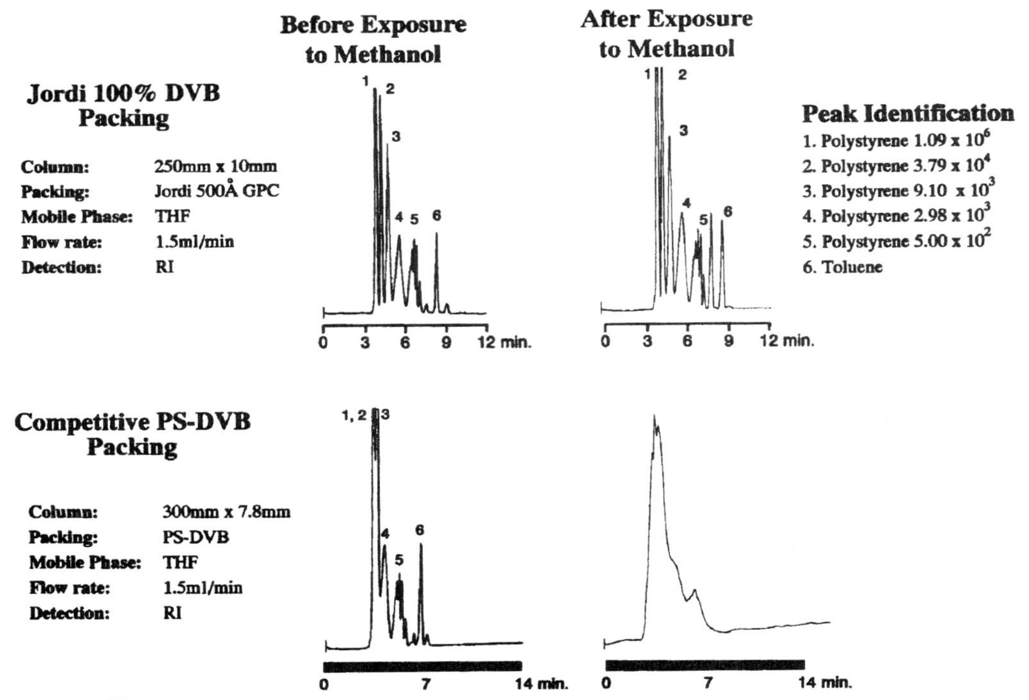

FIGURE 13.1 Initial Chromatograms were obtained using THF as mobile phase. Each column was then stressed by an abrupt change of mobile phase from THF to methanol. Methanol was run through each column for approximately 16 hours, at which time the mobile phase was changed back to THF.

surfaces of PDVB gels in such a way as to make them compatible with aqueous mobile phases and buffered systems that can survive in virtually any pH from 0 to 14 and second by trying to make the modifications in such a way that adsorption of organics is reduced greatly or eliminated.

Subsequent sections describe and illustrate in general terms how PDVB gels are produced as well as how some of the bonded phases have been developed at Jordi Associates.

II. THE BASIC POLYDIVINYLBENZENE PACKING

Because most widely used methods used to prepare classical styrene/divinylbenzene copolymers have always been based on suspension polymerization, it seemed logical that a series of porous PDVB gels using similar methodologies could be developed. In suspension polymerization, divinylbenzene is suspended as a dispersion of small droplets in a continuous phase of water and polymerized by classical free radical initiation. This process produces the spherical beads routinely observed. The suspension is stabilized by the addition of one or more suspending agents, depending on the particular gel being manufactured. Surfactants are also used to add a surface charge to the forming gel beads, which also aids in stabilization of the suspension.

Gels made in this way have virtually no usable porosity and are called Jordi solid bead packings. They can be used in the production of low surface area reverse phase packings for fast protein analysis and in the manufacture of hydrodynamic volume columns as well as solid supports for solid-phase syntheses reactions. An example of a hydrodynamic volume column separation is shown in Fig. 13.2 and its calibration plot is shown in Fig. 13.3. The major advantage of this type of column is its ability to resolve very high molecular weight polymer samples successfully.

For the production of permanently porous gels used in the manufacture of 10^5-, 10^4-, 10^3-, and 500-Å gels it is also necessary to add one or more inert diluents to the suspension polymerization mixture. These diluents must have three properties: they must be miscible with the divinylbenzene monomer, they must not react during the polymerization, and they need to be removable/extractable from the gels formed. The production of large pores requires the use of diluents that are good solvents for the monomer, in this case divinylbenzene, and poor solvents for the forming polymer. This causes large pools of diluent to be trapped in the forming polymer, which is later extracted, yielding the desired large pores. Examples include isooctane, 1- or 2-butanol, and t-amyl alcohol. However, the production of small pores requires the use of diluents that are good solvents for both the divinylbenzene monomer and the forming polymer. Because of this, no large pools of diluent are formed in the gel and the pores tend to be quite small. Examples here would be toluene,

Hydrodynamic Volume Separation
Polystyrene Standards

DESCRIPTION
Packing Material: (4)Jordi Gel DVB Solid Bead
Length: 500mm ID: 10mm

TEST CONDITIONS
Mobile Phase: Chloroform
Flow Rate: 1.5ml/min
Detector: UV 254nm
Temperature: 25°C
Pressure: 2900 PSIG
Sensitivity: 0.2 AUFS

Sample: PS **Inj. vol.: 10µl**

Component	Conc (mg/ml)
1. 8420K	1.0
2. 1090K	1.0
3. 354K	1.0
4. 43K	1.0
5. 2.8K	1.0

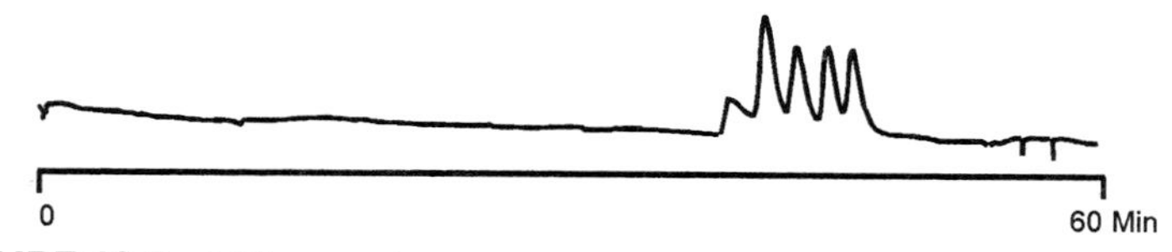

FIGURE 13.2 NPR separation.

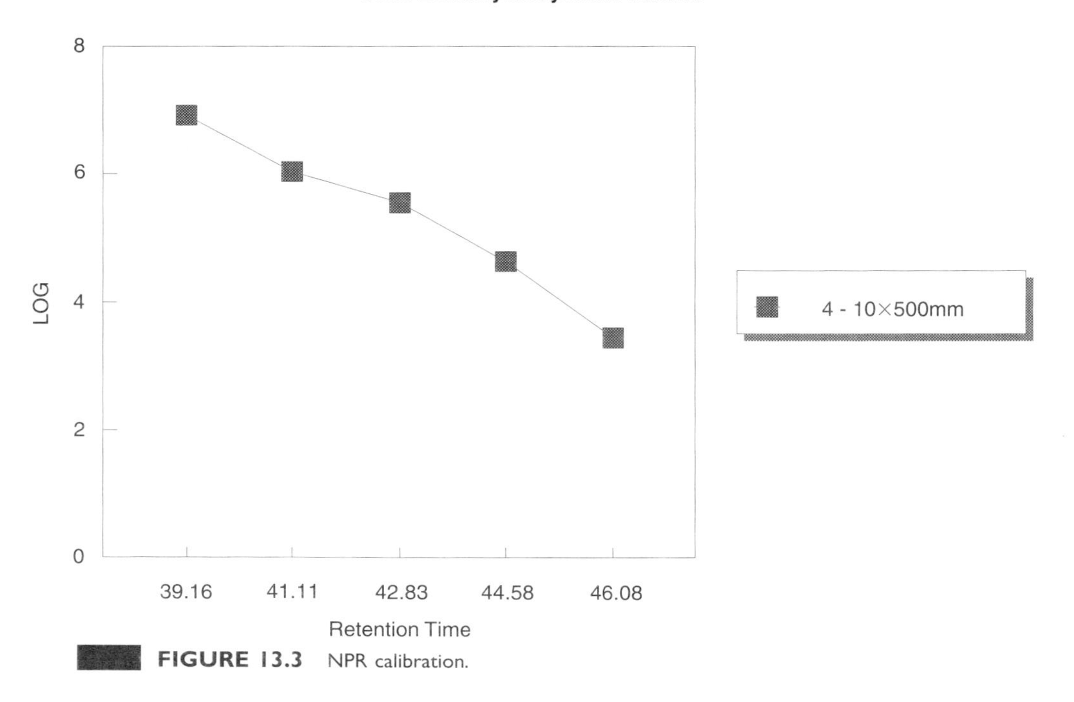

FIGURE 13.3 NPR calibration.

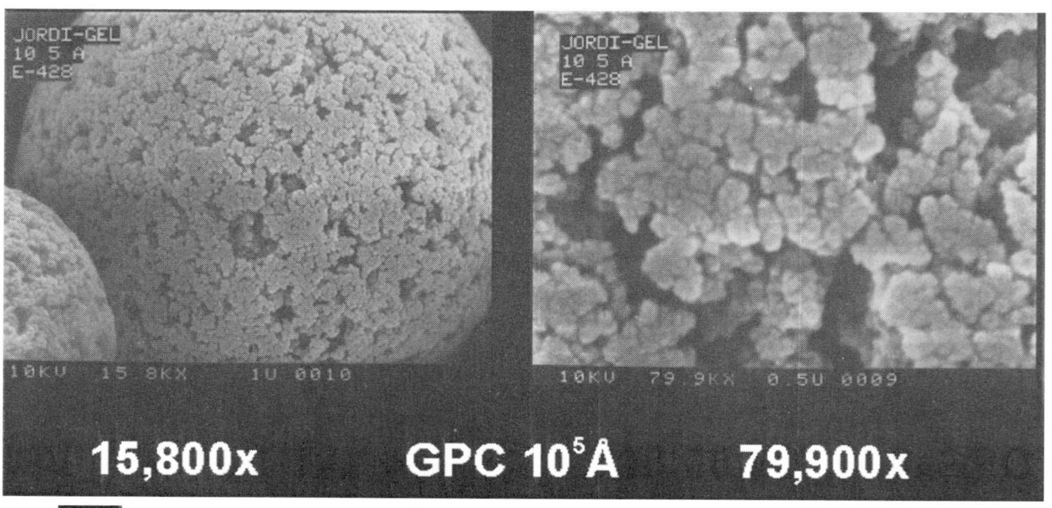

FIGURE 13.4 SEM of 10^3Å gel.

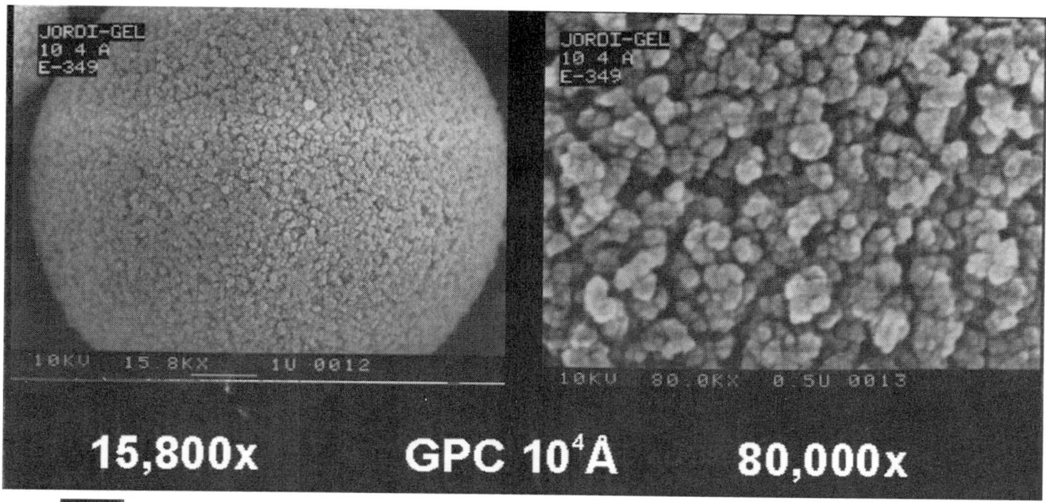

FIGURE 13.5 SEM of 10^4Å gel.

ethylbenzene, diethyl benzene, and so on. These diluents are used in the manufacture of the small pore gels, i.e., 500 and 10^3 Å.

Figures 13.4–13.7 are scanning electron micrographs of 10^5-, 10^4-, 10^3-, and 500-Å Jordi gels, respectively, produced as described earlier. Generally, 100-Å polydivinylbenzene gels are not used because the porosities obtainable are very low, and hence a 67% pore volume 500-Å column will actually do

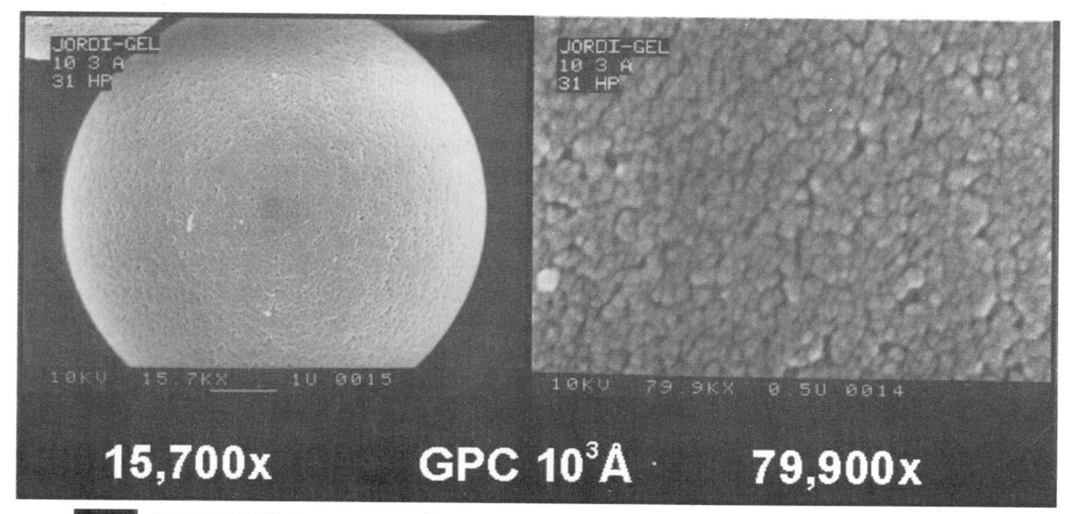

FIGURE 13.6 SEM of 10^3Å gel.

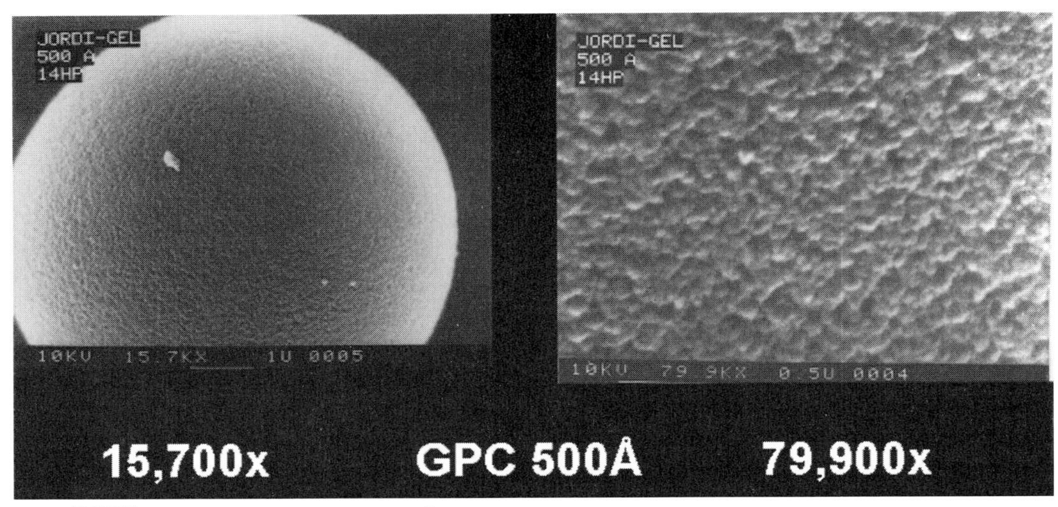

FIGURE 13.7 SEM of 500Å gel.

a better job resolving low molecular weight species than a 30% pore volume 100-Å column.

Gels of various particle sizes can be made by varying the stirring rate, the shape of the stir blades used, the concentration of the suspending agents, and, to a lesser degree, the ratio of the continuous water phase to the discontinuous organic phase.

Figures 13.8 and 13.9 show the separation of polystyrene standards using a typical mixed-bed column and its calibration plot, respectively. The major advantages of using a large i.d. 10-mm column are low back pressure and relatively short run times. As seen in Fig. 13.8, 10 standards from toluene thru 8.4×10^6 MW can be resolved in a mere 21 min. Because of the large 10-mm i.d. columns, 1.5-ml/min flow rates give a linear velocity equivalent to that of only 0.9 ml/min using a 7.6-mm i.d. column. Also, the gel volume contained in one 10 mm i.d. × 500 mm column is 39.3 ml, whereas a 7.6 mm i.d. × 300 mm column contains only 13.6 ml of gel volume. This bulk volume factor, combined with the large pore volumes of gels, obtains essentially the same resolution as that obtained on three standard 7.6 × 300-mm columns in series, but in about one-half the usual time required using the smaller columns.

The exact recipes for these products are of course proprietary, but the development of some of these gels has required years of effort to develop. Since the early 1980s, well over 1000 gel batches have been made and we are always looking for better ways to manufacture our resins. For the reader interested in a more detailed description of gel making, see Potschka and Dubin (2) as well as the excellent older work (1).

III. BASIC AND CHEMICALLY MODIFIED POLYDIVINYLBENZENE PACKINGS

As we began to use the underivatized porous and nonporous PDVB gels, we had a number of early successes but also encountered many sample types that

Jordi Gel DVB Mixed Bed Separation
Polystyrene Standards

DESCRIPTION
Packing Material: Jordi Gel DVB Mixed Bed
Length: 500mm ID: 10mm

TEST CONDITIONS
Mobile Phase: Chloroform
Flow Rate: 1.5ml/min
Detector: UV 254nm
Temperature: Ambient
Pressure: 700 PSIG
Sensitivity: 0.2 AUFS

Sample: PS **Inj. vol.: 15μl**

Component	Conc (mg/ml)
1. 8420K	1.0
2. 1090K	1.0
3. 791K	1.0
4. 354K	1.0
5. 98.9K	1.0
6. 43K	1.0
7. 10.1K	1.0
8. 2.8K	1.0
9. 0.53K	1.0
10. 0.092K	1.0

0 23 Min

FIGURE 13.8 Polystyrene standards.

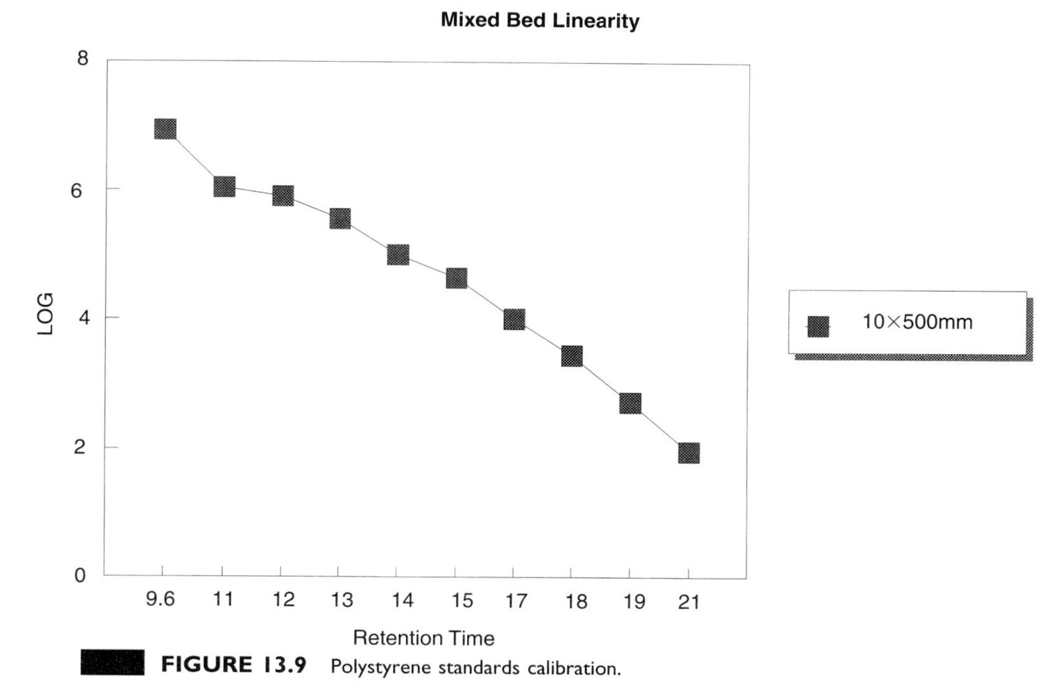

FIGURE 13.9 Polystyrene standards calibration.

were difficult or impossible to run using native PDVB gels. It soon became apparent that to address these difficult to run samples we would need to either develop entirely new gel types or develop bonded phases to minimize the adsorptive effects seen with certain sample types. After much thought it seemed that since basic PDVB gels were so exceptionally durable, the best answer would be to derivatize base PDVB gels and hopefully retain the pH, temperature, and pressure stability enjoyed in the original resins.

The first derivatized gels developed at Jordi Associates were sulfonated PDVB resins. The basic chemical structure of these gels is given in Fig. 13.10.

Unlike earlier sulfonated styrene/divinylbenzene copolymers, these sulfonated gels can be run in virtually any solvent from water and buffers to pure organics as well as most any mixed solvent systems desired. In aqueous systems they absorb water and in organic solvents they stay swollen by imbibing organic solvents.

The next major bonded phase project was the development of the GBR resin, which stands for modified glucose bonded on both the backbone and the ring of basic PDVB gels. The manufacture of this product was ultimately achieved, as outlined later. The gel is first brominated, which places bromine atoms on both tertiary hydrogens of the PDVB. The brominated gel is then reacted with chlorosulfonic acid, and a specially treated reduced D-glucosamine is coupled to the gel. This process has the potential to covalently bond up to three sugar residues to each available divinylbenzene residue in the PDVB polymer. The exact reaction conditions used are proprietary; however, the surface of the finished product is believed to look similar to Figs. 13.11 and 13.12.

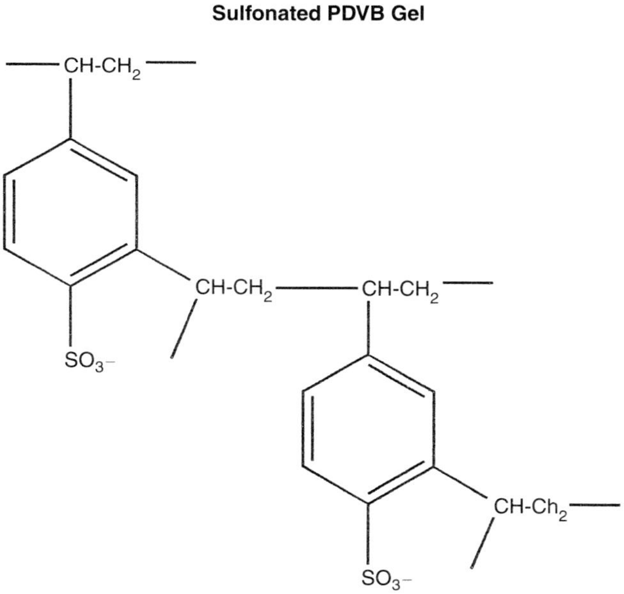

Sulfonated PDVB Gel

FIGURE 13.10 Sulfonated gel.

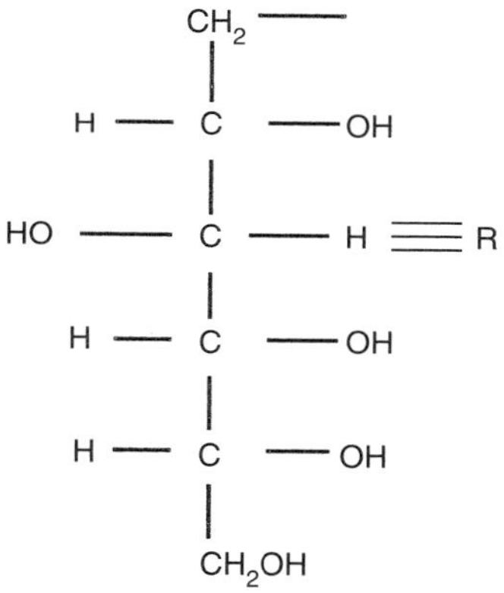

CH_2 —
H — C — OH
HO — C — H ≡ R
H — C — OH
H — C — OH
CH_2OH

FIGURE 13.11 R group for GBR gel.

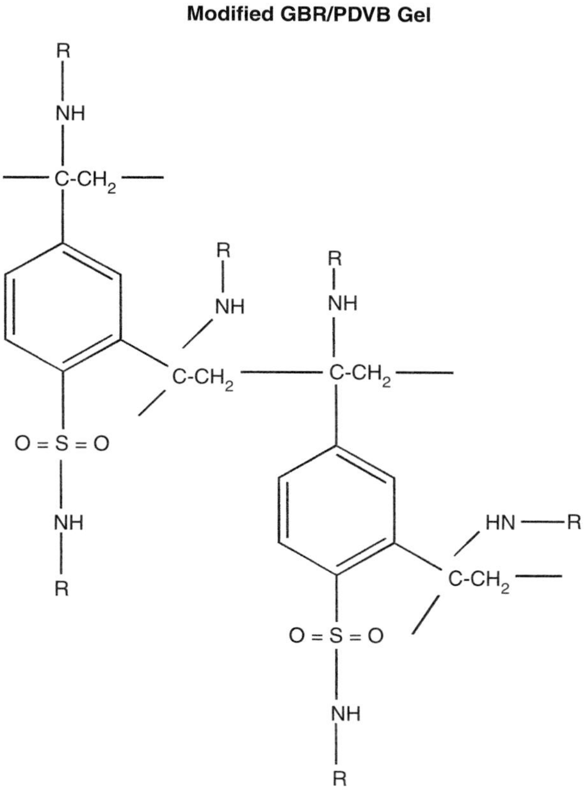

Modified GBR/PDVB Gel

R
NH
— C-CH_2 —
R R
NH NH
C-CH_2 — C-CH_2 —
O = S = O
NH
R
HN — R
C-CH_2 —
O = S = O
NH
R

FIGURE 13.12 GBR gel.

IV. MODIFIED GBR/PDVB GEL

It has been found that both bromination and chlorosulfonation reactions are very useful in coupling a wide range of bonded phases to the gels. It is possible to couple virtually any primary amine to brominated and/or chlorosulfonated PDVB gels. This has allowed the development of many other phases such as polyamino, octadecyl, polyethyleneimine, and quaternary amine. Figures 13.13–13.16 show an assortment of various bonded phases developed since the mid-1980s.

The Jordi PDVB family of columns has very nonpolar packings with a high degree of aromatic character. They are made from the purest DVB available and currently serve as the base material for all other Jordi packings (Fig. 13.13).

The Jordi polyamine column is a polar column for simple sugar and polysaccharide applications. The amine groups are bonded to the DVB backbone and are stable in aqueous mobile phases. This material does not self-hydrolyze as do many silica-based amino packings (Fig. 13.14).

The Jordi C18-DVB column has C18 chains bonded to the DVB backbone. It is a nonpolar reversed-phase or GPC material recommended for applications that require a C18 bonded phase (Fig. 13.15).

The Jordi glucose-DVB column is a highly polar GPC column used for separating polar compounds. Modified glucose units are bonded to the DVB backbone to yield a hydrophilic surface (Fig. 13.16).

The Jordi polyethyleneimine phase has been developed in an attempt to block adsorptive effects of the PDVB surfaces toward cationic polymers such as chitosan and quaternary water-soluble flocculants (Fig. 13.17).

The Jordi quaternary amine phase has been developed to perform ion chromatographic applications and also can be used in a GPC mode to block

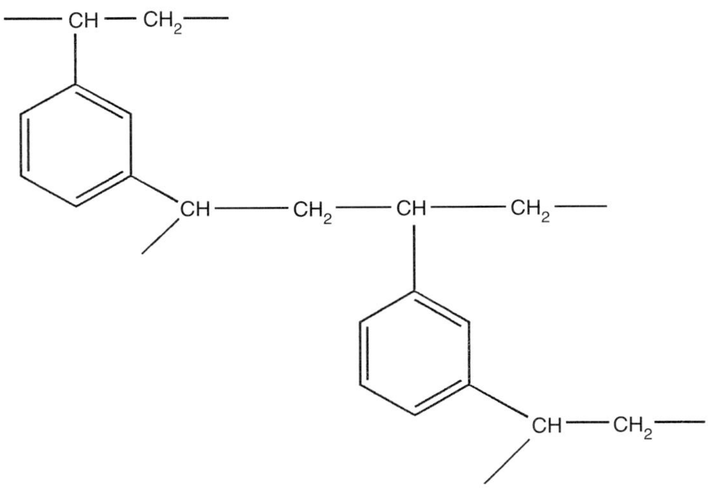

FIGURE 13.13 Base PDVB resin.

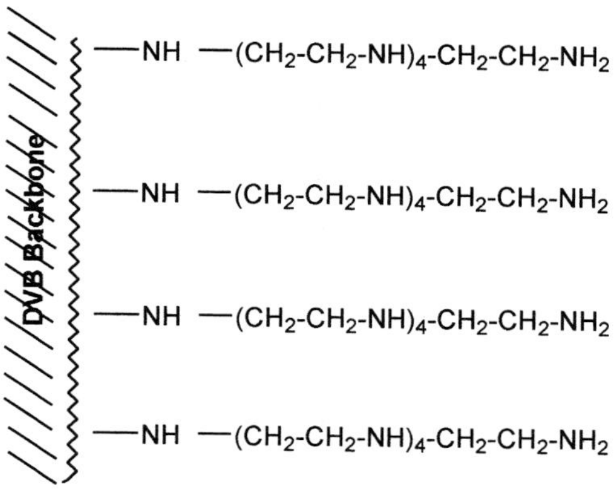

DVB Backbone
—NH —(CH2-CH2-NH)4-CH2-CH2-NH2
—NH —(CH2-CH2-NH)4-CH2-CH2-NH2
—NH —(CH2-CH2-NH)4-CH2-CH2-NH2
—NH —(CH2-CH2-NH)4-CH2-CH2-NH2

FIGURE 13.14 Polyamino phase.

DVB Backbone
—NH-(CH2)17-CH3
—NH-(CH2)17-CH3
—NH-(CH2)17-CH3
—NH-(CH2)17-CH3

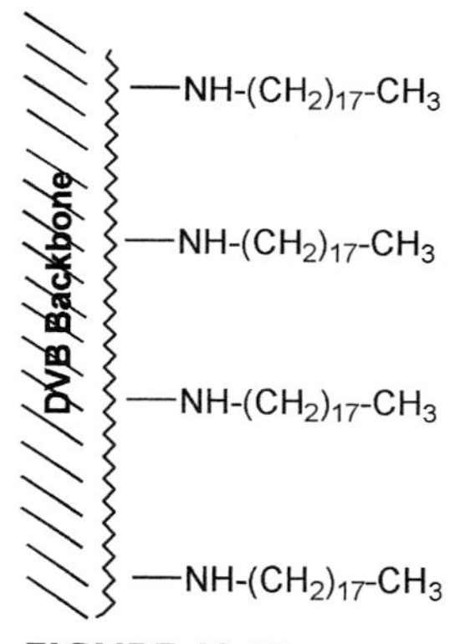

FIGURE 13.15 Octadecyl phase.

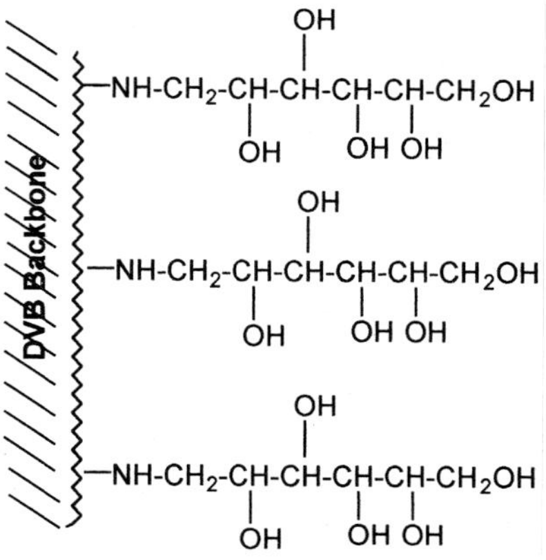

DVB Backbone
OH
—NH-CH2-CH-CH-CH-CH-CH2OH
OH OH OH
OH
—NH-CH2-CH-CH-CH-CH-CH2OH
OH OH OH
OH
—NH-CH2-CH-CH-CH-CH-CH2OH
OH OH OH

FIGURE 13.16 GBR phase.

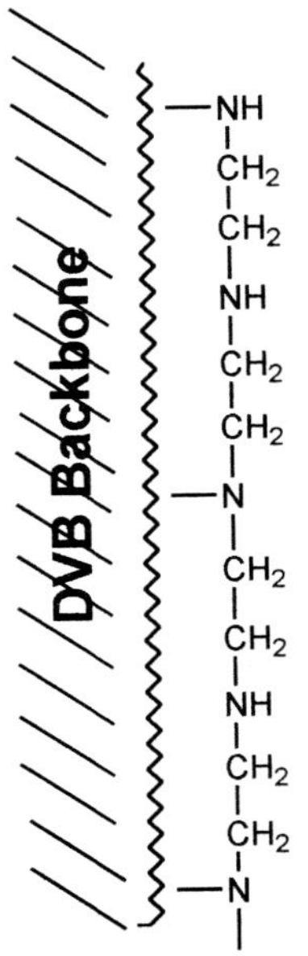

FIGURE 13.17 PEI/WAX phase.

adsorptive effects of the PDVB surfaces toward quaternary water-soluble polymers. The R group is proprietary (Fig. 13.18).

Current ongoing research at Jordi Associates is investigating new ways of bonding hydrophillic groups directly to the aromatic centers of PDVB gels, again in the hope of minimizing adsorptive effects.

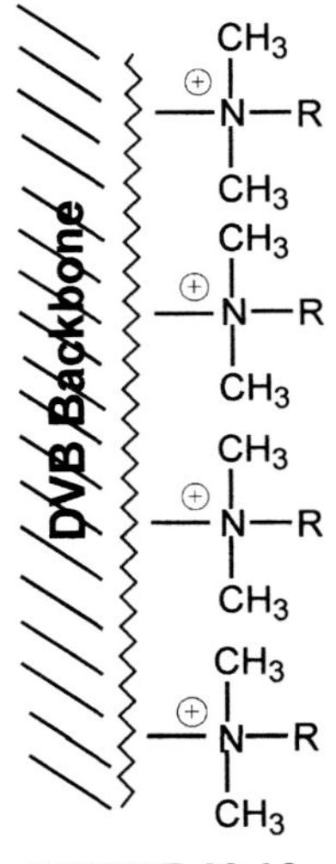

FIGURE 13.18 Quat phase.

V. DISCUSSION OF CHROMATOGRAPHIC APPLICATIONS OF VARIOUS BONDED AND NONBONDED PHASES

Many successful GPC separations have been developed on nonderivatized PDVB gels because of their innate properties such as very high pressure stability, pH stability from 0 to 14, compatibility with virtually all common solvents other than pure water, and very good temperature stability up to 145°C. These properties allowed the use of some very untypical GPC solvents to minimize or eliminate adsorptive effects for many of the difficult sample types such as amines. For example, in order to run low MW GPC separations of amines and surfactants used in the manufacture of nylons, *n*-butylamine at 80°C was used as the mobile phase. Figure 13.19 shows this separation. Nontailing peaks and good resolution of Tween 20, cetylalcohol, stearylamine, isophoronediamine, 1,6-hexanediamine, ϵ-caprolactam, and *m*-xylenediamine were obtained in approximately 80 min using a bank of 5- to 500-Å columns (10 × 500 mm) at a flow rate of 1.5 ml/min. Figure 13.20 shows another way to run many basic sompounds. It turns out that the acetate in sodium acetate acts as the conjugate base and apparently performs much the same function as triethylamine or *n*-butylamine would in deactivating the adsorption sites on PDVB gels.

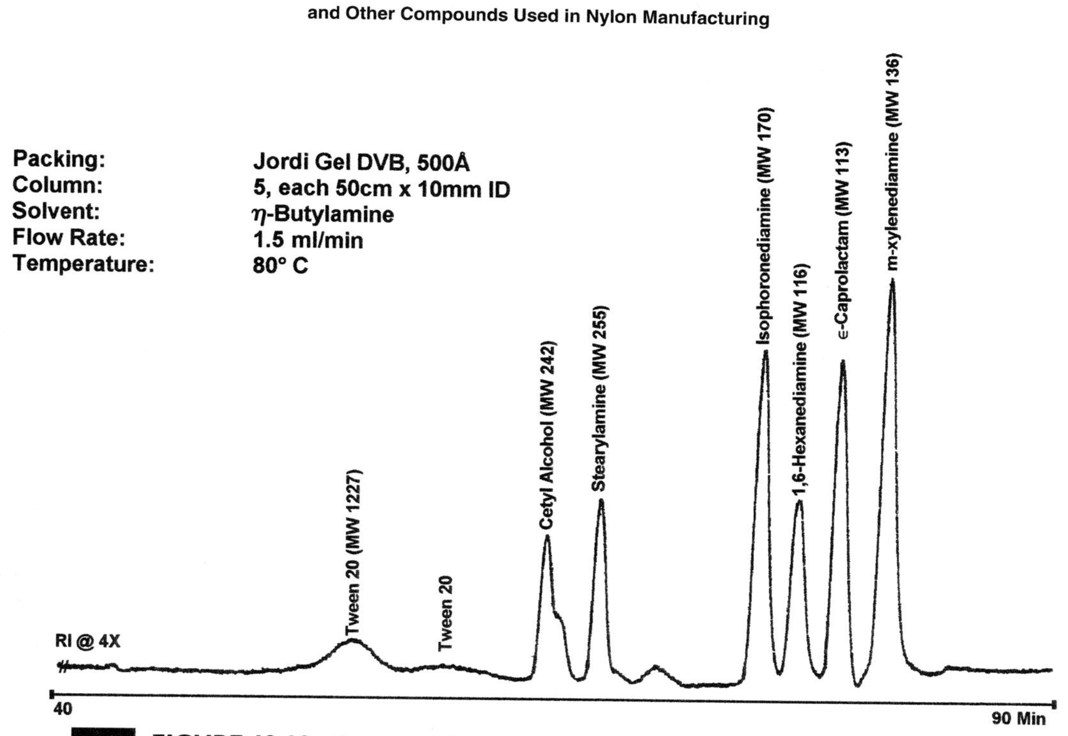

FIGURE 13.19 Separation of amines and surfactants.

Hindered Amine Light Stabilizer Standards
Antioxidant Standards

Packing:	Jordi Gel DVB 500A
Column:	(6) 50cm x 10mm ID
Solvent:	67/33 1-Methoxy-2-Propanol/2,2,4-Trimethylpentane with 0.01M NaAc
Flow Rate:	1.0 ml/min
Temperature:	80°C
Detector:	Waters 990 PDA
Injection:	100μl
Conc:	2mg/ml Chimassorb 944
	1mg/ml Irganox1010, Irgafos 168, BHT

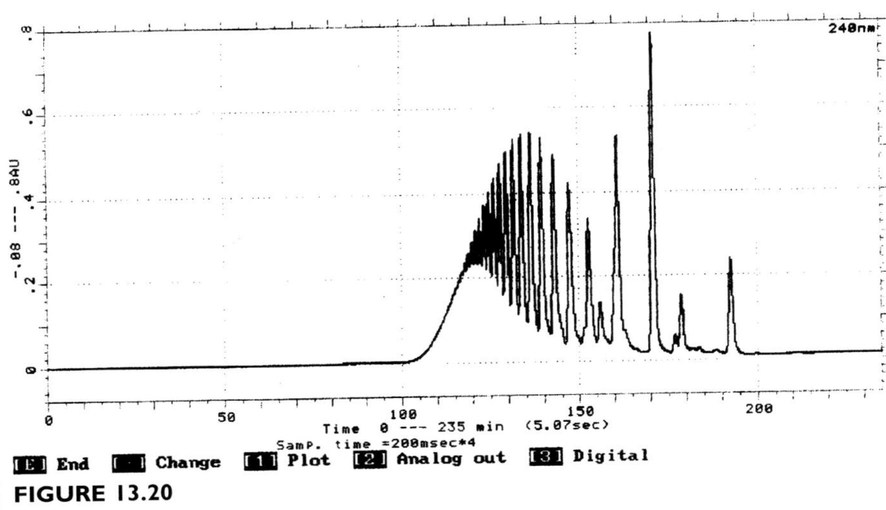

FIGURE 13.20

In the absence of the acetate Chimassorb 944 does not elute. Also, the mobile phase used here is UV transparent down to at least 220 nm, which can aid in the detection of low UV-absorbing analites.

Figure 13.21 shows the resolution of a dozen polymer additives at very high resolution using chloroform as the mobile phase. Tinuvin 622 will elute in pure chloroform whereas Chimassorb 944 and many other hindered amine light stabilizers (HALS) will not. With the addition of 1% triethyl amine to the chloroform, however, virtually all HALS will elute.

Figure 13.22 shows the resolution of the surfactants Tween 80 and SPAN. The high resolution obtained will even allow the individual unreacted ethylene oxide oligomers to be monitored. Figure 13.23 details the resolution of many species in both new and aged cooking oil. Perhaps the most unique high resolution low molecular weight SEC separation we have been able to obtain is shown in Fig. 13.24. Using 1,2,4-trichlorobenzene as the mobile phase at 145°C with a six column 500-Å set in series, we were able to resolve C_5, C_6, C_7, C_8, C_9, C_{10}, and so on hydrocarbons, a separation by size of only a methylene group. Individual ethylene groups were at least partially resolved out to C_{58}. This type of separation should be ideal for complex wax analysis.

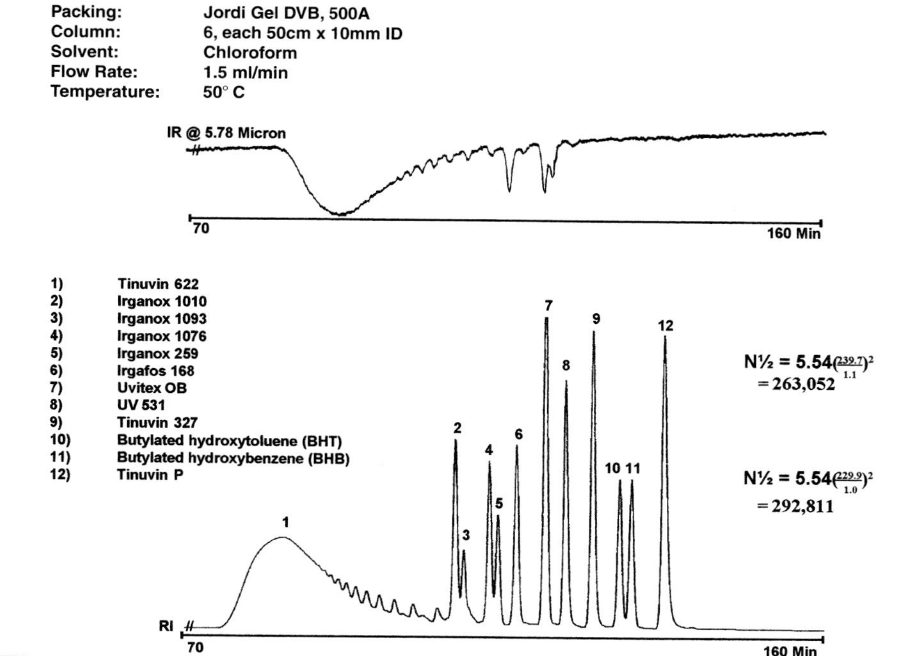

FIGURE 13.21 A series of polymer additives using the infrared detector at 5.78 micron. Efficiencies were calculated using the last peak, Tinuvin P, and a plate count of 290,000 was achieved.

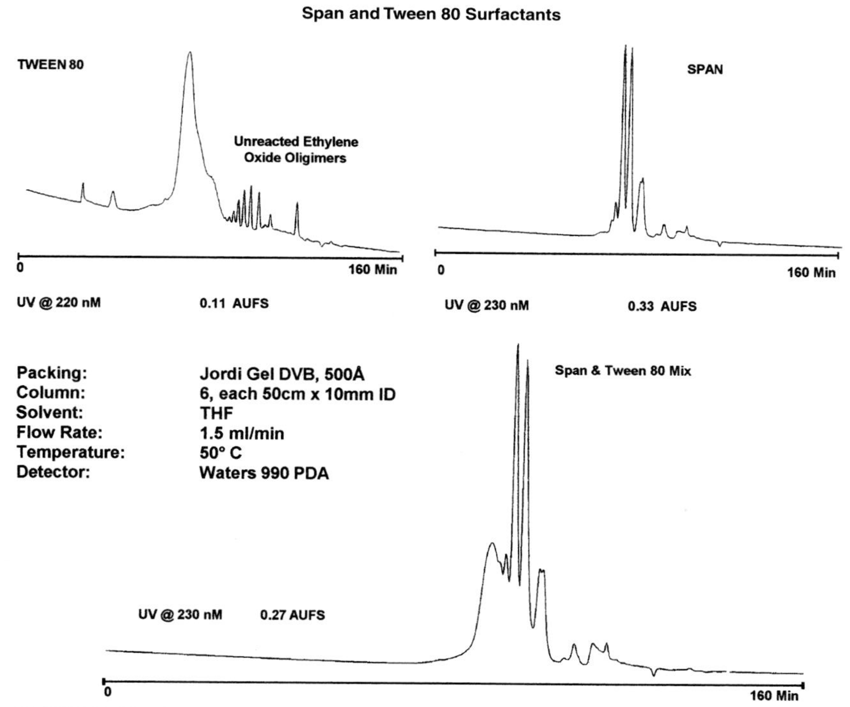

FIGURE 13.22 Surfactants on the 6 column set, using a photodiode array (PDA) detector. The Span and Tween 80 can be distinguished from each other very nicely in a mix. THF was the solvent used at 50°C. (A restrictor after the detector minimizes bubbles.)

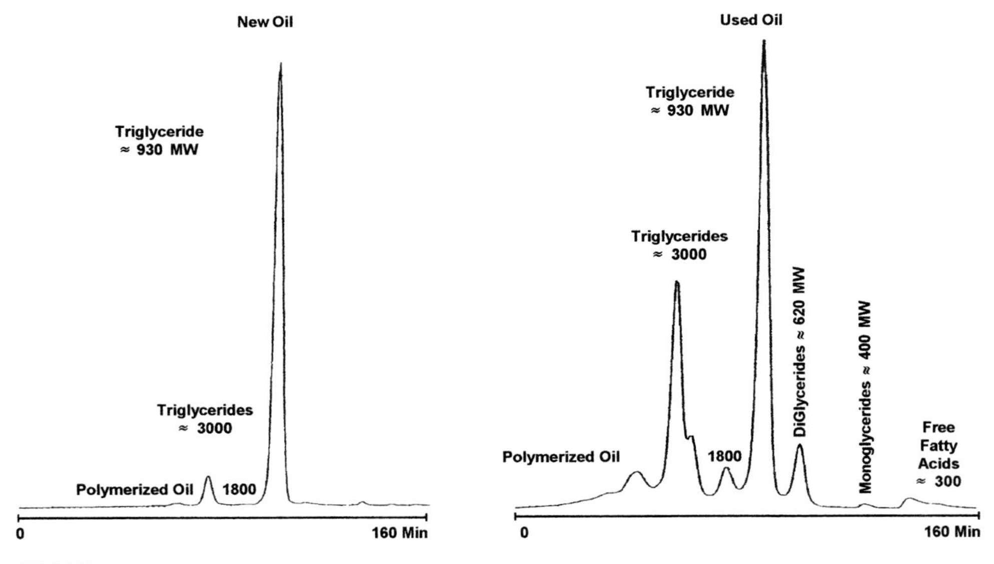

FIGURE 13.23 Using the PDA detector and THF at 50°C, a comparison of soybean cooking oils provides good separations of the glycerides.

Figures 13.25–13.28 show the ultrahigh resolution separations in chloroform of polystyrene standards, polytetramethylene glycol, urethanes and isocyanates, and epoxy resins, respectively. Multiple column sets of anywhere from two to six columns in series have been used for well over a year with no apparent loss of efficiency. The 500- and 10^3-Å gels can easily tolerate 15,000 psi or more. In fact, the limiting factor in the number of columns that can be used in series is generally the pump or injector in the HPLC system. A pump capable of 10,000 psi operation should allow the use of a column bank of 10–12 50-cm columns with a total plate count of 500,000 or more.

As a final example of column durability and solvent resistance in small pore gels we were able to resolve nylon 6 oligomers using a methanol mobile phase and 205-nm UV detection as shown in Figure 13.29. In fact, polar solvents such as acetone, acetonitrile, methanol, and 2-propanol, are used routinely as needed with no ill effects.

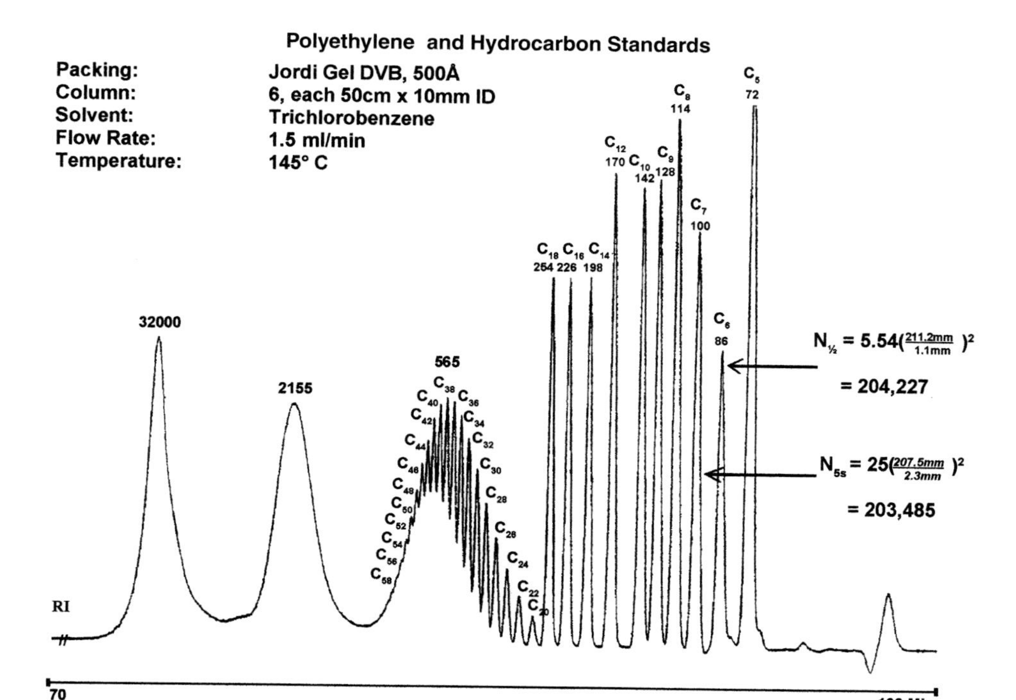

FIGURE 13.24 A mixture of hydrocarbons and some polyethylene standards at 145°C. Column backpressure was approximately 5500 psi. Plate counts calculated on the hexane and heptane peaks yield 204,000 plates.

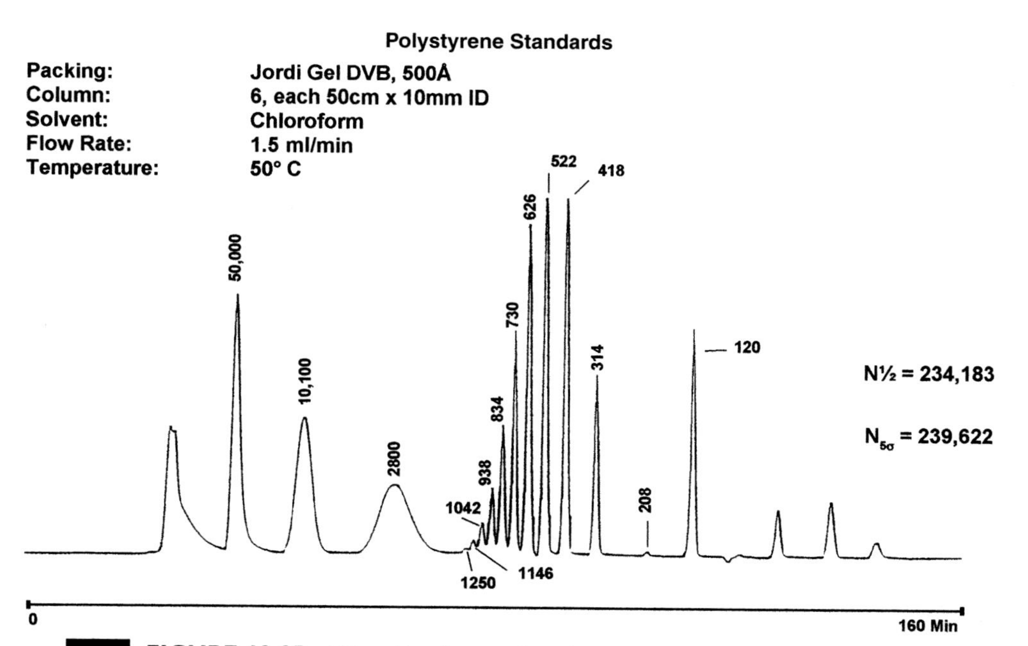

FIGURE 13.25 Using chloroform as the solvent, a mixture of polystyrene standards were nicely separated on the 3-m set of columns. Run times here were 160 min. Plate count for toluene was calculated at 240,000 plates. The 500 MW Standard is separated nicely into its oligomers.

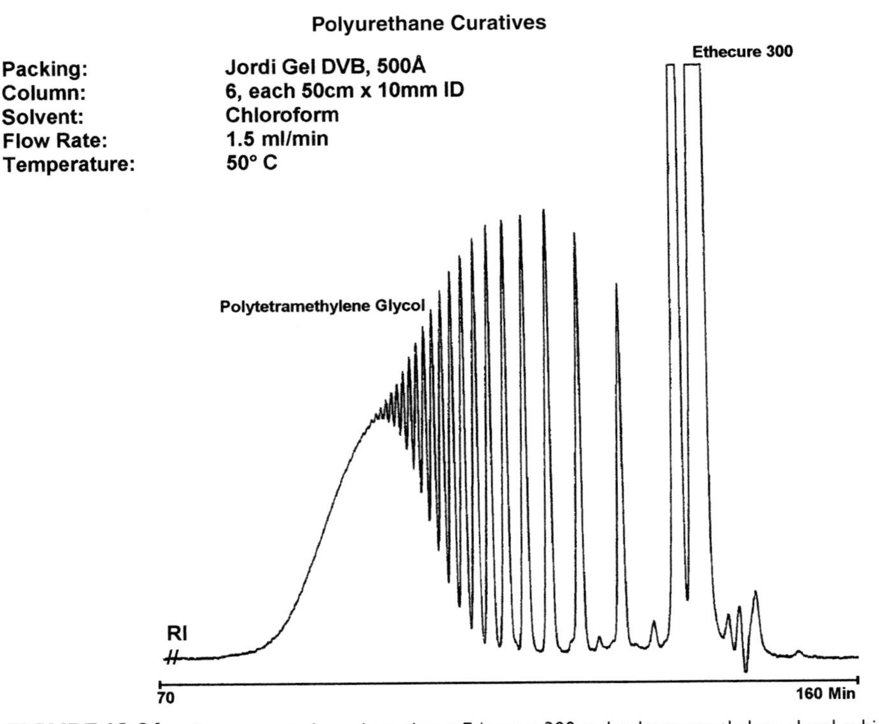

FIGURE 13.26 Curing agents for polyurethane: Ethecure 300 and polytetramethylene glycol, which separates very nicely into its oligomers up to molecular weights of ~1000 before the characteristic Gaussian distribution predominates.

As the porosities of PDVB gels increase above 10^3 Å, the pressure limits drop, with 2500 psi being the maximum usable pressure for 10^4 Å, 10^5 Å, and mixed-bed columns. Because the normal operating pressures in most solvents for these columns tend to be in the range of 1000 psi or less for a 10×500-mm column, there is seldom an operational problem. Figure 13.8 shows the resolution of a typical mixed-bed column run in chloroform at 1.5 ml min yielding a back pressure of 700 psi and running polystyrene standards.

Figures 13.30–13.53 demonstrate the use of various mobile phases for polymer SEC using standard mixed-bed DVB columns. Once again these applications demonstrate that PDVB gels will easily tolerate virtually any solvent or mixed solvent system.

Figures 13.54 thru 13.56 have been singled for special mention because of the unusual nature of the applications presented. Because peptides are also polyamides and because we routinely run polyamides using hexafluoro-2-propanol (HFIP) as solvent, a modified polypeptide, poly(leucine-comethyl glutamate), was run in HFIP. Figure 13.54 shows the results and suggests that modified HFIP might well prove to be a good solvent for at least some polypeptide purifications/characterizations. Figure 13.55 demonstrates the use of modi-

Urethane and Isocyanates

Packing:	**Jordi Gel DVB, 500Å**
Column:	**6, each 50cm x 10mm ID**
Solvent:	**Chloroform**
Flow Rate:	**1.5 ml/min**
Temperature:	**50° C**

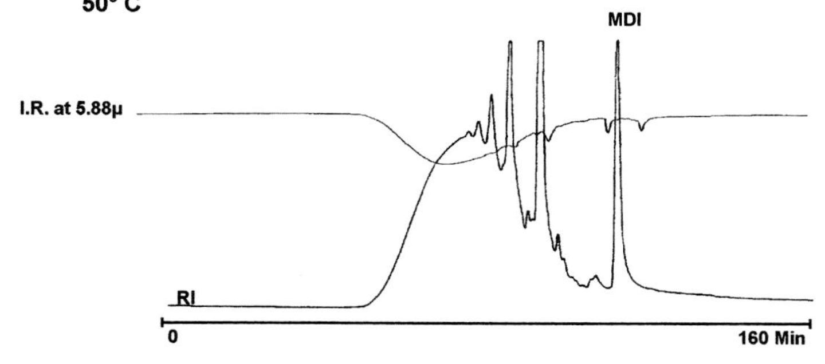

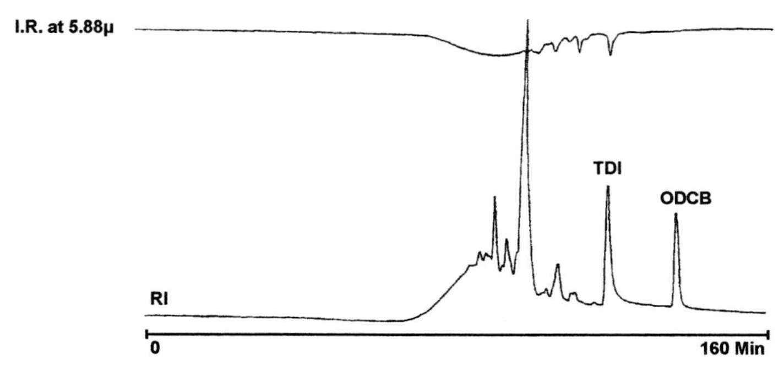

FIGURE 13.27 Urethane and some isocyanates on the same system, but including an infrared detector in series at 5.88 micron. The MDI and TDI have different retentions so they can be distinguished from each other in a urethane.

fied 1,1,2-trichloro-1,2,2-trifluoroethane for the analysis of perfluoroether Z-Dol 2000 lubricant. Figure 13.56 demonstrates that the PDVB resin can even tolerate dichloroacetic acid containing mobile phases. Using 3 : 1 $CHCl_3$: dichloroacetic acid as the mobile phase, we were able to analyze polyether ether ketone or PEEK.

Figure 13.57 discusses in some detail the use of mobile-phase modifiers to prevent adsorption on PDVB resins. These concepts are very valuable in developing methods. For example, note how the observed column efficiencies improve for paraben analysis in the order of methanol < acetonitrile < 50/50 methanol/acetonitrile < THF. Furthermore, when THF is used the chromato-

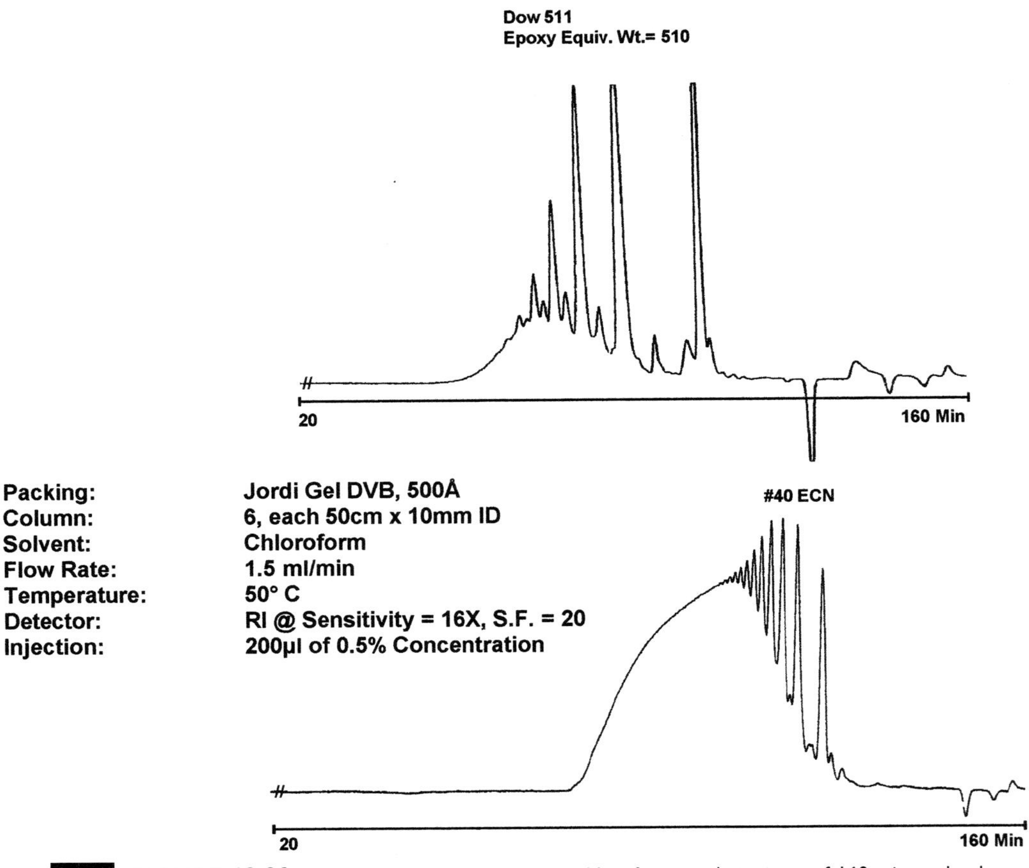

FIGURE 13.28 Different epoxy resins run in chloroform, analysis times of 160 min, and column temperature maintained at 50°C.

graphic mode also changes from reverse phase to SEC. The use of triethylamine or *n*-butylamine in the 1- to 10% (v/v) range with other solvents will often allow the SEC analysis of otherwise adsorbing polymers. Figure 13.46 is a typical example.

Several other examples of modified mobile phases are given in Figs. 13.58 and 13.59 using 90/5/5 THF/MeOH/ACN and 95/5 chloroform/*n*-butylamine for the SEC analysis of poloxamer and nitrile-butadiene rubber samples, respectively.

Even with mobile-phase modifiers, however, certain polymer types cannot be run due to their lack of solubility in organic solvents. In order to run aqueous or mixed aqueous/organic mobile phases, Jordi Associates has developed several polar-bonded phase versions of the PDVB gels as discussed earlier. Figures 13.60 thru 13.99 detail examples of some polar and ionic polymers that we have been able to run SEC analysis of using the newer bonded PDVB resins.

Caprolactam
(Nylon 6 Oligomer Quantitation)

Packing: 2-Jordi Gel DVB 500Å
Column: 50cm x 10mm ID
Solvent: CH_3OH
Flow Rate: 1.5ml/min
Temp: 35°C
Detector: Waters 990 PDA
Inj: 400µl
Conc: 2mg/ml

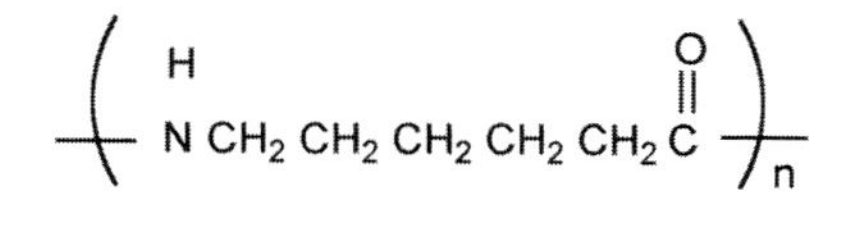

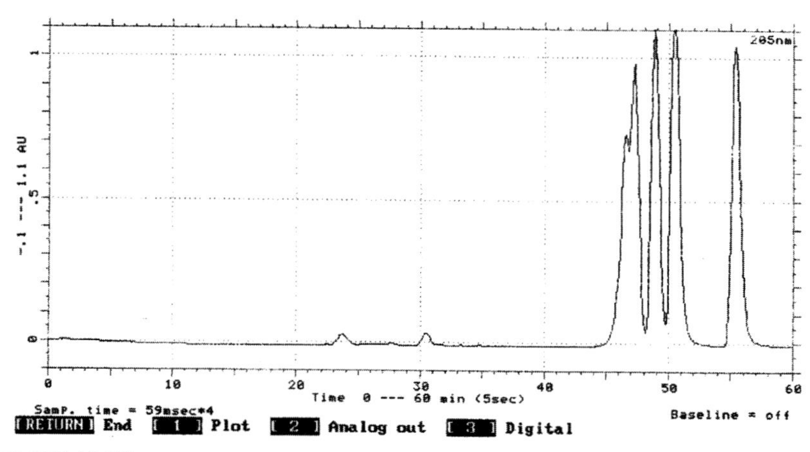

FIGURE 13.29 Nylon oligomer separation.

Hot Melt Adhesive Sample

Packing: (1) Jordi Gel DVB Mixed Bed
Column: 500mm x 10mm ID
Solvent: THF
Sample: Hot Melt Adhesive Sample
Flow Rate: 1.2 ml/min
Temp: 35°C
Detector: Waters 410; 16x, 20
Inj: 200µl
Conc: 0.25% (w/v)

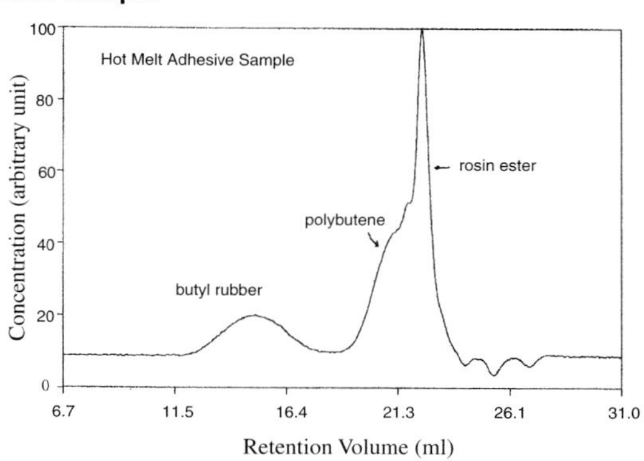

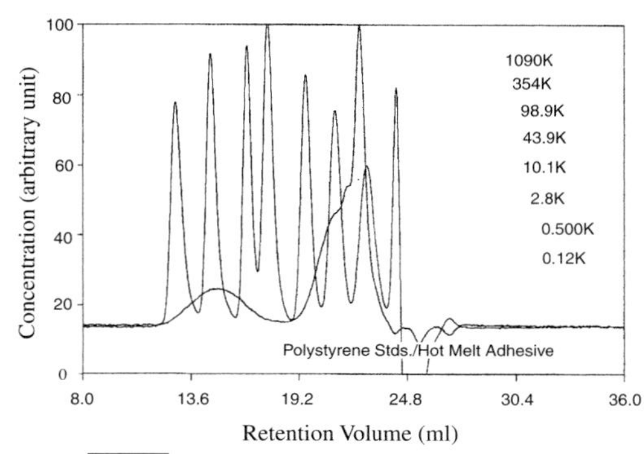

FIGURE 13.30 Hot melt adhesive.

Nitrocellulose

Packing:	(2) Jordi Gel DVB Mixed Bed
Column:	250mm x 10mm ID
Solvent:	DMF/0.05M LiBr
Sample:	Nitrocellulose
Flow Rate:	1.0 ml/min
Temp:	35°C
Detector:	Waters 410; 16x, 20/Fitted with a PD2000 Light-scattering Detector @ 90°
Inj:	200µl
Conc:	0.25% (w/v)

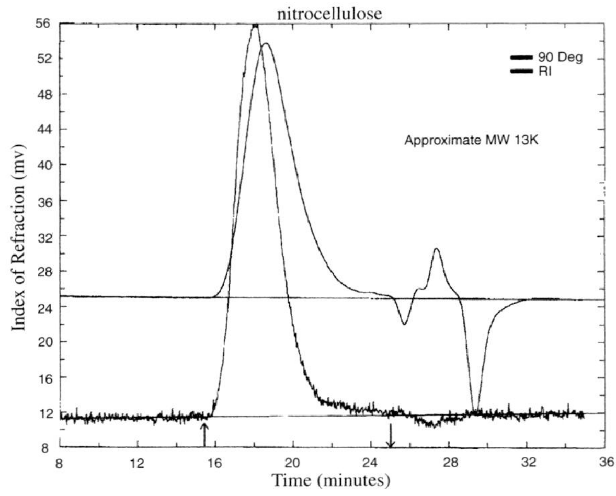

FIGURE 13.31 Nitrocellulose separation.

Ethyl Acrylate Water Based Emulsion

Packing:	(1) Jordi Gel DVB Mixed Bed
Column:	500mm x 10mm ID
Solvent:	THF
Sample:	Ethyl Acrylate Water Based Emulsion
Flow Rate:	1.2 ml/min
Temp:	35°C
Detector:	Waters 410; 8x, 20
Inj:	300µl
Conc:	0.25% (w/v)

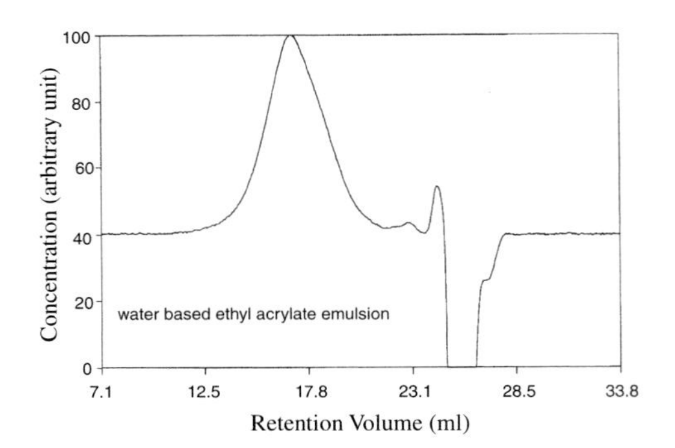

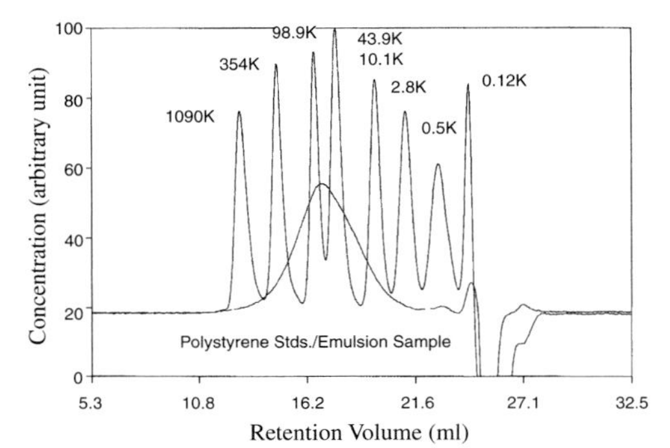

FIGURE 13.32 Ethyl acrylate emulsion.

Hydroxyethyl Starch

Packing:	(2) Jordi Gel DVB Mixed Bed
Column:	500mm x 10mm ID
Solvent:	Dimethylacetamide (DMAC)/0.1M LiBr
Sample:	Hydroxyethyl Starch
Flow Rate:	1.0 ml/min
Temp:	35°C
Detector:	Waters 410; 16x, 20/Fitted with a PD2000 Light-scattering Detector @ 15°, 90°
Inj:	100µl
Conc:	0.28% (w/v)

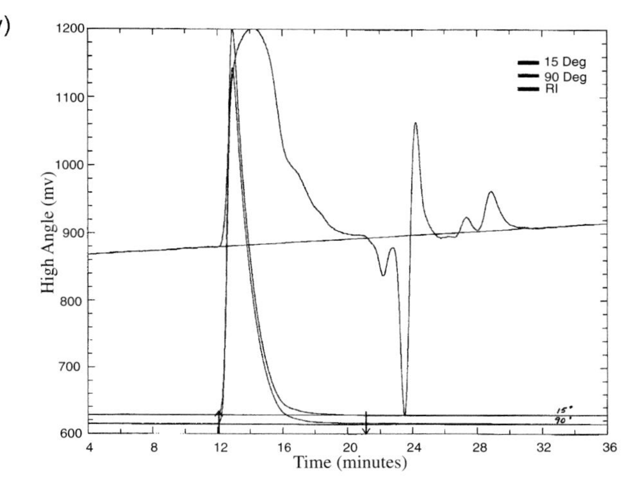

FIGURE 13.33

Poly(Tetramethylene Teraphthalate)

Packing:	(1) Jordi Gel DVB Mixed Bed
Column:	500mm x 10mm ID
Solvent:	Dichloromethane
Sample:	Poly (Tetramethylene Teraphthalate)
Flow Rate:	1.2 ml/min
Temp:	35°C
Detector:	Waters 410; 8x, 20
Inj:	250µl
Conc:	1.0% (w/v)
	PS Stds. 30µl, 1mg/ml [0.1% (w/v)]

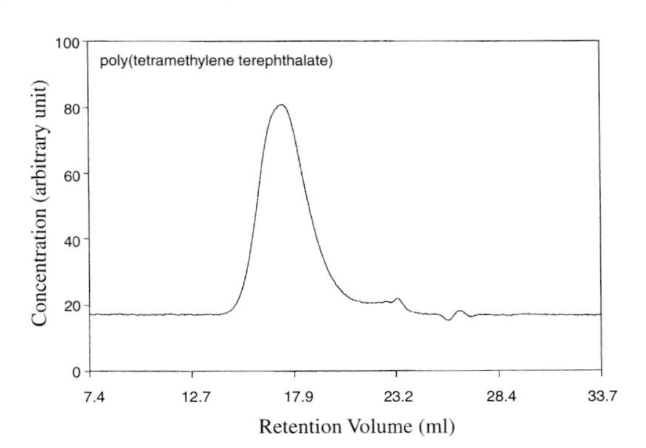

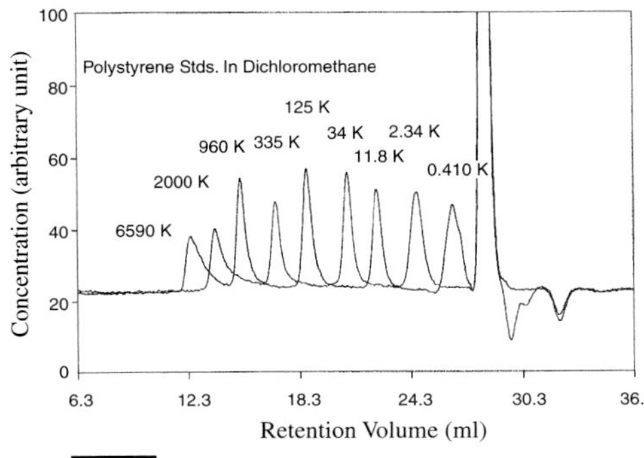

FIGURE 13.34

Polyvinylchloride (PVC)

Packing:	(1) Jordi Gel DVB Mixed Bed + (1) Jordi Gel DVB 500Å
Column:	250mm x 10mm ID + 500mm x 10mm ID
Solvent:	THF
Sample:	Polyvinylchloride (PVC)
Flow Rate:	1.3 ml/min
Temp:	35°C
Detector:	RI, Waters 410
Inj:	350µl
Conc:	0.25% (w/v)

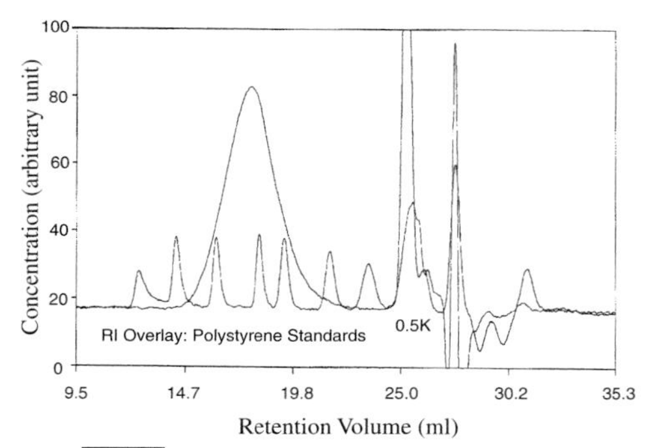

FIGURE 13.35 Polyvinylchloride (PVC) resin.

Amidoamine

Packing:	(1) Jordi Gel DVB Mixed Bed
Column:	250mm x 10mm ID
Solvent:	HFIP/0.01M NATFAT
Sample:	Amidoamine
Flow Rate:	0.6 ml/min
Temp:	Ambient
Detector:	RI, Waters 410, 8x
Inj:	90µl
Conc:	0.2% (w/v)

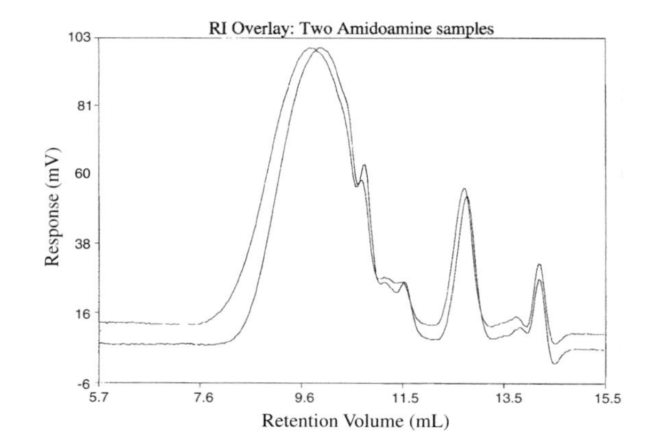

FIGURE 13.36

Nitrile Rubber Samples

Packing:	(1) Jordi Gel DVB Mixed Bed
Column:	500mm x 10mm ID
Solvent:	THF
Sample:	Nitrile Rubber Samples
Flow Rate:	1.3 ml/min
Temp:	35°C
Detector:	RI, Waters 410, 8x, 20
Inj:	250µl
Conc:	0.25% (w/v)

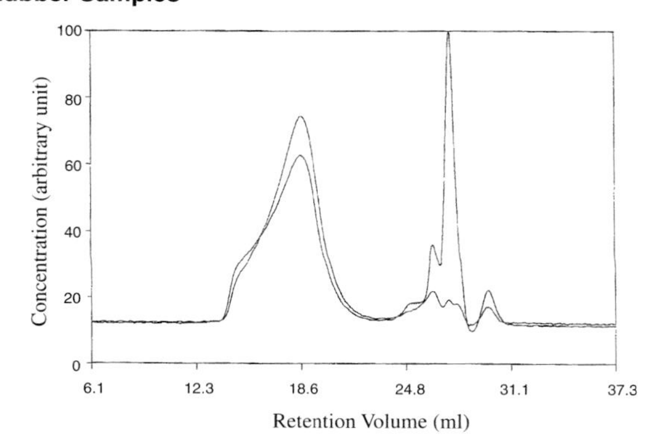

FIGURE 13.37

Alkyd Resins

Packing:	(1) Jordi Gel DVB Mixed Bed
Column:	500mm x 10mm ID
Solvent:	$CHCl_3$
Sample:	Alkyd Resins
Flow Rate:	1.3 ml/min
Temp:	35°C
Detector:	RI, Waters 410, 8x, 20
Inj:	400µl
Conc:	0.2% (w/v)

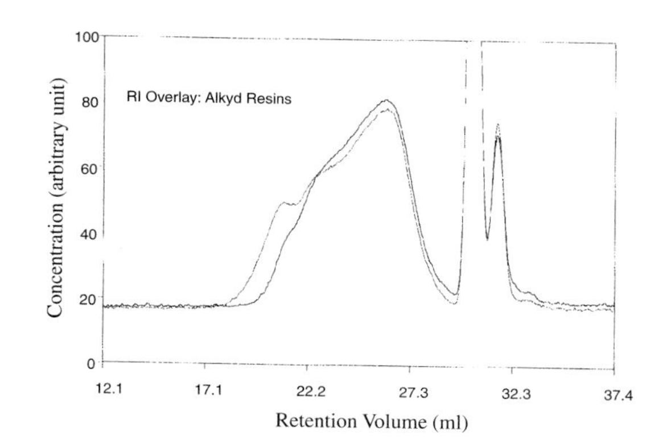

FIGURE 13.38

Cariflex SBR Block Copolymer

Packing:	(2) Jordi Gel DVB Mixed Bed
Column:	500mm x 10mm ID
Solvent:	THF
Sample:	Cariflex SBR Block Copolymer
Flow Rate:	1.2 ml/min
Temp:	35°C
Detector:	Waters 150C
Inj:	75µl
Conc:	2.5mg/ml

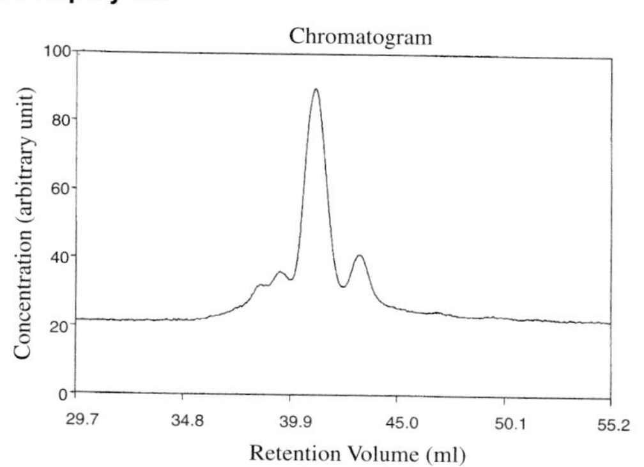

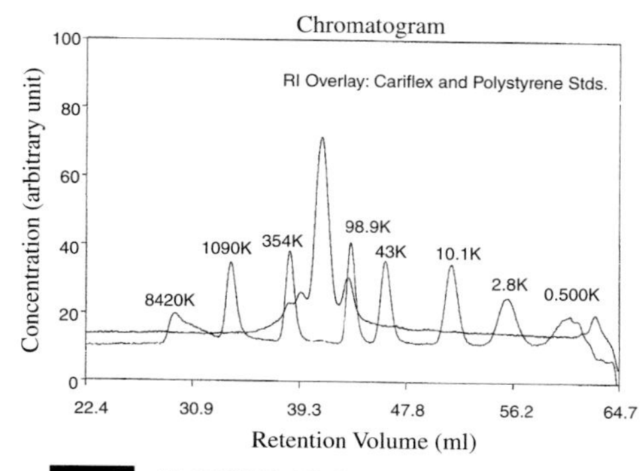

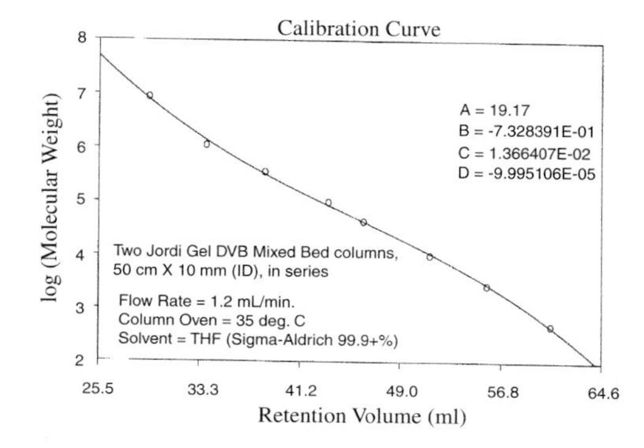

FIGURE 13.39 SBR copolymer.

Poly(Vinylidene Fluoride)

Packing:	**Jordi DVB, Mixed Bed Linear**
Column:	**50cm X 10mm I.D.**
Solvent:	**DMAC**
Flow Rate:	**1.2 ml/min**
Temp:	**100°C**
Detector:	**RI**
Injection:	**250µl**
Conc:	**0.25% W/V**

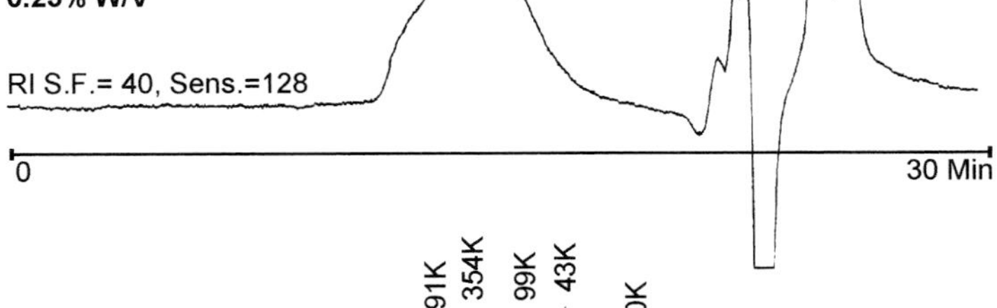

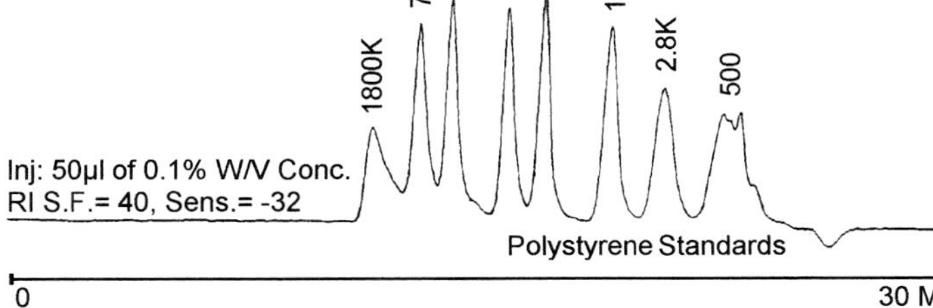

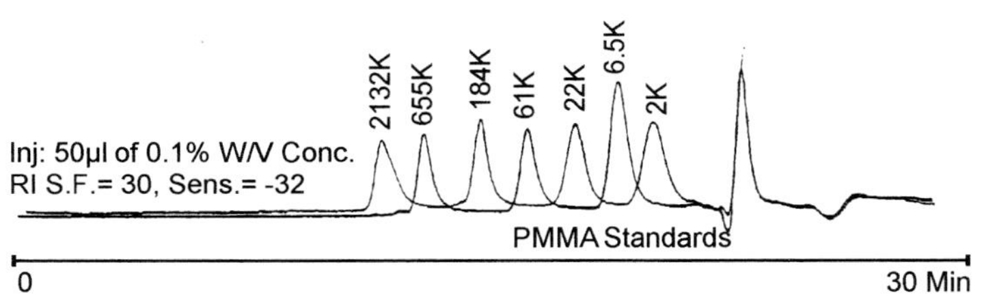

FIGURE 13.40

PET(Polyethylene Terephthalate)

Packing:	**Jordi DVB, Mixed Bed Linear**
Column:	**50cm X 10mm I.D.**
Solvent:	**75/25 V/V $CHCl_3$/Dichloroacetic Acid**
Flow Rate:	**1.2 ml/min**
Temp:	**25°C**
Detector:	**RI**
Injection:	**200µl**
Conc:	**0.3% W/V**

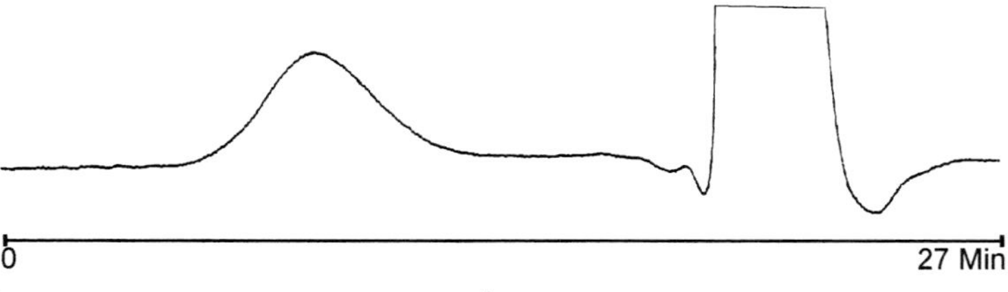

FIGURE 13.41

Polypropylene Oxide

Packing:	Jordi DVB, Mixed Bed Linear
Column:	50cm X 10mm I.D.
Solvent:	90% V/V DMAC with 9% V/V PEG
Flow Rate:	1.2 ml/min
Temp:	140°C
Detector:	RI
Injection:	200µl
Conc:	0.45% W/V

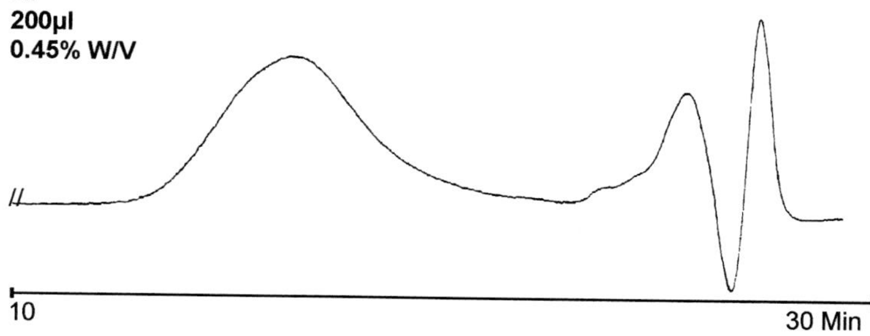

FIGURE 13.42

Polyether Sulfone

Packing:	Jordi DVB, Mixed Bed Linear
Column:	50cm X 10mm I.D.
Solvent:	N-Methylpyrrolidone (NMP)
Flow Rate:	1.2 ml/min
Temp:	80°C
Detector:	RI
Injection:	150µl
Conc:	0.25% W/V

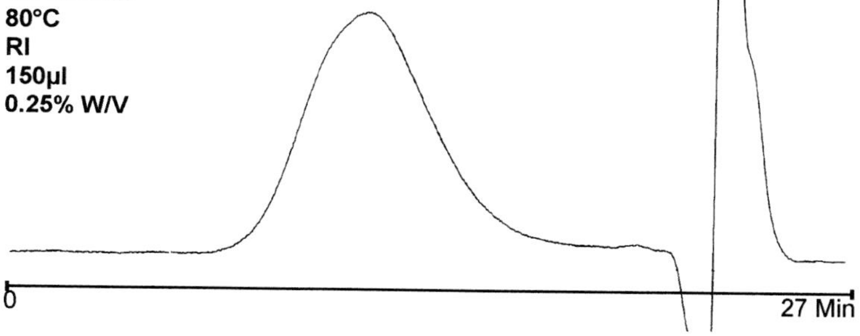

Poly Sulfone

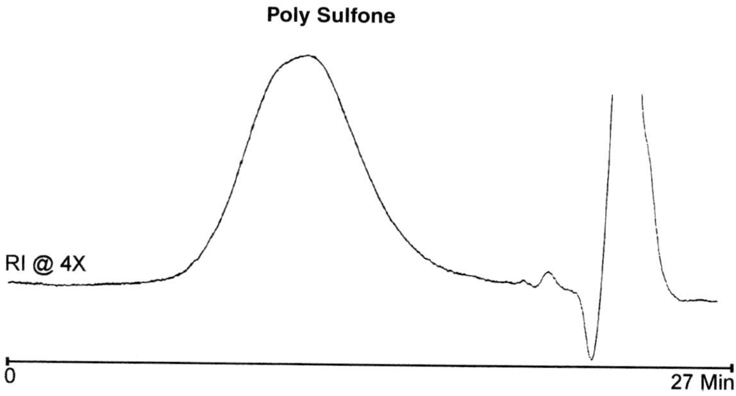

FIGURE 13.43 Polysulfones.

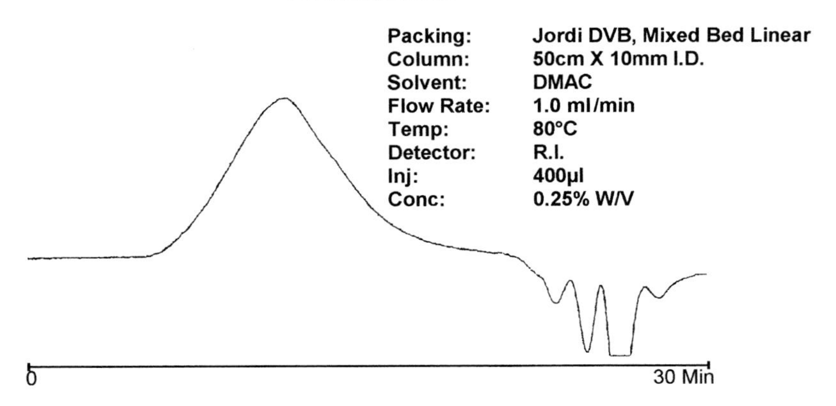

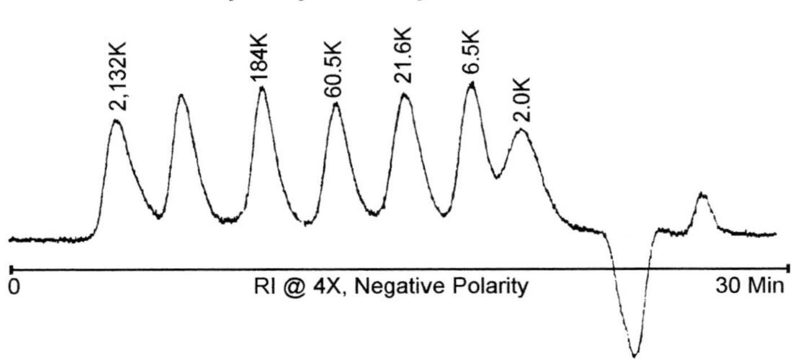

FIGURE 13.44

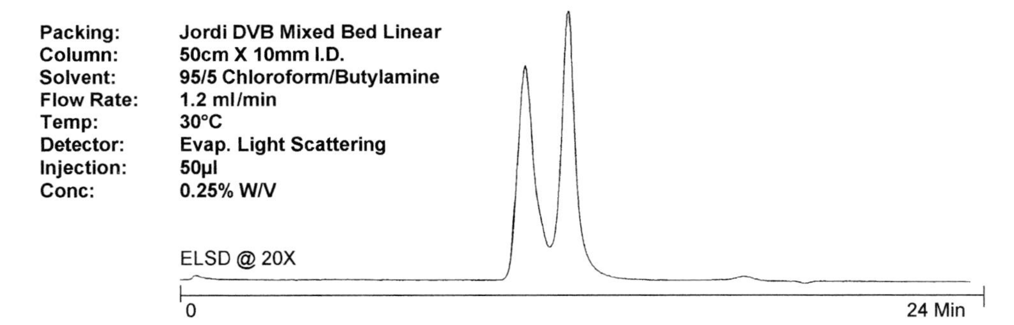

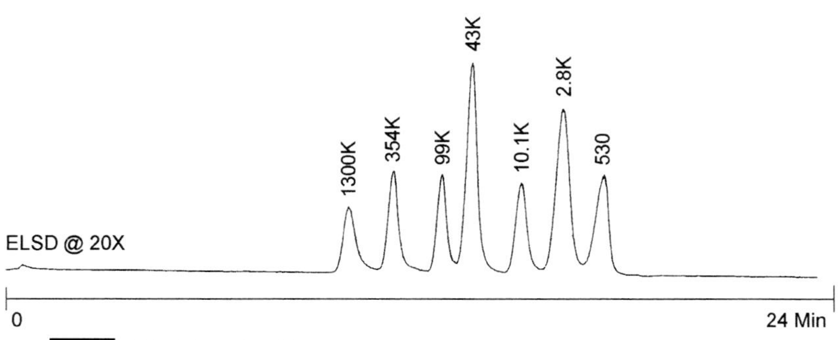

FIGURE 13.45

Poly-4-Vinyl Pyridine

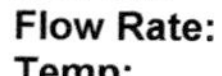

Packing: Jordi DVB Mixed Bed Linear
Column: 50cm X 10mm I.D.
Solvent: 97/3 DMF/Butylamine
Flow Rate: 1.0 ml/min
Temp: 60°C
Detector: RI
Injection: 300µl
Conc: 0.25% W/V

FIGURE 13.46

Polyether Sulfone
Peso Resin

Packing: Jordi Gel DVB Mixed Bed Linear
Column: 50cm X 10mm I.D.
Solvent: DMF with 0.5M LiBr
Flow Rate: 1.2 ml/min
Temp: 60°C
Injection: 150µl

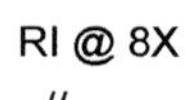

FIGURE 13.47

Polyether Glycols
Polytetramethylene Ether Glycol

Packing: Jordi Gel DVB Mixed Bed Linear
Column: 50cm X 10mm I.D.
Solvent: 90/10 THF/Acetone
Flow Rate: 1.3 ml/min
Temp: 40°C

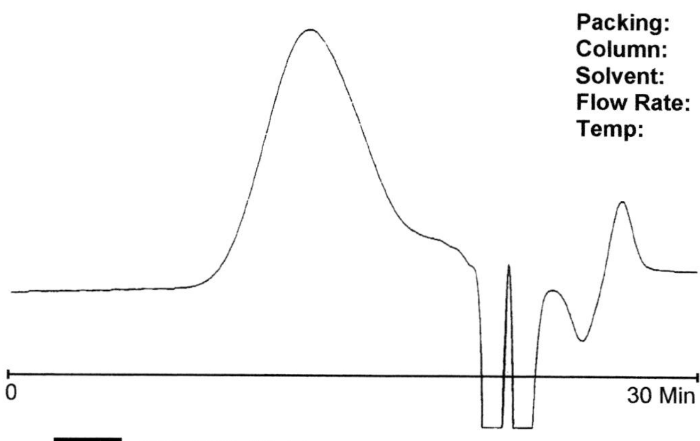

FIGURE 13.48

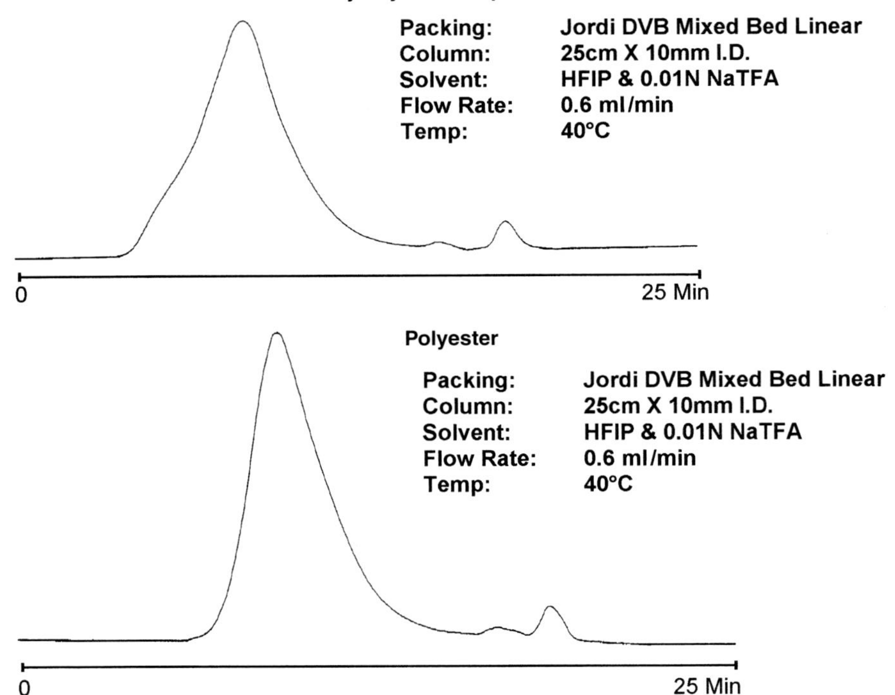

FIGURE 13.49 Polyesters.

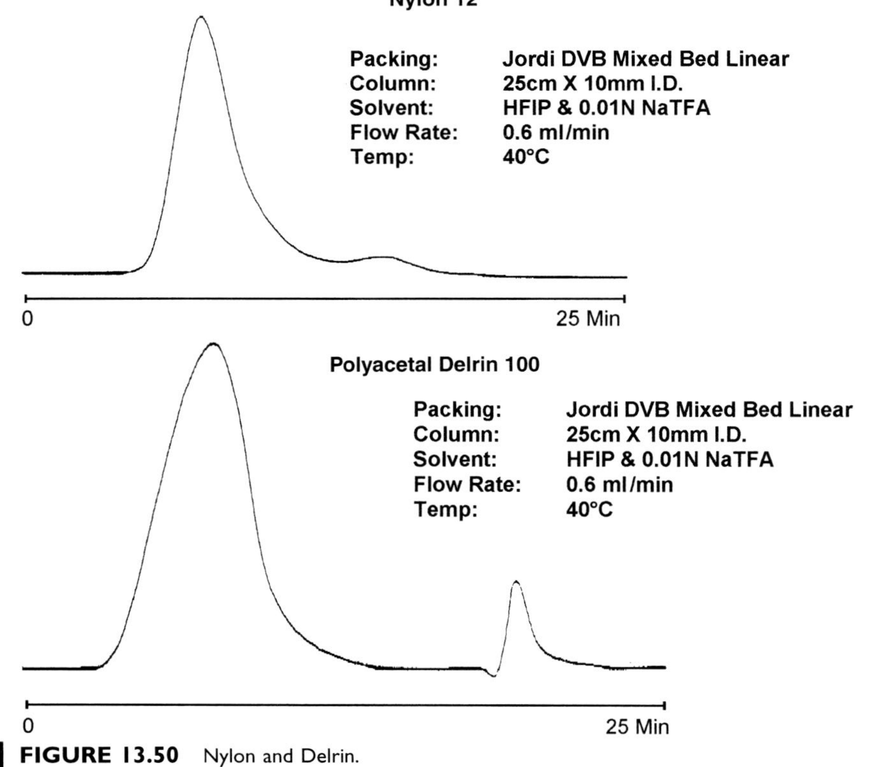

FIGURE 13.50 Nylon and Delrin.

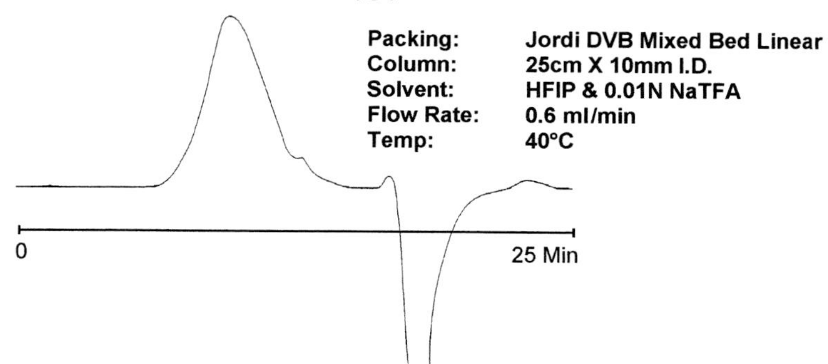

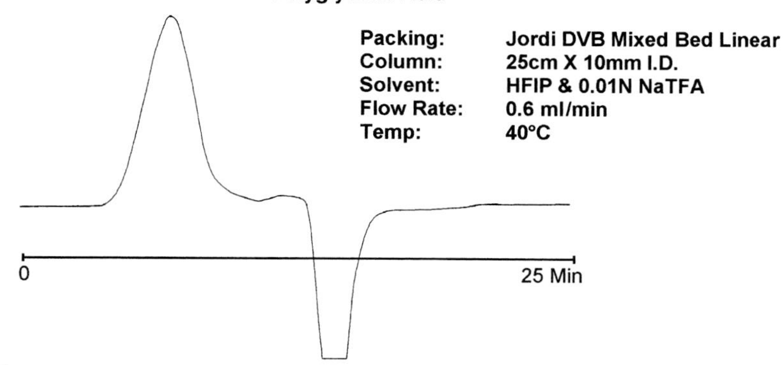

FIGURE 13.51 Polyglycolides.

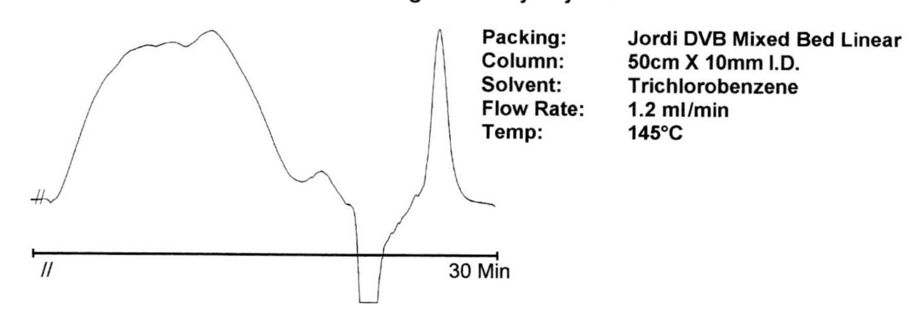

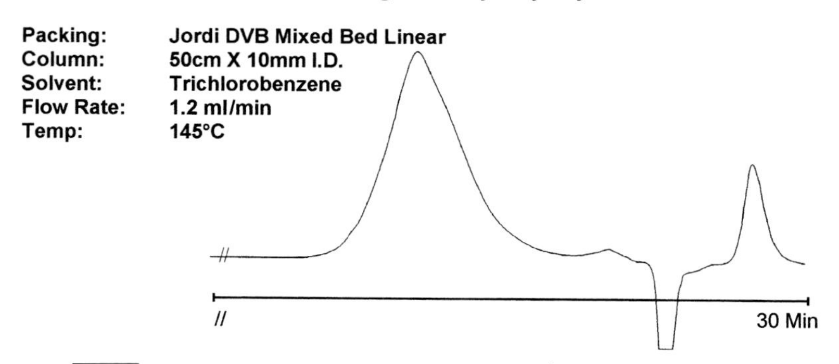

FIGURE 13.52 High temperature GPC of polyethylenes.

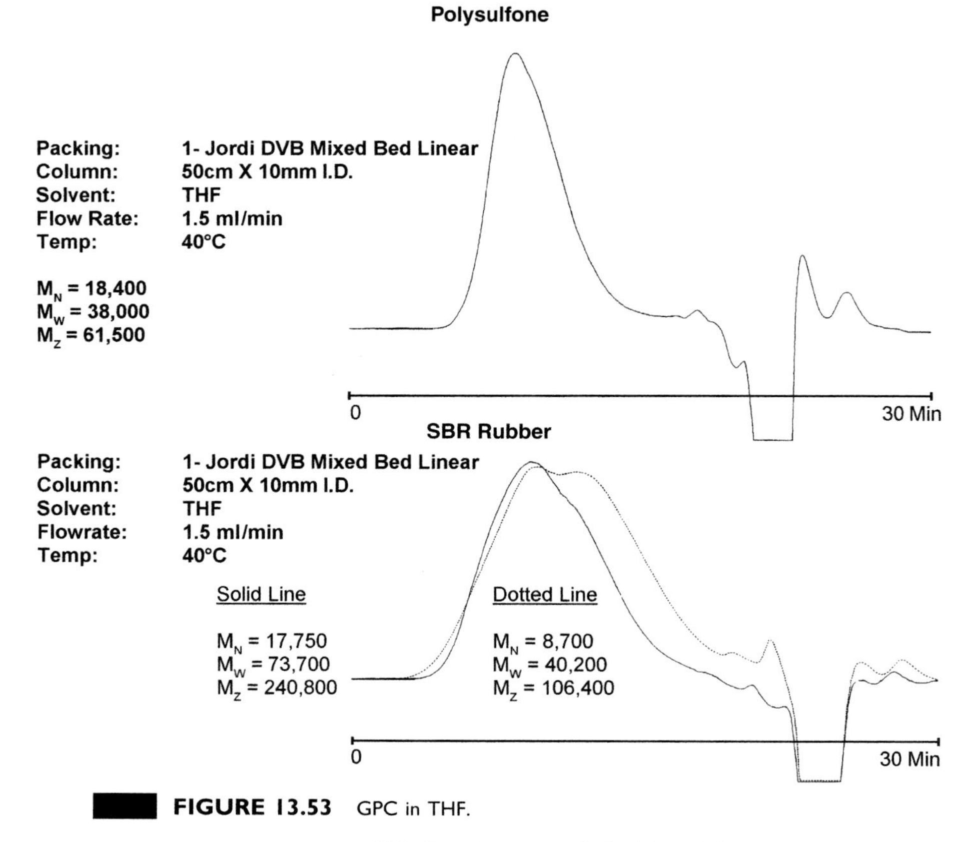

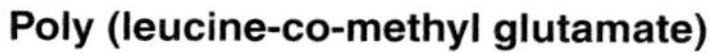

FIGURE 13.53 GPC in THF.

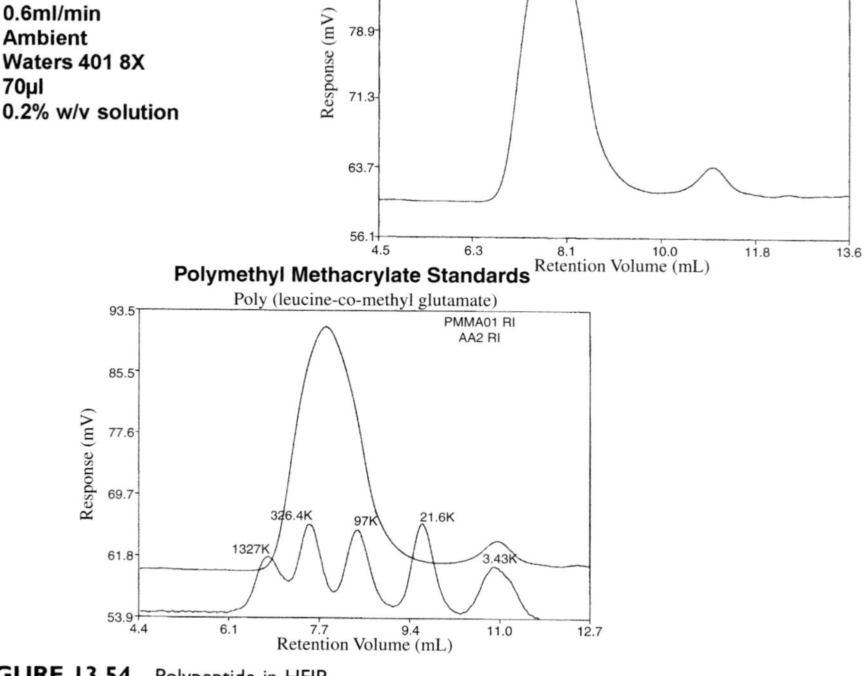

FIGURE 13.54 Polypeptide in HFIP.

Perfluoroether Z-Dol 2000 Lubricant

Packing:	**(1) Jordi Gel DVB Mixed Bed**
Column:	**50cm x 10mm ID**
Solvent:	**1, 1, 2 Trichloro-1, 2, 2, Trifluoroethane with modifiers**
Sample:	**Perfluoroether Z-Dol 2000 Lubricant**
Flow Rate:	**1.3 ml/min**
Temp:	**27°C**
Detector:	**Waters 410; 8x, 20**
Inj:	**100µl**
Conc:	**10 mg/ml**

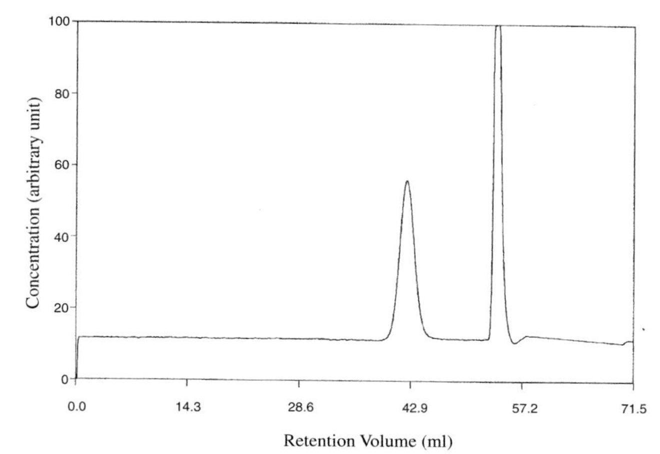

FIGURE 13.55 GPC in freon of perfluoroether Z-Dol 2000 lubricant.

PEEK (Polyether Ether Ketone)

Packing:	**Jordi DVB, Mixed Bed Linear**
Column:	**50cm X 10mm I.D.**
Solvent:	**3:1 $CHCl_3$:Dichloroacetic Acid**
Flow Rate:	**1.2 ml/min**
Temp:	**25°C**
Detector:	**RI**
Injection:	**200µl**
Conc:	**0.3% W/V**

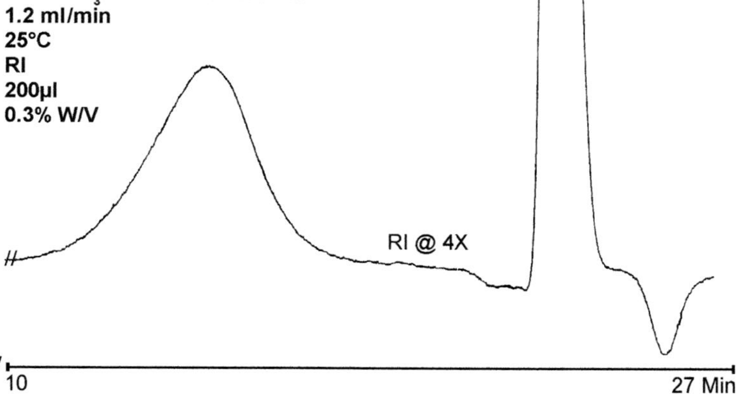

FIGURE 13.56

Parabens by GPC and LC

Strongest Deactivator **Intermediate Deactivator** **Weakest Deactivator**

Packing: (2) Jordi Gel DVB, 500Å
Column: 50cm X 10mm I.D.
Flow Rate: 1.5 ml/min
Temp: 25°C

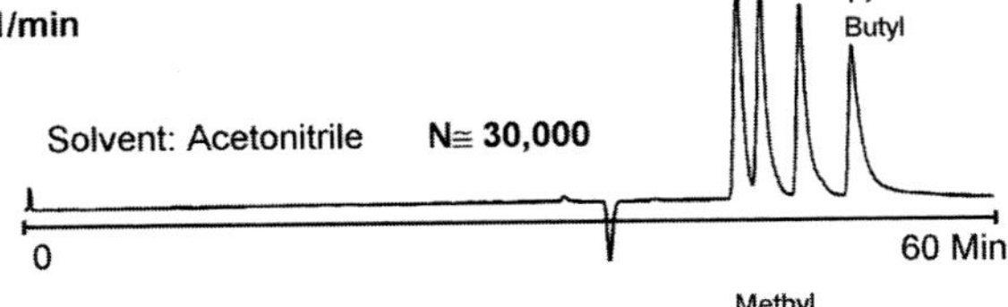

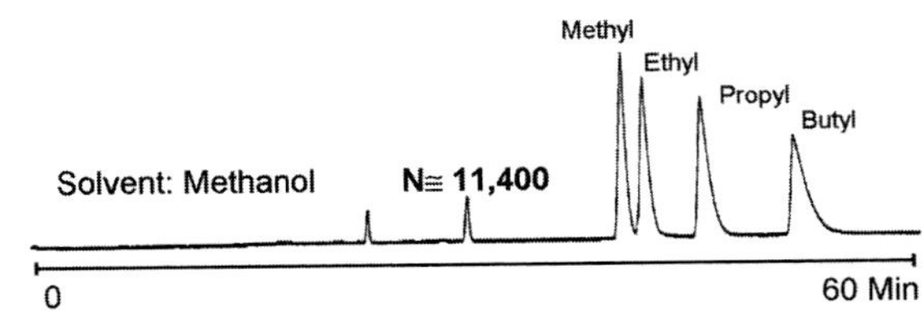

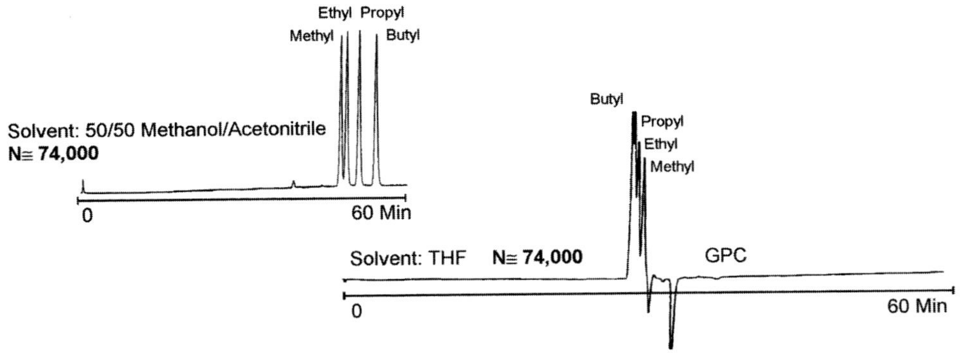

FIGURE 13.57 NOTE: The importance of Solvent/column interaction using Jordi DVB columns *cannot* be over emphasized. We have found that a 50/50 mix of MeOH/ACN for the strong solvent is adequate for many reverse phase separations and is better than either alone. We have now observed that the use of THF/ACN as strong solvent is often better than MeOH/ACN. In general Lewis bases (electron donor solvents) deactivate the aromatic rings and often dramatically increase column efficiencies.

The GBR resin works well for nonionic and certain ionic polymers such as various native and derivatized starches, including sodium carboxymethylcellulose, methylcellulose, dextrans, carrageenans, hydroxypropyl methylcellulose, cellulose sulfate, and pullulans. GBR columns can be used in virtually any solvent or mixture of solvents from hexane to 1 *M* NaOH as long as they are miscible. Using sulfonated PDVB gels, mixtures of methanol and 0.1 *M* Na acetate will run many polar ionic-type polymers such as poly-2-acrylamido-2-methyl-1-propanesulfonic acid, polystyrene sulfonic acids, and poly aniline/polystyrene sulfonic acid. Sulfonated columns can also be used with water: glacial acetic acid mixtures, typically 90/10 (v/v). Polyacrylic acids run well on sulfonated gels in 0.2 *M* NaAc, pH 7.75.

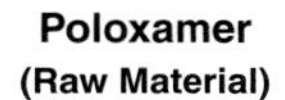

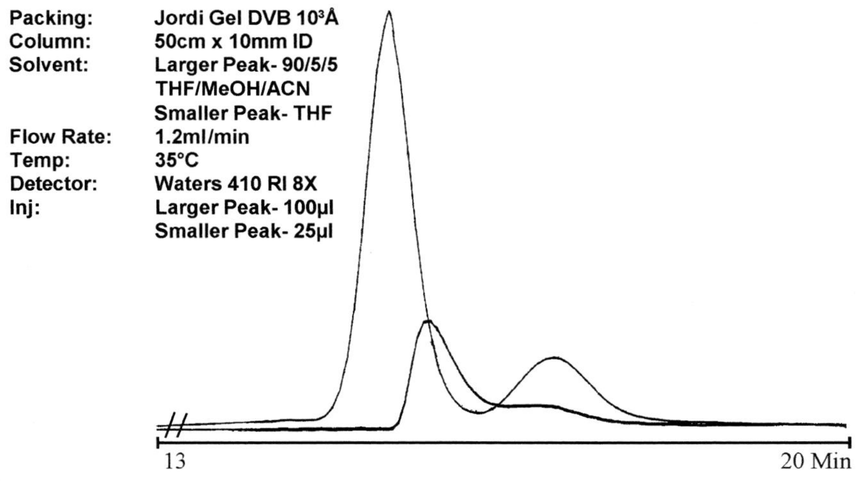

FIGURE 13.58 Poloxamer.

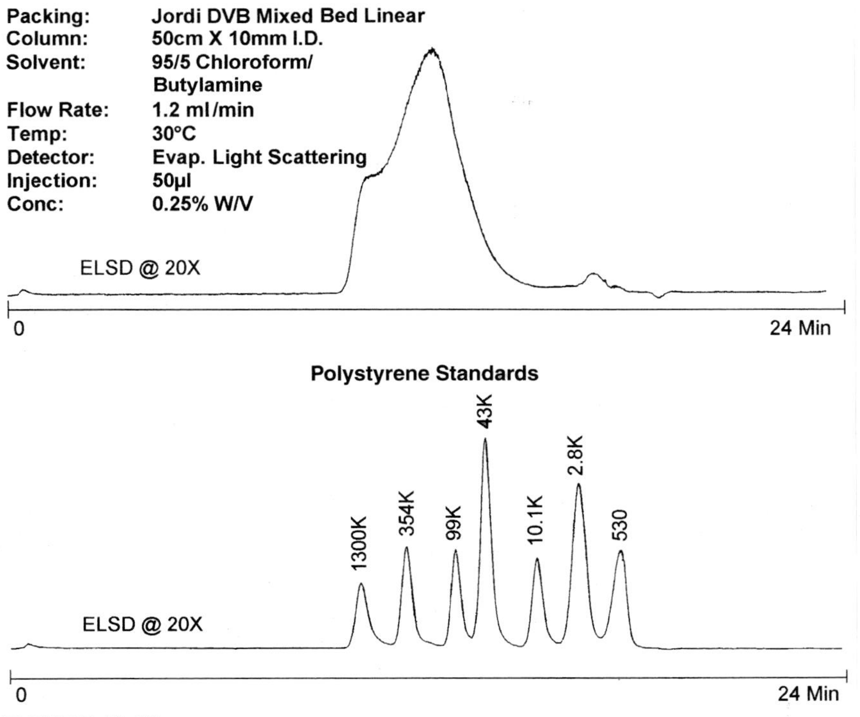

FIGURE 13.59

Poly-2-Acrylamido-2-Methyl-1-Propanesulfonic Acid

Packing:	**(2) Jordi Gel DVB Mixed Bed Sulfonated**
Column:	**25cm x 10mm ID**
Solvent:	**80/20 (v/v) 0.1M Na Acetate/Methanol**
Sample:	**Poly-2-Acrylamido-2-Methyl-1-Propanesulfonic Acid**
Flow Rate:	**1.0 ml /min**
Temp:	**35°C**
Detector:	**Waters 410; 8x, 20**
Inj:	**100µl**
Conc:	**0.25% (w/v)**

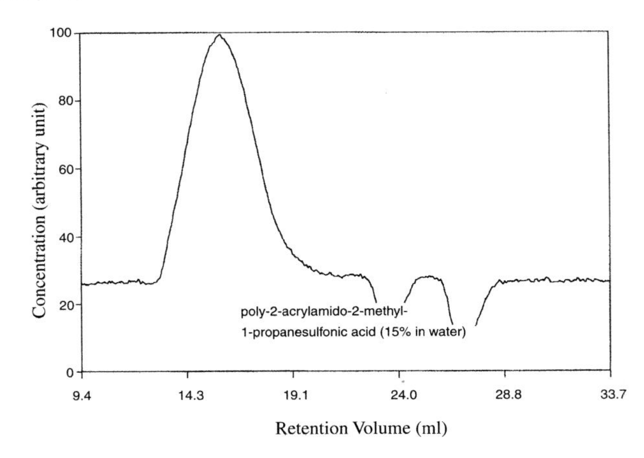

FIGURE 13.60

Poly(styrene sulfonic acid)

Packing:	**(2) Jordi Gel DVB Mixed Bed Sulfonated**
Column:	**25cm x 10mm ID**
Solvent:	**80/20 (v/v) 0.1M Na Acetate/Methanol**
Sample:	**Poly(styrene sulfonic acid)**
Flow Rate:	**1.0 ml/min**
Temp:	**35°C**
Detector:	**Waters 410; 8x, 20**
Inj:	**100µl**
Conc:	**0.25% (w/v)**

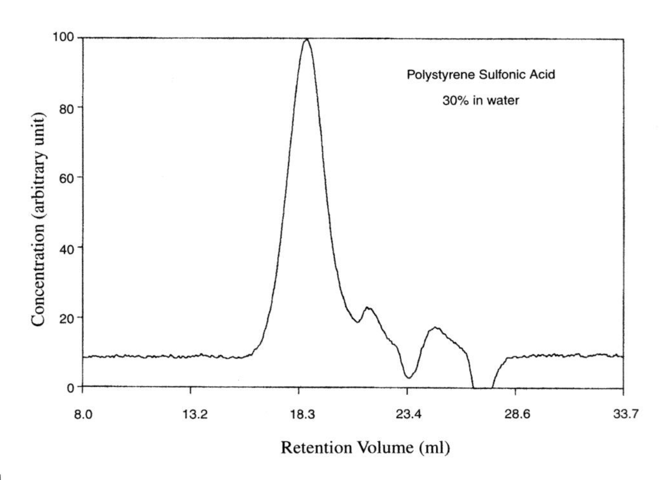

FIGURE 13.61

Polyaniline/Polyacrylamido-2-Methyl-1-Propanesulfonic Acid

Packing: (2) Jordi Gel DVB Mixed Bed Sulfonated
Column: 25cm x 10mm ID
Solvent: 80/20 (v/v) 0.1M Na Acetate/Methanol
Sample: Polyaniline/Polyacrylamido-2-Methyl-1-Propanesulfonic Acid
Flow Rate: 1.0 ml/min
Temp: 35°C
Detector: Waters 410; 8x, 20
Inj: 250µl
Conc: 0.25% (w/v)

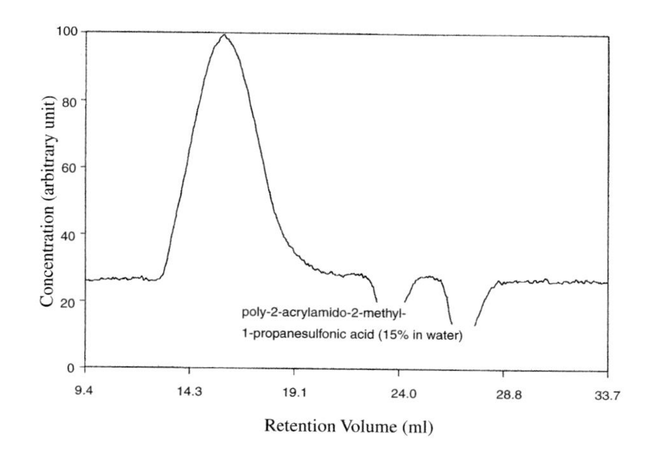

FIGURE 13.62

Polyaniline/Polystyrene Sulfonic Acid

Packing: (2) Jordi Gel DVB Mixed Bed Sulfonated
Column: 25cm x 10mm ID
Solvent: 80/20 (v/v) 0.1M Na Acetate/Methanol
Sample: Polyaniline/Poly(styrene sulfonic acid)
Flow Rate: 1.0 ml/min
Temp: 35°C
Detector: Waters 410; 8x, 20
Inj: 250µl
Conc: 0.25% (w/v)

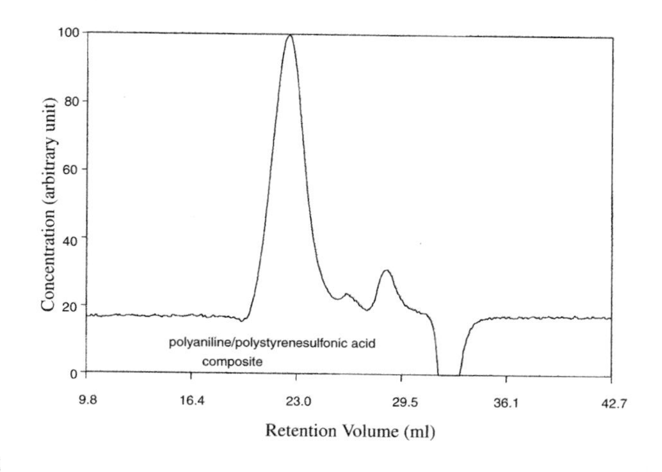

FIGURE 13.63

Phospholipid EFA Sample

Packing:	**(1) Jordi Gel PEI WAX 10³Å**
Column:	**25cm x 10mm ID**
Solvent:	**90/10 (v/v) H_2O/Glacial Acetic Acid**
Sample:	**Phospholipid EFA Sample**
Flow Rate:	**1.0 ml/min**
Temp:	**35°C**
Detector:	**Waters 410; 8x, 20**
Inj:	**100µl**
Conc:	**Sample 0.85% (w/v); Polymer 0.255% (w/v)**

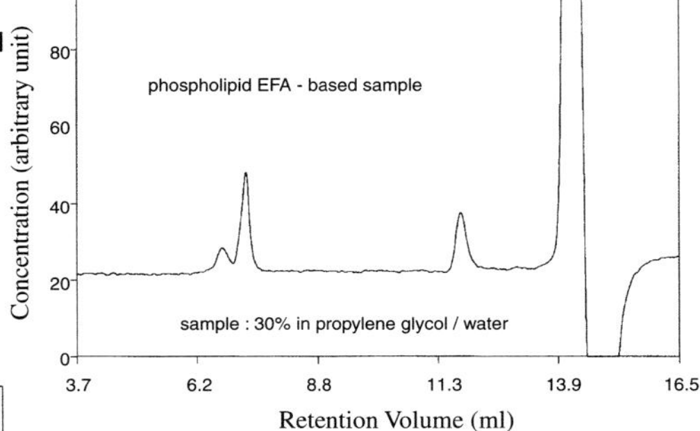

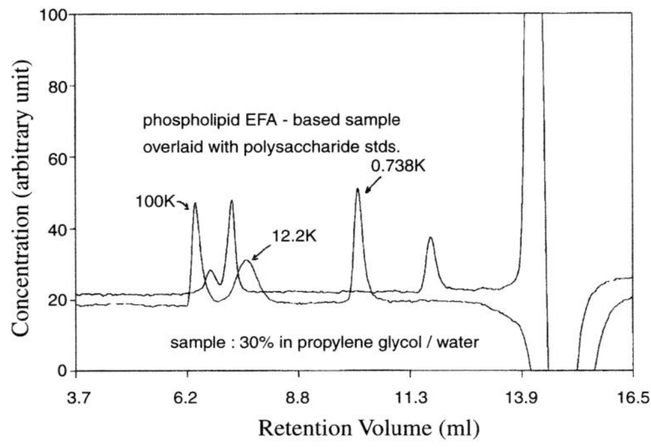

FIGURE 13.64

Polycarboxylate/Polyacrylic-Itaconic Acid

Packing:	**(1) Jordi Gel DVB Mixed Bed Sulfonated**
Column:	**50cm x 10mm ID**
Solvent:	**90/10 (v/v) Water: Glacial Acetic Acid**
Sample:	**Polycarboxylate/ Polyacrylic-Itatonic Acid**
Flow Rate:	**1.2 ml/min**
Temp:	**45°C**
Detector:	**Waters 410 DRI; 4x, 20**
Inj:	**300µl**
Conc:	**0.5% (w/v)**

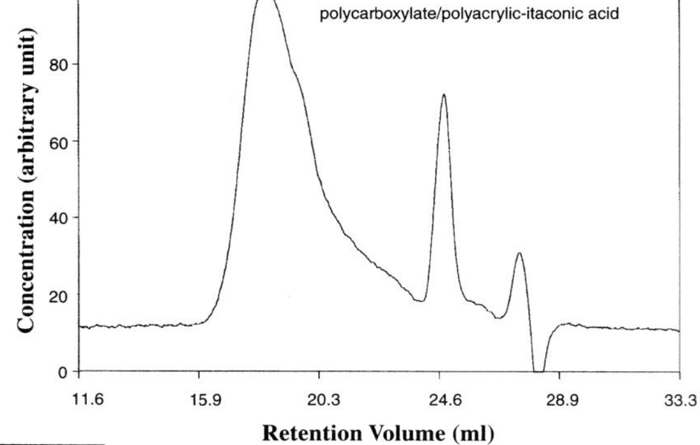

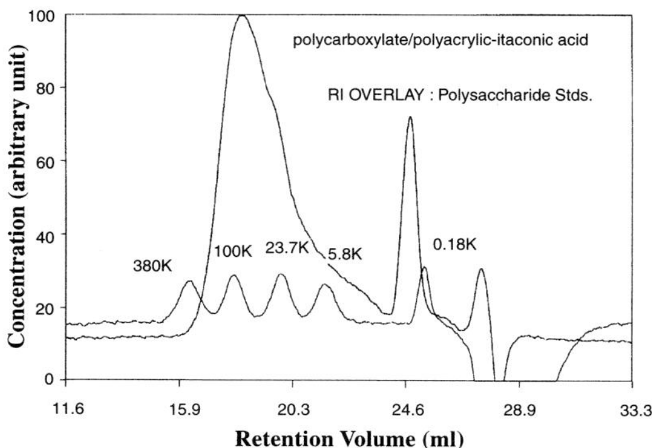

FIGURE 13.65

Polyethyleneimine Pullulan Standards

Packing: (2) Jordi Gel PEI WAX Mixed Bed
Column: 24cm x 10mm ID
Solvent: 95/5 (v/v) Water:Glacial Acetic Acid
Sample: Polyethyleneimine
Flow Rate: 1.0ml/min
Temp: 45°C
Detector: Waters 410 DRI; 8x, 20
Inj: 75μl
Conc: 0.1% (w/v)

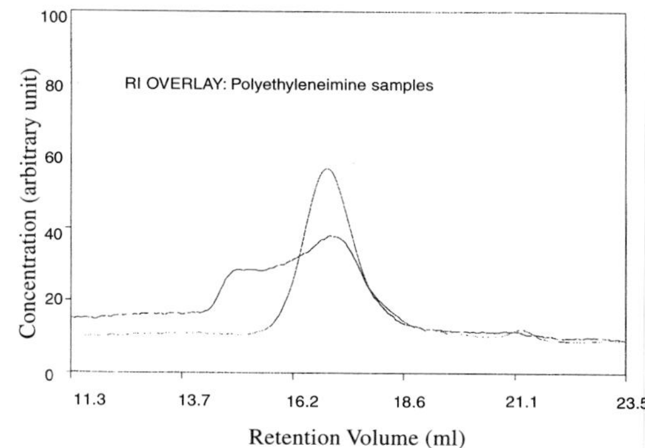

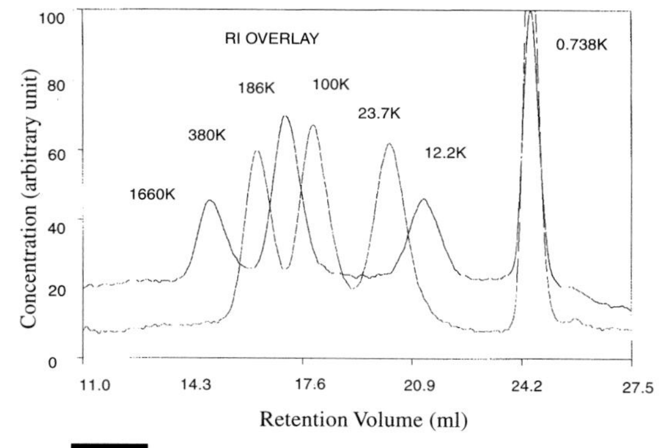

FIGURE 13.66 Polyethyleneimine.

Water-Soluble Flocculant

Packing: (1) Jordi Gel DVB 10^5Å Polar Pac WAX
Column: 25cm x 10mm ID
Solvent: 90/10 (v/v) Water:Glacial Acetic Acid
Sample: Water-soluble Flocculant
Flow Rate: 1.0ml/min
Temp: Ambient
Detector: RI, Waters 410, 4x
Inj: 100μl
Conc: 5 mg/ml

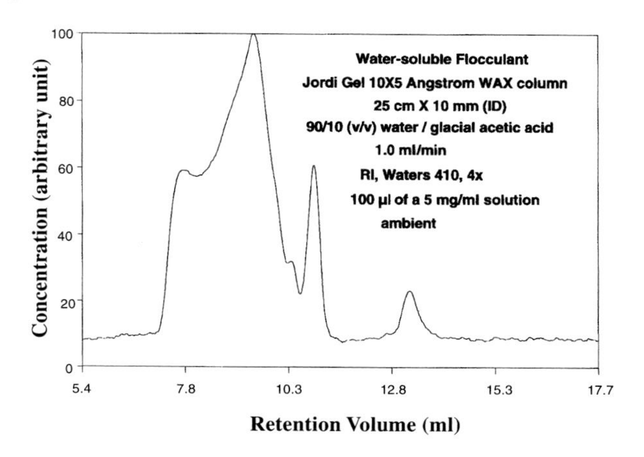

FIGURE 13.67

Analysis of Corn Syrup

Packing:	**(1) Jordi Gel DVB 500Å Sulfonated Polar Pac SCX**
Column:	**25cm x 10mm ID**
Solvent:	**Water**
Sample:	**Corn Syrup**
Flow Rate:	**1.0ml/min**
Temp:	**35°C**
Detector:	**RI**
Inj:	**100μl**
Conc:	**5 mg/ml**

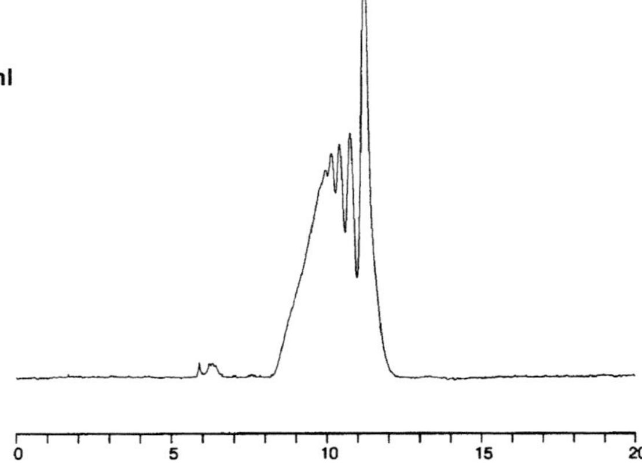

FIGURE 13.68

Methacrylamidopropyltrimethyl Ammonium Chloride-Based Polymer

Packing:	**(2) Jordi Gel Polar Pac WAX Mixed Bed DVB**
Column:	**25cm x 10mm ID**
Solvent:	**90/10 Water/Glacial Acetic Acid**
Flow Rate:	**1.2ml/min**
Temp:	**35°C**
Detector:	**Waters 410 RI 8X, Scale Factor 20**
Inj:	**400μl**
Conc:	**0.15% (w/v)**

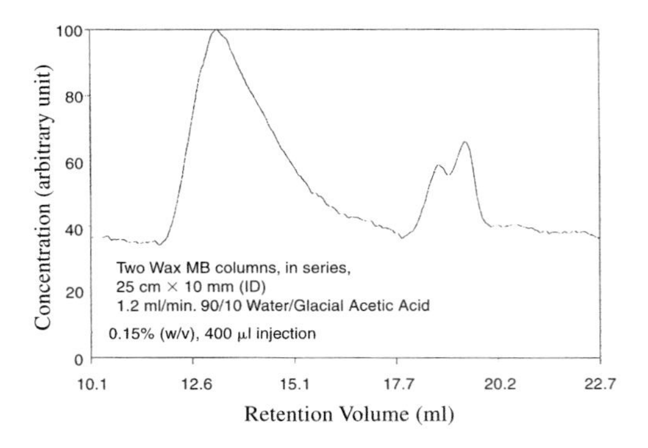

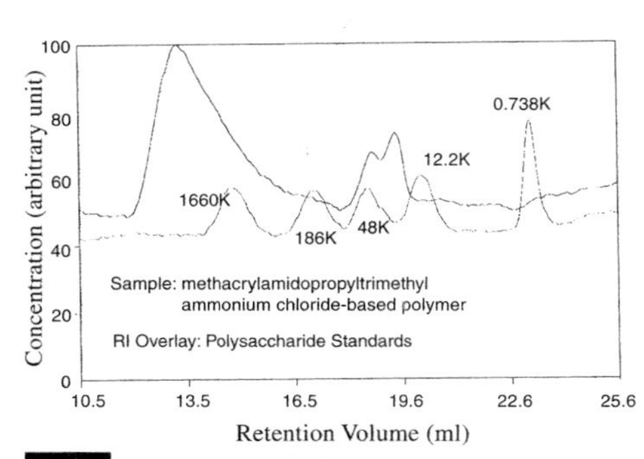

FIGURE 13.69

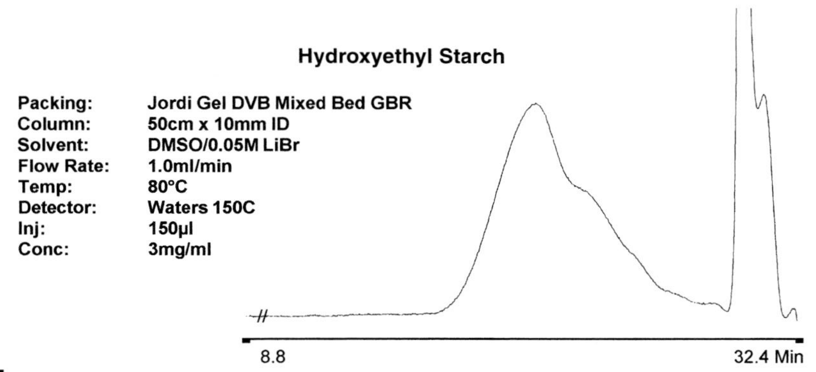

FIGURE 13.70

One of the more difficult sample types to run has traditionally been the positively charged polymers. They seem to adsorb even to GBR gels and not elute at all or elute in a mixed mode way even with modifiers present. Another bonded phase—the PEI or WAX phase on PDVB gels—has been developed to eliminate the unwanted adsorption of polyquaternary-type polymers. By using acids as mobile phases, the surface of the column packing becomes positively charged and tends to repel positively charged samples, thus eliminating unwanted adsorption. Figures 13.66, 13.67, 13.69, 13.72, and 13.74 show

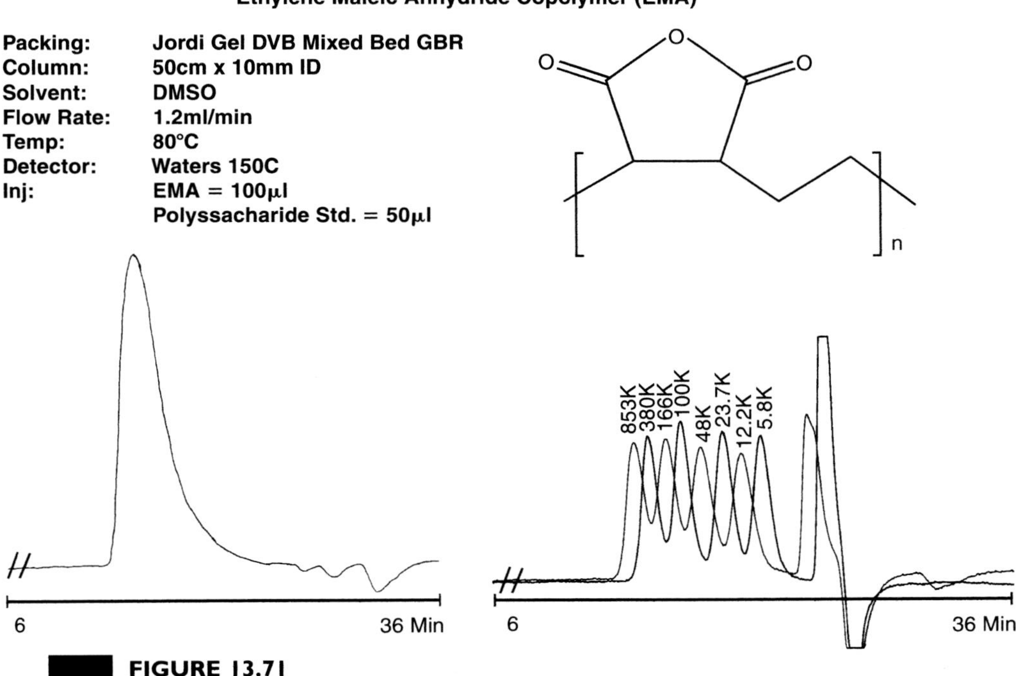

FIGURE 13.71

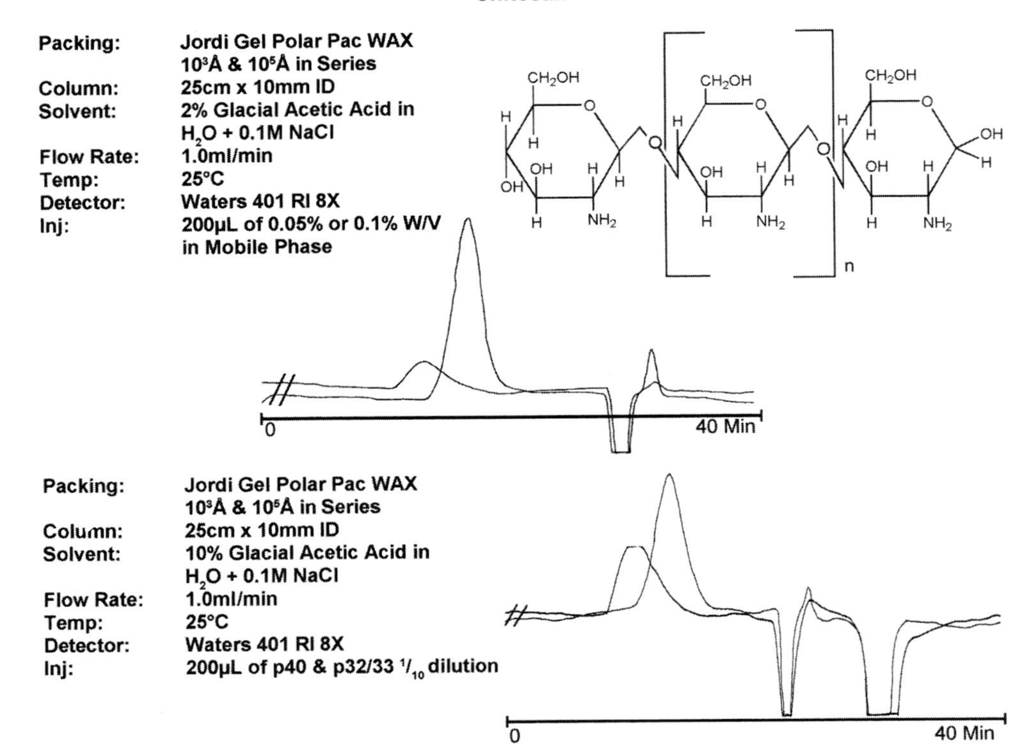

FIGURE 13.72

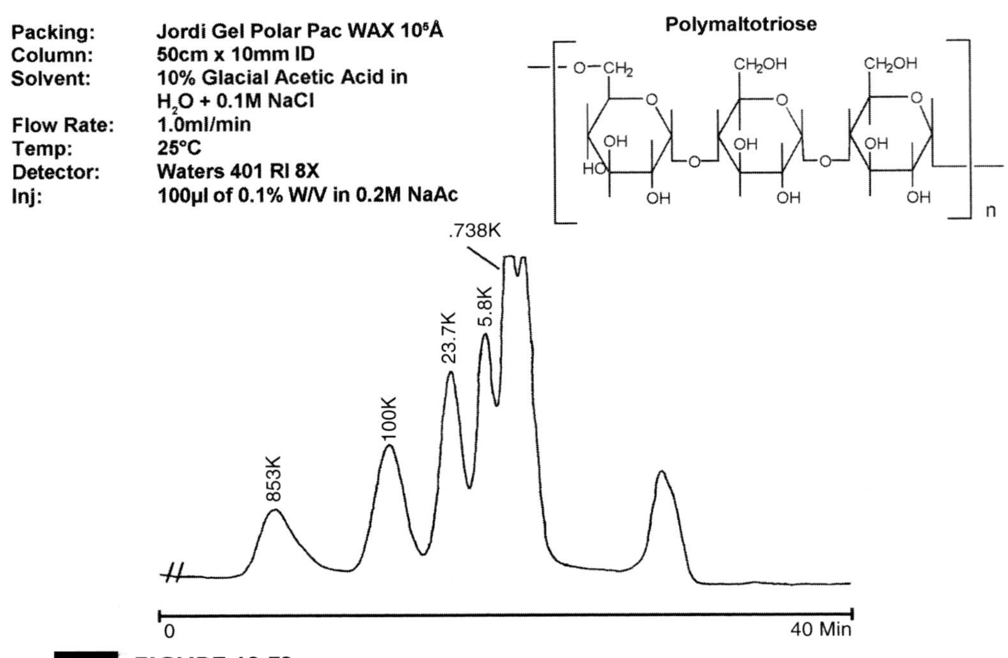

FIGURE 13.73

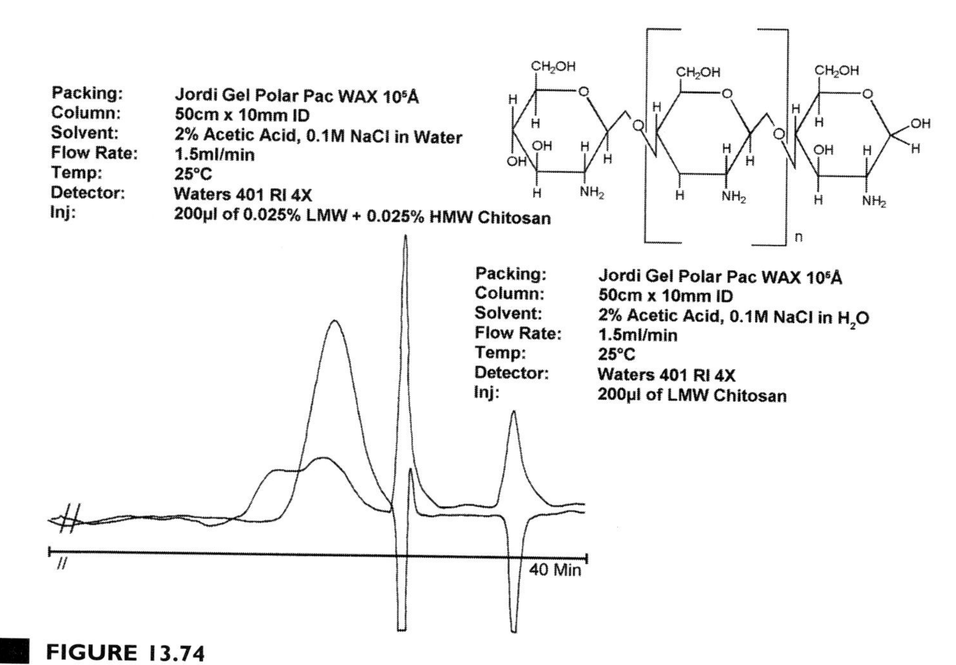

FIGURE 13.74

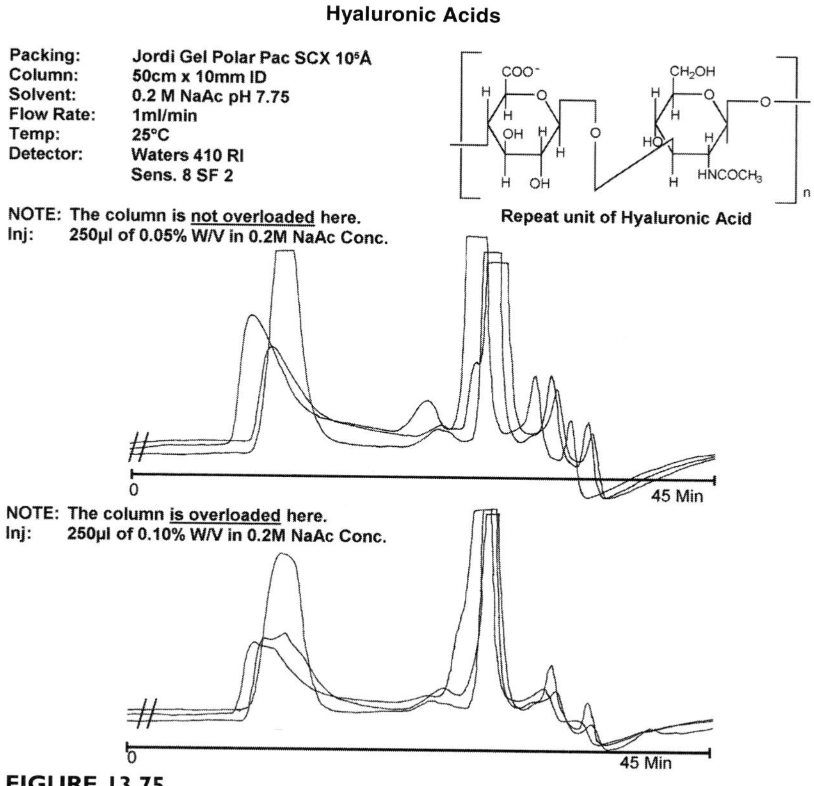

FIGURE 13.75

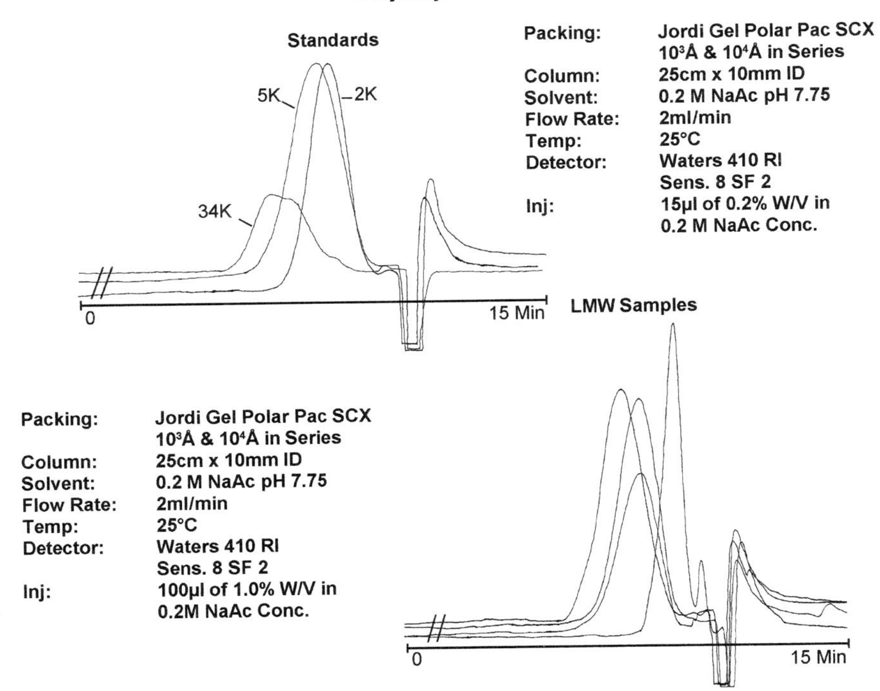

FIGURE 13.76

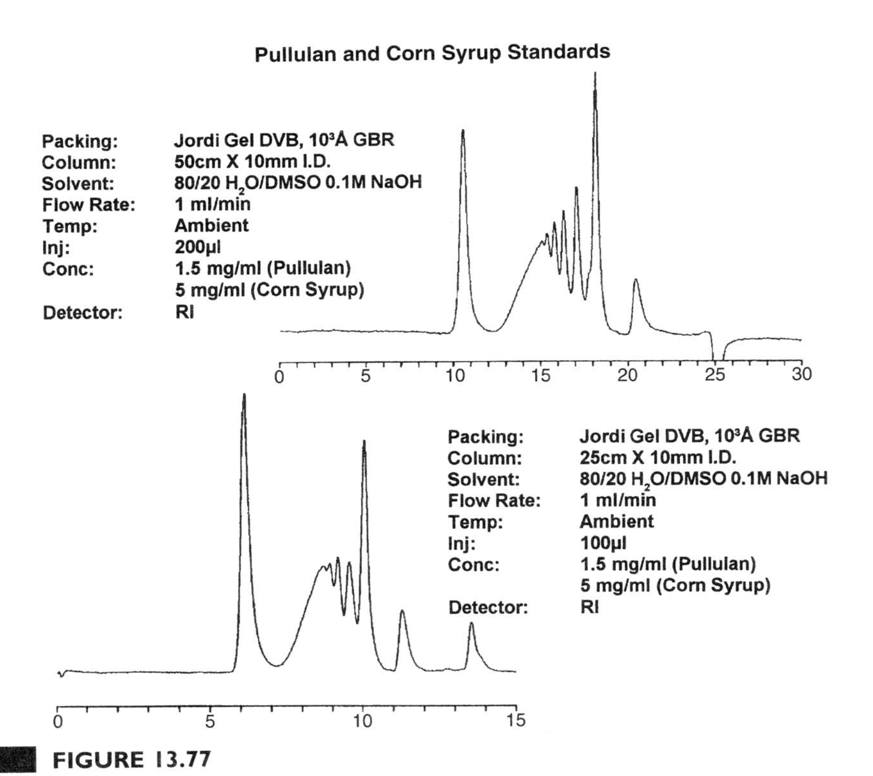

FIGURE 13.77

Hydroxypropyl Methyl Cellulose and Dextran Mix

Packing:	2-Jordi DVB Glucose BR, Mixed Bed Linear
Column:	25cm X 10mm I.D.
Solvent:	DMSO
Flow Rate:	1.0 ml/min
Temp:	80°C
Detector:	RI
Injection:	200µl
Conc:	0.1% W/V

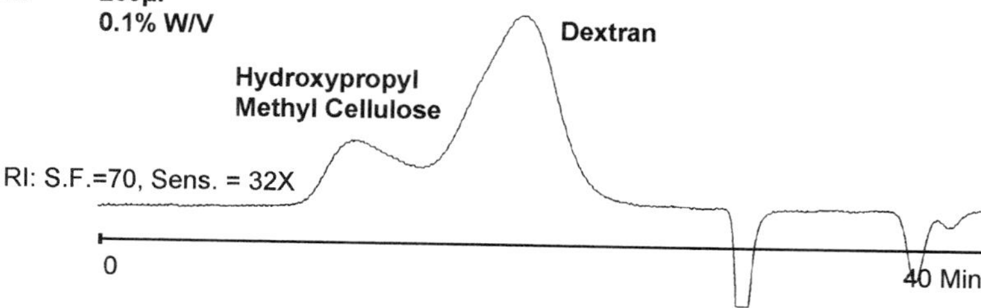

Polysaccharide Standards

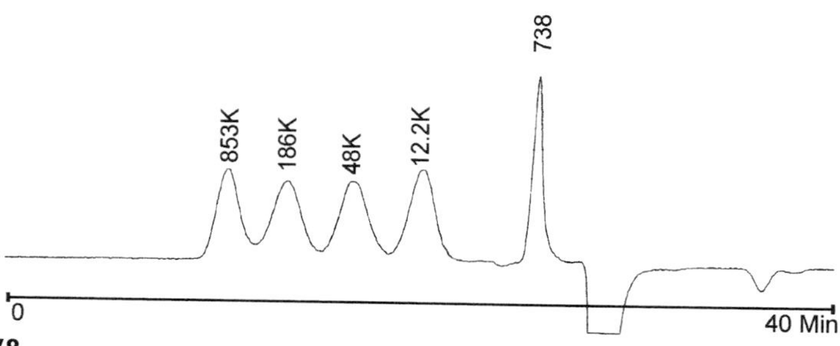

FIGURE 13.78

Cellulose Sulfate

Packing:	Jordi DVB Glucose BR, Mixed Bed Linear
Column:	50cm X 10mm I.D.
Solvent:	80:20 V/V 0.1M NaOH/DMSO
Flow Rate:	1.0 ml/min
Temp:	60°C
Detector:	RI
Injection:	500µl
Conc:	0.05% W/V

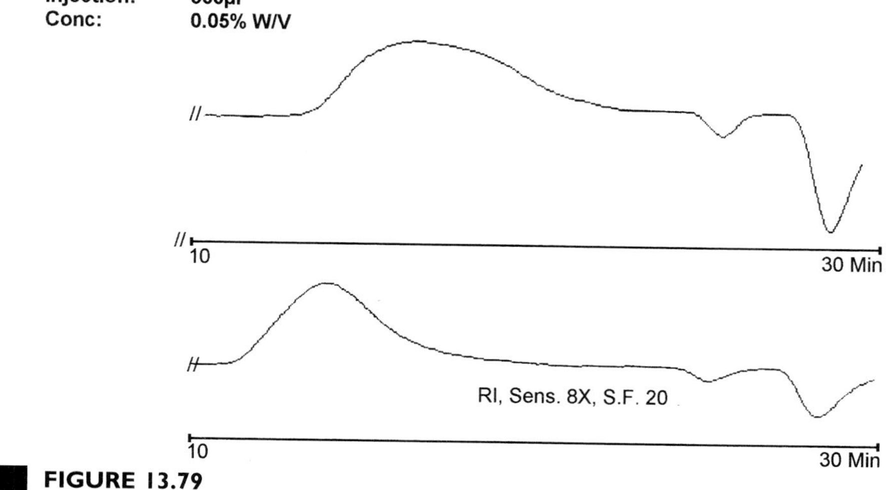

FIGURE 13.79

Gelatins

Packing: 2-Jordi DVB Glucose BR Mixed Bed Linear
Column: 25cm X 10mm I.D.
Solvent: 90%[80:20 V/V H_2O/DMSO] + 10% V/V Butylamine
Flow Rate: 1.2 ml/min
Temp: 80°C
Detector: RI
Injection: 75µl
Conc: 1.0% W/V

RI @ 8X
6 min
24 Min

Polysaccharide Standards
853K
100K
12.2K
0
24 Min

FIGURE 13.80

Poly (Ethylvinyl Alcohol/Maleic Anhydride)

Packing: Jordi DVB Glucose BR Mixed Bed Linear
Column: 50cm X 10mm I.D.
Solvent: DMSO
Flow Rate: 1.0 ml/min
Temp: 80°C
Detector: RI
Injection: 200µl
Conc: 0.1% W/V

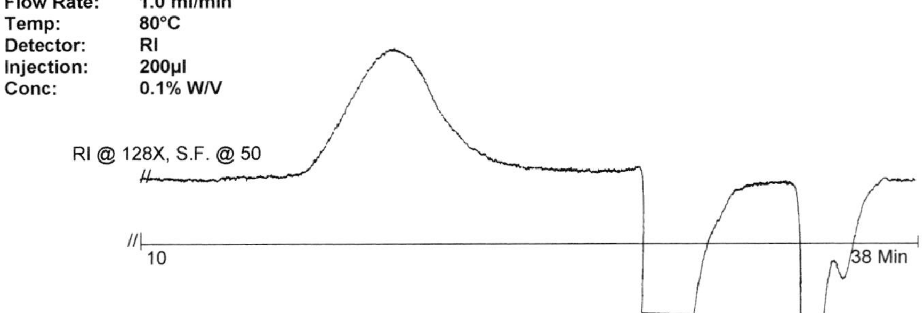

Hydroxylpropyl Methylcellulose

Packing: Jordi DVB Glucose BR Mixed Bed Linear
Column: 50cm X 10mm I.D.
Solvent: DMSO
Flow Rate: 1.0 ml/min
Temp: 80°C
Detector: RI
Injection: 250µl
Conc: 0.1% W/V

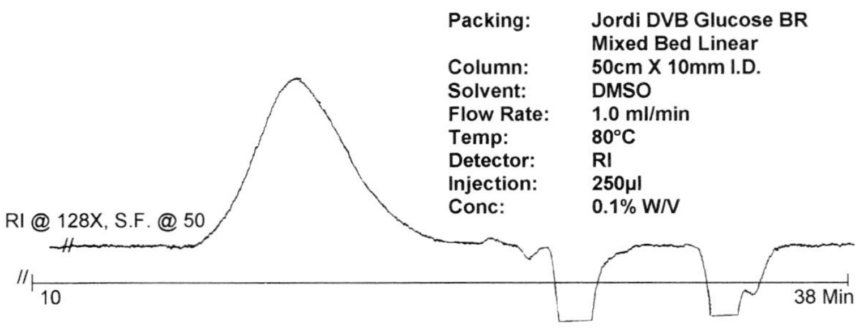

FIGURE 13.81

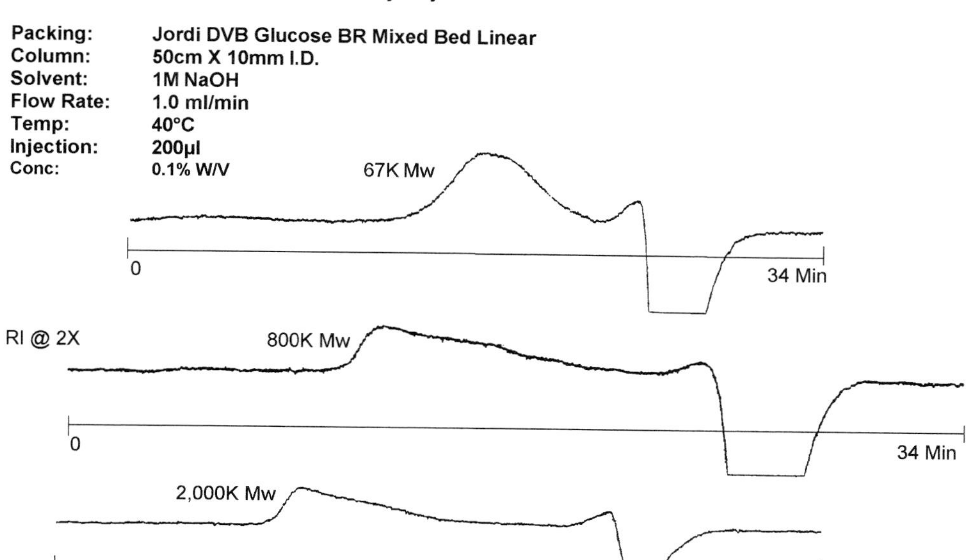

FIGURE 13.82

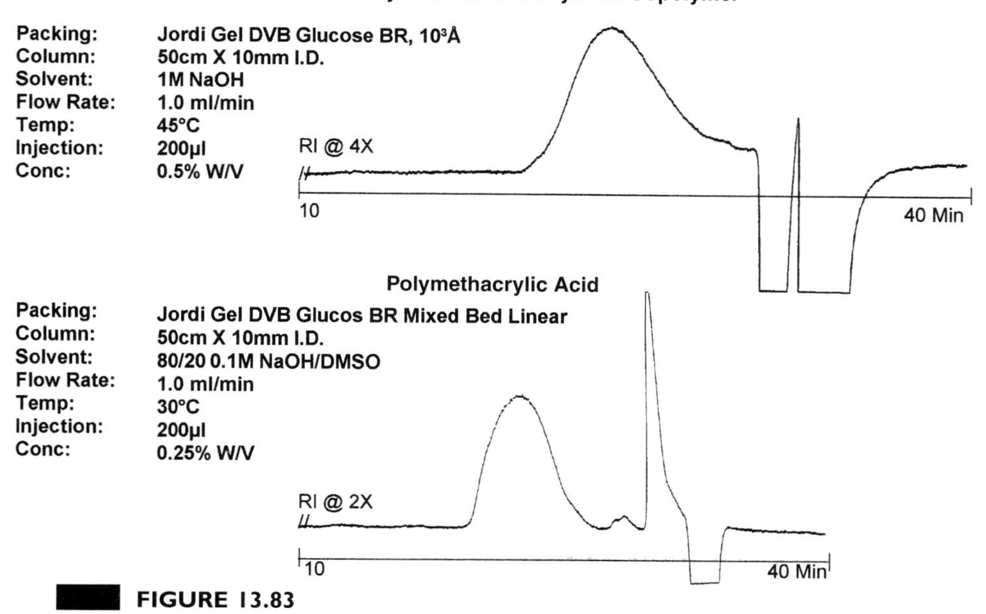

FIGURE 13.83

Polyacrylic Acid

Packing: 2-Jordi Gel DVB Glucose BR, 10^3Å
Column: 50cm X 10mm I.D.
Solvent: 1M NaOH
Flow Rate: 1.5 ml/min
Temp: 60°C
Injection: 150µl
Conc: 1% W/V

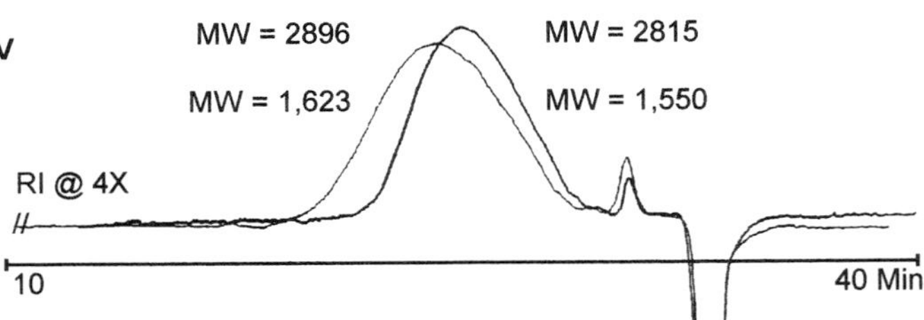

FIGURE 13.84

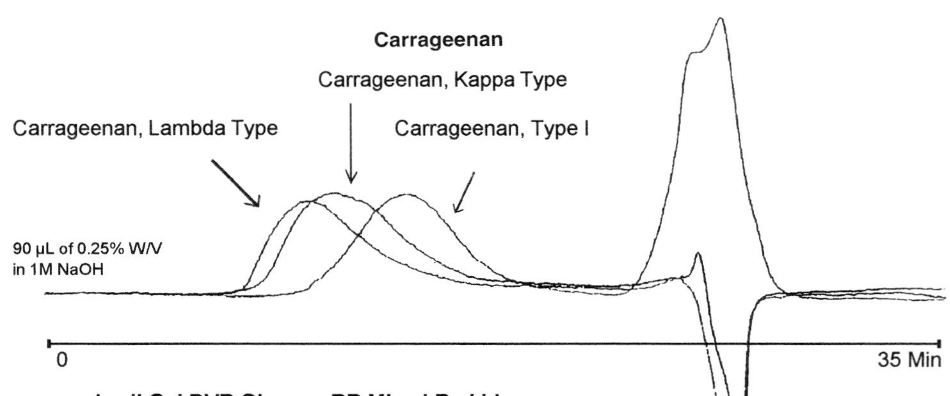

Packing: Jordi Gel DVB Glucose BR Mixed Bed Linear
Column: 50cm X 10mm I.D.
Solvent: 1M NaOH
Flow Rate: 1.0 ml/min
Temp: 40°C

Carrageenan

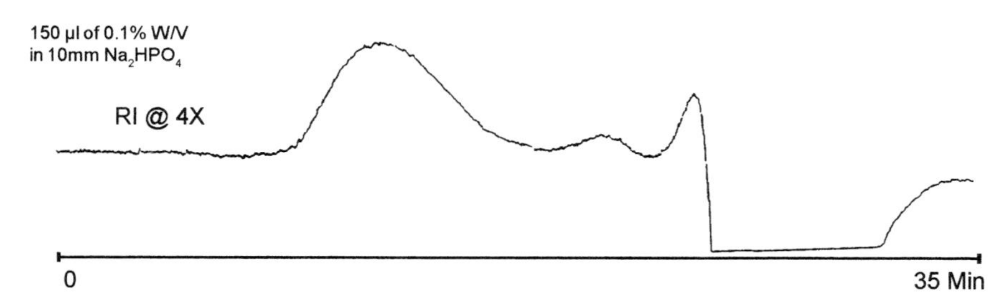

FIGURE 13.85 Carrageenans.

Gum Arabic

Packing:	**Jordi Gel DVB Glucose BR Mixed Bed Linear**
Column:	**50cm X 10mm I.D.**
Solvent:	**80/20 0.1M NaOH/DMSO**
Flow Rate:	**1.0 ml/min**
Temp:	**45°C**
Injection:	**300µl**
Conc:	**0.1% W/V in H_20**

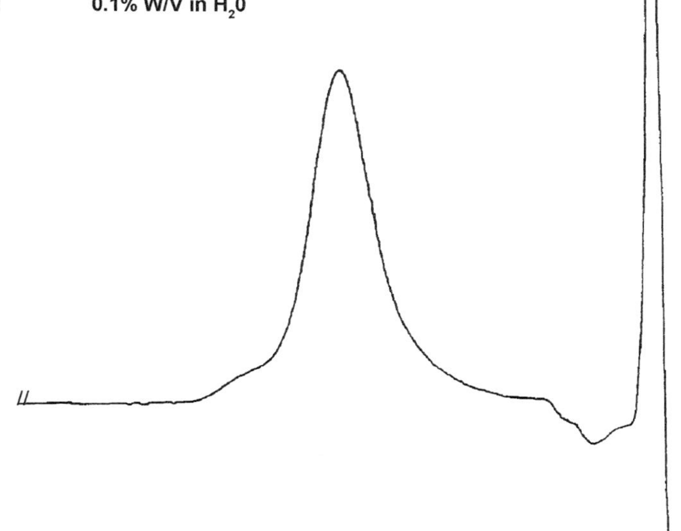

FIGURE 13.86

Alginate A

Packing:	**Jordi Gel DVB Glucose BR, 10^5Å**
Column:	**50cm X 10mm I.D.**
Solvent:	**1M NaOH**
Flow Rate:	**1.0 ml/min**
Temp:	**60°C**
Injection:	**350µl**
Conc:	**0.1% W/V in 1M NaOH**

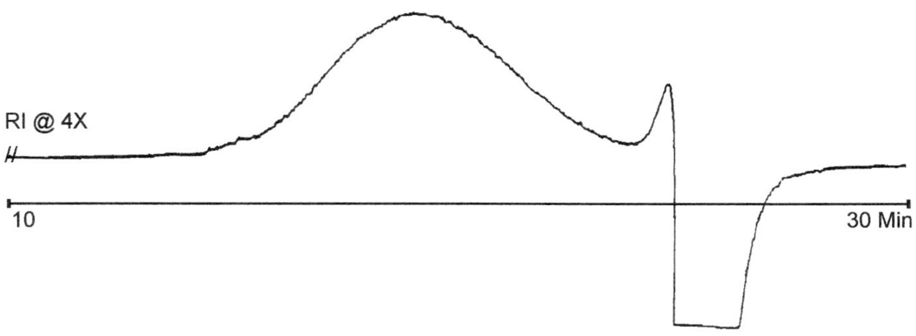

FIGURE 13.87

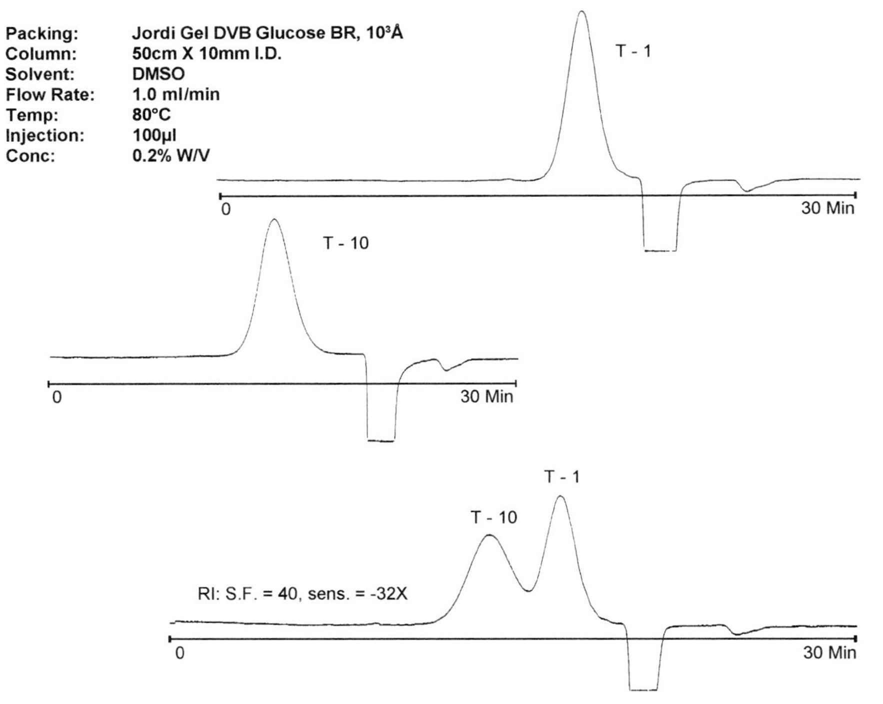

FIGURE 13.88

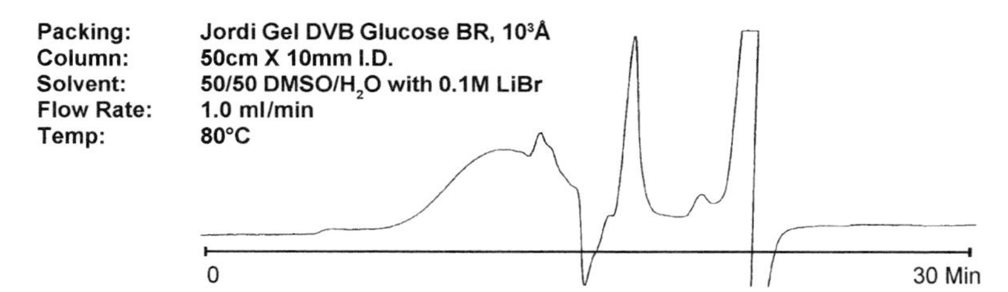

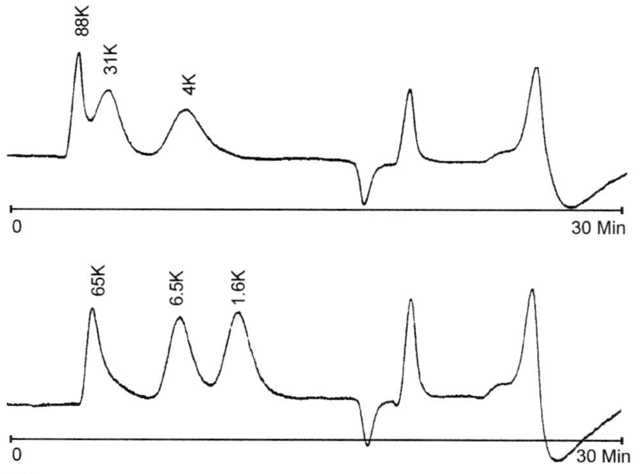

FIGURE 13.89

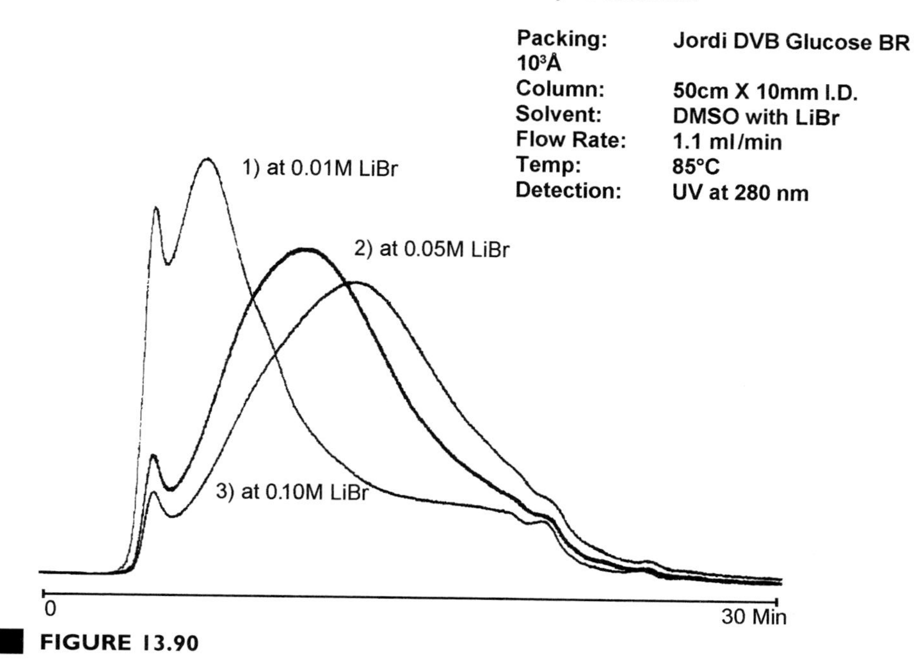

FIGURE 13.90

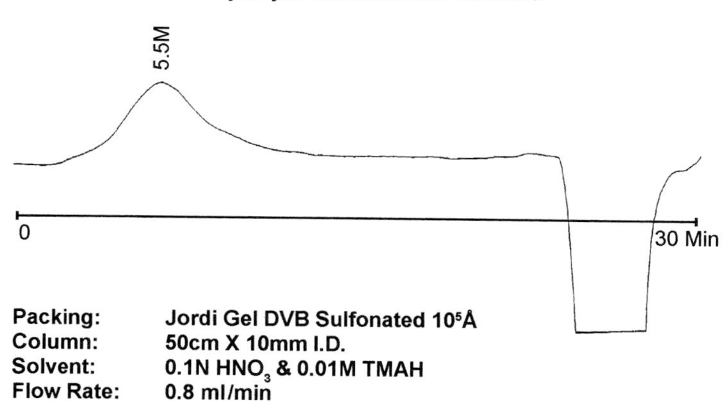

Packing: Jordi Gel DVB Sulfonated 10⁵Å
Column: 50cm X 10mm I.D.
Solvent: 0.1N HNO_3 & 0.01M TMAH
Flow Rate: 0.8 ml/min

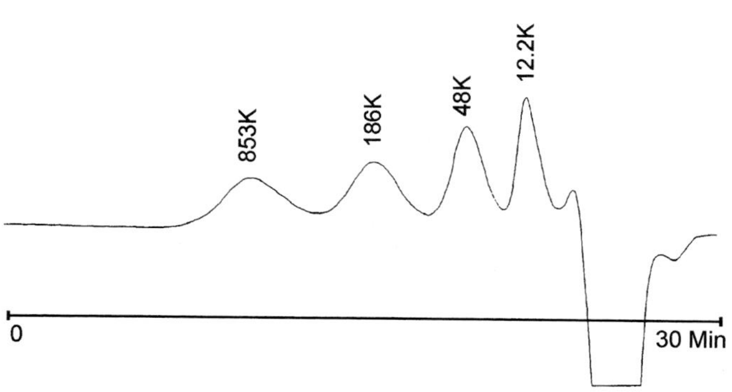

FIGURE 13.91

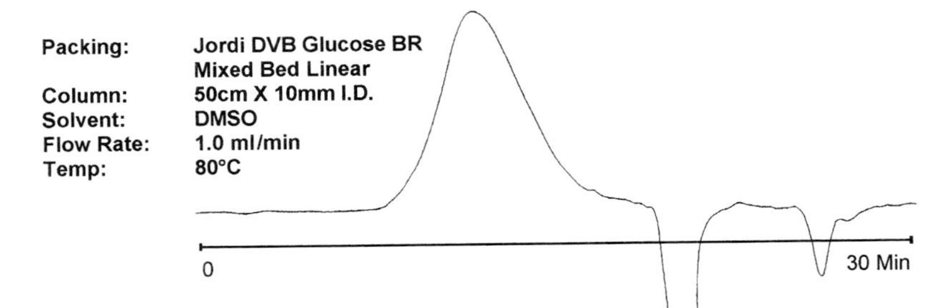

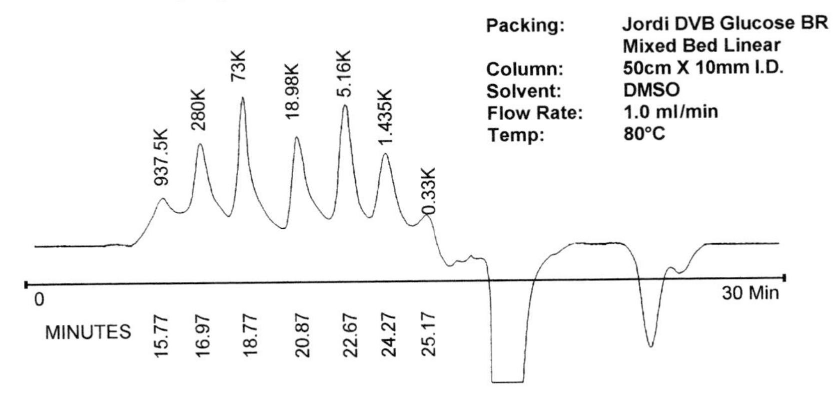

FIGURE 13.92

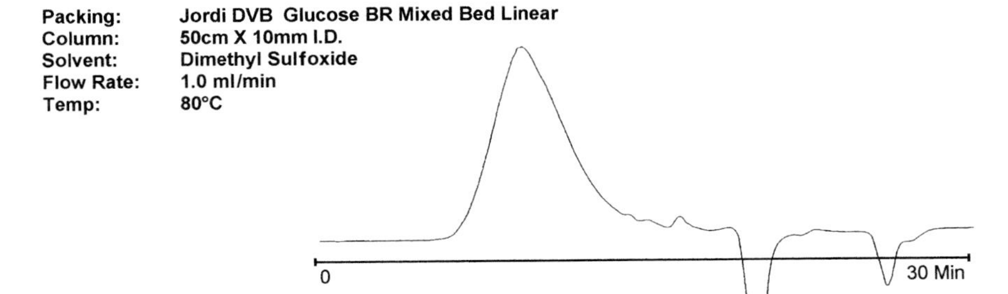

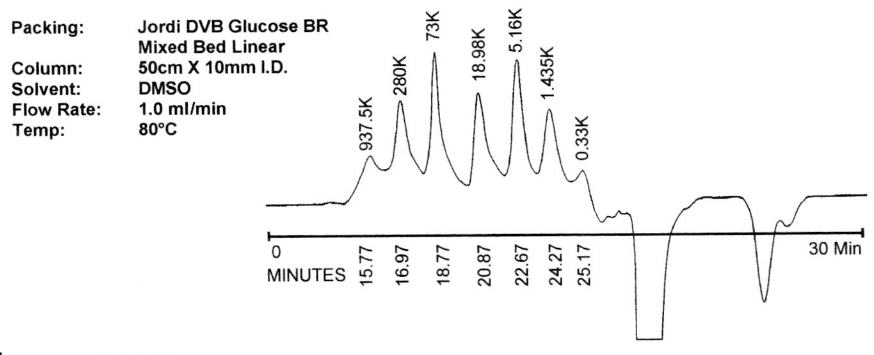

FIGURE 13.93

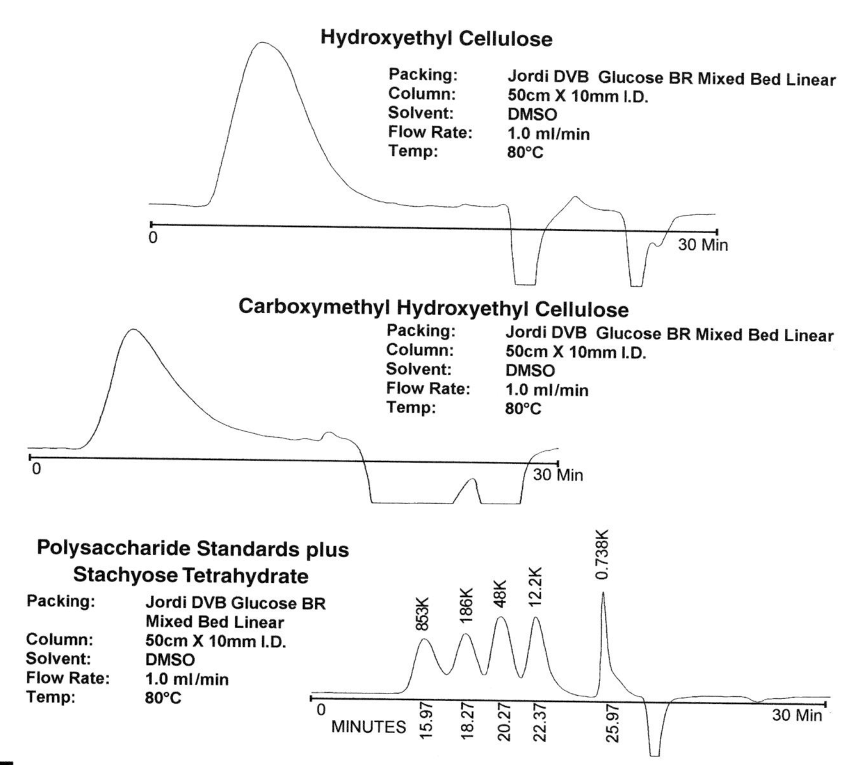

FIGURE 13.94 Various derivatized celluloses.

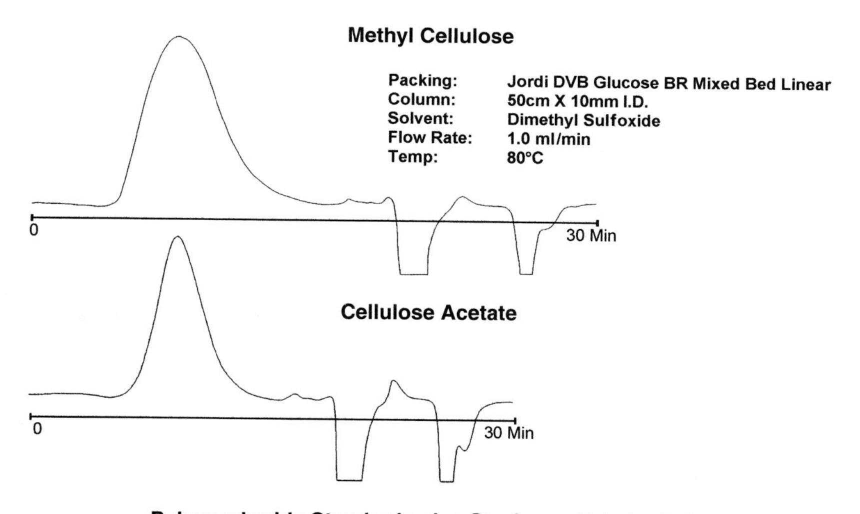

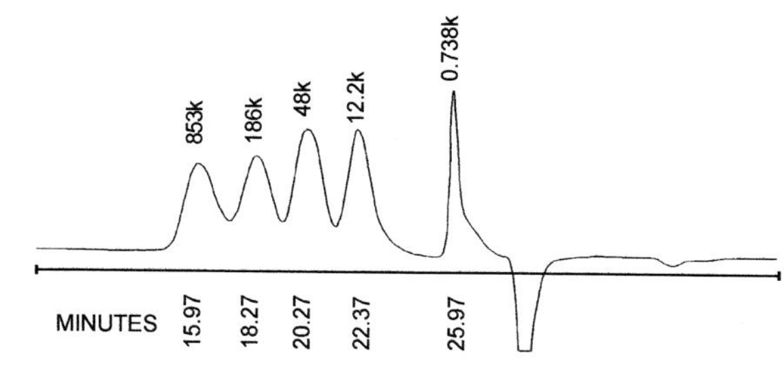

FIGURE 13.95 Various derivatized celluloses.

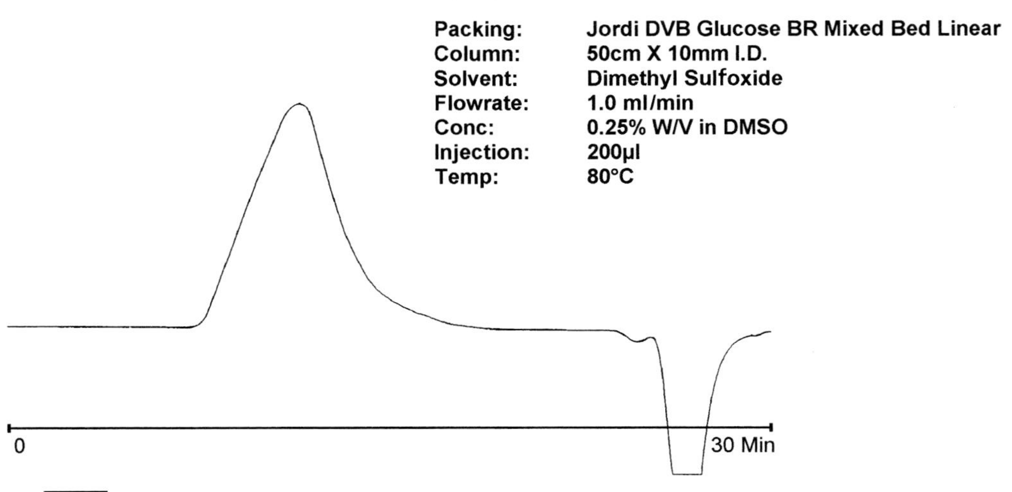

FIGURE 13.96

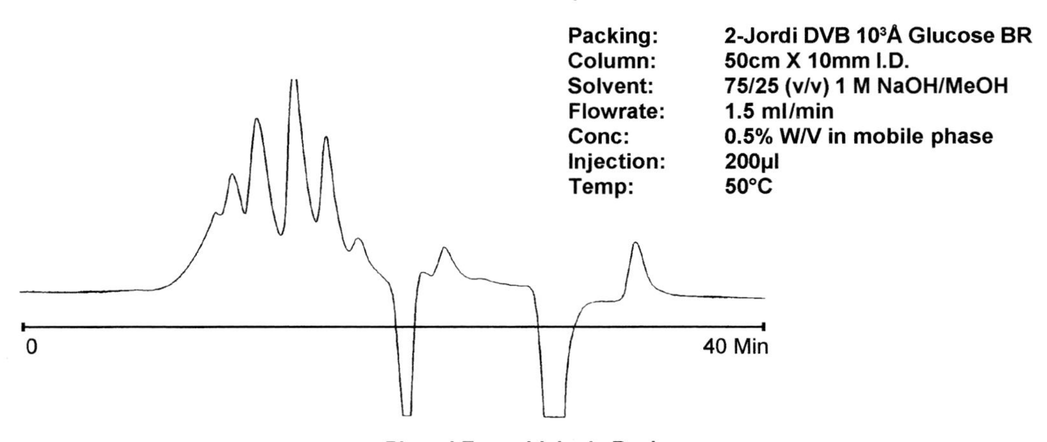

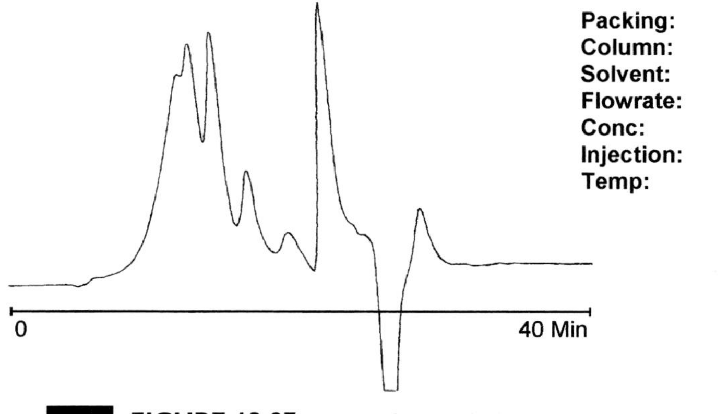

FIGURE 13.97 Phenol formaldehyde resins.

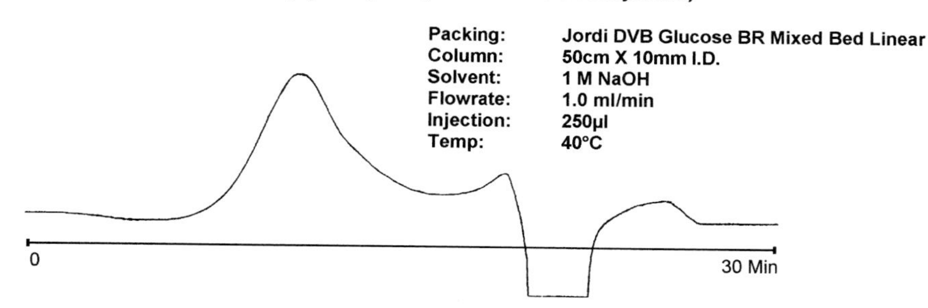

Sodium Carboxymethyl Cellulose

Packing: Jordi Gel - Glucose BR Mixed Bed Linear
Column: 50cm X 10mm I.D.
Solvent: 1 M NaOH
Flowrate: 1.0 ml/min
Injection: 100µl
Temp: 40°C

0 30 Min

FIGURE 13.98 Aqueous GPC in 1 *M* base.

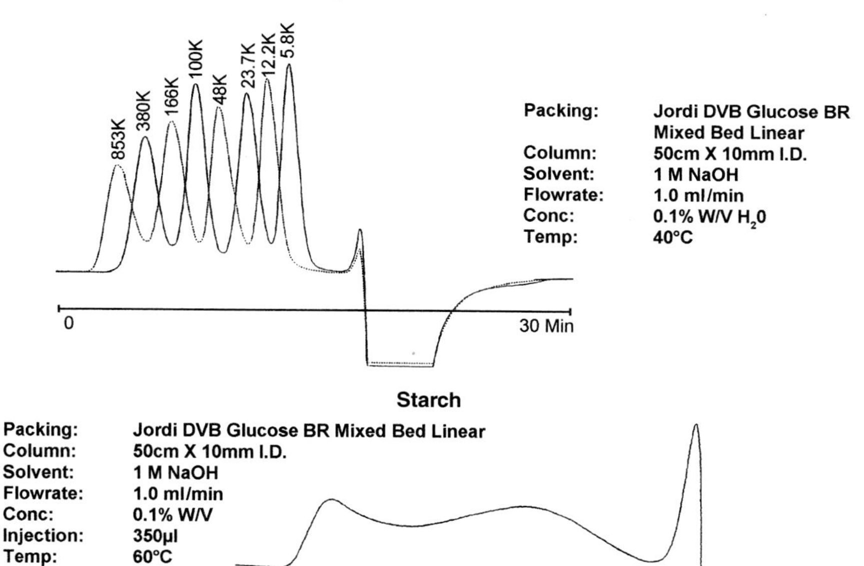

Packing: Jordi DVB Glucose BR Mixed Bed Linear
Column: 50cm X 10mm I.D.
Solvent: 1 M NaOH
Flowrate: 1.0 ml/min
Conc: 0.1% W/V
Injection: 350µl
Temp: 60°C

0 30 Min

FIGURE 13.99 Starch analysis.

examples of this type of SEC analysis for polyethyleneimines, a quaternized water-soluble flocculant, methacrylamidopropyltrimethyl ammonium chloride-based polymer, and chitosans, respectively.

In all Jordi gels the end user is free to use virtually any solvent system required to accomplish the job at hand. This can frequently serve as a very large advantage. If one mobile phase does not work, just try another. In doing this we have more or less accidentally discovered that GBR gels, for example, will run virtually anything runable on nonderivatized PDVB gel, as well as the aqueous soluble type polymers the phase was originally developed for. Figures 13.100 thru 13.104 give several examples of organic soluble polymers run on the GBR phase. Figure 13.103 is an example of a separation of an isobutylene/isoprene copolymer run on a GBR mixed-bed column using a solvent of 1/99 (v/v) 2-propanol/heptane, which also allowed for the UV detection of unsaturation. While more traditional mobile phases such as THF could also be used, they would not allow for effective low UV monitoring. In the absence of the 1% 2-propanol the polymer would not elute. Sometimes even water can be used to improve an organic SEC separation on GBR columns. Figure 13.102

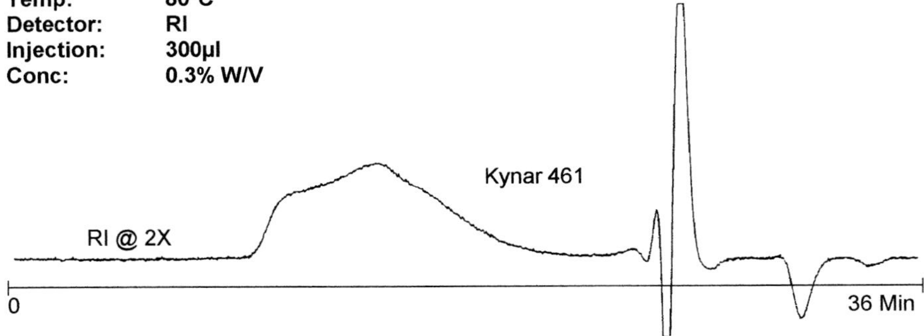

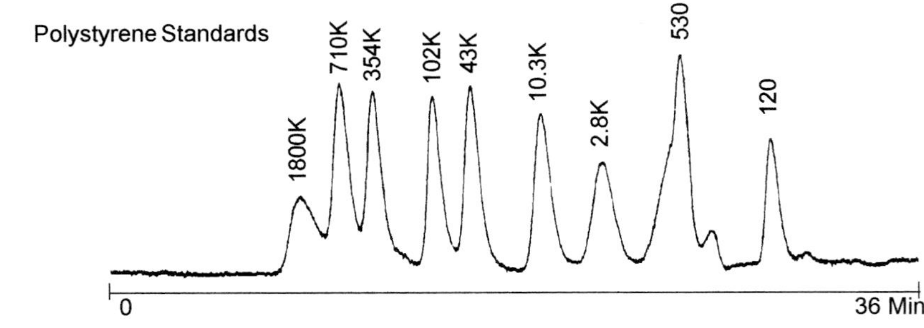

FIGURE 13.100

Acrylonitrile-Butadiene-Styrene (ABS)

Packing:	Jordi DVB Glucose BR Mixed Bed Linear
Column:	50cm X 10mm I.D.
Solvent:	Chloroform
Flow Rate:	1.5 ml/min
Temp:	35°C
Detector:	RI
Injection:	200µl
Conc:	0.25% W/V

RI @ 16X, S.F. @ 20

10 26 Min

FIGURE 13.101

Cellulose Acetate

Packing:	Jordi Gel DVB GBR Mixed Bed Linear
Column:	50cm X 10mm I.D.
Solvent:	Acetone/Water (95/5, V/V)
Flow Rate:	1.0 ml/min
Temp:	Ambient

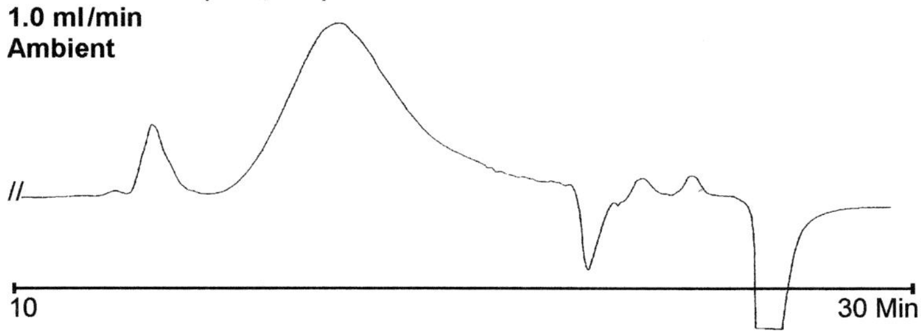

Polymethyl Methacrylate Standards

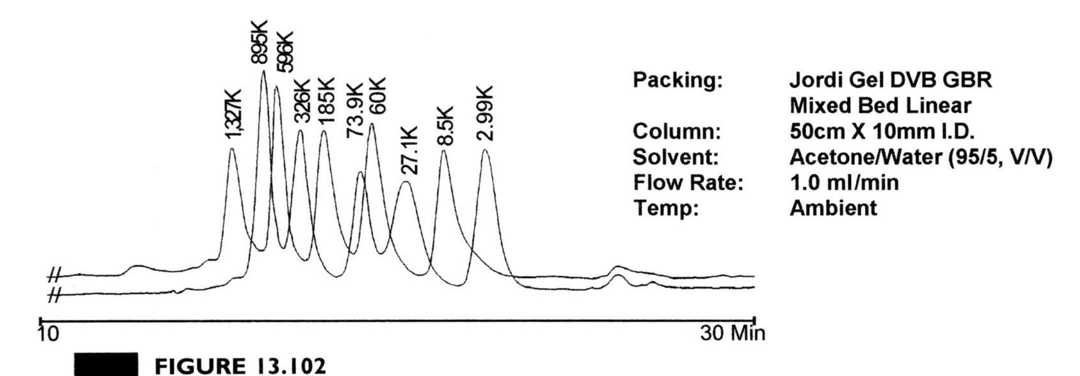

FIGURE 13.102

Butyl Rubber Copolymer
97% Iso Butylene: 3% Isoprene

Packing:	Jordi Gel DVB Glucose BR Mixed Bed Linear
Column:	50cm X 10mm I.D.
Solvent:	1% V/V IPOH in Heptane
Flow Rate:	1.0 ml/min
Temp:	30°C
Inj:	200µl
Conc:	0.2% W/V

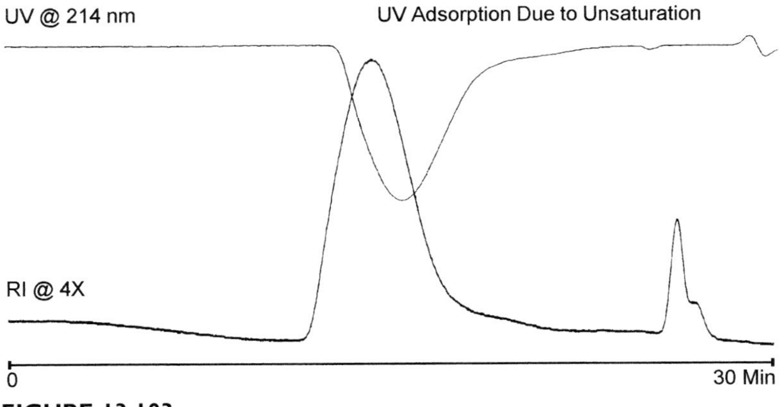

FIGURE 13.103

Gutta Percha
(Trans-1,4-Polyisoprene)

Packing:	Jordi Gel DVB Glucose BR Mixed Bed Linear
Column:	50cm X 10mm I.D.
Solvent:	Chloroform
Flow Rate:	1.5 ml/min
Temp:	30°C
Injection:	250 µl
Conc:	0.25% W/V

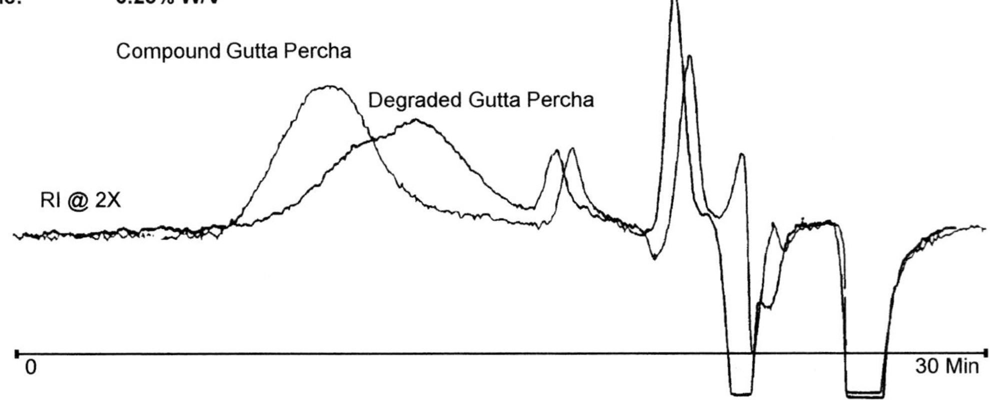

FIGURE 13.104

TABLE 13.1 Charge Types of Each Phase

Phase	Charge	Best for polymer charge of	Solvent allowed
Native PDVB	Neutral	Neutral	Any except dimethyl sulfoxide and water
Glucose BR	Neutral	Neutral or negatively charged	Any
Sulfonated	Negative	Neutral or negatively charged	Any
WAX	Neutral to positive	Neutral or positively charged	Any

details the use of 95/5 acetone/water as the mobile phase for the analysis of cellulose acetate. In the absence of the 5% water, the high molecular weight material does not elute.

As described earlier, Jordi Associates provides columns for both neutral and charged polymers. Table 13.1 lists the charge types of each phase. This will allow for planning a coherent strategy for method development.

REFERENCES

1. Cantow, M. J. R. (ed.) (1967). "Polymer Fractionation." Academic Press, New York.
2. Potschka, M., and Dubin, P. L. (eds.) (1996). "Strategies in Size Exclusion Chromatography." *Am. Chem. Soc.*, Washington DC.

II

CONTRIBUTIONS FROM USERS OF SIZE EXCLUSION CHROMATOGRAPHY AND GEL-PERMEATION CHROMATOGRAPHY

14

GENERAL CHARACTERIZATION OF GEL-PERMEATION CHROMATOGRAPHY COLUMNS

Rudolf Bruessau

BASF Aktiengesellschaft, ZKM/A-B1, Kunstofflaboratorium, D-67056 Ludwigshafen, Germany

I. INTRODUCTION

The selection of the right gel-permeation chromatography (GPC) column or column combination for a special task is a multistage procedure. First, a decision about the type of column has to be made according to the polymer samples in question:

• **Molar Mass Interval.** GPC columns are offered for different molar mass intervals; for larger intervals it is possible to combine some columns of different pore size types or to combine a few so-called "mixed bed" or "linear" columns. Both possibilities have their own special advantages and disadvantages: mixed bed columns with a linear separation range of more than four molar mass decades are suitable to quickly get an overall view of a new sample, whereas a column set, carefully selected from different pore size types, often has a much better separation efficiency in a limited mass interval (for details, see Sections III and IV).

• The low molecular weight range should not be too narrow: often it is very important to sufficiently separate the oligomer range of the sample from the elution area of system peaks (also called "impurity peaks", "salt peaks", etc.).

• The high molar mass range has to be considered in the same way. Normally the M_w value lies at the high molecular side of the chromatographic maximum, but the high end of the molar mass distribution can be larger than one decade. For a sufficient fractionation the high molecular end of the analyzed sample should not lie outside the linear range of the calibration curve.

• **Eluent.** The solubility of the sample determines the elution solvent for the GPC experiment: the better the solubility the lower the danger of undesirable

Column Handbook for Size Exclusion Chromatography

interactions between sample coils and the surface of the separation material. The affinity between eluent and sample should be greater than between the sample and the surface of the porous material, as otherwise nonsize exclusion interactions between surface and injected sample can appear (see Section V).

Distinctions have to be made among the solvent for shipment, eluent for the chromatographic experiment, and the solvent for long-term storage.

Changing the eluent in a ready-made column is sometimes combined with a loss of packing quality. The manufacturers try to hold the number of specified solvents for one packing type as low as possible, but some eluents have to be nominated at order time. Any replacement has to be performed strictly according to the instructions of the manufacturers. Not every eluent is suitable for the long-term storage of a used column (e.g., reactive decomposition products of tetrahydrofuran or corrosion by some aqueous buffers) and have to be replaced.

• **Column Temperature.** Some manufacturers offer special columns for GPC experiments at higher temperatures. A slight raise of column temperature over room temperature is mostly useless and leads to more experimental difficulties than advantages: The effect on a decrease of eluent viscosity, an increase of diffusion rate, or prevention of adsorption is low; using a differential refractometer at low T differences against room temperature, the disturbances by T fluctuations (e.g., of the connecting capillaries) are more difficult to control than at higher temperatures. Higher temperatures are justified for only two reasons:

1. if the viscosity of an eluent is higher than 0.6 mPa, the motion of the polymer molecules into and out of the pores is hindered: the peak broadening increases and the fractionation effectivity decreases. In such cases the viscosity has to be lowered by an elevated temperature (e.g., dimethylformamide, dimethylacetamide, benzylalcohol, trichlorobenzene, *N*-methylpyrrolidone, *m*-cresol).
2. sometimes the polymer is insoluble in the eluent at lower temperatures.

The lifetime for GPC columns used under high temperatures is lower than at room temperature. If higher temperatures are necessary, the user should look for special offers of high temperature columns or check which temperature interval is permitted for the selected column type. For a maximum lifetime a few precautions are advantageous:

- the rate of heating should be low and under a reduced flow rate
- the operating temperature should be changed as seldom as possible; intermediate cooling to room temperature must be avoided
- if a column is not in use for shorter periods, the working temperature has to be maintained but the flow rate should be reduced
- the flow rate (but this is a more general rule) should be changed only in small steps

These three criteria—molar mass interval, eluent, and working temperature—are fixed by the group of samples to be analyzed and considerably restrict the number of suitable columns. The selection has to be done from current lists of the manufucturers. It is useless to collect these data here, as such tables would be antiquated before this book is printed. This chapter deals with the quality of the selected columns. At this stage, columns of the same application profile are compared. The most important properties are (1) the number of theoretical plates. This is the most important measure of the column quality according to the packing and to the construction of the fittings; this property goes down continuously by using the column. (2) The asymmetry factor shows the effect of an uneven packing or of an unsufficient inlet or outlet on the chromatogram shape and can suddenly get worse for a damaged column. (3) Separation efficiency is a measure for the separation of the molecules according to their molar mass and is mostly constant during the lifetime of a column, but can suddenly go down by polluted separation material.

The difficulties start at this point: the manufacturers mostly report only the first number, but the extent of the warranty is low ("typical properties" instead of "guaranteed values" or low, easy to fulfill limits). The characterization of the column packing by an asymmetry factor is an exception, but if the interval is too broad, results are meaningless. The description of the third criterion is usually done only by figures with typical calibration curves. These graphics are suitable only for a rough comparison of the different pore size types of one manufacturer, as comparisons of different producers are not possible by these very schematic curves. Calibration curves do not immediately give information about the relation between the total pore volume and the matrix of a separating material, but these attached figures are often too small, vague, and incomplete for a calculation of the separation efficiency and the apparent pore size distribution according to Sections III and IV. Therefore, it is nearly impossible to make a rational decision by catalogue. Additional information such as the specification of the bead size or printed application examples, is not enough. This chapter gives a detailed description of the measurement and interpretation of these three properties. Unfortunately, these tests are only possible if a column is bought and present in your laboratory.

Further aspects, which come from special demands (e.g., multiple detection or samples with reactive functional groups), are also possible; these are discussed later.

II. DETERMINATION OF THE NUMBER OF THEORETICAL PLATES

The separation of a sample in a chromatographic column can be described as a stepwise process analogous to the consecutive discrete plates of a distillation column. In this model the injected sample solution is diluted beyond the injection volume and the concentration profile takes on the shape of an ideal symmetric Gaussian distribution. Assuming the correctness of this model, the column performance can be characterized analogously by a plate number; the model results in a relation between the peak broadening and the elution volume. The "number of theoretical plates" N [or "plate number," see Ettre (1)] is defined by the following equation between the retention volume V_r of a test sample and the standard deviation σ of the resulting peak:

$$N = (V_r/\sigma)^2, \tag{1}$$

where the determination of N consists of the injection of a low molecular weight test sample, measurement of the retention volume at the peak maximum, and the determination of σ from the peak shape. The different equations used in practice for calculating N differ in the method to get σ of the real peak assumed to have the shape of an ideal Gaussian distribution:

• **Tangents Method.** The tangents on either side of the peak are drawn through the inflection points until the baseline. For an ideal Gaussian peak the resulting base line interval W_b is equal to 4σ. Equation (1) becomes with the width (W_b):

$$N = 16^*(V_r/W_b)^2. \quad (2)$$

The determination of the tangents on the chart strip is difficult, as even little mistakes lead to significant deviations in the calculated N. This fact and the disappearance of the chart recorders in modern GPC units indicate another characterization procedure.

• **Half-Height Method.** For an ideal Gaussian distribution the peak width W_h at the half of the maximal height is 2.355 σ. If the width at the half-height is inserted in Eq. (1), we get

$$N = 5.545^*(V_r/W_h)^2. \quad (3)$$

This method is the most popular procedure, as it can be used without problems by both manual and computer calculations (Fig. 14.1).

• **5σ Method.** The peak width of an ideal Gaussian distribution is 5σ at 4.4% of the maximal height, Eq. (1) changes to

$$N = 25^*(V_r/W_{0.044})^2. \quad (4)$$

This result is strongly influenced by any tailing of the peak. This sensitivity is too high for daily problems and therefore this equation is seldom used.

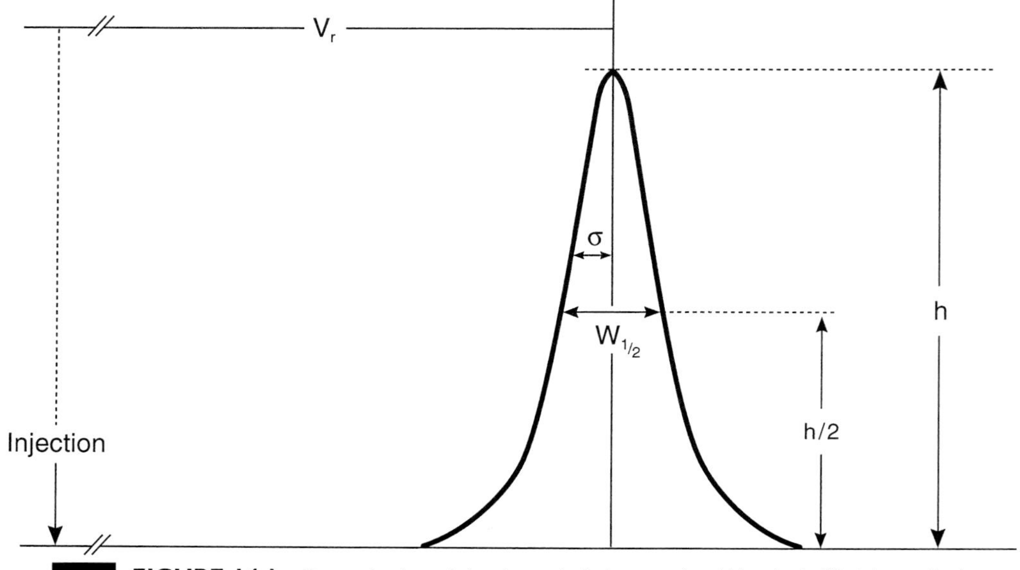

FIGURE 14.1 Determination of the theoretical plate number N by the half-height method.

The chromatogram of the test sample for plate number determination normally is not a Gaussian-shaped peak, which means that all three equations result in different N values. Therefore, it is important to specify which method is used for the calculation of plate number. If the method used deviates from the normal practice, it must be justified. This is especially true for a column manufacturer.

The results of a N determination are very sensitive to experimental conditions and to a careful and accurate execution of the method. The control measurements of an incoming new column have to be carried out strictly under the conditions fixed by the manufacturer, even if the conditions are not an optimum in your opinion or are not typically practiced in your laboratory (e.g., using the carcinogenic benzene as the testing sample). Otherwise any complaint can be rejected by the supplier. This chapter does not give general recommendations for the best method, as the exact choice depends on the type of columns, testing a new column for warranty or a used column for an internal quality control, quality of the available eluent (occurrence of system peaks), additives in the eluent, and testing a single column or a column combination.

The influence of the σ calculation was mentioned earlier. The results for N are additionally influenced by:

1. Test Sample

The plate number measured by a low molecular weight sample gives information about the packing quality, but the resulting N value is not the effective plate number for the separation of polymers. This value is lower. A detailed inspection shows that the plate numbers for one column measured by different low molecular weight samples can also be significantly different. When a test sample is selected, it is important that the sample peak does not overlay any system or impurity peak, especially in the case of negative peaks measured by a differential refractometer. This is critical because the reduced size of the positive test sample peak simulates a higher N value.

2. Eluent

The plate number for the same column also depends on the eluent, e.g., a permitted operation for some styrene–divinylbenzene columns is to change the eluent from tetrahydrofuran (THF) to dimethylacetamide (DMAC) and then return to THF. The plate number in DMAC is considerably lower than in THF. After the replacement of DMAC by THF the old N value is obtained again.

3. Injection Volume and Concentration

The peak width increases with injection volume. Therefore this parameter has to be fixed for comparative measurements. It has become the custom to inject low molecular weight test samples in very small volumes at very high concentrations, occasionally even as pure compounds. This extreme is not recommended as it is more important to inject a constant sample amount, reproducibly, in a precisely kept volume. Typical GPC injections are between 50 and 200 μl. It is better to inject a larger volume of a lower concentration polymer solution. GPC units are often not designed for injection volumes lower than 20 μl. An injection volume between 20 and 30 μl is recommended for plate number determinations. If lower volumes are prescribed for a new column by the manufacturer, then it has to be determined that such low volumes can be applied reproducibly on the given equipment.

4. Flow Rate

Because test samples for the *N* determination are low molecular weight compounds, the influence of flow rate is usually marginal. Sometimes, however, very high concentrations are injected, so it is better to fix the flow rate too.

5. Instrument Band Spreading

The peak broadening measured for a plate number characterization is the sum of the variances (σ^2) for the column and the chromatographic equipment used:

$$\sigma^2_{\text{total}} = \sigma^2_{\text{column}} + \sigma^2_{\text{equipment}}. \tag{5}$$

The contribution of the equipment between injection unit and detector cell should be negligable in relation to the column for a sufficient column characterization; short connections with narrow capillaries and zero dead volume unions are the precondition for reliable plate numbers. Every end fitting of a column causes additional band broadening. In the past a column type was offered that could be directly combined without any capillary links; unfortunately, it has disappeared from the market.

The Waters company recommends a system check of the chromatographic equipment that is used for plate number determination and analyses (2): the columns in the GPC unit used are replaced by a zero dead volume union. Then the test sample is injected under the same conditions such as a plate number determination. The 5σ peak width measured on a suitable recorded peak is evaluated: this 5σ width of a 20-μl injection should be lower than 150 μl.

6. Data Sampling

The results of an experiment for the determination of a plate number are very sensitive to the collection and representation of the experimental data, e.g., test chromatograms have to be large enough for manual evaluation, and that a measuring error in the size of the line width does not play a significant part in the calculated plate number. For evaluation by computer, the density of measuring points has to be so high that interpolation errors can be neglected. Therefore the ISO standard method for GPC analyses (3) prescribes: ". . . Determine the peak width at half-height either electronically from at least 30 data points per peak or manually on a chromatogram where the peak is at least 2 cm wide at half-height and at least 15 cm high at the peak maximum . . . ." These requirements are the minimum in our experience.

7. Discarding a Column

A GPC column is like an alive individual: every column is unique and is getting older with his own curriculum. Therefore the question arises, when does a column need to be discarded?

There are different reasons to discard a column: a column can be damaged by irreversible adsorption of reactive polymer samples. Small amounts of styrene oligomers are known to permanently elute from styrene–divinylbenzene materials with tetrahydrofuran as the eluent, which means a continuous shear degradation of the separation material and consequently a decrease of the packing quality; this observation is very important if fractions are collected and used for further analyses, e.g., for the determination of infrared (IR) spectra. One can presume that similar effects are present with other organic materials too.

Waters recommends (2) replacing a column if the plate number is decreased 30% below the original value. This is a very strict limit for an expensive tool as a column. Such a decision should be made more accordingly to the special purpose of given equipment, for instance: will the performance of a column (combination of columns) be sufficient to distinguish between different polymer samples for continued use? The ISO standard method (3) requires 20,000 plates per meter for interlaboratory experiments independent of original values. Significant changes in the shape of the test peak (asymmetry, tailing) plays an important part too.

III. ASYMMETRY FACTOR

The peak measured for a plate number determination contains additional information about the packing quality of a column. The same peak may also be used to quantify information about the shape as well. The peak width on both sides of the perpendicular through the peak maximum is measured at a height of 10% of the maximum height (see Fig. 14.2). The quotient of the back by the front part of the peak is defined as the asymmetry factor (AF):

$$\mathrm{AF} = W_b/W_f. \tag{6}$$

The value of AF should be as close to 1.0 as possible; the typical range is between 0.7 and 1.2. An upper limit of 1.6 as in the specification of the TosoHaas company is too high.

Most manufacturers do not specify the asymmetry factor. Therefore this parameter can serve only for the observation of the column performance during its use. For interpretation, see the remarks about discarding a column.

IV. SEPARATION EFFICIENCY

One of the most important properties of a chromatographic column is the separation efficiency. A measure of this parameter could be the difference of the retention volume for two different compounds. The result of a GPC analysis is usually, however, only one large peak, and a separation into consecutive molar mass species is not possible. Additionally there is no standard for higher molar masses consisting only of a species that is truly monodisperse. Therefore, the application of the equation to the chromatographic resolution of low

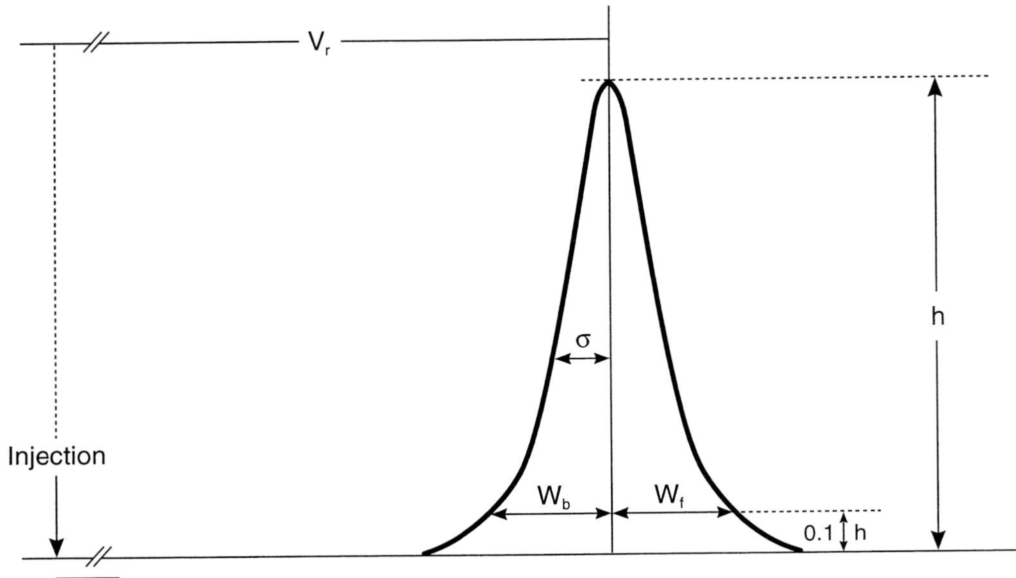

FIGURE 14.2 Determination of the asymmetry factor.

molecular weight samples is only possible with vague assumptions. So we have to look for a method that can help us with the specific conditions of GPC.

Normally a calibration curve—molar mass against the total retention volume—exists for every GPC column or column combination. As a measure of the separation efficiency of a given column (set) the difference in the retention of two molar masses can be determined from this calibration curve. The same eluent and the same type of calibration standards have to be used for the comparison of different columns or sets. However, this volume difference is not in itself sufficient. In a first approximation the cross section area does not contribute to the separation. Dividing the retention difference by the cross section area normalizes the retention volume for different diameters of columns. The ISO standard method (3) contains such an equation

$$\text{separation performance} = \frac{V_{\text{r,Mx}} - V_{\text{r,(10*Mx)}}}{\text{column cross-sectional area}} > 6, \tag{7}$$

where $V_{\text{r,Mx}}$ is the retention volume (in cm^3) for polystyrene of molar mass M_x and $V_{\text{r,(10*Mx)}}$ is the retention volume for 10 times the molar mass M_x; column cross-section area, in cm^2.

The ISO method prescribes polystyrene standards with tetrahydrofuran as the eluent, but this equation can also be used with other narrow distribution standards, provided the same elution solvent and the same standards are used for a comparison. Further, the ISO method requires the result to be greater than 6 for one decade of the molar mass. Because calibration curves are usually not linear, this decade should lie nearly symmetrically around the peak maxima of the samples in question. The required value of 6 is easy to fulfill, as results of 10 or more are usual with modern columns. If so-named "linear" or "mixed bed" columns with extended separation ranges are used, then it is sometimes difficult to meet this demand of 6.

The result of this equation describes the quality of the separation on the basis of an ideal size exclusion mechanism with a given pore volume distribution. The quality of the packing is deliberately excluded from this consideration. This parameter should be measured separately and judged by the plate number. The ASTM standard method for HPSEC of polystyrene (4) contains the following equation for resolution (R_s):

$$R_s = 2\ (V_{R2} - V_{R1})/[(W_1 + W_2)\log_{10}(M_1/M_2)], \tag{8}$$

where V_{R1}, V_{R2} are the peak elution volumes measured at the peak maximum of polymer standards 1 and 2, W_1, W_2 are the peak widths of standards 1 and 2 measured in volume such as V_R, and M_1, M_2 are the peak molar masses of standards 1 and 2.

This equation is based on experience with liquid chromatography of low molecular weight samples displaying single peaks. Its application for the GPC of polymers, however, contains a disadvantage, as it mixes two inseparable properties: the retention difference for the separation and the peak width for the contrary effect of band broadening. Such a procedure is acceptable if both effects are accessible for an experimental examination. For the GPC experiment, we do not possess polymer standards, consisting of molecules that are truly monodisperse. Therefore, we cannot determine the real peak width necessary for a reliable and reproducible peak resolution R_s. This equation then is not qualified for a sufficient characterization of a GPC column.

V. APPARENT PORE SIZE DISTRIBUTION

The maximum difference ΔV_{max} of possible retention volumes is limited to

$$\Delta V_{max} = V_{penetr} - V_{excl} \tag{9}$$

in a GPC experiment with an ideal size exclusion mechanism, where V_{penetr} is the elution volume at the penetration limit and V_{excl} is the elution volume at the exclusion limit.

In the classical model of the size exclusion mechanism this difference stands for the effective pore volume of the separating model. Any elution of samples or fractions outside this interval always means a perturbation by a different mechanism. Such conditions have to be avoided. It is not possible to expand this elution difference ΔV_{max} significantly for a given column. For this reason, GPC column sets are considerably longer than LC columns for other mechanisms.

With these facts in mind, it seems reasonable to calculate the pore volume from the calibration curve that is accessible for a certain molar mass interval of the calibration polymer. A diagram of these differences in elution volume for constant M or ΔM intervals looks like a pore size distribution, but it is not [see the excellent review of Hagel *et al.* (5)]. Absolute measurements of pore volume (e.g., by mercury porosimetry) show that there is a difference on principle. Contrary to the absolute pore size distribution, the distribution calculated from the calibration curve is influenced by accessibility given by the pore structure; the apparent reduction of the pore volume by the radius of the polymer coil itself (the so-called "wall effect") and is the reason that molar mass separations are also possible with uniform pore size materials; and the influence of the diffusion speed of polymer molecules on the calibration curve, which decreases with increasing molar mass. This flow dependence can be observed very clearly with narrowly distributed high molecular standards.

Nevertheless, the interpretation of the calibration curve in the form of a "selectivity curve" is an excellent tool for the comparison of the separation effectivity for columns of different origin.

The eluent and standards have to be the same for such comparative experiments. In a first approximation the flow rate should be the same as well. It is, however, possible that the flow rate can be normalized according to the same linear velocity or to the same retention time of the penetration limit. The procedure for the calculation of the apparent pore size distribution is easy: for the series M = 100, 200, 400, . . . 102,400, 204,800 . . . the retention volumes of a polystyrene calibration curve are determined, and the differences of neighboring values are normalized with the volume of the empty column (set). By this analogy, the volumes for penetration and exclusion limit are treated. The resulting series of numbers is a fast available empirical measure for the accessible pore volume and its distribution of the different M intervals. The normalization enables comparisons with columns of different dimensions. With modern columns, significant deviations can also be observed between columns of different manufactures, which result in different separation efficiencies, despite similar high plate numbers. Beyond this, it is possible that the series of volume fractions of single columns add and simulate in this way a column combination. Such simulations are helpful with respect to designing a column combination for a given task. This is especially true if the gap between the oligomer range and the impurity peaks were to be increased or the separation of the very high molecular interval were to be improved. Then the addition of single columns can increase the pore volume in narrow M intervals. The result is a kink in the calibration curve and artificial shoulders in the chromatogram. In most cases the available regression models for calibration curves are not flexible enough to follow such sharp bends. As a result, the artificial shape of the chromatogram remains and falsifies the calculated differential distribution curve. Simulations with apparent pore size distributions can help to smooth or avoid such abrupt transitions in advance.

In summary, such series of normalized volumes are more informative than the typical calibration curves in the leaflets of the manufacturers, especially in the classification of the so-called "mixed bed columns" for extended separation intervals. Such tables were found only once (6), and unfortunately these descriptions were never repeated in later brochures.

VI. SURFACE INTERACTIONS

A typical description of a porous material for the GPC separation indicates that it is made from styrene and divinylbenzene or from hydroxyethylmethacrylate

(HEMA) or similar. Such a description is correct, but not complete. For the suspension polymerization of these little beads, some auxiliary components are used that cannot be completely removed by washing the beads after the polymerization. Traces of these compounds remain on the surface of the separating material, possibly bound even by covalent bonds. The result is interactions with injected polymer samples, which are not expected for the main polymer network, e.g., polyisocyanates on the base of dimethyldiphenyldiisocyanate can be analyzed in tetrahydrofuran on styrene–divinylbenzene materials. Such columns are later useless for polymers with free amino or hydroxylic groups (e.g., polyetherpolyols), resulting in unspecific, broad, and retarded chromatograms. How can these observations be interpreted? For the suspension polymerization, colloid stabilizers are often used on the base of vinylacetate–vinylalcohol copolymers. If fractions of this auxiliary component remain on the bead surface, then the first injected polyisocyanate reacts partially with the hydroxylic groups of the vinylalcohol unit so that now surplus isocyanate groups are on the surface. These groups can react with suitable functionalities of samples injected later.

We also have to count on variations of the main monomer unit as well. For example, HEMA materials always contain traces of unesterified free carboxylic groups. These acidic groups interact with polycations. As result, retardation is observed until total adsorption of the injected sample. Consequently, polycations can be analyzed on such a material only in acidic eluents, where the dissociation of the $-COOH-$ group is suppressed.

The extent of such surface interaction also often varies for the same general packing type between different manufacturers. If there are such difficulties with a group of samples, sometimes changing to columns of the same nominal type but of a different manufacture solves the problem. Unfortunately, most manufacturers refuse to say anything about the existence and the amount of functional groups on the surface of their separating material. Additionally, such properties change without any announcement. This means that publications in the literature, especially about conditions for the elution behavior of cationic polyelectrolytes, are often useless and have to be proven by elution experiments with currently available columns. We can only recommend analyzing samples that show interactions on columns of different origins either on borrowed columns or in the application laboratorium of a manufacturer. The chromatograms then have to be checked and compared very closely for delayed polymer peaks, e.g., for overlapping of polymer and system peaks, for tailing beyond the system peaks, or for other unexpected peak shapes.

VII. SEPARATION OF THE OLIGOMER RANGE FROM SYSTEM PEAKS

As stated in Section I, columns should be selected so the low molar mass portions of the samples in question can be sufficiently separated from the elution interval of the system peaks. This task cannot always be accomplished, e.g., dimethylacetamide often replaces dimethylformamide as a GPC eluent; the analyzed, mostly polar, samples require a neutral salt (e.g., LiBr) (7). The calibration is usually carried out with poly(methylmethacrylate) standards

(PMMA). Under these conditions, some inevitable peaks are observed in the low molar mass range. This area starts for an elution volume of M_{PMMA} of about 600. Enclosing a further column for this low molecular size range generally improves the separation, but the elution sequence of oligomer and disturbing peaks remains constant.

Water-soluble polymers obtained through a radical polymerization [e.g., poly(acrylic acid) PAA] often contain sodium sulfate Na_2SO_4 as a decomposition product of the initiator. The peak of Na_2SO_4 is eluted before the dimer. In the interpretation of the chromatogram, a typical GPC program has to be truncated before the Na_2SO_4 peak, or at a M_{PAA} value of about 200. The calibration curve in this region can be flattened by an additive small pore column as well, but the principle problem remains unsolved.

The reason for such difficulties is the GPC mechanism itself. We do not separate by molar mass but by the size of the solvated molecules. Different solvation of chemical unlike molecules results in breaking the M sequence of the calibration curve; this becomes visible especially in the low molar mass range. Sometimes such difficulties can be circumvented if a specific detector is used, e.g., if the sample absorbs in the ultraviolet (UV) range and the disturbing peaks are UV transparent.

VIII. COLUMN DIMENSIONS

The latest trend is to smaller beads in smaller columns, as this saves eluent and shortens the time for a chromatographic analysis. This argument can be correct if only one suitable detector is used. However, these modern small columns are not optimal for a combination of detectors. So-called "multiple detection" is a combination of some detectors with different measurement principles (differential refractometer, spectral photometer, light-scattering detector, on-line viscometer) behind the last column, mostly in series, seldom in a branched ("parallel") order. In this way, the tedious preparative fractionation of a polymer sample can often be avoided.

The detector cells normally are connected by a capillary. For the interpretation of the detector signals the volume of this connection must be known: data of the following detectors have to be shifted for the delay toward the first cell. Usually this dead volume cannot be measured immediately, as it has to be determined indirectly by test analyses, e.g.,

- difference of peak maxima of low molar mass samples; if one of the detectors is a light-scattering detector or a viscometer, then this test sample has to be strictly monomolecular
- dead volume between concentration detectors: iterative variation of the dead volume until the same calibration curve results for both detectors the same molar mass averages.

Experience shows that such methods can result in significantly different dead volumes for the same GPC unit. There are at least two reasons for these observations: (1) the peak shape is changed by the laminar flow with a parabolic flow rate distribution in the connecting capillaries (8) and (2) it seems important if a detector is sensitive to a radial concentration distribution (like a differential refractometer) or not (e.g., photometers).

Such effects principally cannot be observed in multiband detectors such as a UV diode array detector or a Fourier transform infrared (FTIR) detector because all wavelengths are measured under the same geometry. For all other types of detectors, in principle, it is not possible to totally remove these effects of the laminar flow. Experiments and theoretical calculations show (8) that these disturbances can only be diminished by lowering the concentration gradient per volume unit in the effluent, which means that larger column diameters are essential for multiple detection or that narrow-bore columns are unsuitable for detector combinations. Disregarding these limitations can lead to serious misinterpretations of GPC results of multiple detector measurements. Such effects are a justification for thick columns of 8–10 mm diameter.

IX. CONCLUSION

If we consider the properties of a GPC column, discussed in the preceding sections, then two questions arise: (1) Which information does the user need when buying a new GPC column? (2) What should a user do to characterize a column?

The first question is a request to the manufacturer and the second results in a checklist for the start in the laboratory. The following lists contain recommendations, but they cannot comprise all possibilities and not all points are always necessary. The order of points is more according to technical reasons, it does not represent the importance.

A. Characterization of a GPC Column by the Manufacturer

1. Chemical Composition of the Separation Material

- basis polymer
- irreversibly bonded residues of auxiliary components
- quantitative declaration of functional groups on the surface of separation material, able to interact with reactive polymer samples or polyelectrolytes

2. Physical Characterization and Permitted Chromatographic Conditions

- column dimensions, characterization of the in-line filters
- guaranteed plate number
- guaranteed asymmetry
- bead size distribution
- maximum flow rate
- permitted maximum back pressure
- designated temperature interval
- preferential eluents

- possibilities and measures for a safe change of the eluent
- recommended solvents for unused columns in stock

3. Separation Properties

- calibration curve: data shall be given in a form that can be used for further calculations. It must be possible quantitatively to compare calibration curves of columns offered by different manufacturers. The small figures given in most leaflets at present are not suitable for a reliable assessment of GPC columns.
- exclusion limit, extension of linear range
- apparent pore size distribution: here the same requirements have to be fulfilled as the calibration curve.
- relation between volume of matrix material and total pore volume, usable by GPC (very important for the fractionation of ultrahigh molar mass samples and for high performance separations of, e.g., anionically initiated blockcopolymers)
- recommended chromatographic conditions (among others: injection for calibration and for broad distributed samples)

B. Testing a GPC Column before Using It

1. Single Column

- plate number
- asymmetry of the plate number peak
- back pressure under the designated conditions: flow rate, eluent, and temperature

2. Column Combination with the Final Eluent

- plate number of the column combination as a base for regular maintenance
- determination of the calibration curve
- inspection of the calibration curve for the low molar mass cutoff, caused by system peaks; exclusion limit; extension of the linear range, suitable for the planned samples; and for irregularities caused by an uneven pore size distribution ("kinks").
- analyses of few polymers of the designated samples and inspection of the chromatograms for nonsize exclusion effects: changing of the chromatograms for consecutive repeated injections of the same sample according to size, shape, and elution volume; retardation of the polymer peak into the range of system peaks; long tailing beyond the system peaks; unusual low slope of the calibration curve measured with a light-scattering detector compared with the standard calibration curve; and total adsorption of the injected sample.

The order and the extent of testing a column is not constant, it depends on the task: testing a newly bought column for daily work or looking for the best choice for a new polymer type. This list is more a proposal of points worth considering. Often the simple injection of a typical sample gives a lot of hints for the right selection of a column.

REFERENCES

1. Ettre, L. S. (1993). *Pure Appl. Chem.* **65/4,** 819–872, especially §3.10.03.
2. "Waters Styragel Column, Care and Use Manual," Waters Milford MA 01757, brochure #044491TP, Revision 1, March 1994.
3. ISO Committee draft ISO/CD 13885 (state of 1997): "GPC Using THF as Eluent."
4. ASTM D 5296 - 92 (1994). "Standard Test Method for Molecular Weight Averages and Molecular Weight Distribution of Polystyrene by High Performance Size-Exclusion Chromatography." Annual Book of ASTM Standards, Vol.08.03, pp. 419–431.
5. Hagel, L., Oestberg, M., and Andersson, T. (1996). *J. Chromatogr. A* **743,** 33–42.
6. Waters Associates (1970). "Chromatography Packings Components Instruments Services," pp. 53–55.
7. DIN 55672-2 "Gel Permeation Chromatography, 2. *N,N*-Dimethylacetamide as Eluent," first publication as draft in November 1997.
8. Bruessau, R. J. (1980). *Chromatogr. Sci. Ser.* **13,** 73–93.

15

INTERACTIVE PROPERTIES OF POLYSTYRENE/DIVINYLBENZENE AND DIVINYLBENZENE-BASED COMMERCIAL SIZE EXCLUSION CHROMATOGRAPHY COLUMNS

DUŠAN BEREK

Polymer Institute of the Slovak Academy of Sciences, 842 36 Bratislava, Slovakia

I. INTRODUCTION

Most size exclusion chromatography (SEC) practitioners select their columns primarily to cover the molar mass area of interest and to ensure compatibility with the mobile phase(s) applied. A further parameter to judge is the column efficiency expressed, e.g., by the theoretical plate count or related values, which are measured by appropriate low molar mass probes. It follows the apparent linearity of the calibration dependence and the attainable selectivity of separation: the latter parameter is in turn connected with the width of the molar mass range covered by the column and depends on both the pore size distribution and the pore volume of the packing material. Other important column parameters are the column production repeatability, availability, and price. Unfortunately, the interactive properties of SEC columns are often overlooked.

As known, SEC separates molecules and particles according to their hydrodynamic volume in solution. In an ideal case, the SEC separation is based solely on entropy changes and is not accompanied with any enthalpic processes. In real systems, however, enthalpic interactions among components of the chromatographic system often play a nonnegligible role and affect the corresponding retention volumes (V_R) of samples. This is clearly evident from the elution behavior of small molecules, which depends rather strongly on their chemical nature and on the properties of eluent used. This is the case even for

Column Handbook for Size Exclusion Chromatography

so-called inactive column packings such as polystyrene/divinylbenzene (PS/DVB) heterogeneously cross-linked materials (1,2).

Due to their interaction with the column packing surface, the shifts of retention volumes of separated macromolecules were also observed in numerous systems already in the initial stage of SEC development, in the mid-1960s (2). Both exclusion of macromolecules and their interaction with column packing are afftected by the eluent applied. The thermodynamic quality of the eluent for macromolecules determines the sizes of polymer coils in solution. In contrast, the affinity of eluent molecules to the column packing surface controls the interaction of packing with macromolecules. In the first approximation one can say that if the interaction between the packing surface and eluent molecules is stronger than that between packing and segments of macromolecules, the interactive polymer retention is suppressed. To evaluate interactions between packing and macromolecules, the effect of eluent–polymer interactions on the size of coils of macromolecules has to be eliminated. This became possible after Benoit *et al.* (3) discovered the SEC universal calibration dependence, i.e., the plot of $\log M\,[\eta]$ vs V_R, where M is polymer molar mass and $[\eta]$ its limiting viscosity number in the eluent. The mutual shifts of universal calibration curves measured with the same SEC column or column system for different polymers and/or for different eluents could be interpreted as a consequence of enthalpic interactions between packing and macromolecules, which is influenced not only by the eluent nature but in some cases also by temperature (2,4–6). Attempts were made to quantitatively interpret these shifts. For example, the basic relation among retention volumes of macromolecules in size exclusion chromatography and interstitial volume V_0, pore volume V_p, and the SEC distribution coefficient K_D

$$V_r = V_0 + K_D\,V_p \tag{1}$$

can be modified by introducing a further distribution coefficient, K_p, which relates to interactions (mainly to adsorption) between macromolecules and column packing (5). We then have

$$V_R = V_0 + K_D\,K_p\,V_p\,. \tag{2}$$

Unfortunately, K_p often depends on polymer molar mass, makes the use of Eq. (2) and similar expressions (7) impractical.

Alternatively, one can write for SEC retention volume

$$V_R = V_0 + K_D\,V_p + \sum K_i\,V_{pi}\,, \tag{3}$$

where K_i can be represented by the distribution coefficients of adsorption, partition, electrostatic, and so on secondary separation mechanisms and V_{pi} are the corresponding "effective pore volumes" (8). Equation (3) allows for estimating role of various interactive mechanisms (9); however, the unanimous definition of K_i and V_{pi} parameters has not yet been accomplished, as even an exact determination of effective pore diameters for the ideal size exclusion mechanism is hardly possible (10).

Concentration effects, i.e., the dependences of V_R on injected polymer concentration (c_i), represent a further complicating factor in the quantitative evaluation of SEC polymer–packing interactions. The slopes of linear plots V_R vs c_i depend on the thermodynamic quality of the eluent for the polymer (11). V_R values must be measured at several c_i and extrapolated to $c_i = 0$ to obtain a $V_{R,0}$ value for each eluent; these are further considered in exact process analyses (4,9,10). One can summarize that

(i) the nonexclusion separation mechanisms may be present in many SEC systems. These are mainly caused by adsorption of macromolecules on the surface of column packing, but partitioning and incompatibility (repulsion between separated macromolecules and polymer matrix) can appear as well. In the case of charged macromolecules, ion exchange, ion exclusion, and ion inclusion may be operative. Even if the overall effect of interaction is evaluated quantitatively, it is not possible to identify and discriminate particular contributions that may exhibit an opposite influence on the resulting retention of macromolecules. Eventually, one has to consider a series of so-called side and parasitic effects that may influence the behavior of macromolecules in SEC (2,6), as well. Consequently,

(ii) the quantitative evaluation and a priori description of nonexclusion mechanisms are not possible and only semiempirical expressions were adopted.

(iii) The existence of nonexclusion separation mechanisms affects SEC retention volumes and decreases the precision of measurements, especially if the column is calibrated with a polymer that differs from the investigated one. Data corrections may help to some extent, but they require numerous additional measurements and the correction factors obtained usually cannot be generalized.

(iv) The most important nonexclusion mechanism is the adsorption of macromolecules, which enhances their retention volumes. Adsorption effects may also cause a decrease in polymer recovery that is caused by a selective or overall persistent attachment of macromolecules onto column packing. The extent of adsorption usually depends on the molar mass of polymers. This means that a polymer of a specific molar mass can be retained selectively within the column, causing important errors in the molecular characteristics determined by size exclusion chromatography.

For these reasons, it is very important to effectively suppress the enthalpic interactions in SEC so that Benoit's plot is valid for both calibration standards and characterized polymers. This is done by choosing non(inter)active column packings, adjusting the temperature of the experiment, and applying interaction (mainly adsorption) suppressing liquids (single or mixed) as eluents.

Nonaqueous SEC (gel-permeation chromatography) is presently dominated with column packings based on heterogeneously cross-linked PS/DVB copolymers or DVB homopolymers. Alternatively, silica gels, bare or bonded with appropriate organic phases are used. Lower interactive activity of PS/DVB and DVB-based SEC column packings in comparison with silica gels is considered the main advantage of the former materials.

The elution behavior of various polymers near their critical adsorption point with silica gel packings and various eluents has been studied (12). It was of interest to apply "hybrid" column systems composed of active ("critical") packings (silica gels) in combination with nonactive (nonadsorptive) PS/DVB and DVB-based gels. Some PS/DVB and DVB gels exhibited rather strong adsorptive interactions with some polymers using particular eluents. Moreover, large differences in the adsorptive properties of columns produced from different companies were observed. Therefore, we decided to study the adsorptive properties of various commercial PS/DVB and DVB columns for SEC more systematically. This chapter presents the first sets of such investigations.

II. EXPERIMENTAL

A. Columns

The author decided to investigate the popular linear PS/DVB and DVB SEC columns covering a broad range of molar masses from a few hundred up to over 10^6 g·mol^{-1}. Eight important PS/DVB and DVB size exclusion chromatographic column producers were contacted. Six of them answered and kindly provided us with their columns. These were (alphabetically): Jordi Associates, (Bellingham, MA), Polymer Laboratories (Church Stretton, UK), Polymer Standards Service (Mainz, Germany), Shodex (Tokyo, Japan), Tosoh Co. (Shinnanyo, Japan), and Waters Associates (Milford, MA). The six columns, used in the study are designated 1–6, although this sequence *does not* correspond with the alphabetical order of companies. The author definitively does not intend to make a positive or negative advertisment about any producer in this contribution. Moreover, nothing is known about the batch-to-batch repeatability of the interactive properties of columns produced by particular producers.

New, unused columns were used to minimize the "memory effects" known not only with silica gel "active" packings, but also with PS/DVB materials (13). One single column of each kind (7.5, 7.8, 8, or 10 mm in diameter and 250 or 300 mm in length) was used in each series of experiments.

B. Polymers

In this stage of the investigation, poly(methyl methacrylates) (PMMAs) were selected as the polymeric probes of intermediate polarity. Polymers of medium broad molar mass distribution and of low tacticity (14) were a gift of Dr. W. Wunderlich of Röhm Co., Darmstadt, Germany. Their molar masses ranged from 1.6×10^4 to 6.13×10^5 g·mol^{-1}. For some comparative tests, narrow polystyrene standards from Pressure Co. (Pittsburgh, PA) were used.

C. Solvents

Because we wanted to suppress the effects of thermodynamic quality of the eluent toward the polymer probes, we therefore looked for liquids that would be thermodynamically good solvents for PMMA. At the same time, one solvent should promote polymer adsorption whereas the others should promote desorption.

Tetrahydrofuran (THF) is known as a liquid that exhibits high affinity toward various adsorbents and therefore effectively suppresses the adsorption of many polymers. Still, THF could not fully suppress adsorption, e.g., of

PMMA, on the unmodified porous glass and silica gel, and the universal calibration curves for polystyrenes and poly(methyl methacrylates) did not coincide (10,12,19).

Chloroform has lower affinity toward silica gel-based adsorbents than THF. PMMAs were fully retained from this solvent on the nonmodified silica gel even in the presence of a 1% ethanol stabilizer (14). It has been shown that PMMAs were not retained even within the most interactive PS/DVB column using chloroform stabilized with 1% ethanol. Chloroform is considered a weak displacer.

Toluene was found to have low affinity toward silica gel and strongly promoted the adsorption of PMMA (12,14) on this material. Preliminary experiments revealed that toluene at least moderately promoted the adsorption of PMMA on several PS/DVB and DVB SEC gels.

Stock of solvents were prepared so that their properties changed only little in the course of experiments. All mixed eluent compositions are given in weight %.

D. Apparatus

Waters pumping systems Models 501 and 510 and a Knauer pump Model FR 30 (Berlin, Germany) were used in the course of study. They were set at 1 ml·min^{-1} flow rate and the actual volume of effluent was occasionally controlled with a burette. Rather large flow rate variations were found with the FR 30 pump. The resettability of all pumping systems was, unfortunately, not perfect. Samples were injected from a 50-μl loop using a Rheodyne (Cotati, CA) valve. Columns were kept at 30°C in an column oven from Chroma (Graz, Austria). The detector was an evaporative light-scattering device, Model DDL-21 (Eurosep, Cergy St.-Pontoise, France) in combination with an on-line Knauer RI detector, which was used to monitor the system peaks and peaks of low molecular probes, *n*-hexane and toluene. Data were collected with a PC and processed using Chroma software.

E. Data Evaluation

As mentioned in Section I, the evaluation of interactions of macromolecules with SEC column packings on a strictly theoretical base is hardly possible. The SEC column interactive properties can be judged quantitatively if the concentration of displacing liquid (THF, chloroform) in the adsorption promoting eluent (toluene) is adjusted so that the resulting universal calibration dependence just coincides with that monitored in pure displacer. This approach would require large series of measurements supported with determinations of corresponding polymer limiting viscosity numbers in eluents. Moreover, a "perfect" displacer, i.e., an eluent fully suppressing the polymer–gel interaction, would have to be identified. Therefore, only conventional plots log M vs V_R were compared: the aim of this study was to show the qualitative differences among various columns rather than exactly quantifying their extent and to explain their sources.

III. RESULTS

A. Column No. 1

The results are depicted in Figs. 15.1 and 15.2. This column was one of the most interactive of the set investigated. The behavior of this column has in fact initiated our present study. PMMA dissolved in toluene and injected into the toluene eluent exhibited liquid adsorption elution mechanism: retention volumes increased with polymer molar mass (Fig. 15.1). The polymer recovery dropped sharply over M equal to 5×10^4 and peaks were broadened extensively. PMMAs with molar masses of 1.6×10^5 and higher were totally retained within the column. PMMAs dissolved in THF and injected into the toluene eluent eluted independently of their molar mass. This behavior is typical for liquid chromatography (LC) under limiting conditions of adsorption (LCA) (18). Limiting conditions of adsorption are only rarely reached with an (single) eluent. LC LCA may be utilized for the discrimination of various complex polymers. For example, if a blend of PMMA with a less polar polymer such as polystyrene were injected into the present system, PS would be eluted in the SEC mode without interference from PMMA, which would leave the column separately. PMMA fraction could be forwarded to an on-line noninteractive SEC column for separation according to its molar mass.

Eight percent of THF added to toluene (Fig. 15.1) displaced the calibration curve to the SEC-like mode, but the retention volumes are still strongly shifted

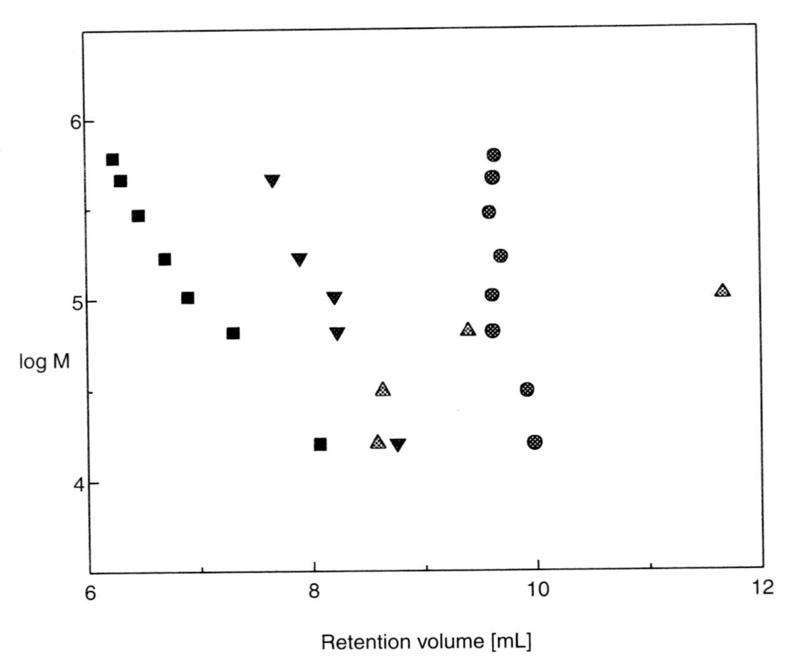

FIGURE 15.1 Column No. 1. Calibration dependences for PMMA dissolved in THF using eluent THF (■), toluene (●), and THF/toluene 8/92 (▼), as well as for PMMA dissolved in toluene in toluene eluent (▲).

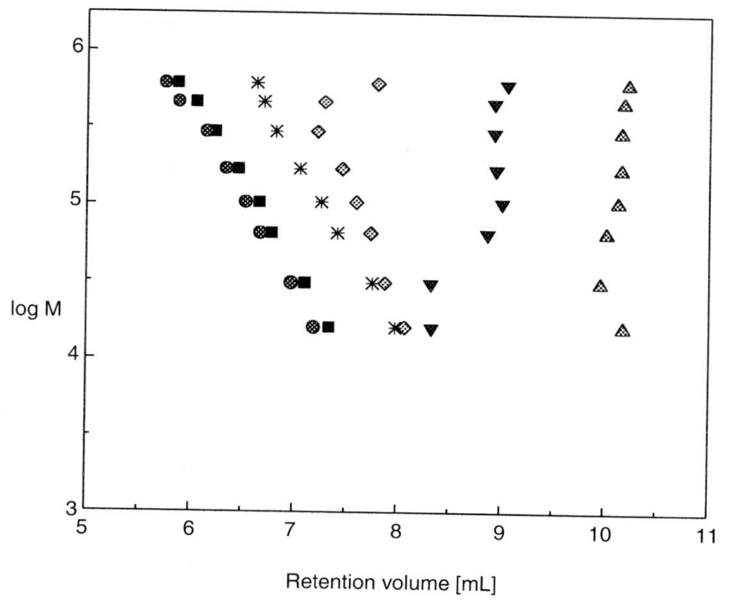

FIGURE 15.2 Column No. 1. Calibration dependences for PMMA dissolved and eluted in THF (■) and chloroform (●), respectively, as well as for PMMA dissolved in $CHCl_3$ and eluted in mixed eluents toluene/chloroform 90/10 (▲), 73/27 (▼), 65/35 (◆), and 50/50 (*).

when compared to that in the THF eluent. This shift is certainly larger than one would expect as a result of hydrodynamic volume changes and indicates the presence of some residual interactive polymer retention. Retention volumes of PMMA in chloroform are only slightly higher than those in THF and the difference hardly exceeds the anticipated hydrodynamic volume differences and the experimental errors. The LC LCA behavior was identified with an eluent containing 10% of chloroform in eluent chloroform/toluene for the entire PMMA molar mass range investigated. Reasonable SEC elution was reached in eluent chloroform/toluene containing 50% of chloroform, but the resulting retention volumes are still rather pronouncedly shifted toward high values. All these results again support our conclusion about the strong interactive properties of column No. 1.

B. Column No. 2

Calibration curves for PS and PMMA are shown in Figs. 15.3–15.5. The slight differences in courses of calibration curves for PS in THF, chloroform, and toluene, as well as the curve for PMMA in THF (Fig. 15.3), can be explained by the flow rate variations for different pumping systems and by the hydrodynamic volume effects, respectively. The calibration curves for PMMA in mixed eluents THF/toluene are shown in Fig. 15.4. Three percent of THF in toluene assured a reasonable SEC elution of PMMA. However, more chloroform was needed to obtain a good SEC elution of PMMA in mixed eluent chloroform/toluene

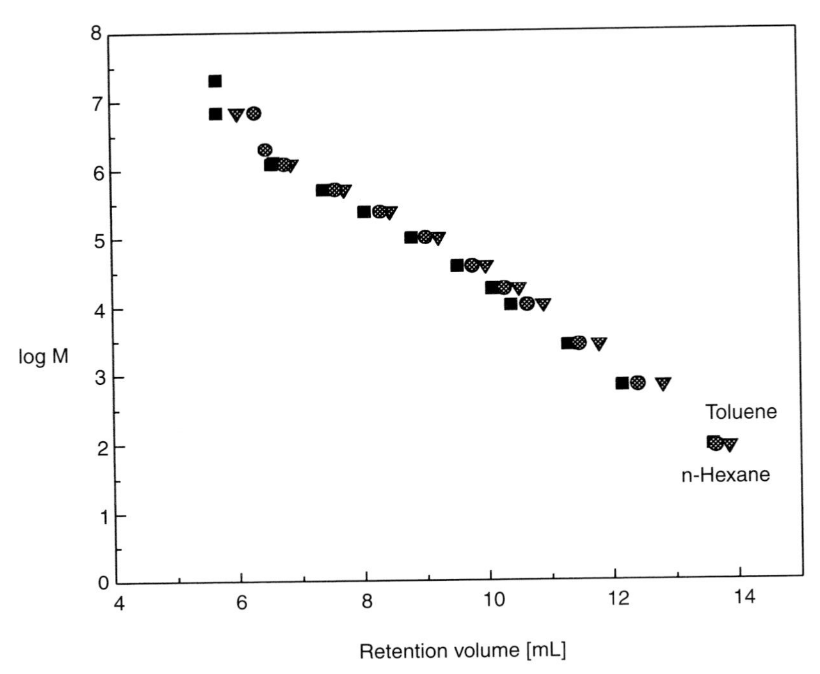

FIGURE 15.3 Column No. 2. Calibration dependences for polystyrenes in THF (■), chloroform (●), and toluene (▼) eluents. Samples were dissolved in eluent.

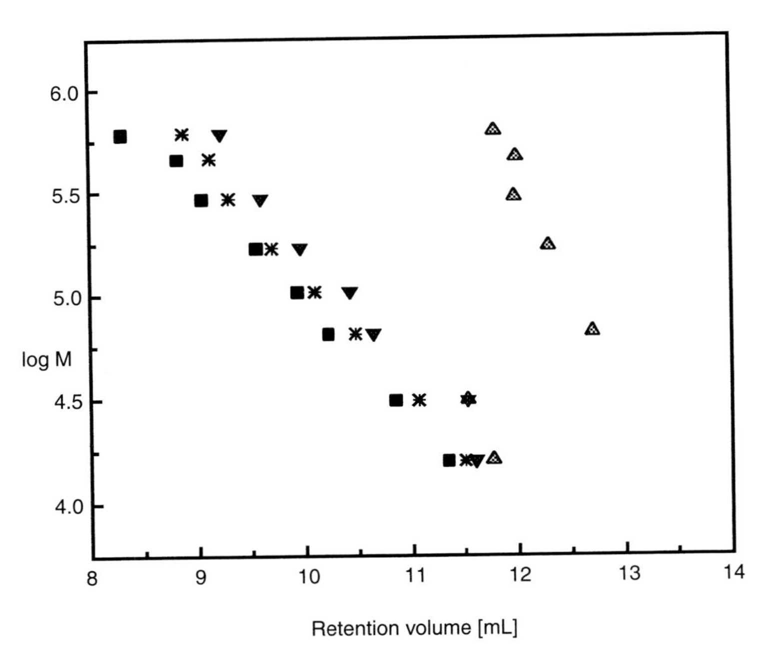

FIGURE 15.4 Column No. 2. Calibration dependences for PMMA dissolved in THF. Eluents were THF (■), THF/toluene 5/95 (*), THF/toluene 3/97 (▼), and THF/toluene 1/99 (▲).

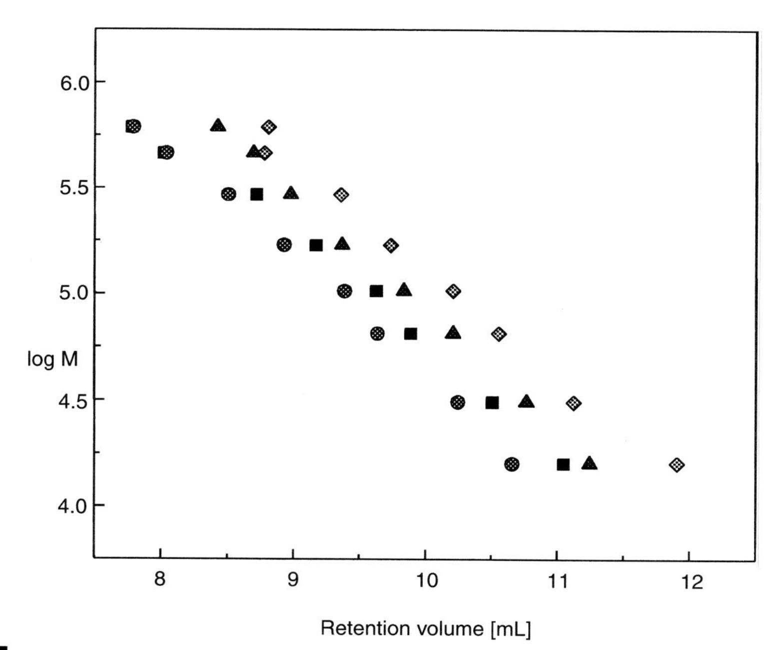

FIGURE 15.5 Column No. 2. Calibration dependences for PMMA dissolved in chloroform and eluted in eluents THF (■) (pumping system Waters 510), chloroform (●), chloroform/toluene 20/80 (▲), and chloroform/toluene 5/95 (◆) (all using the pumping system from Knauer).

(Fig. 15.5). PMMAs dissolved in toluene and injected into toluene eluent were fully retained within the column. Column No. 2 is also very interactive.

C. Column No. 3

Figure 15.6 indicates that this is a relatively noninteractive column. PMMA is eluted in toluene, yet the calibration dependences surprisingly diverge in the area of high molar masses. It has the opposite tendency to that observed for various polymers with PS/DVB columns by Dawkins (5,20) and also contradicts the finding for columns No. 5 and 6.

D. Column No. 4

This is a rather noninteractive column as well (Fig. 15.7). The calibration curves converge slightly in the area of high molar masses.

E. Column No. 5

According to measurements with the PMMA–toluene system, this is a noninteractive column (Fig. 15.8).

F. Column No. 6

This column exhibits a retention behavior for PMMA in toluene that is very similar to column No. 4, it is as if they contained identical column packings

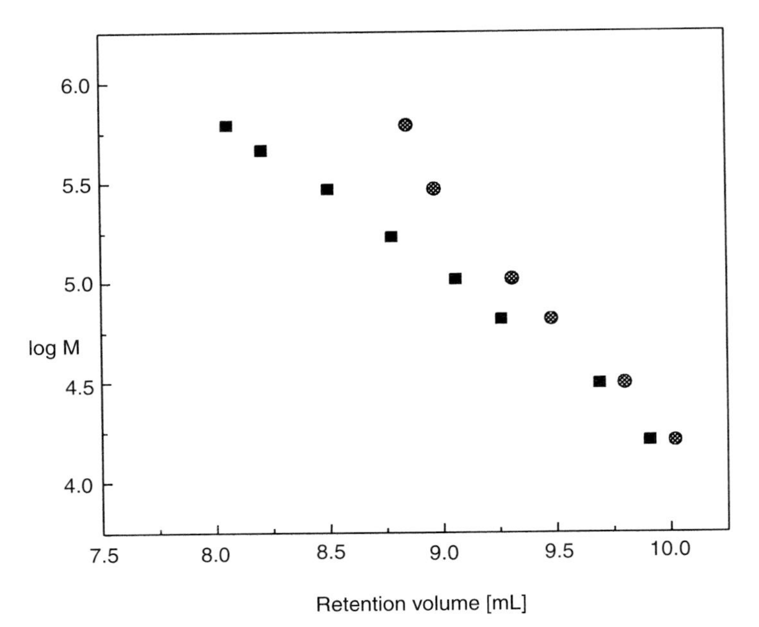

FIGURE 15.6 Column No. 3. Calibrations dependences for PMMA dissolved and eluted in THF (■), as well as dissolved and eluted in toluene (●).

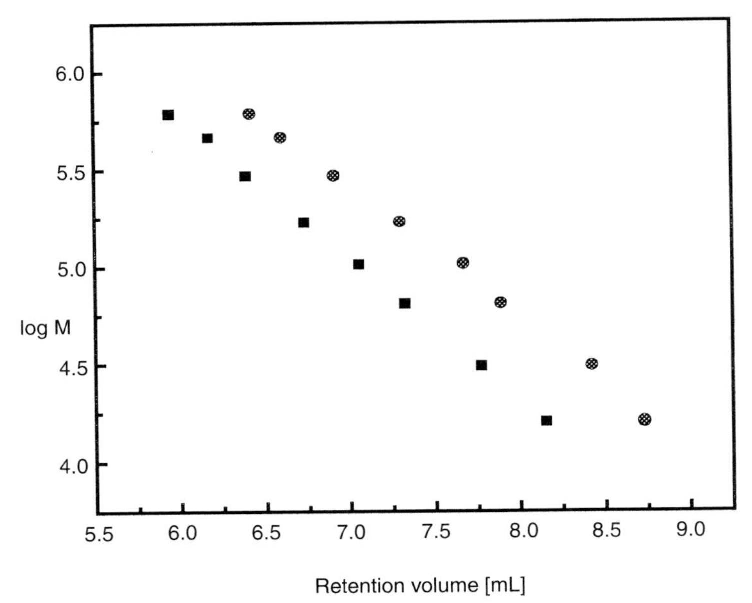

FIGURE 15.7 Column No. 4. Calibration dependences for PMMA dissolved and eluted in THF (■), as well as dissolved and eluted in toluene (●).

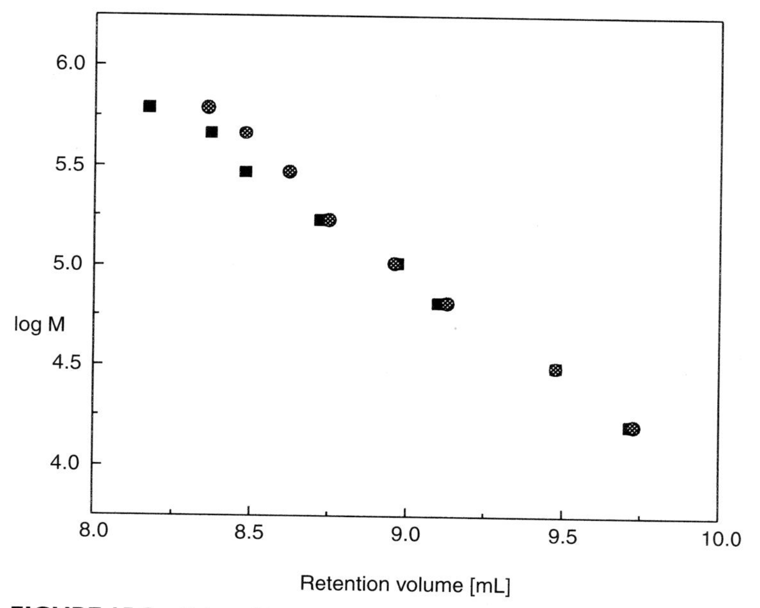

FIGURE 15.8 Column No. 5. Calibration dependences for PMMA dissolved and eluted in THF (■) as well as dissolved and eluted in toluene (●).

(Fig. 15.9). It would be interesting to compare the polymerization systems of both producers; maybe this contribution will help in surmounting the barriers of competition. It is interesting to note that the addition of 50% chloroform shifted the calibration dependence to practically coincide with that for the pure THF eluent.

IV. DISCUSSION

It is evident from these results that the interactive properties of the investigated SEC PS/DVB or DVB gels are very different. Because polar electroneutral macromolecules of PMMA were more retained from a nonpolar solvent (toluene) than from polar ones (THF, chloroform), we conclude that the dipol–dipol interactions were operative. Columns No. 1 and No. 2 were very interactive and can be applied successfully to LC techniques that combine exclusion and interaction (adsorption) mechanisms. These emerging techniques are LC at the critical adsorption point (18), the already mentioned LC under limiting conditions of adsorption (15,18), and LC under limiting conditions of desorption (16). In these cases, the adsorptivity of the SEC columns may even be advantageous. In most conventional SEC applications, however, the interactive properties of columns may cause important problems. In any case, interactive properties of SEC columns should be considered when applying the universal calibration, especially for medium polar and polar polymers. It is therefore advisable to check the elution properties of SEC columns before use with the

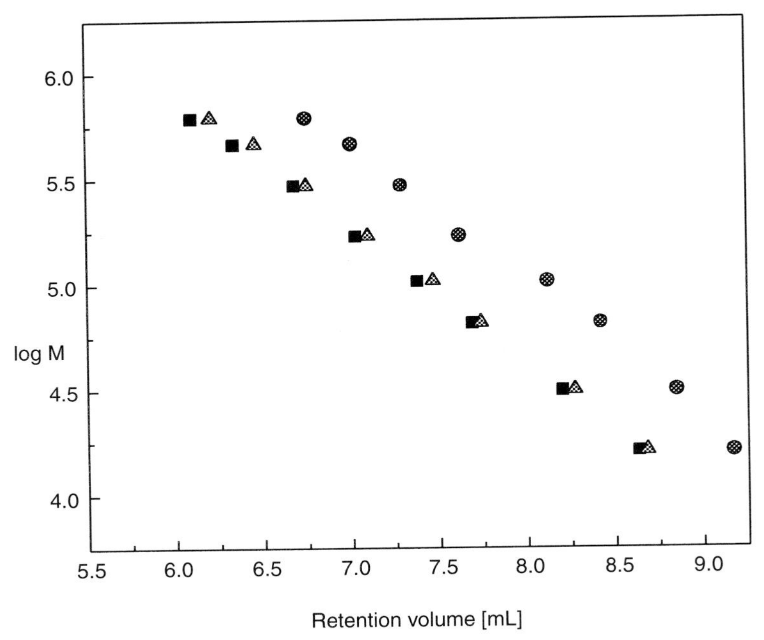

FIGURE 15.9 Column No. 6. Calibration dependences for PMMA dissolved and eluted in THF (■), dissolved and eluted in toluene (●), and dissolved in chloroform and eluted in chloroform/toluene 50/50 (▲).

help of appropriate probes such as PMMA in combination with an adsorption promoting eluent.

So far, it is not clear what is responsible for the interactive activity of PS/DVB and DVB columns for size exclusion chromatography and why the differences are so large. The auxiliary components of polymerization/cross-linking system (e.g., initiators, transfer agents, porogenes, and mainly detergents) may be built into the gel matrix during polyreaction. This, however, does not necessarily mean that the most active gels contain the most "impurities." For example, the molecules of some detergents attached to the gel surface may suppress the adsorptive interaction of PMMA macromolecules and thus decrease their retention volumes, provided their nonpolar parts protrude from the packing surface. The opposite alternative in which the polar parts of detergent molecules jut out into the pore volume seems more probable.

The producer of column No. 6 tried to supplementary remove admixtures from the gel matrix by applying liquid extractions. He revealed that the extraction process was very slow and that the apparently "clean" material started to "bleed" again after some time or when the temperature of extraction was raised. This result indicates that the retention properties of SEC columns may change in the course of their use as a result of "cleaning" their surface. Maybe the recipes for the gel synthesis will have to be modified to suppress the effects of additives. It seems that the producer of column No. 5 is not far from the ideal situation, at least for the PMMA–toluene system. We cannot exclude that the interactive properties would be adverse with this column for other polymers and other adsorption promoting eluents. Further study is needed to better understand the interactive properties of SEC gels. The experiments with rather nonpolar polystyrene and highly polar poly-2-vinyl pyridine polymers as probes are currently in progress in the author's laboratory.

It is hoped that these recommendations and even warnings for SEC column users will also give an important signal for the column producers to introduce further progress in this area.

ACKNOWLEDGMENTS

The author thanks the previously mentioned companies for the columns and to Dr. W. Wunderlich for PMMA samples. This work was supported in part by the Slovak Grant Agency VEGA, Project 2/4012/97, and by Slovak–USA Scientific–technical cooperation project 007/95. The technical assistance of Mrs. J. Tarbajovska is acknowledged.

REFERENCES

1. Saitoh, K., Ozaki, O., and Suzuki, N. (1982). *J. Chromatogr.* **239**, 661.
2. Berek, D., Kubin, M., Marcinka, K., and Dressler, M. (1983). Gélová chromatografia (in Slovak) Veda Bratislava.
3. Grubisic, Z., Rempp, P., and Benoit, H. (1967). *J. Polym. Sci.* **B5**, 753.
4. Berek, D., Bakoš, D., Bleha, T., and Šoltés, L. (1975). *Macromol. Chem.* **176**, 391.
5. Dawkins, J. V., and Hemming, M. (1975). *Macromol. Chem.* **176**, 1795.
6. Berek, D., and Marcinka, K. (1984). Gel chromatography. *In:* "Separation Methods" (Z. Deyl, ed.), p. 271. Elsevier, Amsterdam.
7. Campos, A., Soria, V., and Figueruelo, J. E. (1979). *Macromol. Chem.* **180**, 1961.
8. Hjertén, S. (1970). *J. Chromatogr.* **50**, 189.
9. Bakoš, D., Bleha, T., Ozimá, A., Berek, B. (1979). *J. Appl. Polym. Sci.* **23**, 2233.
10. Bleha, T., Spychaj, T., Vondra, R., and Berek, D. (1983). *J. Polym. Sci. Polym. Phys. Ed.* **21**, 1903.
11. Berek, D., Bakoš, D., Bleha, T., and Šoltés, L. (1974). *J. Polym. Sci. Polym. Lett. Ed.* **12**, 277.
12. Berek, D., Jančo, M., and Meira, G. (1998). *J. Polym. Sci. A Polym. Chem.* **36**, 1363.
13. Bruessau, R. (1996). *Macromol. Symp.* **110**, 15.
14. Berek, D., Jančo, M., Hatada, K., Kitayama, T., and Fujimoto, N. (1997). *Polym. J.* **29**, 1029.
15. Berek, D. (1997). *ACS PMSE* **77**, 20.
16. Berek, D. (1998). *Macromolecules.* Submitted for publication.
17. Petro, M., and Berek, D. (1993). *Chromatographia* **37**, 549.
18. Berek, D. (1998). *Macromol. Symp.* Submitted for publication.
19. Chiantore, O. (1984). *J. Liq. Chromatogr.* **7**, 1.
20. Dawkins, J. (1979). *Pure Appl. Chem.* **51**, 1473.

16 ANALYTICAL AND PREPARATIVE COLUMNS FOR AQUEOUS SIZE EXCLUSION CHROMATOGRAPHY OF POLYSACCHARIDES

ANTON HUBER

Institut für Physikalische Chemie, KF-Universität, A-8010 Graz, Austria

WERNER PRAZNIK

IFA-Tulln, Analytikzentrum, A-3430 Tulln, Austria, and Institut für Chemie, Universität für Bodenkultur, A-1190 Vienna, Austria

I. INTRODUCTION

Regarding the total amount of assimilated biomass, carbohydrates/polysaccharides represent the dominant class with an estimated amount of $1\text{–}4 \times 10^{11}$ tons of dry matter produced by photosynthesis annually:

- carbohydrates/polysaccharides: ~75% of total biomass
- lignin: ~20% of total biomass
- lipids: ~2% of total biomass
- proteins: ~2% of total biomass
- others (e.g., DNA)

The ranking of carbohydrates/polysaccharides according to mass percentages of subclasses reads:

- nonbranched $\beta(1\rightarrow4)$ glycosidically linked glucan cellulose: ~45%
- hemicellulose: ~20–25% of total carbohydrates/polysaccharides
- $\alpha(1\rightarrow4)$ linked and $\alpha(1\rightarrow6)$ branched glucan starch: ~2–5%
- others (e.g., mannan, galactan, fructose, ribose)

Although carbohydrates/polysaccharides exist in such huge amounts, their industrial processing is expensive due to enormous quality fluctuations of succeeding raw material batches. The reason for these fluctuations is a high variability on the molecular level, particularly in the degree of polymerization distribution, in branching characteristics, and in complex interactive properties.

Column Handbook for Size Exclusion Chromatography

From the organochemical point of view, carbohydrates/polysaccharides are more or less substituted polyhydroxy aldehydes (e.g., glucose→glucans) or polyhydroxy ketons (e.g., fructose→fructans). From the physicochemical point of view, an enormous heterogeneity also exists in

- the composition of repeating units
- polymer-elongating glycosidic linkages
- branching characteristics
- degree of polymerization distribution (including average values)
- polymer coil conformation (coil packing density)
- occupied volumina (excluded volumina)
- the kind of intra- and intermolecular interactions

Chromatographic separation of polysaccharide components, particularly separation by means of size exclusion chromatography (SEC) according to different occupied volumina, is an appropriate analytical, semipreparative, and even preparative tool for obtaining detailed information about characteristics and contributions of individual fractions to the total mixture. Because of the high number of hydroxyl groups, carbohydrates/polysaccharides are quite polar and thus, at least theoretically, well soluble in aqueous media. However, these hydroxyl groups form intra- and intermolecular H bonds, and true aqueous solutions hardly can be achieved by simply dissolving a powder in H_2O. Highly alkaline pH or nonaqueous polar solvents such as dimethy sulfoxide (DMSO) are necessary in most cases, but high ionic strength or DMSO often causes problems for further processing and/or analyses. For analytical purposes, aqueous SEC can be applied to remove interfering solvent components and/or to change solvent completely: the small particles of the initial solvent are shifted to the total exclusion volume, and in the course of separation more and more sample components become dissolved in the eluent. As a result of the obtained mass distribution of constituting components, SEC elution profiles contain a wide range of physicochemical information:

- Characteristics due to chemical functionalities (e.g., carboxyl groups) of sample components that control solubility of the sample in aqueous media, viscosity of carbohydrate/polysaccharide solutions, and stability of obtained solutions.
- Interactions in the ternary system: sample components eluent/SEC matrix.
- Stability of dissolved polymers with respect to degradation, precipitation (adsorption), and aggregation.
- Range of molecular weight distribution (mass/molar).

Based on this background, a selected number of preparative, semipreparative, and analytical aqueous SEC systems for the analysis of polysaccharides, summarized in Table 16.1, are presented and discussed briefly in this chapter.

A. Data Processing

Data in this chapter have been processed with CPCwin (Colloid and Polymer Characterization for Windows; A.H group, Austria; e-mail: anton.huber@kfuni), a software package that has been developed to handle SEC data. The

TABLE 16.1 SEC Systems and Correlated Applications

SEC system	Application
Preparative Sephacryl S-1000	Fractionation of potato starch glucans
Semiprepartive Sephacryl system S-200 / S-400 / S-500 / S-1000	Fractionation/characterization of nb/lcb fraction of potato starch glucan
Semipreparative/analytical Sephacryl system S-500 / S-1000	Long-time reproducibility of elution profiles; broad standard calibration with dextran T-500; transformation of a scb-type calibration function into nb/lcb-type via universal calibration; dp of synthetic glucans in the presence of significant amounts of monomer; mass and molar degree of polymerization of Triticale (hybride) starch
Preparative Sephacryl system S-200 S-1000	Fractionation of a suspected to be nb/lcb glucan sample for detailed analysis
Fractogel TSK HW 40(s)/HW 50(s)/ Superose 6 (prep grade)	Analysis of nb/lcb glucan fractions obtained from a preparative S-200/S-1000 system
Sepharose CL 2B Sepharose S-1000	Comparison of separation performance for quantification of a high dp scb glucan fraction in a nb/lcb glucan sample
Sepharose CL 4B/2B	Separation of potato starch and two fractions (nb/lcb, scb) of potato starch
Analytical Superose 6	Peak position calibration with nb glucans Differences between molar and mass degree of polymerization distribution of a partially hydrolized starch
Analytical Superose 6	Peak position calibration with nb glucans Differences between molar and mass degree of polymerization distribution of a partially hydrolized starch
Biogel P-6 / Sephacryl S-200	Degree of polymerization distribution of a plant fructan (inulin) at increasing physiological age of the source; remarkable performance of S-200 in the low dp range; degree of polymerization distribution obtained from bad (P-6) and good (S-200 / P-6) resolution of high dp components
High-performance TSK pre/PW6000/ 5000/3000	Absolute system calibration with a set of narrow distributed nb glucans (pullulans) by means of dual detection of mass and scattering intensity applying a mixed peak position/ broad standard calibration
High-performance TSK pre/PWM/6000/ 5000/4000/3000	On-line changing of solvent from DMSO to H_2O for a starch glucan; establishing an absolute calibration function; characteristics of mass and molar degree of polymerization distribution for a broad distributed starch sample

CPCwin concept includes step-by-step developing and visualization of intermediate and final results in terms of data lines/functions and parameters/parameter values. Data lines/functions utilized in this chapter are listed in Table 16.2.

Among the variety of CPCwin-supported features, there are

- import/export of ASCII-formatted data lines/functions; communication with text and spread sheet editors such as MS Winword and MS Excel

TABLE 16.2 Characteristics of Data Lines and Functions of SEC Experiments as Handled by Software Package CPCwin

Dataline/function	Characteristics
raw_mass	Elution profile of fraction masses: either obtained from on-line detection via mass detection or reconstructed from off-line determined fraction masses
raw_LS_5	Elution profile of scattering intensity at a scattering angle 5°
mass_ev	Normalized mass chromatogram: area = 1.0 within the selective separation range
LS_5_EV	Normalized scattering chromatogram for scattering angle 5°; normalization with respect to initial laser intensity and applied sample concentration: as a result the area within the selective separation range numerically equals the weight average molecular weight (M_w) of the investigated sample
raw_MWV	Raw data for SEC calibration • from peak position calibration with narrow distributed standards • from broad standard calibration • from universal calibration • established without external references (absolute) as the ratio of LS_5_EV/mass_ev
m_MWD_d	Differential molecular weight distribution: mass fractions normalized to area = 1.0 • M_w: weight average molecular weight • M_n: number average molecular weight • M_w/M_n: polydispersity
m_dpD_d	Differential degree of polymerization distribution: mass fractions normalized to area = 1.0 • dp_w: weight average degree of polymerization • dp_n: number average degree of polymerization • dp_w/dp_n: polydispersity
n_MWD_d	Differential molecular weight distribution: molar fractions normalized to area = 1.0
n_dpD_d	Differential degree of polymerization distribution: molar fractions normalized to area = 1.0

- computation of intermediate and final results (data lines/functions and average values) starting from raw data via operator-customizable assistants
- several approaches used to establish SEC-calibration:
  - peak position calibration with a set of narrow distributed standards
  - broad standard calibration (linear and/or nonlinear mode with one single broad distributed standard)
  - universal calibration approach: transformation of an already existing calibration for a certain polymer coil conformation [e.g., of short chain branched (scb) glucans] into a calibration for another conformation [e.g., of nonbranched/long chain branced (nb/lcb) glucans]
- handling of multiple detection data
- procedures to estimate dimensions of SEC-separated polymers in terms of occupied volumina and sphere-equivalent particle radii
- algorithms to estimate branching characteristics of SEC-separated polymers

• several options to test obtained molecular weight/degree of polymerization distributions on differences/identity with "most-probable"distributions (Flory, Schulz)

B. Broad Standard Size Exclusion Chromatography (SEC) Calibration

Establishing a calibration function with one single broad distributed sample is an alternative to traditional peak postion calibration of SEC systems with a set of narrow distributed standards. An obvious advantage of this technique is time: for peak position calibration elution profiles for the set of standards need to be determined; for broad standard calibration the elution profile of one sample needs to be determined only. Establishing a linear calibration function with a broad distributed standard includes startup information [M_w(true), M_n(true)] and an iterative (repeat . . . until) algorithm:

- M_w(true): known weight average molecular weight of broad distributed standard
- M_n(true): known number average molecular weight of broad distributed standard
- mass__ev: experimentally determined normalized elution profile of the standard
- arbitrary linear SEC calibration function raw__MWV with slope "a_1" and intercept "a_0"

  repeat

  calculate M_n(i) and M_w(i) for mass__ev utilizing initial/modified raw__MWV

  check distance of M_n(i) / M_n(true) ratio from 1.0

  check distance of M_w(i) / M_w(true) ratio from 1.0

  modify "a_1" and "a_0" of raw__MWV to approach 1.0 in the next cycle;

  until $M_w(i) = M_w(\text{true}) \pm 0.1\%$ **and** $M_n(i) = M_n(\text{true}) \pm 0.1\%$
- establish broad standard SEC calibration function (raw__MWV) with resulting polynomial coefficients: $\log(M)_i = a_0 + a_1.V_{ret}$

If the startup calibration function (raw__MWV) was already reasonable, such an algorithm typically needs 5–15 iterations to establish a linear SEC calibration function, which yields M_w(true) ± 0.1% and M_n(true) ± 0.1% for the eluogram of the broad distributed standard. If nonlinear calibration needs to be applied (e.g., to establish a sigmoid calibration function), molecular weight averages (M_n, M_w) have to be substituted by molecular weight distribution (m__MWD__d), and the number of polynomial coefficients (a_x) may be increased compared to the linear mode.

C. Universal SEC Calibration

Universal SEC calibration reflects differences in the excluded volume of polymer molecules with identical molecular weight caused by varying coil conformation, coil geometry, and interactive properties. Intrinsic viscosity, in the notation of Staudinger/ Mark/Houwink power law ($[\eta]=K.M^a$), summarizes these phenomena with K and a as material constants for given conditions. As occupied volume is the separation criterion in SEC, a calibration considering the excluded volume of individual components may be assumed to be universal, i.e., to fit for any coil conformation over a wide range of interactive characteristics. Consequently, calibration of a SEC system for a polymer sample of known geometry and interaction characteristics ($[\eta] \rightarrow K_{\text{sample}}$, a_{sample}, M_{sample}) can be achieved easily by transformation of an existing calibration function (Std) with K_{Std} and a_{Std} values according to

$$\log(M)_{\text{sample}} = \frac{1}{1 + a_{\text{sample}}} log\left(\frac{K_{\text{Std}}}{K_{\text{sample}}}\right) + \frac{1 + a_{\text{Std}}}{1 + a_{\text{sample}}} \log(M)_{\text{Std}}. \tag{1}$$

D. Molecular Weight/Degree of Polymerization: Average Values and Distributions

$$\text{Number average molecular weight: } \overline{\text{M}}_{\text{n}} = \frac{\sum \text{n}_i \text{M}_i}{\sum \text{n}_i} \tag{2}$$

$$\text{Weight average molecular weight: } \overline{\text{M}}_{\text{w}} = \frac{\sum \text{n}_i \text{M}_i^2}{\sum n_i M_i} \tag{3}$$

$$\text{dp (degree of polymerization): dp}_n = \frac{\text{M}_{\text{i}}}{\text{M}_{\text{o}}} \tag{4}$$

$$\text{Number average degree of polymerization: } \overline{\text{dp}}_{\text{n}} = \frac{\sum \text{n}_{\text{i}} \frac{\text{M}_i}{\text{M}_{\text{o}}}}{\sum_{n_{\text{i}}}} \tag{5}$$

$$\text{Weight average degree of polymerization: } \overline{\text{dp}}_{\text{w}} = \frac{\sum \text{n}_i \left(\frac{\text{M}_i}{\text{M}_{\text{o}}}\right)^2}{\sum \text{n}_{\text{i}} \frac{\text{M}_i}{\text{M}_{\text{o}}}} \tag{6}$$

m__MWD__d: differential molecular weight distribution: mass fractions

$$\text{m(M)}_i = \frac{\text{d m(M)}}{\text{d M}} \text{ with } \int_0^\infty \text{m(M) d M} = \text{m\_\_MMD\_\_d} = 1.0 \tag{7}$$

m__dpD__d: differential degree of polymerization distribution: mass fractions

$$\text{m(dp)}_i = \frac{\text{d m(M)}}{\text{d M}} \text{ with } \int_0^\infty \text{m(dp) d dp} = \text{m\_\_dpdD\_\_d} = 1.0 \tag{8}$$

n__MWD__d: differential molecular weight distribution: molar fractions

$$n(M)_i = \frac{m(M)_i/M_i}{\int_0^\infty \left[m(M)_i/M_i \right] d\,M} \quad \text{with} \int_0^\infty n(M)\, d\,M = n\_MWD\_d = 1.0 \qquad (9)$$

n__dpD__d: differential degree of polymerization distribution: molar fractions

$$n(dp)_i = \frac{m(dp)_i/dp_i}{\int_0^\infty \left[m(dp)_i/dp_i \right] d\,dp} \quad \text{with} \int_0^\infty n(dp)\, d\,M = n\_dpD\_d = 1.0 \qquad (10)$$

where i is the ith component, n is the mole fraction, m is the mass fraction, M is the molecular weight, and M_0 is the molecular weight of constituting unimer (monomer).

II. DEXTRAN GELS

Dextran gels have been utilized since the late 1950s (1) for the separation of biopolymers. First attempts on Sephadex (2–5) and Sephadex/Sepharose (6–8) systems are documented for hydrolyzed and native starch glucans. Up until now, particularly for the preparative and semipreparative separation of polysaccharides, a range of efficient and mechanically stable Sephacryl gels (9–14) have been developped.

The performance of several Sephacryl gel combinations is illustrated by results achieved for glucans from different types of starch granules. The applied Sephacryl gels of Pharmacia Biotech (15) are cross-linked copolymers of allyl dextran and *N,N'*-methylene bisacrylamide. The hydrophilic matrix minimizes nonspecific adsorption and thus guarantees maximum recovery. Depending on the pore size of the beads, ranging between 25 and 75 μm in diameter, aqueous dissolved biopolymers up to particle diameters of 400 nm can be handled.

Chemical stability:

- stable with all commonly used aqueous buffers
- 0.2 *M* NaOH, 0.1 *M* HCl
- long-term pH stability: 3–11
- 1 *M* acetic acid,
- 8 *M* urea, 6 *M* guanidine–HCl, 1% sodium dodecyl sulfate
- 2 *M* NaCl
- 24% ethanol, 30% propanol, 30% acetonitrile

Separation ranges for Sephacryl (S-type) gels are listed in Table 16.3.

Sephacryl gels were preswelled in H_2O(dest) + 0.01% Merthiolate. The diluted and degassed gels were then packed at medium pressure assisted by a pump-sucking eluent with a flow rate of approx 0.4 ml/min.

TABLE 16.3 Producers (Pharmacia Biotech) Specification of Fractionation Ranges of Cross-Linked Allyl Dextran/*N,N'*-Methylene Bisacrylamide Copolymer-Based Sephacryl Gels for Dextrans

	S-200	S-300	S-400	S-500	S-1000
From (g/M)	1×10^3	2×10^3	1×10^4	4×10^4	5×10^5
To (g/M)	8×10^4	1×10^5	2×10^6	2×10^7	$x \times 10^8$ [a]

[a] x = 1–9.

A. Preparative Sephacryl S-1000 and Semipreparative Sephacryl System S-200/S-400/S-500/S-1000

Glucans of starch granules from potatoe species *Ostara* were isolated to investigate their structural characteristics. From an aqueous suspension small (<35 μm in diameter) and large granules (60–120 μm in diameter) were pooled by sedimentation fractionation. Glucans of both fractions were dissolved in DMSO and splitted in a pool of nb/lcb (amylose-type) glucans and a pool of scb (amylopectin-type) glucans by *n*-butanol precipitation and subsequent methanol precipitation from the remaining supernatant, respectively. The obtained aqueous redissolved glucans were then separated on a Sephacryl S-1000 to investigate molecular characteristics of high and low dp fractions (Scheme 16.1).

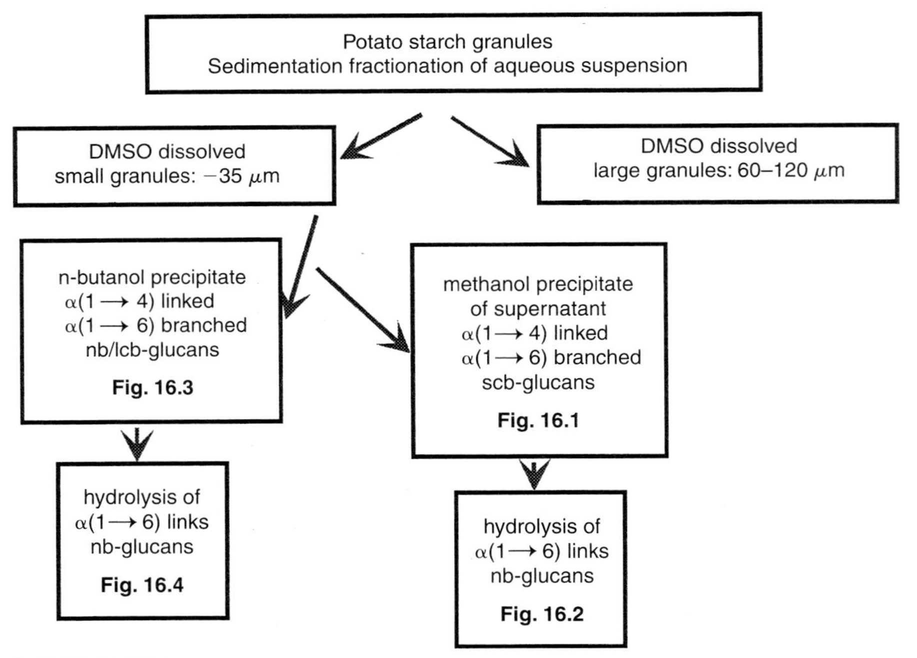

SCHEME 16.1 Analytical strategy for investigating potato starch glucans.

Preswelled Sephacryl S-1000 was prepared in a K26/100 column (88 × 2.6 cm). Equilibration with 0.005 *M* NaOH containing 0.002% NaN_3 at a flow rate of 0.67 ml/min was achieved after 20 hr. Sample solutions were applied with a 5-ml injection loop. The mass and iodine-complexing potential of separated glucan components was determined off-line for each of the subsequently eluted 5-ml fractions. Based on the determined mass of carbohydrate for each of the fractions, elution profiles such as Fig. 16.1 were constructed.

Figure 16.1 illustrates the pooling of scb glucans of small starch granules in a high dp (fraction 1) and a low dp fraction (fraction 2) for following detailed investigations of their branching characteristics.

Therefore the $\alpha(1\rightarrow6)$ branching linkages were hydrolized selectively. The resulting constituting chain-length distribution was determined on a semipreparative Sephacryl S-200/S-400/S-500/S-1000 (12 + 55 + 66 + 135 × 1.6 cm) system. Normalized eluograms (mass_ev) of high- and low dp glucans show significant differences: whereas high dp fraction 1 consists of two glucan chain populations, fraction 2 comes up with three dp populations (Fig. 16.2). Obviously, fraction 1 is a fraction of scb glucans, whereas fraction 2 may be

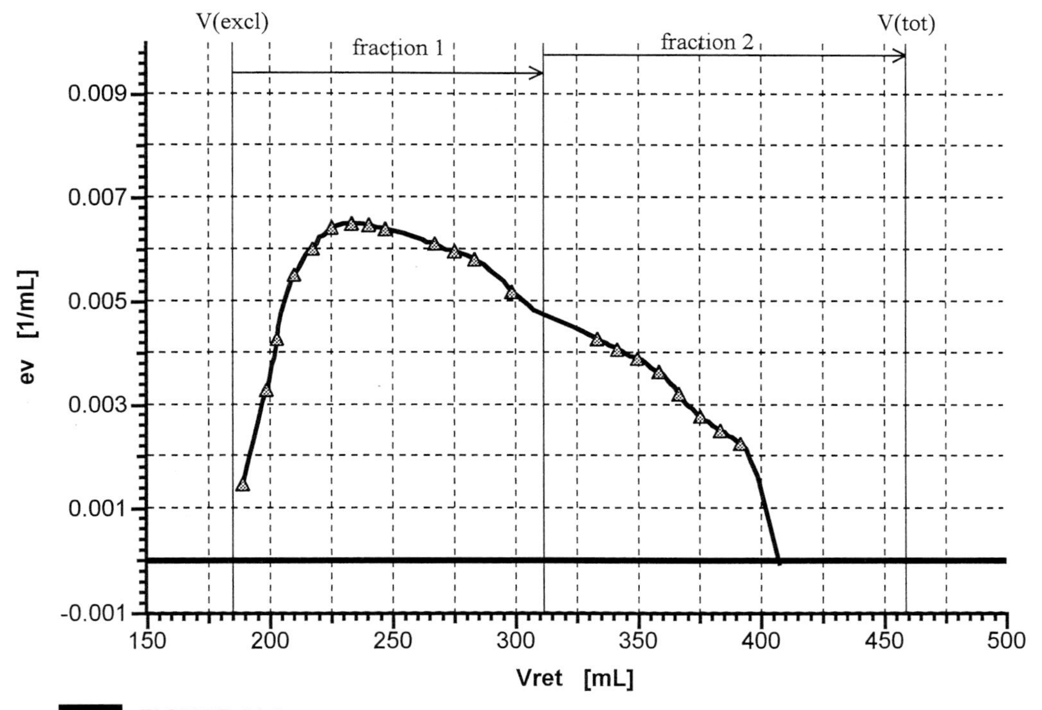

FIGURE 16.1 Preparative SEC of short-chain (scb) branched glucans of "small" (<35 μm) starch granules of potato species *Ostara* separated on Sephacryl S-1000 (88 × 2.6 cm); eluent: 0.005 *M* NaOH; the normalized chromatogram (area = 1.0) was constructed from an off-line determined carbohydrate content of succeeding 5-ml fractions; flow rate: 0.67 ml/min; V_{excl} = 185 ml, V_{tot} = 460 ml; fraction 1: high dp fraction; fraction 2: low dp fraction.

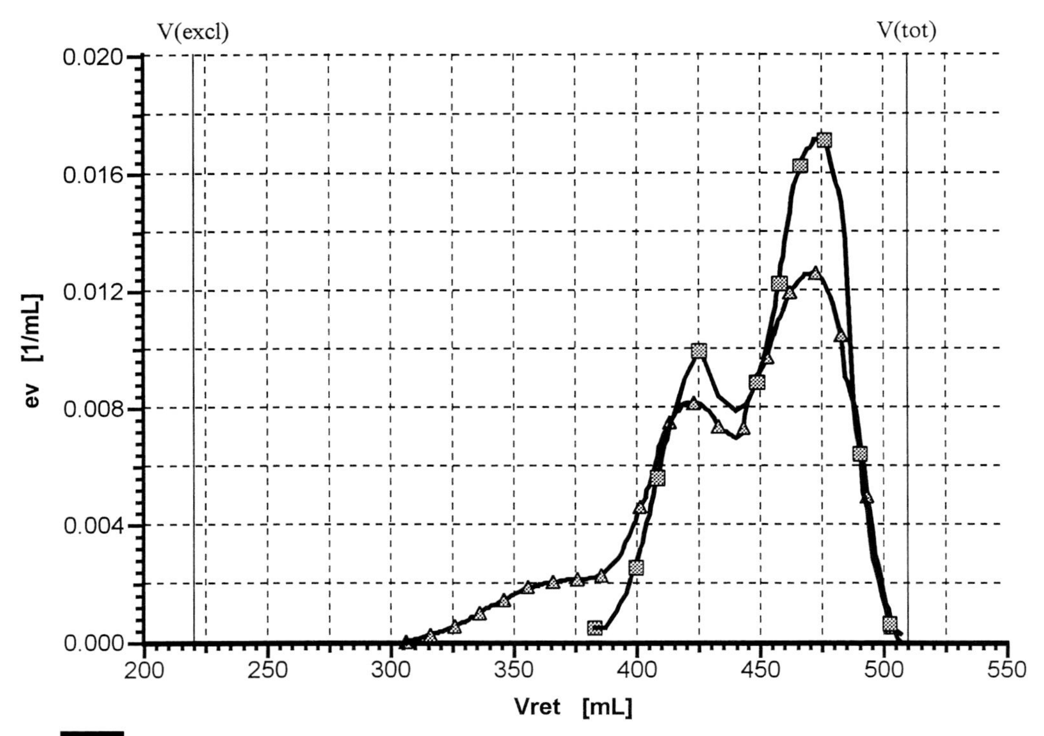

FIGURE 16.2 Chromatograms of totally debranched glucans of fraction 1 (□) and fraction 2 (△) of scb glucans (Fig. 16.1) separated on Sephacryl system S-200/S400/S-500/S-1000 (12 + 55 + 66 + 135 × 1.6 cm); V_{excl} = 220 ml; V_{tot} = 510 ml.

judged as a transition-type lcb→scb glucans. Such differences in branching characteristics result in significantly different packing densities of the polymer molecules (high density for scb, lower density for lcb) and different interactive properties. Such differences in branching characteristics may result in different percentages of amorphous (scb) and/or semicrystaline (lcb) domains in starch granules, and, of course, technological starch quality is influenced strongly by the kind of branching pattern.

Separation of nb/lcb glucans of the small starch granules is illustrated in Fig. 16.3. Once again a high dp (fraction 1) and a low dp fraction (fraction 2) were collected for further analysis. The constituting chain-length distribution was determined for both fractions after selective hydrolysis of $\alpha(1\rightarrow6)$-branching linkages. Although dp of the constituting components was found in the same range for both fractions, the percentage of long and short chains was quite different (Fig. 16.4). The low dp glucans of fraction 2 contain a high percentage of long chains, whereas the high dp glucans of fraction 1 are constructed primarily of short glucan chains.

These results strongly favor a concept of initially formed nb glucans that become lcb branched in a first step and increasingly scb branched with increasing physiological age and functional requirements.

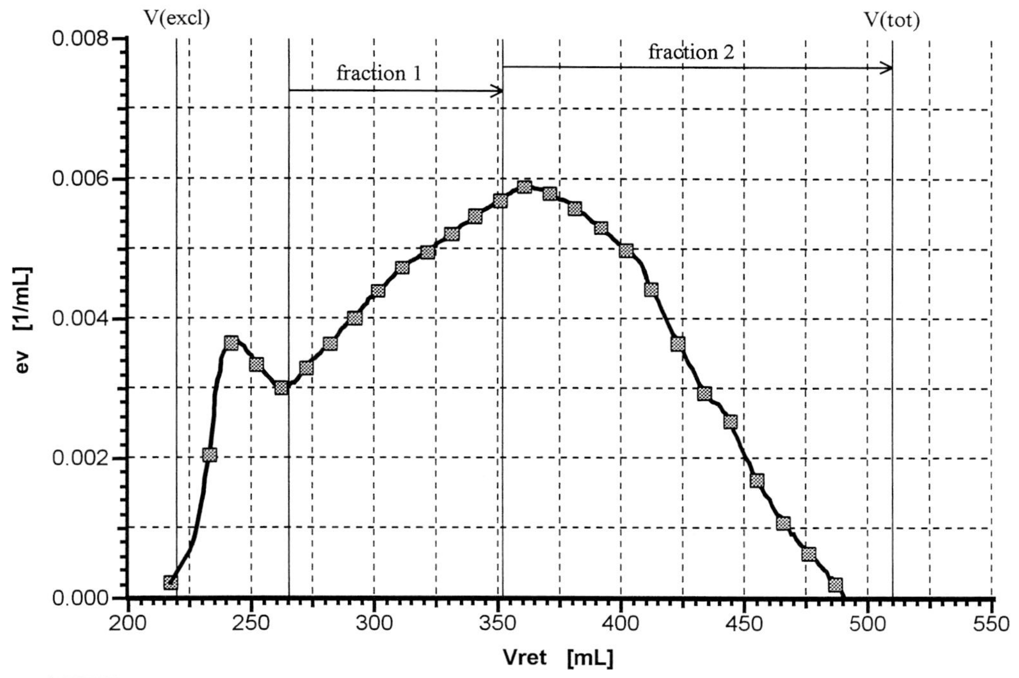

FIGURE 16.3 Preparative SEC of nonbranched/long-chain (nb/lcb) branched glucans of "small" (<35 μm) starch granules of potato species *Ostara* separated on Sephacryl S-200/S400/S-500/S-1000 (12 + 55 + 66 + 135 × 1.6 cm); eluent: 0.005 *M* NaOH; the normalized chromatogram (area = 1.0) was constructed from an off-line determined carbohydrate content of succeeding 5-ml fractions; flow rate: 0.67 ml/min; V_{excl} = 220 ml; V_{tot} = 510 ml; fraction 1: high dp fraction; fraction 2: low dp fraction.

B. Semipreparative Sephacryl S-1000

In accordance with manufacturer specifications, Sephacryl S-1000 can be judged as a suitable gel for very high molecular weight polysaccharides. A mixture of high-molecular Blue Dextran, broad-distributed Dextran T-500 and glucose was utilzed for the calibration of S-1000. The resulting linear broad standard calibration based on the characteristics of Dextran T-500 and the limits of selective separation (exclusion limit V_{excl} (Blue Dextran), total permeation limit V_{tot} (glucose) is illustrated in Fig. 16.5. Absolute molecular weight averages M_w and M_n (dp_n and dp_n, respectively) of Dextran T-500, which were needed for the construction of the linear calibration function, have been obtained from independent SEC differential refractive index (DRI) low angle laser light-scattering (LALLS) experiments.

Components of a highly short chain branched (scb) waxymaize and a more long chain branched (lcb) amylomaize were separated on the semipreparative Sephacryl S-1000 system. Both samples contained high dp components that eluted in the exclusion volume, but the percentage of these components was quite different: 90% for the scb waxymaize starch and approximately 10% for the lcb amylomaize starch (Fig. 16.6). The degree of polymerization averages for these samples was determined utilizing the previously established linear

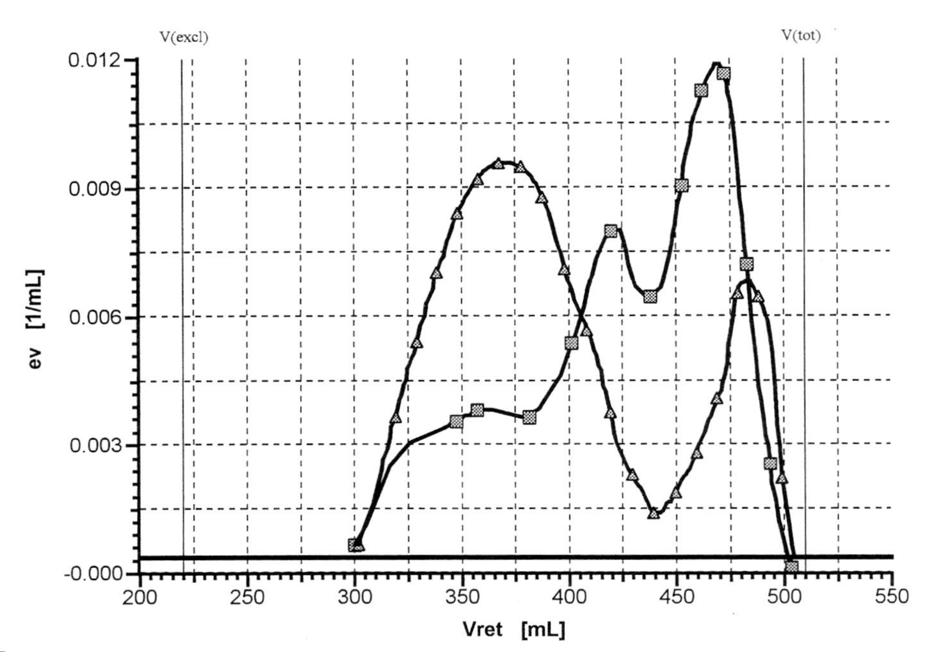

FIGURE 16.4 Chromatograms of totally debranched fraction 1 (□) and fraction 2 (▲) of lcb glucans (Fig. 16.3) separated on Sephacryl S-200/S-400/S-500/S-1000 (12 + 55 + 66 + 135 × 1.6 cm); flow rate: 0.40 ml/min; V_{excl} = 220 ml; V_{tot} = 510 ml.

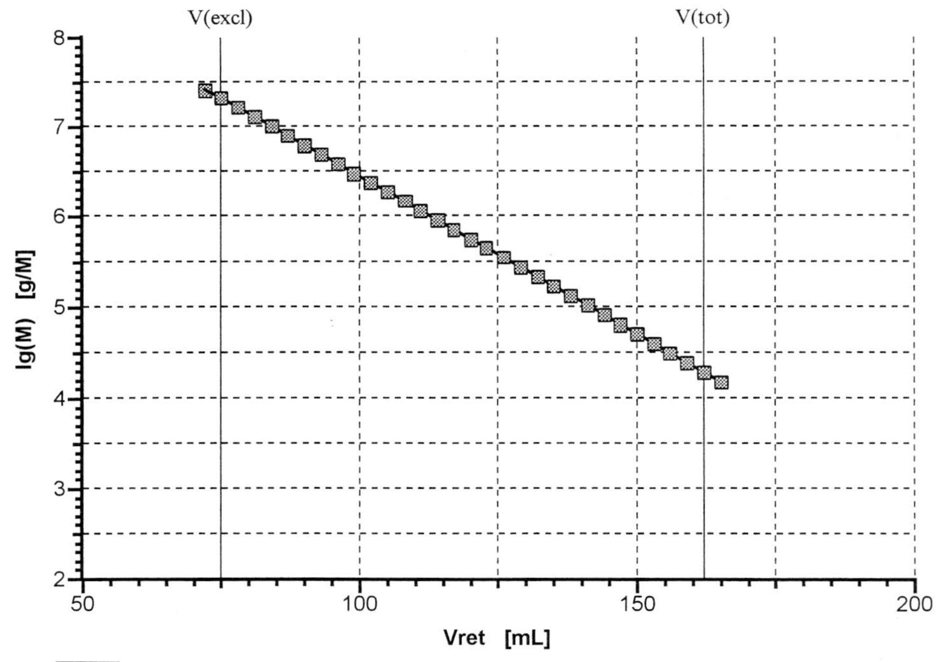

FIGURE 16.5 Broad standard calibration (linear mode) of a semipreparative Sephacryl S-1000 system (95 × 1.6 cm) with an aqueous mixture of Blue Dextran, Dextran T-500, and glucose; eluent: 0.005 *M* NaOH; V_{excl} = 75 ml, V_{tot} = 162 ml.

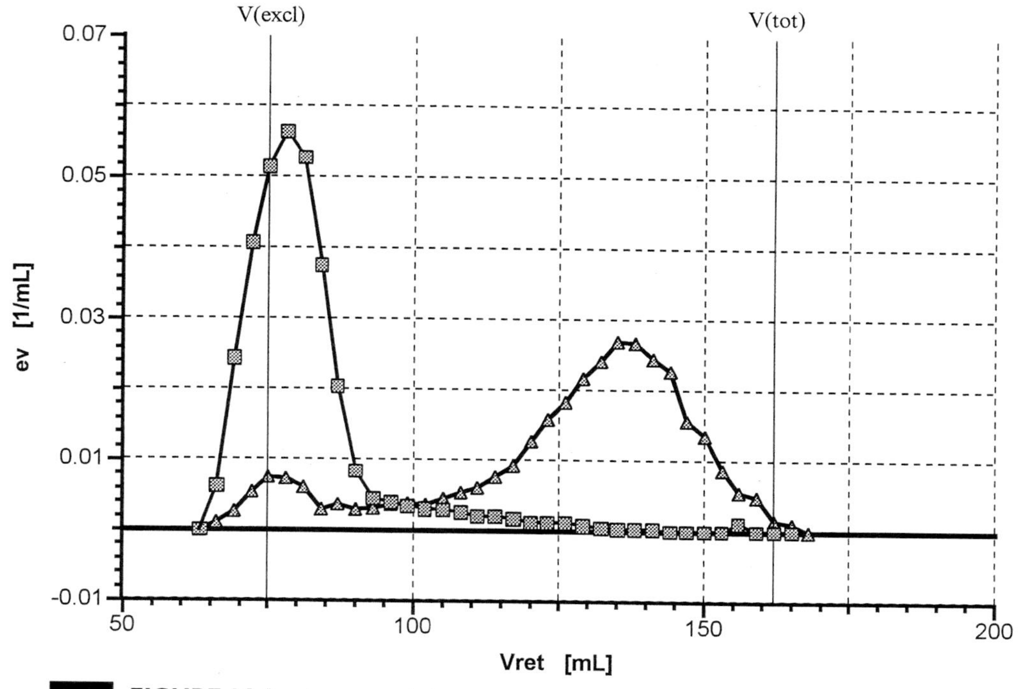

FIGURE 16.6 Waxymaize (□) and amylomaize (▲) separated on Sephacryl S-1000 (95 × 1.6 cm); pooled in 3-ml fractions for further analysis; normalized (area = 1.0) eluogram profiles (ev) constructed from an off-line determined mass of carbohydrates for each of the 3 ml fractions; flow rate: 0.41 ml/min; V_{excl} = 75 ml, V_{tot} = 162 ml; eluent: 0.005 *M* NaOH.

broad standard calibration: dp_w = 106,000 glucose units for the waxymaize and dp_w = 16,300 glucose units for the amylomaize.

Another native starch (a mixture of high dp scb glucans and of midrange dp lcb glucans) and individual fractions of nb/lcb and scb glucans of this starch were separated on Sephacryl S-1000 to investigate correlations between degree of polymerization and braching characteristics (Fig. 16.7).

For identical conditions, different branching patterns of scb and lcb glucans cause differences in molecular conformation and thus differences in SEC separation characteristics. Compared to lcb glucans, scb glucans are packed more compactly: a scb molecule with an identical excluded volume as a lcb glucan contains more molar mass and is higher in dp. Thus, application of dextran-based broad standard calibration will yield too high values for lcb glucans and too low values for scb glucans. However, independent of such systematic deviations, as well weight average degree of polymerization (dp_w = 8300 glucose units for the lcb fraction; dp_w = 80,700 glucose units for the scb fraction) as the profiles of the mass distributions clearly indicate the lcb fraction to be midrange/low dp and the scb fraction to be high dp glucans.

In summary, semipreparative Sephacryl S-1000 proved to be an appropriate system for the separation of glucan components of any kind of branching

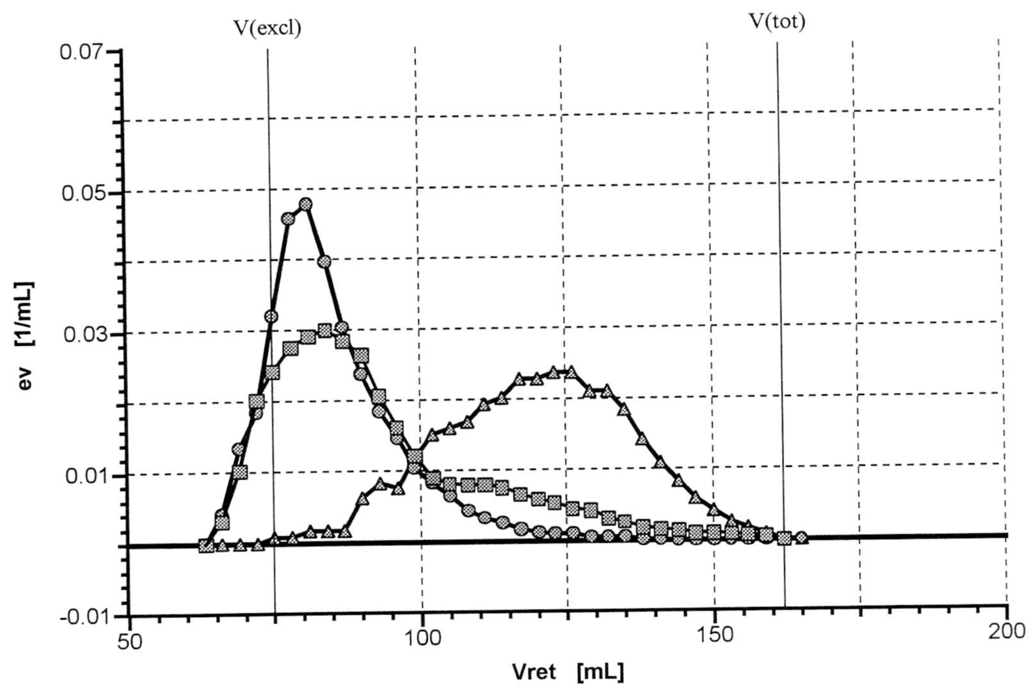

FIGURE 16.7 Native starch (□) and fractions of native starch differing in their branching characteristics (nb/lcb "amylose"-type fraction: ▲; scb "amylopectin"-type fraction: ●) separated on semipreparative Sephacryl S-1000 (95 × 1.6 cm); 3-ml fractions were pooled for further analysis; normalized (area = 1.0) eluogram profiles (ev) constructed from an off-line determined mass of carbohydrates for each of the pooled fractions; flow rate: 0.40 ml/min; V_{excl} = 75 ml, V_{tot} = 162 ml; eluent: 0.005 *M* NaOH.

pattern with the degree of polymerization between some ten up to some hundred thousand glucose units.

C. Semipreparative Sephacryl System S-500/S-1000

Reproducibility of separation for a Dextran T-500 sample was tested on a semipreparative Sephacryl system S-500/S-1000 (65 + 95x1.6 cm) over a period of 6 months. The elution profiles of Dextran T-500 could be superimposed with deviations in the elution axis of ±3 ml (±1 fraction), and deviations in carbohydrate content within ±5% referring to the maximum value at V_{ret} = 213 ml (Fig. 16.8).

Absolute degree of polymerization distribution (m__dpD__d) and degree of polymerization averages of Dextran T-500 have been determined from data of independent SEC DRI/LALLS experiments (Fig. 16.9). With these values for weight and number average degree of polymerization (dp_n = 1174 glucose units; dp_w = 3055 glucose units) and the elution profile for Dextran T-500 on the S-500/S-1000 system, a linear calibration function was established (Fig. 16.10).

To improve interpretation of nb/lcb glucan profiles from the S-500/S-1000 system, the initially obtained broad standard calibration function was

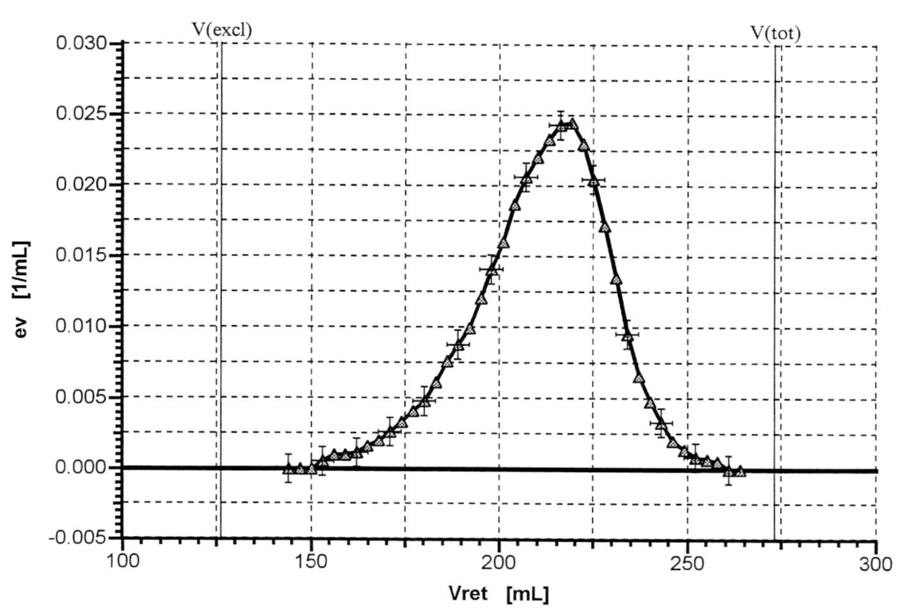

FIGURE 16.8 Dextran T-500 (▲) separated on Sephacryl S-500/S-1000 (60 + 95 × 1.6 cm); sampled in 3-ml fractions; normalized (area = 1.0) eluogram profiles (ev) constructed from an off-line determined mass of carbohydrates within each of the fractions; flow rate: 0.42 ml/min; V_{excl} = 126 ml, V_{tot} = 273 ml; eluent: 0.005 *M* NaOH; reproducibility of results over a period of 6 months: ±5% of ev maximum and ±3 ml (±1 fraction).

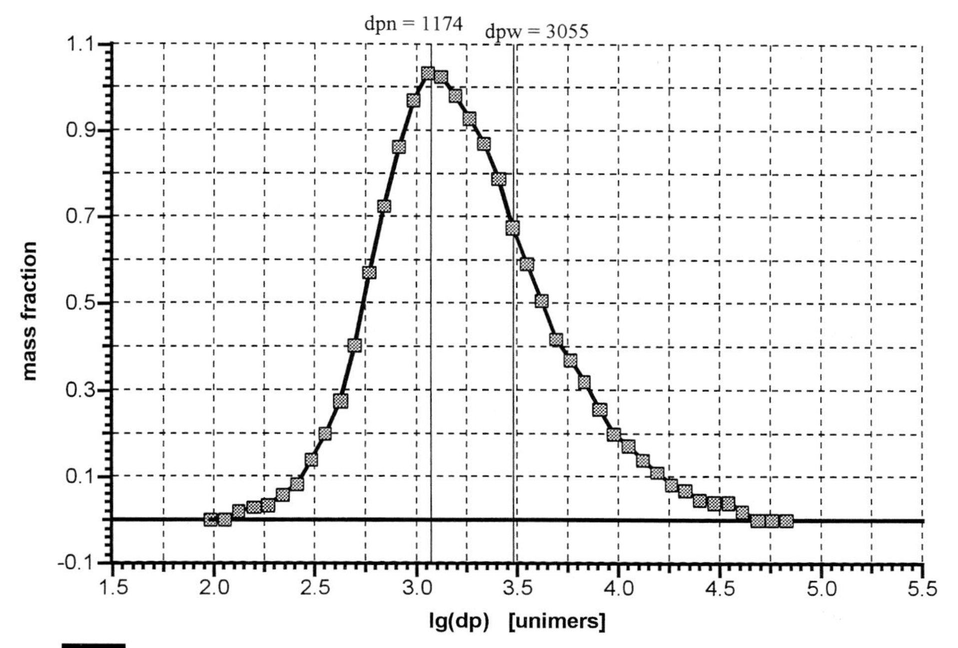

FIGURE 16.9 Degree of polymerization distribution of Dextran T-500 (□) separated on Sephacryl S-500/S-1000 (60 + 95 × 1.6 cm) achieved from broad standard calibration with known number average degree of polymerization dp_n = 1174 [glucose units] and weight average degree of polymerization dp_w = 3055 [glucose units]; differential degree of polymerization distribution normalized to area = 1.0.

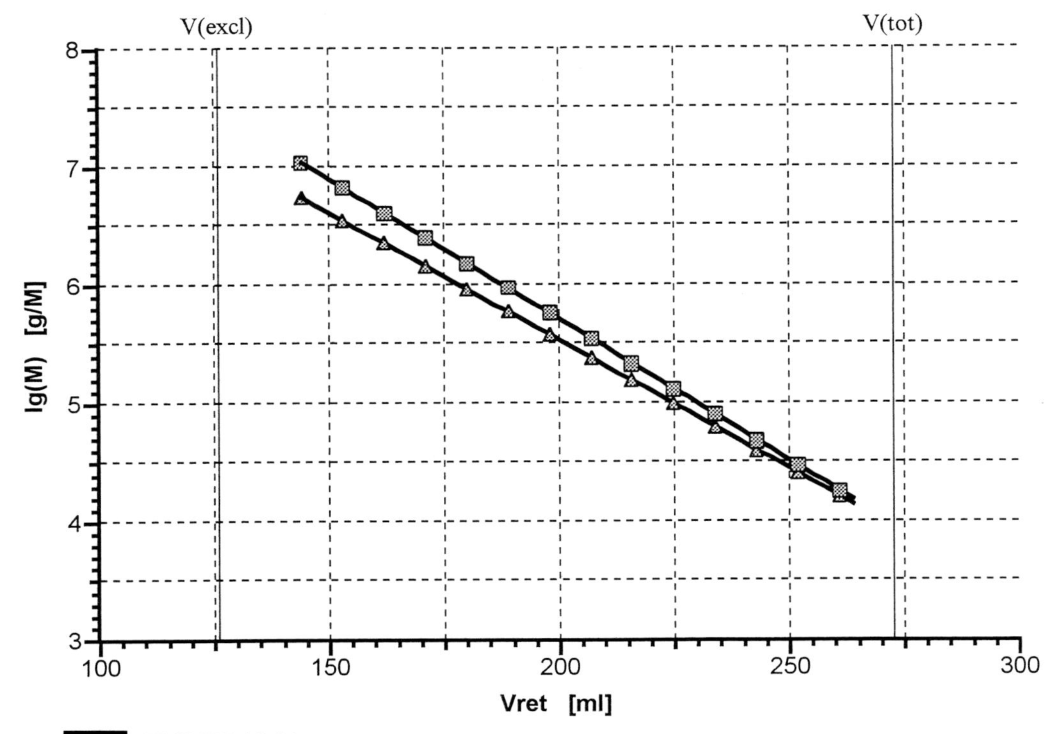

FIGURE 16.10 Intial dextran calibration (□) and resulting nb/lcb glucan calibration (△) for the Sephacryl S-500/S-1000 (60 + 95 × 1.6 cm) system achieved from broad standard calibration with Dextran T-500 and universal calibration, respectively; Staudinger/Mark/Houwink constants (dextran: K_{Std} = 0.0978 ml M g^{-2}, a_{Std} = 0.50; nb/lcb ≡ amylose ≡ pullulan: K_{sample} = 0.0268 ml M g^{-2}, a_{sample} = 0.65).

transformed from a scb-type calibration function into a lcb-type calibration function applying universal SEC calibration. The resulting nb/lcb-type calibration of S-500/S-1000 obtained with SMH constants for dextran (K_{Std} = 0.0978 ml M g^{-2} and a_{Std} = 0.50) and SMH constants for nb/lcb glucans (K_{sample} = 0.0268 ml M g^{-2} and a_{sample} = 0.65) is illustrated in Fig. 16.10.

The selective separation range of the S-500/S 1000-system for glucans is shown for hybrid starch Triticale T22. This mixture of scb and nb/lcb glucans contains components in the range between approximately dp 50–300,000 glucose monomers (Fig. 16.11). The degree of polymerization distribution obtained from dextran-based calibration was computed as well in terms of mass fractions (m__dpD__d) as in terms of molar fractions (n__dpD__d). Average values dp_w = 32,000 [glucose units], dp_n = 900 [glucose units], and the polydispersity of dp_w/dp_n = 36 show a broad distribution with the interesting fact that components exceeding dp 3000 represent approximately 3–5% of molar concentration only, whereas mass contribution of these glucans is approximately 60%.

Sephacryl S-500/S-1000 is an appropriate system to control the extent of enzymatic synthesis of nonbranched $\alpha(1\rightarrow4)$-linked glucans. The degree of

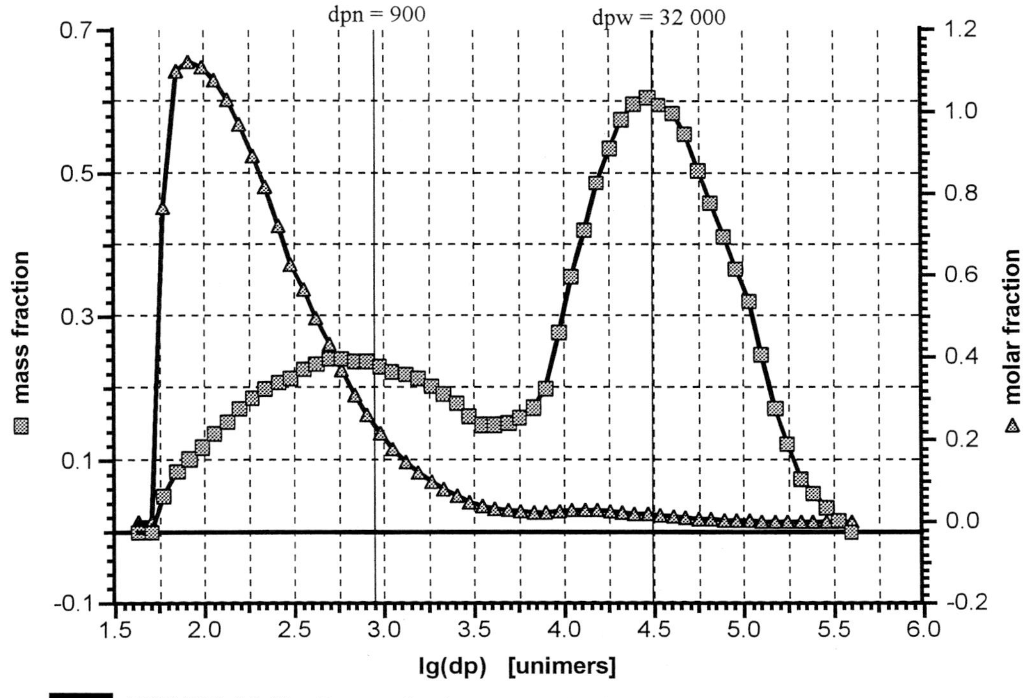

FIGURE 16.11 Degree of polymerization distributions: mass (□; m_dpD_d) and molar (△; n_dpD_d) fractions for Triticale starch separated on Sephacryl S-500/S-1000 (60 + 95 × 1.6 cm); dp_w = 32,000 [glucose units] and dp_n = 900 [glucose units] (polydispersity dp_w/dp_n = 36).

polymerization of enzymatically generated products hardly can be determined by means of nondestructive techniques due to interferences caused by significant monomer concentration and the high ionic strength of utilized buffers. Application of the reaction mixture on the S-500/S-1000 systems separates the high/midrange dp glucans from low dp components (monomer, salts) (Fig. 16.12) and easily enables the determination of degree of polymerization distribution (m__dpD__d) and average values for the degree of polymerization (dp_w, dp_n) for the synthesized glucans. As the reaction scheme predicts the formation of nb glucans only, computation of the degree of polymerization distribution was achieved with a nb/lcb calibration of S-500/S-1000 (Fig. 16.13).

D. Preparative Sephacryl System S-200/S-1000

A range of preparative and semipreparative soft gel systems with an improved mechanical stability and thus the chance to run them with increased flow rates were tested for their potential on the separation of starch glucans. For each of these systems a Sephacryl S-200 precolumn proved to be a perfect shock absorber for sample application, improved reproducibility of separations, and increased lifetime of soft gel systems.

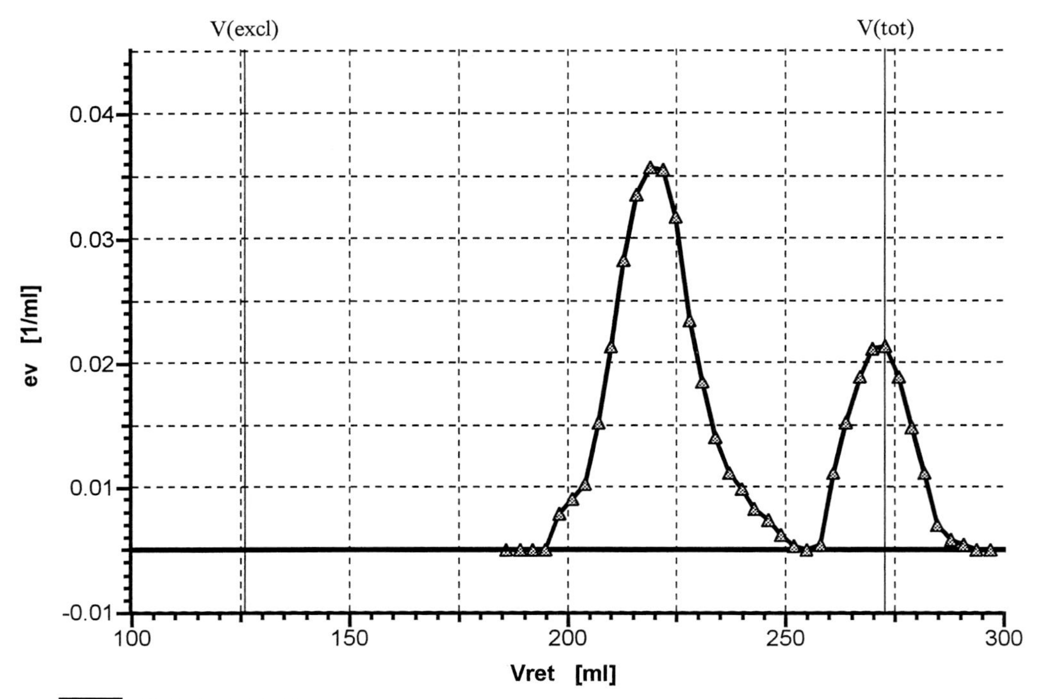

FIGURE 16.12 Enzymatically synthesized "amylose"-type nb/lcb glucans (▲) with a significant amount of the substrate glucose-1-PO_4 separated on Sephacryl S-500/S-1000 (60 + 95 × 1.6 cm); 3-ml fractions were collected for further analysis; normalized (area = 1.0) eluogram profiles (ev) constructed from an off-line determined mass of carbohydrates for each of the pooled fractions; flow rate: 0.42 ml/min; V_{excl} = 126 ml, V_{tot} = 273 ml; eluent: 0.005 *M* NaOH.

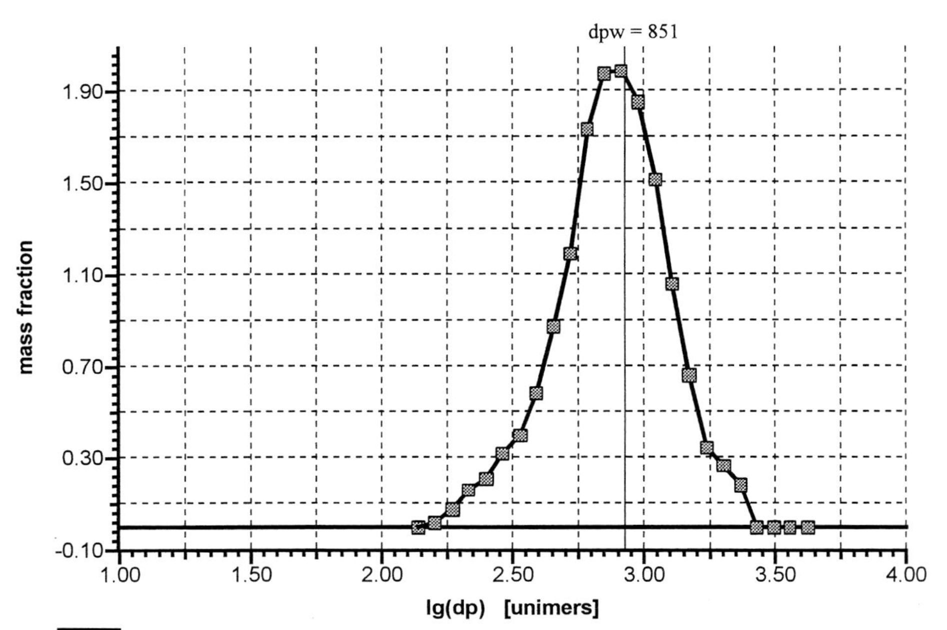

FIGURE 16.13 Degree of polymerization distribution (m_dpD_d) for synthesized "amylose"-type nb/lcb glucans calculated from dextran-calibrated → amylose-converted calibration of S-500/S-1000; dp_w = 851 [glucose units].

Preparative Sephacryl S-200/S-1000 was utilized to obtain three pools of glucans from a sample expected to be nb/lcb glucans as they have been obtained by *n*-butanol precipitation from aqueous dissolved potato starch (Fig. 16.14).

Each of the three fractions was applied to a combined SEC system of Fractogel/Superose to investigate their molecular characteristics in detail. These high resolution and mechanically stable gels allow the application of eluents with increased ionic strengths (e.g., 0.05 *M* KCl) at a reasonable resolution of sample components (Fig. 16.15).

III. COMBINED SYSTEM: FRACTOGEL TSK HW40 (s)/HW50 (s)/SUPEROSE 6 (PREP GRADE)

Fractogel TSK HW of Merck (16) is a packing material based on spherical polyethylene glycol dimethylacrylate (PGM) particles. The manufacturer claims that this gel is particularly suitable for aqueous systems, including diluted HCl (pH 1) and diluted NaOH (pH 14), but it even stands organic solvents. Compared to dextran/agarose gels, PGM Fractogel is highly temperature stable:

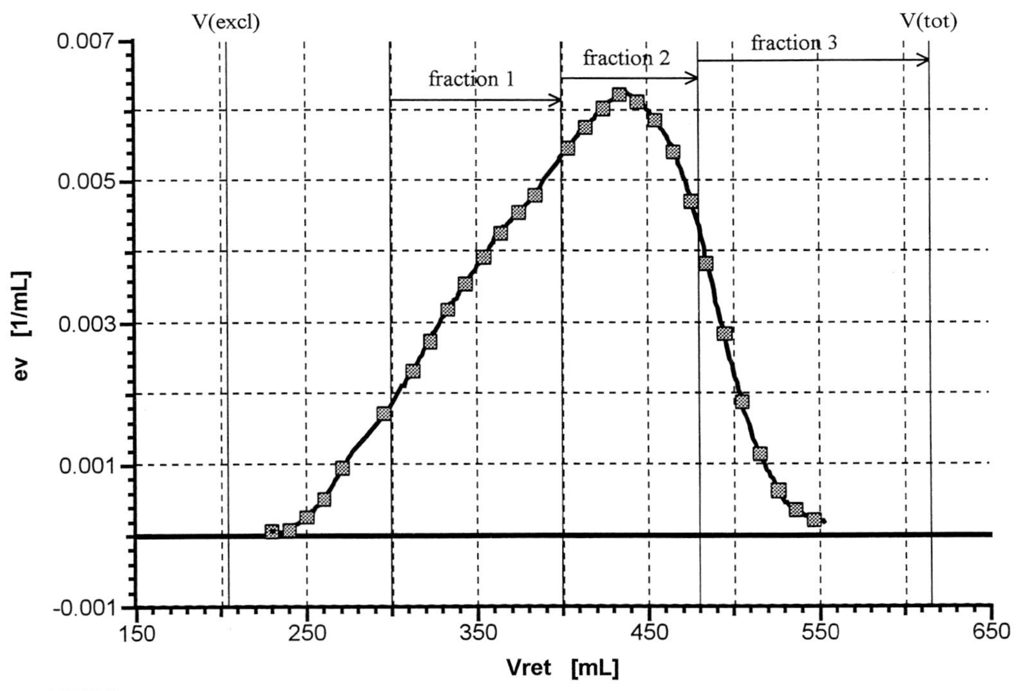

FIGURE 16.14 Nonbranched/long chain branched glucans separated on preparative Sephacryl SEC system S-200/S-1000 (31 + 85 × 2.6 cm); pools: fraction 1: 35% of nb/lcb glucans, high excluded particle volume; fraction 2: 45% of nb/lcb glucans, midrange excluded particle volume; 3: 10% of nb/lcb glucans, small excluded particle volume; applied volume of sample solution: 5 ml of a 15-mg/ml solution; normalized (area = 1.0) eluogram profiles (ev) constructed from an off-line determined mass of carbohydrates for each of the subsequentely pooled 10-ml fraction; flow rate: 0.65 ml/min; V_{excl} = 204 ml, V_{tot} = 616 ml; eluent: 0.005 *M* NaOH.

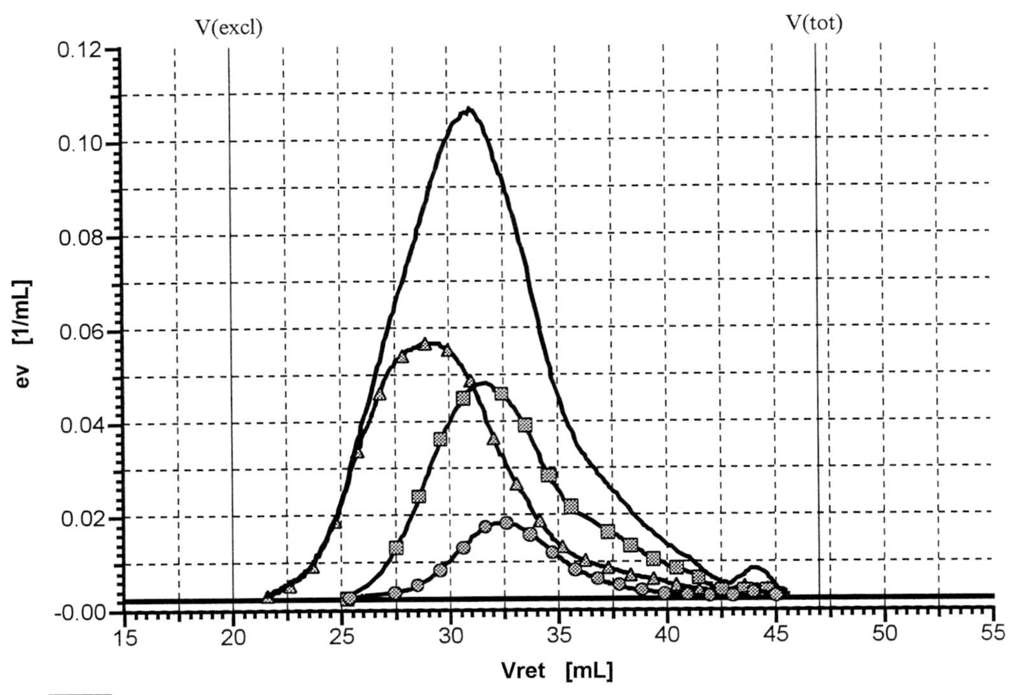

FIGURE 16.15 Three nb/lcb glucan fractions from preparative Sephacryl SEC system S-200/S-1000 (31 + 85 × 2.6 cm) (Fig. 16.14), separated on analytical system Fractogel TSK HW40(S)/HW50(S)/ Superose 6(prep grade) (29 + 29 + 29 × 1.0 cm); fraction 1 (▲), fraction 2 (□), fraction 3 (●), and arithmetic sum of fractions 1(50%), 2(40%), and 3(10%) (—); applied sample volume: 500 μl of 10-mg/ml solutions; normalized (area = 1.0) DRI eluogram profiles (ev); flow rate: 0.70 ml/min; V_{excl} = 20 ml, V_{tot} = 47 ml; eluent: 0.05 *M* KCl + 0.02% NaN_3.

up to 120°C. HW-type gels are available in four bead sizes: C (coarse grade: 50–100 μm), M (medium grade: 45–90 μm), F (fine grade: 32–63 μm), and S (superfine grade: 25–40 μm). S-type gels yield better resolution than F-type gels at identical flow rates, and F-type gels can be run at higher flow rates (system pressure) with acceptable results.

Fractogel TSK HW is mechanically stable, withstanding pressure up to several bars without significant deformation. Fractogel TSK HW swells in polar organic solvents and solvent mixtures; however, the bed volume decreases with decreasing polarity of the solvent, but swells more (~10%) in dimethylformamide and DMSO than in water. Separation ranges for Fractogel HW gels are listed in Table 16.4.

Superose gel material of Pharmacia Biotech is a highly epichloro-hydrine cross-linked agarose matrix that has a pH range of 3–12 (short term: 1–14). Hydrophilic interactions may be noticeable for lipids, peptides, and small aromatic compounds, but such interactions might even improve resolution. Superose medium is available in two different porosities: Superose 6 HR 10/30 (bead size 13 ± 2 μm; maximum pressure: 1.5 MPa) and Superose 12 HR 10/30 (bead size 10 ± 2 μm; maximum pressure: 3.0 MPa).

TABLE 16.4 Producers (Merck) Specification of Fractionation Ranges for Polyethylene Glycol Dimethylacrylate-Based Fractogel HW Gels

	Fractogel TSK HW 40	Fractogel TSK HW 50
From (g/M)	10^2	5×10^2
To (g/M)	5×10^3	2×10^4

Superose gels are also available as "prep-grade" materials with almost identical selectivity as Superose HR, but with larger particle size (Superose 6 and 12 prep-grade: 20–40 μm; maximum pressure: 0.4 and 0.7 Mpa, respectively) and thus the extent of interaction is less than with prepacked Superose HR gels. Separation ranges for Superose gels are listed in Table 16.5.

With the HW40(s)/HW50(s)/Superose 6 (prep grade) system, even viscous solutions of polysaccharides with molecular weights exceeding 1×10^6 [g/M] can be separated without problems. Compared to the Sephacryl systems, additional advantages are increased reproducibility and fast (~2 hr) separation.

Figure 16.15 shows the resulting chromatograms for the three glucan fractions obtained by previous preparative separation on Sephacryl S-200/S-1000 (Fig. 16.14). From the normalized fraction chromatograms, the elution profile of the initial mixture has been reconstructed by mixing 50% fraction 1, 40% fraction 2, and 10% fraction 3. Compared to the chromatogram of the preparative Sephacryl S-200/S-1000 system, separation with the TSK/Superose system yields improved resolution in the low dp (high V_{ret}) domain.

IV. AGAROSE GELS

Agarose gels have been used for more than two decades to separate polysaccharides (17–22). In particular, Sepharose CL 2B is widely used (6–8) to separate native starch, but continuously improved mechanical and chemical stability made all of the Sepharose CL gels perfect systems for the analysis of high molecular and broad distributed polysaccharides (23–28).

Sepharose CL gels of Pharmacia Biotech are cross-linked agarose derivatives with cross-links formed by 2,3-dibromopropanol. Compared to pure agarose, the cross-linking improves chemical and physical resistance of the

TABLE 16.5 Producers (Pharmacia Biotech) Specification of Fractionation Ranges of Epichloro-Hydrine Cross-Linked Agarose-Based Superose Gels

	Superose 6 HR 10/30	Superose 12 HR 10/30
From (g/M)	$\times 10^3$	$\times 10^2$
To (g/M)	$\times 10^6$	$\times 10^5$

matrix. Sepharose CL is available with different percentages of cross-linked agarose: 2%→CL 2B, 4%→CL 4B, and 6%→CL 6B. Cross-linked agarose decreases matrix porosity, increases rigidity of the gel, and modifies the selective separation range. The Sepharose matrix ranges in wet bead sizes between 60–200 μm for CL 2B and 45–165 μm for CL 4B and CL 6B. All of these gels show minimum nonspecific adsorption phenomena with pH stability between 3 and 13 (short term: 2–14). Fractionation ranges for Sepharose CL gels are listed in Table 16.6.

A. Sepharose CL 2B

The molecular uniformity of constituting components of a nb/lcb glucan fraction of potato starch was investigated with Sepharose CL 2B (Fig. 16.16) as well as with Sephacryl S-1000 (Fig. 16.17). Therefore, each of the subsequently eluted 3-ml fractions was analyzed on their potential to form inclusion complexes with iodine, a sensitive test for the presence of nb/lcb glucans. Results are shown in Fig. 16.17 in terms of "branching index," the ratio of extinction of pure iodine solution and of nb/lcb glucan/iodine complex: the higher the index, the more pronounced the nb/lcb characteristics.

In accordance with expectations, the branching index classifies the major partition of the investigated sample as nb/lcb glucans. On CL 2B, as well as on S-1000, the major percentage of components was eluted in the selective separation range, and a minor amount of high dp scb glucans in both systems was eluted at the exclusion limit.

Briefly summarized, S-1000 shows better resolution in the high dp range, whereas the CL 2B system resolves low dp components better. Due to these differences in separation performance, the degree of polymerization distribution and degree of polymerization average values for the same sample obtained from broad scb→nb/lcb-transformed dextran calibration yielded dp_w = 29,900 glucose units for the S-1000 system and dp_w = 21,100 glucose units for the CL 2B system (Fig. 16.18).

B. Sepharose CL 4B/2B

Combining Sepharose CL 2B and CL 4B singnificantly increases resolution in the low dp range of broad distributed samples. As an example, wild-type potato starch and two fractions of this sample differing in their branching

TABLE 16.6 Producers (Pharmacia Biotech) Specification of Fractionation Ranges of 2,3-Dibromopropanol Cross-Linked Agarose Derivative-Based Sepharose CL Gels

	Sepharose CL 2B	Sepharose CL 4B	Sepharose CL 6B
From (g/M)	1×10^5	3×10^4	1×10^4
To (g/M)	2×10^7	5×10^6	1×10^6

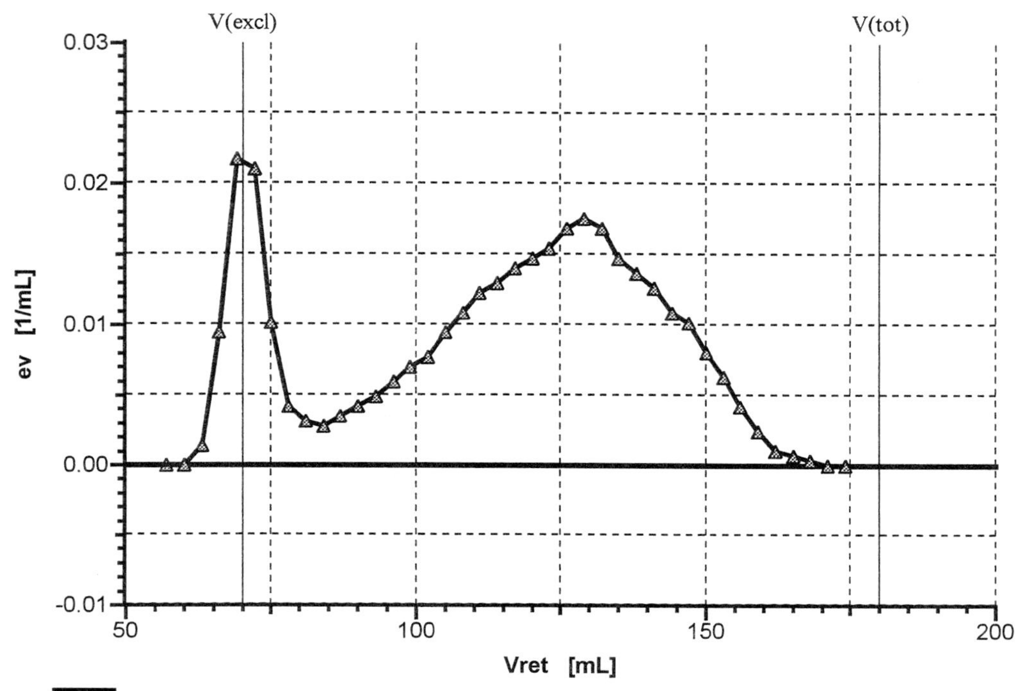

FIGURE 16.16 Nonbranched/long chain branched glucans of potato starch dissolved in hot water–steam and 0.1 *M* NaOH; 1.2 ml of the 18-mg/ml solution was separated on Sepharose CL 2B (88 × 1.6 cm); 3-ml fractions were collected for further analysis; normalized (area = 1.0) eluogram profiles (ev) constructed from an off-line determined mass of carbohydrates of each of the fractions; flow rate: 0.15 ml/min; V_{excl} = 70 ml, V_{tot} = 180 ml; eluent: 0.01 *M* NaOH.

pattern, nb/lcb and scb glucans, were separated on a combined Sepharose CL 4B/CL 2B system. The degree of polymerization distribution and average values of degree of polymerization were obtained applying a broad standard calibration (Fig. 16.19).

Components of both samples, nb/lcb and scb glucans, cover the identical dp range, but whereas the major percentage of high dp components are of the scb type, the midrange/low dp fraction are dominantly nb/lcb glucans. These results typically illustrate the separation range of the CL 4B/CL 2B system, ranging between some x up to some x 000 000 glucose monomers ($x = 1$–9) for starch glucans. As particularly low dp components are resolved, perfectly, this system may be recommended strongly for broad distributed samples with significant percentages of midrange/low dp components.

C. Analytical Superose 6 HR 10/30

Prepacked columns with cross-linked high-resolution (HR) agarose gels provide a high number of theoretical plates and fast separations (29,30).The Superose gel material of Pharmacia Biotech is a highly epichloro-hydrine cross-linked

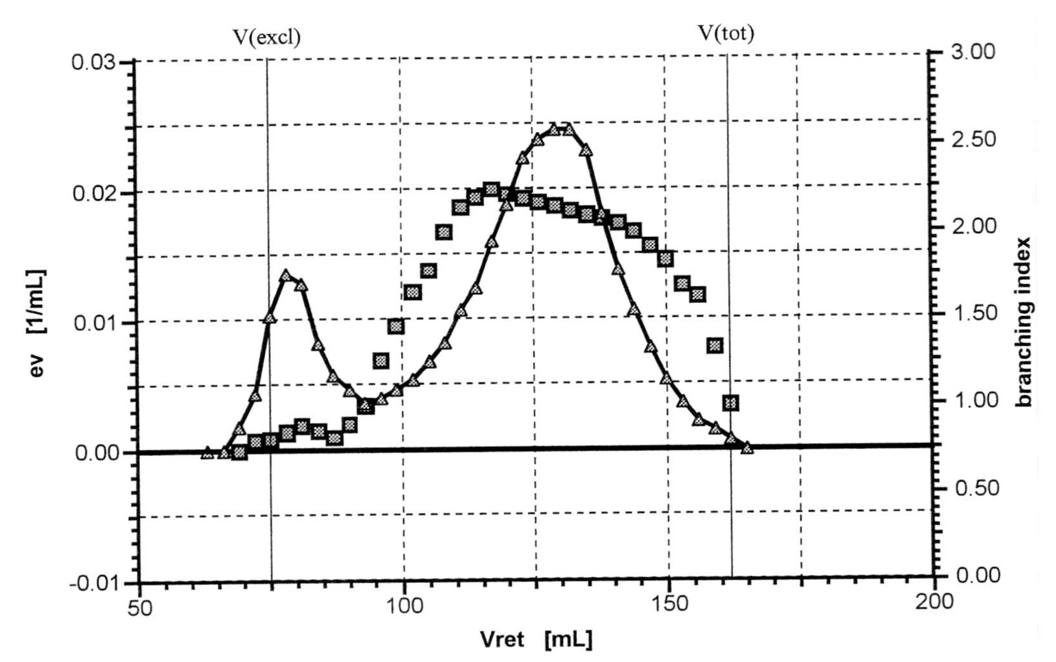

FIGURE 16.17 Nonbranched/long chain branched glucans of potato starch dissolved in hot water–steam and 0.1 *M* NaOH; 1.2 ml of the 18-mg/ml solution was separated on Sephacryl S-1000 (95 × 1.6 cm); 3-ml fractions were collected for further analysis; normalized (area = 1.0) eluogram profiles (ev) constructed from an off-line determined mass of carbohydrates for each of the fractions; branching index (□) determined from iodine-complexing potential of individual 3-ml fractions; flow rate: 0.40 ml/min; V_{excl} = 75 ml, V_{tot} = 162 ml; eluent: 0.005 *M* NaOH.

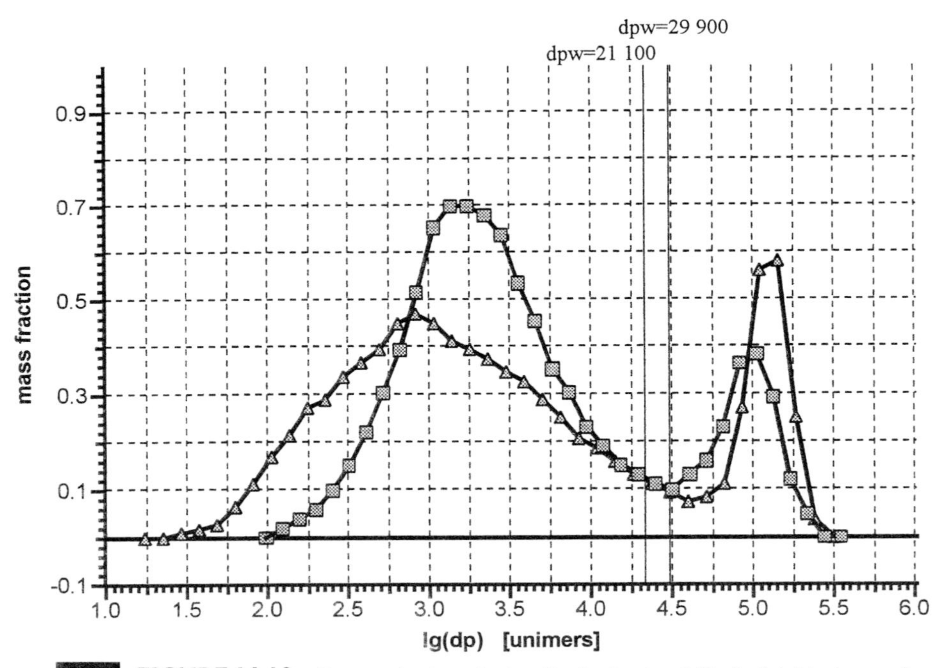

FIGURE 16.18 Degree of polymerization distribution (m_dpD_d) of nb/lcb glucnas of potato starch separated on Sepharose CL 2B (□) and on Sepharose S-1000 (▲); degree of polymerization distribution is normalized (area = 1.0) with indicated weight averages of degree of polymerization (CL 2B: 29,900 glucose units; S-1000: dp_w = 21,100 glucose units).

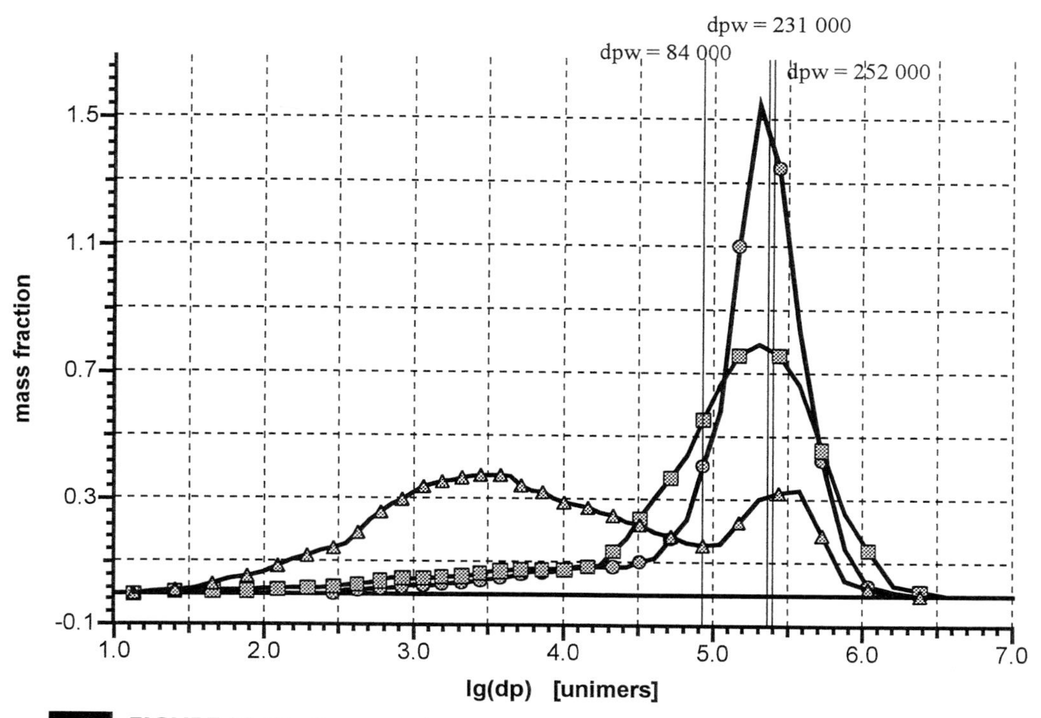

FIGURE 16.19 Degree of polymerization distribution (m_dpD_d) of wild-type potatoe starch (□), a nb/lcb fraction ("amylose"-type: ▲), and a scb fraction ("amylopectin"-type: ●) of the native starch; mass distributions are normalized (area = 1.0) with indicated weight averages of degree of polymerization (dp_w = 252,000 [glucose units] for the native starch, dp_w = 84,000 [glucose units] for the nb/lcb fraction; dp_w = 231,000 [glucose units] for the scb fraction).

agarose matrix that stands pH 3–12 (short term: 1–14). Particularly for lipids, peptides, and small aromatic compounds, hydrophilic interactions may be noteworthy, but such interactions might even improve resolution. The Superose medium is available in two different porosities: Superose 6 HR 10/30 (bead size 13 ± 2 μm; maximum pressure: 1.5 MPa), preferably for broad distributed samples, and Superose 12 HR 10/30 (bead size 10 ± 2 μm; maximum pressure: 3.0 MPa) for high dp samples.

Components of a partially hydrolized starch glucan were separated on the Superose 6 system. The mass (m__dpD__d) and molar (n__dpD__d) degree of polymerization distribution and average values of degree of polymerization (Fig. 16.21) were computed applying the previously with a set of narrow distributed standards (Table 16.7) established calibration function (Fig. 16.20).

The comparison of mass and molar distribution of hydrolized starch glucans shows high molar concentrations of low dp components and the negligible percentage of components exceeding dp 100. However, this molar minority of high dp glucans represents the dominant mass contribution and is highly important as these components control bulk properties and thus technological material qualities of starch-based materials.

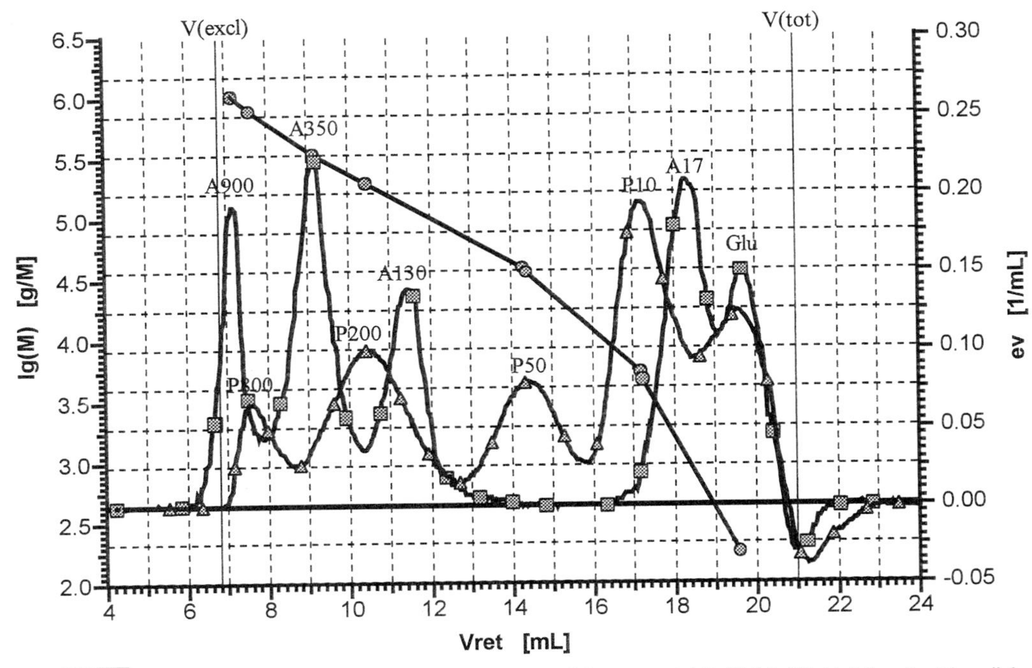

FIGURE 16.20 Peak position calibration of Superose 6 HR 30/16 (29 × 1.0 cm) with pullulan standards (P800, P200, P50, P10) and synthetic amyloses (nb glucans: A900, A350, A130, A17) and glucose; injected volume of sample solutions: 50 μl of 20-mg/ml aqueous solutions; flow rate: 0.6 ml/min; eluent: 0.05 *M* KCl.

As one separtion is achieved in approximately 40 min at a flow rate of 0.6 ml/min, Superose 6 may be classified as a fast and efficient system of broad distributed samples indeed.

V. COMBINED SYSTEM: BIO-GEL P-6/SEPHACRYL S-200

Bio-Gel P materials of Bio-Rad are polyacrylamide beads (45-90 μm in diameter) prepared from copolymerization of acrylamide and *N,N'*-methylenebis-

TABLE 16.7 Molecular Weight and Polydispersity of Standards for Peak Position Calibration of Superose 6 HR 10/30

Synthetic amylose	$M_w \times 10^{-4}$ (g/M)	M_w/M_n	Pullulan	$M_w \times 10^{-4}$ (g/M)	M_w/M_n
A-900	94.7	1.09	P-800	85.3	1.14
A-350	36.8	1.08	P-200	18.6	1.13
A-130	13.4	1.01	P-50	4.80	1.09
A-17	0.24	1.13	P-10	1.22	1.06

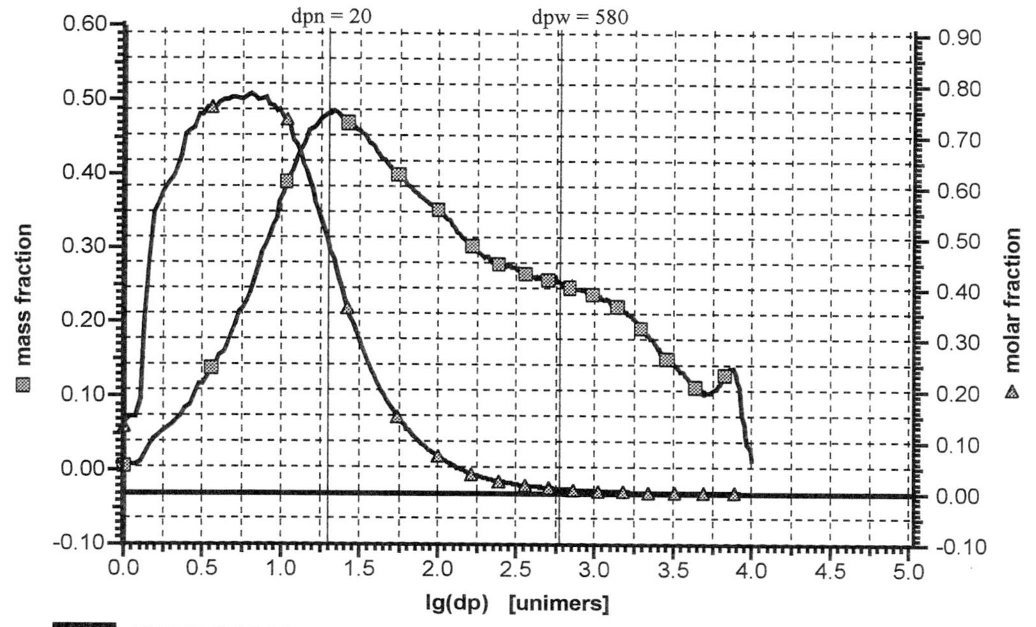

FIGURE 16.21 Degree of polymerization distributions of a broad distributed partially hydrolized starch glucan separated on Superose 6; mass distribution (□),molar distribution (▲), and weight average degree op polymerization dp_w = 580 [glucose units] and number average degree of polymerization dp_n = 20 [glucose units]; polydispersity dp_w/dp_n = 29.

acrylamide. Bio-Gel P resins are available with different packing densities (P-2, P-4, P-6, P-10, P-30, P60, P-100) and correspondingly different separation ranges (Table 16.8). Bio-Gel P resins are extremely hydrophilic and contain no charges. At room temperature they can withstand pH 2–10, 8 *M* urea, detergents, and organic solvents.

Bio-Gel P-6 and a combined system of P-6/S-200 were utilized for investigations of inulin-type $\beta(2\rightarrow1)$-linked nb fructans. With a flow rate of 0.33 ml/min, each separation of a sample volume of 1 ml of a 20- to 30-mg/ml concentrated solution typically lasted 20 hr, i.e., one run per day. Both systems (P-6 and P-6/S-200) maintained stability for approximately 1 year, equivalent to approximately 100 runs.

TABLE 16.8 Producers (Bio-Rad) Specification of Fractionation Ranges of Polyacrylamide-Based Bio-Gel P Gels

	Bio-Gel P-4	Bio-Gel P-6	Bio-Gel P-10
From (g/M)	$x \times 10^{2}$ [a]	$x \times 10^{2}$ [a]	$x \times 10^{2}$ [a]
To (g/M)	4×10^3	6×10^3	$x \times 10^{4}$ [a]

[a] x = 1–9.

Degassed and preswelled Bio-Gel P-6 and Sephacryl S-200 were packed in self-made glass columns (70 × 1.5 cm; 140 × 1.5 cm) and equilibrated for 20 hr with H_2O(dest.) + 0.002% NaN_3 to prevent microbial growth. The mass of eluted fractions was detected with a differential refractive index detector (Waters 403 RI, sensitivity: 8).

The selective separation range of P-6/S-200 was determined with Blue Dextran (V_{excl}, exclusion limit) and fructose (V_{tot}, total permeation limit). Molecular weight (degree of polymerization) calibration (Fig. 16.22) was established with dextran standards and low dp pullulans (dp 3, 6, 9, 12, 15, 18) formed by the controlled hydrolysis of high dp pullulan.

Aqueous dissolved inulin, isolated from plant tubers of increasing physiological age (small = young → midrange → large = old), was separated on P-6. Obviously, the mass contribution of early eluting (high dp) components increases with increasing tuber size, i.e., increasing physiological age of the tubers (Fig. 16.23). Results achieved by application of narrow distributed standard calibration (Fig. 16.22) supported these qualitative results with increasing weight averages of degree of polymerization for inulin of small (dp_w = 16 fructose units), midrange (dp_w = 18 fructose units), and large (dp_w = 22 fructose units) tubers.

The remarkable performance of P-6 in the separation of oligosaccharides, particularly of glucans and fructans, is well known (31,32) and is illustrated

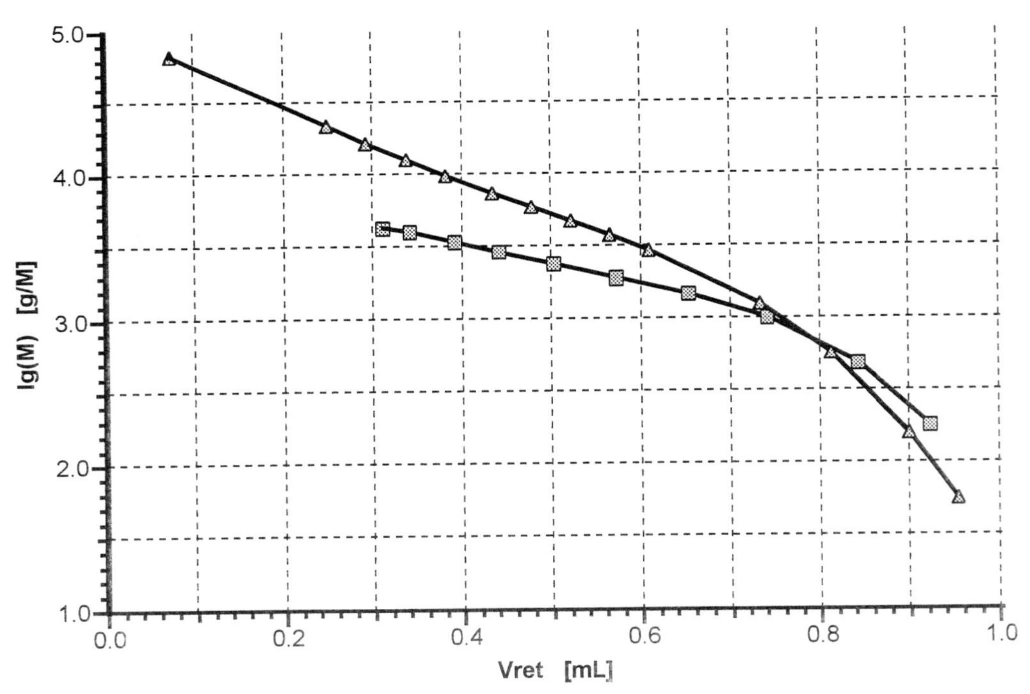

FIGURE 16.22 Peak position calibration of Bio-Gel P-6 (□) and Bio-Gel P-6/Sephacryl (▲) S-200 with dextran standards (T10, T40) and low dp pullulans.

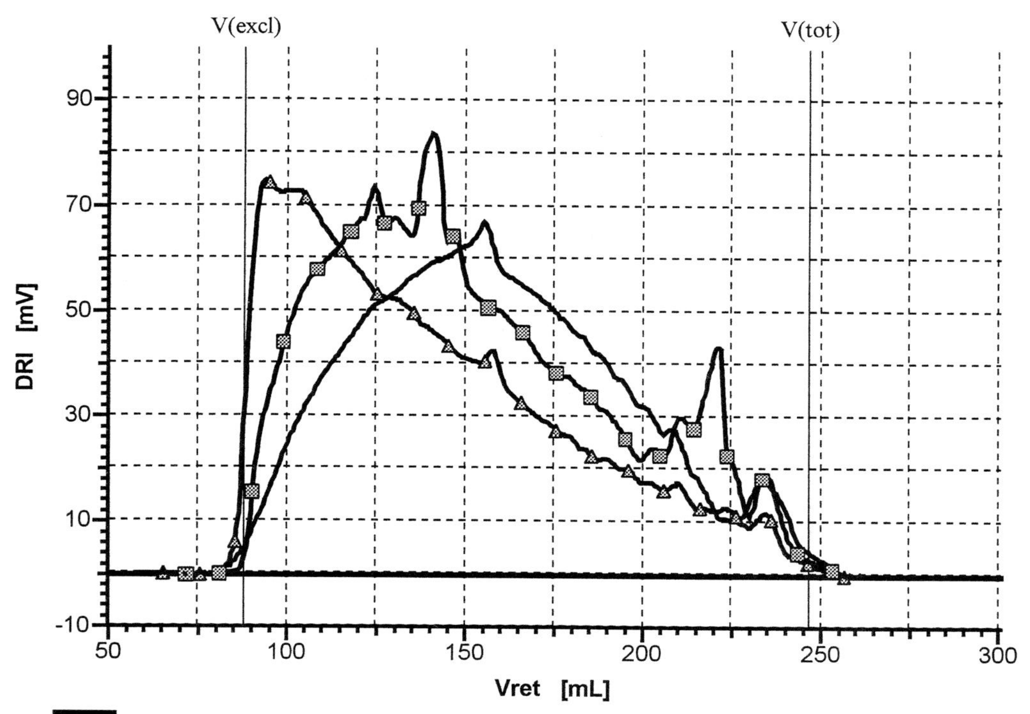

FIGURE 16.23 Inulin isolated from small (—), medium (▢) and large (▲) tubers separated on P-6; (140 × 1.5 cm); flow rate: 0.33 ml/min; eluent: H_2O(dest) + 0.002% NaN_3; mass detection: Waters 403 R differential refractive index detector, sensitivity 8×; applied sample solution volume: 1 ml of a 20-mg/ml aqueous inulin solution.

for $\beta(2\rightarrow1)$-linked nb fructan inulin, isolated from mature Topinambur tubers (Fig. 16.24). Although not baseline separated, individual oligomer components of this sample may be identified easily for elution ranges exceeding $V_{ret} = 190$ ml.

For samples with significant percentages of high dp components (up to 60,000 [g/M]), a supplementary separation system to P-6 is needed. Tests with Sephadex gels, such as G-25 or G-50, or Bio-Gels P-10 and P-100 were negative, but a combined system of P-6 and S-200 worked perfectly.

As an illustration for the improved performance of P-6/S-200 compared to P-6 solely, highly purified inulin was separated on both systems. Obviously, the high dp inulin components could not be resolved on P-6 (Fig. 16.25), whereas all of the inulin components were eluted within the selective separation range of the P-6/S-200 system (Fig. 16.26).

Consequently, inappropriate (P-6) and appropriate (P-6/S-200) separations yield significantly different results for the degree of polymerization distribution and average values of degree of polymerization (Fig. 16.27). In a first and qualitative evaluation, dp distribution achieved from P-6/S-200 differs significantly in symmetry from the dp distribution of P-6. In particular, differences in separation performance become obvious for high dp components. In a

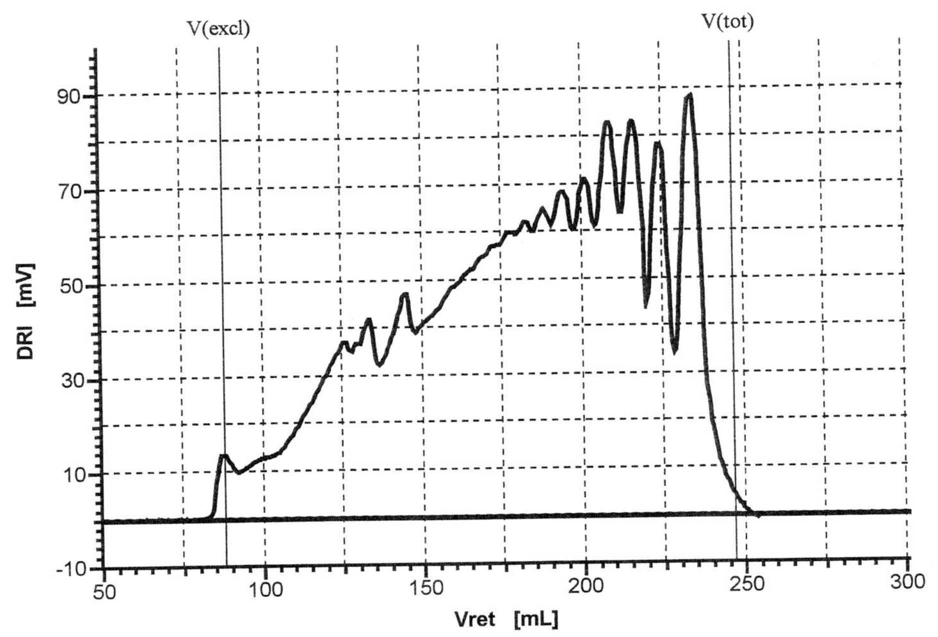

FIGURE 16.24 Inulin isolated from mature Topinambur tubers sparated on Bio-Gel P-6 (140 × 1.5cm); flow rate: 0.33 ml/min; eluent: H_2O(dest) + 0.002% NaN_3; detection: Waters 403 R differential refractive index detector, sensitivity 8×; applied sample: 1 ml of a 20-mg/ml aqueous solution; weight average degree of polymerization $dp_w = 9$; number average degree op polymerization $dp_n = 2$; polydispersity $dp_w/dp_n = 4.5$.

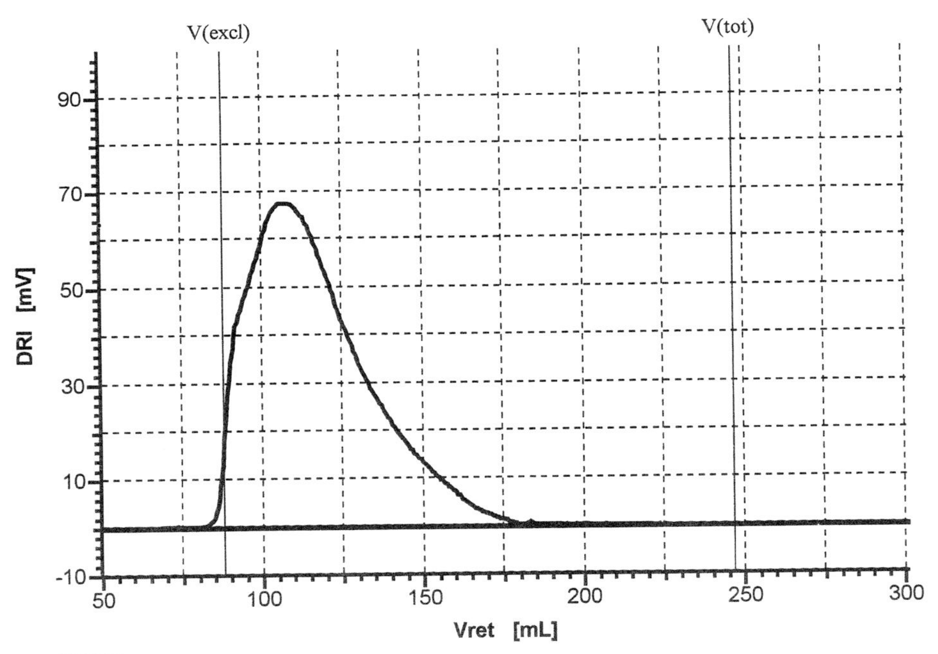

FIGURE 16.25 Highly purified inulin separated on Bio-Gel P-6 (140 × 1.5 cm); eluent: H_2O(dest) + 0.002% NaN_3; flow rate: 0.33 ml/min; mass detection: Waters 403 R differential refractive index detector; applied sample volume: 1 ml of a 20-mg/ml aqueous solution.

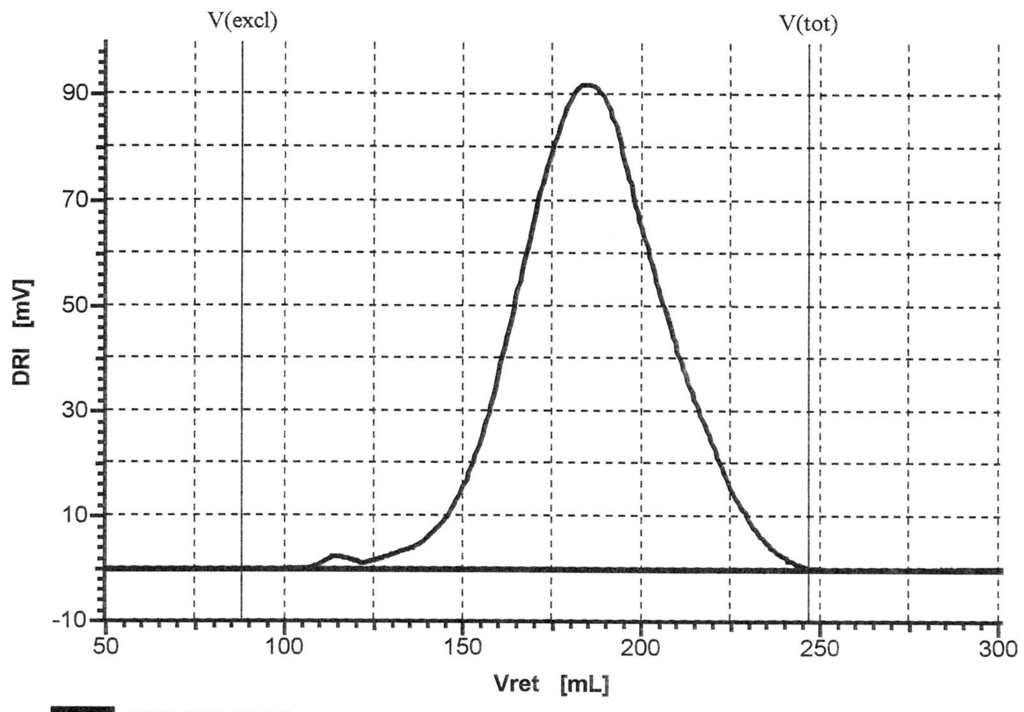

FIGURE 16.26 Highly purified inulin separated on Bio-Gel P-6 / Sephacryl S-200 (140 + 70 × 1.5 cm); eluent: H_2O(dest) + 0.002% NaN_3; flow rate: 0.33 ml/min; mass detection: Waters 403 R differential refractive index detector; applied sample volume: 1 ml of a 20-mg/ml aqueous solution.

more quantitative comparison, even values for the weight average degree of polymerization obtained from chromatograms of both systems differ significantly: in the logarithmic scaling for 23% [log(dp_w) 2.274 instead 1.843] and much more drastically in normal scaling: 170% (dp 38 instead of 70).

VI. PREPACKED SYSTEM: TSK-GEL PW COLUMNS

TSK-GEL PW column packing from TosoHaas (33), basically copolymers of ethylene glycol and methacrylate, are hydrophilic, rigid, spherical, porous polymeric beads with excellent chemical and mechanical stability. They are able to withstand eluents in the range of pH 2–12 and take up to 50% organic solvents and temperatures up to 80°C. Although most TSK-GEL PW packings bear a slight negative charge (5–18 μeq/ml), the separation of nonionic, anionic, and most cationic samples is not affected if small molarities of (neutral) salts are added to the eluent.

TSK-GEL PW resins are available in a wide range of pore sizes and are packed in 30 or 60 × 0.75-cm columns with particle sizes of either 10 or 17

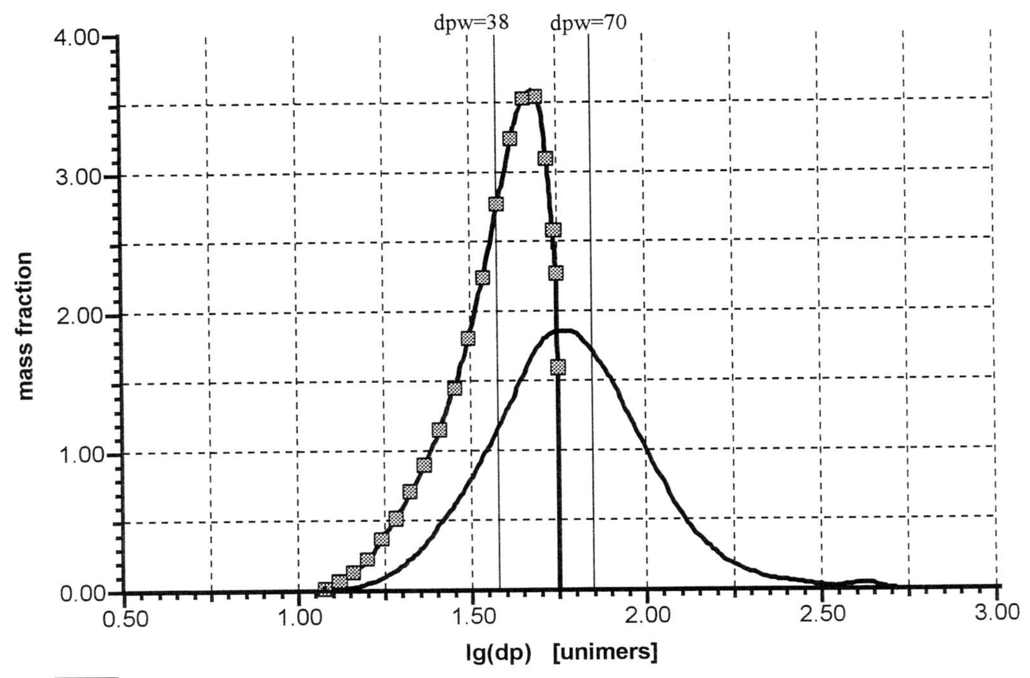

FIGURE 16.27 Degree of polymerization distribution (m_dpD_d) computed utilizing system calibrations established with dextrans and low dp pullulans (Fig. 16.22) for a highly purified inulin separated on Bio-Gel P-6 (□) (140 × 1.5 cm) and on a combined system of Bio-Gel P-6/Sephacryl S-200 (70 + 70 × 1.5 cm); dp_w, weight average degree of polymerization.

μm and efficiency rates between 3000 and 5000 theoretical plates per 30 cm. Separation ranges for TSK PW gels are listed in Table 16.9.

Although each of the TSK PW matrices may be used individually, a combination of two or more columns yields optimum results (34–37). The dual detection of mass and scattering intensity of narrow distributed nb glucans ranging between 5×10^3 and 1×10^6 [g/M] utilizing a combination of precolumn + PW6000 + 5000 + 3000 (5 + 60 + 60 + 60 × 0.75 cm) is illustrated in Figs. 16.28 and 16.29.

In these experiments, simultaneous DRI (mass) and LALLS detection is applied to obtain absolute information about molecular weight (degree of polymerization) distribution and average values of degree of polymerization. Absolute system calibration is achieved from normalized DRI (mass: mass__ev) (Fig. 16.28) and normalized LALLS (scattering intensity: LS__5__EV) chromatograms (Fig. 16.29) of a set of narrow distributed pullulan standards:

- mass__ev-area=1.0 for each chromatogram
- LS__5__EV chromatogram normalized as well with respect to applied laser intensity as to sample concentration yielding an area that numerically equals M_w (weight average molecular weight) of the sample.

TABLE 16.9 Producers (TosoHaas) Specification of Fractionation Ranges of Ethylene Glycol/Methacrylate Copolymer-Based TSK PW Gels

	G1000PW	G2000PW	G2500PW	G3000PW	G4000PW	G5000PW	G6000PW	GMPW
Pore size (nm)	<10	12.5	<20	20	50	100	>100	100–1000
From (g/M)					1×10^3	5×10^4	5×10^5	
To (g/M)	$x \times 10^3$ [a]	$x \times 10^3$ [a]	$x \times 10^4$ [a]	6×10^4	7×10^5	7×10^6	5×10^7	$<5 \times 10^7$

[a] $x = 1–9$.

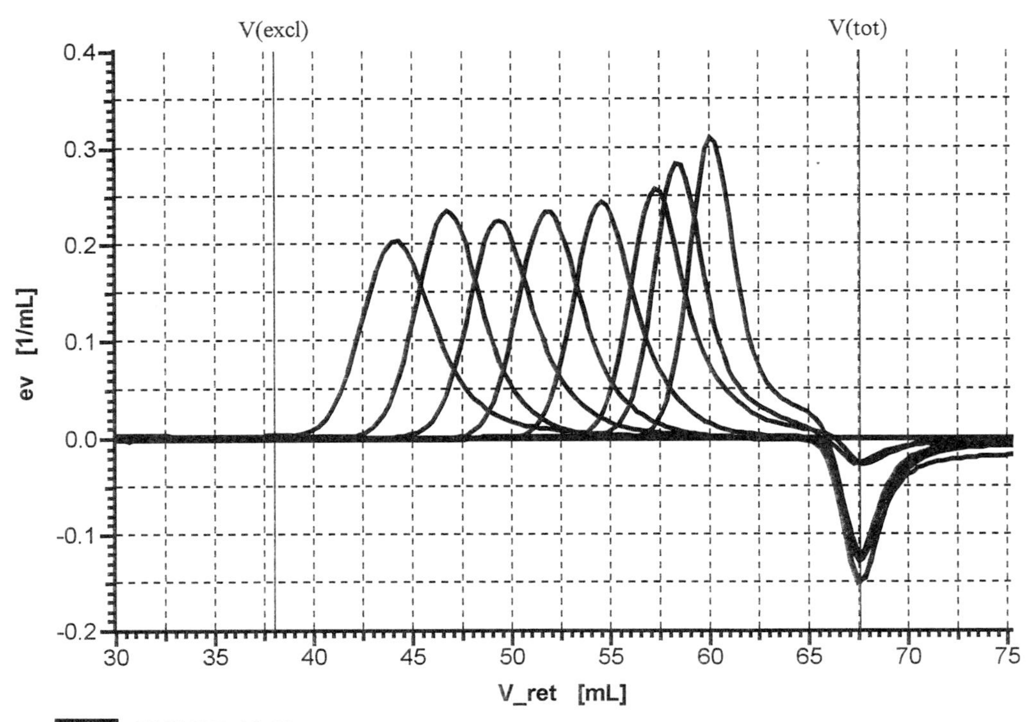

FIGURE 16.28 Normalized mass chromatograms (mass_ev) of nb glucan (pullulan) standards separated on TSK PW precolumn + PWG6000 + 5000 + 3000 (5 + 60 + 60 + 60 × 0.75 cm); eluent: 0.05 *M* NaCl; flow rate: 0.71 ml/min; detection: Optilab 903 interferometric differential refractometer; applied sample mass/volume: 200 μl of 4-mg/ml aqueous solutions; V_{excl} = 38 ml; V_{tot} = 67.5 ml.

- Then, simple summing of mass__ev functions (Σmass__ev)
- and of LS__5__EV functions (ΣLS__5__EV) over the selective separation range
- and computation of decadic logarithm of (ΣLS__5__EV/Σmass__ev) ratio
- yields raw data SEC calibration (raw__MWV) in terms of logarithmic molecular weight [log(M)] versus retention volume (V_{ret})

A fit to these raw data finally provides the absolute SEC calibration (Fig. 16.30).

Another TSK combination (precolumn + PWM + 6000 + 5000 + 4000 + 3000) was tested on differences in separation performance between individual narrow distributed samples and mixtures of several narrow distributed samples. The result is summarized in Fig. 16.31: within experimental error the summed chromatograms (theory) of four narrow distributed glucans (dextran) match perfectly with the experimentally determined chromatogram of the mixture. The (theory/experimental) ratio, plotted for quantification of the match, in-

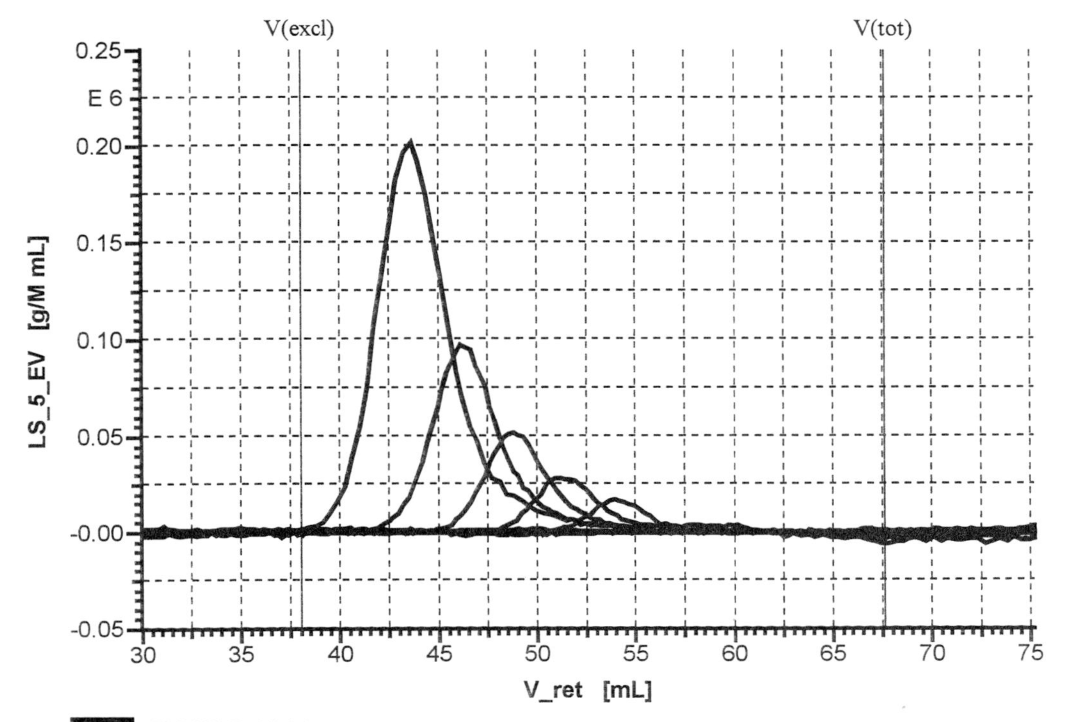

FIGURE 16.29 Normalized LALLS chromatograms (LS_5_EV) of nb glucan (pullulan) standards separated on TSK PW precolumn + PWG6000 + 5000 + 3000 (5 + 60 + 60 + 60 × 0.75 cm); eluent: 0.05 *M* NaCl; flow rate: 0.71 ml/min; detection: low-angle laser light scattering; applied sample mass/volume: 200 μl of 4-mg/ml aqueous solutions; V_{excl} = 38 ml; V_{tot} = 67.5 ml.

forms about a slight shift to increased retention volumina for high dp samples if they are components of a broad distributed mixture (ratio: 1.1→0.9 with increasing V_{ret}). The separation behavior of low dp glucans was found to be independent whether they occur as individual narrow distributed samples or as components of a broad distributed mixture (ratio = ±1.0). As an important consequence of these results, axial dispersion phenomena, which become more and more problematic the more narrow distributed the samples are, can be judged to be of neglectible for the utilized TSK column system.

Because polysaccharides contain a huge number of polar groups, H bond-breaking solvents are often applied to obtain true solutions. One such solvent is NaOH, but high pH and/or high ionic strength (due to following neutralization) often causes problems. DMSO, another solvent, avoids such problems but comes with the disadvantage of nonaqueous solutions.

In Fig. 16.32, application of a TSK PW SEC system consisting of a combination of precolumn + PWM + 6000 + 5000 + 4000 + 3000 demonstrates a possibility for analytical purposes to change from DMSO-dissolved glucans to an aqueous solution. An initially DMSO-dissolved potato starch sample was applied to the TSK PW system and because separated with an aqueous

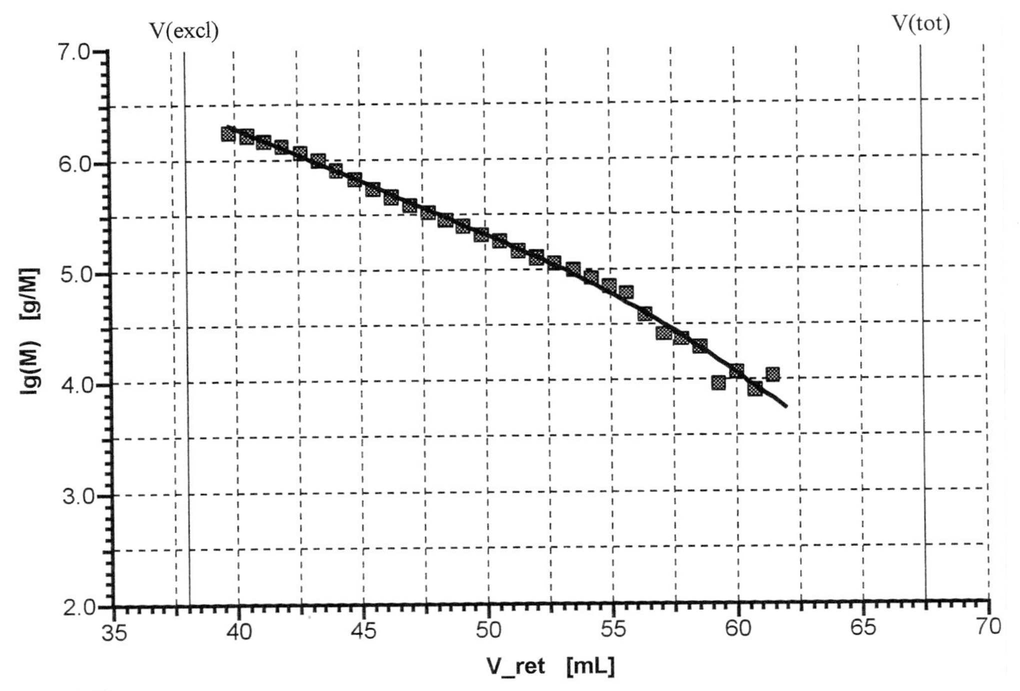

FIGURE 16.30 Raw_data calibration and final absolute system calibration achieved from a fit to the raw_data for nb glucan (pullulan) standards for the system TSK PW precolumn + PWG6000 + 5000 + 3000 (5 + 60 + 60 + 60 × 0.75 cm).

0.05 M NaCl eluent. Because of the good performance of the TSK PW system in the low dp range, DMSO was shifted outside the selective separation range ($V_{tot} > 50$ ml) and aqueous eluent-dissolved starch components could be analyzed in terms of absolute degree of polymerization distributions (mass: m__dpD__d; molar: n__dpD__d) from mass- and light-scattering signals. Changing to an aqueous solvent in particular is important for absolute techniques such as SEC DRI/LALLS, as optical constants and material characteristics typically are known pure (aqueous or DMSO) solutions and not for mixed systems. Based on dual detection signals, an absolute linear calibration function was found for the investigated broad distributed sample (Fig. 16.33). Application of this calibration function yielded a mass (m__dpd__D) and molar (n__dpD__d) degree of polymerization distribution and average values of degree of polymerization (dp_w, dp_n) as illustrated in Fig. 16.34.

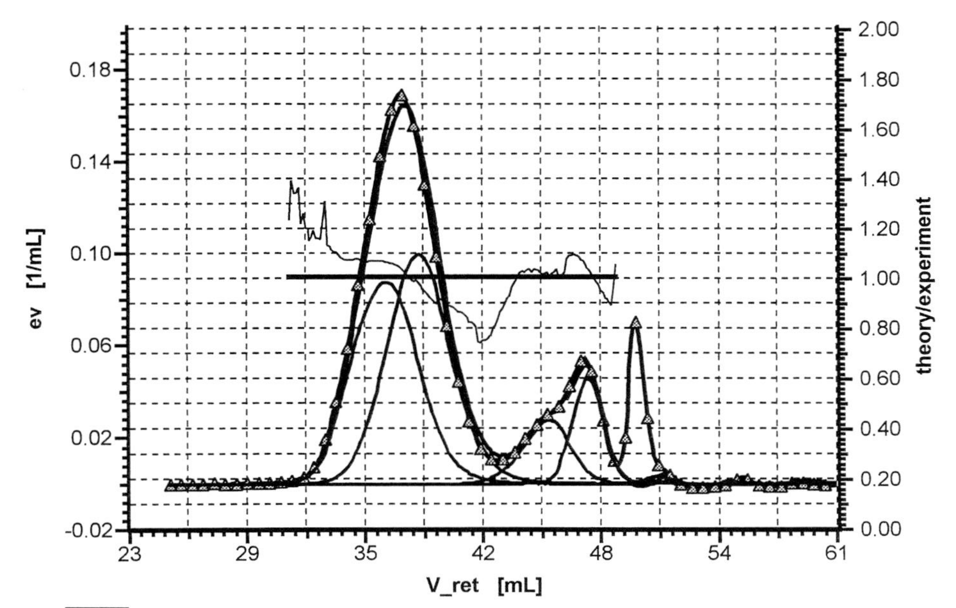

FIGURE 16.31 Four individual dextran standards (—) (39:43:9:9) and a mixture of four dextran standards (experimental: ▲) separated on TSK PW pre/PWM/G6000/5000/4000/3000 (5 + 30 + 30 + 30 + 30 + 30 × 0.75 cm); V_{excl} = 28 ml; V_{tot} = 50 ml; eluent: 0.05 *M* NaCl; flow rate: 0.80 ml/min; detection: Optilab 903 interferometric differential refractometer; applied sample mass/volume: 200 μl of 2-mg/ml aqueous solutions; sum of individual chromatograms (theory: —) and (theory/experimental) ratio (—) plotted for quantification of deviations in separation performance between narrow distributed samples and broad distributed samples.

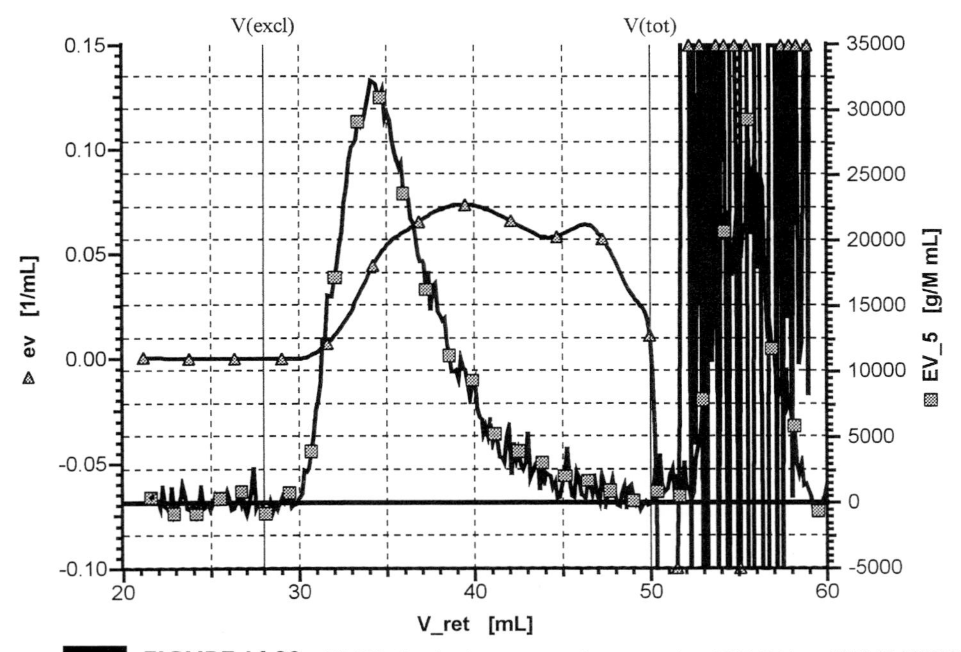

FIGURE 16.32 DMSO-dissolved potato starch separated on TSK PW pre/PWM/6000/5000/4000/3000 (5 + 30 + 30 + 30 + 30 + 30 × 0.75 cm); normalized mass chromatogram (ev: ▲) with DMSO shifted outside beyond V_{tot}; normalized LALLS chromatogram (EV_5: □); eluent: 0.05 *M* NaCl; V_{excl} = 28 ml, V_{tot} = 50 ml.

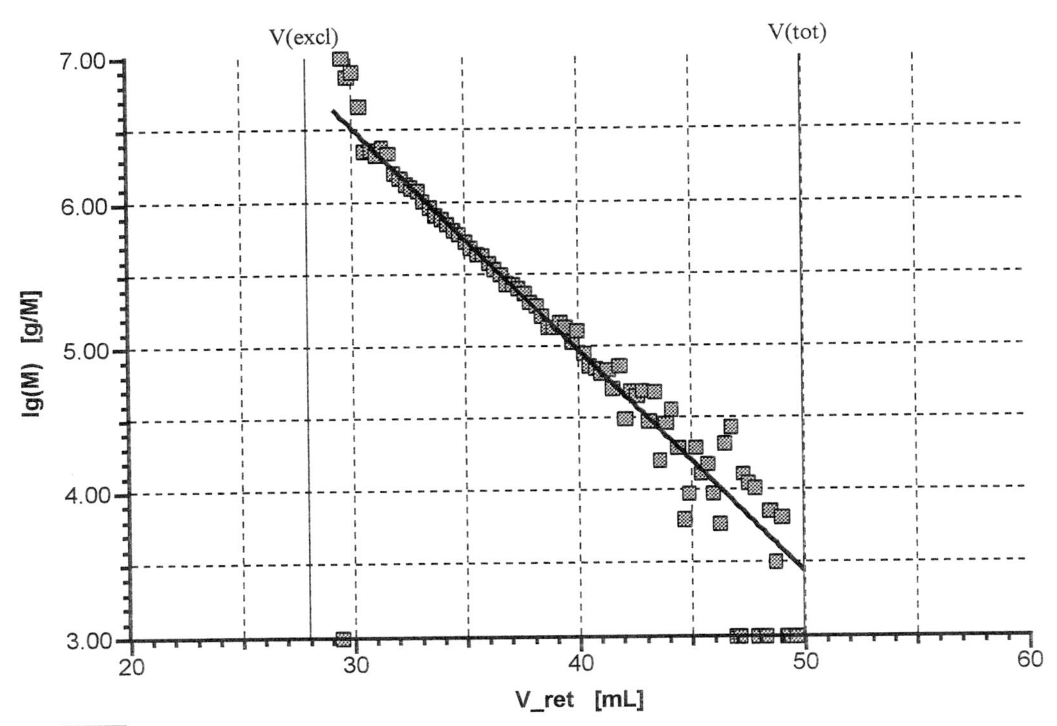

FIGURE 16.33 Absolute calibration function (raw_data: ▨ ▨ ▨; fit to the raw_data: ———) for DMSO-dissolved potato starch separated on TSK PW pre/PWM/6000/5000/4000/3000 (5 + 30 + 30 + 30 + 30 + 30 × 0.75 cm); eluent: 0.05 *M* NaCl; V_{excl} = 28 ml, V_{tot} = 50 ml.

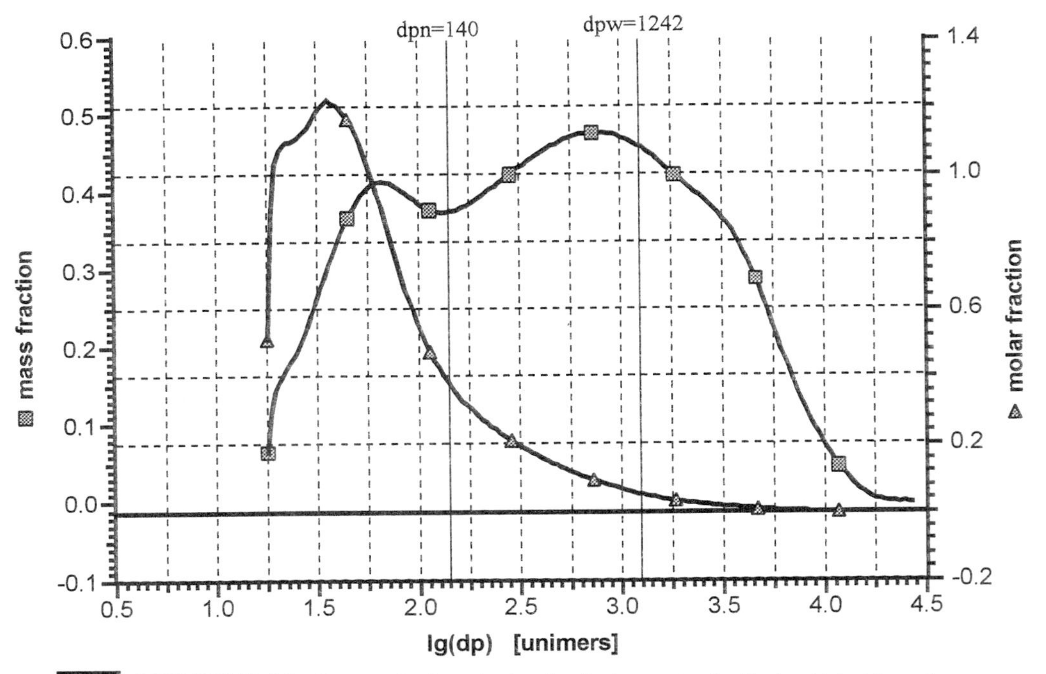

FIGURE 16.34 Degree of polymerization distribution: molar distribution (▲) with number average degree of polymerization (dp_n = 140 glucose unimers); mass distribution (▨) with weight average degree of polymerization dp_w = 1242 glucose unimers.

ACKNOWLEDGMENT

This work was supported by the Austrian FWF 'Fonds zur Foerderung wissenschaftlicher Forschung' project P-12498-CHE.

REFERENCES

1. Porath, J., and Flodin, P. (1959). *Nature* **183, 1657.**
2. Robin, J. P., Mercie, R., Charbonniere, R., and Guilbot, A. (1974). *Cereal Chem.* **51, 389.**
3. Robin, J. P., Mercier, R., Duprat, F., Charbonniere, R., and Guilbot, A. (1975). *Starch* **27**, 36.
4. Kayisu, K., and Hood, L. F. (1981). *J. Food Sci.* **46**, 1894.
5. Ikawa, Y., Glover, D. F., Sugimoto, Y., and Fuwa, H. (1981). *Starch* **33**, 9.
6. Ebermann, R., and Praznik, W. (1975). *Starch* **329,**
7. Ebermann, R., and Schwarz, R. (1975). *Starch* **27**, 361.
8. Praznik, W., Schmidt, S., and Ebermann, R., (1983). *Starch* **35**, 58.
9. Praznik, W., Burdicek, G., and Beck, R. H. F. (1986). *J. Chromatogr.* **357**, 216.
10. Praznik, W., Burdicek, G., Beck, R. H. F. (1986). *Starch* **38**, 181.
11. Praznik, W. (1986). **38**, 392.
12. Praznik, W., Schillinger, H., and Beck, R. H. F. (1987). *Starch* **39**, 183.
13. Praznik, W. (1990). *Nutrition* **14**, 427.
14. Praznik, W., Huber, A., Watzinger, S., and Beck, R. H. F. (1994). *Starch* **46**, 88.
15. Pharmacia Biotech Consumables Catalogue: Chromatography, Electrophoresis, Molecular and Cell Biology.
16. Fractogel TSK: Hydrophilic Gel Media—Tools for Chromatographic Techniques in Biotechnology, Merck, Darmstadt, FRG.
17. Matheson, N. K. (1971). *Phytochemistry* **10**, 3213.
18. Yamada, T., and Taki, M. (1976). *Starch* **28**, 374.
19. Biliaderis, C. G., Grant, D. R., and Vose, J. R. (1979). *Cereal Chem.* **56**, 475.
20. Manners, D. J., and Matheson, N. K. (1981). *Carbohydr. Res.* **90**, 99.
21. Schweizer, T. F., and Reimann, S. (1982). *Z. Lebensm. Unters. Forsch.* **174**, 23.
22. Craig, S. A. S., and Stark, J. R. (1984). *Starch* **36**, 127.
23. Sargeant, J. G. (1982). **34**, 89.
24. Praznik, W., and Ebermann, R. (1979). **31**, 288.
25. Bruun, H., and Henriksnäs, H. (1979). *Starch* **29**, 122.
26. Boyer, C. D., Damewood, P. A., and Simpson, E. K. G. (1981). **33**, 125.
27. Atwell, W. A., Hoseney, R. C., and Lineback, D. R., (1980). *Cereal Chem.* **57**, 12.
28. Baba, T., Arai, Y., Yamamoto, T., and Itho, T. (1982). *Phytochemistry* **21**, 2291.
29. Beck, R. H. F., and Praznik, W. (1986). *J. Chromatogr.* **369**, 208.
30. Praznik, W., Beck, R. H. F., and Eigner, W.-D. (1987). *J. Chromatogr.* **387**, 467.
31. John, M., Schmidt, J., Wandrey, C., and Sahm, H. (1982). *J. Chromatogr.* **247**, 281.
32. Sabbagh, N. K., and Fagerson, I. S. (1976). *J. Cromatogr.* **86**, 184.
33. TosoHaas: The Bioseparation Specialist—TSK-Gel, Toyo Pearl, Amberchrom.
34. Huber, A., and Praznik, W. (1996). *In* "ACS Symposium Series No. 635. Strategies in Size-Exclusion Chromatography" (M. Potschka and P. L. Dubin, eds.), Chapter 19, p. 351.
35. Huber, A., and Praznik, W. (1994). *J. Liq. Chromatogr.* **17**(18), 4031.
36. Huber, A. (1992). *In* "Analysis of Polymers/Molar-Mass and Molar-Mass Distribution of Polymers, Polyelectrolytes and Latices" (W.-M. Kulicke, ed.), Hüthig & Wepf Verlag, 61, 248.
37. Huber, A. (1991). *J. Appl. Polym. Sci. Appl. Polym. Symp. Ed.* **48**, 95.

17 COMPARISON OF FOUR COMMERCIAL LINEAR AQUEOUS SIZE EXCLUSION COLUMNS AND FOUR SETS OF COMMERCIAL POLYETHYLENE OXIDE (PEO) STANDARDS FOR AQUEOUS SIZE EXCLUSION CHROMATOGRAPHY OF POLYVINYLPYRROLIDONE AND PEO

CHI-SAN WU, LARRY SENAK, DONNA OSBORNE, AND TOM M. H. CHENG

International Specialty Products, Wayne, New Jersey 07470

INTRODUCTION

Polymers with narrow, broad, or bimodal molecular weight distribution designed to deliver value added products have been introduced to the marketplace at a fast pace. Size exclusion chromatography (SEC) is the most practical method to determine the molecular weight (MW) and molecular weight distribution (MWD) of polymers. It is also the simplest method to obtain an overlay of MWD of two polymers. For SEC to be used as an effective and rugged quality assurance method, the accuracy and precision of SEC must be evaluated thoroughly. Column and calibration standards are two important factors in determining the quality of the MW and MWD obtained from SEC.

TSK PW columns and Shodex OH-pak columns are made respectively by Toyo Soda and Showa Denko and have become very popular aqueous SEC columns for synthetic water-soluble polymers since the early 1980s. Linear or mixed-bed columns are used widely these days due to the high resolution for a very broad molecular weight range (typically three orders of magnitude or more) for these columns. Therefore, this study evaluates only the linear columns.

The particle size of the TSK GM-PW_{XL} column is smaller than the PW column. Shodex OH-pak has gone through several change from KB to SB to SBHQ with reported compatibility with polar organic solvents (1). However, the basic chemical composition of these columns has not been disclosed adequately by the manufacturers, and the descriptions in the literature for these columns have not been consistent. For example, Shodex OH-pak packing is described as polyhydroxymethacrylate (2) and TSK GM-PW packing is described as "hydrophilic" (3) in the current official product catalogs. A paper published by Toyo Soda described the PW column as a hydrophilic polymer gel with -$CH_2CHOHCH_2O$- as the main constituent component (4). However, another report on PW columns from Toyo Soda indicated that columns with a small pore size (G1000PW and G2000PW) are different in chemical nature from PW columns with a large pore size (G3000PW and G6000PW) (5). An earlier paper described the Shodex OH-pak as a methacrylate glycerol copolymer and PW columns as a hydroxylated gel (6).

Column manufacturers normally provide basic information about their columns, such as plate count, particle size, exclusion limit, and calibration curve. This information is necessary and fundamental, however, it is not sufficient to allow users to make an intelligent decision about a column for a specific application. For example, separation efficiency, the dependence of separation efficiency on the mobile phase, the ability to separate the system peaks from the polymer peak, the symmetry of the polymer peak, and the possible interaction with polymers are seldom provided.

Narrow distribution polystyrene is the most widely used standard for polymers soluble in organic solvent. Narrow distribution polystyrene standards have been evaluated for the determination of weight average molecular weight of polystyrene by SEC and as long as calibration was constructed by passing through most data points smoothly, no difference in calculated MW among standards from different vendors was noticed (7). Narrow distribution polyethylene oxide (PEO) is one of the most widely used calibration standard for water-soluble polymers and is available in sets of 7 to 10 standards from several vendors. However, the quality of these standards has not been evaluated adequately. NIST has introduced two narrow distribution PEO standards. In this study, the NIST standards are used as a bench mark to evaluate commercial PEO standards.

The purpose of this study is twofold: to compare four linear aqueous SEC columns made by Tosoh and Showa Denko in terms of composition and performance and to evaluate the effect of commercial PEO standards on the accuracy and precision of the MW and MWD of polyvinylpyrrolidone (PVP) and NIST PEO standards. In terms of performance, emphasis will be placed on factors not commonly covered by column manufacturers. Successful SEC conditions for PVP in water, in water/methanol, and in dimethylformamide can be found in the literature (8,9,10). This study deals mainly with the effects of column, mobile phase, and PEO standards on the MW and MWD of PVP.

II. EXPERIMENTAL

Commercial grades of PVP, K-15, K-30, K-90, and K-120 and the quaternized copolymer of vinylpyrrolidone and dimthylaminoethylmethacrylate (poly-VP/DMAEMA) made by International Specialty Products (ISP) were used in this study. PEO standard calibration kits were purchased from Polymer Laboratories Ltd. (PL), American Polymer Standards Corporation (APSC), Polymer Standards Service (PSS), and Tosoh Corporation (TSK). In addition, two narrow NIST standards, 1923 and 1924, were used to evaluate commercial PEO standards. Deionized, filtered water, and high-performance liquid chromatography grade methanol purchased from Aldrich or Fischer Scientific were used in this study. Lithium nitrate ($LiNO_3$) from Aldrich was the salt added to the mobile phases to control for polyelectrolyte effects.

The weight average molecular weights (M_w) and molecular weights at peak evolution volume (M_p) of the PEO standards provided by the suppliers are shown in Table 17.1. All four sets of PEO standards cover a similar molecular weight range, about 10,000 to 1,000,000.

Two linear columns from Showa Denko, Shodex SB-806M and Shodex SB-806MHQ, and two linear columns from TosoHaas, TSK GM-PW$_{XL}$ and TSK GM-PW, were evaluated. Prior to the evaluation, the number of theoretical plates for Shodex SB-806MHQ, SB-806M, PW$_{XL}$, and PW was determined to be 15,100, 15,700, 11,390, and 4710, respectively, as per manufacturer inspection. The lower plate count of the TSK PW column is due to the larger particle size of this column. Two mobile phases, water with 0.1 *M* $LiNO_3$ and 50:50 methanol/water (v/v) with 0.1 *M* $LiNO_3$, were used for each of the four columns. These four columns were new and only PEO and PVP were analyzed with these columns in this study. Waters Ultrahydrogel columns have also been used in this laboratory. However, Ultrahydrogel columns are exactly the same as the TSK GM-PW$_{XL}$ columns based on the calibrations curves supplied by the manufacturers and by the pyrolysis GC data discussed later.

TABLE 17.1 Molecular Weights of PEO Standards Provided by Suppliers

TSK		PSS		PL		APSC	
M_w	M_w (GPC)	M_w	M_p	M_w	M_p	M_w	M_p
920,000	917,000	780,000	825,000	896,000	963,000	913,000	878,000
510,000	630,000	438,000	448,000	649,000	646,000	860,000	800,000
340,000	400,000	218,000	217,000	284,000	288,000	510,000	498,000
170,000	190,000	155,000	155,000	209,000	205,000	438,000	448,000
95,000	101,000	108,000	111,000	122,000	120,000	250,000	245,000
46,000	42,000	42,900	44,400	72,300	73,400	170,000	162,000
21,000	14,000	21,400	22,100	49,300	50,100	108,000	111,000
				33,000	32,600	95,000	92,900
				18,200	18,300	42,300	44,400
				10000	10,000	26,000	24,000

SEC measurements were made using a Waters Alliance 2690 separation module with a 410 differential refractometer. Typical chromatographic conditions were 30°C, a 0.5-ml/min flow rate, and a detector sensitivity at 4 with a sample injection volume of 80 μl, respectively, for a sample concentration of 0.075%. All or a combination of PEO standards at 0.05% concentration each were used to generate a linear first-order polynomial fit for each run throughout this work. Polymer Laboratories Caliber GPC/SEC software version 6.0 was used for all SEC collection, analysis, and molecular weight distribution overlays.

Pyrolysis GC/MS experiments were performed on packing materials recovered from retired columns that had been washed thoroughly. Packing materials near the entrance and exit of the retired column were discarded and not included in the pyrolysis GC experiment. GC/MS used in this study is HP and the conditions are:

Pyrolyzer: CDS Model Pyroprobe 2000 with coil sample probe and quartz boat sample holder
Pyrolysis temperature: about 650°C for 20 sec
Pyrolysis sample size: 0.2–0.3 mg
GC: Hewlett Packard Model 5890
GC column: J&W Scientific DB-5, 30 m, 0.53 mm i.d. and 1.5-μm micron film
Oven temperature: 50°C for 1 min; 10°C/min to 250°C and hold 10 min
Detector: FID at 270°C
Injector: split injector at 240°C
Carrier gas: helium at 8 ml/min
Split ratio: 10
GC/MS: Hewlett Packard HP5890 connected to HP5971 mass detector with a jet separator

III. RESULTS AND DISCUSSION

A. Precision of Size Exclusion Chromatography (SEC)

The precision of SEC must be established before a comparison of columns and calibration standards can be made. Consistency in flow rate or elution time is the first requirement to obtain precision in SEC. Consistency in flow rate or elution time can be monitored by the elution time of the PEO standards, which are run before and after the samples. Elution time or flow rate can be considered consistent if the elution times of the PEO standards before and after the samples agree within 0.1 min.

Table 17.2 shows good agreement between the retention times from the TSK PEO standards analyzed in groups of two or three and the retention times of the TSK PEO standards analyzed individually. ASTM-D5296 requires that for standards to be run as a group, the molecular weight must differ by a factor of 10 (11). The results in Table 17.2 showed that a difference in molecular weight by a factor of 6 is adequate to obtain consistent flow times for standards for the modern linear columns.

TABLE 17.2 Retention Times (RT) of TSK PEO Standards Using Shodex SB-806MHQ in Water with 0.1 *M* $LiNO_3$

MW (group no.)	RT before sample as groups A,B,C	RT after sample as groups A,B,C	RT as individual PEO
920,000 (A)	13.42	13.42	13.43
510,000 (B)	14.17	14.15	14.20
340,000 (C)	14.62	14.62	14.63
170,000 (C)	15.35	15.32	15.35
95,000 (B)	15.95	15.93	15.92
46,000 (C)	16.67	16.65	16.67
21,000 (A)	17.40	17.40	17.40

Determination of MW and MWD by SEC using commercial narrow molecular weight distribution polystyrene as calibration standards is an ASTM-D5296 standard method for polystyrene (11). However, no data on precision are included in the 1997 edition of the ASTM method. In the ASTM-D3536 method for gel-permeation chromatography from seven replicates, the M_w of a polystyrene is 263,000 ± 30,000 (11.4%) for a single determination within the 95% confidence level (12). A relative standard deviation of 3.9% was reported for a cooperative determination of M_w of polystyrene by SEC (7). In another cooperative study, a 11.3% relative standard deviation in M_w of polystyrene by GPC was reported (13).

In this study from six replicates the relative M_w of the NIST-1924 polyethylene oxide standards for a single determination within the 95% confidence level was 123,000 ± 3900 (or 3.2% of 123,000) and 124,000 ± 3250 (or 2.6% of 124,000) on 2 different days. Similarly, the relative M_w of a polyvinylpyrrolidone sample was 570,000 ± 32,000 (or 5.6% of 570,000) and 548,000 ± 6400 (or 1.2% of 548,000) on 2 different days. The variability of 3% for NIST PEO standards and 5% for polyvinylpyrrolidone will be used to evaluate the significance of difference in M_w later in this chapter.

B. Comparison of PEO Calibration Standards

M_w of NIST-1923 and -1924 PEO standards were calculated from the respective linear calibration curves generated from the TSK, PL, PSS, and APSC PEO standards. M_p of the PL standards and M_w from light scattering of the TSK, PSS, and APSC standards are used to set up the respective linear calibration curves because they provide the best agreement with NIST values. The results are shown in Tables 17.3–17.6. NIST-1923 and -1924 are certified to have a M_w of $26.9 \times 10^3 \pm 2.2 \times 10^3$ g/mol with a polydispersity of 1.06 and $120.9 \times 10^3 \pm 9.0 \times 10^3$ g/mol with a polydispersity of 1.04, respectively. In Tables 17.3–17.6, a plus sign is used to indicate if the calculated M_w agrees with the NIST value within the 3% variability.

TABLE 17.3 Molecular Weight and Polydispersity of PVP and PEO for the TSK PW Column

Column	Mobile phase	Polymer	PEO standard	MW	Polydispersity
TSK-PW	Water/methanol 50:50, 0.1 *M* lithium nitrate	PVP K-90	TSK	539,000	5.5
			APSC	580,000	6.0
			PL	600,000	7.6
			PSS	611,000	5.8
		PVP K-30	TSK	33,200	2.9
			APSC	33,400	3.1
			PL	28,500	3.6
			PSS	35,800	3.0
		NIST 1923 MW 24,700–29,100	TSK	32,600	1.4
			APSC	32,500	1.4
			PL (+)	27,000	1.5
			PSS	34,900	1.4
		NIST 1924 MW 111,900–129,900	TSK	142,000	1.3
			APSC	147,000	1.3
			PL (+)	134,000	1.3
			PSS	156,000	1.3
TSK-PW	Water, 0.1 *M* lithium nitrate	PVP K-90	TSK	522,000	4.5
				590,000	5.1
			PL	625,000	7.1
			PSS	644,000	5.5
		PVP K-30	TSK	31,900	3.2
			APSC	32,100	3.5
			PL	25,100	4.5
			PSS	32,500	3.7
		NIST 1923 MW 24,700–29,100	TSK	31,100	1.2
			APSC	30,800	1.3
			PL	23,000	1.3
			PSS	30,900	1.3
		NIST 1924 MW 111,900–129,900	TSK	144,000	1.2
			APSC	152,000	1.2
			PL	132,000	1.3
			PSS	158,000	1.2

For the NIST PEO standards, it is obvious from Tables 17.3–17.6 that the agreement between the calculated M_w and the NIST certified values depends strongly on the PEO calibration standards as well as the columns used. The agreement is much better for Shodex columns than for TSK columns, no matter which set of PEO standards is used. For the PW_{XL} column the poor agreement in M_w for the high molecular weight NIST PEO standard (1924) is most likely due to the lower exclusion limit of this column (see Fig. 17.5). The agreement in M_w for NIST PEO standards for the TSK PW column is poorer than the other three columns in both mobile phases. For Shodex columns, the agreement for the SB-806MHQ column is about the same as for the SB-806 column. The agreement is about the same in water or in water/methanol for these four columns. Despite the fact that M_w based on the PL standards are in general lower than the other three sets of PEO standards, the agreement is about the same for TSK, PL, and APSC PEO standards, no matter which column is used.

TABLE 17.4 Molecular Weight and Polydispersity of PVP and PEO for the TSK PW_{XL} Column

Column	Mobile phase	Polymer	PEO standard	MW	Polydispersity
TSK PW_{XL}	Water/methanol 50:50, 0.1 *M* lithium nitrate	PVP K-90	TSK	452,000	4.9
			APSC	478,000	5.5
			PL	446,000	5.4
			PSS	451,000	4.3
		PVP K-30	TSK	31,000	4.9
			APSC	30,500	5.4
			PL	28,800	5.3
			PSS	34,100	4.3
		NIST 1923 MW 24,700–29,100	TSK (+)	28,100	1.3
			APSC (+)	27,100	1.3
			PL (+)	25,800	1.3
			PSS	31,800	1.3
		NIST 1924 MW 111,900–129,900	TSK	158,000	1.2
			APSC	161,000	1.2
			PL (+)	151,000	1.2
			PSS	167,000	1.2
TSK PW_{XL}	Water, 0.1 *M* lithium nitrate	PVP K-90	TSK	424,000	4.5
			APSC	455,000	5.1
			PL	440,000	5.3
			PSS	455,000	4.7
		PVP K-30	TSK	28,400	5.3
			APSC	27,400	6.1
			PL	24,100	6.9
			PSS	29,100	5.6
		NIST 1923 MW 24,700–29,100	TSK (+)	27,200	1.4
			APSC (+)	25,700	1.4
			PL	22,100	1.5
			PSS (+)	27,700	1.4
		NIST 1924 MW 111,900–129,900	TSK	151,000	1.3
			APSC (+)	133,000	1.3
			PL	141,000	1.4
			PSS	159,000	1.3

However, the agreement for the PSS PEO standards is not as good as the TSK, PL, and APSC PEO standards.

C. Effect of PEO Calibration Standards and Columns on the Molecular Weight and Polydispersity of Polyvinylpyrrolidone (PVP)

The M_w of a PVP K-90 sample and a PVP K-30 sample were determined with four linear columns with four sets of PEO standards to study the dependency of M_w of PVP on the column and the PEO standard used. M_w for PVP K-90 calculated from the TSK GM-PW column are in general higher than the other three columns in this study for all four sets of PEO standards. This is discussed further in Section III,G,2. Similar to the case of the M_w of NIST PEO standards calculated from PL PEO standards, M_w for PVP K-30 calculated from the PL PEO standards are lower than M_w from TSK, APSC, and PSS PEO standards, no matter which column or mobile phase is used. However, M_w for PVP K-

TABLE 17.5 Molecular Weight and Polydispersity of PVP and PEO for the Shodex SB-806MHQ Column

Column	Mobile phase	Polymer	PEO standard	MW	Polydispersity
Shodex SB-806MHQ	Water/methanol 50:50, 0.1 *M* lithium nitrate	PVP K-90	TSK	474,000	5.0
			APSC	514,000	5.6
			PL	542,000	7.3
			PSS	546,000	5.3
		PVP K-30	TSK	29,200	2.5
			APSC	29,400	2.6
			PL	23,700	3.1
			PSS	31,400	2.6
		NIST 1923 MW 24,700–29,100	TSK	30,100	1.4
			APSC (+)	30,000	1.4
			PL (+)	25,500	1.4
			PSS	32,100	1.4
		NIST 1924 MW 111,900–129,900	TSK (+)	124,000	1.7
			APSC (+)	128,000	1.8
			PL (+)	114,000	2.0
			PSS	136,000	1.7
Shodex SB-806MHQ	Water, 0.1 *M* lithium nitrate	PVP K-90	TSK	457,000	5.5
			APSC	504,000	7.1
			PL	512,000	10.2
			PSS	521,000	6.2
		PVP K-30	TSK	25,600	3.4
			APSC	25,700	3.4
			PL	19,800	4.3
			PSS	26,100	3.4
		NIST 1923 MW 24,700–29,100	TSK (+)	27,400	1.3
			APSC (+)	26,800	1.5
			PL	20,100	1.6
			PSS (+)	27,400	1.4
		NIST 1924 MW 111,900–129,900	TSK (+)	112,000	1.5
			APSC (+)	117,000	1.4
			PL	99,900	1.6
			PSS (+)	113,000	1.5

90 calculated from PL PEO standards are similar to the M_w from the other three sets of PEO standards. Except for the TSK GM-PW_{XL} column, the polydispersity for PVP K-90 calculated from the PL PEO standards are higher than TSK, APSC, and PSS PEO standards.

It is surprising to note that for all four columns there is good agreement between the M_w determined in water and in water/methanol, despite the fact that PEO and polyvinylpyrrolidone have different hydrodynamic volumes in these two mobile phases and that the column packings may swell or shrink differently in these two mobile phases.

D. Comparison of Linear Calibration Curves of Shodex and TSK Columns

Due to the limitation of space, only the linear calibration curves for TSK , PSS, APSC, and PL PEO standards in water/methanol for the Shodex SB-806M

TABLE 17.6 Molecular Weight and Polydispersity of PVP and PEO for the Shodex SB-806M Column

Column	Mobile phase	Polymer	PEO standard	MW	Polydispersity
Shodex SB-806M	Water/methanol 50:50, 0.1 *M* lithium nitrate	PVP K-90	TSK	462,000	4.7
			APSC	449,000	4.9
			PL	536,000	6.4
			PSS	529,000	4.5
		PVP K-30	TSK	29,700	2.3
			APSC	31,000	2.3
			PL	26,100	2.7
			PSS	32,700	2.3
		NIST 1923 MW 24,700–29,100	TSK (+)	29,800	1.2
			APSC	31,300	1.2
			PL (+)	25,500	1.2
			PSS	32,900	1.2
		NIST 1924 MW 111,900–129,900	TSK (+)	126,000	1.2
			APSC (+)	133,000	1.2
			PL (+)	123,000	1.2
			PSS	141,000	1.2
Shodex SB-806M	Water, 0.1 *M* lithium nitrate	PVP K-90	TSK	470,000	4.8
			APSC	510,000	5.1
			PL	534,000	7.2
			PSS	543,000	5.3
		PVP K-30	TSK	29,200	2.3
			APSC	30,300	2.4
			PL	28,700	2.8
			PSS	30,700	2.4
		NIST 1923 MW 24,700–29,100	TSK	31,200	1.2
			APSC	32,200	1.2
			PL (+)	24,500	1.2
			PSS	32,600	1.2
		NIST 1924 MW 111,900–129,900	TSK (+)	128,000	1.1
			APSC	135,000	1.2
			PL (+)	118,000	1.2
			PSS (+)	114,000	1.2

column are shown in Figs. 17.1–17.4. All four sets of PEO standards cover the same molecular weight range, about 20,000 to 1,000,000. The curves in water are very similar to the curves in water/methanol for all four columns. Coefficient of determination (R^2) for the linear calibration curves for these four linear columns in water and in water/methanol from all four sets of PEO standards are better than 0.99, except for the PL PEO standards and TSK GM-PW$_{XL}$ column in water/methanol, as shown in Table 17.7. The coefficient of determination for the TSK GM-PW$_{XL}$ column in general is not as good as the other three linear columns. The coefficient of determination for the TSK PEO standards showed the least dependency on columns and mobile phases. The TSK GM-PW$_{XL}$ column has a lower exclusion limit than the TSK GM-PW, Shodex SB-806, and SB-806MHQ columns, as evidenced by the deviation of the highest molecular weight PL standard from the calibration curve in Fig. 17.5. Even

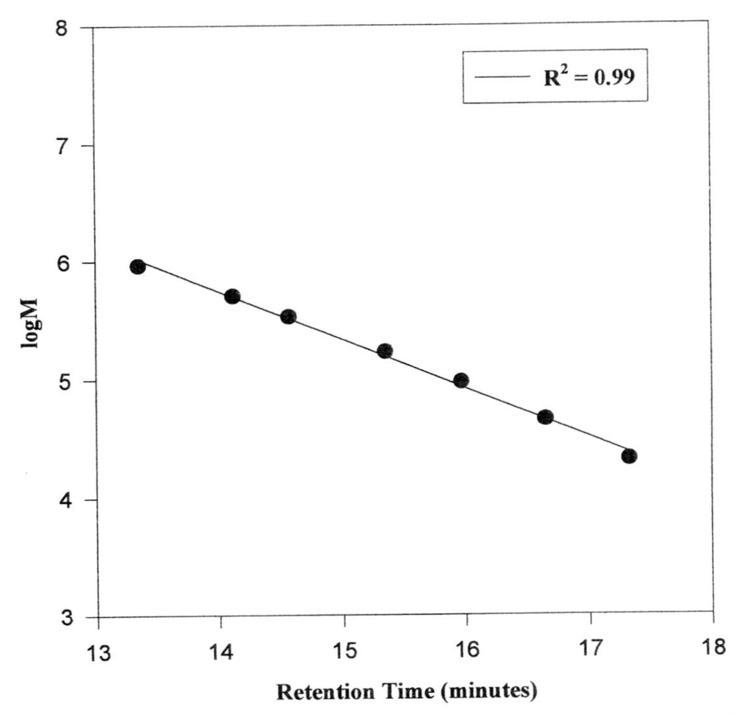

FIGURE 17.1 TSK PEO Calibration in 50 : 50 (v/v) methanol/water with 0.1 *M* $LiNO_3$ for a Shodex SB-806M column.

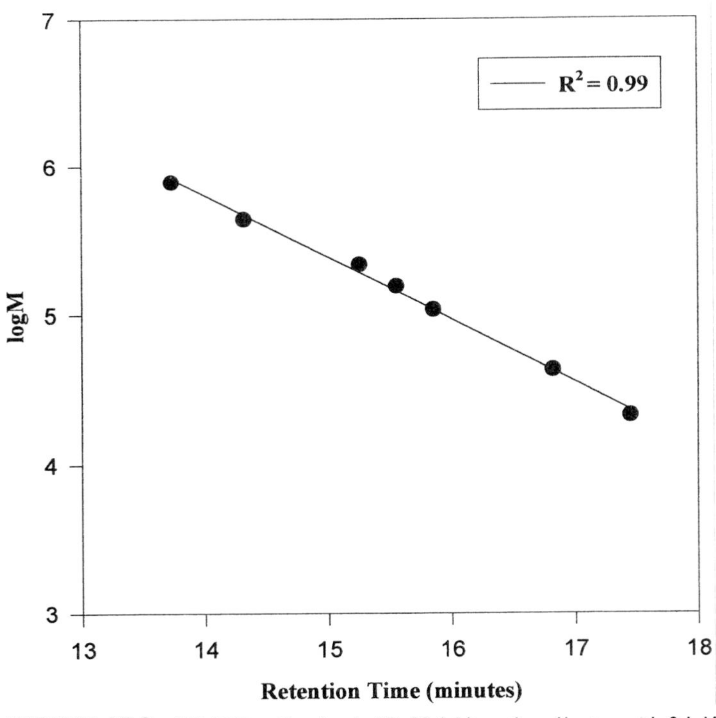

FIGURE 17.2 PSS PEO calibration in 50 : 50 (v/v) methanol/water with 0.1 *M* $LiNO_3$ for a Shodex SB-806M column.

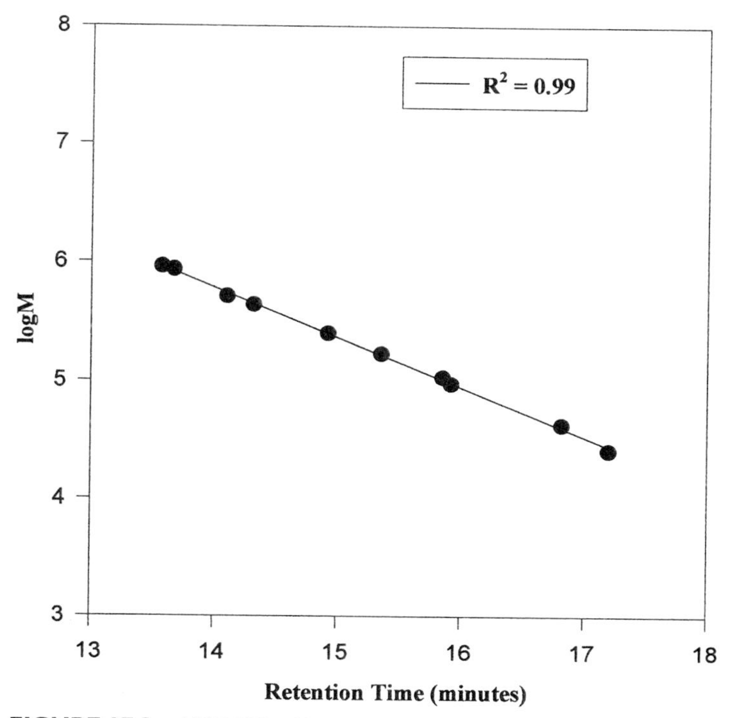

FIGURE 17.3 APSC PEO calibration in 50 : 50 (v/v) methanol/water with 0.1 *M* $LiNO_3$ for a Shodex SB-806M column.

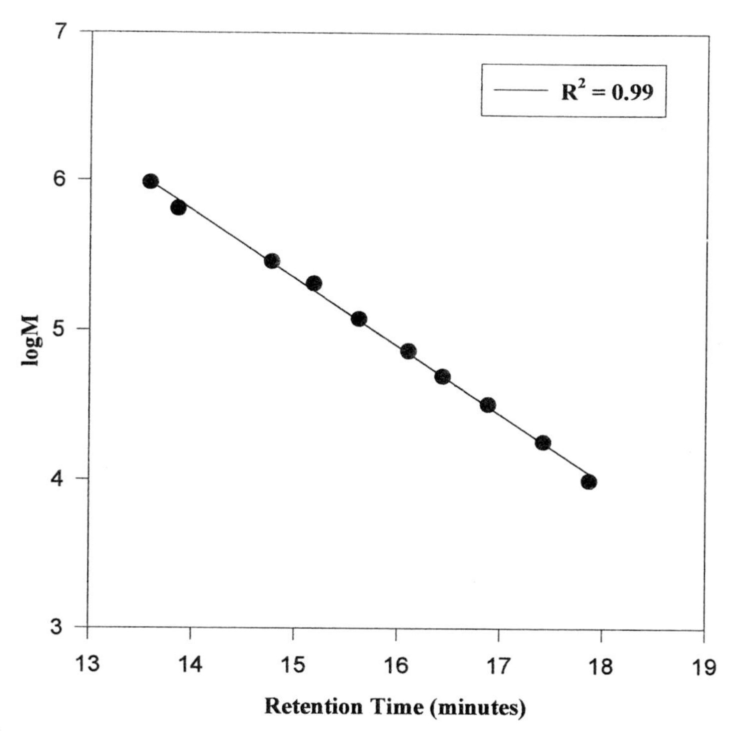

FIGURE 17.4 PL PEO calibration in 50 : 50 (v/v) methanol/water with 0.1 *M* $LiNO_3$ for a Shodex SB-806M column.

TABLE 17.7 Coefficient of Determination (R^2) for Linear Calibration Curves for Four Linear Columns and Four Sets of PEO Standards

Column	PEO	Mobile phase	R^2
TSK GM-PW	TSK	Water	0.9964
	PSS		0.9971
	APSC		0.9993
	PL		0.9993
	TSK	Water/methanol	0.9956
	PSS		0.9961
	APSC		0.9989
	PL		0.9954
TSK GM-PW_{XL}	TSK	Water	0.9969
	PSS		0.9970
	APSC		0.9913
	PL		0.9921
	TSK	Water/methanol	0.9945
	PSS		0.9904
	APSC		0.9989
	PL		0.9869
Shodex SB-806M	TSK	Water	0.9949
	PSS		0.9941
	APSC		0.9978
	PL		0.9970
	TSK	Water/methanol	0.9955
	PSS		0.9962
	APSC		0.9986
	PL		0.9977
Shodex SB-806MHQ	TSK	Water	0.9954
	PSS		0.9937
	APSC		0.9977
	PL		0.9967
	TSK	Water/methanol	0.9944
	PSS		0.9966
	APSC		0.9984
	PL		0.9970

though the coefficient of determination for the TSK PW column is about as good as the Shodex columns, the separation efficiency is not (see the next section), due to the larger particle size and lower plate count for this column.

E. Comparison of the Separation Efficiency of Commercial Aqueous SEC Columns in Water and in Water/Methanol

Comparison of the separation efficiency between two columns in the same mobile phase or one column in two mobile phases is based on the extent of resolution of the peaks of the PEO standards in the respective chromatograms of the PEO A, B, and C group. Due to the limitation of space, only the TSK PEO A chromatograms for the four columns in water and water/methanol are

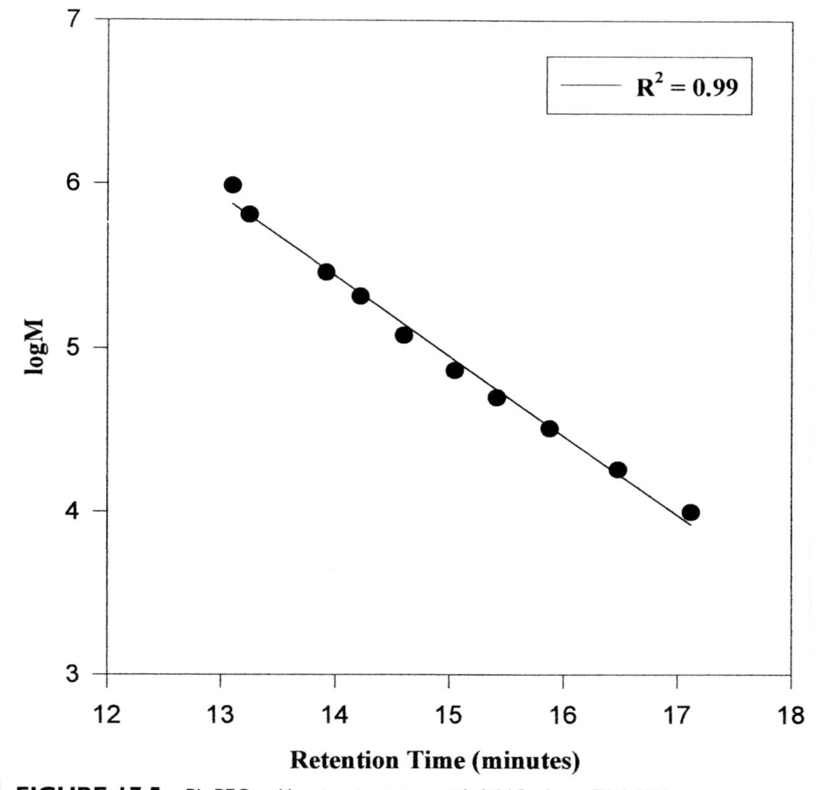

FIGURE 17.5 PL PEO calibration in water with $LiNO_3$ for a TSK PW_{XL} column.

shown in Figs. 17.6–17.13. From the PEO A chromatogram the resolution between the 920,000 PEO and 170,000 PEO (R_a) and the resolution between the 170,000 PEO and 21,000 PEO (R_b) were calculated according to the resolution equation defined in the ASTM D5296 method for the four columns in water and water/methanol (8). According to ASTM5296, a column should provide resolution greater than 1.7 for polymer standards with molecular weights differing by a factor of 10.

$$R = 2(V_{R2} - V_{R1})/[(W_1 + W_2)\log(M_1/M_2)],$$

where V_{R1} and V_{R2} are peak elution volumes or times measured at the peak maximum of polymer standards 1 and 2; W_1 and W_2 are peak widths of standards 1 and 2 measured in elution volumes or times; and M_1 and M_2 are peak molecular weights of standards 1 and 2.

R_a and R_b in Table 17.8 are less than 1.7 because the molecular weights of the standards differ by a factor less than 10. They will be used later to compare the separation efficiency in the high molecular weight range and low molecular weight range for these four linear columns.

The results in Table 17.8 indicate that for polyethylene oxide standards Shodex columns have better separation efficiency than the TSK GM-PW and

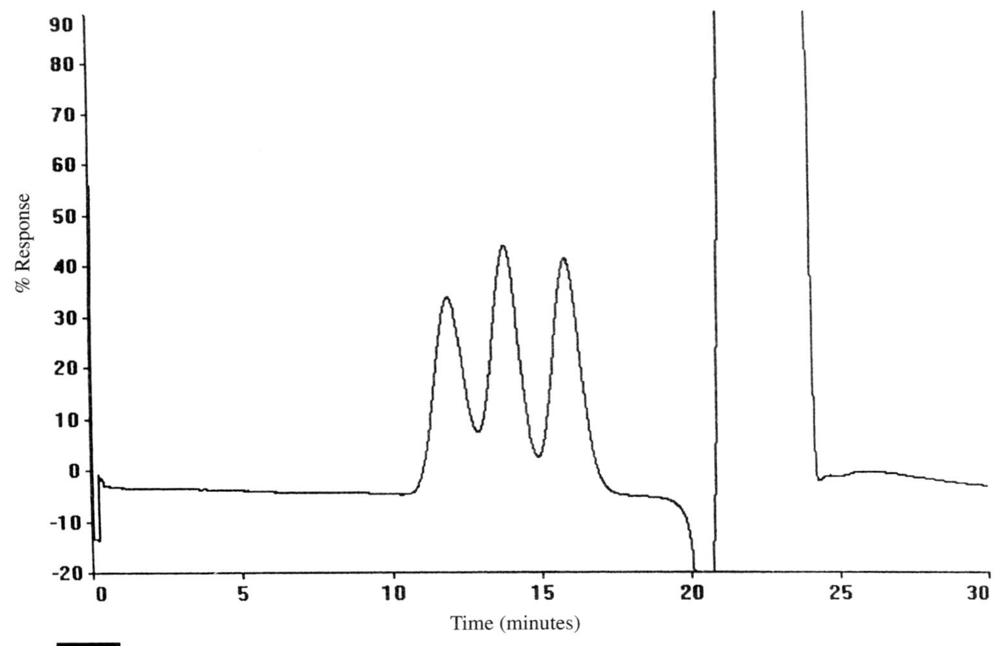

FIGURE 17.6 TSK PEO A standards in water with 0.1 *M* $LiNO_3$ for a TSK PW column.

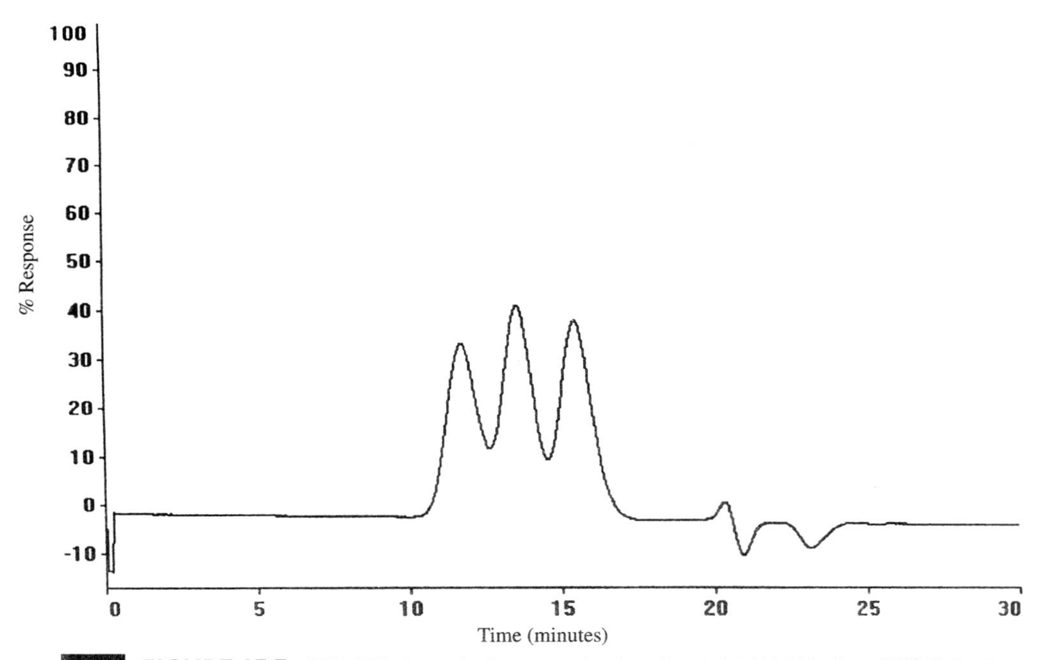

FIGURE 17.7 TSK PEO A standards in water/methanol with 0.1 *M* $LiNO_3$ for a TSK PW column.

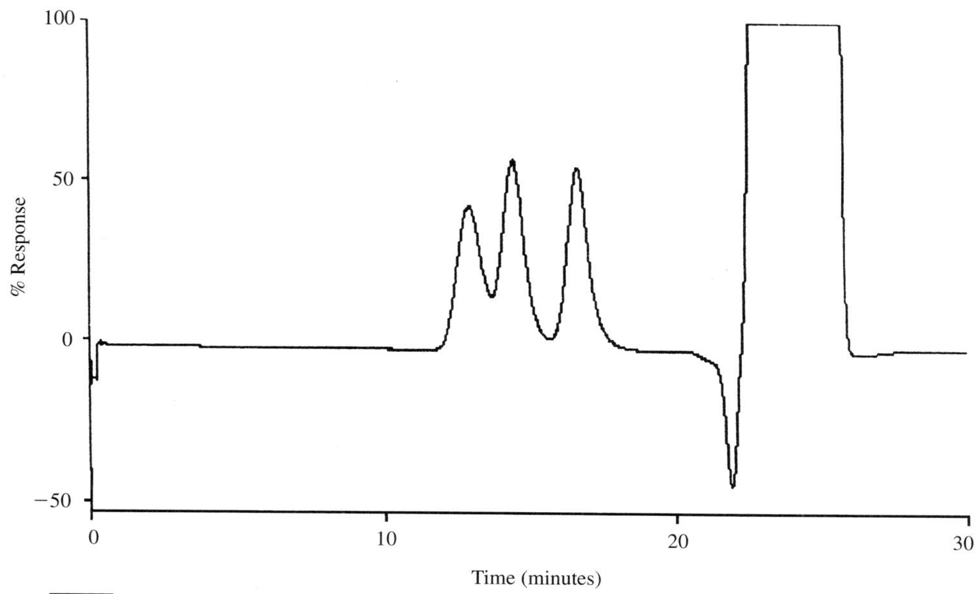

FIGURE 17.8 TSK PEO A standards in water with 0.1 *M* $LiNO_3$ for a TSK PW_{XL} column.

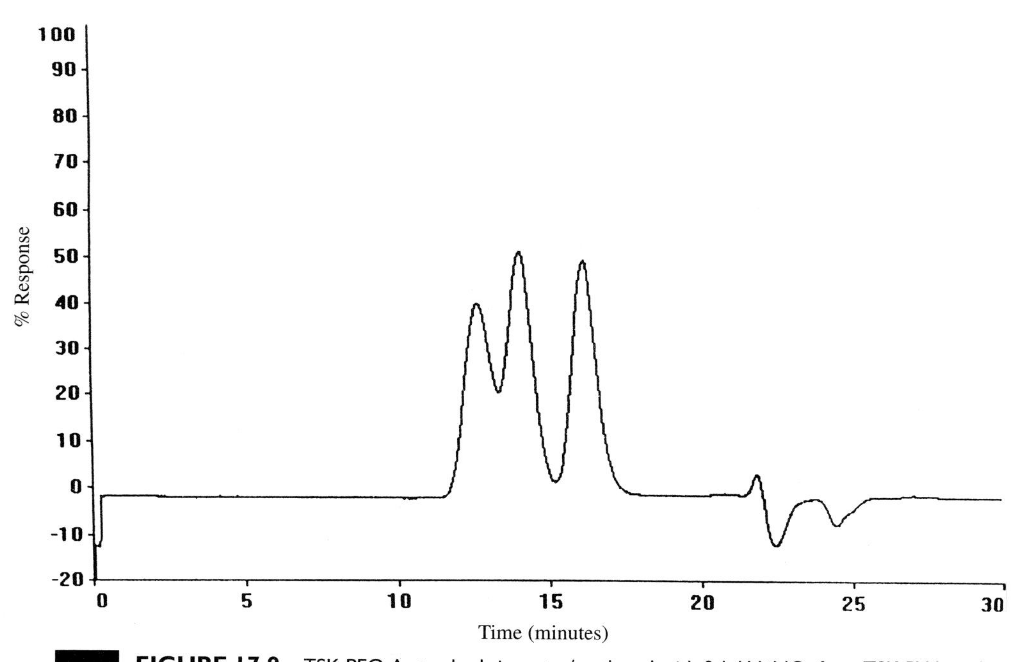

FIGURE 17.9 TSK PEO A standards in water/methanol with 0.1 *M* $LiNO_3$ for a TSK PW_{XL} column.

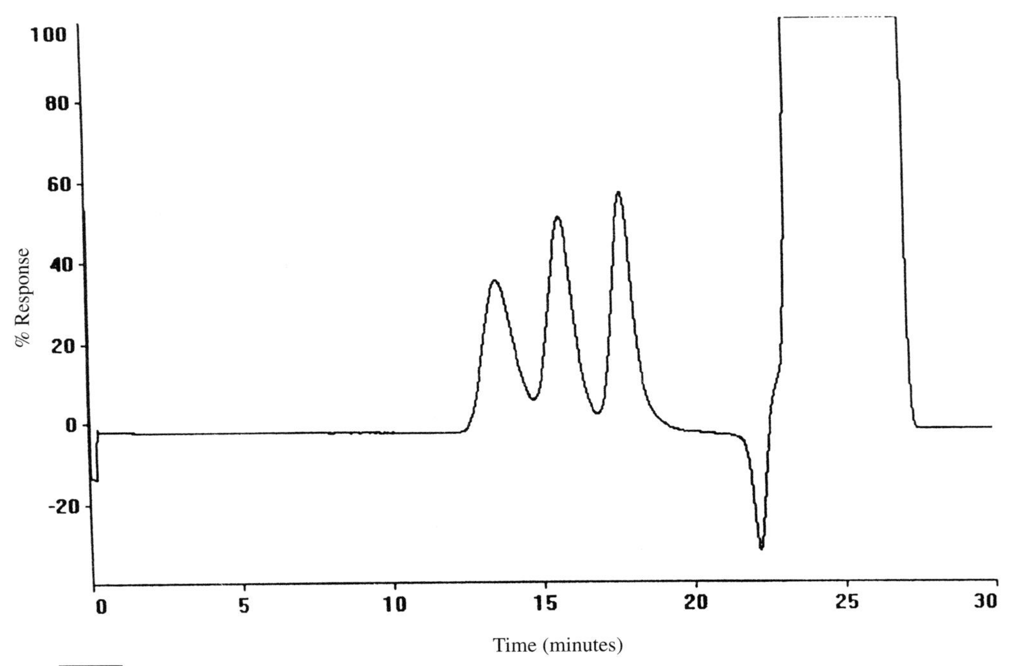

FIGURE 17.10 TSK PEO A standards in water with 0.1 *M* $LiNO_3$ for a Shodex SB-806M column.

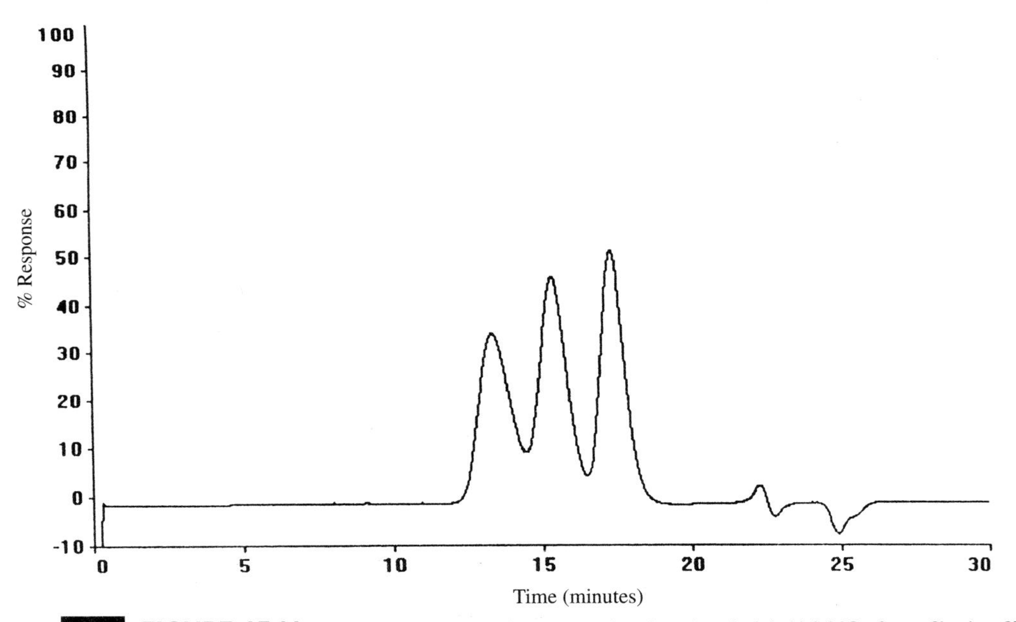

FIGURE 17.11 TSK PEO A standards in water/methanol with 0.1 *M* $LiNO_3$ for a Shodex SB-806M Column.

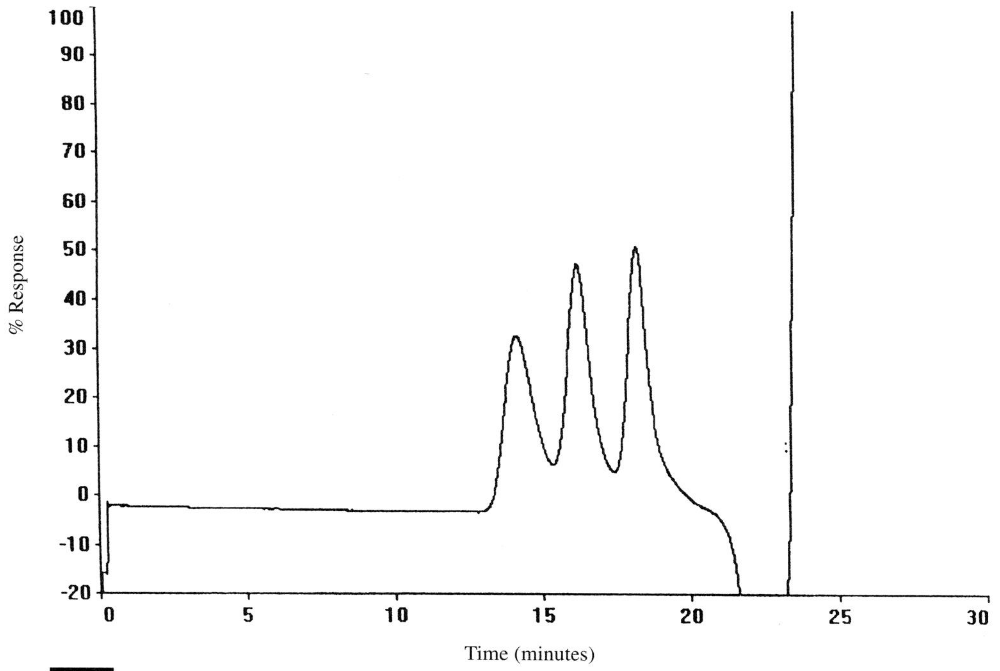

FIGURE 17.12 TSK PEO A standards in water with 0.1 *M* $LiNO_3$ for a Shodex SB-806MHQ column.

PW_{XL} columns in the high molecular weight range and the TSK GM-PW column in the low molecular weight range in water and water/methanol. Among the four columns, TSK GM-PW_{XL} has the best separation efficiency at the low molecular weight end in water and the worst separation efficiency in the high molecular weight end in water/methanol.

The ranking of the four columns for separation efficiency in water and in water/methanol is SB806M = SB806MHQ > PW = PW_{XL} (for high molecular weight range in water); SB806M = SB806MHQ > PW > PW_{XL} (for high molecular weight range in water/methanol); and PW_{XL} > SB806M > SB806-MHQ > PW (for low molecular weight range in water and in water/methanol).

F. Comparison of the Void Volume of Shodex and TSK Aqueous Columns in Water and in Water/Methanol

The difference in retention times between the 920,000 PEO and the 21,000 PEO in Table 17.9 can be used as a measure of the void or pore volume that effectively provides the linear separation range for these columns in water and in a water/methanol mixture. The better separation efficiency of the Shodex columns over the TSK columns is partially related to the larger void volumes of the Shodex columns than the TSK columns. The difference in void volumes for the Shodex, TSK GM-PW, and TSK GM-PW_{XL} columns is partially attributed to the difference in the inner diameters of the three columns, which are 8 (Shodex), 7.8 (TSK GM-PW) and 7.5 (TSK GM-PW_{XL}) mm. Table 17.9 also

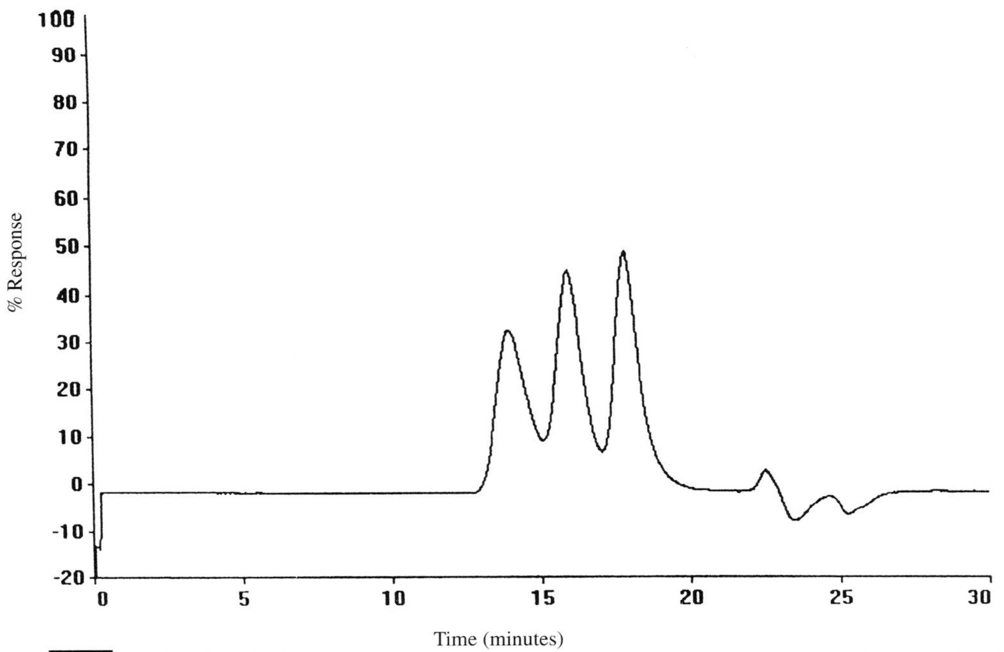

FIGURE 17.13 TSK PEO A standards in water/methanol with 0.1 *M* $LiNO_3$ for a Shodex SB-806MHQ column.

TABLE 17.8 Separation Efficiency of Four Linear Columns in Water and Water/Methanol for Polyethylene Oxide Standards

Column	Mobile phase	Resolution (R_a) 920,000; 170,000	Resolution (R_b) 170,000; 21,000
TSK GM-PW	Water	1.198	1.162
TSK GM-PW	Water/methanol	1.058	0.9773
TSK GM-PW_{XL}	Water	1.151	1.609
TSK GM-PW_{XL}	Water/methanol	0.9310	1.387
Shodex SB-806M	Water	1.455	1.474
Shodex SB-806M	Water/methanol	1.262	1.233
Shodex SB-806MHQ	Water	1.453	1.438
Shodex SB-806MHQ	Water/methanol	1.313	1.206

TABLE 17.9 Retention Time and Void Volume of Columns in Water and in Water/Methanol

Column	PEO MW	Retention time: Water	Retention time: Water/ methanol	Retention time: Difference	Void volume: Water	Void volume: Water/ methanol	Void volume: Difference
Shodex SB-806MHQ	920,000	14.15	13.97	0.18	4.07	3.91	0.16
	510,000	14.97	14.80	0.17			
	340,000	15.42	15.22	0.20			
	170,000	16.22	15.95	0.27			
	95,000	16.78	16.53	0.25			
	46,000	17.48	17.23	0.25			
	21,000	18.22	17.88	0.34			
Shodex SB-806M	920,000	13.55	13.35	0.20	4.15	3.98	0.17
	510,000	14.35	14.12	0.23			
	340,000	14.87	14.57	0.30			
	170,000	15.65	15.35	0.30			
	95,000	16.27	15.97	0.30			
	46,000	17.02	16.65	0.37			
	21,000	17.70	17.33	0.37			
TSK PW	920,000	11.92	11.73	0.19	3.93	3.74	0.19
	510,000	12.75	12.50	0.25			
	340,000	13.15	12.88	0.27			
	170,000	13.82	13.58	0.24			
	95,000	14.33	14.07	0.26			
	46,000	15.07	14.65	0.42			
	21,000	15.85	15.47	0.38			
TSK PW_{XL}	920,000	12.90	12.67	0.23	3.7	3.48	0.22
	510,000	13.48	13.17	0.31			
	340,000	13.83	13.45	0.38			
	170,000	14.40	14.03	0.37			
	95,000	14.95	14.53	0.42			
	46,000	15.73	15.30	0.43			
	21,000	16.60	16.15	0.45			

shows that void volumes for all four columns in water are higher than in the water/methanol mixture and interestingly by about the same amount.

Solvent can affect separation in two different ways. Because water is a better solvent for these four columns than water/methanol, based on the swelling or void volume of the columns in Table 17.9, the separation should be better in water than in water/methanol. The relative viscosity of a 0.5% PEO standard from Aldrich (Lot No. 0021kz, MW 100,000) in water and in water/methanol with 0.1 *M* lithium nitrate is 1.645 and 1.713, respectively. This indicates that the hydrodynamic volume of PEO in water is smaller than in water/methanol. The difference in hydrodynamic volume between two PEO standards should also be larger in water/methanol than in water. Hence, the separation for PEO should be better in water/methanol than in water. The results in Table 17.8 indicate that separation efficiency is better in water than in water/methanol for all four columns. This indicates that the better swelling of the columns in water more than compensates for the larger difference in hydrodynamic volumes of PEO in the water/methanol mixture.

Several factors can contribute to the difference in retention times for PEO in different mobile phases: the viscosity of a mobile phase, the hydrodynamic volume of a PEO, and the swelling or void volume of a column. Shodex and TSK columns should swell more in water than in water/methanol, and PEO should therefore come out later in water than in water/methanol. PEO should also elute later in water than in water/methanol because water/methanol is a better solvent for PEO than water. The viscosity of the 50:50 water/methanol mobile phase is higher than the viscosity of water. PEO should therefore elute later in water/methanol than in water due to the difference in viscosity. The results in Table 17.9 indicate that the difference in retention time for PEO in water and in water/methanol depends more on the swelling of columns and the hydrodynamic volumes of PEO than the viscosities of mobile phases.

The higher the molecular weight of a PEO, the larger the difference in hydrodynamic volumes of this PEO in water and in water/methanol. Therefore, the difference in retention time for the respective PEO standard in water and in water/methanol should decrease with decreasing molecular weight due to the diminishing hydrodynamic volume factor. However, the difference in retention time for a PEO in water and in water/methanol should increase with decreasing molecular weight due to the column swelling or void volume factor, as the small molecule can take advantage of the increase in void volume more efficiently than the large molecule. Table 17.9 shows that the retention time difference for PEO in water and in water/methanol increases with decreasing molecular weight. For TSK PW columns and Shodex OH-pak columns in water and in water/methanol, this indicates that the change in column swelling or void volume more than offsets the change in hydrodynamic volume in determining the retention time for a PEO.

G. Comparison of the Performance of Four Aqueous SEC Columns in Water and Water/Methanol Mixture for SEC of PVP

Different grades of PVP have been analyzed successfully by Shodex OH-pak and TSK GM-PW columns (9). In the present study, Shodex SB-806M, SB-806MH, TSK GM-PW, and PW_{XL} columns were used to analyze PVP K-15, 30, 90, and 120 in comparing the differences in performance in water and in 50:50 methanol/water (v/v) with 0.1 *M* $LiNO_3$. Overlays of the SEC curves of each grade of PVP from the four columns are shown in Figures 17.14–17.21.

1. For Low Molecular Weight Grades PVP K-15 and K-30

For low molecular weight polymers the separation of the system peaks from the low molecular weight end of the polymer peak is very critical in obtaining accurate MWD and the percentage of low molecular weight materials in the polymer. The water/methanol mixture is a better solvent for PVP than water. PVP K-15 and K-30 should be better separated from the system peaks in the water/methanol mixture than in water because the difference in hydrodynamic volumes between PVP K-15 or K-30 and system peaks is larger in

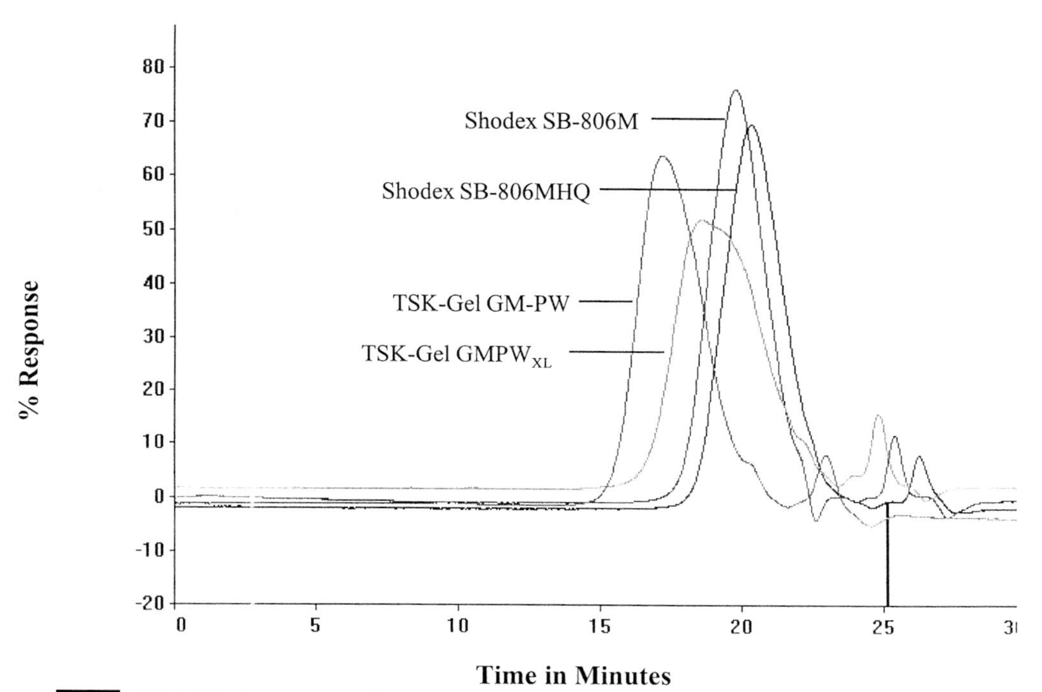

FIGURE 17.14 SEC chromatographic overlays of PVP K-15 for various columns in water with 0.1 *M* $LiNO_3$.

water/methanol than in water. In addition to the hydrodynamic volume factor, separation of the system peaks from the polymer peak also depends on the chemistry of the packing materials.

PVP K-15 and K-30 peaks are symmetric in water and water/methanol, except for the TSK GM-PW$_{XL}$ column in water. This suggests an interaction between PVP K-15 and K-30 with the TSK GM-PW$_{XL}$ column in water. System peaks overlap with the low molecular weight tails of the PVP K-15 and K-30 peaks for all four columns in water. In water/methanol the separation of the system peaks from the polymer peaks is much better for all four columns.

It is interesting to note that the ranking of separation efficiency for PEO standards for these four linear columns in the low molecular weight range discussed in the previous section can not be used to predict which column will give the best separation of PVP K-15 and K-30 in water and in water/methanol. For example, the GM-PW$_{XL}$ column with the best separation efficiency among the four columns for low molecular weight PEO cannot be used for PVP K-15 and K-30 in water due to interaction with the polymer, whereas the GM-PW column with the worst separation efficiency for low molecular weight PEO provides as good as (if not better) separation efficiency for PVP K-15 and K-30 due to its ability to separate the system peaks from the polymer peak. The SB-806MHQ cannot separate the system peak from the PVP K-15 and K-30 peaks as well as the SB-806M and TSK GM-PW columns, even though it has

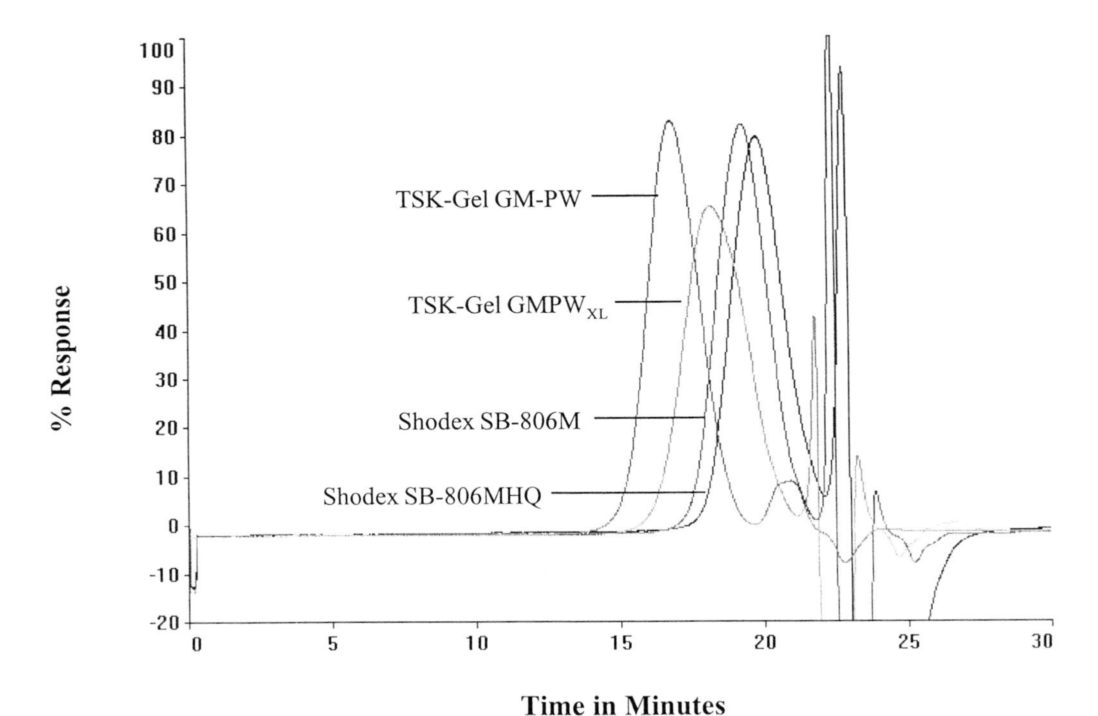

FIGURE 17.15 SEC chromatographic overlays of PVP K-15 for various columns in 50 : 50 water/methanol with 0.1 *M* $LiNO_3$.

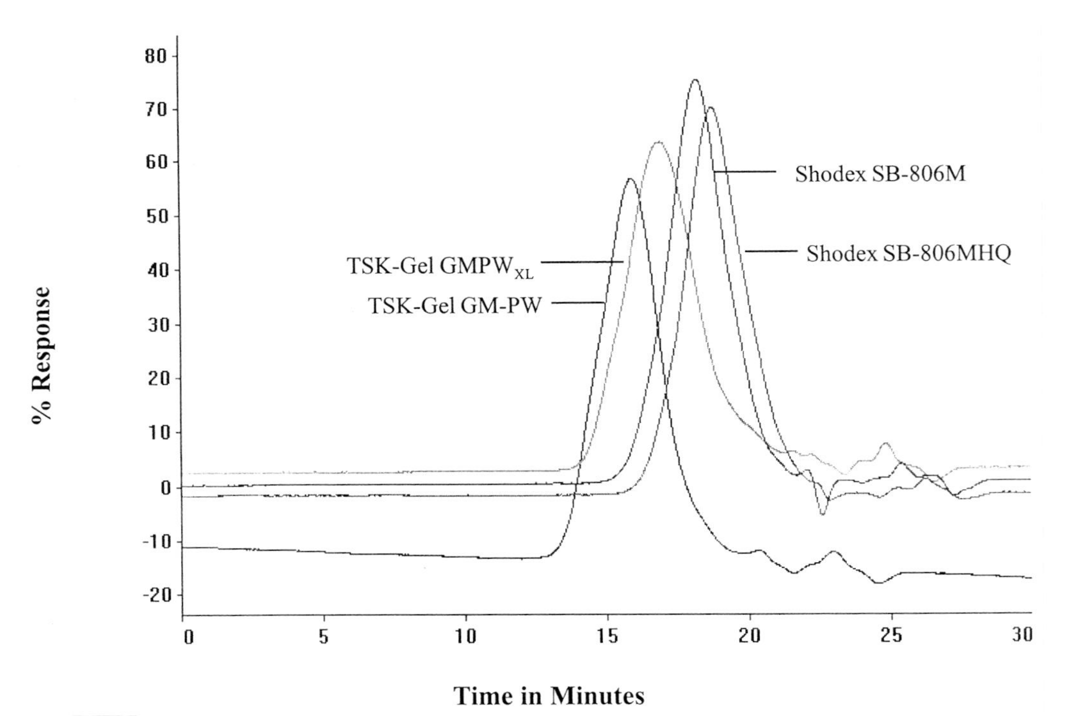

FIGURE 17.16 SEC chromatographic overlays of PVP K-30 for various columns in water with 0.1 *M* $LiNO_3$.

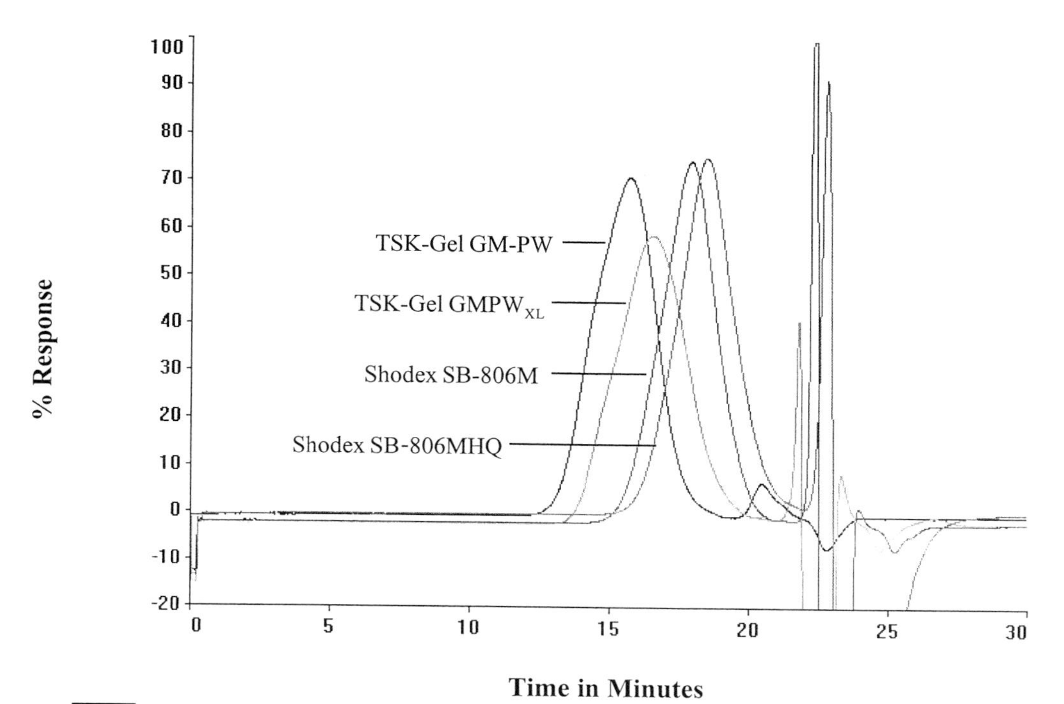

FIGURE 17.17 SEC chromatographic overlays of PVP K-30 for various columns in 50 : 50 water/methanol with 0.1 *M* $LiNO_3$.

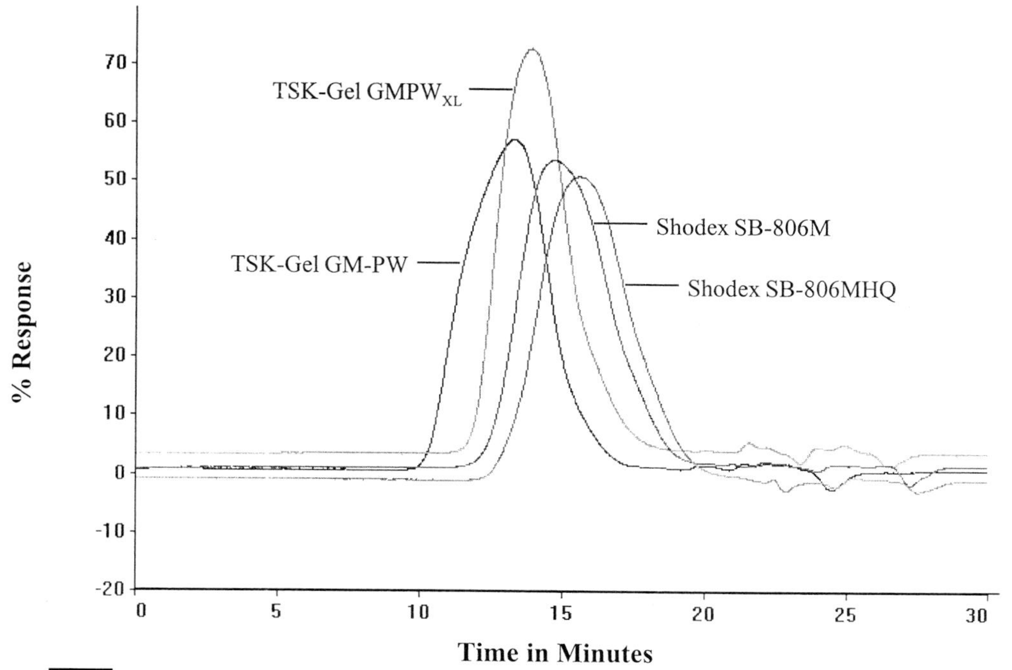

FIGURE 17.18 SEC chromatographic overlays of PVP K-90 for various columns in water with 0.1 *M* $LiNO_3$.

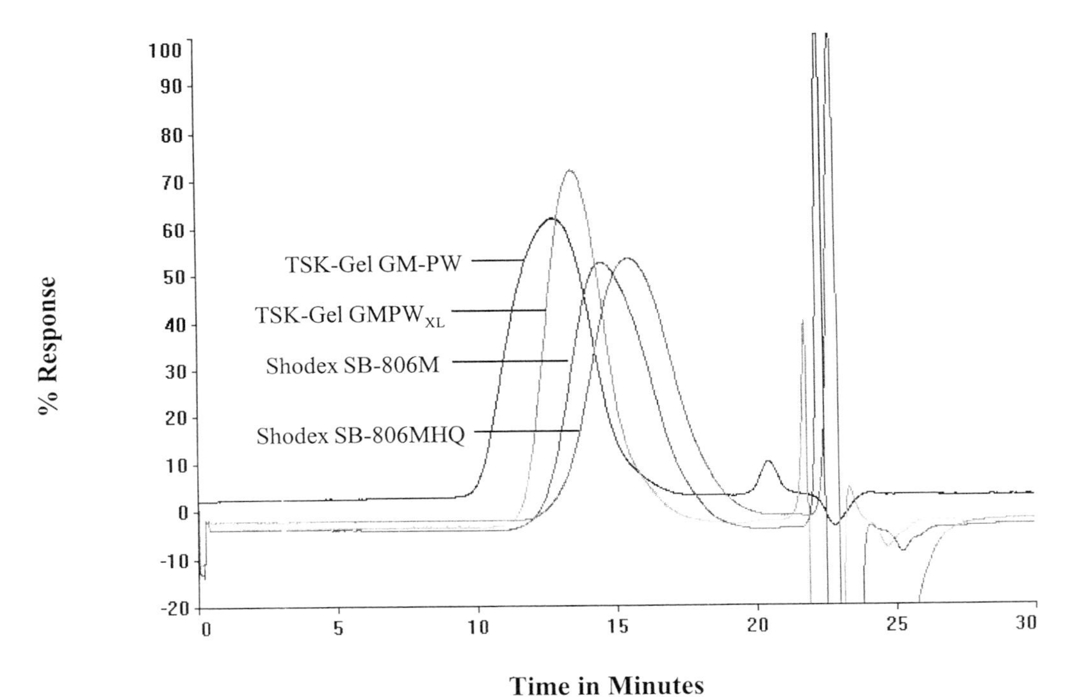

FIGURE 17.19 SEC chromatographic overlays of PVP K-90 for various columns in 50:50 water/methanol with 0.1 *M* $LiNO_3$.

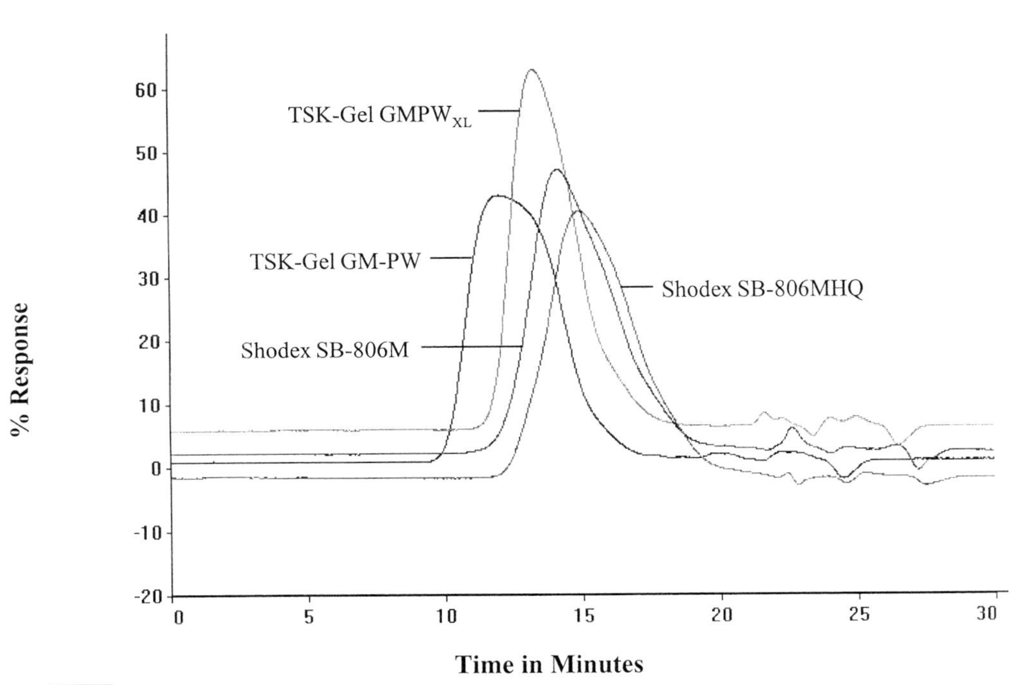

FIGURE 17.20 SEC chromatographic overlays of PVP K-120 for various columns in water with 0.1 *M* $LiNO_3$.

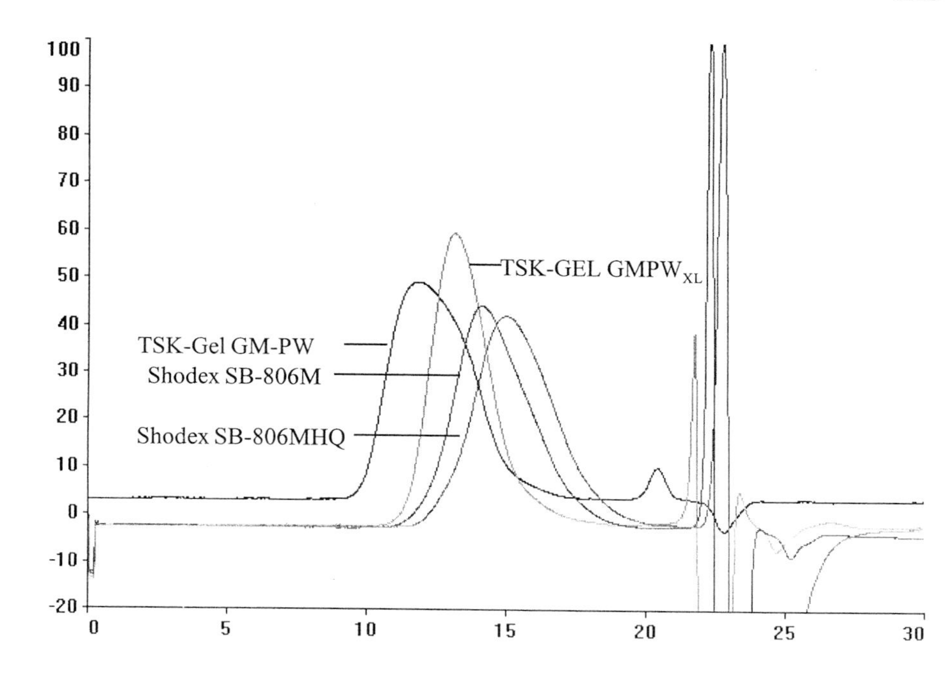

FIGURE 17.21 SEC chromatographic overlays of PVP K-120 for various columns in 50 : 50 water/methanol with 0.1 *M* $LiNO_3$.

better separation efficiency for low molecular weight PEO than the latter two columns. Overall, the best separation for PVP K-15 is achieved using the TSK GM-PW, TSK GM-PW$_{XL}$, or SB-806M column in water/methanol.

2. For High Molecular Weight Grades PVP K-90 and K-120

PVP K-120 peaks are distorted at the high molecular weight end for the TSK-PW column in water and in water/methanol. The distortion is more severe in water than in water/methanol. If the distortion is caused by poor separation of the TSK GM-PW column, then the distortion should be more severe in water/methanol than in water, as it was shown earlier that the resolution should be better in water than in water/methanol. The distortion is probably caused by interaction between PVP K-120 and the TSK GM-PW column. There is a high molecular weight shoulder in the PVP K-90 peak for the TSK GM-PW column, especially in water. This may be the reason why M_w for PVP K-90 determined with the TSK GM-PW column are higher than from the other columns, as shown in Tables 17.3–17.6.

The PVP K-90 and K-120 peaks are not distorted at the high molecular weight end for the GM-PW$_{XL}$ column, even though the GM-PW$_{XL}$ column has less separation efficiency for high molecular weight PEO standards than the GM-PW column. This is another example that the separation efficiency for these four columns for PEO standards cannot always be used to predict which column will provide the best separation for PVP. Similar to the other grades of PVP, the low molecular weight tails for PVP K-90 are more prominent in water than in the water/methanol mixture. Overall, the best separation for PVP K-90 and 120 is achieved with the SB-806M, SB-806MHQ, or TSK-PW_{XL} column in water/methanol.

In summary, methanol as a mobile-phase modifier has a significant effect on the separation of PVP in aqueous SEC with these four linear columns. The best separation of all PVP grades can be achieved with the SB-806M column in 50:50 water/methanol with 0.1 *M* lithium nitrate. It is interesting to note that despite the improvements reported by the manufacturers for the newer columns (SB-806MHQ and PW_{XL}), the newer columns do not necessarily perform better than the older columns (SB-806 and PW) for aqueous SEC of PVP.

H. Comparison of Shodex and TSK Columns for SEC of a Cationic Polymer

The quaternized copolymer of vinylpyrrolidone and dimethylaminoethylmethacrylate (poly-VP/DMAEMA) has been analyzed successfully with Ultrahydrogel columns and a mobile phase of a 0.1 *M* Tris pH 7 buffer with 0.3 or 0.5 *M* lithium nitrate (14). In this study, poor recovery of a poly-VP/DMAEMA sample was noticed when 0.2 *M* lithium nitrate was used for KB-80M, SB806-MHQ, and TSK GM-PW_{XL} columns. Good recovery was achieved with 0.4 *M* lithium nitrate, and M_w of the poly-VP/DMAEMA were found to be 290,000, 300,000, and 320,000 for the respective columns. This demonstrates the equivalence of these columns for SEC of cationic polymers.

I. Comparison of Shodex and TSK Columns by Pyrolysis GC and GC/MS

The pyrolysis GC chromatograms of the Shodex OH-pak columns and TSK PW columns are shown in Figs. 17.22–17.26. The assignment of the peaks (Table 17.10) is done by pyrolysis GC/MS. The TSK G2500 PW_{XL} column, TSK G3000 PW column, and Ultrahydrogel column are very similar in composition. Shodex OH-pak KB-80M is only slightly different from these three columns in composition. The major peaks are methacrylic acid, hydroxyethylmethacrylate, and ethylene glycol dimethacrylate (cross-linker). Two minor peaks are unidentified methacrylates. In contrast to the other four columns, peaks in the pyrolysis GC chromatogram for the Shodex SB-80M column are too weak to be identified by GC/MS. It is obvious from the pyrolysis chromatogram that the Shodex OH-pak SB-80M column is very different in composition from the other four columns with methacrylate as the only identifiable peak. The cross-linker for the Shodex SB-80M column cannot be identified either. Pyrolysis GC will be performed again on the retired Shodex SB family of columns in the future. It is very unfortunate that column manufacturers have not been forthcoming in discussing the basic chemistry of their columns.

IV. CONCLUSIONS

1. The TSK G2500 PW_{XL} column, TSK G3000 PW column, and Ultrahydrogel column are very similar in composition. Shodex OH-pak KB-80M is

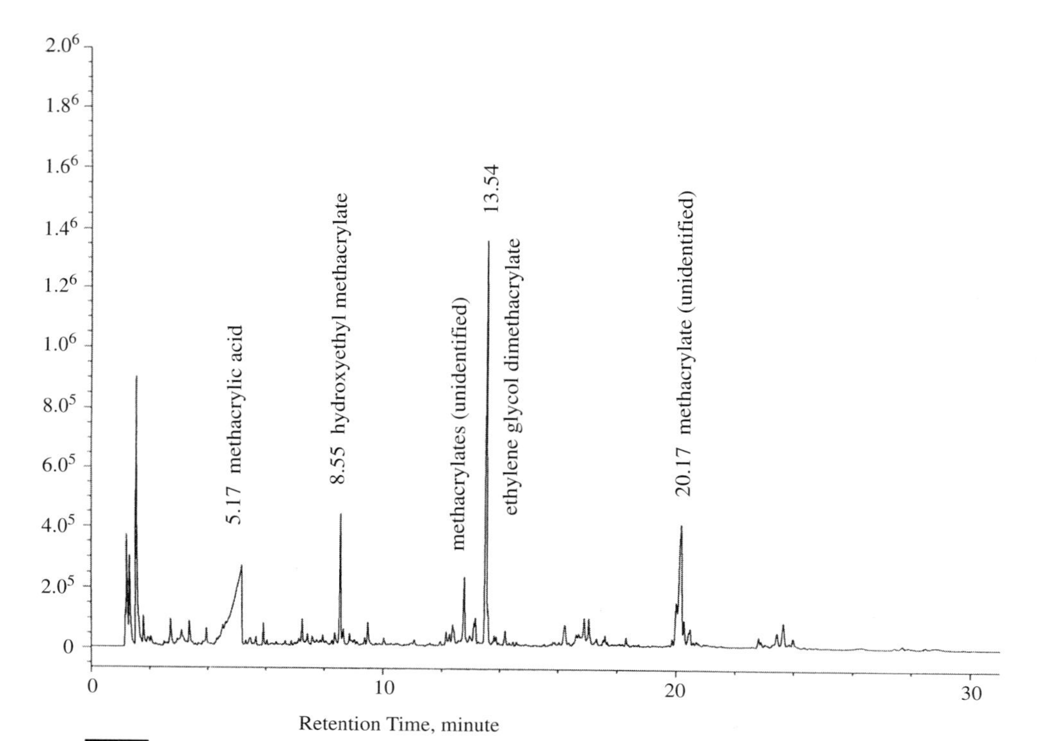

FIGURE 17.22 Pyrolysis GC chromatogram of TSK G2500 PW_{XL} column.

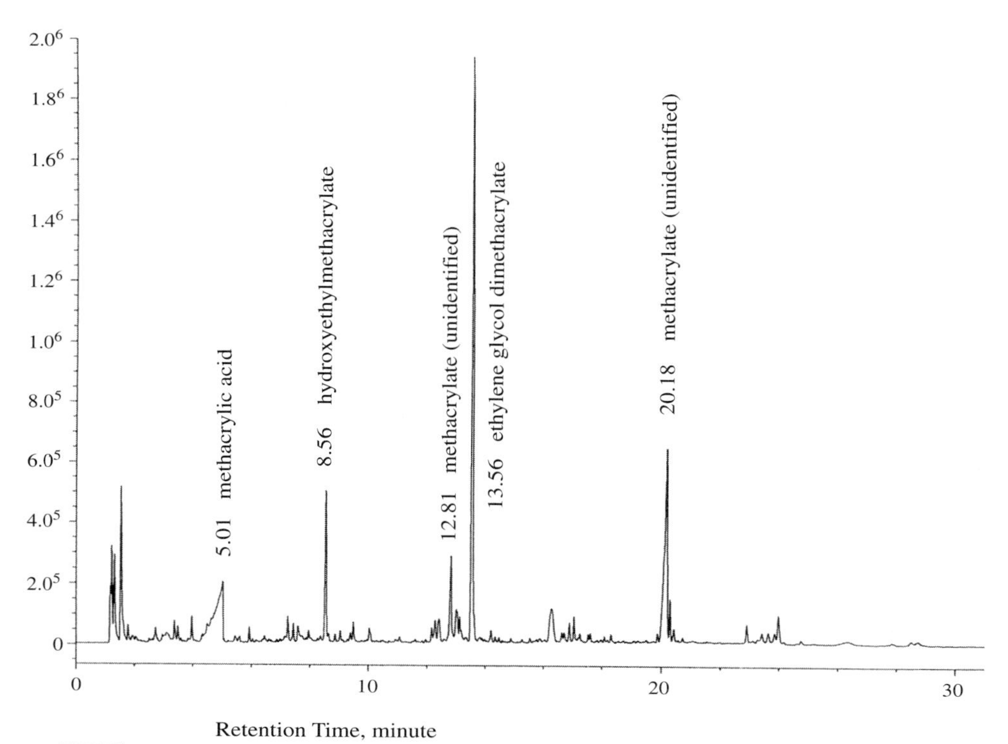

FIGURE 17.23 Pyrolysis GC chromatogram of TSK G3000 PW column.

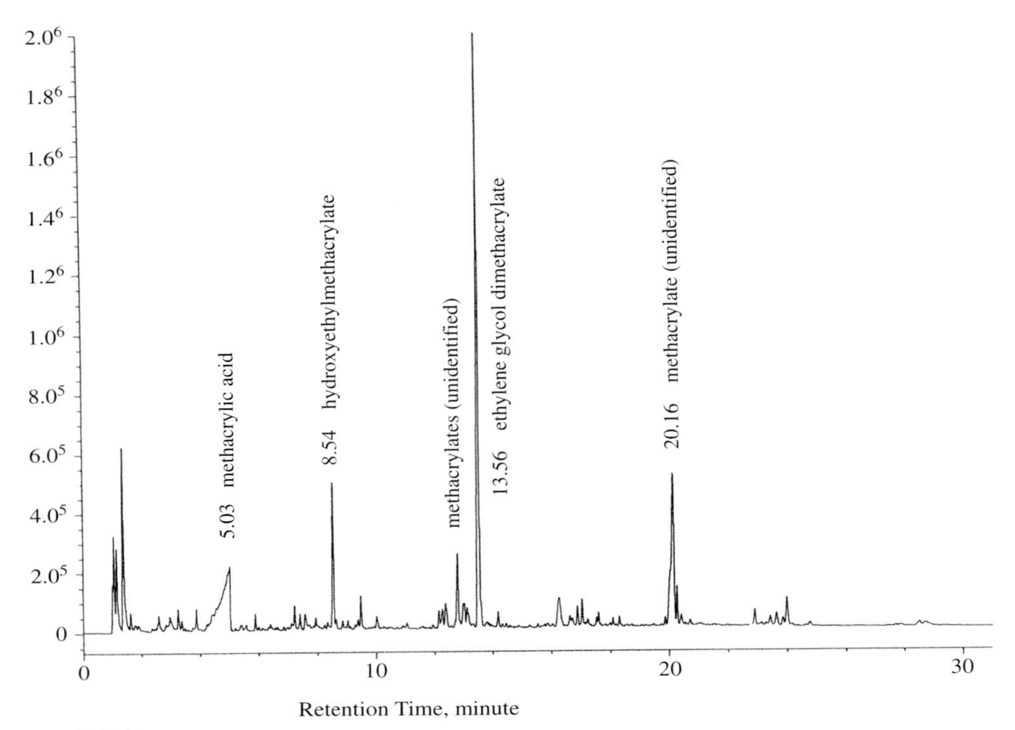

FIGURE 17.24 Pyrolysis GC chromatogram of Ultrahydrogel linear column.

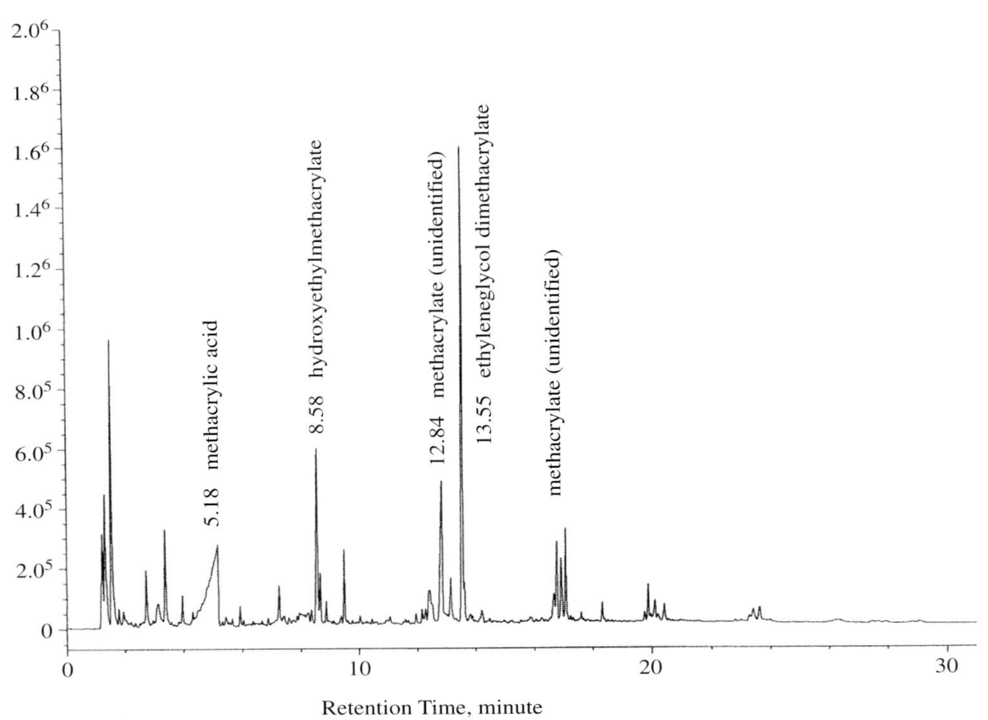

FIGURE 17.25 Pyrolysis GC chromatogram of Shodex OH-pak KB-80M column.

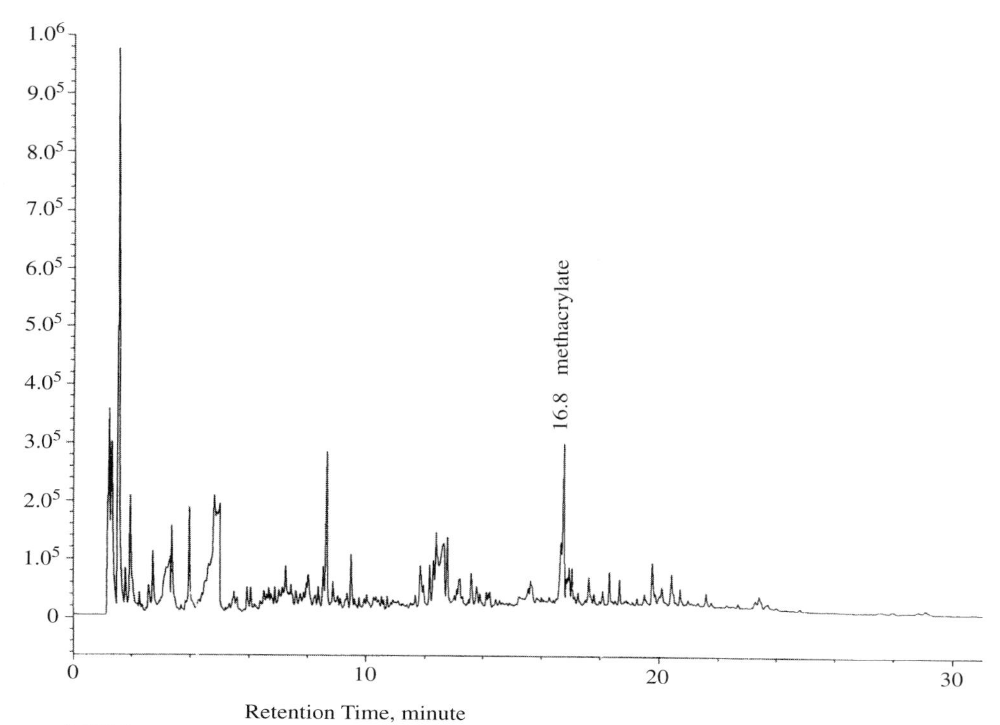

FIGURE 17.26 Pyrolysis GC chromatogram of Shodex OH-pak SB-80M column.

TABLE 17.10 Comparison of TSK and Shodex Columns by Pyrolysis GC and GC/MS

Retention time of major peaks (min)	Identification	TSK GM-PW$_{XL}$ G3000	TSK GM-PW	Ultrahydrogel linear column	Shodex OH-pak, KB-80M
1.1–1.6	Hydrocarbons	17.6%	12.7%	13.4%	18.8%
5.1	Methacrylic acid	18.3%	10.7%	13.1%	17.6%
8.5	Hydroxyethyl methacrylate	4.3%	4.9%	4.8%	6.5%
9.5	Unidentified methacrylates	NA	NA	NA	1.9%
12.8	Unidentified methacrylates	2.9%	3.8%	3.1%	7.4%
13.5	Ethyleneglycol dimethacrylate	13.3%	21%	22.6%	14.8%
16.6–17.0	Unidentified methacrylates	5.0%	4.8%	4.1%	9.0%
20–20.1	Unidentified methacrylates	9.6%	14.1%	11.6%	1.4%
	Others	29%	28%	27%	23%

only slightly different from these three columns in composition. The major peaks in the pyrolysis GC chromatogram are methacrylic acid, hydroxyethylmethacrylate, and ethylene glycol dimethacrylate (cross-linker). However, the Shodex OH-pak SB-80M column is very different in composition from the other four columns with only one identifiable peak (methacrylate) due to the much lower peak intensity for the SB-80M column in the pyrolysis GC chromatogram. The information provided by the manufacturers on the chemistry of these columns is not consistent with the pyrolysis GC and GC/MS results.

2. For all four sets of PEO standards the coefficient of determination (R^2) for the linear calibration curves for the four linear columns in water and in water/methanol are better than 0.99, except for the PL PEO standards and the TSK GM-PW_{XL} column in water/methanol. The coefficient of determination for the TSK GM-PW_{XL} column in general is not as good as the other three linear columns. The coefficient of determination for the TSK PEO standards showed the least dependency on columns and mobile phases. The TSK GM-PW_{XL} column has a lower exclusion limit in the high molecular weight range than TSK GM-PW, Shodex SB-806, and SB-806MHQ columns.

3. The ranking of the four columns for separation efficiency for PEO standards in water and in water/methanol are as follows: SB806M = SB806-MHQ > PW = PW_{XL} (for high molecular weight range in water); SB806M = SB806MHQ > PW>PW_{XL} (for high molecular weight range in water/methanol); and PW_{XL} > SB806M > SB806MHQ > PW (for low molecular weight range in water and in water/methanol).

4. For NIST PEO standards the agreement in M_w between the NIST-certified values and the values determined by SEC with the commercial PEO standards is better for the Shodex SB806M and 806MHQ columns than for the TSK GM-PW and PW_{XL} columns, no matter which set of commercial PEO standards is used. The agreement for TSK, PL, and APSC standards is better than for PSS standards, no matter which column is used. The agreement is about the same in water or in water/methanol for all columns and PEO standards.

5. For the purpose of setting up a calibration curve, PEO standards can be run as a group if the molecular weights of the PEO standards differ by a factor of 6.

6. Methanol as a mobile-phase modifier has a significant effect on the separation of PVP in aqueous SEC with these four linear columns. The best separation of all PVP grades can be achieved with the SB-806M column in 50:50 water/methanol with 0.1 *M* lithium nitrate. The separation of the system peaks from the PVP peak is better in water/methanol than in water. Columns that give the best separation efficiency based on PEO standards at the high and low molecular weight ends are not necessarily the columns that will give the best separation for PVP in water and in water/methanol. Despite the improvements reported by the manufacturers for newer columns (SB-806MHQ and PW_{XL}), the newer columns do not necessarily perform better than the older columns (SB-806 and PW) for the aqueous SEC of PVP.

7. Shodex SB-806M and 806MHQ columns have larger void volumes than the TSK GM-PW and PW_{XL} columns due to the larger diameters of the

Shodex columns. The TSK GM-PW column have a larger void volume than the TSK GM-PW_{XL} column for the same reason.

8. TSK PW_{XL}, Shodex OH-pak KB-80M, and SB806MHQ columns are equivalent for SEC of a quaternized poly-(VP/DMAEMA) in a pH 7 mobile phase.

REFERENCES

1. Macfarlane, J. D., Tokuda, T., Mori, K., Yamada, T., and Moriguchi, S. Enhanced Organic GPC Analysis Employing Aqueous GPC HPLC Column, presented at the 1995 Pittsburgh Conference.
2. Shodex Packed Columns for HPLC and Shodex Instruments for HPLC, '95/96, p. 10, Showa Denko.
3. TosoHaas, The Separation Specialist, p. 18, TosoHaas.
4. Hashimoto, T., Sasaki, H., Aiura, M., and Kato, Y. (1978). *J. Polym. Sci. Polym. Phys. Ed.* **18**, 1789–1800.
5. Barth, H. G. (1980). *J. Chromatogr. Sci.* **18**, 409–428.
6. Sasaki, H., Natsuda, T., Ishikawa, O., Takamatsu, T., Tanaka, K., Kato, Y., and Hashimoto, T. (1985). "New Series of TSK-GEL PW Type for High Performance Gel Filtration Chromatography." Scientific Report of Toyo Soda, 29(1).
7. Mori, S., Takayama, S., Goto, Y., Nagata, M., Kinugawa, A., Housaki, T., Yabe, M., Takada, K., and Shimidzu, M. (1996). *Bunseki Kagaku* **45**(5), 447–453.
8. Herman, D. P., and Field, L. R. (1979). *J. Chromatogr.* **185**, 305–319.
9. Wu, C. S., Curry, J. F., Malawer, E. G., and Senak, L. (1995). "Handbook of Size Exclusion Chromatography," p. 311, 318–321. Dekker, New York.
10. Mori, S. (1983). *Anal. Chem.* **55**, 2414–2416.
11. ASTM D-5296-92, Standard Test Method for Molecular Weight Averages and Molecular Weight Distribution of Polystyrene by High Performance Size-Exclusion Chromatography. ASTM Annual Book of ASTM Standards, Vol.08.03, pp. 425–438.
12. ASTM D-3536-91, Standard Test Method for Molecular Weight Averages and Molecular Weight Distribution by Liquid Exclusion Chromatography (Gel Permeation Chromatography-GPC), ASTM Annual Books of ASTM Standards, Vol. 08.02, P. 349–359. This method was deleted in the 1997 edition of the ASTM book.
13. Aida, H., Matsuo, T., Hashiya, S., and Urushisaki, M. (1991). *Kobunshi Ronbunshu* **48**(8), 507–515.
14. Wu, C., Senak, L., and Malawer, E. G. (1990). *J. Lig. Chromatogr.* **13**(5) 851–861.

18 APPLICATION OF SIZE EXCLUSION–HIGH-PERFORMANCE LIQUID CHROMATOGRAPHY FOR BIOPHARMACEUTICAL PROTEIN AND PEPTIDE THERAPEUTICS

DAVID P. ALLEN

Biopharmaceutical Product Development, Eli Lilly and Company, Indianapolis, Indiana 46285

I. SCOPE

This chapter reviews the applications of size exclusion–high-performance liquid chromatography (SE-HPLC) for the analysis of protein and peptide therapeutics in the biopharmaceutical industry. The chapter begins with the basic theory and primary application of SE-HPLC for the assessment of size heterogeneity by the SE-HPLC measurement of apparent molecular weight. SE-HPLC assays for insulin and recombinant human growth hormone (hGH) are presented as examples. The basic elements and considerations for assay development, validation, and specification assignment are reviewed along with citations to the relevant industry guidance documents. Several specialized uses of SE-HPLC for the physical characterization of protein and peptides are also described briefly. Finally, a survey of additional technologies used to measure the physical property of the size of the protein and peptide molecules is also presented.

II. THEORY AND PRIMARY APPLICATIONS

Size exclusion HPLC has many other common names, such as gel permeation, gel filtration, steric exclusion, molecular sieve chromatography, or gel chromatography. These names all reflect the theoretical mode of action for this type

Column Handbook for Size Exclusion Chromatography

of chromatography. A review article by Barth *et al.* (1) provides an excellent source of detailed information and literature citations for size exclusion chromatography. The mode of SE-HPLC separation is an equilibrium distribution of molecular species between the mobile phase and the porous stationary phase. The resolution of SE-HPLC is the relationship of the molecule's hydrodynamic volume relative to the pore size of the stationary phase.

Hydrodynamic volume refers to the combined physical properties of size and shape. Molecules of larger volume have a limited ability to enter the pores and elute the fastest. A molecule larger than the stationary phase pore volume elutes first and defines the column's void volume (V_0). In contrast, intermediate and smaller volume molecules may enter the pores and therefore elute later. As a measure of hydrodynamic volume (size and shape), SE-HPLC provides an approximation of a molecule's apparent molecular weight. For further descriptions of theoretical models and mathematical equations relating to SE-HPLC, the reader is referred to Refs. 2–5.

The primary application of analytical SE-HPLC is to quantitatively determine the apparent molecular heterogeneity of samples. As a purity method, resolved molecular species are quantitated by comparing the individual areas to the summed areas of all species. Data are typically reported as a percentage of the resolved species. The naming of the resolved molecular species typically relates to the measured molecular size, such as monomer. Species of higher apparent molecular weight are generally labeled as higher molecular weight polymers (HMWP), but may also be named for the specific higher order self-associated forms such as dimer and trimer. Molecular species smaller than the monomer form are usually called fragment species. A second application is the determination of protein concentration using either the total or the main peak response. The quantitation is accomplished by comparison to a well-characterized international reference standard or to an in-house characterized reference standard material. The specificity, precision, accuracy, and sensitivity (mcg limit of quantitation) along with the convenience of automation make SE-HPLC a good method for protein quantitation.

III. EXAMPLES

The first example is the SE-HPLC analysis of insulin. The term insulin refers to a variety of marketed products that includes both animal and recombinant bulk drug sources and a series of solutions and crystalline formulation strengths. SE-HPLC plays a key role in establishing the purity and quality of insulin bulk and formulated finished products. SE-HPLC assays are described in several compendial monographs for insulin. The formation of high molecular weight polymers is one of the known degradation mechanisms of insulin. As an example, a typical elution profile for a protamine-containing insulin suspension is shown in Fig. 18.1. The SE-HPLC elution profile of insulin has three regions of interest. The first region is the HMWP species and is defined as the sum of all insulin polymers present. It can include one or all of the following: covalent insulin dimer (CID), covalent insulin–protamine polymer (CIPP), and higher molecular weight insulin polymers. The second region is the main peak, which

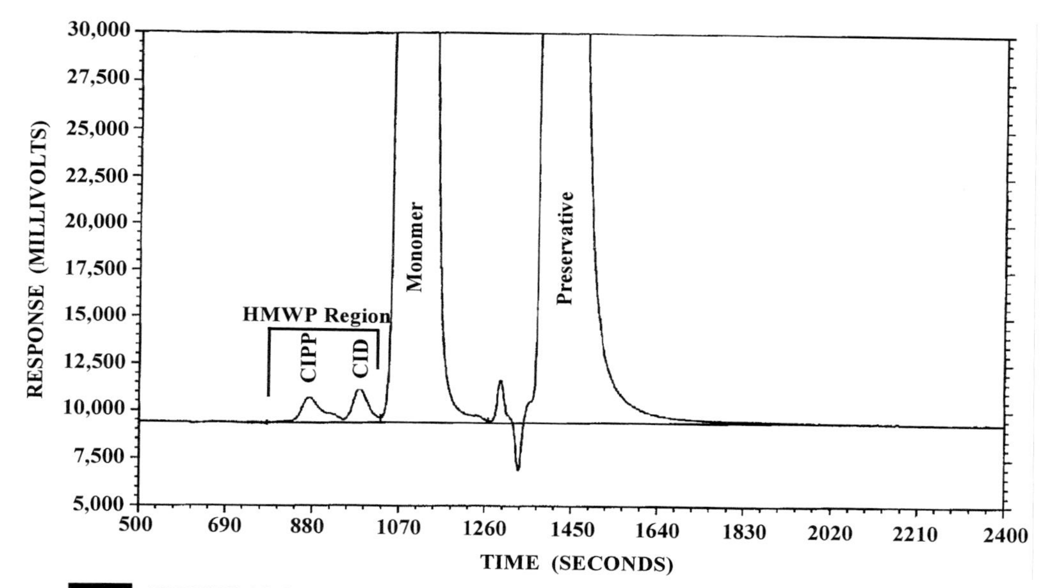

FIGURE 18.1 Typical chromatogram (baseline view) of a protamine containing insulin drug product sample integration. HMWP is defined as the sum of all insulin polymers present and can include one or all of the following: covalent insulin dimer (CID), covalent insulin–protamine polymer (CIPP), and higher molecular weight insulin polymers.

is monomeric insulin. The third region is late eluting preservative compounds, which approximates the column volume (V_t). HMWP in insulin bulk and formulated finished product samples are determined on a HMWP peak area versus total insulin peak area basis. As an example, the compendial limits for insulin bulk and finished product(s) are listed in Table 18.1.

The second example is the SE-HPLC analysis of recombinant hGH. In this example, SE-HPLC is used for both a purity and a protein concentration method for bulk and formulated finished products. This method selectively separates both low molecular weight excipient materials and high molecular weight dimer and aggregate forms of hGH from monomeric hGH, as shown

TABLE 18.1 Compendial Specifications

Compound	% HMWP[a]
Insulin bulk	1.0%
Insulin finished product	2.0% (nonprotamine) 3.0% (protamine containing)
hGH	4%
hGH finished product	6%

[a]References: Insulin bulk Ph Eur monograph, insulin finished product proposed monograph, and hGH proposed USP monograph.

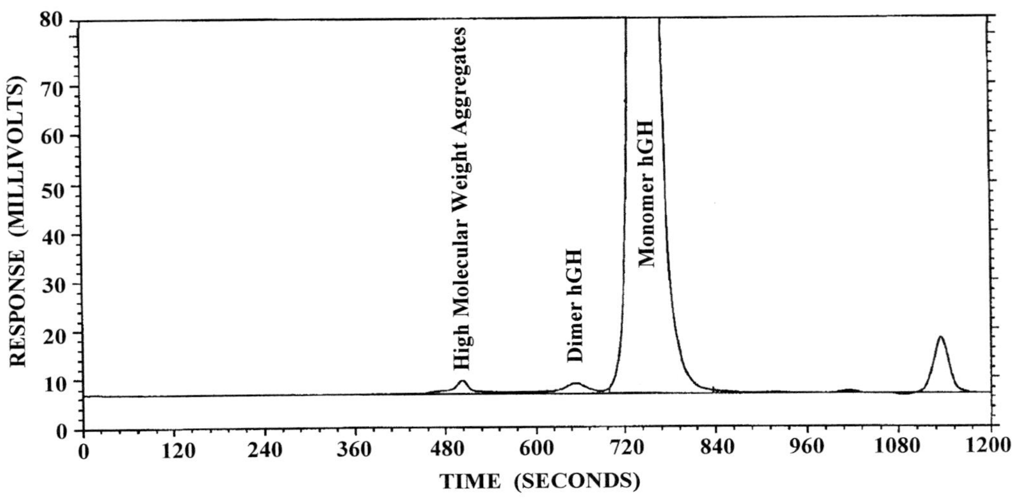

FIGURE 18.2 Typical chromatogram (baseline view) of finished product recombinant hGH formulation. This method is performed with a TSK GEL G3000SW column.

in Fig. 18.2. Compendial limits for hGH HMWP are shown in Table 18.1. The protein concentration is determined by the main monomer response with respect to a characterized reference standard material. The results may be reported as milligrams per milligram of solids, milligrams per milliliter of a solution, or milligrams per vial.

IV. ASSAY DEVELOPMENT AND VALIDATION

The basic elements and considerations for assay development, validation, and specification assignment are reviewed briefly. Assay development produces a method that requires validation for the analysis and release of materials (bulk or formulated finished product) for use in clinical development. The cumulative analysis of materials and stability considerations is then used to established specifications for internal and regulatory submission.

In the development of a SE-HPLC method the variables that may be manipulated and optimized are the column (matrix type, particle and pore size, and physical dimension), buffer system (type and ionic strength), pH, and solubility additives (e.g., organic solvents, detergents). Once a column and mobile phase system have been selected the system parameters of protein load (amount of material and volume) and flow rate should also be optimized. A beneficial approach to the development of a SE-HPLC method is to optimize the multiple variables by the use of statistical experimental design. Also, information about the physical and chemical properties such as pH or ionic strength, solubility, and especially conditions that promote aggregation can be applied to the development of a SE-HPLC assay. Typical problems encountered during the development of a SE-HPLC assay are protein insolubility and column stationary phase effects. These issues are best addressed by the manipulation and optimization of the column matrix type and mobile-phase parameters.

Assay validation is a process used to establish if an analytical method is acceptable for its intended use (6). The validation of an analytical method evolves as the development of a therapeutic compound progresses. The industrial guidelines that describe assay validation include compendia and regulatory documents which include recent International Conference on Harmonization initiatives (6–9). Following these guidance documents, assay validation protocols for SE-HPLC analysis for purity and protein concentration are easily developed. As a part of validation, the important method criteria of the system suitability test must also be addressed. The United States Pharmacopoeia has described in detail HPLC system suitability testing (10). Standard chromatographic parameters such as resolution, tailing, and precision are typically specified for SE-HPLC assays. An important and final aspect of the assay validation process is documentation.

The specification development process is a data-driven activity that requires a validated analytical method. The levels of data needed include assay precision, replicate process results (process precision), and real-time stability profiles. A statistical analysis of these data is critical in setting a realistic specification. Most often, aggregation and fragmentation degradation mechanisms are common to protein and peptide therapeutics. Therefore, the SE-HPLC method provides a critical quality parameter that would need to be controlled by a specification limit.

V. SPECIALIZED APPLICATIONS

This section briefly describes two examples of additional protein characterization applications for the use of size exclusion chromatography. This is not an exhaustive review of these applications, but provides several examples that may allow the reader to see broader applications of this technique. The measurement of apparent molecular weight has been used to study refolding to of a recombinant protein (11) and to measure protein interactions (12–15). The refolding of *Escherichia coli* produced recombinant protein requires the denaturation and refolding of the protein of interest. A major factor in decreasing the final yield of a biologically active refolded protein is the formation of irreversible aggregates. In this example, the recombinant lysozyme was successfully refolded from denaturing conditions using SE-HPLC for controlled buffer exchange (11). As a second application, SE-HPLC is used to measure protein/protein interactions. In this application the interaction of the fibronectin receptor with talin is measured by mixing the proteins followed by analysis using SE-HPLC that has been preequilibrated with talin (12). This technique, termed high-performance receptor-binding chromatography, has also been applied as a novel bioassay for hGH with an isolated soluble hGH receptor (13). As a final protein interaction application, the use of SE-HPLC to identify and characterize the avidity of antibody epitopes has been described (14,15). The central theme of these specialized applications is to utilize and take advantage of changes in hydrodynamic volume.

VI. ALTERNATIVE METHODS FOR SIZE MEASUREMENTS

In addition to SE-HPLC, the following methods, sodium dodecyl sulfate–polyacrylamide gel electrophoresis (SDS-PAGE), capillary electrophoresis (CE), ultracentrifugation (UC), and mass spectrometry, are also used to measure the molecular weight or apparent molecular weight of protein or peptide therapeutics. Again, this is not an exhaustive review of these techniques, but provides the reader with an introduction to the techniques. The literature citations provided are to assist in seeking more detailed information. The first three techniques, SDS-PAGE, CE, and UC, all result in apparent molecular weight determinations by the measurement of hydrodynamic volume. SDS-PAGE is a high-resolution qualitative method that measures the size heterogeneity. The apparent molecular weight is determined by comparing the migration of the SDS–protein complex to that of molecular weight standards. Capillary electrophoresis systems for size determination include SDS-PAGE filled, polymer filled, and sieving medium (16–19). These systems provide the resolution of electrophoresis and the quantitative detection capability of HPLC. The precision characteristics (repeatability, intermediate precision, and reproducibility) of CE are not yet sufficiently reliable and have limited its routine use for quantitative measurements. Analytical ultracentrifugation has returned as a routine characterization method to examine self-association in protein formulation systems (20,21). The final technique to present is protein mass spectroscopy, which measures the absolute mass of a protein or peptide. Advances in mass spectroscopy instrumentation have allowed for the direct mass assignment and verification of mass by direct protein sequencing (22–25). These orthogonal methods of measuring size heterogeneity are typically used in parallel with SE-HPLC as part of a well-characterized examination of protein and peptide bulk and final finished products.

ACKNOWLEDGMENTS

The TSK-GEL column is a trademark of Tosoh Haas Co. Ltd and the Waters "Insulin HMWP" column is a trademark of Waters Corporation. The author thanks the book's editor Chi-san Wu, the assigned peer reviewer, and the Eli Lilly and Company reviwers John Towns, Ralph Riggin, and James Kelley for their assistance, time, and professional comments.

REFERENCES

1. Barth, H. G., Boyes, B. E., and Jackson C. (1996). *Anal. Chem.* **68**, 445R–466R.
2. Degoulet, C., Busnel, J. P., and Tassin, J. F. (1994). *Polymer* **35**, 1957–1965.
3. McCoy, B. J. (1995). *J. Chromatogr. A* **697**, 533–540.
4. Harlan, J. E., Picot, D., Loll, P. J., and Garavito, R. M. (1995). *Anal. Biochem.* **224**, 557–563.
5. Kuntz, M. A., Dubin, P. L., Kaplan, J. L., and Metha, M. S. (1994). *J. Phys. Chem.* **98**, 7063–7067.
6. International Conference on Harmonization (Step 5, 1996), Q24A: Validation of Analytical Procedures, Definitions and Terms.
7. United States Pharmacopeia 23, Validation of Compendial Methods <1225>.

19

COLUMN SELECTION AND RELATED ISSUES FOR ACRYLIC ACID AND ACRYLATE ESTER POLYMERS

MICHAEL T. BENDER AND DANIEL A. SAUCY

Analytical Research Department, Rohm and Haas Company,
Spring House, Pennsylvania 19477

I. INTRODUCTION

Polyacrylates are an industrially important class of polymers. The name "polyacrylate" is variously used to refer to polymers of acrylate esters [e.g., poly(methyl methacrylate)] as well as polymers of acrylic acids [e.g., poly(methacrylic acid)]. Because the former is organic soluble while the latter is not, chromatographic analysis of these two requires quite different conditions. This chapter discusses both types of polymers, separating their consideration when necessary. We will refer to both types of polymers as "polyacrylates," letting the context indicate whether we are referring to an ester or to an acid polymer.

This chapter makes no distinction between gel-permeation chromatography (GPC) and size exclusion chromatography (SEC). We make mention of specific analysis conditions wherever possible. We have attempted to include a variety of conditions but by no means should this chapter be considered a comprehensive review of conditions for analyzing polyacrylates. We have drawn extensively from our own experience in selecting examples.

A. Polymers of Acrylate Esters

Acrylic esters can be polymerized by a number of routes. Anionic polymerization gives the narrow standards used primarily for calibration, but is not used on an industrial/commercial scale. Free-radical polymerization is the dominant mode of polymerization for making these polymers on an industrial scale. Significant volumes of polymer are made by both solution polymerization

Column Handbook for Size Exclusion Chromatography

and emulsion polymerization. From the viewpoint of the analyst interested in molecular weight, the mode of polymerization is of secondary importance. The examples provided and discussion given apply to any acrylic polymer, regardless of how it was prepared.

Tetrahydrofuran (THF) is usually the solvent of choice for poly(acrylates). It is an excellent thermodynamic as well as kinetic solvent, its only drawback being its volatility and flammability.

GPC analysis of homo- and copolymers of acrylic esters ("acrylic" will be used throughout this chapter to refer to both acrylic and methacrylic) is quite straightforward. A representative set of conditions is

Solvent: THF
Polymer concentration: 0.1 to 0.5%
Flow rate: 1 ml/min
Injection volume: 100 μl
Column set: 2 TosoHaas 30 cm × 7.8 mm H-series
Detection: Refractive index

TosoHaas columns are styrene cross-linked with divinylbenzene (DVB). Columns of similar composition are available from Polymer Laboratories (PL Gel), Waters (Ultrastyragel), Shodex, Jordi (JordiGel), and many others. Columns based on derivatized silica are also available, but are less widely used.

Molecular weight calibration from a monomer to several million daltons can be carried out by a variety of techniques. Because narrow standards of p(methyl methacrylate) (pMMA) are available, these are often used. Narrow standards of p(styrene) (pSty) are also available and can be used. Using the Mark-Houwink-Sakurada equation and the parameters for pSty and pMMA, a system calibrated with pSty can give pMMA-equivalent values, and vice versa.

Solvents other than THF can also be employed. These include toluene, dimethylformamide (DMF) (pSty is insoluble), acetone, chloroform, and dimethyl sulfoxide (DMSO). The specific recommendations of the column manufacturer must be followed closely with regard to changeover from one solvent to another.

B. Polymers of Acrylic Acid

Polyacrylic acid (pAA) homopolymers and related copolymers have become a commercially important class of water-soluble polymers. Acrylic acid polymers can range in molecular mass from less than 1000 Da to greater than 1,000,000 Da. A representative set of analysis conditions is

Solvent: 50 m*M* sodium phosphate buffer, pH 7
Polymer concentration: 0.1 to 0.5%
Flow rate: 1 ml/min
Injection volume: 100 μl
Column set: 2 TosoHaas 30 cm × 7.8 mm PW series
Detection: Refractive index

A wide array of columns can be used for the homopolymers and related copolymers. These can be based on derivatized silica (e.g., TosoHaas TSK

SW series and Alltech Macrosphere series), hydroxylated methacrylate (e.g., TosoHaas TSK PW series, Waters Ultrahydrostyragel, Shodex OHPak), functionalized styrene/DVB (Polymer Laboratories PL aquagel-OH), poly(vinyl alcohol) (Shodex OHPak), and others.

Commercial narrow standards [such as poly(ethylene glycol) (pEG), polystyrene sulfonate, pAA, poly *n*-vinyl pyrrolidinone, dextrans] are available from American Polymer Standards Corporation, Polymer Laboratories, Polymer Standards Service USA, Toyo Soda, and others. While these standards are often not as narrow as pSty or pMMA that has been anionically polymerized, they are acceptable for narrow standard calibrations.

II. SOLUBILITY AND ADSORPTION

A. Solubility

For the GPC separation mechanism to strictly apply, there must be no adsorption of the polymer onto the stationary phase. Such adsorption would delay elution of the polymer, thereby resulting in the calculation of too low a molecular weight for the polymer. The considerable variety of undesirable interactions between polymers and column stationary phases has been well reviewed for GPC by Barth (1) and this useful reference is recommended to the reader. Thus, the primary requirement for ideal GPC is that the solvent–polymer interaction be strongly thermodynamically favored over the polymer–stationary phase interaction.

On inspecting a polymer/solvent pair for solubility, the analyst is often judging solubility based on observing a clear and homogeneous solution. Cases can arise, however, where a sample solution appears clear and homogeneous to the analyst yet will not filter through even a 1-μm filter. In many of those cases, the culprit is a swollen polymer. A highly swollen polymer may have a refractive index so close to that of the solvent that little or no light scattering occurs. This misleading situation is more likely to occur when working with solvents of higher refractive index, such as DMF and DMSO. In the extreme limit, if the solvent and polymer refractive indices are the same, a totally insoluble polymer will appear to be soluble.

If intrinsic viscosity data are available or can be estimated for the polymer/solvent pair, then the magnitude of the exponent of the Mark-Houwink-Sakurada (MHS) fit for that pair can be used to judge the solvent suitability for GPC analysis of that polymer. For random coil polymers, the MHS exponent usually ranges from 0.5 to 0.8, with 0.5 corresponding to Θ temperature conditions and 0.8 corresponding to highly solvated (large excluded volume) conditions. One typically finds that if the polymer is (1) coil-like, (2) not a polyelectrolyte, and (3) has no specific functionality that would indicate a likely stationary-phase interaction, then any polymer/solvent pair with an MHS exponent of greater than 0.55 to 0.60 is a good choice for an ideal GPC analysis solvent of that polymer. This estimate must be modified somewhat to consider significantly branched polymers (which typically have lower MHS exponents than their unbranched coil analogs) or rod-like polymers (which have higher

MHS exponents than their coil analogs) , but it is a reasonable rule of thumb for many polymers. MHS exponents are conveniently found in references such as Du *et al.* (2) or Grulke (3).

B. Adsorption

The analyst must remember that solubility of a polymer in the chosen eluant is a necessary, but not sufficient, requirement for ideal GPC separations. Once injected on the column, the polymer has a choice of partitioning onto the stationary phase or remaining in the solvent. It is imperative that the analyst choose solvent and column conditions such that the ideal, nonadsorptive, GPC mechanism can occur.

It is tempting to believe that if one works at, or slightly above, the Θ temperature of a given polymer–solvent pair that no strong interactions will occur between the polymer and the stationary phase. Polymers at their Θ temperature assume theoretically calculable, free-flight chain conformations, but that does not mean that they are strongly solvated in that solvent at that temperature. One way to think of a Θ solvent for a polymer is that it is a thermodynamically poor solvent in which polymer chain segments attract one another just enough to compensate for the effect of the physically occupied volume of the chain (4). This means that to obtain ideal GPC separation mechanisms, analyses may need to be performed well above the Θ temperature so as to avoid any potential polymer–stationary-phase interactions.

Data in Fig. 19.1 illustrate this point with a set of three anionically polymerized pMMA standards, analyzed at two different sets of conditions. The three pMMA standards were made up in room temperature acetonitrile and easily dissolved into that solvent. The samples were injected onto a GPC column (Polymer Laboratories PLgel, 20-μm mixed bed A, 7.8 $\times$ 30 cm) pumped with THF at 35°C. The chromatogram shows the expected resolution of the three standards (and also the large peak at the total permeation volume of the column resulting from the injected acetonitrile). The very same samples were injected onto the same column (but pumped with acetonitrile at 65°C, a temperature more than 30°C above the reported Θ temperature of the pMMA/acetonitrile system). The chromatogram shows that no peaks eluted prior to the total permeation volume of the column (approximately 10 min). Thus, this example shows that even working at 30°C above the Θ temperature may not be sufficient to prevent strong partitioning of the polymer onto the stationary phase of the GPC column.

It is well known that pMMA and pSty in THF follow ideal GPC behavior on many common GPC columns. However, many commercially important acrylate polymers contain a wide array of other monomers. In general, acrylic polymers composed of monomers that do not contain polar groups will yield well-behaved polymers, giving ideal GPC separations. Monomers that contain polar groups should prompt the analyst to carefully evaluate the possibility of adsorption of the analyte onto the column. The most common functionalities of concern are hydroxyl groups, amine groups, ethylene oxide units, and carboxylic acids. In many cases, such monomers can be tolerated. However, the acceptable level can vary considerably with even apparently minor changes in

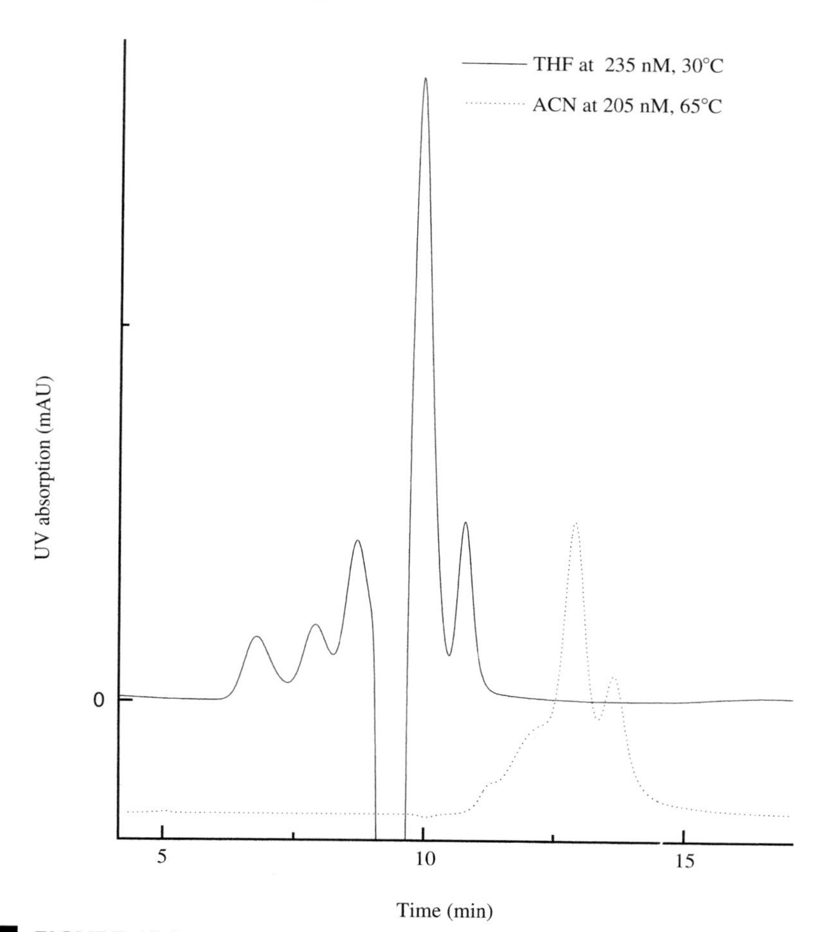

FIGURE 19.1 Theta condition effects on pMMA elution. Upper trace: acetonitrile 65°C. Lower trace: THF 35°C.

structure. For example, acrylic acid in an acrylate polymer is not tolerated as well as methacrylic acid. For acrylate polymers in THF, approximate levels of some polar monomers are shown in Table 19.1.

TABLE 19.1 Acceptable Levels of Polar Monomers

Functional group	Maximum % by weight
Acid	10–20
Hydroxyl	1–5
Amine	1–5
Blocks of ethylene oxide	Depends on length

Because the stationary-phase surface plays a key role in adsorption, one can find large changes in adsorption behavior when switching from one column manufacturer to another, even though both claim to have a common stationary-phase chemistry (e.g., styrene/divinylbenzene). We have found that we can analyze up to 100,000-Da copolymers containing multiple blocks of 8000-Da pEG on TosoHaas ProGel TSK G-series columns, but those same materials adsorb severely when analyzed on Waters Ultrastyragel columns. However, we have also seen cases of columns from different lots from the *same* manufacturer exhibiting different behavior with troublesome systems such as these.

In certain instances, it may be difficult or impossible to assess the extent of adsorption and/or eliminate it. If there is no choice but to live with adsorption, then the analyst should pay particular attention to the analysis. As the column ages, its surface character may change, which may, in turn, change the elution times of the polymer. This elution time change would appear as a change in molecular weight. If a new column is installed to replace that column, its adsorption characteristics may be very different from the in-service column, thereby again causing a change in the apparent molecular weight of the sample. These two problems can be reduced by using a broad standard calibration approach based on a standard of the same composition as the samples themselves.

However, the nature of the adsorption process is such that adsorption is often molecular weight dependent and thus not easily corrected for. This is demonstrated by Fig. 19.2. Figure 19.2A shows a series of pEG standards run in inhibited THF on TosoHaas ProGel TSK G-series columns. While the 73 and 19k samples look fine, the 288k sample is clearly broadened and tailed. Figure 19.2B demonstrates the need to check for adsorption with standards of the same chemical composition as the unknown. Although the pEG sample is clearly adsorbing, the p(Sty) standard that nearly coelutes with it shows no evidence of any irregularity.

In summary, while it may be possible to successfully run a method where adsorption is taking place, it is a method with a high potential for problems.

III. GENERAL GPC PRACTICES

A. Column Quality Indicators

It is difficult to decide what should serve as adequate column quality parameters for describing the performance of a set of GPC columns. The two most common measures are plate count and resolution. While both of these can be useful for monitoring the performance of a column set over time, it is not generally possible to a priori specify the performance needed for a specific analysis. This will depend on the nature of the polymer itself, as well as the other matrix components.

Most GPC columns are provided with vendor estimates of the plate count of the column and a chromatogram of a series of test peaks. These plate count estimates are usually obtained using small molecule analytes that elute at the total permeation volume (V_p) of the column. The Gaussian peak shape model

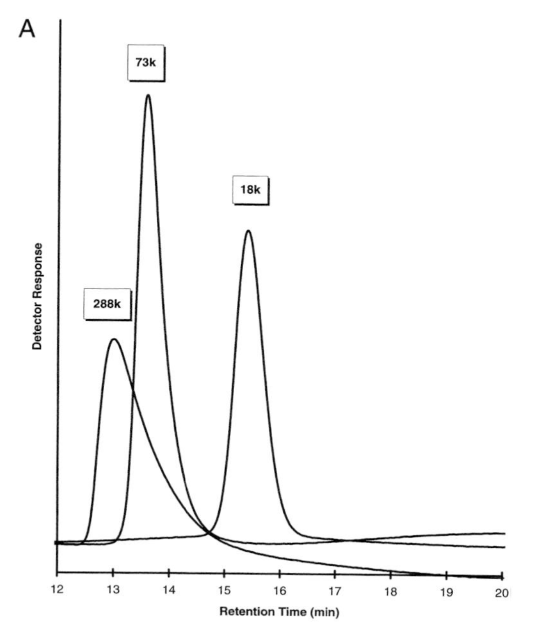

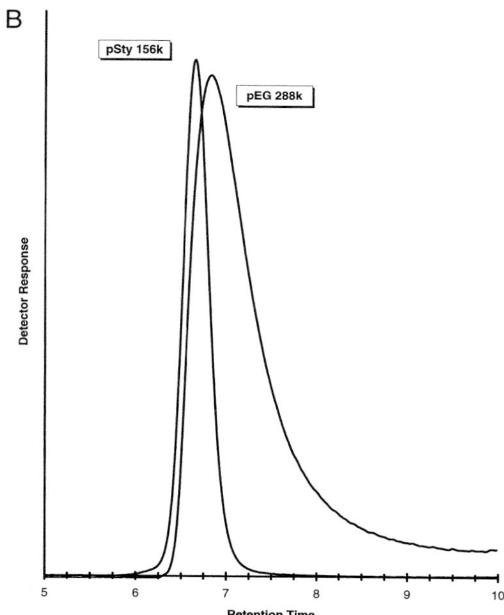

FIGURE 19.2 (A) pEG 18, 73, and 288k in THF on TosoHaas G3000Hxl + G4000Hxl columns. (B) Comparison of pEG 288k and pSty 156k on TosoHaas G-series columns.

is then used to determine the plate count of the column. It would be much more desirable to calculate the plate count using a sample that elutes in the working range of the column, but it is difficult to find a monodisperse material that can be used to test the plate count in this region. It is quite likely that the plate counts found by such a monodisperse standard eluting in the column working range would be lower than those currently reported by pure solvents at V_p, but they would perhaps be more indicative of the quality of separation available for polymers on that column.

Because the polydispersity of the polymer is reflected in the width of the chromatographic peak, we require that the column band broadening contribution to the peak width be minor compared to that from the polymer itself. This criterion cannot always be met.

Care must be taken when comparing data between columns sets when the effects of column band broadening, e.g., resolution, have not been accounted for. Figure 19.3 shows a set of narrow pSty standards analyzed on a 10-μm column set (Waters HT-6, -5, -4, and -2) and a 20-μm column set (two Polymer Laboratories Mixed A plus a Polymer Laboratories guard column). The narrow standards clearly demonstrate the difference in resolution of these two column sets, arising both from the difference in particle size and from the difference in bed volume. The molecular weight polydispersities calculated for one of these standards were 1.06 and 1.34 versus a claimed polydispersity for the sample of 1.03. When a solution, free-radical polymerized polyacrylate was analyzed on the columns sets described earlier, without accounting for the difference in resolution, the respective molecular weight polydispersities observed were 1.88 and 2.20. Thus, one must always be aware of the influence of column resolution on the observed polydispersity.

For acrylic polymers produced via emulsion polymerization, a set of two or more 30-cm-long columns with 10-μm or less packing material will usually ensure that the observed polydispersities are minimally influenced by column band broadening.

Substantial disagreement exists in test method recommendations as to what plate counts are required for a "good" set of GPC columns. ASTM D5296-92 recommends (5) that columns have greater than 13,100 plates/meter for them to be considered an adequate set for the analysis of pSty in THF. The Organization for Economic Cooperation and Development (OECD) has no recommended (6) plate count per meter for a column set whereas the DIN 55672-1:1995-02 standard (7) is 20,000 plates/meter. This issue becomes more important in choosing column sets when an analyst is trying to assemble a regulatory package on a polymer and needs to conform to the standards required by a particular nation or collection of nations.

A greater disparity is found in terms of the resolution requirements that are placed by various organizations on GPC column sets. A common resolution requirement is to measure the resolution R_s as defined in Eq. (1):

$$R_s = 2 \times \frac{(V_{r2} - V_{r1})}{(w_1 + w_2)} \times \log_{10}\left(\frac{M_1}{M_2}\right), \tag{1}$$

where V_{ri} are the elution volumes of a pair of peaks, w_i is the width of those respective peaks, and M_i is the molecular weight of the respective probe molecules. The parameter R_s is typically measured with narrow polymers (polydispersity less than 1.07) that have a molecular weight difference of a factor of

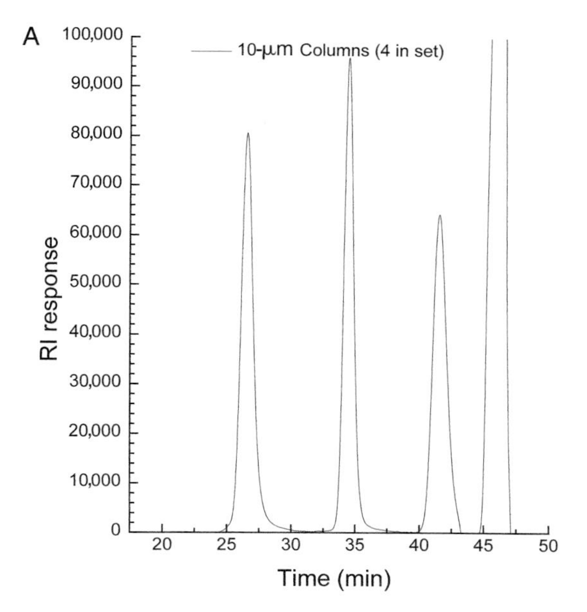

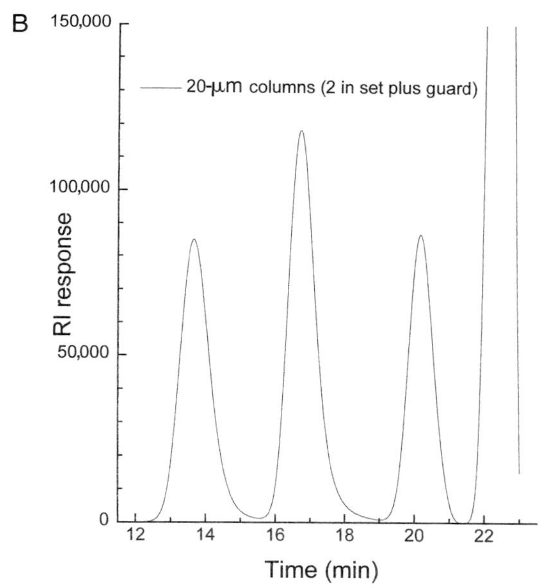

FIGURE 19.3 Comparison of pSty standards on Polymer Laboratories PLgel mixed A 20-μm and Waters HT 10-μm columns.

10. ASTM D5296-92 requires this R_s greater than 1.7 whereas the DIN 55672-1:1995-02 standard requires R_s to be greater than 2.5. The OECD was uncertain on this point in 1994 and provisionally selected an R_s of 2.5 but the OECD also recommended this value for further review by the OECD member nations. This resolution criterion in Eq. (1) is a useful one for judging column set quality

and the disagreement about the minimal values for R_s should not dissuade the reader from making use of this column quality parameter.

One final quality parameter that is sometimes specified in test methods is the peak capacity of the column set. In the literature of the OECD, this peak capacity is required to have a value of 6.0 or greater. This peak capacity is defined by Eq. (2):

$$\frac{(V_{eM_1} - V_{eM_2})}{\pi^* r^2} > 6.0 \text{ cm}^3/\text{cm}^2, \tag{2}$$

where V_{eMi} is the elution volume of a polymer of mass M_i, r is the internal radius of the column, and $M_1/M_2 = 10$. The OECD has specified that the test use polystyrene standards of narrowness sufficient to satisfy the previously referenced DIN standard. This column criterion is also a valuable one for defining the initial suitability of a column set for a particular analysis.

It should be noted that the values for the previously mentioned quality criteria (plate count, R_s, and peak capacity) have been defined by the various testing groups with higher resolution, i.e., smaller particle diameter, column sets in mind. There are instances where it is profitable, or even necessary, to use larger bead diameter columns. In these circumstances, the plate count and resolution of the column set may fall below the stated standard recommendations. This situation may arise when larger bead columns are required to elute microgel or insolubles from the GPC samples or to obtain larger pore sizes for analyzing higher molecular weight materials.

B. Guard Column Economics

One component that is often advertised for GPC systems is a guard column. Some vendors market guard columns with little justification for them other than that one might want to make their GPC system look analogous to a standard high-performance liquid chromatography (HPLC) system. It benefits the GPC analyst to consider closely the scientific and economic reasons for or against guard columns in their GPC system.

One vendor's sales literature that was reviewed while writing this chapter gave the sole justification that guard columns were used to ". . . filter solvent and sample particles . . ." from the GPC system. When looking at the price of guard columns, one sees that careful sample filtration plus the use of effective, yet inexpensive, in-line filters are a much more economic way of solving particulate problems than purchasing and using guard columns. Also, the in-line filter does not add to the run time for the GPC experiment whereas the guard column inevitably increases the GPC run time.

A more justifiable reason for the added cost and analysis time that a guard column brings is to avoid a problem with adsorption of sample matrix components on the stationary phase of the analytical GPC columns. In many industrial laboratories, the usual GPC experiment deals with fairly well-controlled sample matrices that may not have absorbable components and therefore rarely require a guard column. In situations where slow adsorption of matrix components may occur on GPC columns, it may make better economic sense to periodically replace the first column of a set. Nevertheless, in cases where adsorption of matrix components is severe, a guard column may be appropriate.

C. Flow Markers

The use of flow markers in GPC is a fairly simple, yet important, means of making first-order compensations for the molecular weight errors that result from small magnitude variations in the flow rate delivered by the GPC system pump. It must be remembered that the flow marker reflects only the *average* flow rate during the run. Thus, the flow marker is only capable of correcting for a consistent error in flow rate due, for example, to temperature-induced changes in the viscosity of the solvent. Use of a flow marker can not compensate for a flow rate that changes significantly over the course of a run.

In most cases these flow markers are species that are mixed with the sample and coinjected with the analyte onto the GPC column. The retention time of this marker is used to adjust the time axis to compensate for any moderate pump variability during the running of the standards and the samples.

Flow markers are often chosen to be chemically pure small molecules that can fully permeate the GPC packing and elute as a sharp peak at the total permeation volume (V_p) of the column. Examples of a few common flow markers reported in the literature for nonaqueous GPC include xylene, dioctyl phthalate, ethylbenzene, and sulfur. The flow marker must in no way perturb the chromatography of the analyte, either by coeluting with the analyte peak of interest or by influencing the retention of the analyte. In all cases it is essential that the flow marker experience no adsorption on the stationary phase of the column. The variability that occurs in a flow marker when it experiences differences in how it adsorbs to a column is more than sufficient to obscure the flow rate deviations that one is trying to monitor and correct for.

A more difficult criterion to meet with flow markers is that the polymer samples not contain interferents that coelute with or very near the flow marker and either affect its retention time or the ability of the analyst to reproducibly identify the retention time of the peak. Water is a ubiquitous problem in nonaqueous GPC and, when using a refractive index detector, it can cause a variable magnitude, negative area peak that may coelute with certain choices of totally permeated flow markers. This variable area negative peak may alter the apparent position of the flow marker when the flow rate has actually been invariant, thereby causing the user to falsely adjust data to compensate for the "flow error." Similar problems can occur with the elution of positive peaks that are not exactly identical in elution to the totally permeated flow marker. Species that often contribute to these problems are residual monomer, reactants, surfactants, by-products, or buffers from the synthesis of the polymer.

The use of totally permeated flow markers in aqueous GPC offers similar advantages along with many of the same shortcomings that one finds in non-aqueous GPC. One problem commonly found in aqueous GPC is that salt peaks due to the on-column ion exchange of counter ions of a polyelectrolyte with dissimilar ions in the GPC mobile phase will occur at or near the total permeation volume of the column. These salt peaks will often obscure the flow marker used in the analysis. Short of preconditioning the sample to exchange away the offending counter ions, there is little to do in this situation except to change the flow marker system in some way so as to avoid the coeluting salt peak.

Some GPC analysts use totally excluded, rather than totally permeated, flow markers to make flow rate corrections. Most of the previously mentioned requirements for totally permeated flow marker selection still are requirements for a totally excluded flow marker. Coelution effects can often be avoided in this approach. It must be pointed out that species eluting at the excluded volume of a column set are not immune to adsorption problems and may even have variability issues arising from viscosity effects of these necessarily higher molecular weight species from the column.

It is not common practice, but quite viable, to inject the flow marker at a predetermined volume offset from the polymer injection. This can be accomplished either with a second injection at a predetermined time into the run or by having two coordinated injection valves separated by a fixed volume of tubing. This approach can avoid many of the pitfalls described earlier. However, the mathematics of this correction is slightly different from that for a coinjected marker. The proper correction for the delayed volume injection is shown in Eq. (3):

$$t'' = \left[(t - d) \times \frac{(f^0 - d)}{(f - d)} \right] + d, \tag{3}$$

where t'' is the corrected analyte retention time, t is the observed analyte retention time, f^0 is the expected marker retention time, f is the observed marker retention time, and d is the injection delay time. Unfortunately, most data systems have only the coinjection correction programmed into them. That correction is shown in Eq. (4):

$$t'' = t \times \frac{f^0}{f}, \tag{4}$$

The difference between these two equations dictates that there will be an error if the coinjection equation is used for a delayed injection marker. Table 19.2 shows several examples and the magnitude of the errors generated.

In evaluating the effect of these errors, it must be remembered that these values are errors in retention time. The error in molecular weight will depend on the slope of the calibration curve, but will usually be considerably larger.

TABLE 19.2 Errors Generated by Improper Flow Marker Correction

System parameters						
Observed sample retention time	15	15.15	15.45	15.45	15.45	15.75
% flow rate shift	0%	1%	3%	3%	3%	5%
Time delay	0	1	1	2	3	3
Corrected sample retention times						
Delayed injection calculation	15.00	15.00	15.00	15.00	15.00	15.00
Coinjection calculation	15.00	15.00	15.01	15.03	15.04	15.07
% difference	0.00%	0.03%	0.09%	0.18%	0.27%	0.43%

IV. NONROUTINE USES OF GPC COLUMNS

A. Semiprepping

In industry and academia the need often arises to isolate portions of a polymer sample, whether it be to separate low molecular weight material from a sample or to actually fractionate the polymer across its molecular weight distribution. If gram quantities of isolated polymer are needed, true preparative chromatography equipment and techniques are usually necessary.

However, for quantities substantially less than this level, 7- to 10-mm i.d. analytical columns can often be used in a semipreparative mode. By repeatedly injecting 300 to 500 μl of up to 1% polymer, reasonable quantities of polymer can be isolated. An autosampler and automated fraction collector can be setup to perform such injections around the clock. Although the larger injections and higher concentrations will lead to a loss of resolution, in some situations the result is quite acceptable, with a considerable savings in time being realized over other means of trying to make the same fractionation.

Inhibited THF is problematic for semipreparative separations. Because small quantities of polymer are being collected along with larger volumes of solvent, more inhibitor, usually butylated hydroxytoluene (BHT), than sample is often collected in each fraction. Thus, one must carefully consider if the BHT will cause a problem in the subsequent analysis of the isolated fractions. If it does, uninhibited THF or other alternate solvents should be used. It must be remember that if uninhibited THF is used, the analyst must pay careful attention to the inevitable peroxide formation in the solvent/fractions.

Aqueous GPC can also be semiprepped in manner just like nonaqueous GPC. In this case one must consider carefully the buffers, salts, and biocides used in the eluant. If the fractions are destined for nuclear magnetic resonance experiments it will be imperative to either reduce the salt concentration in the eluant or remove salt after the initial fractionation. Likewise, if the collected samples are destined for infrared (IR) analysis, it is important to choose salts and buffers that have good IR transparency in the wavenumber ranges of interest.

B. Injection of Insolubles

The particle diameter of the GPC column stationary phase plays a role not just in determining the resolution of the column, but also in determining how well the column elutes insolubles or microgels that may be present. It is usually advisable to avoid the injection of insolubles or microgels that might block the frits or interstices of any GPC column, but in some instances the analysis of these materials by GPC is possible and even desirable.

The approximate radius of insoluble material that may pass through a GPC column without being screened out by the physical sieving processes on the column can be approximated by R_c, defined in Eq. (5):

$$R_c = d_p \times \frac{\varepsilon}{3 \times (1 - \varepsilon)}, \tag{5}$$

where ε is the column porosity and d_p is the diameter of the column packing material (8).

In preparing commercially important acrylic polymers by emulsion polymerization, a significant amount of lightly cross-linked polymer may be formed. Because this material may contribute to the performance of the polymer, its quantitation may be of interest. GPC can sometimes handle this need. Figure 19.4 shows two GPC chromatograms of the same emulsion polymer. When analyzed on a Polymer Laboratories PLgel 5-μm particle size column, one peak, at approximately 12 min, is observed. However, when the very same sample is analyzed on a PLgel 20-μm particle size column, two peaks are seen. The second peak, at approximately 16 min, appears, from its width, to be the soluble polymer, and thus the same as the single peak seen with the 5-μm column. The sharp peak at 7 min is a polymer that is totally excluded, either because of its molecular weight or because it is cross-linked and swollen. If this material were excluded because of its high molecular weight, we would

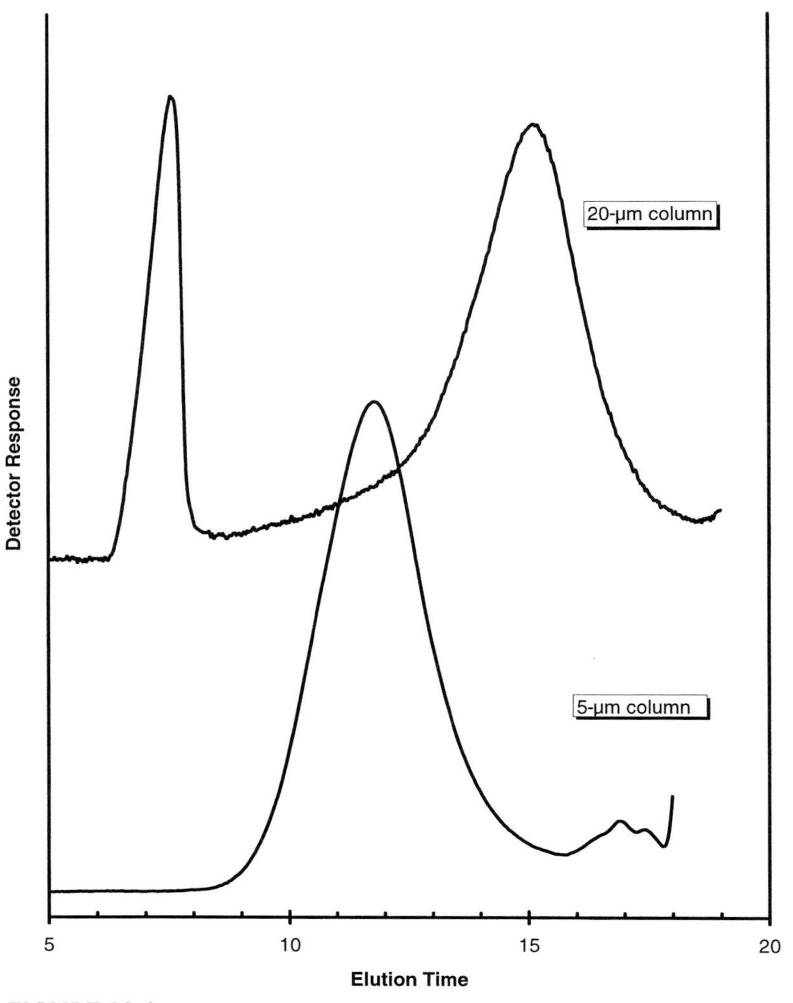

FIGURE 19.4 Elution of cross-linked polymer from 5- and 20-μm particle size columns.

also expect to see it excluded on the 5-μm column. The fact that we do not leads us to suspect that it is a cross-linked polymer that is "filtered" out by the smaller passages of the 5-μm column. Not only does this process prevent us from making the measurement on the 5-μm column, but in the long run it will destroy the column.

The elution of such gels is an example not of size exclusion but rather of hydrodynamic fractionation (HDF). However, it must be remembered that merely being able to physically fit an insoluble material through the column interstices is not the only criterion for whether the GPC/HDF analysis of an insoluble material will be successful. A well-designed HDF packing and eluant combination will often elute up to the estimated radius in Eq. (5) , but adsorption can drastically limit this upper analysis radius. For example, work in our laboratory using an 8-mm-bead-diameter Polymer Laboratories aqueous GPC column for HDF found that that column could not elute 204 n*M* pSty particles, even though Eq. (5) estimates a critical radius of ~1.5 μm.

These combined HDF and GPC separations require the use of detectors such as static light scattering or viscometers to help sort out the convoluted elution profiles seen in those type of experiments. It should also be remembered in these situations that the typical refractive index or ultraviolet detector responses may not be representative of the actual mass fraction of insolubles eluting from the column because of the significant light scattering that can occur with those large particles in the detector cell.

V. GPC OF ACRYLATES USING SOLVENTS OTHER THAN TETRAHYDROFURAN

While THF is the solvent of choice for ordinary acrylate ester polymers, there are numerous monomers that can be incorporated into those acrylic polymers that cause problems either with solubility or with adsorption onto the stationary phase. In some cases, these problems can be overcome by switching to a solvent other than THF.

For acrylate polymers with higher levels of carboxylic acids, THF can be modified by the addition of acids such as acetic, phosphoric, or trifluoroacetic. Levels as high as 10% acetic acid are considered acceptable by most manufacturers for their styrene/DVB columns. If such a modified mobile phase is used, it may need to be premixed rather than "generated" using a dynamic mixing HPLC pump because on-line mixing often leads to much noisier baselines, particularly when using a refractive index detector.

For amine-containing polymers, DMF is often a good choice of solvent. DMF can also be a good choice for polymers of higher carboxylic acid content. However, DMF does present some experimental difficulties. It must be run at an elevated temperature, typically 60°C, because of its viscosity. Also, because most polymers have a much lower refractive index response in DMF, the signal-to-noise ratio for a polymer in this solvent is diminished versus the same ratio for common acrylates in THF.

The importance of adding some ionic strength to the DMF, typically with LiBr, has been reported in the literature (9). This can be important for ionizable monomers, but can also be a factor even if only the end group is ionic. The addition of salt is useful in working with DMF, but with fatty side chain acrylates one quickly finds that adsorption problems start to occur in DMF/salt systems.

Several manufacturers supply columns packed in DMF. Careful attention must be paid to the manufacturer's recommendations for solvent changeover. If possible, one should purchase the columns packed in DMF and reserve them exclusively for use in DMF.

VI. POLYACRYLIC ACID AND RELATED POLYMERS

A. Polyelectrolyte Considerations

Because the monomer units in p(acrylic acid) can be ionized, the possibility of electrostatic interaction between groups within the same chain, as well as between the chain and the stationary phase, needs to be considered.

1. Intrachain Electrostatic Repulsion

Electrostatic repulsion between charged groups within the same polymer chain can and does affect the hydrodynamic size of the polymer, and hence its elution. The magnitude of the effect is influenced by both the pH and the ionic strength of the mobile phase. The pH affects the degree of ionization and the ionic strength influences the dielectric constant and hence the strength of a given interaction.

These effects dictate that analyses are most reproducible if done in a system buffered at a pH 1.5 to 2 units away from the pK_a of the polymer (approximately 4.3–5.5 for pAA homopolymer, depending on the ionic strength of the mobile phase). Such pH's ensure that slight changes in pH from run to run, week to week, do not greatly change the ionization state of the polymer. Such changes can lead to significant changes in hydrodynamic size and hence elution characteristics. We generally find that a pH on the high side of the pK_a provides better chromatography than one on the low side.

Because the dielectric constant of the mobile phase affects the degree of repulsion, the ionic strength of the mobile phase also needs to be considered. As the ionic strength, I, increases, the hydrodynamic size of the polymer will decrease because of reduced repulsion, leading to later elution. Thus, the mobile phase should have a reproducible ionic strength. Although changes in I can be calibrated away if the calibrants are the same composition as the analyte, this is not usually the case. Thus, it becomes important that ionic strength be reproducible. Generally, an I in the range from 0.01 to about 1.0 works well. Simple inorganic salts such as sodium nitrate can be used to raise I, although the buffer itself often provides enough ionic strength.

There is usually an ionic strength above which there is no more effect on hydrodynamic size or, worse yet, there are hydrophobic interactions of the polymer with the stationary phase. Thus, the optimum I is usually at the low I end of the plateau of size vs I. This concentration will minimize ionic strength effects while also minimizing wear on the pump seals and pistons.

These ionic strength effects are illustrated for p(acrylic acid) homopolymer in Fig. 19.5. With no ionic strength, the polymer is excluded. The addition of even 10 m*M* sodium nitrate has a marked effect, but once approximately 50 m*M* is reached, no further changes are seen.

2. Ion Exclusion

A second electrostatic effect that can occur is ionic exclusion. The hydrophilic methacrylate polymer used as the stationary phase for many of the columns available (e.g., TosoHaas PW series, Waters Ultrahydrostyragel, Polymer Laboratories aquagel-OH) contains some carboxylic acid groups. These may be ionized at the pH of the analysis, giving the surface of the gel a negative charge. If the polymer also has a negative charge at this pH, there will be electrostatic repulsion between the polymer and the surface of the gel. The net effect is to decrease the effective pore size of the packing, leading to an earlier elution, by several percent, of the polymer. One effect of ion exclusion is that the observed exclusion volume of the column set will often be dependent on the chemical composition of the polymer.

This ion exclusion effect can sometimes be exploited beneficially. For example, by purposefully choosing a column with some carboxyl groups and a pH that ionizes them (greater than approximately 6.5), it may allow separation of a charged and an uncharged polymer that have the same hydrodynamic size. Alternatively, one may be able to fine-tune elution of a polymer by adjusting pH.

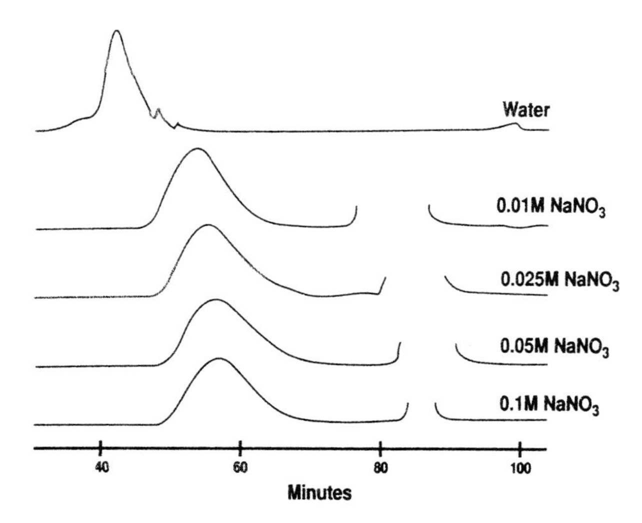

Column: TSKgel GMPW, two 17μm, 7.5mm x 60cm columns in series
Sample: 0.5ml of 0.05-0.1% of the sodium salt of polyacrylic acid, an anionic polymer
Elution: water, 0.01M, 0.025M, 0.05M or 0.1M $NaNO_3$ in water
Flow rate: 0.5ml/min
Detection: RI

FIGURE 19.5 Effect of ionic strength on elution of p(acrylic acid) from TosoHaas GMPW columns. (Courtesy of TosoHaas.)

B. Adsorption

Problems with adsorption onto the packing material are more common in aqueous GPC than in organic solvents. Adsorption onto the stationary phase can occur even for materials that are well soluble in water if there are specific interactions between the analyte and the surface. A common example of such an interaction is the analysis of pEG on a silica-based column. Because of residual silanols on the silica surface, hydrogen bonding can occur and pEG cannot be chromatographed reliably on silica-based columns. Likewise, difficulties are often encountered with polystyrenesulfonate on methacrylate-based columns.

Many acrylic acid-based polymers contain significant amounts of hydrophobic monomers. Examples include various types of associative thickeners, surfactants, scale inhibitors, and corrosion inhibitors. By design, these polymers associate with each other or with surfaces. Thus, they are prime candidates for adsorption onto the packing material, even though they appear to be completely soluble in water. In some cases, adsorption can be overcome by adding organic modifiers to the mobile phase. Most packing materials can tolerate up to 20% organic modifiers such as acetonitrile or methanol. Changeover, however, must be done slowly because of significant differences in the swelling of the packing material.

Adsorption can also occur because of electrostatic effects. A positively charged polymer can adsorb strongly on a negatively charged stationary phase. Thus, amine-containing polymers chromatographed on methacrylate columns often adsorb because of the carboxylic acid sites present on the packing. Fortunately, this situation can often be avoided by the proper choice of pH and/or ionic strength. Figure 19.6 shows the effect of ionic strength on the elution of

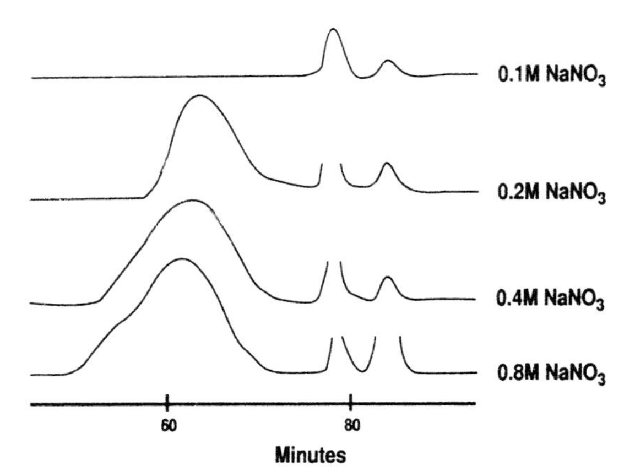

Column: TSKgel GMPW, two 17μm, 7.5mm x 60cm columns in series
Sample: 0.5ml of 0.05-0.1% of the cationic polymer DEAE-dextran
Elution: 0.1M, 0.2M, 0.4M or 0.8M $NaNO_3$ in water
Flow rate: 0.5ml/min
Detection: RI

FIGURE 19.6 Effect of ionic strength on elution of cationic polymer, DEAE-dextran, from TosoHaas GMPW columns. (Courtesy of TosoHaas.)

a cationic polymer, DEAE-dextran, on a methacrylate column. At low ionic strength, the polymer completely adsorbs onto the column. Only with the addition of considerable ionic strength can the electrostatic attraction between the analyte and the stationary phase be screened out so that the sample elutes.

C. Calibration Strategy

Electrostatic and adsorption effects conspire to make aqueous GPC more likely to be nonideal than organic solvent GPC. Thus, universal calibration is often not obeyed in aqueous systems. Hence, it is much more critical that the standard chosen for calibration share with the polymer being analyzed chemical characteristics that affect these interactions. Because standards that meet this criterion are often not available, it is prudent to include in each analysis set a sample of a secondary standard of the same composition and molecular weight as the sample. Thus, changes in the chromatography of the analyte relative to the standards will be detected.

Alternatively, either the Hamielec or the integral broad standard calibration procedure can be used. For this approach, the broad standard would be initially analyzed using a calibration based on some convenient standard that may be quite dissimilar in chemistry from the standard. From that point forward, the system is calibrated with the secondary standard. Thus, changes that affect the sample will be compensated for in the calibration procedure.

VII. CLOSING REMARKS

GPC has many uses and is a powerful analysis technique for acrylate polymers. With care in selecting solvents and stationary phases, one finds that many polymers can be analyzed successfully. Opportunities always exist to use analytical GPC columns in nonstandard ways (semiprep, HDC, pseudo-HPLC combined with GPC) to the benefit of the analyst, but the analyst must always be keenly aware of which mode of operation is dominating when practicing such nonroutine analyses.

REFERENCES

1. Barth, H. G. (1987). *In* "ACS Symposium Series 352, Detection and Data Analysis in Size Exclusion Chromatography" (T. Provder, ed.), pp. 29–46. American Chemical Society, Washington, DC.
2. Du, Y., Xue, Y., and Frisch, H. L. (1996). *In* "Physical Properties of Polymers Handbook" (J. E. Mark, ed.), pp. 227–241. AIP Press, Woodbury, N.Y.
3. Grulke, E. A. (1989). *In* "Polymer Handbook" (J. Brandrup and E. H. Immergut, eds.), pp. VII519–VII561. Wiley-Interscience, New York.
4. Morawetz, H. (1975). *In* "Macromolecules in Solution," 2nd Ed., p. 128. Wiley-Interscience, New York.
5. ASTM D 5296-92, "Standard Test Method for Molecular Weight Averages and Molecular Weight Distribution of Polystyrene by High Performance Size-Exclusion Chromatography." ASTM, Philadelphia, PA, 1992.

6. OECD Guideline for the Testing of Chemicals, "Determination of the Number-Average Molecular Weight and the Molecular Weight Distribution of Polymers using Gel Permeation Chromatography," Draft Proposal, May 1994, Paris, France.
7. DIN 55672-1: 1995-02, Gelpermeationschromatographie (GPC) Teil 1: Tetrahydrofuran (THF) als Elutionsmittel, February 1995.
8. Venema, E., Krakk, J. C., Poppe, H., and Tijssen, R. (1996). *J. Chromatogr. A* **740,** 159–167.
9. Dias, M., Mano, E., and Azuma, C. (1997). *Eur. Polym J.* **33**(4), 559–564.

20 APPLICATIONS AND USES OF COLUMNS FOR AQUEOUS SIZE EXCLUSION CHROMATOGRAPHY OF WATER-SOLUBLE POLYMERS

DENNIS J. NAGY

Analytical Technology Center, Air Products and Chemicals, Inc., Allentown, Pennsylvania 18195

I. INTRODUCTION

A. Size Exclusion Chromatography (SEC) of Water-Soluble Polymers

Water-soluble polymers comprise a major class of polymeric materials and are used in a wide variety of applications. Synthetic water-soluble polymers include poly(vinyl alcohol), poly(acrylamide), poly(acrylic acid), poly(ethylene oxide), poly(vinyl pyrrolidone), cellulosics, and many copolymers of these types. Their end uses are quite varied and their applications depend mainly on their viscosifying, rheological, and surface-active properties (1). For example, poly(vinyl alcohol) is used in adhesives, fibers, textile and paper sizing, packaging, as a stabilizer for emulsion polymerization, and as a precursor for the manufacture of poly(vinyl butyral), which is used in automotive windshields. Poly(vinyl alcohol) is also the world's largest volume, commodity, water-soluble polymer.

The chemistry of water-soluble polymers can take various forms. These polymers can be anionic, cationic, or nonionic. Their polymer backbone can contain hydrophilic and hydrophobic pendant groups. Branching and polymer stereoregularity also play a role in the physical behavior of these materials.

The end-use applications of water-soluble polymers require accurate means to characterize the molecular weight distribution (MWD) and to provide a better understanding of product performance. The molecular weight affects many physical properties such as solution viscosity, tensile strength, block resistance, water and solvent resistance, adhesive strength, and dispersing power. Commercially available polymers such as poly(vinyl alcohol),

Column Handbook for Size Exclusion Chromatography

poly(acrylamide), and poly(vinyl pyrrolidone) come in a wide range of molecular weights and polydispersities. Thus, it is critical that sensitive and reliable methods be employed for molecular weight and molecular weight distribution analysis. Size exclusion chromatography (SEC) is commonly used for these purposes.

The characterization of water-soluble polymers by SEC has closely followed the advances in column and detection technology since the 1960s. Aqueous SEC of water-soluble polymers for the determination of molecular weight and molecular weight distribution is, however, often more challenging than the analysis of polymers under organic-based conditions. Several mechanisms that compete with the size exclusion process can easily complicate the characterization process. These include such phenomena as ion exchange, ion exclusion, ion inclusion, adsorption, and viscous "fingering." Ideally, one wants only size exclusion as the operable mechanism when characterizing water-soluble polymers for molecular weight. The composition of the mobile phase must be carefully chosen to prevent enthalpic interactions between polymer and packing. For polyelectrolytes, polymer-packing interactions can be caused by electrostatic as well as hydrophobic forces. For nonionic polymers, hydrophobic forces as well as hydrogen bonding can lead to adsorption. Thus, mobile-phase composition and column chemistry play a vital role in the utilization of an effective SEC process for polymer separation. These effects have been reviewed in detail by Barth (1).

In addition to competing, nonsize exclusion effects, the detection system used in aqueous SEC can also present additional challenges. On-line, differential viscometry detection requires the use of polymer standards and the obeyance of universal calibration for the determination of molecular weights. Low-angle laser light scattering (LALLS) and multiangle laser light scattering (MALLS) require a particulate-free mobile phase to eliminate excessive background scatter. Prior knowledge of the specific refractive index increment values of the polymer under the conditions of analysis are also required for these type of light-scattering measurements.

B. Scope

This chapter describes the use of three commercially available SEC column types for the characterization of nonionic, anionic, and cationic, synthetic water-soluble polymers. These include TSK-PW, Synchropak, and CATSEC columns. Specific examples and experimental procedures are discussed for each type of column. However, the major emphasis is on the use of TSK-PW columns due to their broad applicability for a variety of water-soluble polymers.

The general theme of this chapter is to illustrate the use of the previously discussed columns for specific applications related to molecular weight characterization. Other chapters in this book describe in more detail the physical properties of the packing materials, calibration procedures, column packing techniques, flow rate effects, and resolution issues concerning TSK-PW, Synchropak, and CATSEC columns. Discussions and examples in this chapter emphasize the utilization of these columns for measuring absolute molecular weight (M_w = weight-average molecular weight, M_n = number-average molecular weight), radius of gyration, conformational coefficients from light scattering, and an examination of universal calibration and the measurement of Mark-Houwink constants from differential viscometry. These properties are very dependent on the proper choice of column packing, quality of the packing material, and mobile-phase chemistry. TSK-PW, Synchropak, and CATSEC columns are well suited to these tasks. The applications described in this chapter are those from the successful experiences of this author utilizing these SEC columns.

Specific polymers discussed in this chapter and the type of column used for their characterization are summarized in Table 20.1. The polymers are categorized as nonionic, anionic, or cationic. The nomenclature (acronyms) used for the different polymer types are also listed in Table 20.1.

II. AQUEOUS SEC USING TSK-PW COLUMNS

A. Description and Experimental Procedures

TSK-PW columns belong to the family of polymeric-based, high-performance packing materials. They first became available commercially in 1978 and are produced by Tosoh Corporation, Japan, and are sold through a number of distributors in North America. TSK-PW columns contain a hydrophilic, polyether gel packing of $(\text{-}CH_2CHOHCH_2O\text{-})_n$ with residual carboxylate functionality on the packing surface (2,3). They are available in a wide variety of pore sizes (40 to 1000 Å). The molecular weight exclusion limits of each pore size are summarized in Table 20.2 (3). Typical plate counts range from 7,000 to 10,000 plates per column (30 cm length). TSK-PW columns provide extraordinary performance for characterizing water-soluble polymers. They exhibit excellent resolution, durability, stability over a pH range from 2 to 12, and are useful for a wide variety of polymer types. Typical solvents compatible with these columns include water, aqueous salt solutions, buffer solutions, and

TABLE 20.1 Polymer Nomenclature and Column Types

Type	Polymer nomenclature	TSK-PW	Synchropak	CATSEC
Nonionic	Poly(vinyl alcohol) (PVA)	X	X	
	Poly(ethylene glycol) (PEG)	X	X	
	Poly(ethylene oxide) (PEO)	X	X	
	Poly(saccharide) (PSC)	X	X	X
	Poly(acrylamide) (PAAM)	X	X	
	Poly(ethenylformamide) (PEF)	X	X	
Anionic	Poly(styrene) sulfonated (PSS)		X	
	Poly(acrylic acid) (PAA)		X	
Cationic	Poly(2-vinyl pyridine) (PVP)			X
	Poly(vinylamine) (PVAm)			X
	Poly(vinyl alcohol-vinylamine) (PVA-VAm-HCl)			X

TABLE 20.2 Molecular Weight Exclusion Limits for TSK-PW Columns[a]

TSK-PW column	Pore size (Å)	Molecular weight exclusion limit
G1000	40	1,000
G2000	50	5,000
G3000	200	50,000
G4000	400	300,000
G5000	1000	800,000
G6000	—	8,000,000

[a]Based on PEG/PEO.

aqueous solutions with organic modifiers such as methanol or acetonitrile. The columns can be operated up to 50°C. With care, TSK-PW columns can easily be used for 1 to 2 years before significant deterioration of the packing occurs.

A summary of typical experimental conditions used with TSK-PW columns for nonionic polymers is described in Table 20.3. A common mobile phase is an aqueous solution of 0.05 *N* sodium nitrate. A salt solution of sodium nitrate is a good choice because it is not as corrosive as a solution of sodium chloride. For the descriptions and examples that follow, a bank of either five or six TSK-PW columns in series (G1000–G5000 or G1000–G6000) was used for the aqueous SEC work. These configurations allow for molecular mass characterization from less than 1,000 Da to 1,000,000 Da or greater.

Synthetic, nonionic polymers generally elute with little or no adsorption on TSK-PW columns. Characterization of these polymers has been demonstrated successfully using four types of on-line detectors. These include differential refractive index (DRI), differential viscometry (DV), LALLS, and MALLS detection (4–8). Absolute molecular weight, root mean square (RMS) radius of gyration, conformational coefficients, and intrinsic viscosity distributions have

TABLE 20.3 Summary of Experimental Conditions Used with TSK-PW Columns

Columns:	TSK-PW: G1000, G2000, G3000, G4000, G5000, G6000, each 30 cm × 7.5 mm/i.d. (Tosoh Corporation)
Mobile phase:	0.05–0.10 *N* sodium nitrate; acetonitrile/0.05 sodium nitrate (20/80)
Flow rate:	1.0 ml/min (nominal)
Detectors:	Model 410 differential refractive index (Waters Corporation) Model 100 differential viscometer (Viscotek Corporation) Dawn-F multiangle laser light-scattering photometer (Wyatt Technology)
Standards:	PEG and PEO (American Polymer Standards Corporation) PSC (Polymer Laboratories)
Temperature:	35°C
Injection volume:	0.200 ml (samples and standards)

all been measured employing these type of detectors with these columns. The superb characteristics (described earlier) of TSK-PW columns enable the investigation and measurement of these polymer properties.

B. SEC of Water-Soluble Standards

PEG and PEO monodisperse standards are often used to calibrate TSK-PW columns. SEC chromatograms of a PEG (M_w 4950 Da) and a PEO (M_w 160,000 Da) standard are illustrated in Fig. 20.1 using a Waters Model 410 DRI (Waters Corporation, Milford, MA). Generally, a polymer concentration

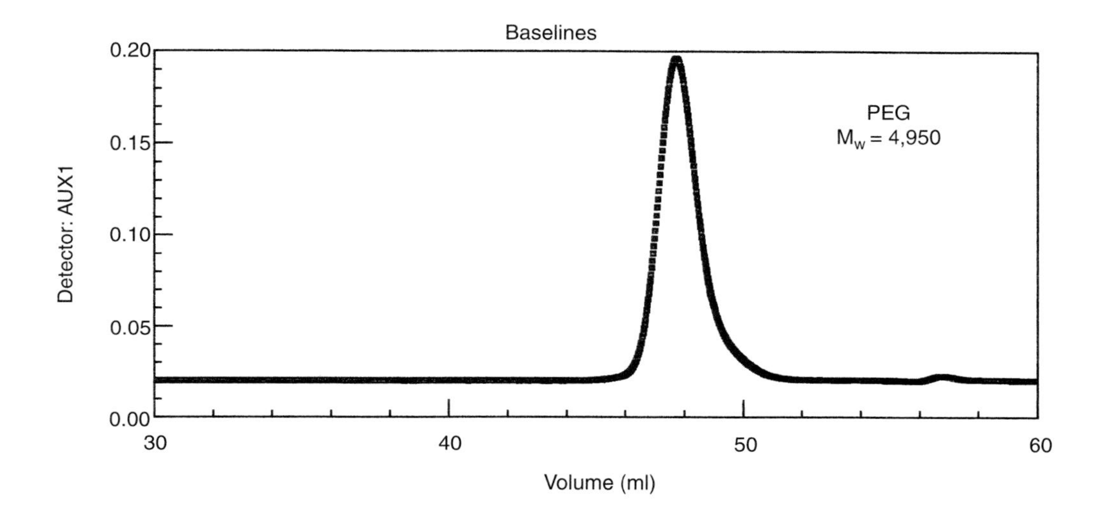

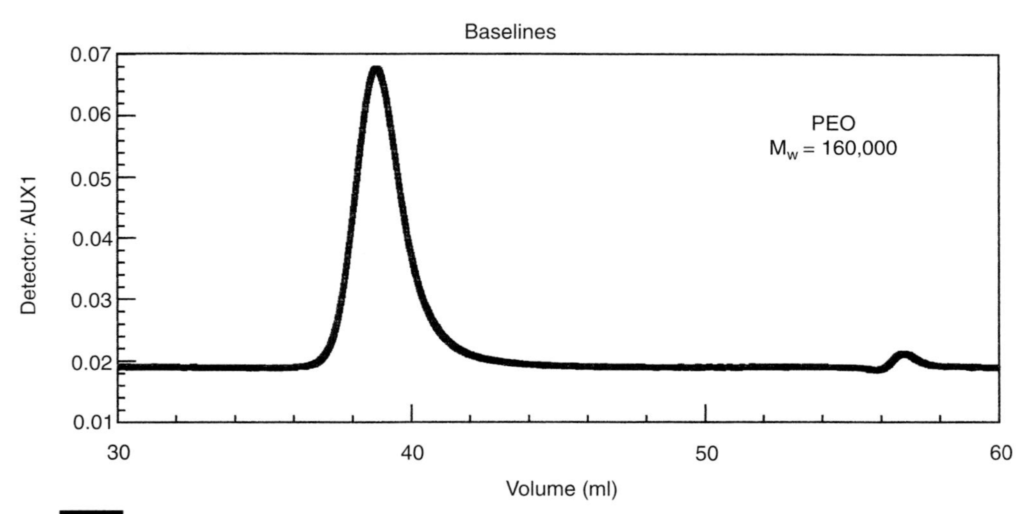

FIGURE 20.1 DRI chromatograms of PEG and PEO standards using TSK-PW columns.

of 0.2 weight % works well under these conditions. PEG standards range from 975 to 17,500 Da and PEO standards range from 43,500 to 800,000 Da. PEG and PEO are chemically identical with -(CH_2CH_2O)- repeat units. Usually, molecular weights less than 20,000 are referred to as PEG and those over 20,000 as PEO.

PSC standards are also used for the calibration of TSK-PW columns. DRI chromatograms of two different molecular weight PSC standards (M_w 23,700 Da and M_w 186,000 Da) are shown in Fig. 20.2. PSC standards are available in the molecular mass range from approximately 5000 to 850,000 Da.

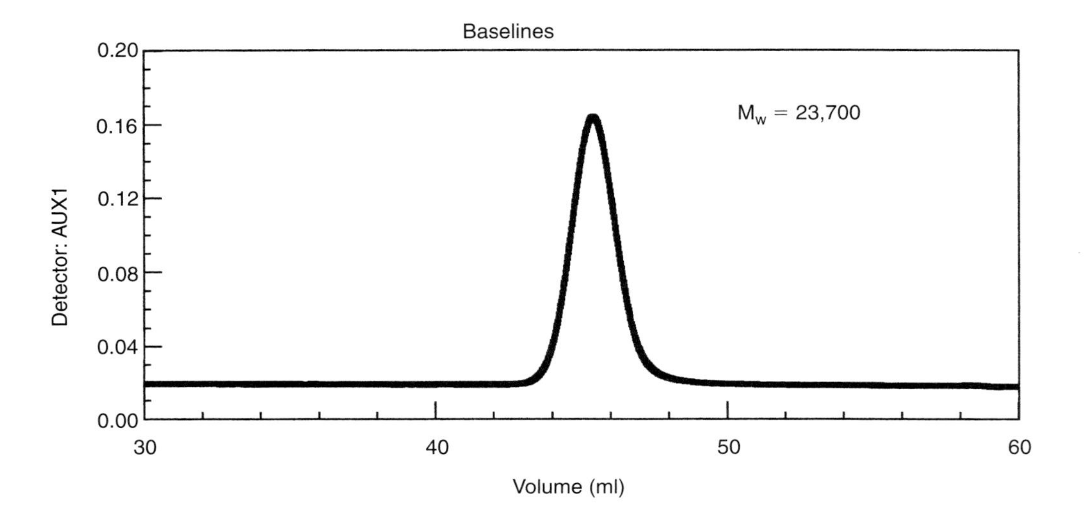

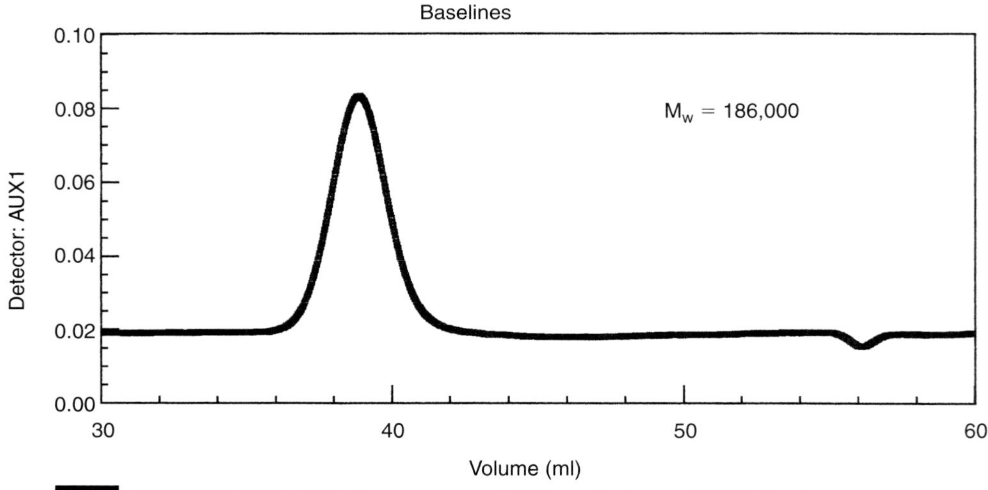

FIGURE 20.2 DRI chromatograms of PSC standards using TSK-PW columns.

C. Universal Calibration and Mark-Houwink Constants

In SEC, universal calibration is often utilized to characterize a molecular weight distribution. For a universal calibration curve, one must determine the product of log(intrinsic viscosity · molecular weight), or log($[\eta] \cdot$ M). The universal calibration method originally described by Benoit *et al.* (9) employs the hydrodynamic radius or volume, the product of $[\eta] \cdot$ M as the separation parameter. The calibration curves for a variety of polymers will converge toward a single curve when plotted as log($[\eta] \cdot$ M) versus elution volume (V_r), rather than plotted the conventional way as log(M) versus V_r (5). Universal calibration behavior is highly dependent on the absence of any secondary separation effects. Most failures of universal calibration are normally due to the absence of a pure size exclusion mechanism.

To use universal calibration, intrinsic viscosity must be measured. An online, DV detector can measure specific viscosity, η_{sp}, which is related to intrinsic viscosity by the expression

$$[\eta] = \lim_{c \to 0} (\eta_{sp}/c), \quad (1)$$

where c is the concentration at a particular increment within an SEC chromatogram. Thus, a DV detector (in tandem with a DRI concentration detector) can be used to measure intrinsic viscosity of a series of well-characterized polymer standards to generate a universal calibration curve.

The relationship between molecular weight, M, and intrinsic viscosity, $[\eta]$, of a polymer is given by the widely used Mark-Houwink expression

$$[\eta] = K(M)^a, \quad (2)$$

where K and a are the empirical Mark-Houwink constants for a particular polymer/solvent system. The importance of this equation is its relationship between molecular weight and $[\eta]$, the volume per unit mass (intrinsic viscosity is usually reported in units of dl/g). Mark-Houwink constants are usually determined from a log–log plot of intrinsic viscosity versus molecular weight.

A universal calibration curve using PEG, PEO, and PSC standards in a mobile phase of 0.10 N $NaNO_3$ at 35°C with TSK-PW columns is shown in Fig. 20.3. This curve was obtained using a Viscotek Model 100 DV detector (Viscotek Corporation, Houston, TX). The plot of log($[\eta] \cdot$ M) versus V_r shows that these polymers fall on a single calibration curve. The importance of these results is that universal calibration is obeyed under aqueous conditions using these columns and other appropriate experimental conditions. The absence of nonexclusion effects occurs during the chromatography of these hydrophilic, water-soluble polymers. Even high molecular weight PEO and PSC exhibit this response in lieu of any adsorption effects. Universal calibration has also been demonstrated for these same polymer standards on TSK-PW columns using a mobile phase of 20% acetonitrile in aqueous 0.10 N $NaNO_3$ (5).

Mark-Houwink constants for PEG, PEO, and PSC are summarized in Table 20.4. These were measured in either an aqueous mobile phase of 0.10 N $NaNO_3$ or a mobile phase of 20% acetonitrile in aqueous 0.10 N $NaNO_3$ (which also exhibits universal calibration behavior). The values for a fall within

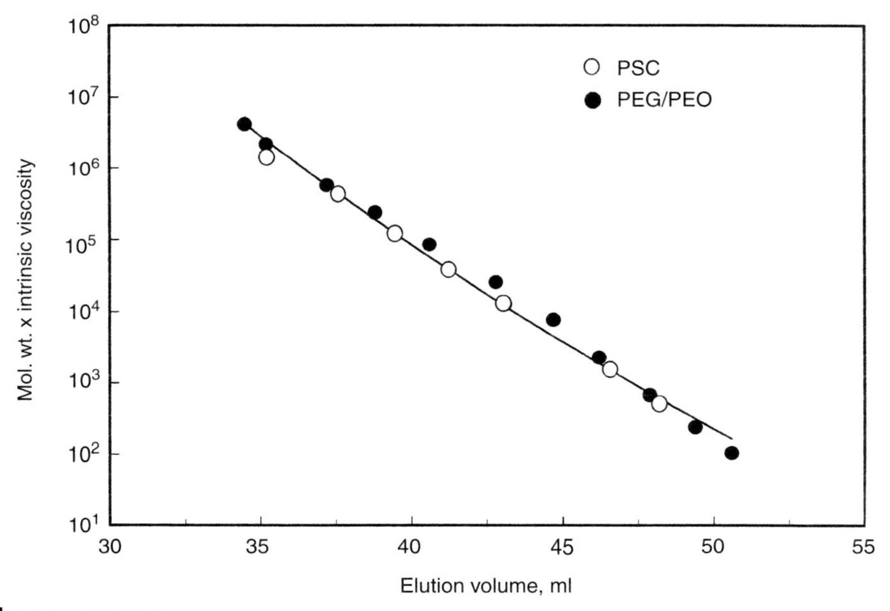

FIGURE 20.3 Universal calibration curve for TSK-PW columns with an aqueous mobile phase of 0.10 *N* sodium nitrate.

the 0.5 to 0.8 range expected for random-coiled polymers. It is important to note that the optimized SEC separation mechanism within the TSK-PW columns allows the measurement of the small difference in the Mark-Houwink *a* value between PEO and PEG (molecular mass less than 20,000 Da). Similar behavior has been observed for PSS standards under aqueous conditions and poly(styrene) standards in tetrahydrofuran (5).

In addition to polymer standards, a number of broad distribution, water-soluble polymers can be characterized on TSK-PW columns using universal calibration. These include both fully and partially hydrolyzed PVA, PAAM, PEF, and dextran. PVA, the world's largest-volume, synthetic, water-soluble polymer, was first successfully separated on TSK-PW columns by Hashimota *et al.* (10). In the 1980s, the use of low-angle, laser light-scattering detection

TABLE 20.4 Mark-Houwink Constants for PEG, PEO, and PSC from SEC/Viscometry Using TSK-PW Columns

	0.10 *N* $NaNO_3$		20/80 acetonitrile/ 0.10 *N* $NaNO_3$	
	a	*K*	*a*	*K*
PEG	0.58	1.01×10^{-3}	0.60	8.73×10^{-3}
PEO	0.70	3.47×10^{-4}	0.73	2.45×10^{-4}
PSC	0.66	2.14×10^{-4}	0.59	3.72×10^{-4}

was reported for PVA using TSK-PW columns (11,12). Soon after, on-line viscometry detection was reported for PVA (13). An SEC–viscometry chromatogram of a medium molecular weight, fully hydrolyzed PVA (98% hydrolysis type, Air Products and Chemicals, Inc.) in a mobile phase of 0.10 *N* $NaNO_3$ is shown in Fig. 20.4. This illustrates the excellent signal response associated with viscometry detection when used with these columns.

On-line viscometry measurements allow for calculation of the Mark-Houwink constants under the conditions used for the analysis. The fact that fully hydrolyzed PVA exhibits pure size exclusion on TSK-PW columns allows the K and a values to be determined over a broad range of molecular weight. The Mark-Houwink constants for different molecular weight grades of fully hydrolyzed PVA are summarized in Table 20.5. For comparative purposes, other K and a values for PVA reported in the literature are also included (14). The average values based on all the determinations are: $K = 1.332 \times 10^{-3}$ and $a = 0.57$. The a value for fully hydrolyzed PVA is in the expected range of 0.5–0.8 for random-coil polymers, but is slightly lower than other published values (15–17). This may be due to the fact that previously reported values were taken from PVA determinations in water. The addition of sodium nitrate, although a weak salting-out species for PVA, contributes to a somewhat greater degree of chain compression and coil contraction of PVA in a salt solution than pure water (14,18).

The significance of knowing the K and a values of fully hydrolyzed PVA is that molecular weight distribution data can be directly calculated using two methodologies. The first is the Mark-Houwink method, which requires prior knowledge of K and a values for fully hydrolyzed PVA and calibration standards such as PEG, PEO, or PSC. The second method is the intrinsic viscosity method. This method utilizes a simple ratio of the concentration signal to the specific

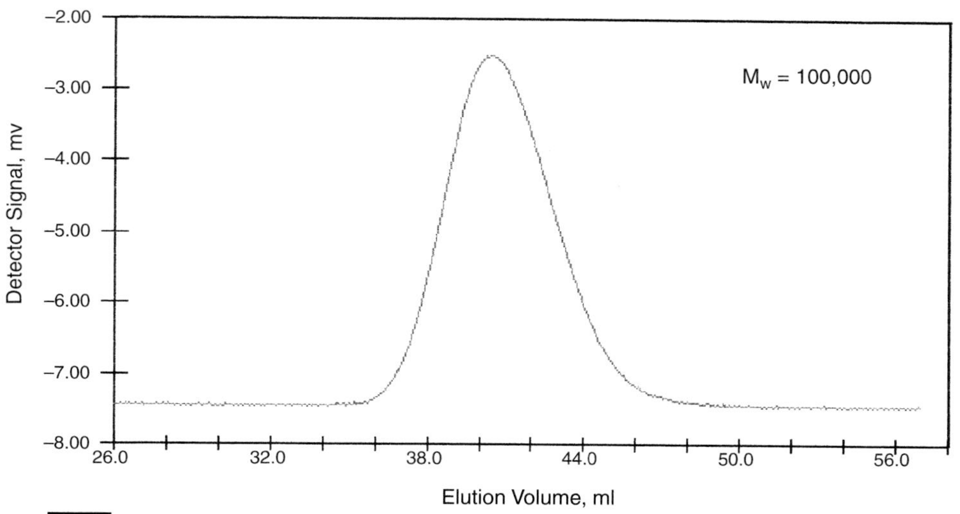

FIGURE 20.4 Differential viscometry chromatogram of fully hydrolyzed PVA using TSK-PW columns.

TABLE 20.5 Comparison of Mark-Houwink Constants for PVA, Using TSK-PW Columns

a	*K*	Reference
0.570 (M_w 249,000)	1.332×10^{-3}	14
0.567 (M_w 143,000)	1.356×10^{-3}	14
0.560 (M_w 98,400)	1.376×10^{-3}	14
0.569 (M_w 31,600)	1.281×10^{-3}	14
0.61	0.690×10^{-3}	15
0.62	0.887×10^{-3}	16
0.64	0.750×10^{-3}	17

viscosity signal. By inputting the Mark-Houwink values for PVA, molecular weights can then be calculated using the Mark-Houwink relationship (6). This method is not limited to PVA and can be applied to other water-soluble polymers.

D. SEC/Multiangle Laser Light Scattering

The use of multiangle laser light-scattering detection for water-soluble polymers has added a new dimension to characterization because it provides absolute molecular weight, size, and conformational information. MALLS is considered a primary technique because it can determine molecular weight and size in solution, independent of elution volume and without the need for column calibration. It is the unique ability of MALLS to measure the angular dependence of scattered light that distinguishes the technique from a single, angular measurement such as LALLS, which provides only molecular weight and no size or conformational information. LALLS measurements also require extreme care and patience due to the inherent difficulty with measurements at low angles, especially in aqueous media. The use of an aqueous-based mobile phase for measuring light scattering can prove quite challenging as water is a notorious scatterer of contaminants and dust, especially at low angles. In this regard, light-scattering measurements in organic solvents such as toluene or tetrahydrofuran are considerably easier. The use of good column support is critical for optimized LALLS and MALLS measurements in aqueous media. Columns that shed particulates due to deterioration of the packing material make light-scattering measurements almost impossible. TSK-PW columns have proven to be an excellent type of support to use due to their inherent stability and long life. A clean, prefiltered mobile phase is just as important. With these two criteria met, one can obtain conditions as good as those from organic-based SEC-MALLS (7).

A convenient way to examine the light-scattering response from each angle of detection (detector signal versus elution volume) is shown in Fig. 20.5 for a fully hydrolyzed PVA. The chromatograms from all 15 angles of detection from a Dawn-F (Wyatt Technology Corporation, Santa Barbara, CA) are

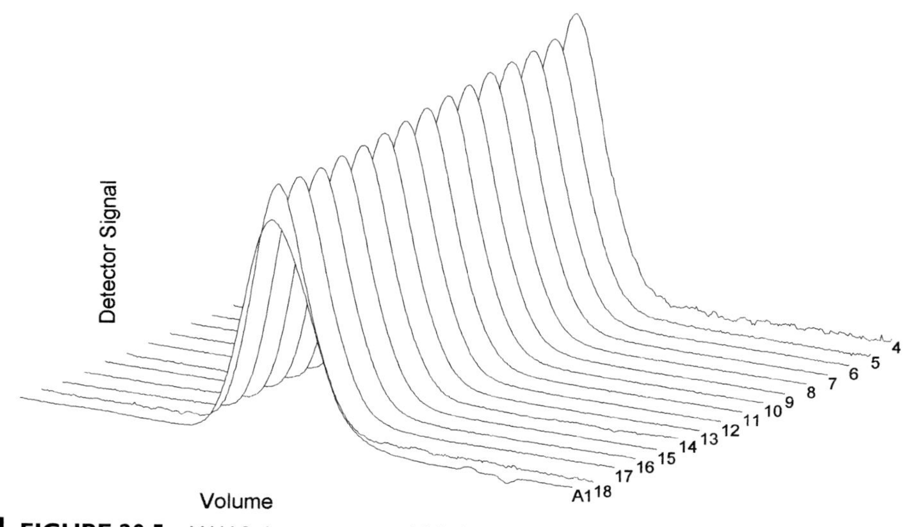

FIGURE 20.5 MALLS chromatograms of fully hydrolyzed PVA using TSK-PW columns. A1 = DRI, 4 = 35.5°, 11 = 90°, 18 = 144.5°

overlaid as a function of detector angle. It is important to note how the chromatograms exhibit a gradual increase in intensity with decreasing detector angle (decreasing detector number). It is this angular variation that enables determination of the RMS radius. Also note that detector noise due to excess scatter is present at the lowest angle, but not to a significant degree (7).

The presence of secondary, nonexclusion effects must be avoided when determining molecular weight using aqueous SEC-MALLS. As an example, fully hydrolyzed PVA generally exhibits excellent separation in a mobile phase of 0.05 *N* $NaNO_3$ on TSK-PW columns. However, partially hydrolyzed PVA (88% hydrolysis type, Air Products and Chemicals, Inc.) presents more of a problem due to its increased hydrophobicity. It is known that secondary, nonsize exclusion effects result in a slight retardation of partially hydrolyzed PVA on TSK-PW columns when fractionated using the same conditions as for fully hydrolyzed PVA (6). Figure 20.6 exhibits DRI chromatograms for a partially hydrolyzed PVA that was fractionated using two different mobile phases: 0.05 *N* $NaNO_3$ and 20/80 (v/v) acetonitrile/0.05 *N* $NaNO_3$. Less broadening of the chromatogram (especially at low molecular weight and higher elution volumes) is observed with the 20/80 acetonitrile/0.05 *N* $NaNO_3$ mobile phase. This organically modified mobile phase does a better job eliminating the hydrophobic interactions between the polymer and column packing. This is an important issue to consider because the presence of secondary effects can adversely affect molecular weight data obtained from MALLS, resulting in M_w/M_n ratios biased low (7).

Molecular weights for fully hydrolyzed and partially hydrolyzed grades of PVA (expressed in terms of 4% solution viscosity) from aqueous SEC-MALLS are summarized in Table 20.6. M_w values range from less than 20,000 Da for a 3 centipoise (cP) viscosity grade to over 230,000 Da for a super-high viscosity

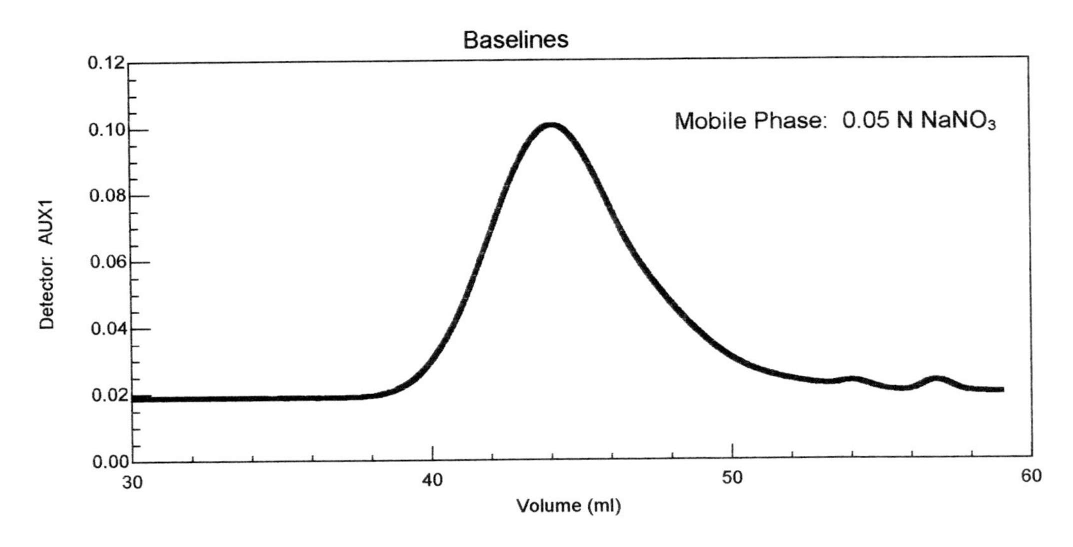

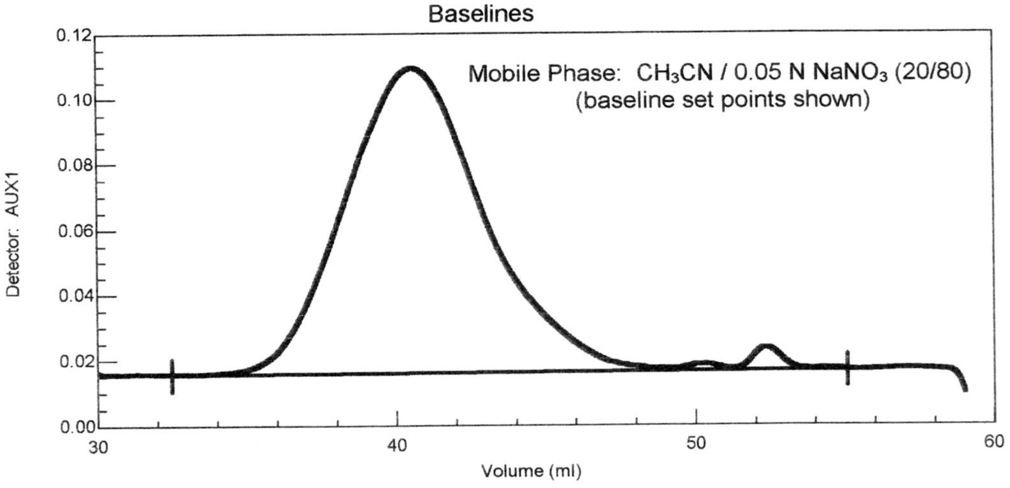

FIGURE 20.6 DRI chromatograms of partially hydrolyzed PVA (low molecular weight type) using TSK-PW columns.

grade (234 cP). Polydispersity values are generally around 2.0. RMS radius values for fully and partially hydrolyzed PVA grades are also summarized in Table 20.6. Data show the expected trend in which the RMS radius decreases with decreasing molecular weight.

The measurement of molecular weight and RMS radius provides the means to examine the conformational characteristics of a polymer using the relationship

$$\mathrm{RMS} = \beta\, \mathrm{M}^{\alpha}, \tag{3}$$

where α is the conformational coefficient and β is a constant. It is known that $\alpha = 1.0$ for a rigid rod, 0.5 for random coils in a Θ solvent, and 0.33 for a

TABLE 20.6 Molecular Weight of PVA from Aqueous SEC-MALLS Using TSK-PW Columns

4% solution viscosity (cp)[a]	M_w	M_n	M_w/M_n	RMS radius (nm)
Fully hydrolyzed				
3	19,100	11,600	1.6	6.8
7	32,000	16,400	1.9	7.7
21	74,400	43,000	1.7	14.7
25	95,100	55,300	1.7	16.1
50	138,000	58,100	2.4	19.4
234	233,000	110,000	2.1	31.9
Partially hydrolyzed				
3	24,100	13,900	1.7	—
5	38,600	21,300	1.8	11.7
23	109,000	56,700	1.9	17.1
40	143,000	54,000	2.6	21.6

[a]In water, 20°C.

sphere. Flory (19) has shown that the value for a of a polymer in a good solvent is 0.55–0.60. The conformational coefficient is determined from the slope of a log–log plot of RMS radius versus molecular weight. A summary of polymer conformation data for PVA and several other common, water-soluble polymers is shown in Table 20.7. Partially hydrolyzed PVA, PEO, PEF, and PSC exhibit values typical for random coil polymers. The a value of 0.38 for the commercial dextran is consistent with the fact that this material is highly branched. Similar values have also been reported using aqueous SEC-MALLS (20).

III. AQUEOUS SEC USING SYNCHROPAK COLUMNS

A. Description and Experimental Procedures

The use of nonpolymeric, column supports has also been a popular way to characterize water-soluble polymers. Synchropak SEC columns from Micra

TABLE 20.7 Conformational Coefficients for Water-Soluble Polymers in 0.05 *N* Sodium Nitrate with TSK-PW Columns

Polymer	α
Fully hydrolyzed PVA	0.50
Partially hydrolyzed PVA	0.48
Poly(ethylene oxide)	0.49
Poly(saccharide)	0.56
Poly(ethenylformamide)	0.35–0.54
Dextran	0.38

Scientific (Northbrook, IL) contain a silica support with a γ-glycidoxypropylsilane-bonded phase to minimize interaction with anionic and neutral polymers. The columns come in five different pore sizes ranging from 100 to 4000 Å. The packing material has a diameter from 5 to 10 μm and yields in excess of 10,000 plate counts. With a rigid silica packing material, the columns can withstand high pressure (maximum of 3000 psi) and can be used under a variety of salt and/or buffered conditions. A mobile phase above pH 8, however, will dissolve the silica support of the column (21). A summary of the experimental conditions used for Synchropak columns is described in Table 20.8.

B. Anionic Polymers

Anionic, water-soluble polymers can be characterized easily using Synchropak columns in an aqueous mobile phase of 0.10 *N* $NaNO_3$. This is due to the fact that the bonded γ-glycidoxypropylsilane layer on the packing surface eliminates most, if not all, secondary interactions with the polymer. Thus, anionic polymers are separated according to size, as is the case for nonionic polymers using TSK-PW columns. DRI chromatograms of a sulfonated poly(styrene) standard and poly(acrylic acid) standard using a bank of Synchropak columns of 100-, 300-, 1000-, and 4000-Å pore sizes are shown in Fig. 20.7. These two anionic, water-soluble polymers are much more difficult to characterize using TSK-PW columns, even under high pH conditions. It is the author's experience that Synchropak columns provide a much better means for the separation of anionic polymers (22).

C. Nonionic Polymers

Nonionic polymers such as PAAM and fully hydrolyzed PVA can also be characterized on Synchropak columns. Chromatograms of a PAAM and a fully hydrolyzed PVA using a three-angle MALLS detector (Mini-Dawn, Wyatt Technology) are shown in Figs. 20.8 and 20.9. Detector signal versus elution volume are shown for the DRI response and the three angles from the Mini-Dawn. As with the TSK-PW columns, the light-scattering chromatograms are very clean, with little or no spiking due to the presence of particulates. Synchropak columns tend not to shed particles and are ideally suited for this type of

TABLE 20.8 Summary of Experimental Conditions Used with Synchropak Columns

Columns:	Synchropak 4000, 1000, 300, 100 Å, each 25 × 4.6 cm (Micra Scientific)
Mobile phase:	0.10 *N* sodium nitrate
Flow rate:	0.40 ml/min (nominal)
Detectors:	Differential refractive index (Waters), Mini-Dawn multiangle laser light-scattering photometer (Wyatt Technology)
Temperature:	35°C (columns and detectors)
Injection volume:	0.050 ml (samples and standards)

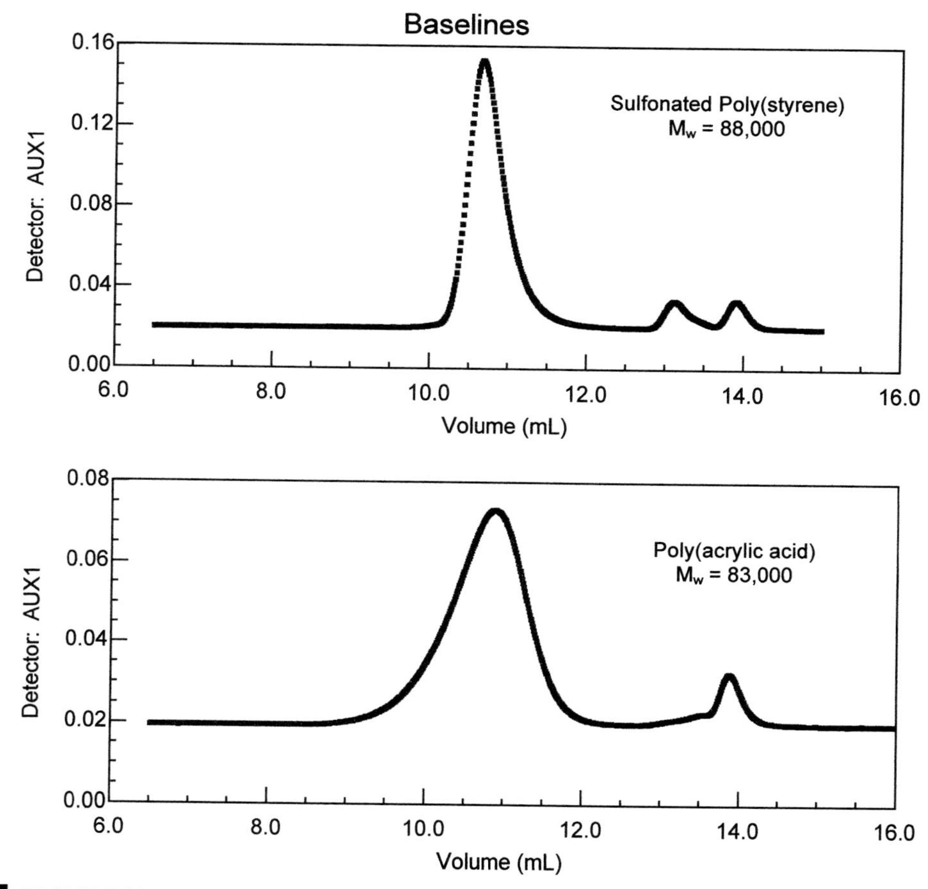

FIGURE 20.7 DRI chromatograms of anionic polymers using Synchropak columns with an aqueous mobile phase of 0.10 *N* sodium nitrate.

on-line SEC detection. Other hydrophilic, water-soluble polymers such as PEG, PEO, PEF, poly(vinyl pyrrolidone), and dextran also work well on these columns. More hydrophobic type polymers, however, such as partially hydrolyzed PVA, tend to tail significantly due to some type of nonexclusion interaction with the bonded packing phase.

IV. AQUEOUS SEC USING CATIONIC COLUMNS

A. Description and Experimental Procedures

The use of bonded, silica column supports has also become a useful way to characterize cationic, water-soluble polymers. CATSEC SEC columns from Micra Scientific contain a silica support with a polymerized polyamine-bonded phase. This imparts a cationic surface charge on the packing that can be

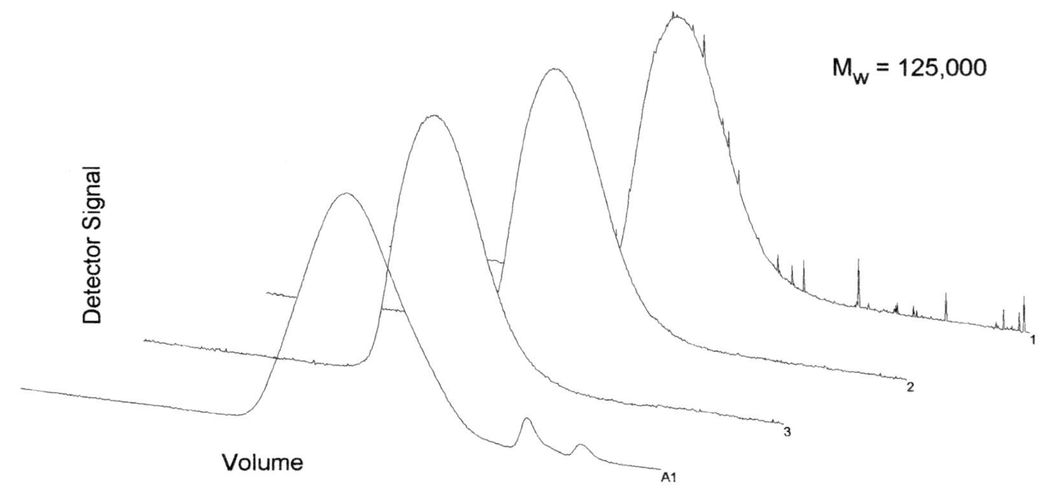

FIGURE 20.8 MALLS and DRI chromatograms of PAAM using Synchropak columns. A1 = DRI, 1 = 41.5°, 2 = 90°, 3 = 138.5°

"controlled" with the proper level of salt in the mobile phase (8,23). The bonded phase also neutralizes the surface silanols, enabling the successful elution of cationic polymers (24). Cationic polyelectrolytes will elute without adsorption using the proper mobile-phase ionic strength and pH.

CATSEC columns come in four different pore sizes ranging from 100 to 4000 Å. The packing material has a diameter from 5 to 10 μm and yields in excess of 10,000 plate counts. With a rigid silica packing material, the columns can withstand high pressure (maximum of 3000 psi) and be used under a

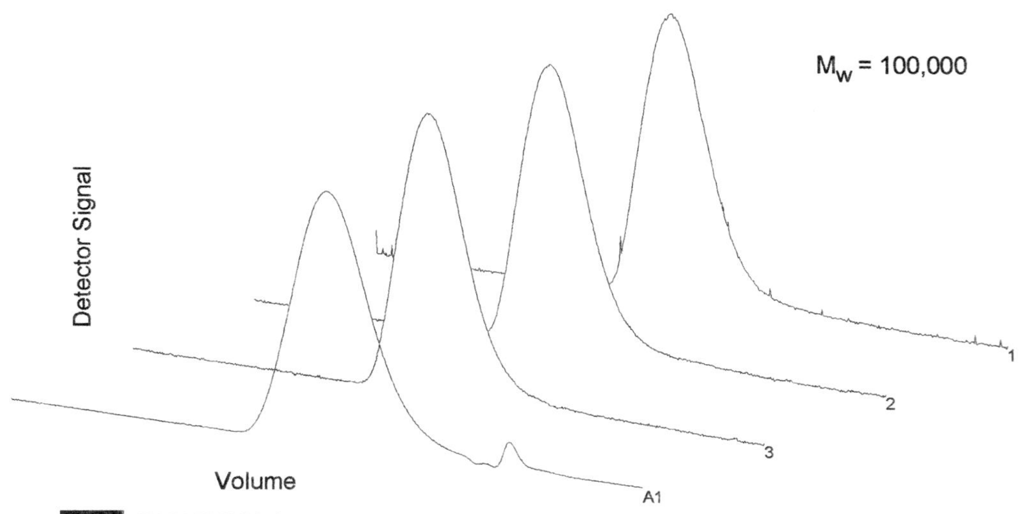

FIGURE 20.9 MALLS and DRI chromatograms of fully hydrolyzed PVA using Synchropak columns. A1 = DRI, 1 = 41.5°, 2 = 90°, 3 = 138.5°

variety of salt and/or buffered conditions. To avoid the corrosive effects of NaCl solutions, aqueous solutions of $NaNO_3$ in the range from 0.05 to 0.20 *N* have been used with these columns (23). Cationic polymers usually require a mobile phase with a low pH (≤3.0). A 0.1% concentration of trifluoroacetic acid (TFA) is generally added to the mobile phase for this purpose. As with Synchropak GPC columns, a mobile phase above pH 8 will dissolve the silica support of the column. A summary of the experimental conditions used for CATSEC columns is listed in Table 20.9.

B. Calibration with Cationic Standards

A commercially available cationic standard that can be used for the calibration of CATSEC columns is poly(2-vinyl pyridine), or PVP. Cationic PVP can be characterized easily on CATSEC columns over a broad range of molecular weight. DRI chromatograms of two cationic PVP standards using a bank of CATSEC columns (100-, 300-, 1000-, and 4000-Å pore size) and a mobile phase of 0.05 *N* $NaNO_3$/0.1% TFA are shown in Fig. 20.10.

A universal calibration curve, using PVP standards and DV detection, is shown in Fig. 20.11 for the set of CATSEC columns described earlier. The mobile phase used was 0.20 *N* $NaNO_3$/0.1% TFA. Intrinsic viscosity values as measured by DV detection ranged from 0.08 dl/g for the PVP standard with M_w = 3300 Da to 5.5 dl/g for the PVP standard with M_w = 1,000,000 Da (23).

C. SEC-MALLS of Cationic Polymers

Weight-average molecular weights of PVP calculated using a Mini-Dawn (three-angle) MALLS detector with CATSEC columns generally show good agreement with the vendor-supplied molecular weights as summarized in Table 20.10. For standards less than 10,000 Da, the agreement is not as good. This may be a consequence of the lower limit of detection of MALLS, which is dependent on the specific refractive index increment (dn/dc) of the polymer, the injected mass, and the laser wavelength (8). For standards greater than 10,000 Da, M_w values from MALLS and those supplied by the vendor agree quite well (within 8%).

TABLE 20.9 Summary of Experimental Conditions Used with CATSEC Columns

Columns:	CATSEC 4000, 1000, 300, 100 Å, each 25 × 4.6 cm each (Micra Scientific)
Mobile phase:	0.05–0.20 *N* sodium nitrate/0.1% trifluoroacetic acid
Flow rate:	0.40 ml/min (nominal)
Detectors:	Model 410 differential refractive index (Waters Corporation), Mini-Dawn multiangle laser light-scattering photometer (Wyatt Technology)
Standards:	PVP (American Polymer Standards Corporation)
Temperature:	35°C (columns and detectors)
Injection volume:	0.050 ml (samples and standards)

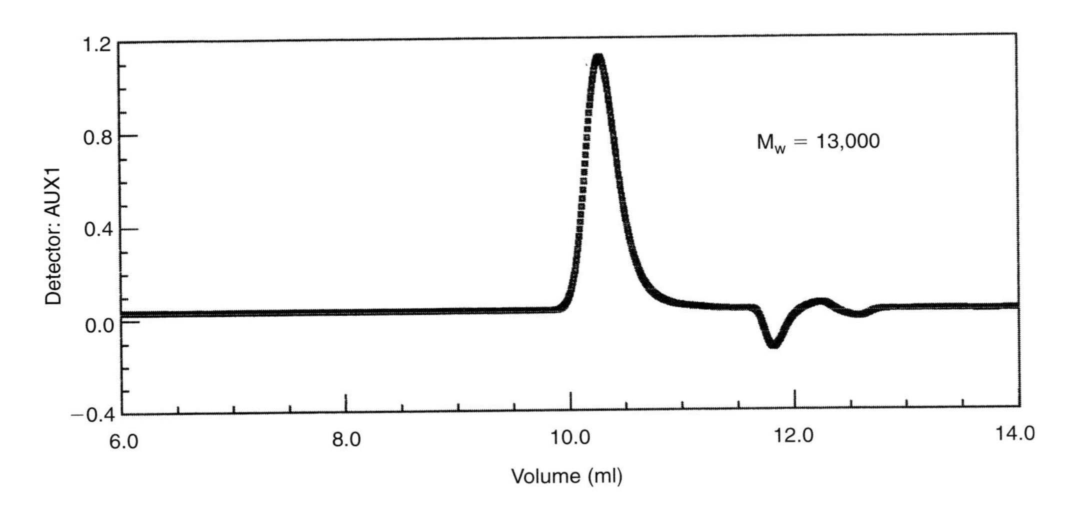

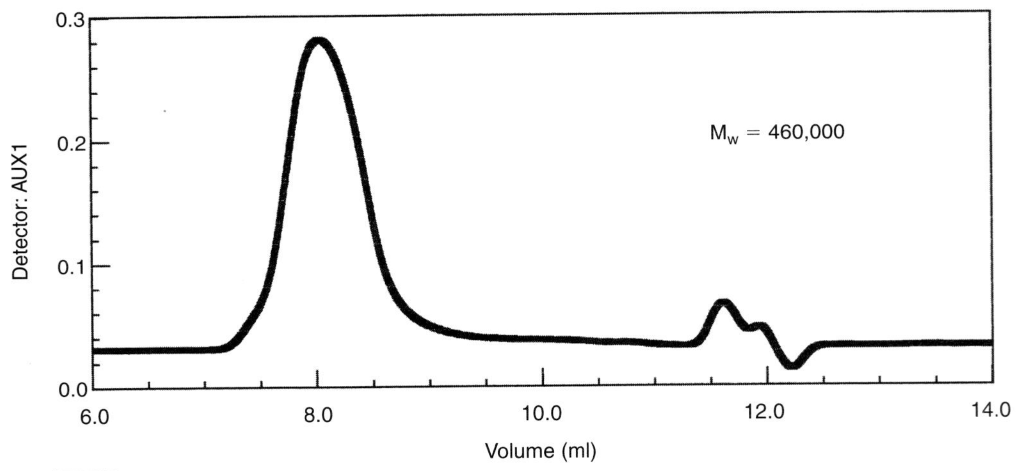

FIGURE 20.10 DRI chromatograms of cationic PVP using CATSEC columns.

In addition to cationic standards, broad distribution cationic polymers can be characterized using CATSEC columns. These include polymers of the hydrochloride salts of poly(allylamine), poly(vinylamine), and poly(vinyl alcohol-vinylamine), to name a few (free base analogs of these polymers are not discussed here). Chromatograms from the Mini-Dawn MALLS detector of a poly(vinylamine)-HCl homopolymer (PVAm-HCl) with M_w = 104,000 Da in a mobile phase of 0.05 *M* $NaNO_3$/0.1% TFA are shown in Fig. 20.12 (samples from Air Products and Chemicals, Inc.). As is the case with Synchropak columns, there is virtually noparticulate spiking or baseline scatter present in these chromatograms. CATSEC columns tend not to shed particles and are ideally suited for this type of on-line SEC detection. This type of behavior is a prerequisite for the accurate calculation of molecular weight and RMS radius values

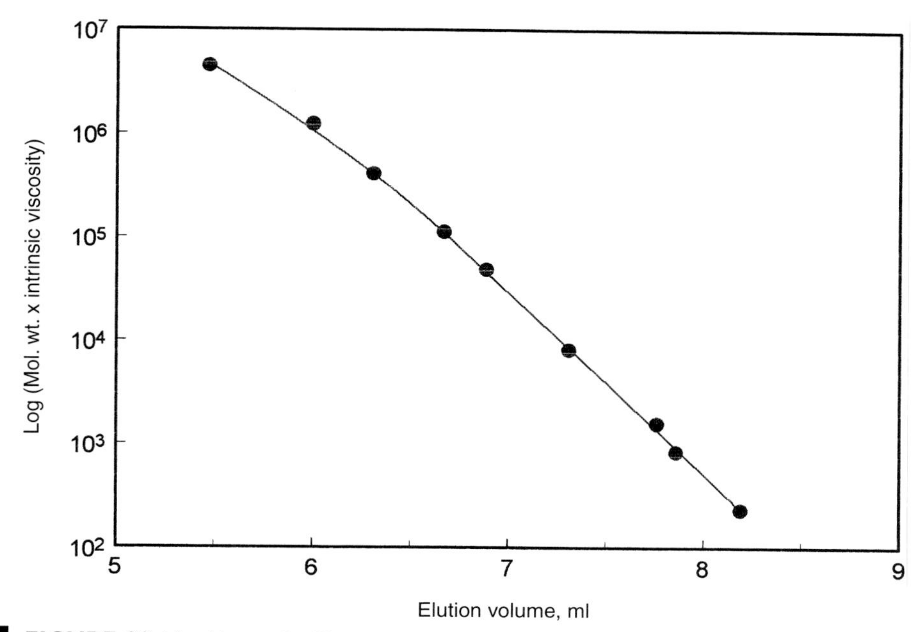

FIGURE 20.11 Universal calibration curve of PVP using CATSEC columns in an aqueous mobile phase of 0.20 *N* sodium nitrate/0.1% trifluoroacetic acid.

under these conditions. Figure 20.13 shows Mini-Dawn MALLS chromatograms of a cationic poly(vinyl alcohol-vinylamine) copolymer (PVA-VAm-HCl) on CATSEC columns in the same mobile phase. Similar behavior is observed as for the PVAm-HCl sample shown in Fig. 20.12.

The ionic strength of the mobile phase can directly affect the elution volume and the RMS radius of a cationic polymer. Figure 20.14 is a plot of molecular weight (calculated with the Mini-Dawn detector) versus elution volume for a PVAm-HCl homopolymer (M_w = 95,000 Da) for three ionic strength levels

TABLE 20.10 SEC-MALLS Data for Cationic Poly(2-vinyl pyridine) Standards Using MALLS and CATSEC Columns[a]

M_w vendor	M_w MALLS	M_n MALLS	M_w/M_n
3,300	4,600	4,000	1.2
8,000	9,900	8,700	1.1
13,100	13,900	11,600	1.2
34,700	36,200	23,600	1.5
105,000	100,000	85,200	1.2
159,000	153,000	41,000	1.1
309,000	285,000	216,000	1.3
460,000	469,000	391,000	1.2

[a]dn/dc = 0.250 ml/g.

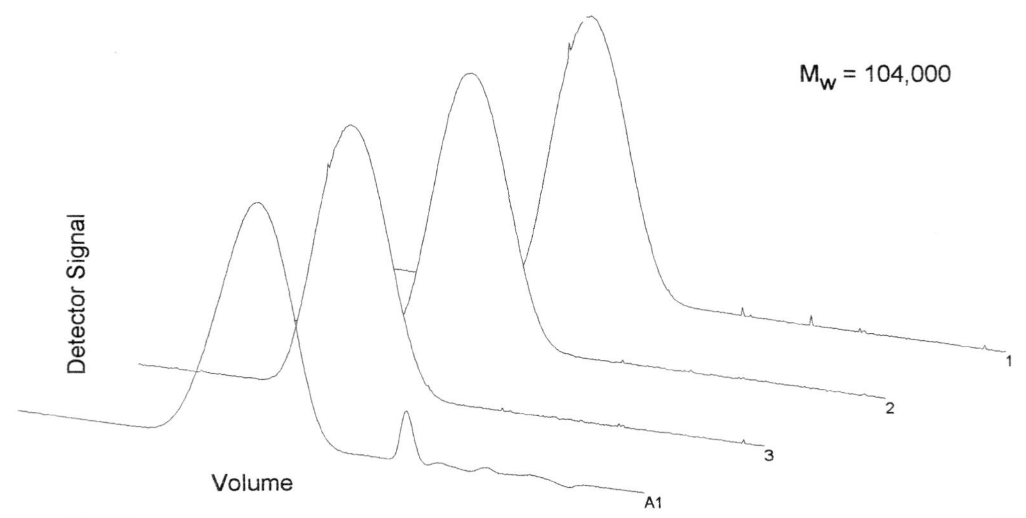

FIGURE 20.12 MALLS and DRI chromatograms of PVAm-HCl using CATSEC columns. A1 = DRI, 1 = 41.5°, 2 = 90°, 3 = 138.5°

(with 0.1% TFA) in the mobile phase. The same set of CATSEC columns was used as described earlier. These ionic strengths include no sodium nitrate, 0.02 *M* sodium nitrate, and 0.20 *M* sodium nitrate. As the ionic strength of the mobile phase increases, the polycation coil size decreases, and further suppression of electrostatic double layer effects between the polymer and the column packing results in an increase in elution volume as more pore volume

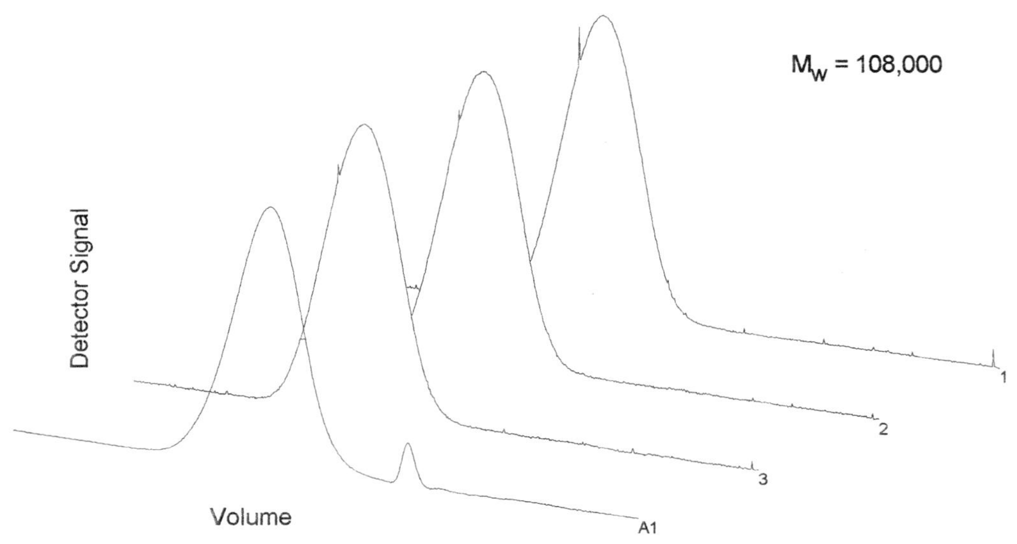

FIGURE 20.13 MALLS and DRI chromatograms of a PVA-VAm-HCl copolymer using CATSEC columns. A1 = DRI, 1 = 41.5°, 2 = 90°, 3 = 138.5°

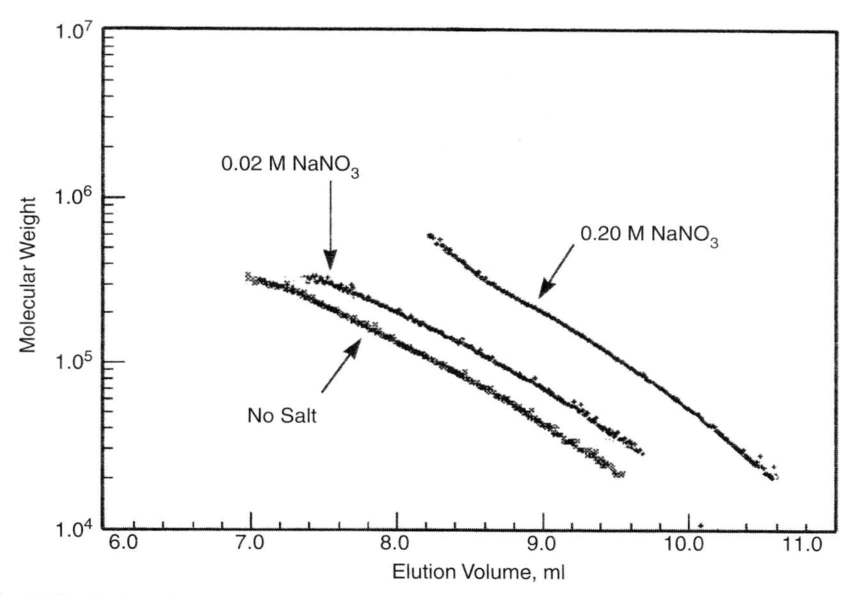

FIGURE 20.14 Molecular weight versus elution time as a function of ionic strength for PVAm-HCl using CATSEC columns.

in the SEC packing becomes accessible to the polymer. At lower ionic strengths, the increased level of electrostatic repulsion between packing and polymer contributes to less accessible pore volume and lower elution volumes. This type of effect has been measured previously using on-line differential viscometry detection for cationic polymers (25).

RMS radius values determined from the Mini-Dawn can be used to ascertain ionic strength effects. Figure 20.15 shows a plot of RMS radius versus concentration of sodium nitrate for the PVAm-HCl homopolymer from Fig. 20.12 and the PVA-VAm-HCl copolymer from Fig. 20.13 and a fully hydrolyzed PVA homopolymer (nonionic). The RMS radius for the PVAm-HCl and PVA-VAm-HCl exhibits the expected decrease with increasing ionic strength. The increased salt concentration of the mobile phase contributes to the suppression of intramolecular screening and repulsive effects of the cationic PVAm-HCl, which results in a decreased hydrodynamic size of the PVAm-HCl. The PVA-VAm-HCl copolymer, which contains approximately 6 mole percent of vinylamine present as the hydrochloride salt, decreases less than the PVAm-HCl due to the lower amount of cationic charge along the polymer backbone. PVA exhibits virtually no change in RMS radius due to its nonionic character. These data demonstrate the sensitivity of on-line MALLS detection to changes in hydrodynamic size (8). The excellent stability and resolution of CATSEC columns help enable these types of measurements to be made.

V. SUMMARY

The use of TSK-PW, Synchropak, and CATSEC columns can prove very effective for the characterization of a broad range of water-soluble polymers: non-

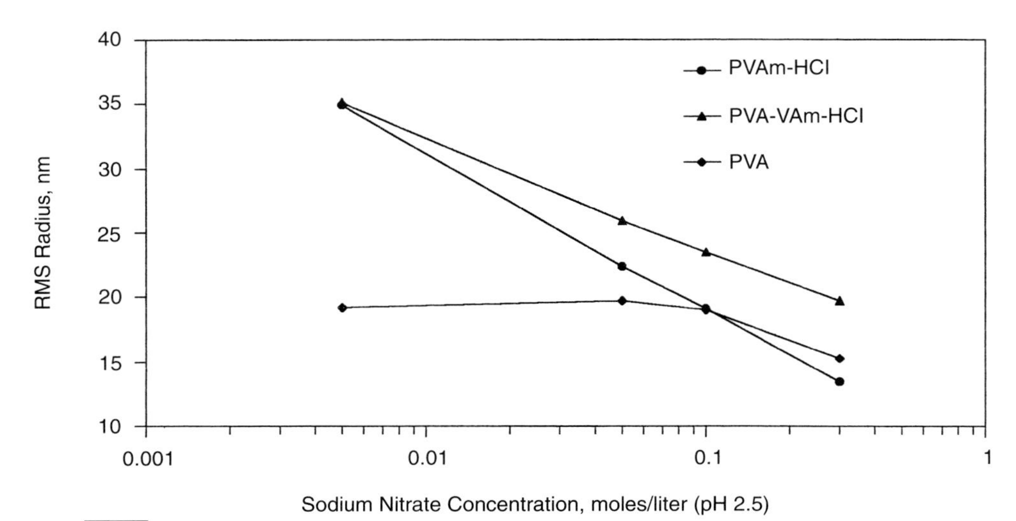

FIGURE 20.15 RMS radius versus ionic strength of cationic polymers and fully hydrolyzed PVA using CATSEC columns.

ionic, anionic, and cationic. TSK-PW columns are a suitable platform for the characterization of nonionic polymers over a broad range of molecular weights due to their stability and excellent resolution. On-line light scattering and differential viscometry detectors interface and work well with TSK-PW columns. These combinations have proven to be extremely valuable for the characterization of molecular weight, RMS radius, conformation, intrinsic viscosity, and Mark-Houwink coefficients for synthetic, water-soluble polymers. The durability and lack of secondary, nonsize exclusion effects of TSK-PW columns have enabled the measurement of these types of physical properties using differential viscometry and multiangle laser light scattering.

Synchropak columns are very useful for characterizing hydrophilic, anionic, and nonionic, water-soluble polymers. CATSEC columns work best for characterizing cationic polymers utilizing both light scattering and/or differential viscometry detection over a wide range of molecular weights.

REFERENCES

1. Barth, H. J. (1986). *In* "Water-Soluble Polymers: Beauty with Performance, ACS Advances in Chemistry Series" (J. E. Glass, ed.), No. 213, p. 31. Am. Chem. Society, Washington.
2. Hashimoto, T., Sasaki, H., Aiura, M., and Kato, Y. (1978). *J. Polym. Sci. Polym. Phys. Ed.* **16**, 1789.
3. TSK-Gel Product Literature, Phenomenex, Torrance, CA, 1995.
4. Nagy, D. J. (1986). *J. Appl. Polym. Sci. C, Polym. Lett.* **24**, 87.
5. Nagy, D. J. (1990). *J. Liq. Chromatogr.* **13**, 4.
6. Nagy, D. J. (1995). *In* "Handbook of Size Exclusion Chromatography, Chromatographic Science Series" (C. W. Wu, ed.), Vol. 69, p. 279. Dekker, New York.
7. Nagy, D. J. (1995). *Am. Lab.* **27**, 4.
8. Nagy, D. J. (1996). *J. Appl. Polym. Sci.* **59**, 1479.

9. Benoit, H., Grubsic, Z., and Rempp, R. (1987). *J. Polym. Sci, B* **5**, 753.
10. Hashimota, T., Sasaki, H., Aiura, M., and Kato, Y. (1978). *J. Polym. Sci. Polym. Phys.* **16**, 1789.
11. Gebben, B., van den Berg, H. W. A., Bargeman, D., and Smolders, C. A. (1985). *Polymer* **26**, 1737.
12. Nagy, D. J. (1986). *J. Appl. Polym. Sci. C Polym. Lett.* **24**, 87.
13. Nagy, D. J. (1987). "Proceedings of the 1987 International GPC Symposium." Waters/Millipore Corporation, Chicago.
14. Nagy, D. J. (1993). *J. Liq. Chromatogr.* **16**, 3041.
15. Beresniewicz, A. (1959). *J. Polym. Sci.* **39**, 63.
16. Staudinger, H., and Schneider, J. (1939). *Liebigs Ann.* **541**, 151.
17. Sakurada, I. (1944). *Kogyo Kagaku Zasshi* **47**, 137.
18. Finch, C. A. (ed.) (1992). "Poly(vinyl alcohol): Developments," p. 166. Wiley, New York.
19. Flory, P. (1951). "Principles of Polymer Chemistry." Cornell Univ. Press, Ithaca, NY.
20. Bahray, M. S., and Hogan, M. P. (1994). International GPC Symposium, Orlando, FL.
21. Synchropak Product Literature, Micra Scientific, Lafayette, Indiana, 1996.
22. Nagy, D. J. (1994). First International GPC-Viscometry Symposium, Houston, TX.
23. Nagy, D. J., and Terwilliger, D. A. (1989). *J. Liq. Chromatogr.* **12**, 1431.
24. CATSEC Product Literature, Micra Scientific, Lafayette, Indiana, 1996.
25. Nagy, D. J., Terwilliger, D. A., Lawrey, B. D., and Tiedge, W. F. International GPC Symposium, Boston, October 1989.

21 QUALITY CONTROL OF COLUMNS FOR HIGH-TEMPERATURE GEL-PERMEATION CHROMATOGRAPHY

A. WILLEM deGROOT

The Dow Chemical Company, Freeport, Texas 77541

I. INTRODUCTION

In the mid-1980s the Dow Chemical Company decided to globally standardize polyethylene characterization. This decision was made in part because of the globalization of the polyethylene business and the constant need to compare data between different Dow locations around the world. While for many analytical procedures, such as melt index, density, and additives analysis, accomplishing this task meant simply standardizing a procedure, for high temperature gel-permeation chromatography (GPC) analysis the process was much more complex. For this analytical procedure we examined every step of the process from the dissolution procedure to the software that does the calculations. For polyethylene research in Texas, this was in some ways an eye-opening experience because the methodology that was followed, including both the software and the GPC columns used, were derived from the original work of John Moore (1) and Lu Ho Tung (2). Although many of our laboratories had standardized on commercial software packages and commercially packed GPC columns, Dow's Freeport laboratory was using software written in-house and we were packing our own columns. We had very little exposure to what was commercially available either in software or in columns at the time, mainly because we were satisfied with what we had.

The search for high temperature GPC columns proved to be a much bigger challenge than any of us had envisioned at the time. In our search for commercially available columns, we found immediately that none of them met our expectations (3). This immediately meant to us that we needed to work with

Column Handbook for Size Exclusion Chromatography

a manufacturer to design columns to meet our needs. We decided to work closely with Polymer Laboratories, Ltd. of Shropshire, United Kingdom, based on their current product in the market and their willingness to work with us to design columns to meet our needs. The purpose of this chapter is to relate Dow's experiences in designing columns for high temperature GPC and monitoring the quality of GPC columns for high temperature GPC of polyethylenes in such a way that others can benefit from the experience.

II. EXPERIMENTAL

All column evaluations were done with the Waters 150C GPC using 1,2,4-trichlorobenzene stabilized with 200 ppm butylated hydroxytoluene (BHT) as the solvent. The GPC operating conditions are listed in Table 21.1. Polyethylene samples were prepared by dissolving them at 160°C for 4 hrs at a concentration of 4 mg/ml. Plate counts were determined using eicosane at a concentration of 0.34 mg/ml and an injection volume of 50 μl. The system was calibrated with narrow molecular weight distribution polystyrene standards. These are described in Table 21.2. The narrow polystyrene standards were prepared by dissolving them at 160°C for approximately 30 min with gentle mixing. All data were analyzed with software written in house.

III. COLUMN SELECTION CRITERIA

The first step in deciding what columns to purchase was to determine what criteria, as best as we could, we wanted these columns to meet. We came up with the following list of features that we wanted in the columns.

1. Linearity of Calibration

This simply relates to how linear the relationship between the peak molecular weight of narrow polystyrene standards versus elution volume fits a straight line. This is typically measured with the linear correlation coefficient, r^2.

TABLE 21.1 Operating Conditions for Waters 150C GPC

Injection compartment temperature	140°C
Column compartment temperature	140°C
Solvent flow rate	1.0 ml/min
Injection volume	100 μl
Analysis time	45 min
Detector	Differential refractive index
Sensitivity	128 for polyethylene −128 for polystyrene
Scale factor	4

TABLE 21.2 Polystyrene Standards Used

Polystyrene peak molecular weights	Concentration (mg/ml)
8,400,000	0.15
2,950,000	0.18
1,000,000	0.20
702,000	0.20
320,000	0.40
120,000	0.40
68,000	0.40
34,500	0.40
9,200	0.40
7,600	1.1
3,250	1.1
1,060	1.5
580	1.8

2. Narrow Particle Size Distribution of Gel

This criterion was based on experience in the past with designing our own GPC columns and has been documented in the open literature (4,5). We found that it was very important in obtaining a column that was very efficient and had high resolution to have the beads as narrow in size distribution as possible.

3. Peak Asymmetry

Peak asymmetry or skewing is a well-documented (4,6,7) characteristic of chromatographic peaks and is measured easily by ratioing the peak half widths at 10% height as shown:

$$\text{Peak asymmetry} = (B - P)/(P - F),$$

where B is the peak retention time at the back of the peak at 10% of the peak height, P is the peak retention time, and F is the retention time at the front of the peak at 10% of the peak height. We decided to standardize our calculations and use a peak from eicosane to measure peak asymmetry.

4. Resolution

Traditionally, column efficiency or plate counts in column chromatography were used to quantify how well a column was performing. This does not tell the entire story for GPC, however, because the ability of a column set to separate peaks is dependent on the molecular weight of the molecules one is trying to separate. We, therefore, chose both column efficiency and a parameter that we simply refer to as $D_2\sigma$, where D_2 is the slope of the relationship between the log of the molecular weight of the narrow molecular weight polystyrene standards and the elution volume, and σ is simply the band-broadening parameter (4), i.e., the square root of the peak variance.

a. Column Efficiency or Plate Count

The following equation was standardized for determining the efficiency of the columns:

$$N = \frac{5.54}{L}\left(\frac{V_r}{W_{1/2}}\right)^2,$$

where N is the number of theoretical plates, L is the column set length in meters, V_r is the peak retention volume in milliliters, and $W_{1/2}$ is the peak width at half height in milliliters. Experimentally, we standardized using eicosane and a 50-μl injection volume.

b. Resolving Power

In our early evaluations, three parameters were utilized for the resolving power of the columns (3,4,7). These were the valley-to-peak height ratio, v, the peak separation parameter, P, and the parameter mentioned earlier, $D_2\sigma$. The valley-to-peak height ratio is defined as

$$v = \frac{h_v}{0.5\ (h_1 + h_2)} \times 100\%,$$

where h_v is the height of the valley between two peaks and h_1 and h_2 are the heights of the two peaks. From this definition one can see that the closer v is to zero, the better the separation or resolution.

Another peak separation parameter, P, is defined as

$$P = \frac{f}{(g + f)},$$

where f is the depth of the valley below a straight line connecting the two peak maxima and g is the distance between the valley and the baseline. From this definition it can be seen that the closer P is to 1, the better the separation the column set is performing.

5. Molecular Weight Accuracy

A method for determining the molecular weight accuracy of a column set has been described by Yau *et al.* (5). These relationships are shown as

$$\overline{M}_n^* = e^{-(1/2)(\sigma D_2)^2} - 1$$

$$\overline{M}_w^* = e^{-(1/2)(\sigma D_2)^2} - 1$$

where M^* is defined as the molecular weight accuracy and represents errors caused by band broadening only. Here one can see quite clearly that as the resolution gets better or $D_2\sigma$ gets smaller, the better the predicted molecular weight accuracy will be for a given column set.

6. Column Durability

This is a difficult parameter to measure, particularly on one or two evaluation column sets. It was found that most column problems are actually caused by catastrophic instrument failure or poor filtering of the solvent or sample.

It usually takes many columns to get a good gauge of how durable columns are. Our solvent is distilled and filtered through 0.5-μm filters, but we only filter samples that we suspect might have gels or have an unknown origin. Our expected column lifetime is 6 to 9 months with very heavy usage, not taking into account instrument failures or failures due to plugging. We considered a column set as being unusable when the plate count drops below 20,000 plates/meter or an obvious pressure increase occurs as a sample is passed through it.

7. Guarantee of Supply to All Dow Sites

Because Dow is a global company and many different Dow sites would rely on these columns it was very important for the supplier to be able to guarantee column availability to all Dow locations.

8. Sensitivity to Overloading

Overloading refers to the shift of a peak to higher elution volumes and consequently lower apparent molecular weights. In the analysis of polyethylene resins this is an important factor to consider because polyethylene molecular weights can reach several million and the samples can be very broad in molecular weight distribution. Overloading is believed to originate from two primary causes (4,8,12): a decrease in the dimensions of the molecules due to crowding and "viscous fingering" due to high solution viscosities. Because both these effects cause the polymer to elute later, they can be minimized by looking at the relationship between concentration and the retention volume of the sample.

9. Optimum Flow Rate

Resolution is strongly dependent on flow rate, as indicated by the van Deemter equation (13). This was an important criterion because we wanted to be able to operate our GPCs at a flow rate of about 1.0 ml/min in order to minimize sample run time.

10. Polymer Shear Degradation

Shear degradation is another parameter that is very difficult to quantify and it is not fully understood. Several studies have been made on this subject (16–19) and the topic has been reviewed (20). We have found in our own work (21) that NBS 1476, a low-density polyethylene, is very susceptible to shear degradation in the GPC column and that the amount of degradation that will occur is very dependent on both flow rate and the particle size of the packing.

After coming up with these criteria we evaluated the currently available commercial high temperature GPC columns and did not find any columns that we thought met the previously described criteria as well as the ones we packed ourselves. Therefore, we changed our plan to looking for column manufacturers who would work with us to design a new column set for our use. Toward this end we selected Polymer Laboratories, Inc. based on their current product and on their willingness to work with us to make it meet our needs. After several iterations of gel blending, they came up with a product that met most of our needs very well.

Tables 21.3 and 21.4 show the results of our evaluation on a column set that we felt performed very well. These tables address criteria 1 through 5 described previously. We judged the values listed to be very acceptable for high temperature GPC applications. For room temperature applications, where a smaller particle size column could be used, better values would be expected.

The rest of the criteria listed earlier were not as easily quantifiable. Column lifetime, for example, is very dependent on several things that are not measured easily, such as instrument reliability, solvent purity, and particles in samples. We elected to not filter all the samples we run because of the possibility of exposure of the technicians to hot solvents when performing this operation. We therefore put our column set at a risk whenever we run samples. To partially address this we used a guard column as a filter and installed it in the pump compartment so that it could be changed out without having to cool the column compartment of the Waters' 150C. In the 10 years that we have used the Polymer Laboratories 10-μm columns at our Freeport, Texas facility, the columns last about 4 to 6 months when they are used very heavily. This range includes failures due to instrument malfunction that might result in large temperature changes without flow and failures due to partial plugging from microgels or particulates in a sample. Other Dow locations get longer lifetimes by filtering all of their samples and running fewer samples than we typically run at Dow in Texas.

Polymer Laboratories was able to guarantee column supply by making up the gel in large batches and keeping packed columns in stock. This simple procedure has resulted in our never being without columns over the 10 years that they have been our supplier.

We did not make any direct measurements in trying to optimize either the sensitivity to overloading or the flow rate. Here we relied on knowledge in the art and the manufacturers recommendations.

Finally, the last criterion, polymer shear degradation, was addressed in two ways. First, because polymer shear degradation was difficult to measure at the time, we went with a larger particle size than is typically used in analytical GPC in an effort to try and minimize the effect. Several years later, we did a study to try and understand the relationship among flow rate, particle size of the gel, and shear degradation of polyethylene (18). In doing this study, some polyethylenes that had a high molecular weight tail such as high pressure, low density polyethylene, shear degradation could become very significant with

TABLE 21.3 Column Evaluation Data for an Acceptable High Temperature GPC Column

Batch No.	Particle size (μm)	N^a	Peak symmetry	$D_2 \times 10^{3b}$	r^2	σ^c	$D_2\sigma$	$M_n{}^a$ (%)	$M_w{}^a$ (%)
52	8–10	37,000	1.1	6.15	0.9992	13.2	0.081	0.33	0.33

[a]Plate count using eicosane in units of theoretical plates/meter.
[b]In units of sec^{-1}.
[c]In units of seconds.

TABLE 21.4 Peak Resolution Parameters Valley-to-Peak Ratios, *v*, and Peak Separation Parameters, *P*, for Batch 52 Gel

Resolution parameters	Polystyrene standards used in the evaluation		
	8.4×10^6 and 2.95×10^6	1.0×10^6 and 3.2×10^5	6.4×10^4 and 3.45×10^4
Valley-to-peak ratio, v	33.2	12.5	47.5
Peak separation parameter, P	0.418	0.467	0.345

10-μm columns. However, for typical polyethylenes produced using Ziegler-Natta catalysts, which do not have a high molecular weight tail, shear degradation was not a concern.

IV. QUALITY CONTROL PROCEDURE

With the first part of the project having been completed, two more challenges remained. We first had to decide how we would ensure that the quality of data from each column set remained high. This meant setting up a quality control procedure for each GPC. The second challenge was determining what to do when we ran out of columns from a particular batch of gel from the manufacturer.

To address the first challenge, we set up a system involving routinely performing plate count determinations and setting up a run chart plotting the molecular weights of a commercial polyethylene. The details of the system are as follows: First, every Monday a plate count determination is made. If the plate count is below 20,000 plates per meter the column set is replaced. Second, for each new column set after it is calibrated with polystyrene standards a run chart is set up using the weight average molecular weights of DOWLEX* 2056, a linear low-density polyethylene (LLDPE) produced by Dow. The run chart is set up by injecting the sample 30 times and setting the upper and lower limits of the chart at 3 times the standard deviation of the molecular weights for the 30 injections. Next, every Monday, Wednesday, and Friday a sample of Dowlex 2056 is run and charted. The following rules were followed.

1. If three consecutive samples show a trend of being on either the high or the low side of the average, a fourth sample is run immediately. If this sample shows the same trend, a new calibration is performed and a new run chart is created. In this case the average is created using only 15 injections and the previous standard deviations are used to compute the new upper and lower control limits.

2. If a molecular weight of Dowlex 2056 is outside the upper or the lower control limits, then the standard is rerun immediately to verify the trend. If the molecular weight of the rerun sample is again outside the control limits,

* Trademark of the Dow Chemical Company.

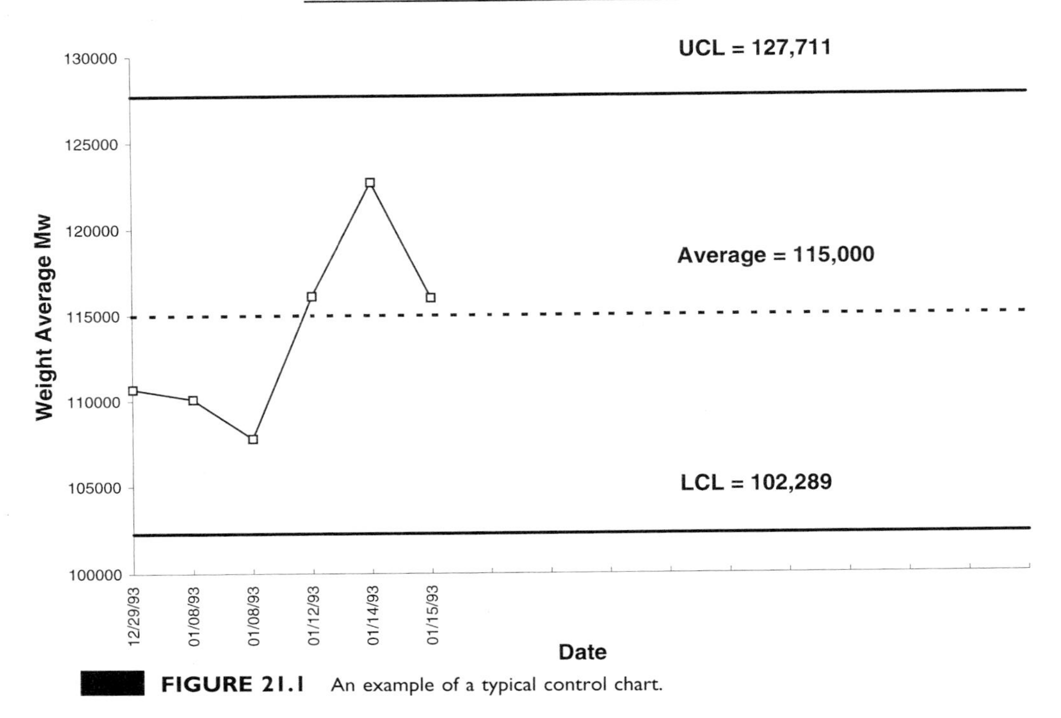

FIGURE 21.1 An example of a typical control chart.

the system is recalibrated and a new average is determined on 15 injections of Dowlex 2056.

Figure 21.1 shows an example of a typical control chart. Here the date is plotted on the *x* axis so the operator can keep track of how the instrument is performing.

TABLE 21.5 Column Evaluation Data

Batch No.	N^a	Peak symmetry	$D_2 \times 10^{3b}$	r^2	σ^c	$D_2\sigma$
52	37,000	1.1	6.15	0.9992	13.2	0.081
58	33,000	1	5.97	0.9996	12	0.072
62	46,630	1.03	5.86	0.9994	12.2	0.071
68	42,700	1	6.03	0.9998	11.8	0.071
74	36,949	—	5.91	0.9993	11.5	0.068
87	39,360	−1.0	5.35	0.9998	12.4	0.067

[a]Plate count using eicosane in units of theoretical plates/meter.
[b]In units of sec^{-1}.
[c]In units of seconds and determined on a 68,000 MW polystyrene standard.

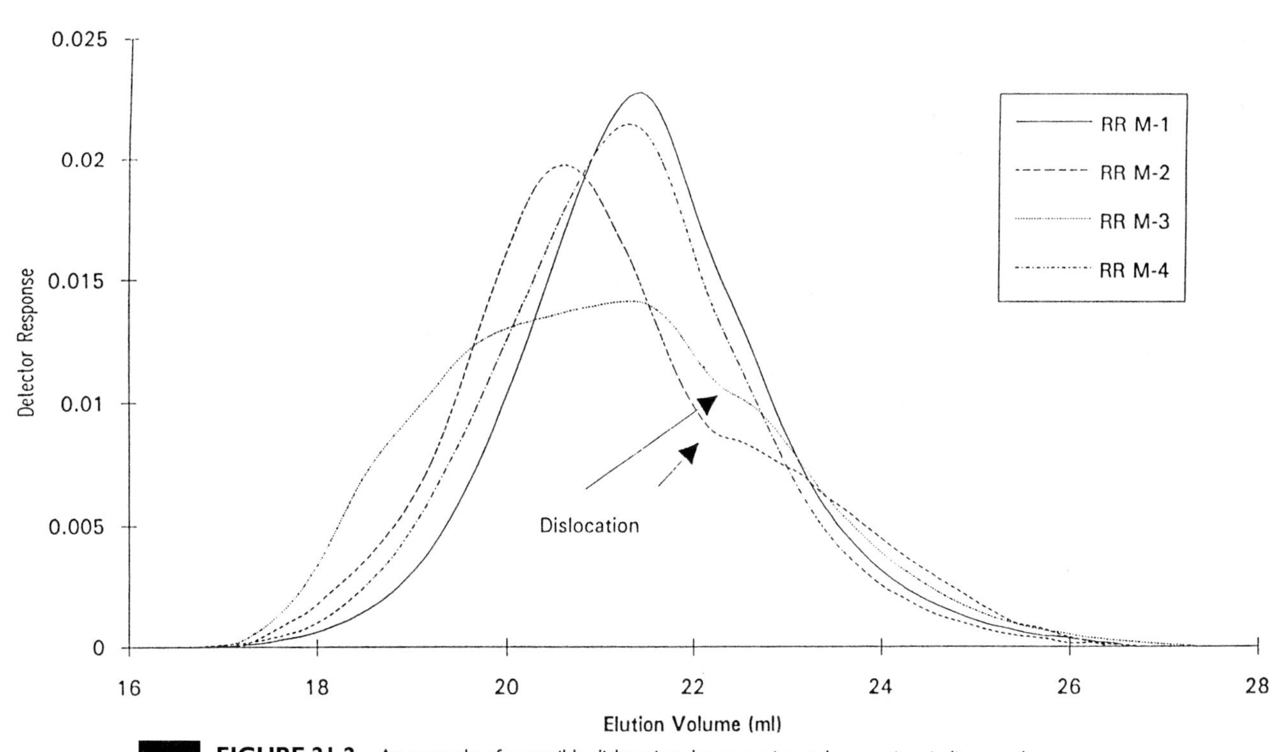

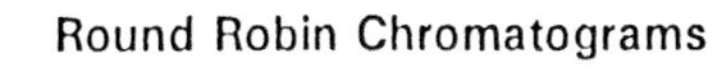
FIGURE 21.2 An example of a possible dislocation due to a mismatch pore sizes in linear columns.

V. NEW GEL BATCH EVALUATION PROCEDURE

The final step in the process of standardizing our columns was to try and maintain the high quality of columns from batch to batch of gel from the manufacturer. This was done by following the basic procedures outlined earlier for the initial column evaluation with two exceptions. First, we did not continue to use the valley-to-peak ratios or the peak separation parameters. We decided that the $D_2\sigma$ values told us enough information. The second modification that we made was to address the issue of discontinuities in the gel pore sizes (18,19). To do this, we selected six different polyethylenes made via five different production processes. These samples are run every time we do an evaluation to look for breaks or discontinuities that might indicate the presence of a gel mismatch. Because the resins were made by several different processes, the presence of a discontinuity in several of these samples would be a strong indication of a problem. Table 21.5 shows the results for several column evaluations that have been performed on different batches of gel over a 10-year period. Table 21.5 shows how the columns made by Polymer Laboratories have improved continuously over this time period. Figure 21.2 shows an example of a discontinuity that was identified in one particular evaluation. These were not accepted and the manufacturer quickly fixed the problem.

VI. CONCLUSIONS

With the move of many large chemical companies in the United States and abroad becoming more and more global, the need to be able to compare high-quality data between various company locations becomes essential. We have addressed part of this issue, i.e., high temperature GPC data, in three ways. First, we have standardized on one type of GPC column. Second, we have implemented a quality control procedure to make sure that data stay at a high quality. Finally, we have a procedure in place to approve future batches of gel to assure that the chromatograms from batch to batch will be very comparable.

REFERENCES

1. Moore, J. C. (1964). *J. Polym. Sci. A* **2**, 835.
2. Tung, L. H. (1966). *J. Appl. Polym. Sci.* **10**, 375.
3. Moldovan, D. G. (1989). "International GPC Symposium Proceedings," 240.
4. Yau, W. W., Kirkland, J. J., and Bly, D. D. (1979). Modern Size-Exclusion Chromatography." Wiley, New York.
5. Yau, W. W., Kirkland, J. J., Bly, D. D., and Stoklosta, H. J. (1976). *J. Chromatogr.* **125**, 219–230.
6. Kirkland, J. J., Yau, W. W., Stoklosa, H. J., and Dilks, Jr., C. H. (1977). *J. Chromatogr. Sci.* **15**, 303.
7. Balke, S. T. (1984). "Quantitative Column Liquid Chromatography A Survey of Chemometric Methods." Elsevier, Amsterdam.
8. Giddings, J. C. (1982). *Adv. Chromatogr.* **20**, 217.
9. Shi, L.-H., Ye, M.-L., Wang, W., and Ding, Y.-K. (1984). *J. Liq. Chromatogr.* **7**, 1851.
10. Chiantore, O., and Guaita, M. (1984). *J. Liq. Chromatogr.* **7**, 1867.

11. Janca, J., Pokorny, S., Zabransky, J., and Bleha, M. (1984). *J. Liq. Chromatogr.* **7**, 1887.
12. Janca, J. (1984). *J. Liq. Chromatogr.* **7**, 1903.
13. van Deemter, J. J., Zuiderweg, F. J., and Klinkenberg, A. (1956). *Chem. Eng. Sci.* **5**, 271.
14. Huber, C., and Lederer, H. (1980). *J. Polym. Sci. Polym. Lett.* **18**, 535.
15. Rooney, J. G., and Verstrate, G. (1981). *In* "Liquid Chromatography of Polymers and Related Materials" (J. Crazes, ed.), Vol. 3. Dekker, New York.
16. Rand, W. G., and Mukherji, A. K. (1982). *J. Polym. Sci. Polym. Lett.* **20**, 501.
17. Wang, P. J., and Glasbrenner, B. S. (1987). *J. Liq. Chromatogr.* **10**, 3047.
18. Barth, H. G., and Carlin Jr., F. J. (1984). *J. Liq. Chromatogr.* **7**, 1717.
19. deGroot, A. W., and Hamre, W. J. (1993). *J. Chromatogr.* **648**, 33.
20. Warner, F. P., Dryzek, Z., and Lloyd, L. L. (1986). *Antec* '86, 456.
21. Warner, F. P., McConville, J. A., Lloyd, L. L., Dryzek, Z., and Brookes, A. P. (1987). GPC Symposium '87, 480.

COLUMNS FOR OTHER RELATED POLYMER SEPARATION OR FRACTIONATION TECHNIQUES

22 MOLECULAR WEIGHT SEPARATION OF MACROMOLECULES BY HYDRODYNAMIC CHROMATOGRAPHY

SHYHCHANG S. HUANG

Advanced Technology Group, The BFGoodrich Company, Brecksville, Ohio 44141

I. INTRODUCTION

Hydrodynamic chromatography (HdC) is a relatively new technique, especially in molecular weight separation. It was first investigated in 1969 by DiMarzio and Guttman (1,2) and was called "separation by flow" (3,4). Small started calling it "hydrodynamic chromatography" in 1974 (5). The application of this technique was first concerned with the separation of particle size. Prud'homme applied it to the molecular weight separation of macromolecules in 1982 (6).

The instrumentation of HdC, including a pump, an injector, a column (set), a detector, and a recorder or computer, is very similar to size exclusion chromatography(SEC). The essence of this technique is the column. There are two types of HdC columns: open microcapillary tubes and a nonporous gel-packed column. This chapter emphasizes column technology and selection and the applications of this technique on the molecular weight analysis of macromolecules.

II. OPEN MICROCAPILLARY TUBULAR HYDRODYNAMIC CHROMATOGRAPHY (OTHdC)

A. Theory

Although the early studies on HdC were done using packed columns, the basic principles of HdC are easily explained by considering the transport of spherical

Column Handbook for Size Exclusion Chromatography

macromolecules in laminar flow through an open microcapillary tube (Fig. 22.1). The solvent velocity in an open microcapillary tube is a parabolic Poiseuille flow. Macromolecules are considered as rigid spheres and are neutrally buoyant. As a result of Brownian motion, macromolecules will diffuse over all possible radial locations in the capillary cross section. Because of their finite sizes, the center of the polymer molecules cannot approach the column wall closer than their own radii, r_p. Due to the monotonic increase of fluid velocity toward the tube centerline, a larger solute molecule travels through the capillary at a greater average velocity than a smaller solute. In other words, the separation of HdC is not due to size exclusion itself, but the faster average flow speed in the area where the macromolecules (or particles) are restricted due to their size.

Tijssen *et al.* (7) showed that the dimensionless retention time of a macromolecule, τ, can be described by a modified quadratic term:

$$\tau = t_p/t_m = (1 + 2\lambda - C\lambda^2)^{-1}, \tag{1}$$

where t_p and t_m are the retention times of a large molecule (or particle) and a very small molecule, respectively; and λ, the aspect ratio, is the ratio of the radii of the large molecule, a, to the capillary tubing, R:

$$\lambda = a/R. \tag{2}$$

A plot of λ versus τ, the calibration curve of OTHdC, is shown in Fig. 22.2. The value of constant C depends on whether the solvent/polymer is free draining (totally permeable), a solid sphere (totally nonpermeable), or in between. In the free-draining model by DiMarzio and Guttman (DG model) (3,4), C has a value of approximately 2.7, whereas in the impermeable hard sphere model by Brenner and Gaydos (BG model) (8), its value is approximately 4.89.

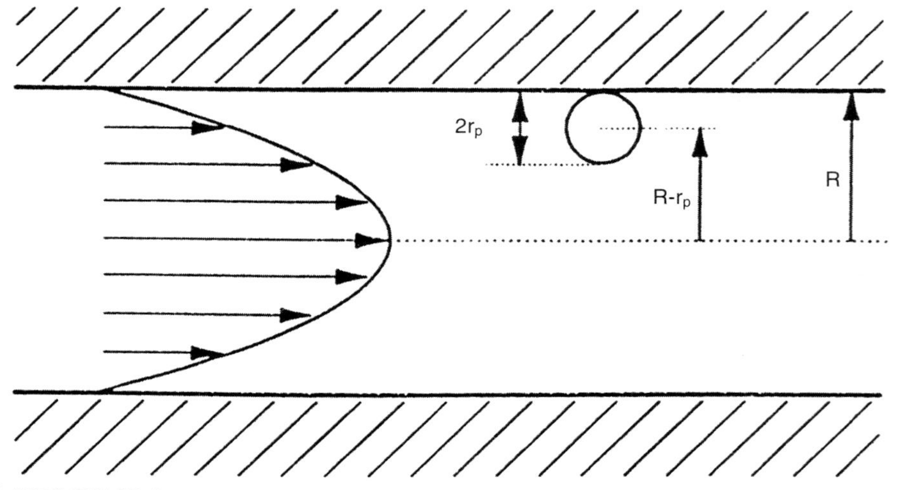

FIGURE 22.1 Transport of a spherical particle immersed in Poiseulle flow through a cylindrical capilliary. (Reprinted from *J. Chromatogr. A*, **657**, 253, Copyright 1993, with permission from Elsevier Science.)

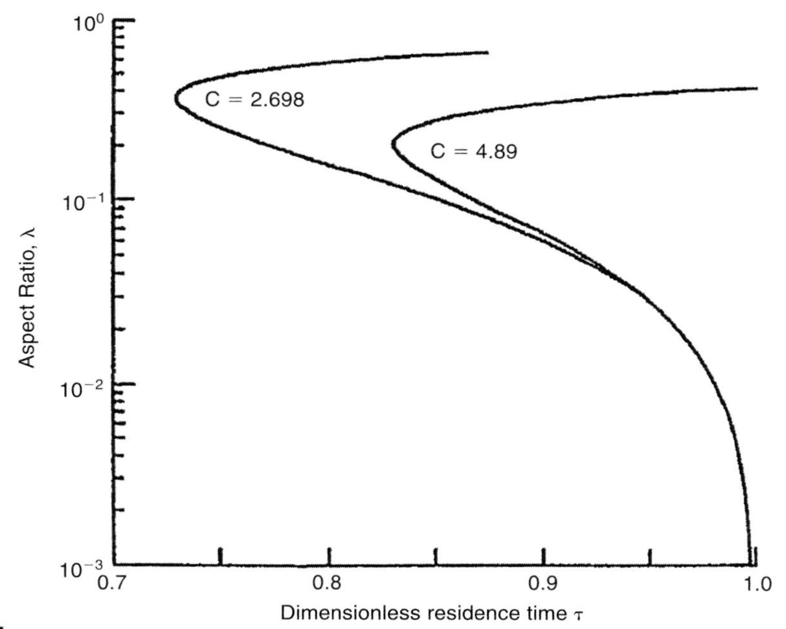

FIGURE 22.2 Calibration curve for OTHdC [Eq. (1), with C = 2.698 for random coils and C = 4.89 for hard spheres]). (Reprinted from *J. Chromatogr. Lib. Ser.*, **56,** 99, Copyright 1995, with permission from Elsevier Science.)

B. OTHdC Columns

Both the radii λ and the retention time τ in Eq. (1) and Fig. 22.2 are dimensionless, suggesting the internal diameter of a capillary chromatographic column can be as large as possible. However, the diffusion rate decreases rapidly when the molecular size increases. For the molecular weight separation of macromolecules, the internal diameter of a microcapillary column is normally limited to 10 μm. Beyond this limit, the number of times a particle moves through the column cross section during the time t of the chromatographic process (Fourier number, F_o) will become too small (9). Bos and Tijssen (9) suggested that F_o should be $\geq$10 to ensure a sufficiently high rate of radial mass transfer. There is also a "tubular pinch effect" (10,11) for larger capillary columns, which is beyond the scope of this chapter.

Figure 22.2 shows the relation of molecular size to the retention time in HdC. However, the correlation of molecular weight to the retention time is needed to apply the technique. Consider all polymer radii to be represented by the general equation

$$a = pM_w^q, \tag{3}$$

which was reported earlier in SEC studies (12,13). In Eq. 3, p and q are polymer- and solvent-dependent constants. Combining eqs. (1), (2), and (3), a theoretical calibration curve in terms of molecular weight versus τ can be expressed:

$$\tau = [\,1 + 2(p\,M_w{}^q/R) - C(p\,M_w{}^q/R)^2]^{-1}. \tag{4}$$

Figure 22.3 shows such theoretical curves as well as experimental points of capillary columns with 0.6- to 1.4-μm radii. The separation ranges of these capillary columns are from 5×10^3 to 10^7. Most of the data points follow the modified DG model. With OTHdC, the molecular hydrodynamic size can be calculated. However, the separation range of a single capillary column is relatively narrow, only about 1.5 order of magnitudes.

For the calculation of molecular weight using Eq. (4), R should be measured accurately, which can be done with a pressure drop-residence time method described in Tijssen *et al.* (7).

C. Applications

The major advantage of the capillary hydrodynamic chromatography is that the mobile phase does not need to have similar solubility parameter as the sample and packing material. (In SEC, nonsize exclusion effects may be observed if the solubility parameter of the sample, packing material, or mobile phase is considerably different.) Therefore, the hydrodynamic size of polymers can be studied in a Θ solvent and even in a solvent that is not compatible with any currently available SEC packing material (9). Figure 22.4 is an example of polystyrene separation in both THF and diethyl malonate. Diethyl malonate is the Θ solvent of polystyrene at 31–36°C.

Another example of OTHdC application is the study of the association/dissociation of copolymer micelles. Using the light-scattering technique, the

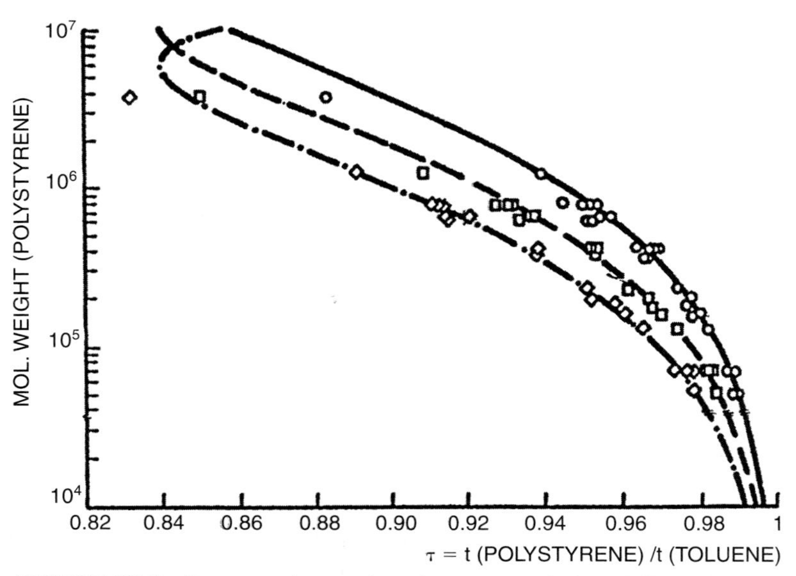

FIGURE 22.3 Experimental points for polystyrene standards in THF and three OTHdC columns (○, 1.342 μm, □, 0.882 μm; and ◇, 0.630 μm) with theoretical curves according to the modified BG model. (Reprinted with permission from Ref. 7. Copyright 1986 American Chemical Society.)

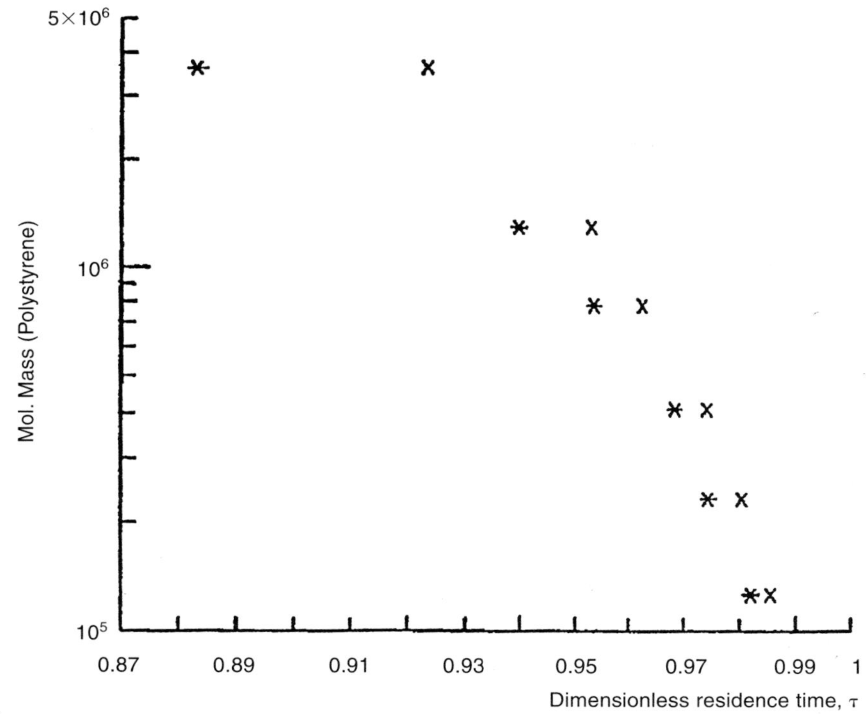

FIGURE 22.4 Comparison of OTHdC experiments on polystyrenes in THF (*) and in diethyl malonate (x). Column diameter 2.68 μm. (Reprinted from *J. Chromatogr. Lib. Ser.*, **56,** 106, Copyright 1995, with permission from Elsevier Science.)

polymer-to-micelle transition may be studied by the average particle/molecular size; however, the OTHdC is able to "observe" the copolymer and micelles, simultaneously. Figure 22.5 shows that a styrene/isoprene diblock copolymer forms micelles from 100 to 85°C in *n*-decane (14).

D. Limitations

Although the OTHdC has several unique applications in polymer analysis, this technique has several limitations. First, it requires the instrumentation of capillary HPLC, especially the injector and detector, which is not as popular as packed column chromatography at this time. Second, as discussed previously, the separation range of a uniform capillary column is rather narrow. Third, it is difficult to couple capillary columns with different sizes together as SEC columns.

III. PACKED COLUMN HdC (PCHdC)

A. Theory

HdC separation also occurs in the interstices of a packed column, although the configuration of channels is not as simple as a microcapillary tube. The

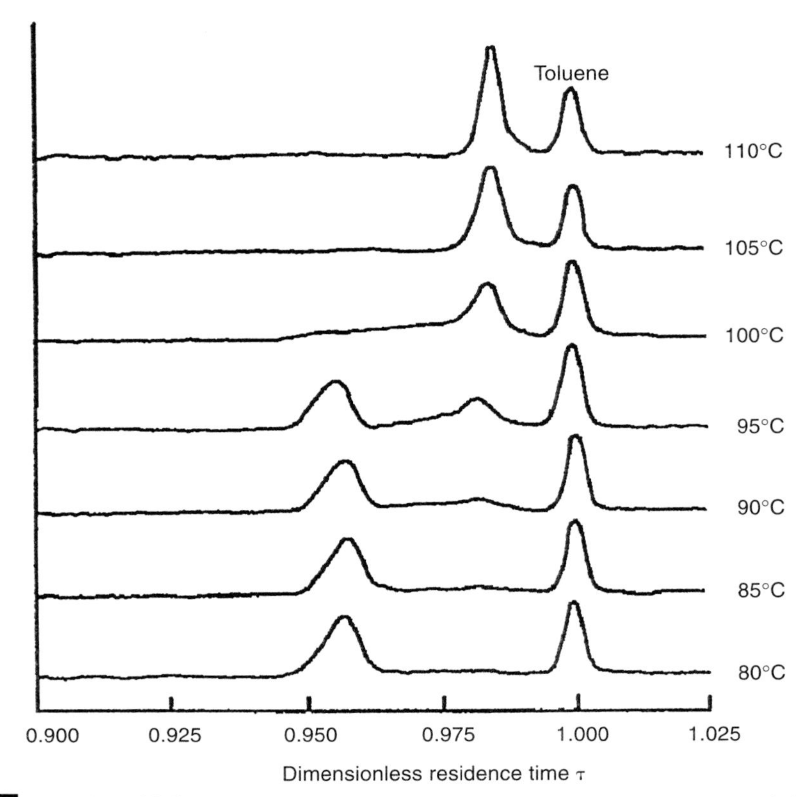

FIGURE 22.5 OTHdC study on the temperature range of the dissociation of the micelle of a styrene–isoprene two-block polymer in *n*-decane. Column: 3.70 μm × 300 cm. (Reprinted with permission from Ref. 14. Copyright 1989 American Chemical Society.)

interstices of a PCHdC column can be envisioned by the packing structure of spheres. The perfect (most compact) packing structure of uniform size spheres is hexagonal closed packing. It can be visualized as being built up as follows: Place a sphere on a flat surface. Surround it with six equal spheres as close as possible in the same plane. Then, form a second layer of equally bunched spheres staggered so that the second-layer spheres nestle into the depressions of the first layer. A third layer can then be added with each sphere directly above a sphere of the first layer. The fourth layer lies directly above a sphere of the second layer, and so forth. The interstices between two layers of the hexagonal packing structure are of two kinds: tetrahedral (four spheres adjacent to it) and octohedral (six spheres adjacent to it), as shown in Fig. 22.6. Mathematically, the volumes of the spheres and interstices of a perfect hexagonal closed-packing structure are 74 and 26% to the total volume, respectively (15). However, the ratio of the interstitial volume of a normal LC/SEC column to the total column volume, the column porosity ε, is approximately 40%. The larger ε is believed to be due to the nonuniform particle size and the nonperfect packing structure.

Because of the three-dimensional void structure of a packed column, the exact form of the velocity profile is not clearly defined as in the microtubular

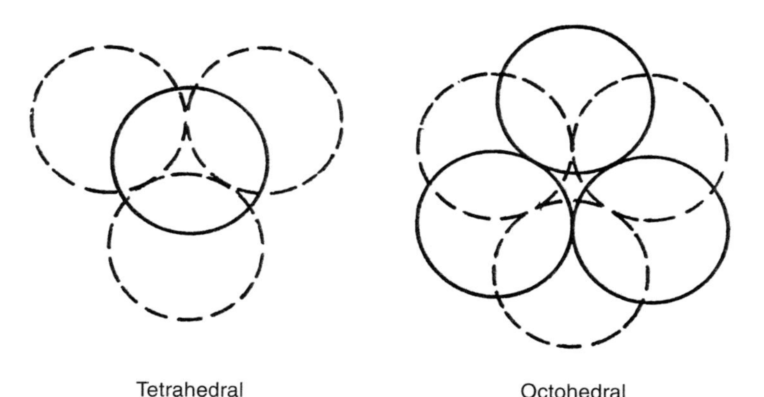

FIGURE 22.6 Two types of interstitial holes between layers of closed-packed spheres.

columns. The mechanism is much more complicated than OTHdC. However, we may consider these voids as a set of equivalent capillaries. Bird *et al.* (16) found that the radius of equivalent capillaries is roughly 0.2 times the mean diameter of the packing beads. Stegeman *et al.*, following earlier work in this field (17,18), concluded that the equivalent capillary radius R_o, the radius of a capillary tube with equal volume-to-surface ratio as the packed bed (19), is

$$R_o = (d_p/3)[\varepsilon/(1 - \varepsilon)], \quad (5)$$

where d_p is the diameter of the packing particles. These simplifying assumptions on flow channel geometry allow for the use of capillary migration models as described earlier. Therefore, the relationship between the retention time and the molecular size of OTHdC, as Eqs. 1 and 4, can also be applied in the PCHdC.

B. PCHdC Columns

Most of the PCHdC columns are packed with nonporous spherical gels, except the work done by Mori *et al.* (20), in which the HdC column was packed with 0.9- to 1.4-μm glass rods.

The particle sizes of spherical gels that were used in the PCHdC studied range from 1.0 to 30 μm (16,21–25). The size of interstitial voids of these columns is, as discussed earlier, equivalent to 0.2- to 6-μm microcapillary tubes. Figure 22.7 shows the calibration curves of PCHdCs with packing diameters of 1.40, 1.91, and 2.69 μm. The results are in good agreement with the theoretical curve, which is in between the free-draining (DG) and solid sphere (BD) models according to the C value, up to around 3×10^6 molecular weight polystyrene.

The larger macromolecules can be separated using larger particle size columns. However, the flow rate should be watched carefully. As the "effective hydrodynamic size" of the macromolecules may be reduced due to the deformation by shear (23). Figure 22.8 shows that the effective hydrodynamic size of a 12–15 $\times$ 10^6 MW polyacrylamide sample will not reach its maximum, or the size without shear, unless the flow rate is reduced to 0.01 ml/min. A

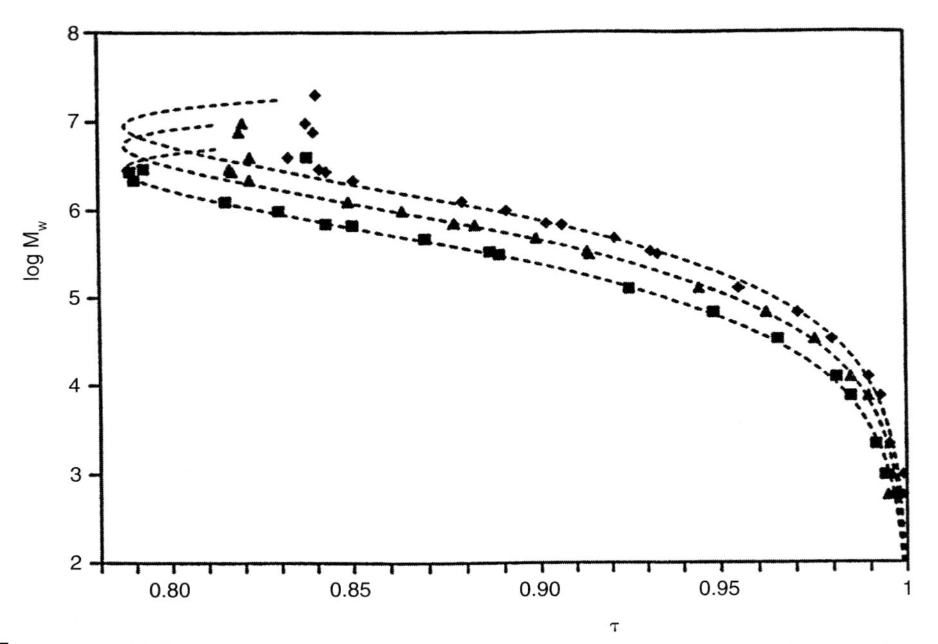

FIGURE 22.7 Elution behavior of polystyrene standards in THF in PCHdC with different packing diameters: ■, 1.40 μm; ▲, 1.91 μm; and ◆, 0.87 μm. Theoretical curves according to Eq. (1), where C = 3.7. (Reprinted from *J. Chromatogr.*, **506**, 554, Copyright 1990, with permission from Elsevier Science.)

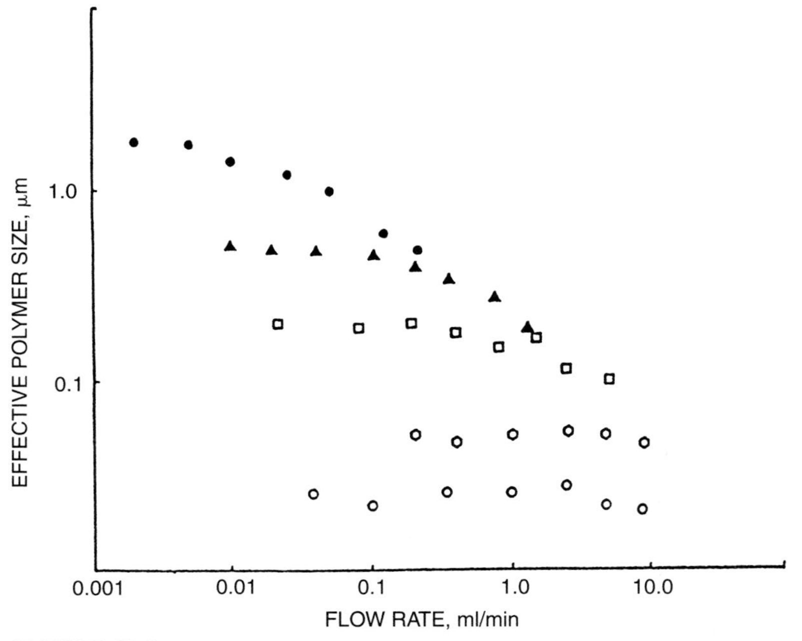

FIGURE 22.8 Flow rate effect on the elution of polyacrylamide. Degradation of polymer during the analysis occurs in each case at flow rates just above those shown. Unmodified MW = 12–15 × 10^6; carboxyl substitution = 9.5%. ●, unmodified; ▲, sheared for 30.0 min; □, sheared for 12 hr; ⬡, sonicated for 1.0 min; ○, sonicated for 5.0 min. (From Ref. 23, Copyright © 1988. Reprinted by permission of John Wiley & Sons, Inc.)

24.8-cm-long column with a 1.0-cm diameter packed with 14.8-μm spherical resins was used in such an experiment. Mechanical shear or ultrasonication reduces the molecular weight and polymer size, allowing the use of higher flow rates for the measurement of the effective hydrodynamic size.

The nonporous spherical gels for PCHdC are often specially prepared for research purposes. However, nonporous polystyrene/divinylbenzene beads, Solid Bead, can be obtained in various particle sizes from Jordi Associates, Inc. (Bellingham, MA). Columns packed with these gels can be used for HdC of the polymers that are currently analyzed using polystyrene/divinylbenzene SEC columns. Fumed silica nanospheres are offered by Cabot (Tuscola, IL) (17), and nonporous silica (NPS) microspheres are offered by Micra Scientific, Inc. (Northbrook, IL). These nonporous silica gels may also be used for HdC.

C. Applications

The major advantage of PCHdC over SEC is that the flow rate in PCHdC can be much greater than in SEC, without sacrificing efficiency or the number of theoretical plates. This is because the HdC separation, like the liquid chromatography of pellicular packing, does not have the "stagnant mobile-phase transfer" (27). Figure 22.9 shows that the reduced plate height ($h = H/d_p$)

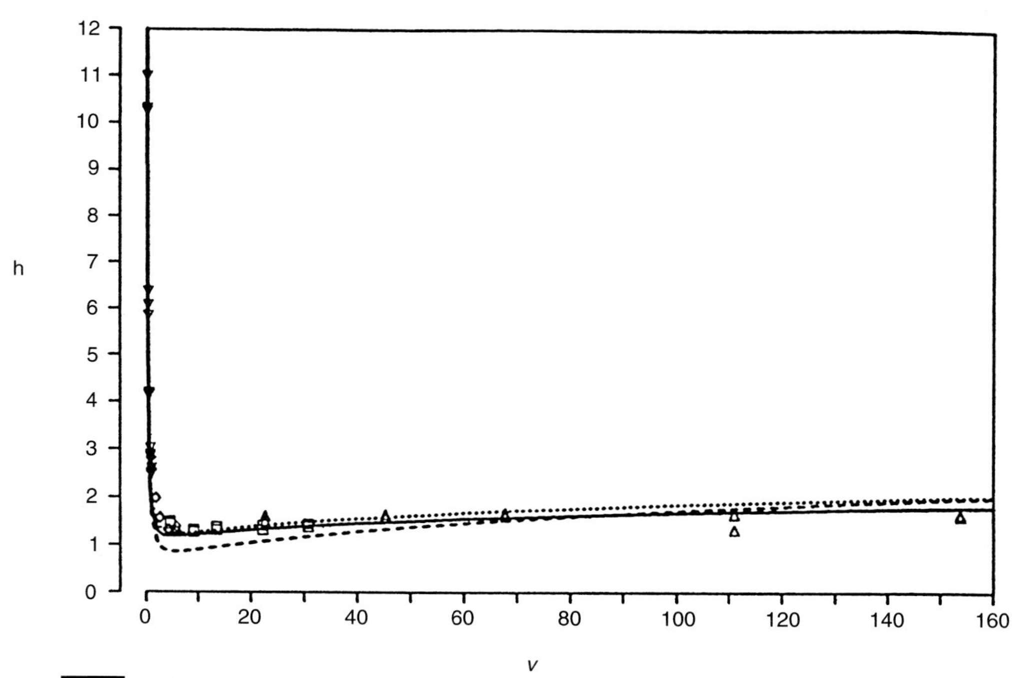

FIGURE 22.9 Reduced plate height versus reduced velocity. Measured data: ▽, toluene; ◇, PS 2200; □, PS 43,900; △, PS 775,000. Theoretical lines: solid lines, Giddings, infinite diameter column; dotted line, Knox, infinite diameter column; dashed line: Knox walled column. (Reprinted from *J. Chromatogr.*, **634,** 154, Copyright 1993, with permission from Elsevier Science.)

does not increase much for wide MW range polystyrene samples up to a reduced velocity ($\nu = vd_p/D_m$, where D_m is the diffusion coefficient in the mobile phase) (28) of 150. Separation of five narrow MW distribution polystyrene samples was achieved in 1.5 min (Fig. 22.10). This fast separation is restricted to relatively low MW samples, i.e., below 10^6. For higher MW polymers, shear-induced deformation may occur under high flow rate.

As in SEC, the surface chemistry of the HdC gels should be similar to that of the mobile phase and the solute. Otherwise, the retention time may increase as with the nonsize exclusion effects. However, the tolerance of PCHdC for a poorer mobile phase is better than SEC. The polymer size under Θ conditions has been studied using PCHdC (19).

D. HdC in Porous Gel Columns

Although most PCHdC studies are conducted using columns packed with nonporous gels, the hydrodynamic separation also occur in SEC columns. This can be easily observed using small pore-size SEC columns (29) as shown in

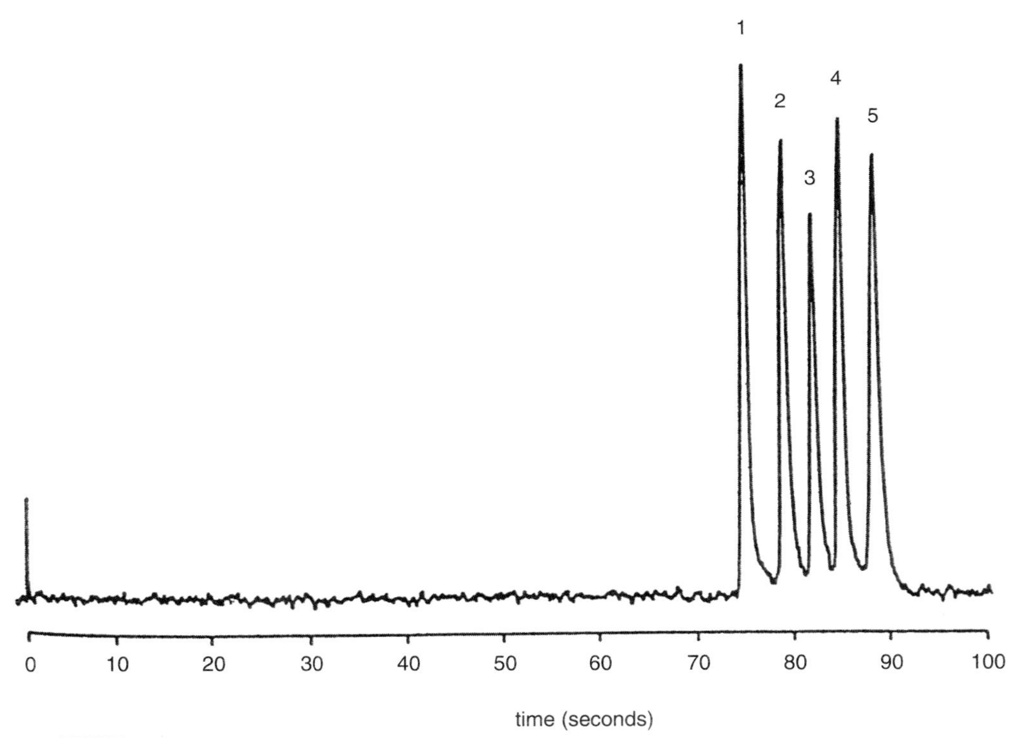

FIGURE 22.10 High-speed PCHdC separation of polystyrenes dissolved in THF. Column, 150 × 4.6 mm; packing, 1.50-μm nonporous silica particles; pressure drop, 200 bar; detection, UV. 1, PS 775,000; 2, PS 336,000; 3, PS 127,000; 4, PS 43,900; and 5, toluene, 0.2 mg/ml each. (Reprinted from *J. Chromatogr. A*, **657**, 255, Copyright 1993, with permission from Elsevier Science.)

Fig. 22.11. Because the HdC separation does not involve the stagnant mobile-phase transfer, the HdC peaks are significantly narrower than the SEC peaks. The HdC separation in columns with large-pore gels will not be obvious because its separation range will overlap and mix with the SEC separation.

The HdC calibration curves of different particle sizes, as shown in Fig. 22.12 (30), are similar to the calibration curves of different pore size columns: the separation ranges of MW due to hydrodynamic chromatography depend on particle size. The larger the particle size, the higher the MW ranges. Stegeman *et al.* (30) proposed that a smooth calibration curve may be achieved by proper ratio of the particle diameter to the pore diameter.

IV. COMPARISON TO OTHER SEPARATION TECHNIQUES

Currently, there are several molecular weight separation techniques, such as OTHdC, PCHdC, SEC, thermal field flow fractionation (ThFFF), and sedimentation field flow fractionation (SdFFF). The molecular weight separation range

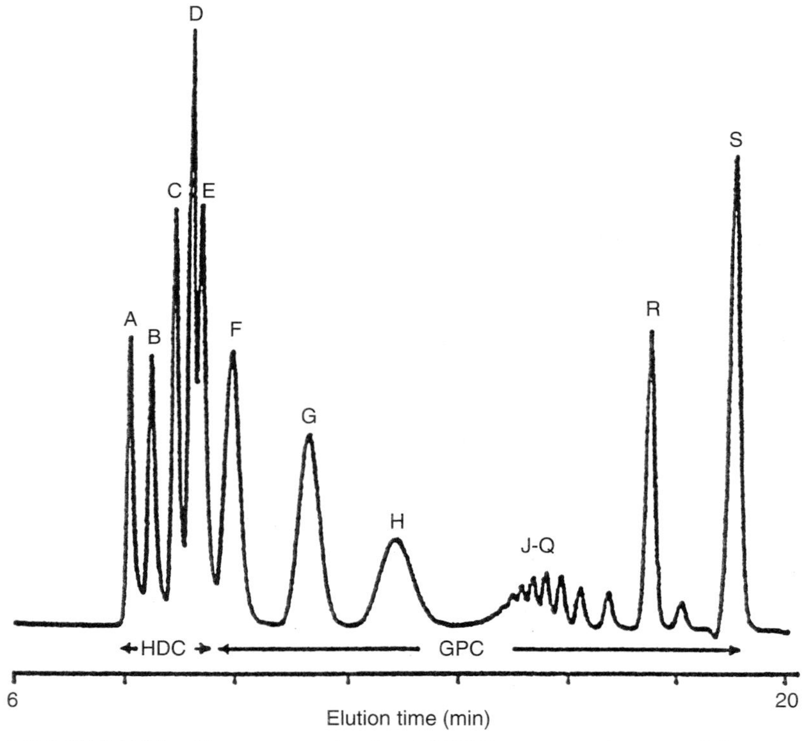

FIGURE 22.11 Chromatogram showing HdC/GPC separation of narrow polydispersity polystyrene standards. Column: two 300 × 7.5-mm PLgel mixed E columns in series; flow rate: 1.0 ml/min. A, 4,000,000; B, 1,550,000; C, 550,800; D, 156,000; E, 66,000; F, 30,300; G, 9,200; H, 3,250; J–Q, oligomers; R, 162; S, toluene. (Reproduced with permission from LC-GC International, Vol. 5, Number 11, November 1992, page 37. Copyright by Advanstar Communications Inc. Advanstar Communications Inc. retains all rights to this article.)

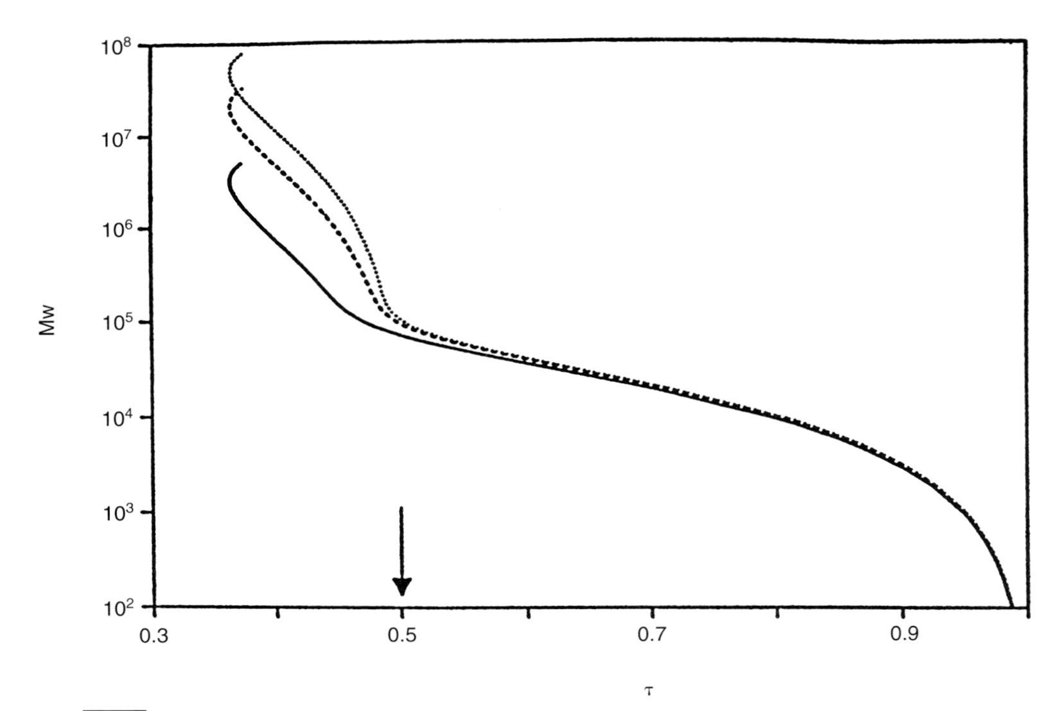

FIGURE 22.12 Theoretical HdC/SEC calibration graphs for different particle diameters. Pore diameter = 10 nm; particle diameters: dotted line, 5 μm; dashed line, 3 μm; solid line, 1 μm. (Reprinted from *J. Chromatogr.*, **550,** 728, Copyright 1991, with permission from Elsevier Science.)

of SdFFF is higher than the interested area for most synthetic polymers. Stegeman *et al.* (30) compared the other four techniques in terms of selectivity, plate height, and minimum separation time, which is of great value to select the best technique for a particular analytical purpose. Revillon (31) also discussed the ThFFF and HdC techniques as alternatives to SEC.

The comparison among these techniques is tabulated in Table 22.1. In summary, HdC is a separation technique with low selectivity; however, the efficiency can be very high. Especially in PCHdC, high analysis speed can be achieved over a wide MW range. ThFFF performs best for the separation of high MW samples. SEC has an intermediate selectivity between HdC and ThFFF. Practicality makes SEC the most suitable method for the common MW range of synthetic polymers. SEC is by far the most commonly used technique for molecular weight distribution determinations. However, HdC is better for the fast analysis purpose.

Another advantage of HdC is its generosity in terms of mobile-phase selection. The polymer size and solution properties of a polymer can be studied using HdC, especially OTHdC, in almost any solvent. In SEC, by comparison, the packing material and mobile phase have to be selected to prevent the nonsize exclusion effect. Because the instrumentation of HdC is similar to SEC, and the packing material and columns have become available commercially, this technique will gain in popularity.

TABLE 22.1 Column Comparison among Four Molecular Weight Separation Techniques[a]

	SEC	OTHdC	PCHdC	ThFFF
Typical column dimensions	i.d.: 7–10 mm Length: 25–30 cm Several columns may be coupled in series	i.d.: 0.1–3 μm Length: 100–500 cm	Similar to SEC	Thin channel Thickness: 50–150 μm Width: 1–2 cm Length: 30–50 cm
Packing material	Spherical porous gels (silica or polymeric)	None	Spherical solid gels (silica or polymeric)	None
MW separation factor	Pore size	Column i.d.	Interstitial void size	Thermal gradient
Resolution factors[b]	Column length (↑) Pore volume (↑)	Column length (↑)	Column length (↑) Void volume (↑)	Column length (↑) Channel thickness (↓)
MW separation range	100–10^7 (shear degradation may occur for MW $>10^6$)	10^4–$>10^7$	10^4–$>10^7$	10^3–$>10^7$
Remarks	Moderate selectivity Moderate separation speed Narrow mobile-phase selection	Low selectivity High-speed separation Most generous mobile-phase selection	Low selectivity High-speed separation Generous mobile-phase selection	Best for high MW polymers Thermal gradient may be programmed for broad MW separation

[a]SEC, size exclusion chromatography; OTHdC, open tubular hydrodynamic chromatography; PCHdC, packed column hydrodynamic chromatography; ThFFF, thermal field flow fractionation.

[b]↑, the resolution will increase when the value increases; ↓, the resolution will increase when the value decreases.

REFERENCES

1. DiMarzio, E. A., and Guttman, C. M. (1969). *J. Polym. Sci. B* **7,** 267.
2. DiMarzio, E. A., and Guttman, C. M. (1969). *Bull. Am. Phys. Soc.* **14,** 424.
3. DiMarzio, E. A., and Guttman, C. M. (1970). *Macromolecules* **3,** 131.
4. Guttman, C. M., and DiMarzio, E. A. (1970). *Macromolecules* **3,** 681.
5. Small, H. (1974). *J. Colloid Interface Sci.* **48,** 147.
6. Prud'homme, R. K., Froiman, G., and Hoagland, D. A. (1982). *Carbohydr. Res.* **106,** 225.
7. Tijssen, R., Bos, J., and van Kreveld, M. E. (1986). *Anal. Chem.* **58,** 3036.
8. Brenner, H., and Gaydos, L. J. (1977). *J. Colloid Interface Sci.* **58,** 312.
9. Bos, J., and Tijssen, R. (1995). *J. Chromatogr. Lib. Ser.* **56,** Chapter 4, p. 95.
10. Segré, G., and Silberberg, A. (1963). *J. Colloid Sci.* **18,** 312.
11. Walz, D., and Grün, F. (1973). *J. Colloid Interface Sci.* **45,** 467.
12. van Kreveld, M. E., and van den Hoed, N. J. (1973). *J. Chromatogr.* **83,** 111.
13. Squire, P. G. (1981). *J. Chromatogr.* **210,** 433.
14. Bos, J., Tijssen, R., and van Kreveld, M. E. (1989). *Anal. Chem.* **61,** 1318.
15. Manas-Zioczower, I., and Tadmor, Z. (1993). "Mixing and Compounding of Polymers: Theory and Practice," p. 561, Hanser-Gardner Pub.
16. Bird, R., Stewart, W. E., and Lightfoot, E. N. (1960). "Transport Phenomena," Wiley, New York.
17. Stegeman, G., Oostervink, R., Kraak, J. C., Poppe, H., and Unger, K. K. (1990). *J. Chromatogr.* **506,** 547.
18. Prieve, D. C., and Hoysan, P. M. (1978). *J. Colloid Interface Sci.* **64,** 201.
19. Stegeman, G., Kraak, J. C., Poppe, H., and Tijssen, R. (1993). *J. Chromatogr. A* **657,** 253.
20. Mori, S., Porter, R. S., and Johnson, J. F. (1974). *Anal. Chem.* **46,** 1599.
21. Lecourtier, J., and Chauveteau, G. (1984). *Macromolecules* **17,** 1340.
22. Langhorst, M. A., Stanley, F. W., Cutie, S. S., Sugarman, J. H., Wilson, L. R., Hoagland, D., and Prud'homme, R. K. (1986). *Anal. Chem.* **58,** 2242.
23. Hoagland, D. A., and Prud'homme, R. K. (1988). *J. Appl. Polym. Sci.* **36,** 935.
24. Hoagland, D. A., and Prud'homme, R. K. (1989). *Macromolecules* **22,** 775.
25. Stegemen, G., Kraak, J. C., and Poppe, H. (1993). *J. Chromatogr.* **634,** 149.
26. Venema, E., Kraak, J. C., Poppe, H., and Tijssen, R. (1996). *J. Chromatogr. A* **740**(2), 159.
27. Snyder, L. R., and Kirkland, J. J. (1979). "Introduction to Modern Liquid Chromatography," 2nd Ed., Chapter 5, Wiley-Interscience, New York.
28. Meehan, E., and Oakley, S. (1992). *LC-GC Intl.* **5**(11), 32.
29. Stegeman, G., Kraak,J. C., and Poppe, H. (1991). *J. Chromatogr.* **550,** 721.
30. Stegeman, G., van Asten, A. C., Kraak, J. C., Poppe, H., and Tijssen, R. (1994). *Anal. Chem.* **66,** 1147.
31. Revillon, A. (1994). *J. Liq. Chromatog.* **17,** 2991.

23 COLUMNS FOR HIGH OSMOTIC PRESSURE CHROMATOGRAPHY

IWAO TERAOKA

Department of Chemical Engineering, Chemistry, and Materials Science, Polytechnic University, Brooklyn, New York 11201

I. WHAT IS HIGH OSMOTIC PRESSURE CHROMATOGRAPHY (HOPC)?

A. History

High osmotic pressure chromatography (HOPC) (1–3) is a method used to prepare fractions with a narrow molecular weight (MW) distribution from a broad-distribution original sample. Developed in 1995, it has been applied to separations of various organic-soluble polymers and water-soluble polymers. The technique allows users to prepare a large amount of narrow-distribution fractions at the minimal expense of solvent. HOPC resembles size exclusion chromatography (SEC) in many respects. For example, both of them use size exclusion by porous materials to separate polymer with respect to MW. The separation principle is, however, different and so are the practice and the columns used. It will be helpful to briefly explain about HOPC in the first two sections before discussing requirements on the columns.

B. Polymer Solution for HOPC

In HOPC, a concentrated solution of polymer is injected. The concentration needs to be sufficiently higher than the overlap concentration c^* at which congestion of polymer chains occurs. The c^* is approximately equal to the reciprocal of the intrinsic viscosity of the polymer. In terms of mass concentration, c^* is quite low. For monodisperse polystyrene, c^* is given as (4)

$$c^*/(\text{g/liter}) = 1330 \times (\text{MW}/1000)^{-0.785}. \quad (1)$$

Column Handbook for Size Exclusion Chromatography

For example, at MW = 4 × 10^5, $c^* \cong$ 12 g/liter, and at MW = 5 × 10^4, $c^* \cong$ 62 g/liter. A polymer solution with concentration $c > c^*$ is called a semidilute solution because mass concentration is low yet repulsive interactions between solutes are strong. Thermodynamics, viscoelasticity, and diffusion properties of semidilute polymer solutions have been studied extensively since the 1960s.

C. HOPC System

HOPC uses a column packed with porous materials that have a pore diameter close to a dimension of the solvated polymer to separate. A concentrated solution of the polymer is injected into the solvent-imbibed column by a high-pressure liquid pump until the polymer is detected at the column outlet. The injection is then switched to the pure solvent, and the eluent is fractionated. A schematic of an HOPC system is illustrated in Fig. 23.1. A large volume injection of a concentrated solution makes HOPC different from conventional SEC.

D. Example of Separation

The following is a typical example of HOPC separation (3). A 25 wt% solution of poly(methyl methacrylate) (PMMA), with M_w = 7.8 × 10^4 and PDI (= M_w/M_n; polydispersity index) = 2.0, in tetrahydrofuran (THF) was injected into a 300 × 7.8-mm column at 0.3 ml/min. Here M_w and M_n are a weight-average MW and a number-average MW, respectively, of the polymer with reference to polystyrene. The packing material was controlled pore glass (CPG) with an average pore diameter of 12.8 nm. By the time the polymer was detected at the column outlet, 5.3 g of the solution was injected. An additional 37 g of THF was injected to collect a total of 28 fractions. Fractions 1 to 20 collected 50 drops each. The number of drops per fraction increased gradually to reach 250 in the last fraction. In early fractions, the eluent was as viscous

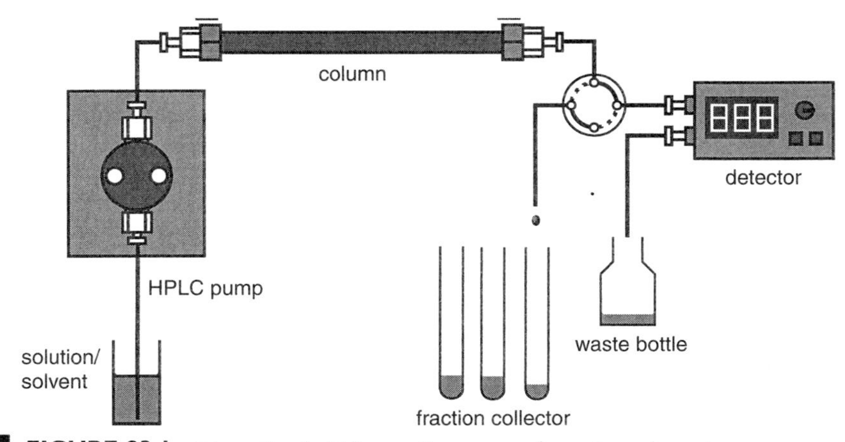

FIGURE 23.1 Schematic of a high osmotic pressure chromatography system.

as the injected solution. Chromatograms for some of the fractions were obtained in analytical SEC (off-line) with a Waters 510 pump (THF mobile phase, 1.0 ml/min), a 410 differential refractometer, and three Phenogel columns (10^3, 10^4, 10^5 Å, Phenomenex). Figure 23.2 shows the chromatograms for the analyzed fractions and for the original PMMA sample. Each chromatogram is normalized by the area under the peak. Initial fractions have a MW distribution centered at the highest end of the MW distribution of the injected material. The PDI narrowed to 1.15, one-fifth power of the PDI of the injected sample. For later fractions, the peak MW decreased and the distribution broadened.

Examples shown in this chapter are for PMMA. Other polymers can be separated as well. The polymers separated so far (1,2) include polystyrene, poly(α-methylstyrene), polycaprolactone, polycarbonate, poly(hexyl isocyanate), polytetrahydrofuran, poly(vinyl methyl ether), and polyvinylpyrrolidone.

E. Features of HOPC

As seen in the example, HOPC has the following features.

a. In early fractions, HOPC separates a sizable amount of high MW components devoid of low MW tail. Late fractions may contain high MW components.
b. Consumption of the solvent is small. To separate a given amount of polymer, HOPC consumes about 1/1,000 of the solvent used in preparative-scale SEC.

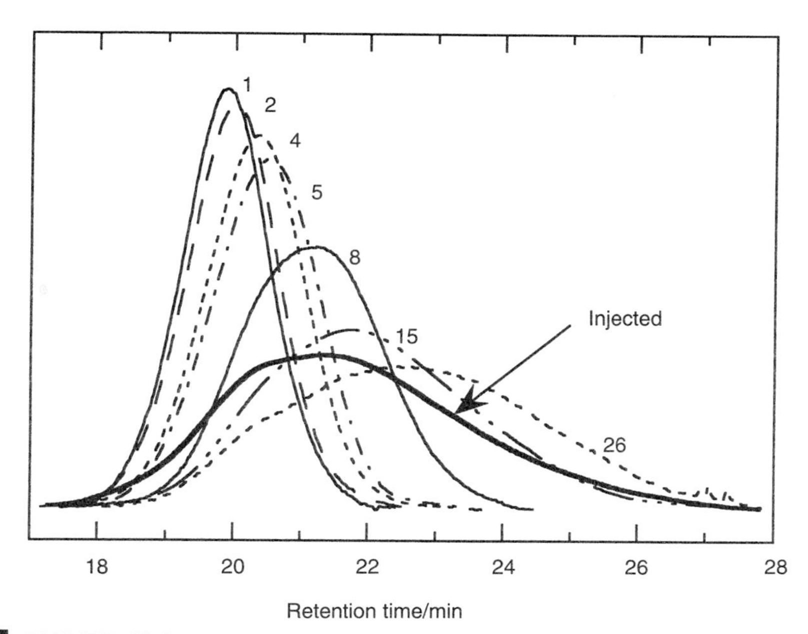

FIGURE 23.2 Example of separation by high osmotic pressure chromatography. The fraction number is indicated adjacent to each chromatogram. (Reprinted from *Tr. Polym. Sci.*, **5**, 258, Copyright 1997, with permission from Elsevier Science.)

c. HOPC is applicable to any polymer that dissolves in a solvent at high concentrations.
d. The eluent is concentrated and therefore can be directly led to the next column to further narrow the MW distribution. Recovery of the solid polymer is easy.
e. HOPC is for preparative or processing purposes (at this moment).
f. At the end of the separation, the column is ready for the next batch.
g. The resolution may well exceed the one typical in the preparative-scale SEC.

II. SEPARATION PRINCIPLE

A. Partitioning of Polymer with a Porous Medium

When a dilute solution of a polymer ($c << c^*$) is equilibrated with a porous medium, some polymer chains are partitioned to the pore channels. The partition coefficient K, defined as the ratio of the polymer concentration in the pore to the one in the exterior solution, decreases with increasing MW of the polymer (7). This size exclusion principle has been used successfully in SEC to characterize the MW distribution of polymer samples (8).

The partitioning principle is different at high concentrations $c > c^*$. Strong repulsions between solvated polymer chains increase the osmotic pressure of the solution to a level much higher when compared to an ideal solution of the same concentration (5). The high osmotic pressure of the solution exterior to the pore drives polymer chains into the pore channels at a higher proportion (4,9). Thus K increases as c increases. For a solution of monodisperse polymer, K approaches unity at sufficiently high concentrations, but never exceeds unity.

Now we consider a polydisperse polymer. The simplest is a bimodal mixture of low and high MW components. When $c << c^*$, $1 > K_L > K_H$, where K_L and K_H are the partition coefficients of low and high MW components, respectively (see Fig. 23.3a). The partition coefficient of each component is the

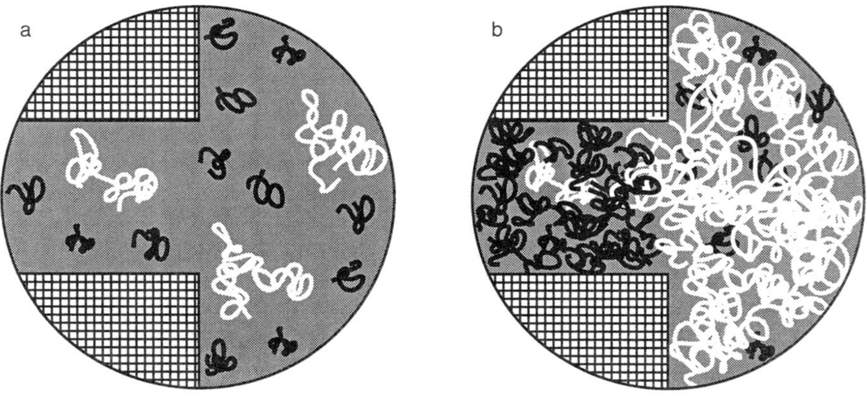

FIGURE 23.3 A solution of a bimodal mixture of low molecular weight components (black) and high weight components (white) in contact with a porous medium: (a) Low and (b) high concentrations.

same as that of a monodisperse polymer of the component. At high concentrations, a phenomenon that never happens for a monodisperse polymer will be seen. When polymer chains are driven into the pore, the size exclusion principle still works, and therefore low MW components dominate the pores. Infusion of low MW components continues until the exterior solution becomes deficient in these components. This means that the exterior solution becomes enriched with high MW components, (see Fig. 23.3b). In terms of partition coefficient, $K_L > 1 > K_H$ (10). The segregation of a polydisperse polymer by MW was formulated into an equilibrium separation technique, termed enhanced partitioning fractionation (11,12).

B. Separation Principle

In HOPC (1), segregation of polymer between the stationary phase (pore channels) and the mobile phase (exterior to the porous medium) is repeated at every plate in the column as the solution of the polymer is transported along the column. The solution front of the mobile phase is enriched further with the high MW components as it approaches the column outlet. The first fraction thus contains the highest MW components present in the injected, original polymer. For the portion of the mobile phase that follows the front end, the mobile phase must be equilibrated with a stationary phase that already contains low MW components. Removal of these components from the mobile phase is not as easy as for the front end of the solution. Thus later fractions decrease the average MW and broaden the distribution. A similar partitioning takes place when the solvent is injected later into the column. Higher MW components in the stationary phase are released into the mobile phase and carried over to the next plate, where the components are exchanged between the two phases to segregate the polymer by MW. Ideally, the highest MW component remaining in the column will be swept out of the column in this way. The high concentration and a large volume injection keep the high concentration condition almost throughout the process.

It is well known that injection of a solution that is not sufficiently dilute distorts the SEC retention curve (overloading). Preparative SEC has been practiced in the low concentration range to minimize complications that arise from overloading, even at the expense of decreased processing capacity. Close examination of the distortion reveals, however, that overloading disperses the rising edge for components of different MWs, whereas it makes the trailing edge overlap. HOPC essentially disregards the tail of the retention curve and makes use of the sharpened front end of the curve. It was shown (3) for a polydisperse polymer that the partition coefficient, which is proportional to the adjusted retention time, depends more sharply on the MW when the concentration is high than at low concentrations. In this respect, the relation of HOPC to SEC is parallel to the relationship of the displacement chromatography to normal-phase and reverse-phase chromatography.

C. Operational Conditions

In HOPC there are several parameters users can adjust for their own needs after a column is selected: concentration of the polymer, injection volume, and flow rate. The dependence of HOPC performance on these parameters has been studied (1). The results are summarized.

a. The concentration of the solution needs to be sufficiently high, at least several times as high as c^*. When a less concentrated solution is injected, narrowing in the MW distribution becomes worse. Injection of a more concentrated solution, in contrast, results in early fractions with a higher MW. The latter fact can be explained as a shift in the boundary MW in the partitioning as an increased osmotic pressure drives more polymer chains into the pore, leaving only very high MW components in the mobile phase. The upper limit of the concentration may only be determined by the system's ability to handle a highly viscous solution.
b. The injection volume needs to be sufficiently large so that most of the column (>50%) is filled with the solution at one time. Injection of a smaller volume degrades the performance. Nonuniform transport of the concentrated solution in the column is a possible cause. Injection of a larger volume broadens the MW distribution in late fractions.
c. The rate of injection needs to be optimized. It was found that a flow rate between 0.1 and 0.2 ml/min gives the best result for a column of 3.9 mm i.d. The performance is, however, not so sensitive to the flow rate as it is to the first two parameters.

D. System Requirements

The requirement for each component of an HOPC system is given.

1. Pump

Most high-performance liquid chromatography (HPLC) pumps can be used in HOPC. The back pressure rating should be at least several thousand pounds per square inch (a few hundred kg/cm^2). A type of pump that does not allow bypassing the pressure transducer or a pulse damper, if it is installed, must not be used. The dead volume should be as small as possible. Pumps with a single plunger are better than those with two plungers.

One of the technical problems in HOPC is the difficulty of injecting a concentrated polymer solution in a large amount against a high back pressure. The solution is highly viscous, or even viscoelastic. Almost all of the HPLC pumps currently available use a reciprocating plunger and a pair of check valves to regulate the flow direction and apply a high pressure. Transport of the solution fails when the ceramic ball (mostly made of ruby) in the check valve does not return to the original seat. A check valve with an improved design, for example, a spring-loaded valve, may prevent backflow. An alternative is to use a syringe pump that guarantees a positive displacement of the liquid. The author has not tested a pump yet and therefore is not certain if those pumps cause other troubles.

2. Detector

A detector is used to determine when to initiate collection of the eluent by a fraction collector. A differential refractometer can be used if the polymer has a refractive index different from that of the solvent. If the polymer has a chromophore, an ultraviolet (UV) absorption detector can be used. Even in the absence of a chromophore, a UV detector may tell if the front of the polymer solution has entered the flow cell in the detector because the concentrated solution disturbs the light path and decreases the intensity of the transmitted light. Visual inspection of the eluent can be used as well; when the eluent is dropped into a nonsolvent of the polymer, a precipitate will be formed as the first polymer comes out of the column. The sensitivity of the detector is not important. Rather, a detector with a small internal fluid volume is recommended to minimize loss of the polymer in the first fraction.

3. Fraction Collector

Any type of fraction collector with a drop counter will do. When the solvent is volatile, attention has to be paid so that the solvent does not dry from the concentrated eluent at the tubing vent.

4. Column Hardware

Columns need to withstand a pressure of at least several thousand pounds per square inch. Stainless-steel columns are recommended.

5. Switching Valves

A switching valve (low pressure) may be used to divert the eluent from the detector to the fraction collector as soon as the polymer is detected. Another switching valve can be used to select the polymer solution or the solvent for introduction into the pump.

6. Tubing and Fittings

From the pump outlet to the column inlet, the pressure is quite high. Stainless-steel tubing, fittings, and end pieces need to be used in this portion.

7. Column

There are no commercial columns designed for HOPC. The user must pack the column with suitable porous materials.

8. Components That Should Not Be Used

Parts that are common in SEC, but should not be used in HOPC are a sample injector (including an autosampler) and a guard column.

III. PACKING MATERIALS

A. Packing Materials for Size Exclusion Chromatography (SEC) and HOPC

HOPC uses a column packed with porous materials. The packing materials are different from those commonly used in SEC in the following three aspects.

1. HOPC uses solid porous materials such as porous silica. Most of the porous gels made of cross-linked polymers, widely used in SEC, will not be appropriate. It is necessary to use porous materials that withstand a high

pressure and do not change the pore size or particle size when immersed in a concentrated polymer solution.

2. A narrow pore size distribution is essential to HOPC. To separate polymer samples with various average molecular weights, users need to prepare columns packed with porous materials of a uniform but different pore size, e.g., 10, 13, 18, and 24 nm. In contrast, a broader pore size distribution is common in a SEC column. A need to analyze a wide range of molecular weights (over many decades) by a single set of columns has spread the use of these columns.

3. The size of the pore used to separate a given polymer in HOPC is smaller compared to the size used in SEC. The former is about one-fourth of the latter (2).

B. Ideal Packing Materials

The porous materials that offer the narrowest possible pore size distribution are those that have cylindrical pores of uniform diameter penetrating the entire medium without branching. Branching gives polymer molecules in the junctions extra conformational entropy. An agglomerate of tiny pieces of these porous materials, interlaced with larger voids (much larger than the pore size), should also be chosen.

Two types of porous materials with cylindrical pores have been developed. One is a nanochannel array glass (13) manufactured in the same way that optical fiber is made. A parent fiber with a cylindrical core and a hexagonal clad is bundled into a larger hexagon. Elongation of the hexagon at a high temperature makes the core thinner while the overall hexagonal shape is still retained. The elongated fiber is cut and bundled together into a larger hexagon. This process is repeated until a desired core size is reached. The bundle is then sliced into thin disks for etching. The core must have an etching rate much greater than that of the clad when soaked in acid. The acid etching leaves uniform cylindrical holes in a regular hexagonal array. In principle, any pore diameter is possible, but pores in the micrometer range are apparently easy to fabricate. So far, 15 nm is the smallest. These nanochannel array glasses are not available commercially, however.

The other type of porous glass that has cylindrical pores is mesoporous silicate (MPS) (14,15). The advantage of MPS is in its feasibility to make a small pore diameter, typically below 10 nm. A columnar-phase liquid crystal, formed from surfactant molecules with a long alkyl chain tail and silicate molecules, is calcined to remove hydrocarbons. At the end, a hexagonal array of straight and uniform cylindrical holes is created in a crystalline order. MPS is not available commercially either.

C. Controlled Pore Glass

All of the currently available porous glasses has a network structure in the pore geometry. They include porous silica such as silica gels, controlled pore glass, and Vycor glass. Silica gels are used widely in normal-phase and reverse-phase liquid chromatography. Unlike nanochannel array glasses and MPS, network-structured porous silica is not free from a pore size distribution or surface roughness. Among the three types of porous silica mentioned earlier, CPG is the most suitable for HOPC because of narrow pore size distribution, a wide range of pore sizes available, and a high mechanical strength.

CPG (16) is produced in spinodal decomposition of sodium borosilicate glass into a bicontinuous structure of a silica-rich phase and a borate-rich phase, followed by acid leaching of the latter phase. Skeletal silica with interconnected tortuous pore space is left. The surface roughness is removed to some degree by etching in a strong base. The pore size and the pore volume can be adjusted by changing the composition of the parent borosilicate glass, the quenching temperature, and the phase-separation time. Compared with several brands of silica gels that claim a narrow pore size distribution, CPG has an even narrower distribution, a smoother surface, and a smaller surface area, yet the mechanical strength of the supporting silica is not sacrificed. The nominal pore size of CPG currently available ranges from 7.5 to 300 nm. CPG has been applied successfully to polymerase chain reaction technology for the synthesis of oligonucleotides.

The profile of the pore size distribution is represented by a plot of the cumulative pore volume (scaled by the total pore volume) as a function of the pore diameter. The pore volume plot, schematically given in Fig. 23.4, is obtained in porosimetry such as nitrogen adsorption/desorption isotherm or mercury intrusion porosimetry (17). The pore diameter is calculated from data of the pore volume and the surface area for a model porous medium that consists of straight cylindrical pores of uniform diameter. In Fig. 23.4, d_x denotes the diameter of the pores below which x % of the total pore volume falls. The median pore diameter, d_m (= d_{50}), is often used to represent the average pore size. It is customary to use d_{90}/d_{10} or $(d_{90}-d_{10})/(2d_m)$ to numerically express the pore size distribution. In various grades of CPG, $(d_{90}-d_{10})/(2d_m)$ is less than 10%. The CPG with the smallest d_m has a slightly broader distribution

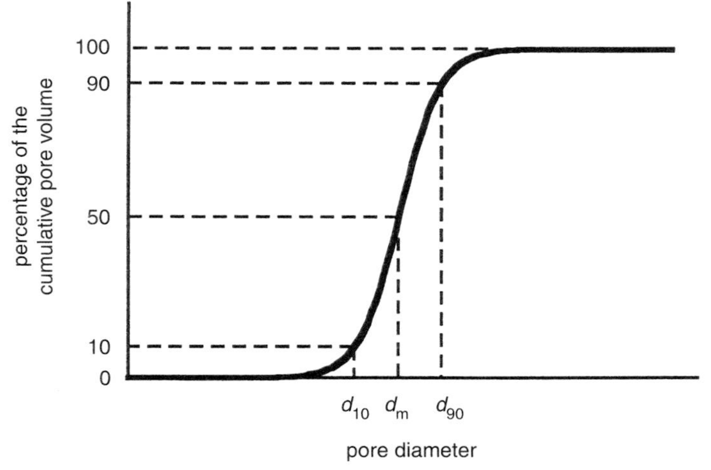

FIGURE 23.4 Percentage of the cumulative pore volume, plotted as a function of pore diameter.

compared with others. In contrast, the same measure of distribution often exceeds 100% in silica gels.

There are two commercial manufactures of CPG in the world. Their addresses are CPG, Inc., 3 Borinski Road, Lincoln Park, NJ 07035, tel: 973-305-8181, fax: 973-305-0884; and Prime Synthesis, Inc., 2 New Road, Suite 126, Aston, PA 19014, tel: 610-558-5920, fax: 610-558-5923.

The porous silica from CPG, Inc. is coded as CPGxxxxxY, where xxxxx is the nominal pore diameter in Å, with leading zero's (target in preparation), and Y is the particle size (A: 80/120, B: 120/200, C: 200/400 in ASTM mesh size). For example, CPG00240C has a d_m of about 240 Å and a mesh size of 200/400. Meshes of 80, 120, 200, and 400 have a sieve opening of 180, 125, 75, and 38 μm, respectively. The nominal pore diameters available are 75, 120, 170, 240, 350, 500, 700, 1000, 1400, 2000, and 3000 Å. For a specific grade of CPG, samples from different lots may be available. There are small variations in d_m and the pore volume, defined as the volume of the pore space per unit mass of CPG.

Prime Synthesis, Inc. offers CPG with pore diameters of 350, 500, 1000, and 3000 Å. Their CPG is coded as Native-xxxxx-CPG, where xxxxx is the median pore diameter.

Vycor glass (Corning, 7930) is porous silica processed in the same way as CPG except that the last step of base etching is missing. Therefore the surface is rough. Only a 40-Å pore is available. When porous silica with a pore size smaller than 75 Å is needed, Vycor glass is the choice. Bulk pieces of Vycor glass are available commercially and need to be crushed into small particles before use.

D. Comparison of Separation by Controlled Pore Glass and Silica Gels

Figure 23.5 shows the importance of using a porous material with a narrow pore size distribution (2). The top chromatograms in Fig. 23.5 were obtained from the separation of PMMA 130K ($M_w = 7.8 \times 10^4$, PDI = 2.0) with silica gels as a separating agent. The bottom chromatograms in Fig. 23.5 are from the separation of the same polymer with CPG. The two separations were carried out in the same condition (25 wt% solution of PMMA in THF, 2 g of the solution was injected into a column of 300 × 3.9 mm at 0.1 ml/min, total 13 fractions collected). The CPG has d_m = 12.8 nm and $(d_{90}-d_{10})/(2d_m)$ = 5.1%. In the silica gels, d_m and the distribution are estimated at 14 nm and 24%, respectively.

The separation by CPG produced fractions with a narrower MW distribution than the separation by silica gels, especially for initial fractions. For fraction 1, PDI dropped to 1.15, about one-fifth power of the PDI of the original sample. In contrast, silica gels decreased the index to 1.27 for fraction 1.

E. Chemical Modification of Pore Surface

Given a polymer to separate, it is important to select an appropriate solvent and appropriate surface chemistry on the pore wall for the optimal separation. In particular, adsorption of the polymer onto the pore surface needs to be prevented. If adsorption occurs, it will favor high MW components, a phenome-

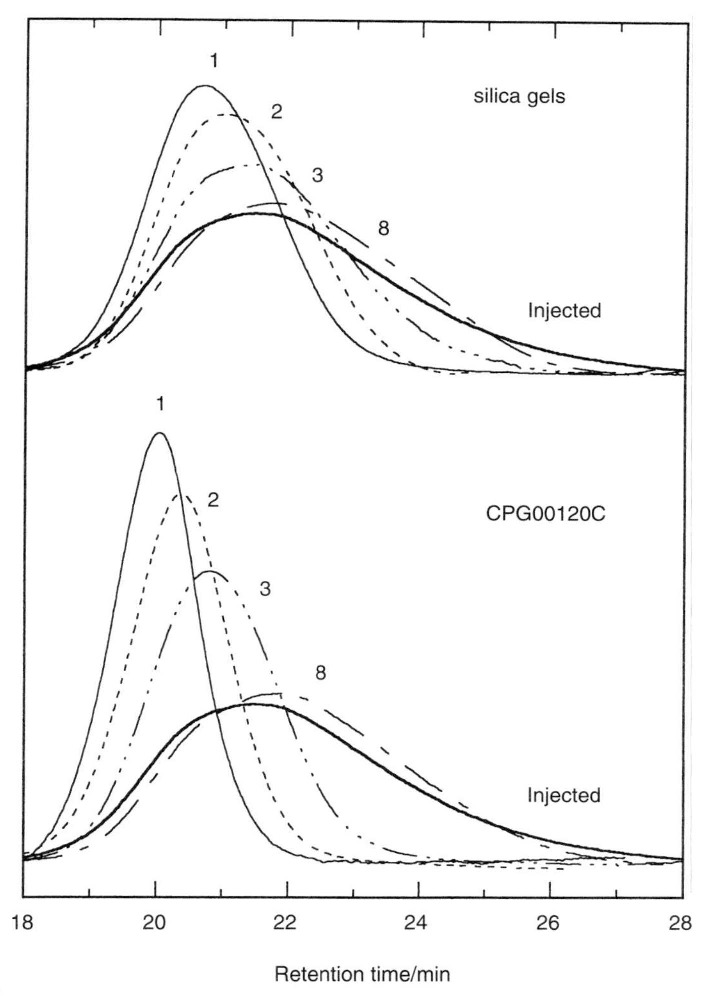

FIGURE 23.5 Comparison of separation by silica gels and separation by CPG that have a similar pore diameter. (Reprinted from *Polymer,* **39,** 891, Copyright 1998, with permission from Elsevier Science.)

non just opposite to the size exclusion. Then the pore surface may accumulate the polymer, resulting in nonreproducible separation with a decreasing resolution, as HOPC batches are repeated on the same column. The requirement to make the surface sufficiently repulsive to the polymer is more stringent in HOPC than it is in SEC. The reason is explained below.

At low concentrations, adsorption is a single-chain phenomenon. The adsorption takes place when the enthalpy gain by the monomer–surface contact with respect to the monomer–solvent contact surpasses the loss of the conformational entropy. In a good solvent the adsorption is not likely unless there is a specific interaction between monomers and the surface. At high concentrations, however, interactions between monomers dominate the free energy of the solution. The adsorption takes place when the enthalpy gain by the monomer–surface contact with respect to the monomer–monomer contact surpasses the loss of the conformational entropy. This situation will happen even in a good solvent when the pore surface's repulsion is not sufficiently strong. To avoid the situation, the surface needs to strongly repel the polymer chains. Needless to say, it is necessary to avoid a solvent that is not sufficiently good. Such a solvent aggregates polymer chains and may precipitate them onto the pore surface.

When porous silica is used as a separating agent, the pore surface needs to be modified depending on the type of the polymer to separate. In aqueous HOPC, native silanols will provide sufficient repulsions to neutral polymers. To separate polyions, other surface chemistry, such as aminopropyl or carboxyl, may be needed, and the pH of the solution needs to be adjusted. In nonaqueous HOPC, organic silanols will prevent adsorption. Many silane-coupling agents are available commercially. They include trimethylchlorosilane $(CH_3)_3ClSi$, hexamethyldisilazane $(CH_3)_3SiNHSi(CH_3)_3$, octyldimethylchlorosilane $C_8H_{17}(CH_3)_2ClSi$, and diphenylmethylchlorosilane $(C_6H_5)_2CH_3ClSi$. In general, the latter two agents will provide sufficiently strong repulsions to most of the polymers.

The silanation procedure is as follows (18):

1. Soak silica beads in concentrated nitric acid at ca. 90°C overnight to remove organic impurities. Rinse the beads thoroughly with deionized water until neutral. If the silica beads are colored, repeat this step until they become colorless.
2. Dry the beads in a clean convection oven at ca. 50°C for about 6 hr and at ca. 90°C for about 24 hr.
3. Prepare a solution of the silanation agent. For trimethylchlorosilane, use toluene as the solvent. The minimum requirement of the silanation agent is calculated by assuming the surface density of silanol to be one functionality per 0.1 nm^2.
4. Transfer the dried beads into a three-neck flask with a thermometer, a nitrogen inlet, and a ventilation stopcock. Add the solution of the silanation agent to the flask with nitrogen sparge.
5. Heat the flask to ca. 60°C. When the temperature has stabilized, close the reaction flask under nitrogen flow. Leave the flask for 3 days. Keep agitating the slurry with a magnetic stirrer.
6. Quench the reaction by adding filtered methanol. Wash the beads with filtered methanol.
7. Dry the beads in a convection oven at about 50°C overnight and at about 100°C in a clean vacuum oven for 1 hr.

If a bulky silanation agent such as octyldimethylchlorosilane is used, there will be unreacted silanols left. These silanols can be end capped by reacting the silica beads with trimethylchlorosilane by repeating steps 4–7.

IV. PORE SIZE AND PARTICLE SIZE

A. Effect of Pore Size

The pore size is the most important parameter that specifies the column used in HOPC. The pore size determines the boundary of MW in the segregation of a polydisperse polymer between stationary and mobile phases. A larger pore diameter shifts the boundary to a higher MW.

An example of three separations is shown in Fig. 23.6 (2). Three columns were packed with CPG00120C, 00240C, and 00350C (d_m = 12.8, 24.2, and 34.3 nm, respectively). About 1.8 g of a 12 wt% solution of PMMA 670K (M_w = 4.7 × 10^5, PDI = 2.8) was injected, and about 14 fractions were collected in each separation. The larger pore produced initial fractions with a shorter peak retention time and a narrower width, but the amount recovered was less. In later fractions, the smallest pore outperformed the others, collecting fractions purer in the low MW components, although the resolution was not good.

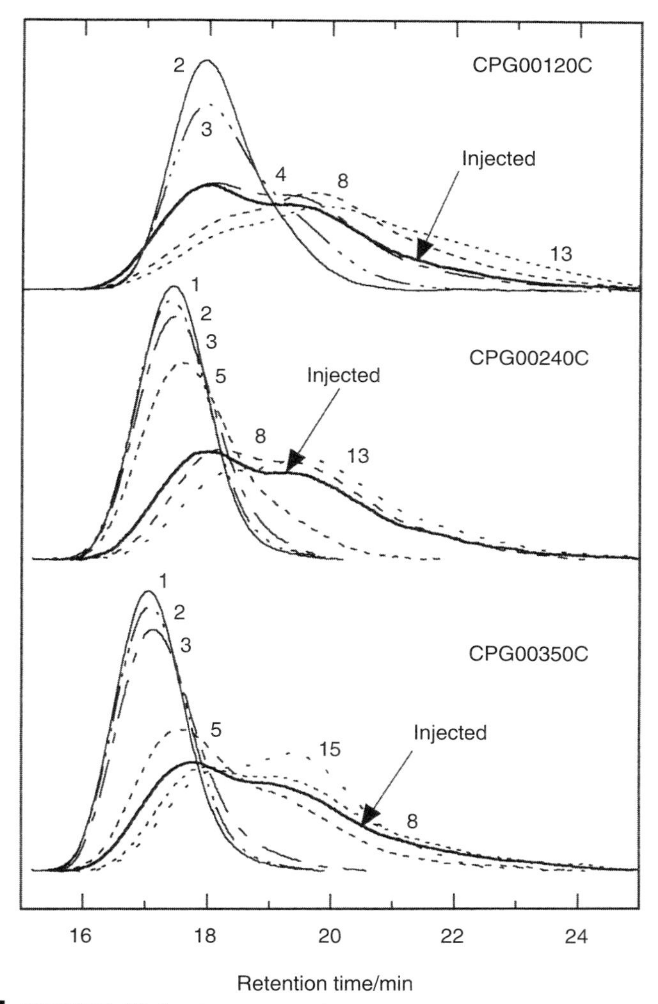

FIGURE 23.6 Comparison of separations by CPG of three different pore diameters. (Reprinted from *Polymer,* **39,** 891, Copyright 1998, with permission from Elsevier Science.)

At M_w of the original PMMA 670K, the radius of gyration R_g of the polymer is estimated to be 30 nm. Here it was assumed that the SEC retention time is a function of R_g, and an empirical formula

$$R_g/\text{nm} = 0.0125 \times M_w^{0.595} \tag{2}$$

obtained (19) for monodisperse polystyrene in a good solvent was used. CPG00120C, 00240C, and 00350C offer the polymer dimension to pore size ratio, $2R_g/d_m$, of 4.6, 2.5, and 1.7, respectively. The separation was poor when the pore was too small.

In contrast to PMMA 670K, PMMA130K ($R_g \cong 10$ nm at M_w) was barely separated by CPG00350C (2). The pore was too large for this polymer. An equally good separation for this PMMA fraction was seen when CPG00075C and CPG00120C were used.

B. Criterion for Pore Size Selection

A criterion for selecting a right pore size to separate a given polydisperse polymer is provided here. To quantify how much the MW distribution narrows for the initial fraction, an exponent α is introduced (2). The exponent is defined by $[\text{PDI}(0)]^\alpha = \text{PDI}(1)$, where PDI(0) and PDI(1) are the polydispersity indices of the original sample and the initial fraction, respectively. A smaller α denotes a better resolution. If $\alpha = 0$, the separation would produce a perfectly monodisperse fraction. Figure 23.7 shows a plot of α as a function of $2R_g/d_m$ (2). Results

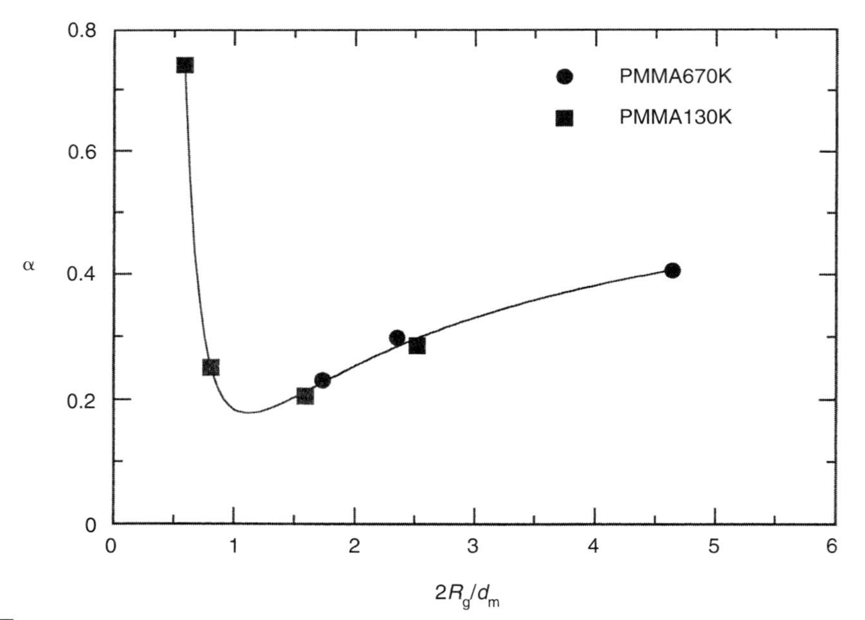

FIGURE 23.7 Exponent α plotted as a function of the ratio of the polymer dimension to the pore diameter, $2R_g/d_m$. (Reprinted from *Polymer,* **39,** 891, Copyright 1998, with permission from Elsevier Science.)

compiled from the separations of PMMA 670K and PMMA 130K with CPGs of four different pore sizes are shown. Apparently, α for a different $2R_g/d_m$ lies on a master curve indicated by a solid line and minimizes in the range of $1 \leq 2R_g/d_m \leq 2$. The selection of pores with pore sizes in that range gives the optimal narrowing for early fractions. From the separation principle of HOPC, it is expected, that the same criterion will be valid for CPGs of other diameters and polymers of other R_gs.

Figure 23.8 shows what range of polystyrene-equivalent M_w of the original polydisperse polymer can be best separated by a CPG of a given d_m indicated adjacent to each bar. The low and high limits of the bar were calculated by using $1 \leq 2R_g/d_m \leq 2$ and Eq. (2). The bar at the bottom may be used for Vycor glass.

In analytical SEC, the optimal pore size for a given polydisperse polymer is approximately $1/4 \leq 2R_g/d_m \leq 1/2$ when a single-pore size column is used (20). In contrast, the pore size needs to be much smaller in HOPC. Small pores are deliberately used in HOPC to exclude nearly all MW components at low concentrations but to allow the entry of low MW components only at high concentrations by the high osmotic pressure (2). Use of the same pore size as used in SEC results in a poor separation. The latter is essentially an overloading in SEC.

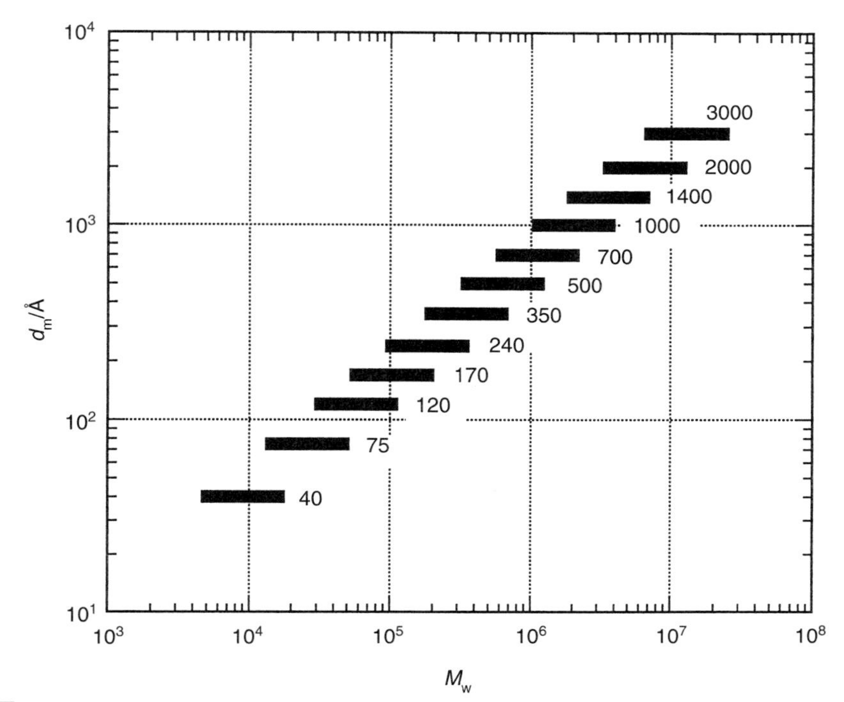

FIGURE 23.8 Range of M_w of the original polydisperse polymer to be separated in high resolution by HOPC using CPG of a given diameter indicated adjacent to each bar.

C. Particle Size

It was shown that the effect of the particle size is not significant in HOPC (1). The experiments were conduced using silica gels of the same pore size but with a different average particle size between 15 and 100 μm. A kinetic effect—enrichment of the mobile phase with high MW components is better at short times before equilibrium is reached—was cited as a possible reason for almost equal quality of separation by large particles. The back-pressure problem was not serious in that range of the particle size.

D. Pore Volume and Packing Density

Other specifications of the porous materials that affect the performance of HOPC include pore volume. A larger pore volume, or equivalently closer packing, of the porous materials increases the ratio of the volume of the stationary phase to the volume of the mobile phase. The difference causes a shift in the segregation boundary in the partitioning and a change in the resolution.

When packing of the porous materials is not sufficiently tight, repeated injection of a viscous solution leads to an increasing packing density, leaving a void space at the column inlet. When the void space was filled with extra porous materials, it was observed (1) that the resolution of HOPC decreased slightly compared with first-time use. When the column ceased to produce the void space, the separation performance became reproducible.

V. COLUMN DIMENSIONS

A. Displacement of a Viscous Liquid

Ideally, the polymer solution should displace the solvent uniformly in the column when the solution is injected. Likewise, the solvent should displace the solution-filled column uniformly when the solvent is injected later. The concentration of the polymer in the stationary phase for a given component should be the same anywhere in a cross section of the column. Similarly, the concentration in the mobile phase should be uniform in the cross section. The large difference in the viscosity between the polymer solution and the pure solvent, however, may cause nonuniform displacement. When a narrow solvent channel is created from the inlet to the outlet of the column in the solvent injection, for example, the incoming solvent will follow the easy path without effectively washing the solution in the pore and the interstitial. The nature of Poiseuille flow of a highly viscous solution may leave the liquid near the column wall unexchanged with the advancing liquid (boundary layer effect). The latter problem is more serious for a longer column and a thinner column.

The nonuniform displacement can occur in the injection of the polymer solution and in the injection of the solvent. The former will be manifested in early fractions. The latter will be seen in the broadening of the distribution in late fractions and in the prolonged time necessary to wash the column.

B. Column Diameter

The performance of HOPC is compared for separation by columns of different diameters (21). A 25 wt % solution of PMMA 130K in THF was injected. Separation of another polymer or of the same polymer at a different concentration results in a different optimal column thickness.

Each of the four columns was packed with CPG00120C (d_m = 13.0 nm). The column dimensions and experimental conditions are listed in Table 23.1. The flow rates (solution and solvent) were set to be proportional to the cross section of the column, whenever possible. The number of drops collected in each test tube was almost proportional to the cross section, especially for the initial fractions that might show a shift in M_w. Figure 23.9 shows chromatograms for some the fractions separated using 2.1-, 3.9-, and 7.8-mm i.d. columns. The result with the 7.8-mm i.d. column is a reproduction of Fig. 23.2 (3). Chromatograms of the fractions obtained from the 1.0-mm i.d. column overlapped with the chromatogram of the injected polymer sample (not shown).

Separation was poor for the two thin columns. Apparently, the transport of the polymer solution through the column was not uniform in the injection, leaving the solvent near the wall unexchanged. The boundary layer effect was serious in these columns. Large amounts of the solution injected and the solvent consumed, relative to the column volumes, are due to the dead volume in the tubing, the end fittings, and the pump head. Compared with these separations, early fractions from the thicker columns (3.9 and 7.8 mm i.d.) collected components of a narrower distribution and a higher MW, with the 7.8-mm i.d. column exhibiting a slightly better separation. The solvent consumed was proportional to the column volume in these two columns. These results show that the thicker columns did not suffer from the boundary layer effect or solvent channels. It is not yet certain whether a column thicker than 8 mm i.d. gives an even better separation.

C. Column Length

A longer column is preferred because of a greater processing capacity and an increased number of plates, as long as the back pressure does not exceed the upper limit and the nonuniform displacement of the solution and the solvent is not serious. The theoretical plate in HOPC is defined as a section in the column in which equivalently full exchange of all of the polymer components

TABLE 23.1 Separation Conditions by Columns of Different Diameters

i.d. (mm)	Length (mm)	Amount of solution injected (g)	Amount of solvent injected (g)	Rate of solution injection (ml/min)
1.0	250	0.71	1.68	0.02
2.1	250	0.98	2.64	0.05
3.9	300	1.94	10.59	0.10
7.8	300	5.30	37.00	0.30

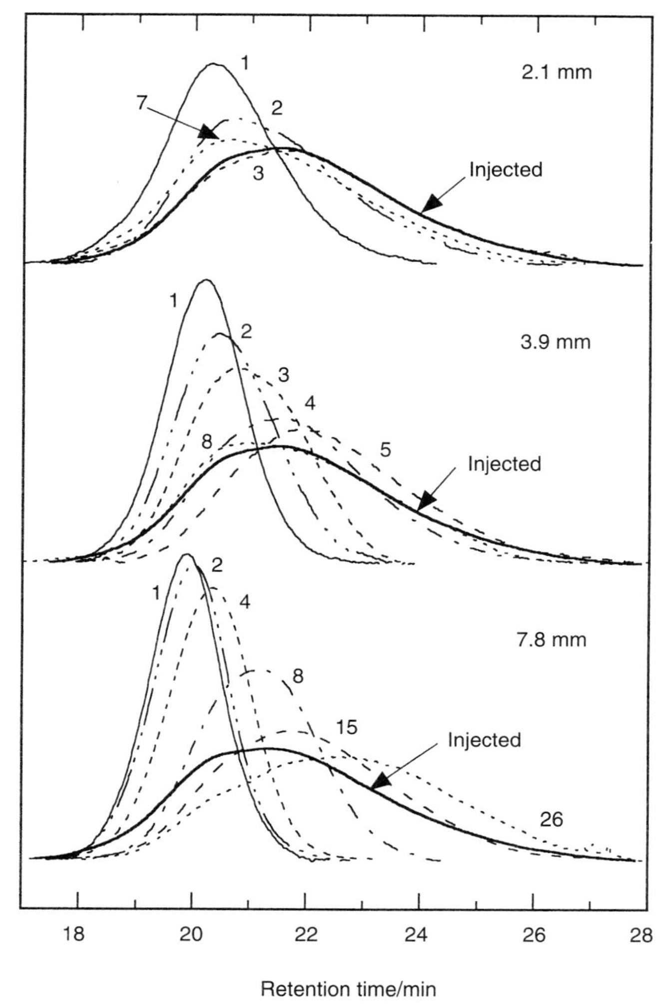

FIGURE 23.9 Comparison of separations by columns of different diameters.

different in MW takes place between the stationary phase and the mobile phase. It is not known how to estimate the height. Cascading two column may be a better alternative to using a longer column.

VI. SUMMARY

HOPC is still at an early stage of development. Not much data have accumulated that tell users which conditions, especially the solvent and the surface chemistry, should be used to separate a given polymer. Nevertheless, a potentially high yield makes it worth while to try to separate the polymer by HOPC.

The following summarizes strategy used in choosing the optimal column and running HOPC.

a. Find the average dimension, R_g, of the polymer to separate. It can be measured in SEC with an in-line multiangle light-scattering detector. A second way to estimate R_g is to calculate the polystyrene-equivalent MW in SEC and use Eq. (2). A third method is to use dynamic light scattering (off-line) to estimate the hydrodynamic radius R_H and to use the relationship $R_g \cong 1.5 \times R_H$.

b. Find the appropriate pore size d_m of CPG using the rule $1 \leq 2R_g/d_m \leq 2$. If a priority is given to securing as large amount as possible at the high MW end of the original distribution, use a smaller pore size. If the priority is to separate the highest MW end at the expense of the recovered amount, use a larger pore.

c. Find a solvent of the polymer that is good to marginal at high concentrations. If the solution is opaque, it means that the solvent gives a near-theta condition to the polymer at that concentration.

d. Choose the surface chemistry that is repulsive to the polymer.

e. Choose a column that has an appropriate d_m and desired surface chemistry.

f. Make a concentrated solution of the polymer for injection. The solution should be at least as viscous as honey.

g. Change the mobile phase to the solvent used to dissolve the polymer and flush the column with the solvent. Run HOPC.

REFERENCES

1. Luo, M., and Teraoka, I. (1996). *Macromolecules* **29**, 4226.
2. Luo, M., and Teraoka, I. (1998). *Polymer* **39**, 891.
3. Teraoka, I., and Luo, M. (1997). *Tr. Polym. Sci.* **5**, 258.
4. Teraoka, I., and Langley, K. H., and Karasz, F. E. (1993). *Macromolecules* **26**, 287.
5. de Gennes, P.-G. (1979). Scaling Concepts in Polymer Physics. Cornell Univ. Press, Ithaca, NY.
6. Doi, M., and Edwards, S. F. (1986). The Theory of Polymer Dynamics. Clarendon Press, Oxford.
7. Casassa, E. F. (1967). *J. Polym. Sci. Poly. Lett. Ed.* **5**, 773.
8. Yau, W. W., Kirkland, J. J., and Bly, D. D. (1979). "Modern Size Exclusion Liquid Chromatography." Wiley, London.
9. Daoud, M., and de Gennes, P. G. (1977). *J. Phys. (Paris)* **38**, 85.
10. Teraoka, I., Zhou, Z., Langley, K. H., and Karasz, F. E. (1993). *Macromolecules* **26**, 3223.
11. Teraoka, I., Zhou, Z., Langley, K. H., and Karasz, F. E. (1993). *Macromolecules* **26**, 6081.
12. Dube, A., and Teraoka, I. (1995). *Macromolecules* **28**, 2592.
13. Tonucci, R. J., Justus, B. L., Campillo, A. J., and Ford, C. E. (1992). *Science* **258**, 783.
14. Kresge, C. T., Leonowicz, M. E., Roth, W. J., Vartuli, J. C., and Beck, J. S. (1992). *Nature* **359**, 710.
15. Beck, J. S., Vartuli, J. C., Roth, W. J., Leonowicz, M. E., Kresge, C. T., Schmitt, K. D., Chu, C. T.-W., Olson, D. H., Sheppard, E. W., McCullen, S. B., Higgins, J. B., and Schlenker, J. L. (1992). *J. Am. Chem. Soc.* **114**, 10834.
16. Haller, W. (1965). *Nature* **206**, 693.
17. Lowell, S., and Shields, J. E. (1991). "*Powder Surface Area and Porosity,*" 3rd Ed. Chapman & Hall, London.
18. "CPG Controlled-Pore Glass." Fluka AG.
19. Huber, K., Bantle, S., Lutz, P., and Burchard, W. (1985). *Macromolecules* **18**, 1461.
20. Haller, W. (1983). "*Application of Controlled Pore Glass in Solid Phase Biochemistry*" (W. Haller, ed.), pp. 535–597. Wiley, New York.

INDEX